Springer Series in Optical Sciences Volume 45

Editor: Peter W. Hawkes

Springer
Berlin
Heidelberg
New York
Barcelona
Budapest
Hong Kong
London
Milan
Paris
Singapore
Tokyo

Springer Series in Optical Sciences

42 **Principles of Phase Conjugation**
By B. Ya. Zel'dovich, N. F. Pilipetsky, and V. V. Shkunov

43 **X-Ray Microscopy**
Editors: G. Schmahl and D. Rudolph

44 **Introduction to Laser Physics**
By K. Shimoda 2nd Edition

45 **Scanning Electron Microscopy**
Physics of Image Formation and Microanalysis
By L. Reimer 2nd Edition

46 **Holography and Deformation Analysis**
By W. Schumann, J.-P. Zürcher, and D. Cuche

47 **Tunable Solid State Lasers**
Editors: P. Hammerling, A. B. Budgor, and A. Pinto

48 **Integrated Optics**
Editors: H. P. Nolting and R. Ulrich

49 **Laser Spectroscopy VII**
Editors: T. W. Hänsch and Y. R. Shen

50 **Laser-Induced Dynamic Gratings**
By H. J. Eichler, P. Günter, and D. W. Pohl

51 **Tunable Solid State Lasers for Remote Sensing** Editors: R. L. Byer, E. K. Gustafson, and R. Trebino

52 **Tunable Solid-State Lasers II**
Editors: A. B. Budgor, L. Esterowitz, and L. G. DeShazer

53 **The CO_2 Laser** By W. J. Witteman

54 **Lasers, Spectroscopy and New Ideas**
A Tribute to Arthur L. Schawlow
Editors: W. M. Yen and M. D. Levenson

55 **Laser Spectroscopy VIII**
Editors: W. Persson and S. Svanberg

56 **X-Ray Microscopy II**
Editors: D. Sayre, M. Howells, J. Kirz, and H. Rarback

57 **Single-Mode Fibers** Fundamentals
By E.-G. Neumann

58 **Photoacoustic and Photothermal Phenomena**
Editors: P. Hess and J. Pelzl

59 **Photorefractive Crystals in Coherent Optical Systems**
By M. P. Petrov, S. I. Stepanov, and A. V. Khomenko

60 **Holographic Interferometry in Experimental Mechanics**
By Yu. I. Ostrovsky, V. P. Shchepinov, and V. V. Yakovlev

61 **Millimetre and Submillimetre Wavelength Lasers** A Handbook of cw Measurements
By N. G. Douglas

62 **Photoacoustic and Photothermal Phenomena II**
Editors: J. C. Murphy, J. W. Maclachlan Spicer, L. C. Aamodt, and B. S. H. Royce

63 **Electron Energy Loss Spectrometers**
The Technology of High Performance
By H. Ibach

64 **Handbook of Nonlinear Optical Crystals**
By V. G . Dmitriev, G. G. Gurzadyan, and D. N. Nikogosyan 2nd Edition

65 **High-Power Dye Lasers**
Editor: F. J. Duarte

66 **Silver Halide Recording Materials for Holography and Their Processing**
By H. I. Bjelkhagen 2nd Edition

67 **X-Ray Microscopy III**
Editors: A. G. Michette, G. R. Morrison, and C. J. Buckley

68 **Holographic Interferometry**
Principles and Methods
Editor: P. K. Rastogi

69 **Photoacoustic and Photothermal Phenomena III**
Editor: D. Bićanic

70 **Electron Holography**
By A. Tonomura 2nd Edition

71 **Energy-Filtering Transmission Electron Microscopy**
Editor: L. Reimer

Volumes 1-41 are listed at the end of the book

Ludwig Reimer

Scanning Electron Microscopy

Physics of Image Formation and Microanalysis

Second Completely Revised and Updated Edition
With 260 Figures

Springer

Professor Dr. Ludwig Reimer

Westfälische Wilhelms-Universität Münster, Physikalisches Institut
Wilhelm-Klemm-Strasse 10, D-48149 Münster, Germany

Library of Congress Cataloging-in-Publication Data. Reimer, Ludwig, 1928– Scanning electron microscopy: physics of image formation and microanalysis / Ludwig Reimer. – 2nd completely rev. and updated ed. p. cm. – (Springer series in optical sciences; v. 45) Includes bibliographical references and index. ISBN 3-540-63976-4 (hardcover; alk. paper) 1. Scanning electron microscopy. I. Title. II. Series. QH212.S3R452 1998 502'.8'25–dc21 98-26178

ISSN 0342-4111
ISBN 3-540-63976-4 2nd Edition Springer-Verlag Berlin Heidelberg New York

ISBN 3-540-13530-8 1st Edition Springer-Verlag Berlin Heidelberg New York

Typesetting: Data conversion by Steingraeber Satztechnik GmbH, Heidelberg
Cover concept by eStudio Calamar Steinen using a background picture from The Optics Project. Courtesy of John T. Foley, Professor, Department of Physics and Astronomy, Mississippi State University, USA
Cover production: *design & production* GmbH, Heidelberg

SPIN 10661468 54/3144-5 4 3 2 1 0 - Printed on acid-free paper

Preface

The aims of this book described in the Preface to the First Edition of 1985 have not changed. In the meantime considerable progress has been made in instrumentation and in imaging and analysing methods, and these developments are considered in this second edition. New aspects of conventional modes of operation mode are also included and the extensive reference list has been brought up to date.

The use of Schottky or field-emitter as an electron source is now standard practice. The resolution can be improved to the order of 1–2 nm especially when using the in-lens mode with the specimen introduced by a side-entry stage between the polepieces of the last probe-forming lens.

It is now possible to work with acceleration voltages below 1 kV, thanks to the development of suitable electron optical configurations. Whereas the conventional SEM worked mainly with electron energies of 5–30 keV, the low-voltage scanning electron microscopy (LVSEM) has expanded the energy range to 0.5–5 keV. Further information about the backscattering and secondary electron emission at low energies is needed to interpret the resulting image contrast. Both the electronics and the image recording of modern SEMs are fully digitalized.

Though the Everhart-Thornley detector for secondary electrons and scintillation or semiconductor detectors for backscattered electrons are still the most important detection systems, the introduction of single-crystal YAG and other materials as efficient scintillators has stimulated the development of new detector designs. With the use of microchannel plates in LVSEM a new detector system without an energy threshold has been introduced.

Equipment for the investigation of hydrated specimens by environmental scanning electron microscopy (ESEM) is now offered by nearly all the manufacturers. The investigation of surface magnetization by spin-polarized secondary electrons has been further developed. Owing to the increasing interest of electron-beam lithography and metrology for the measurement of critical dimensions of integrated circuits short sections on the physical problems of these techniques have been added.

The investigation of crystal structure and orientation by electron backscattering patterns (EBSP) can now be effectively applied with the result that the method of electron channelling has become less important, though still of interest for some applications.

In x-ray microanalysis, the development of Si(Li) energy-dispersive x-ray detectors with ultrathin windows has made possible the routine analysis of elements with characteristic lines corresponding to quantum energies of 0.2–1 keV.

A special acknowledgement is again due to P.W. Hawkes for revising the English text.

Münster, June 1998 *L. Reimer*

Preface to the First Edition

The aim of this book is to outline the physics of image formation, electron-specimen interactions, imaging modes, the interpretation of micrographs and the use of quantitative modes in scanning electron microscopy (SEM). It forms a counterpart to *Transmission Electron Microscopy* (Vol. 36 of this Springer Series in Optical Sciences). The book evolved from lectures delivered at the University of Münster and from a German text entitled *Raster-Elektronenmikroskopie* (Springer-Verlag), published in collaboration with my colleague Gerhard Pfefferkorn.

In the introductory chapter, the principles of the SEM and of electron-specimen interactions are described, the most important imaging modes and their associated contrast are summarized, and general aspects of elemental analysis by x-ray and Auger electron emission are discussed.

The electron gun and electron optics are discussed in Chap. 2 in order to show how an electron probe of small diameter can be formed, how the electron beam can be blanked at high frequencies for time-resolving experiments and what problems have to be taken into account when focsusing.

In Chap. 3, elastic and inelastic scattering is discussed in detail. Usually, the elastic scattering is described as Rutherford scattering at a screened nucleus but we show that Mott's scattering theory is more correct and that large differences can occur; this has an effect on SEM results in which elastic scattering is involved. The inelastic scattering causes deceleration of the electrons and we present the reasoning behind the Bethe continuous-slowing-down approximation in detail because it is the most important model used in diffusion models and in the discussion of specimen damage processes. In Chap. 4, the experimental findings and theoretical models concerning backscattered and secondary electrons and x-ray and Auger electron emission are outlined.

Chapter 5 describes the detector systems employed for secondary and backscattered electrons and their signal-to-noise ratios, electron spectrometers for energy filtering, x-ray spectrometers and light collection systems for cathodoluminescence. The chapter ends with a discussion of the problems of image recording and of analogue and digital image processing.

Chapter 6 presents the typical types of contrast that can be created with secondary and backscattered electrons and demonstrates how the physics of electron-specimen interactions can be used together with an improved detector strategy to make SEM more quantitative. In Chap. 7, the electron-beam-induced current (EBIC) mode for semiconductors and its capacity for measuring semiconductor and device parameters are outlined. A section follows on cathodoluminescence and various other modes, of which the thermal-acoustic mode is likely to attract considerable interest.

The determination of crystal structure and orientation from electron channelling patterns (ECP), electron backscattering pattern (EBSP) and x-ray Kossel patterns is described in Chap. 8, which opens with a brief review of the kinematical and dynamical theories of electron diffraction and of Bloch waves as these are essential for an understanding of these diffraction effects at solid specimens. The final chapter evaluates the correction procedures used to give quantitative information about elemental concentrations by x-ray microanalysis and discusses the problems that can arise when analysing tilted specimens, specimen coatings, particles on substrates and biological tissues. It ends with a survey of different types of x-ray imaging modes.

Electron microscopists who first use a TEM are much more conscious of the problems of image interpretation because the difference between electron and light microscopy is obvious. A SEM produces such beautiful images, which are in some respects comparable to illumination with light, that many SEM users do not give much thought to the origin of the contrast. However, when the contrast needs to be discussed in more detail and the SEM signals are to be used more quantitatively, it becomes necessary to know more about the physics of SEM. It is the aim of this book to provide this knowledge together with ample references which cannot, however, be complete.

Just as for the book about transmission electron microscopy, a special acknowledgement is due to P.W. Hawkes for revising the English text and to K. Brinkmann and Mrs. R. Dingerdissen for preparing the figures.

Münster, November 1984 *L. Reimer*

Contents

1. Introduction

1.1 Principle of the Scanning Electron Microscope

Unlike transmission electron microscopy (TEM) described in the counterpart to this book [1.1], scanning electron microscopy (SEM) can image and analyse bulk specimens [1.2–28]. The principle of SEM is shown in Fig. 1.1. Electrons from a thermionic, Schottky or field-emission cathode are accelerated through a voltage difference between cathode and anode that may be as low as 0.1 keV or as high as 50 keV. The range between 0.1 and 5 keV is called low-voltage SEM (LVSEM). The smallest beam cross-section at the gun – the crossover – with a diameter of the order of 10–50 μm for thermionic emission or the virtual source, with a diameter of 10–100 nm for Schottky or field-emision guns, is demagnified by a two- or three-stage electron lens system, so that an electron probe of diameter 1–10 nm carrying an electron-probe current of 10^{-9}–10^{-12} A is formed at the specimen surface. For modes of operation that need a higher electron-probe current of $\simeq 10^{-8}$ A, the probe diameter increases to $\simeq 0.1$ μm.

When the distance between specimen and lower polepiece, the working distance, of the final probe-forming lens is relatively long, a few millimetres or centimetres for example, the various electrons and quanta emitted can be recorded by detectors inside the specimen chamber and the magnetic field at the specimen is very low. This mode of operation of the probe-forming lens increases the spherical and chromatic aberration and, therefore, the smallest obtainable electron-probe size. Alternatively, the lens excitation can be increased and the specimen placed within the lens, since the aberrations and hence the electron-probe diameter are then smaller. In this case, backscattered and secondary electrons move on spiral trajectories and have to be detected outside the lens field. A disadvantage of the in-lens mode is the small specimen size whereas the conventional mode with long working distances allows us to investigate very large specimens, some centimetres in size for normal specimen stubs and still larger in specially designed specimen chambers.

The electron-probe aperture, that is, the semi-apex angle of the convergent cone of electron trajectories is small. Apertures of the order of ten milliradians are used for routine work and high resolution. Apertures one to two orders of magnitude smaller are necessary to increase the depth of focus and

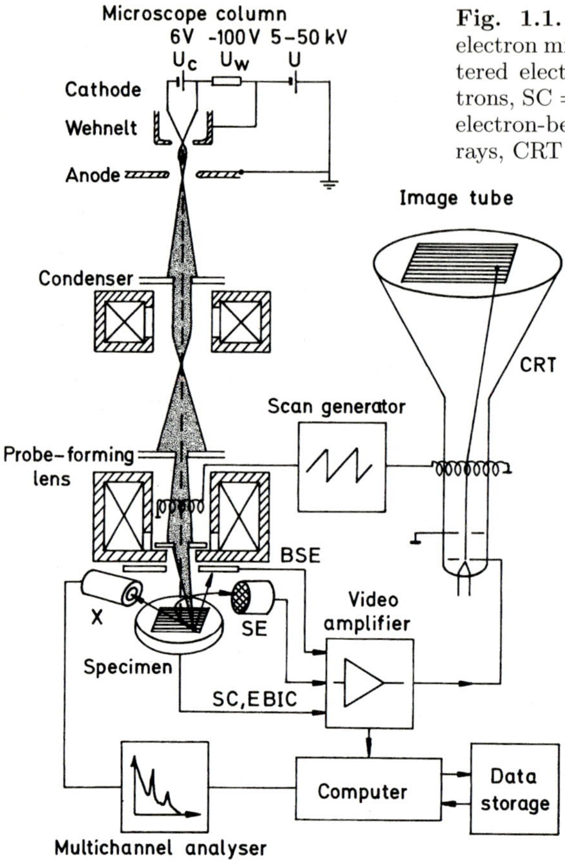

Fig. 1.1. Principle of the scanning electron microscopoe (BSE = backscattered electrons, SE = secondary electrons, SC = specimen current, EBIC = electron-beam-induced current, X = x-rays, CRT = cathode-ray tube)

to improve the angular resolution in electron channelling patterns. Owing to the smallness of the aperture, the depth of focus is much larger than in light microscopy. Specimens with large variations in depth can hence be sharply imaged, even at the lowest magnification of 20–50 times (Fig. 1.2).

Electron-probe current, aperture and probe diameter can all be varied but not independently by changing the excitations of the first condenser lenses and the aperture-limiting diaphragm in the final lens.

A deflection coil system in front of the final lens field scans the electron probe in a raster across the specimen in synchronism with the electron beam of a separate cathode-ray tube (CRT). The intensity of the CRT is modulated by one of the signals recorded (Sect. 1.3) to form an image. It is a mayor advantage of SEM, that, beside the backscattered and secondary electrons, a large variety of electron-specimen interactions can be used to form an image and to furnish qualitative and quantitative information. The magnification can be increased simply by decreasing the scan-coil current and keeping the

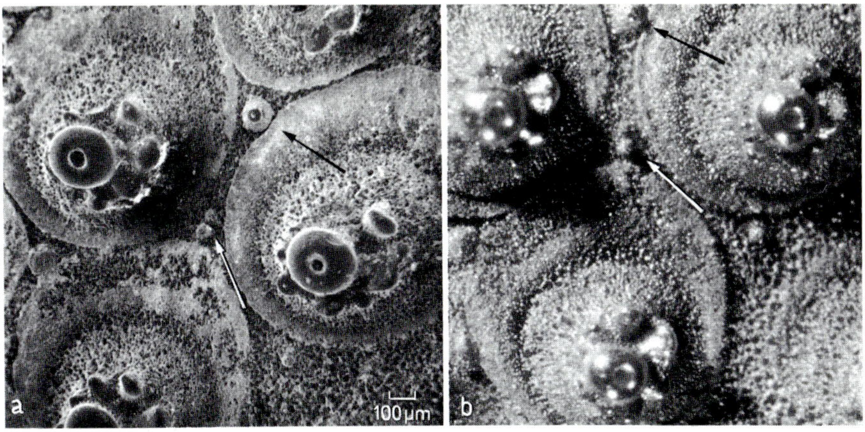

Fig. 1.2. Comparison of (**a**) a SEM micrograph (SE mode) and (**b**) a light-optical image of the same specimen area (see arrows) of a sea-urchin surface

image size, about $10 \times 10\ \mathrm{cm}^2$ and larger on the CRT, constant. Figure 1.3 shows a series of images with increasing magnification.

Other beam-deflection modes involve rocking of the electron beam when the electron probe is at rest and the angle of incidence is raster-scanned to form an electron channelling pattern for crystal analysis; a periodic change of the angle of incidence for recording stereo images at TV frequencies; and periodic blanking or chopping of the beam up to frequencies in the GHz region for stroboscopic modes and time-resolved signals.

1.2 Electron-Specimen Interactions

Elastic and inelastic scattering are the elementary atomic interaction processes, though the final signal used for image formation is, with only a few exceptions, not the result of single scattering processes but of the complete electron diffusion caused by the gradual loss of the electron energy and by lateral spreading caused by multiple elastic large-angle scattering. The consequence of the gradual diminution of electron energy is that the electrons have a finite range R of the order of 10 nm–10 μm depending on electron energy and target density, but the information depth and the lateral extension of the information volume that contribute to each of the possible signals depend on where the corresponding interaction takes place. The information and interaction volumes are not sharply bounded and in some cases, the contribution to the signal decreases exponentially with increasing depth.

Figure 1.4 shows schematically the most important interaction processes and their information volumes. The energy spectrum of the electrons emitted

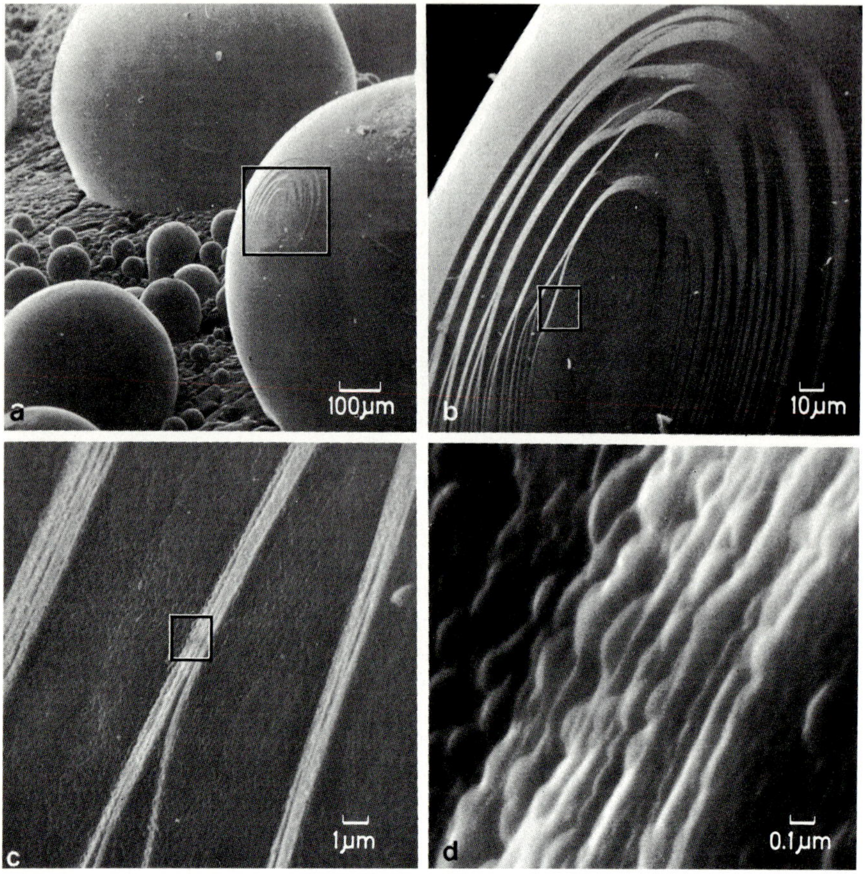

Fig. 1.3. Example of a series of increasing magnification (spherical lead particles imaged in the SE mode)

(Fig. 1.5) consists of contributions from secondary electrons (SE), backscattered electrons (BSE) and Auger electrons (AE). The SE are represented by a peak at low energies with a most probable energy of 2–5 eV. By convention, the limit between SE and BSE is drawn at 50 eV. The SE are generated by inelastic collisions to such high energy levels that the excited electrons can overcome the work function before being decelerated to the Fermi level. The broad spectrum of BSE between 50 eV and the primary electron energy $E = eU$ is caused by the deceleration of electrons that have suffered multiple energy losses and undergone multiple scattering through large angles. Auger electron production is an alternative to characteristic x-ray emission after ionization of an inner shell. The de-excitation energy released when an electron from an upper shell fills the vacancy in the ionized shell can be converted to an x-ray quantum of energy $h\nu = E_2 - E_1$ or the energy may be transferred to another atomic electron, which leaves the specimen as an Auger electron

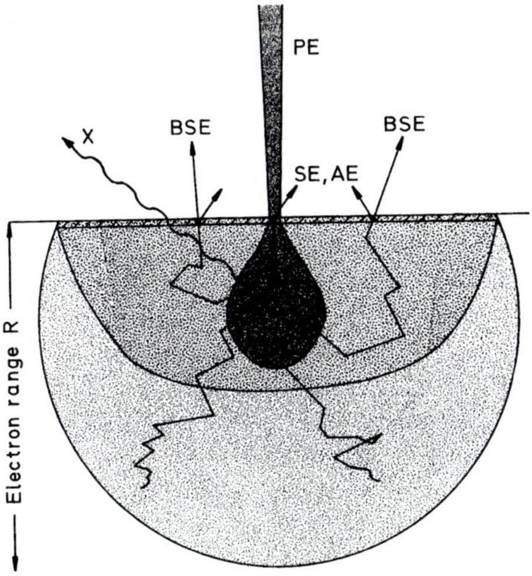

Fig. 1.4. Origin and information depth of secondary electrons (SE), backscattered electrons (BSE), Auger electrons (AE) and x-ray quanta (X) in the diffusion cloud of electron range R for normal incidence of the primary electrons (PE)

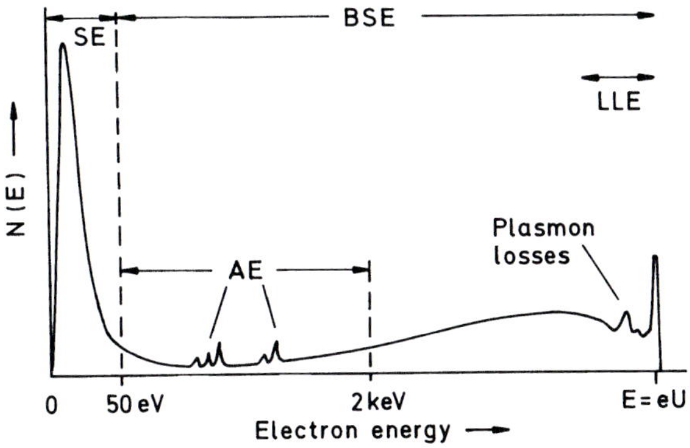

Fig. 1.5. Schematic energy spectrum of electrons emitted consisting of secondary electrons (SE) with $E_{\mathrm{SE}} \leq 50$ eV, low-loss electrons (LLE) with energy losses of a few hundreds of eV, backscattered electrons (BSE) with $E_{\mathrm{BSE}} > 50$ eV and peaks of Auger electrons (AE)

with a characteristic kinetic energy determined by $E_2 - E_1$ less the energy necessary to overcome the ionization energy and the work function.

Secondary and Auger electrons are highly susceptible to elastic and inelastic scattering and can leave the specimen only from a very thin surface layer of the order of a few nanometres thick. They are, however, generated not only by the narrow beam of primary electrons entering this thin surface layer but also by backscattered electrons on their way back through a larger region of the surface.

The most probable energy of the BSE falls in the broad part of the spectrum (Fig. 1.5) but they also show a more or less pronounced elastic peak followed by plasmon losses that depend on the primary energy, the take-off angle and the tilt of the specimen. The continuous-slowing-down approximation assumes that the mean electron energy decreases smoothly with increasing path length of the electron trajectories inside the specimen, though straggling of electron energies also has to be taken into account. In general, electrons with longer path lengths and larger energy losses return to the surface from greater depths. The maximum information depth of BSE is therefore of the order of half the electron range. However, electrons from the elastic peak inside an energy window some 10–100 eV below the primary energy – so-called low-loss electrons (LLE) – can leave the specimen only from a thin surface layer, comparable in extent with that of secondary and Auger electrons. Because BSE are frequently detected by semiconductor or scintillation detectors, which give a signal proportional to the electron energy, BSE with higher energies will make a larger contribution. Thus the energy response of detectors and electron spectrometers has to be taken into account when estimating the information depth.

Characteristic x-rays will only be excited in the volume in which the electron energy exceeds the ionization energy of the inner shell involved. However, x-rays excited in atoms of element 'a' can be absorbed by atoms of element 'b' if the energy of the characteristic quanta of element 'a' is larger than the ionization energy of the atoms of 'b'. The consequence of such absorption is secondary x-ray emission or x-ray fluorescence. The information volume of x-ray fluorescence depends on the absorption of both the primary and the secondary radiation and can exceed the electron range.

Inelastic scattering in semiconductors results in the generation of electron-hole pairs. Charge separation by an applied voltage or, more effectively, by the electric field of a depletion layer results in a charge-collection or electron-beam-induced current (EBIC). The carriers are generated throughout the diffusion cloud. However, the minority carriers, which are responsible for the charge-collection current, can diffuse and their concentration n decreases exponentially with increasing distance x, $n \propto \exp(-x/L)$, L being the diffusion length, which may be of the order of 1–100 μm.

The recombination of electron-hole pairs in an intrinsic semiconductor or in an extrinsic semiconductor via the energy levels of dopant atoms can take place without radiation but may result in the emission of light quanta, an excitation that is known as cathodoluminescence (CL) and can also be excited in some organic fluorescent compounds.

The largest fraction of the primary electron energy that is lost during the cascade of inelastic scattering processes is converted into phonons or heat. In most cases, the thermal conductivity of the material is so high that specimen heating is not a serious problem. Irradiation with a periodically pulsed electron beam creates strongly damped thermal waves, which can generate acoustical waves and these can be picked up by a piezoelectric transducer, for example, to form an image signal.

These examples show that the various electron-specimen interactions generate a great deal of information in the form of emitted particles and quanta and of internally produced signals, all of which can be exploited in the different imaging and operating modes of a SEM.

1.3 Imaging Modes of Scanning Electron Microscopy

1.3.1 Secondary Electrons (SE)

The SE mode is the most important because these electrons can be collected easily by means of a positively biased collector grid placed on one side of the specimen thanks to their low exit energy of a few electronvolts. Behind the collector grid, the SE are accelerated onto a scintillator biased at +10 kV and the light quanta generated are recorded by a photomultiplier. Such an Everhart–Thornley detector is an efficient amplifier, introducing little noise and having a large bandwidth of the order of 10 MHz. In strongly excited electron lenses, the SE move in screw trajectories through the lens field along the lines of the magnetic induction $\boldsymbol{B}$ and can then be collected outside the lens by a scintillation detector.

The dependence of the SE yield on the tilt angle of a surface element, the enhanced emission at edges and small particles and the shadow contrast that results from incomplete collection can all be used to image the surface topography (Figs. 6.2a, 3a, 7a and 11b–c). The low exit depth, of the order of a few nanometres, of the fraction that is produced by the primary beam of small probe size (of the order of 20–50%) allows a resolution of the order of 1–10 nm to be reached. The other fraction of the SE emitted is excited by the backscattered electrons, which have passed through an information volume with a radius of the order of half the electron range (Fig 1.4). Typical BSE contrast is therefore superimposed on every SE micrograph and techniques for separating these two contributions by using the signals from a multidetector system have hence been developed.

The SE are retarded by a positive bias and repelled by a negative bias of the specimen and are influenced by the electrostatic field between regions at different biases. These effects generate voltage contrast, negatively biased areas appearing bright and positively biased regions, dark (Fig. 8.5). When the influence of the electrostatic field is suppressed by pre-acceleration of the SE in a field of a few hundreds volts per millimetre at the surface, the local potential can be measured with high geometrical and voltage resolution by means of an electron spectrometer; this has become an important application in the technology of integrated circuits. Due to the low exit energy, the SE trajectories are also affected by the magnetic stray fields of ferromagnetic domains, recording heads and tapes, and angular selection then creates type-1 magnetic contrast (Fig. 8.2).

1.3.2 Backscattered Electrons (BSE)

Unlike SE, BSE move on straight trajectories and are not affected by electrostatic collection fields, with the result that the detectors have to be mounted with a large solid angle of collection. Scintillators, semiconductors and channel plates are all in use. Since the BSE emission also depends on the surface tilt, the surface topography can be imaged at lower magnifications with a better shadow effect than with SE, owing to the straight trajectories (Figs. 6.2b and 3b) and at higher magnification with a worse resolution, due to the larger information volume and exit area (Fig. 1.4). However the resolution of the BSE image can be increased by filtering the BSE with a spectrometer and using only the low-loss electrons (LLE in Fig. 1.5), which have an exit depth of only a few nanometres. At electron energies below 1 keV, the information volumes of SE and BSE are of much the same order of magnitude.

The most important contrast mechanism of BSE is the dependence of the backscattering coefficient on the mean atomic number $\overline{Z}$, which allows phases with different values of Z to be recognized (Fig. 6.9b). With a detector below the polepiece and high take-off angles, it is mainly this material contrast that is observed and the topographic contrast is suppressed. The SE signal also contains this type of material contrast (Fig 6.9a), thanks to the fraction of the SE signal excited by the BSE.

The backscattering coefficient also depends on the relative orientation of the incident electron beam and the lattice planes. A plane incident electron wave propagates inside the crystal as a Bloch-wave field, which shows anomalous absorption or transmission depending on the angle of incidence relative to the lattice planes. The resulting orientation anisotropy of the backscattering coefficient is very sensitive to small crystal tilts and causes differences in the backscattering coefficient of the order of 1–10%. This allows differently oriented grains in polycrystalline material to be imaged by crystal orientation or channelling contrast (Fig. 6.15). After selecting the LLE by energy filtering, it is possible to observe single lattice defects, dislocations and stacking faults, because the channelling effect contributes to the contrast only in a thin surface layer, the thickness of which is of the order of the absorption length of the primary Bloch-wave field.

Ferromagnetics, which have a high internal magnetic induction $\boldsymbol{B}$, may affect the BSE trajectories inside the specimen with the result that the backscattering coefficient differs by a few tenths of a percent for different orientations of $\boldsymbol{B}$. This allows ferromagnetic domains to be imaged with type-2 magnetic contrast (Fig. 8.3c).

1.3.3 Specimen Current

The specimen current is smaller than the electron-probe current owing to the emission of SE and BSE. When the SE emission is suppressed by a positive bias, the modulation of the specimen current will be complementary to that of the backscattering coefficient. In some cases, this mode can be advantageous. It can easily be put into practice because no additional detector is necessary. However, the resulting low specimen currents cannot be directly amplified with a large bandwidth, so that large probe currents and/or long recording times will be necessary.

1.3.4 Transmitted Electrons

If the specimen is transparent to the incident electrons, the transmitted electrons can easily be recorded by a scintillation or semiconductor detector situated below the specimen support. A specimen that shows sufficient transmission to 10–50 keV electrons will invariably be an excellent specimen for a 100–300 keV TEM, where better resolution is obtainable. The recording of transmitted electrons is merely an additional possibility in SEM of minor interest, useful in a few specialized applications or for comparison with other SEM modes.

1.3.5 Electron-Beam-Induced Current (EBIC)

In semiconductors, a few thousands of electron-hole pairs are created per incident electron. In the depletion layer of a p-n junction, the strong electric field separates the charge carriers and minority carriers can hence reach the junction by diffusion. This results in a charge-collection current or electron-beam-induced current, which can be amplified and used in a quantitative manner to measure the width of the junction and its depth below the surface, the diffusion length and the surface recombination rate of minority carriers, for example. The EBIC signal can also be used to image p-n junctions below the surface (Fig. 7.7), to localize avalanche breakdowns and to image active lattice defects, which influence the recombination rate of minority carriers. The use of Schottky barriers, formed by evaporating a metal layer that forms a non-ohmic contact, allows semiconductor parameters and lattice defects in materials that do not contain a p-n junction to be investigated.

1.3.6 Cathodoluminescence

The emission of ultraviolet, visible or infrared light stimulated by electron bombardment (cathodoluminescence, CL) is exploited in the scintillation detector and the cathode-ray tube but can also be used for imaging. A great many substances, especially semiconductors and minerals but also some organic molecules, show CL, which can be picked up non-dispersively by a light-collecting system and a photomultiplier or dispersively by introducing a spectrometer between specimen and detector. The low intensity of the CL of some substances can become a problem and radiation damage to organic molecules leads to a fading of CL, which also occures in light-optical fluorescence microscopy excited by ultraviolet radiation. Sensitive detector systems for CL must, therefore, work with a large solid angle of collection and the subsequent detection system must be highly transparent. CL contains much analytical information and can reveal material differences that cannot be detected by other methods (Fig. 7.16). A disadvantage is that the dopants that cause CL and the physics of CL in a particular material are often unknown so that the information is only qualitative. However, CL is frequently used in combination with EBIC and the recombination of charge carriers at lattice defects also allows lattice defects to be imaged.

1.3.7 Acoustic Thermal-Wave Microscopy

The acoustic wave excited by periodic thermal expansion at 0.1–5 GHz chopping frequencies of the primary beam depends on the thermal expansion coefficient, on the elastic constants and their anisotropies and on the acoustic wave propagation; the behaviour of this wave thus offers new information, not available from other SEM modes. The acoustic waves can be picked up by a piezoelectric transducer, for example.

1.3.8 Environmental Electron Microscopy

The high vacuum in the specimen chamber has the disadvantage that only dried or dehydrated specimens can be investigated. By increasing the partial pressure of water vapour a few millimetres in front of the specimen by means of a differentially pumped chamber separated by diaphragms from the vacuum of the microscope column, wet specimens and those containing volatile compounds can be investigated under nearly natural conditions. The higher gas pressure causes scattering of the primary electrons but a fraction still remains unscattered. The BSE can be recorded by a scintillator, whereas the SE cannot be collected because the scattering is excessive. However, a voltage of a few hundred volts can be applied to amplify the SE current by a cascade of impact ionizations as in a glow discharge. The generation of positive ions in the gas prevents charging of the specimen.

1.3.9 Imaging with X-Rays

Though x-rays are mainly used for elemental analysis, they can also be used in imaging modes. In x-ray elemental mapping, the CRT displays a bright spot for each characteristic x-ray quantum recorded within a preselected energy window of a multichannel analyser during the scanning of the frame (Fig. 10.40). The concentration of dots is a measure of the concentration of the corresponding elements. Concentration profiles can be recorded by generating a linescan of the x-ray signal. X-ray images can also be produced with the non-dispersive total x-ray signal (TRIX, total rate imaging with x-rays), which comes from a larger information depth than BSE and has a different dependence on atomic number.

In x-ray projection microscopy, a point source of x-rays is produced by the electron beam of the SEM; this point source acts as a projection centre to form an enlarged x-ray absorption image of biological organisms or of thicker metal foils on a photographic emulsion. For single crystals, this can be used to form a topographic image by recording the Bragg-diffracted x-rays.

1.4 Analytical Modes

The term analytical mode is often used to describe all the methods that furnish quantitative information, usually with a stationary electron beam. In this sense, the EBIC mode of semiconductor material and the cathodoluminescent mode are also analytical modes when they are used to measure semiconductor and device parameters. Other analytical modes discussed in this section are crystal analysis by electron diffraction and elemental analysis by x-ray and Auger electron microanalysis.

1.4.1 Crystal Structure and Orientation

The orientation anisotropy of the BSE, which also results in crystal orientation or channelling contrast, can be recorded as an electron channelling pattern (ECP) on the CRT by rocking the primary electron beam in a raster on a single crystal (Figs. 9.23 and 24). This pattern consists of Kikuchi bands, which can be indexed with Miller indices and can be used to determine the crystal symmetry and orientation, the latter because the system of Kikuchi bands or Kossel cones is fixed to the crystal.

The angular exit characteristics of the BSE are also modulated by Kikuchi bands, which can be recorded as an electron backscattering pattern (EBSP) on a fluorescent screen or photographic emulsion inside the specimen chamber (Figs. 9.21 and 26). The advantages of employing this method are that the solid angle is larger ($\pm 25^\circ$) and the electron probe is stationary, whereas the ECP can only be generated with a rocking angle $\pm 5^\circ$ and the beam rocking causes a shift of the electron probe over an area of 1–10 μm in diameter because of the spherical aberration of the final lens.

Like the crystal orientation or channelling contrast, both diffraction patterns are caused by the exponentially damped Bloch-wave field of the incident or backscattered electrons in a thin surface layer only a few nanometres thick and are, therefore, very sensitive to contamination and mechanical distortion of the surface.

For large tilt angles of the specimen, close to 90°, the EBSP becomes a reflection high-energy electron diffraction (RHEED) pattern, which exhibits not only Kikuchi bands but also Bragg diffraction spots and Kikuchi lines.

The x-rays produced in a single crystal are also Bragg reflected and cause defect and excess Kossel cones, with the result that sharp Kossel lines are recorded on a photographic emulsion. Such a Kossel pattern (Fig. 9.31) can also be used to determine lattice constants with an accuracy of $\Delta a/a \simeq 10^{-4}$, which is high enough for local elastic strains to be measured. A disadvantage of this technique is that the characteristic x-rays excited show Bragg reflections only for a limited number of metals. However, characteristic x-rays can be excited in a thin film or particle of another element that does satisfy Bragg's law (pseudo-Kossel technique).

1.4.2 Elemental Microanalysis

Most SEMs are equipped with an energy-dispersive lithium-drifted silicon detector, which allows characteristic x-ray lines to be recorded with a resolution $\Delta E \simeq 100$–200 eV of the x-ray quantum energy. Wavelength-dispersive spectrometers making use of the Bragg reflection at a single crystal are mainly in use in x-ray microprobes, though a few SEMs are also equipped with this type of spectrometer, which has a better resolution, of the order of 5–20 eV, but a lower solid angle of collection. Energy-dispersive spectrometers have the advantage that all quantum energies within 0.2–20 keV are recorded simultaneously and the spectrum is displayed directly by means of a multichannel analyser.

The correction procedures developed for x-ray microprobes and normal incidence can also be used for quantitative analysis in SEM by comparing the number of counts from an element in the specimen with the number of counts from a standard of known composition (normally a pure element). For such a correction to be valid, the specimen must be flat within the diameter of the diffusion cloud and of homogeneous composition. Quantitative analysis becomes more uncertain for inhomogeneous specimens. In particular, additional information from other modes or from the x-ray continuum will be necessary for the quantitative analysis of particles and films on substrates or for biological tissues, for example.

Auger electrons have the advantage that they are excited with a high yield for low-Z elements, for which the x-ray fluoresence yield of characteristic x-ray quanta is low. However, the Auger electrons can leave the specimen without energy loss only from a very thin surface film, of the order of a few monolayers thick and an electron beam of a few keV is needed for optimum excitation. Auger electron spectroscopy is therefore mainly used in specially designed ultrahigh-vacuum Auger electron microprobes. The study of Auger electrons inside the SEM can only be satisfactory when contamination of the specimen is prevented.

2. Electron Optics of a Scanning Electron Microscope

Electrons emitted from thermionic, Schottky or field-emission cathodes are accelerated by a voltage of 0.1–50 keV between cathode and anode. The purpose of the electron optics of a SEM is to produce a small electron probe at the specimen by demagnifying the smallest virtual cross-section of the electron beam near the cathode. For the practical operation of a SEM, it must be possible to vary the electron-probe size, aperture and current; these cannot, however, be varied independently because they are related via the gun brightness. A geometric optical theory of electron-probe formation can be employed when using a thermionic cathode but for a field-emission gun a wave-optical theory is necessary. Electron-beam deflection by transverse electrostatic and magnetic fields is incorporated for scanning the electron probe across the specimen, for tilting the direction of the incident electron beam for stereo-viewing and for recording electron channelling patterns. Deflection systems are further used for blanking and chopping the electron beam up to gigahertz frequencies for stroboscopic modes and for generating time-resolved signals. Owing to the large depth of focus, focusing of SEM images raises no problems but the resolution is limited by the electron-probe size, which decreases with decreasing spherical (C_s) and chromatic (C_c) aberration coefficients of the final probe-forming lens. Though a large working distance between specimen and lower polepiece has the advantage that free space is available for the SE and BSE detectors, a stronger lens excitation and an in-lens position of the specimen decreases the aberration coefficients by one order of magnitude.

2.1 Electron Guns

2.1.1 Thermionic Electron Guns

In thermionic emission, electrons from the Fermi level E_F of the cathode material can overcome the work function ϕ_w by thermionic excitation (Fig. 2.1). The acceleration of the electrons by the electric field $\boldsymbol{E}$ between cathode and anode can be represented by a potential energy $V = -|\boldsymbol{E}|z$. The emission current density j_c is described by the Richardson law

$$j_c = AT_c^2 \exp(-\phi_w/kT_c) \tag{2.1}$$

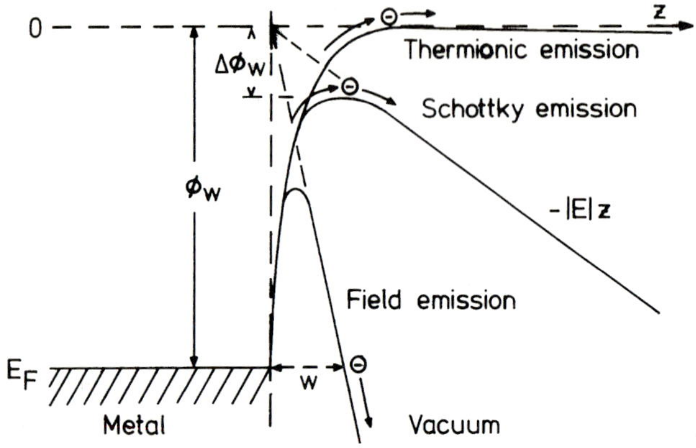

Fig. 2.1. Potential barrier (work function ϕ_w) at the metal–vacuum boundary and decrease of potential energy $V(z)$ with increasing external field $\boldsymbol{E}$ for thermionic, Schottky and field emission

where A is a constant depending on the cathode material, T_c is the temperature of the cathode tip and k is Boltzmann's constant. The current density becomes high enough only just below the melting point T_m, resulting in temperatures T_c = 2500–3000 K for tungsten (ϕ_w = 4.5 eV, T_m = 3650 K). At T_c = 2800 K, a practical value of j_c is of the order of 3 A cm^{-2} [2.1], for example. Lanthanum hexaboride (LaB_6) has a lower work function ϕ_w = 2.7 eV and LaB_6 or CeB_6 cathodes can work at T_c = 1400–2000 K with values of j_c of the order of 20–50 A cm^{-2}.

Tungsten thermionic cathodes are prepared either as a wire bent into a hair-pin or as an indirectly heated tungsten tip (pointed cathode) [2.2] to decrease the emission area. LaB_6 rods with a polished tip are heated indirectly, by squeezing them between carbon electrodes, which are pressed together by a spring of Mo-Re contacts, or by soldering them on refractory metal strips [2.3–9]. The emission from a LaB_6 cathode depends on the crystal orientation of the tip and is found to be ten times higher for a (100)-oriented tip than for a (510)-oriented one [2.10]. Many contradictory results concerning the emissive properties of LaB_6 cathodes can be attributed to differences in the surface structure and the shape of the tip. The cathode has a full cone angle of 90° and its tip is either hemispherical or flat with a radius of 10–200 μm. Below 15 μm the gun brightness is increased but the lifetime is reduced. Further details of LaB_6 cathodes are to be found in [2.11–13].

The lifetime of tungsten cathodes (40–200 h) is limited by evaporation of the cathode material, which results in a break when part of the wire becomes too thin. A vacuum of (1–5)$\times 10^{-3}$ Pa is sufficient to prevent oxidation of the heated cathode (1 Pa = 1 Nm^{-2} = 1.33 $\times$ 10^{-2} Torr). A better vacuum is necessary to increase the lifetime of a pointed thermionic tungsten cathode. LaB_6 cathodes attain lifetimes of a few thousand hours or more but need a vacuum better than 10^{-4} Pa to prevent the formation of volatile oxidation

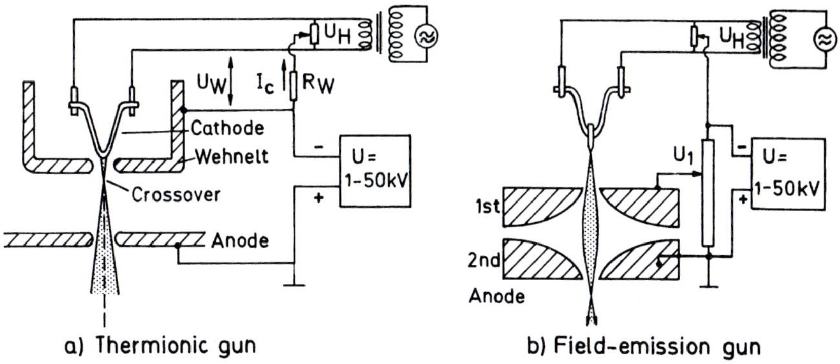

Fig. 2.2. (**a**) Thermionic electron gun consisting of cathode, Wehnelt cup and anode. The Wehnelt bias U_W is provided by the voltage drop of the emission current I_c across R_W. (**b**) Field-emission gun of the Butler type

products. This can be achieved by evacuating the space in front of the cathode with an additional getter ion pump.

The voltage $U = 1$–50 keV between anode and cathode (Fig. 2.2a) accelerates the electrons to a kinetic energy $E = eU$ measured in units of electronvolts (eV). 1 eV = 1.69×10^{-19} Nm is the kinetic energy gained on passing through a potential difference of 1 V. The acceleration results in a relativistic increase of electron mass

$$m = m_0(1 + E/E_0) \,, \tag{2.2}$$

where m_0 denotes the electron rest mass and $E_0 = m_0c^2 = 511$ keV is the rest energy. The dependence of the electron velocity on E is obtained by substituting $m = m_0(1 - v^2/c^2)^{-1/2}$ in (2.2) and solving for v:

$$v = c\left(1 - \frac{1}{(1 + E/E_0)^2}\right)^{1/2} \tag{2.3}$$

and the momentum becomes

$$p = mv = [2m_0E(1 + E/2E_0)]^{1/2} \tag{2.4}$$

which can be used to calculate the de Broglie wavelength $\lambda = h/mv$ (2.22). The influence of relativity on m, v and λ is small for electron energies below 50 keV. Strong relativistic effects can be seen , however, in the elastic scattering cross-sections (Sect. 3.1.3) and the angular characteristics of continuous x-ray quanta (Sect. 10.1.2) because the electrons can be temporarily accelerated in the Coulomb field of an atomic nucleus to much higher energies.

The electron emission is concentrated within a small area of the thermionic cathode tip by means of a negatively biased Wehnelt cup (Fig. 2.2a). This negative bias is generated by auto-biasing, a voltage drop $U_W = I_cR_W$ being created across a 1–10 MΩ resistance R_W in the connection line of the

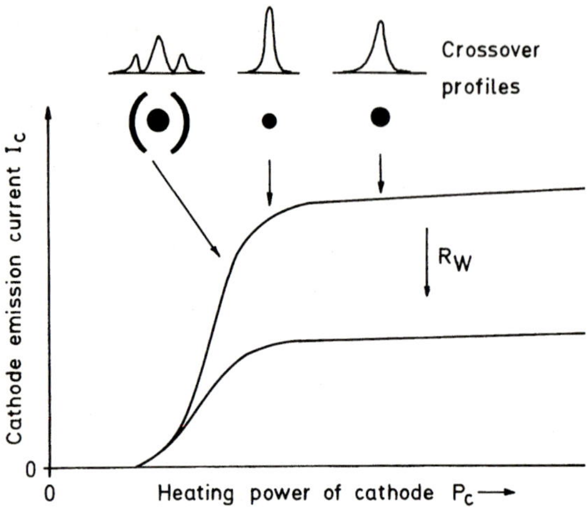

Fig. 2.3

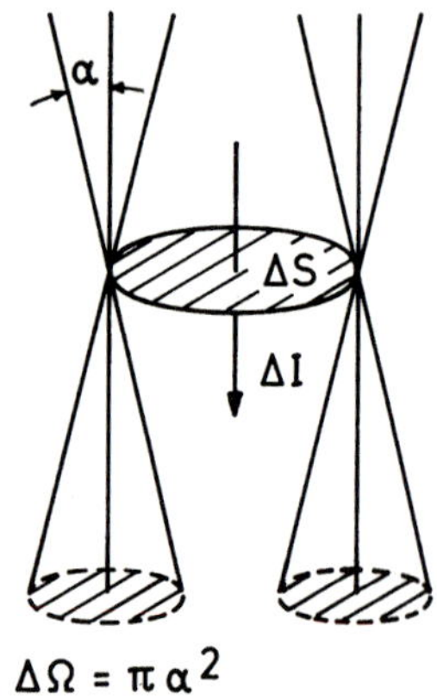

Fig. 2.4

Fig. 2.3. Cathode emission current I_c versus cathode heating power P_c for an auto-biased thermionic electron gun (Fig. 2.2a) and typical profiles of the crossover at three working points

Fig. 2.4. Definition of the gun brightness $\beta = j/\Delta\Omega$ where $j = \Delta I/\Delta S$ is the current density and α the aperture

high-tension supply to the cathode [2.1]. The relation between the resulting beam current I_c and the cathode heating power P_c is plotted in Fig. 2.3. Optimum operation, for which the smallest beam cross-section (crossover) reaches a minimum, occurs at the knee of the I_c vs. P_c diagram of Fig. 2.3 just before saturation sets in. The current density at the crossover shows an approximately Gaussian profile

$$j(r) = j_0 \exp[-(r/r_0)^2] \qquad (2.5)$$

with $r_0 \simeq$ 10–50 μm. Beyond the optimum operation point, an overheated cathode suffers from a reduced lifetime and the angular distribution and the crossover diameter increase with increasing heating power P_c. Below the knee, the cathode is underheated and emission from the outer zone of the cathode tip results in a hollow-beam profile.

Electrons leave the cathode with an initial exit momentum $\boldsymbol{p}$. The normal component p_n is such that the energy distribution is of the Maxwellian type; the energy spread (full width at half maximum) of the accelerated electron beam is of the order of $\Delta E = 2.5\,kT_c$, which is further increased by the Boersch effect [2.14–16] arising from space-charge oscillations near the crossover, so that finally thermionic tungsten cathodes show an energy spread ΔE = 1–3 eV while for LaB_6 cathodes, the value is ΔE = 0.5–2 eV.

Whereas the normal components p_n of the electron exit momentum cause the energy spread, the tangential components p_t result in an angular spread (beam divergence). An important characteristic of an electron gun is the axial gun brightness β [2.17, 18],which is defined to be the current ΔI passing through an area ΔS into a solid angle $\Delta\Omega = \pi\alpha^2$, where α is the beam aperture, the semiapex angle of the incident electron cone (Fig. 2.4):

$$\beta = \frac{\Delta I}{\Delta S \Delta\Omega} = \frac{j}{\pi\alpha^2} = \text{const} \tag{2.6}$$

and $j = \Delta I/\Delta S$ denotes the current density in A/cm^2. The gun brightness β remains constant for all points on the axis of an electron optical system from the cathode tip to the specimen regardless of diaphragms and electron lenses in the electron optics (see [1.1] for details). The theoretically attainable value of β is obtained if we insert the current density j_c at the cathode given by Richardson's law (2.1) and a mean value of $\langle\alpha^2\rangle$ after acceleration to an energy $E = eU$:

$$\beta_{\max} = \frac{j_c}{\pi}\left[1 + \frac{E}{kT_c}(1 + E/2E_0)\right] \simeq \frac{j_c E}{\pi k T_c} \, . \tag{2.7}$$

Typical values for tungsten thermionic cathodes at $E = 25$ keV and $T_c = 2700$ K are $\beta = (0.3\text{–}2)\times10^4$ A cm^{-2} sr^{-1}. Ten to twenty times larger values of β are attainable with LaB_6 cathodes.

Since the electron-probe current at the specimen can be written as $I_p = j_p\pi d_p^2/4$ where d_p denotes the probe diameter, (2.6) shows that the parameters I_p, d_p and α_p which describe the illumination conditions cannot be varied independently. For a fixed value of d_p and α_p, a large value of β will be necessary to get a large probe current I_p. The consequences of this limitation are discussed in Sect. 2.2.4.

The maximum gun brightness (2.7) is proportional to the electron energy E, which hinders the use of thermionic cathodes at low energies of the order of 1 keV in low-voltage SEM. Owing to the lower electron range, however, such low energies are of increasing practical interest and the necessary high gun brightness can be attained only by using Schottky or field-emission cathodes. Regardless of this advantage for LVSEM, these types of cathodes, which are described in the following Sections, will be the electron sources of the future for other electron-optical instruments as well.

2.1.2 Schottky-Emission Guns

When the field strength $\boldsymbol{E}$ at the cathode is increased, the work function

$$\phi_{\text{eff}} = \phi_w - \Delta\phi_w = \phi_w - e\sqrt{\frac{e|\boldsymbol{E}|}{4\pi\epsilon_0}} \, . \tag{2.8}$$

is decreased (Schottky effect, Fig. 2.1). This decrease can be neglected in normal thermionic cathodes. When using a Schottky cathode with a radius $r \leq 1$ μm at the tip, the electric field $|\boldsymbol{E}| \simeq U/r$ at the cathode is strongly increased and a field strength of the order of 10^6 V/cm results in a decrease $\Delta\phi_{\mathrm{w}} \simeq 0.4$ eV.

So-called Schottky emission cathodes are of the ZrO/W(100) type with a tip radius $r \simeq 0.1$–1 μm [2.19, 20]. The work function is lowered by the ZrO coating from $\phi_{\mathrm{w}} = 4.5$ eV (W) to 2.7 eV. This allows the electrons to overcome the work function at a temperature of 1800 K. The cathode is surrounded by a negatively biased suppressor electrode; the tip apex protudes $\simeq$0.3 mm. The electrons are extracted by applying a voltage of 4–8 kV to an extractor electrode; the field strength at the cathode is much higher than for thermionic cathodes but is still about ten times lower than in field-emission sources. This means that the field strength is not sufficient for the wave-mechanical tunneling effect. Although the potential barrier is lowered by $\Delta\phi_{\mathrm{w}}$ (Fig. 2.1), the electrons have to overcome the barrier by their thermal energy. It is therefore confusing to call this type of cathode a field-emission gun. The only common feature of these types of gun is that both are point sources and the Schottky emitter can better be characterized as a field-assisted thermionic emitter. In contrast to a thermionic cathode, this gun shows an energy spread $\Delta E \simeq 0.5$ eV not strongly increased by the Boersch effect; the emission current density $j_{\mathrm{c}} \simeq 500$ A/cm^2 is larger by two orders of magnitude and the gun brightness $\beta \simeq 10^8$ A/cm^2 sr. The size of the virtual source, in which trajectories extrapolated back behind the tip intersect, is $\simeq$ 15 nm, much smaller than the tip radius $r \simeq 0.5$–1 μm. Owing to these properties and the high gun brightness this type of cathode will come into widespread use.

2.1.3 Field-Emission Guns

Field emission from a tungsten tip of radius $r \simeq 0.1$ μm starts when the potential gradient (electric field strength) $|\boldsymbol{E}| \simeq U_1/r$ reaches values larger than 10^7 V/cm; U_1 denotes the potential difference between the cathode tip and the first extraction electrode. Such high fields decrease the width w of the potential wall in front of the cathode to a few nanometres (Fig. 2.1) so that electrons from the Fermi level E_{F} can penetrate the potential barrier by the wave-mechanical tunnelling effect. Field-emission guns (FEG) need an ultra-high vacuum better than 10^{-6} Pa because otherwise the tip radius is destroyed by ion bombardment from the residual gas.

A FEG requires two anodes (Fig. 2.2b). The first regulates the field strength at the tip by the potential difference U_1 and hence the emission current. The second anode accelerates the electrons to the final kinetic energy $E = eU$. A problem with field emission is to concentrate the emission within a small solid angle, so that only a small fraction of the emitted current hits the first anode, because this bombardment can produce ions which in turn damage the tip. FEGs therefore work with a selected crystallographic orientation of the tip. For the (100) orientation in particular, the emission is concentrated within a cone with a semi-apex angle of about 0.1 rad as a result of the dependence of the work function on surface orientation of the tip.

FEG can function at room temperature but often work at $T_c \simeq 1000$–1500 K to avoid gas adsorption at the tip. This lower cathode temperature results in a lower energy spread of the order of $\Delta E = 0.2$–0.3 eV for cold and 0.3–0.5 eV for heated FEG. Cold field emission guns have to be flashed in the evening to ensure the normal operation next morning [2.21]. If need be, fluctuations of the emission current can be compensated by dividing the image signals by a reference current falling on one of the diaphragms in the column [2.22].

The main advantages of FEG are the very substantially higher gun brightness, $\beta \simeq 10^8 - 10^9$ A/cm^{2}sr at $E = 20$ keV, and the smaller diameter of the virtual source of the order of 3–5 nm, so that only one demagnifying lens is needed to get electron-probe diameters below 1 nm. The lifetime of a field-emission cathode can be some years.

2.1.4 Measurement of Gun Parameters

The size and shape of the crossover depend on the height of the cathode tip in the Wehnelt bore and the choice of the working point at the knee of the I_c *vs.* P_c curve of Fig. 2.3. In a TEM, the crossover can be examined by magnifying the latter with the condenser and imaging lenses and this is a standard method of aligning an electron gun on the optic axis. This method is not directly applicable in a SEM, which is primarily designed to demagnify the crossover and normally has no viewing screen. A small magnification of the order of 2–3 at the specimen plane can be produced by weakly exciting only the first condenser lens and the current density in the crossover can then be recorded by scanning this image across a *point detector*. Such a detector may be a small diaphragm and the transmitted electrons then produce secondary electrons at an inclined metal plate [2.23] or are recorded by a Faraday cage or a semiconductor detector [2.24]. Another possibility is to look in the SE mode at an electrolytically sharpened tungsten tip, as used for field emitters, behind a negatively biased diaphragm [2.25]. This device works like a cathode and a Wehnelt cup and concentrates the area from which SE can be collected to 0.2 μm diameter. The source can thus be imaged in the SE mode by scanning the source image across the tip.

The angular emission of an electron gun, which is the emission pattern far from the crossover, can be recorded by scanning the pattern across the diaphragm of the first condenser lens by means of scan coils situated below the anode [2.11, 26]. The pattern is recorded with the usual detector and recording system of the SEM. A 10 μm diaphragm at a distance of 10 cm from the crossover gives an angular resolution of 0.1 mrad. Sharply pointed LaB_6 cathodes show emission patterns with several high-emission maxima (lobes), approximately 0.1 to 1 mrad in width, and one of these is selected to form the electron beam. This method can therefore make an important contribution when aligning a LaB_6 cathode.

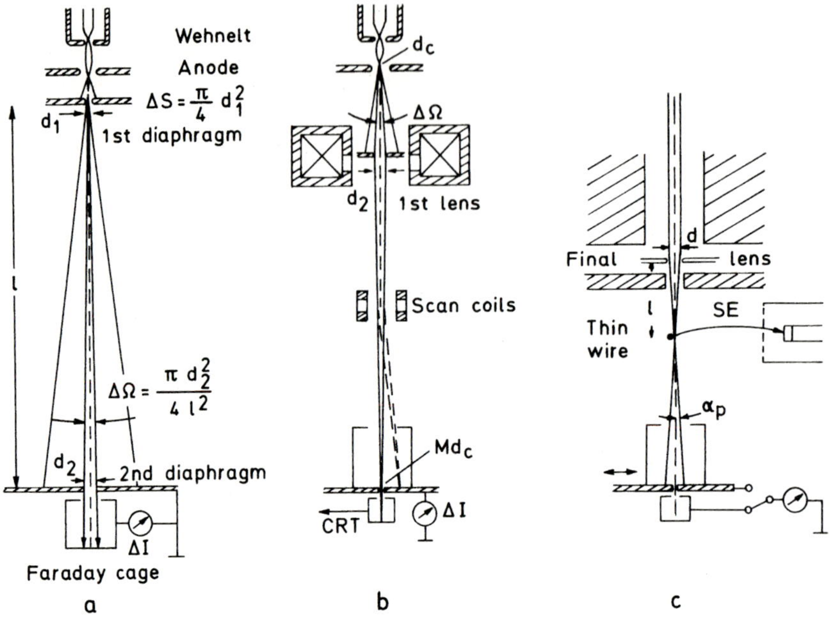

Fig. 2.5. Methods of measuring the gun brightness β by (**a**) the two-diaphragm method, (**b**) the crossover-projection method and (**c**) the electron-probe diameter method

It was shown in Sect. 2.1.1 and 2.1.2 that the gun brightness β (2.6) is an important parameter for characterizing an electron gun. For measurement of β, the quantities ΔI, ΔS and $\Delta\Omega$ have to be known. The following methods can be used in a SEM for measuring β [2.23, 26].

Two-Diaphragm Method. A first diaphragm of diameter d_1 placed just below the anode defines the area $\Delta S = \pi d_1^2/4$ and must be smaller than the diameter of the crossover (Fig. 2.5a). A second diaphragm of diameter d_2 at a distance l = 10–50 cm defines the solid angle $\Delta\Omega = \pi d_2^2/4l^2$. A Faraday cage behind the second diaphragm measures ΔI and

$$\beta = \frac{\Delta I}{\Delta S \Delta\Omega} = \left(\frac{4l}{\pi d_1 d_2}\right)^2 \Delta I \,. \tag{2.9}$$

A test of the accuracy of the measurement is that the resulting value of β should be independent of d_1, which will be the case only if this diaphragm is sufficiently small.

Crossover-Projection Method. The FWHM (full width at half maximum) d_c of the crossover profile is measured by one of the techniques described above and this value is used as d_1 (Fig. 2.5b). The aperture is defined by the lens diaphragm of diameter d_2 and its distance from the crossover. The magnification of the crossover at the point detector and the distance from the crossover to the lens can be determined by placing a grid above the first condenser lens and observing its projection image [2.26].

These two methods cannot be used for Schottky or field-emission guns because the source size is less a few nanometres.

Electron-Probe Method. The focused electron probe is scanned across the edge of a fine tungsten wire or a sharp edge (Fig. 2.5c). The electron-probe diameter can be read from the the line rise profile of the SE signal. The aperture of the electron probe is limited by the diaphragm of the probe forming lens but cannot be calculated exactly by writing $\alpha_p = d/2l$ since the electron trajectories behind the diaphragm may be curved. The aperture can be determined more accurately by scanning the electron beam across a small diaphragm, 5–10 cm below the focal point.

2.2 Electron Optics

2.2.1 Focal Length of an Electron Lens

An electron lens consists of an axial magnetic field with rotational symmetry. The magnetic flux of a coil is concentrated within a small volume by iron polepieces and the stray field at a gap of width S forms the magnetic field (Fig. 2.6), which has a bell-shaped distribution $B(z)$ on the optic axis. The radial component $B_r = -(r/2)\partial B_z/\partial z$ is related to the axial component B_z since div $\boldsymbol{B} = 0$ [1.1]. This relation between B_r and B_z for points near the optic axis explains why the spherical aberration of electron lenses is relatively large. The electrons travel along screw trajectories due to the Lorentz force

$$\boldsymbol{F} = -e(\boldsymbol{E} + v \times \boldsymbol{B}) \tag{2.10}$$

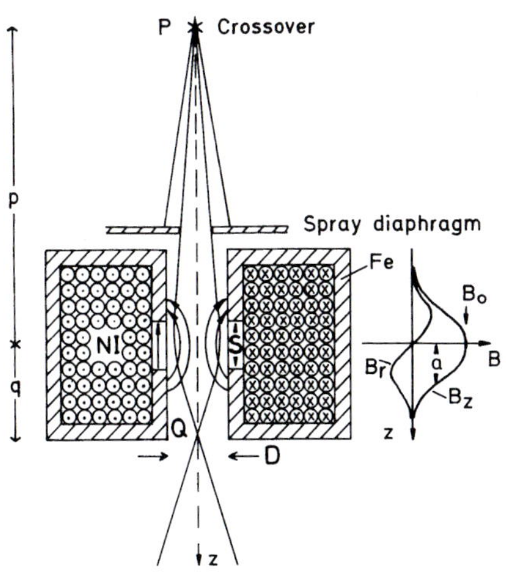

Fig. 2.6. Cross-section of a symmetric magnetic lens, demagnifying the crossover P to Q

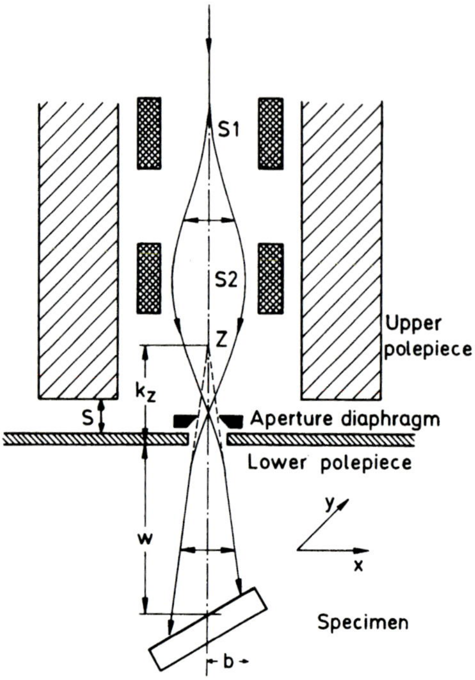

Fig. 2.7. Cross-section of an asymmetric objective lens (S = polepiece gap, S_1 and S_2 scan coils, w = working distance, k_z = position of virtual pivot point Z

and an electron beam diverging from P is focused at Q (Fig. 2.6). In the column of the SEM, the electric field $\boldsymbol{E}$ is zero because all conducting parts are at earth potential. The lenses used in the SEM are normally weak and the lens equation $1/f = 1/p + 1/q$ can be used. The diameter of the crossover at P is demagnified by the factor $M = q/p$. The focal length f is related to $B(z)$ by [2.27]

$$\frac{1}{f} = \frac{e}{8mU} \int_{-\infty}^{+\infty} B_z^2 \mathrm{d}z \,. \tag{2.11}$$

For the symmetric bell-shaped field component B_z of Fig. 2.6, with maximum B_0 and half-width $2a$, we find

$$f = \frac{2a}{\pi k^2} \quad \text{for} \quad k^2 \ll 1 \,, \tag{2.12}$$

where

$$k^2 = \frac{eB_0^2 a^2}{8mU} \tag{2.13}$$

denotes a dimensionless lens parameter, which characterizes the strength of a magnetic lens.

Because the Lorentz force (2.10) is perpendicular to $\boldsymbol{B}$ the electrons move on screw trajectories. Those incident in a meridional plane through the optic axis can be described by a rotation of the meridional plane. The total angle of rotation is found to be

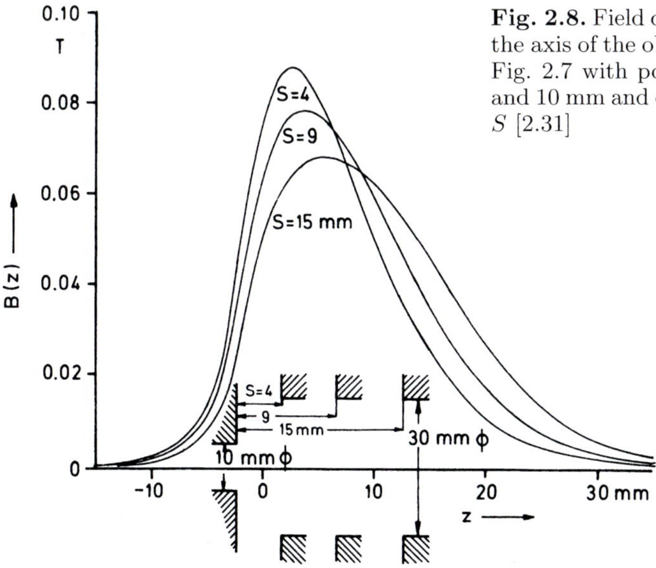

Fig. 2.8. Field distribution $B(z)$ on the axis of the objective lens of type Fig. 2.7 with polepiece bores of 30 and 10 mm and different gap widths S [2.31]

$$\varphi = \sqrt{\frac{e}{8m_0U}} \int_{-\infty}^{+\infty} B_z \mathrm{d}z \; , \tag{2.14}$$

which becomes

$$\varphi = \pi k \quad \text{for} \quad k^2 \ll 1 \tag{2.15}$$

and can be observed as a rotation of the scan direction; the resulting rotation of the image on the cathode-ray tube depends on the lens excitation, though modern SEMs contain a compensation of this rotation.

The last probe-forming lens (Fig. 2.7), also called the objective lens, is normally asymmetric with a small bore in the polepiece plate which faces the specimen. This is to reduce the magnetic field at the specimen. The other polepiece bore is of larger diameter and contains the scan coils S_1 and S_2. Figure 2.8 shows the distribution $B(z)$ of magnetic field on axis for different gaps widths S of the polepieces. Lenses of this type are used for working distances w = 5–20 mm. The polepiece face directed to the specimen is mostly conical with a flat part of 10–20 mm in diameter so that semiconductor or scintillation detectors and channel plates for BSE can be mounted below the polepiece. The conical shape allows larger specimens (semiconductor wafers for example) to be tilted or larger take-off angles to be used for x-ray microanalysis. Other special objective lens designs are discussed in Sect. 2.2.3.

2.2.2 Lens Aberrations

The following lens aberrations are important when discussing electron-probe formation in the SEM.

Spherical Aberration. If parallel rays are incident on an electron lens, those further from the optic axis are focused closer to the lens (Fig. 2.9a). The resulting beam diameter in the Gaussian image plane, which is the focal plane for the paraxial rays, is 2 $C_s\alpha^3$ and in the plane of least confusion [2.28], the diameter is

$$d_s = 0.5\, C_s\alpha^3 \,. \tag{2.16}$$

Here, α is the aperture-limiting semi-angle, which can be altered by changing the diaphragm in the polepiece gap of the objective lens (Fig. 2.7). Strictly speaking, one should distinguish between an *aperture*, which is a semi-angle, and an aperture *diaphragm* and not use 'aperture' for both. Weak lenses used in the SEM for large working distances have large spherical aberrations coefficients C_s = 20–100 mm and C_s increases as f^3 with increasing focal length since

$$C_s \simeq 3f^3/4a^2 \tag{2.17}$$

for a bell-shaped field and $k^2 \ll 1$ [2.29]. Calculations and measurements of C_s for objective lenses are published in [2.30, 31] (see Fig. 2.10, for example). The spherical aberration coefficient C_s can be obtained by measuring the image shift when the objective diaphragm is moved off-axis [2.31]. In more strongly excited objective lenses with in-lens position of the specimen (Sect. 2.2.3), C_s can be reduced to a few millimetres.

Chromatic Aberration. The focal length depends on the electron energy or acceleration voltage (2.11). Since the electrons from the gun have an energy spread ΔE (Sect. 2.1), they are focused into a disc of least confusion (Fig. 2.10b) with diameter

$$d_c = C_c \frac{\Delta E}{E} \alpha \,, \tag{2.18}$$

where C_c is the chromatic aberration coefficient (Fig. 2.10), which is of the order of the focal length for $k^2 \ll 1$. Fluctuations of the acceleration voltage and of lens currents also cause variations of the focal length

$$\frac{\Delta f}{f} = \frac{\Delta U}{U} + 2\frac{\Delta I}{I} \,, \tag{2.19}$$

so that $\Delta U/U$ and $2\,\Delta I/I$ have to be kept smaller than $\Delta E/E \simeq 10^{-4}$ caused by the energy spread. The chromatic aberration coefficient C_c can be found by measuring the image shift when the acceleration voltage is changed and the objective diaphragm is not centred on the axis [2.31].

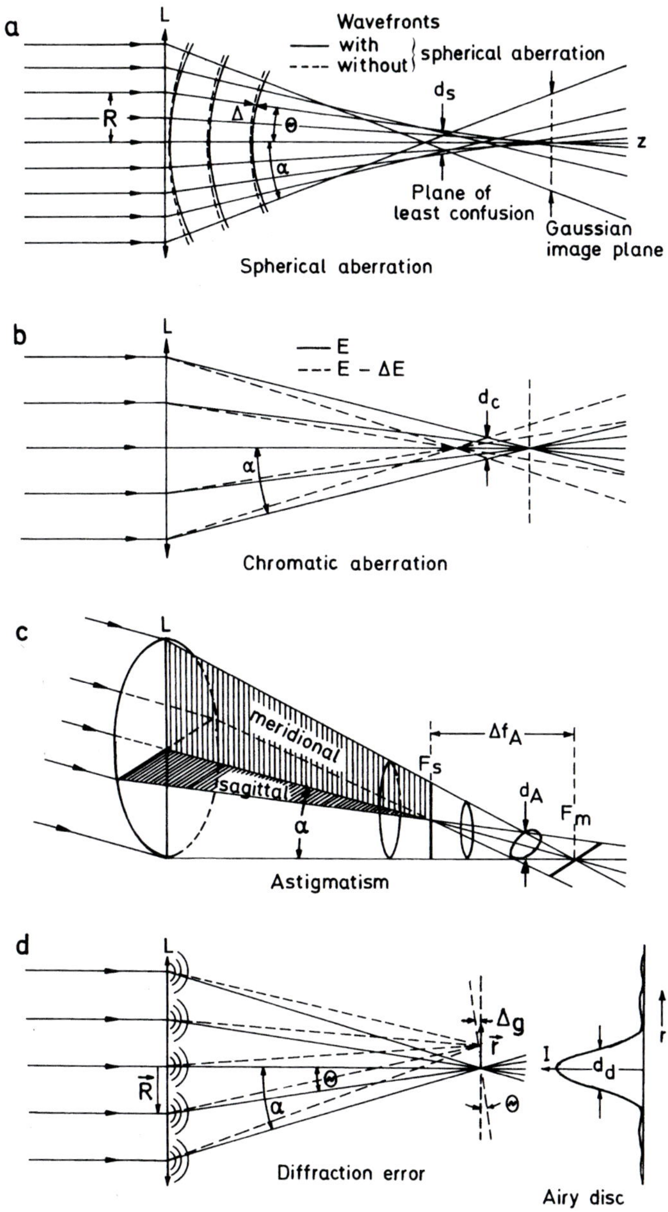

Fig. 2.9. Lens aberrations of an electron lens: (**a**) spherical and (**b**) chromatic aberration, (**c**) axial astigmatism and (**d**) diffraction error disc

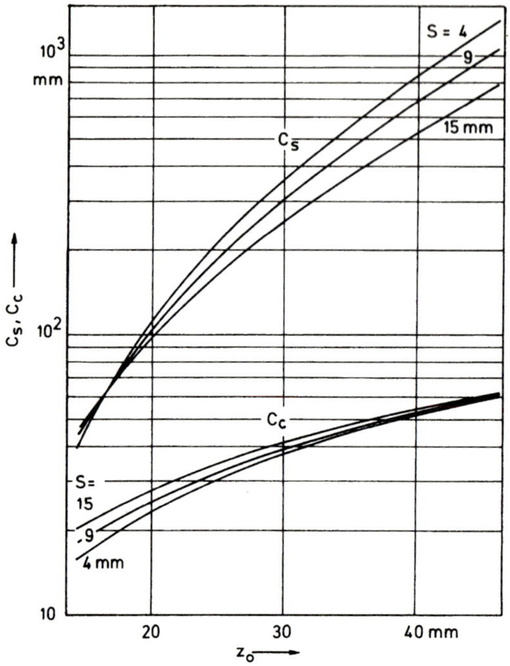

Fig. 2.10. Measured values of the spherical and chromatic aberration coefficients C_s and C_c as a function of distance z_0 of the electron probe from the lens centre for three gap widths S (Fig. 2.8) between upper and lower polepieces of diameters 10 and 30 mm [2.31]; the intermediate image of the crossover is 252 mm in front of the lens

Because $\Delta E/E$ increases with decreasing energy, the electron-probe size will be limited in LVSEM by the chromatic aberration, which can be compensated by a series of electric and magnetic quadrupole lenses, as explained below.

Axial Astigmatism. Magnetic inhomogenities of the polepieces, elliptical polepiece bores and charging effects in the bore or at the aperture diaphragm cause asymmetry in the focusing field. Electrons that are incident in sagittal and meridional planes are focused at different distances from the lens, forming focal lines F_s and F_m (Fig. 2.9c) separated by a distance Δf_A; the diameter of the disc of least confusion is

$$d_A = \Delta f_A \alpha \, . \tag{2.20}$$

The effect of axial astigmatism is similar to that of adding a cylindrical lens to a spherical lens in light optics. This aberration can therefore be compensated by introducimg an additional cylindrical lens, rotated through 90° relative to the field distortion that causes the astigmatism. Such a stigmator takes the the form of a pair of quadrupole lenses, situated near the polepiece gap. It must be variable in strength and azimuth. Methods of detecting astigmatism will be described in Sects. 2.4.1 and 2.4.4.

Diffraction Error. The diaphragm in the objective lens, which limits the aperture to keep the spherical and chromatic aberrations acceptably small, causes a Fraunhofer diffraction pattern (Airy disc) to be formed at the focal plane (Fig. 2.9d) (see also Sect. 2.2.4) with a half-width

$$d_\mathrm{d} = 0.6\lambda/\alpha \,, \tag{2.21}$$

where λ denotes the de Broglie wavelength of the electrons. From (2.4), we have

$$\lambda = h/mv = hc(2EE_0 + E^2)^{-1/2} \tag{2.22}$$

or

$$\lambda = \frac{1.226}{[U(1 + 0.9788 \times 10^{-6}U)]^{1/2}} \tag{2.23}$$

with λ in nm and U in V ($\lambda = 38.8$ pm for $E = 1$ keV and 6.98 pm for 30 keV).

Correction of Lens Aberrations. Other lens aberrations such as coma, distortion and non-axial errors can be avoided by on-axis alignment of the electron beam and by the use of small apertures $\alpha \leq 20$ mrad.

The discussion of electron-probe formation in Sects. 2.2.4 and 5 will show that correction of the spherical and chromatic aberration can considerably decrease the probe size and hence increase the resolution. This is very important for LVSEM where the resolution is mainly limited by the chromatic aberration since the factor $\Delta E/E$ increases with decreasing E.

Scherzer [2.32] showed that the spherical and chromatic aberration can be corrected by a series of electric and/or magnetic multipole lenses. A multipole corrector [2.33, 34] has been incorporated in a low-voltage SEM [2.35, 2.36] in order to correct C_s and C_c. The improvement in resolution was demonstrated by resolving 2–5 nm colloidal gold at 1 keV.

2.2.3 Special Designs of Objective Lenses

In-Lens Operation. An improvement in the electron optical properties is achieved by placing the specimen between the polepieces (*in-lens operation*). A strong excitation of the lens increases the lens parameter k^2 (2.13) to a value of $\simeq$3. The focus then lies at the centre (maximum) of the axial magnetic field between the upper and lower polepieces [2.37–40]. The aberration coefficients C_s and C_c can be decreased to values of 2–4 mm and 0.5–2 mm, respectively, which allows electron probe diameters of $\leq$1 nm to be reached at high electron energies and of 2–3 nm, at 1 keV [2.38]. This mode also makes it possible to work in the scanning transmission electron microscope (STEM) mode and the low-loss electron mode [2.41], whereas SE can only be extracted by spiraling along the magnetic field lines [2.42] (Sect. 5.4.1).

A disadvantage of in-lens operation is the small specimen size. A compromise that allows large specimens to be studied while retaining a higher lens excitation to reduce the aberrations is the semi-in-lens mode [2.43], in which the specimen is situated below the lower polepiece but the working

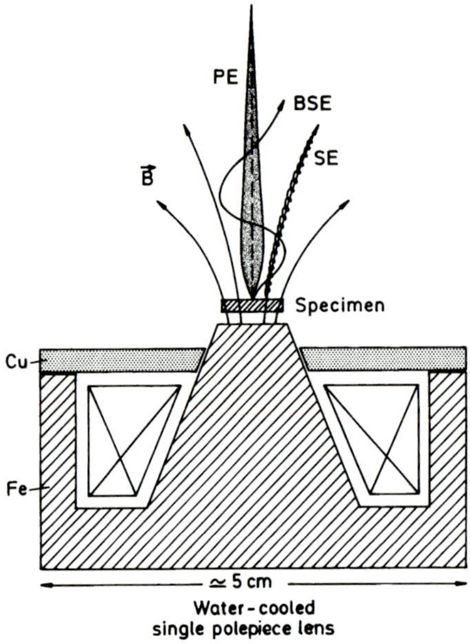

Fig. 2.11. Single polepiece lens used as probe-forming lens behind the specimen with SE and BSE trajectories inside the magnetic front field

distances are small, typically 1–2 mm. This means that the magnetic field at the specimen is high and, as in the in-lens mode, the secondary electrons are collected by a small Everhart–Thornley detectors inside the polepiece gap [2.44].

Single-Polepiece Lenses. Conventional electron lenses as shown in Figs. 2.6 and 2.7 work with a nearly closed magnetic yoke the axial magnetic field being confined within a gap of a few millimetres between the upper and lower polepieces. In contrast, the magnetic field in *single-polepiece lenses* [2.45] is concentrated at a conical polepiece, and the magnetic flux returns to the yoke over a distance 1–10 cm (Fig. 2.11). This type of lens has the advantage that the spherical and chromatic aberration coefficients are lower. A version with a pierced polepiece above the specimen [2.46] can be used to investigate tilted wafers at low working distances w. A compact conical polepiece can also be placed a few millimetres behind the specimen [2.47–49]. In both versions the specimen is immersed in the magnetic lens field with a large extension in one direction.

Combinations with Electrostatic Fields. The presence of magnetic and electrostatic retarding fields allows higher electron energies to be used in the column of a LVSEM, thereby decreasing the influence of magnetic and electrostatic stray fields, together with a deceleration inside or behind the objective lens so that the electrons hit the specimen with a decreased landing energy [2.50–52]. As an example, Fig. 2.12 shows the combination of a conical magnetic lens and an electrostatic lens [2.53–55]. The latter consists of an inner electrode at the acceleration potential and an outer electrode at earth

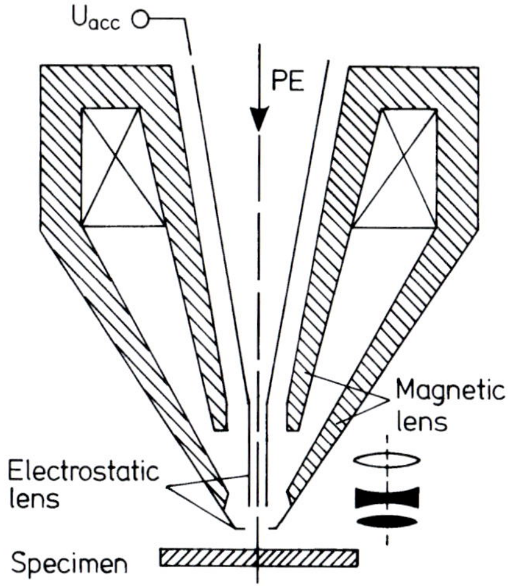

Fig. 2.12. Magnetic-electrostatic compound objective lens with the inner electrostatic electrode at the potential of the acceleration voltage and the outer electrode at earth potential [2.55]

potential. In the gap of the electrostatic lens, the electrons are decelerated to their final energy. The SE are collected by the electric field at the specimen and accelerated to the potential of the inner electrode. They can be detected by a scintillator situated above the lens system. Such a combination can even have lower aberration coefficients [2.53]. A similar *electrostatic detector–objective lens* with conical electrodes is used in a configuration designed to correct C_s and C_c by means of multipole elements [2.35] (Sect. 2.2.2).

The cathode-lens principle [2.56] has been employed in very low energy microscopy to retard the electron beam from 10–15 keV to landing energies of 0–500 eV with the specimen close to the cathode potential [2.57]. It is then also possible to operate in the electron mirror microscopy mode (Sect. 8.5). This technique is similar to low-energy electron microscopy (LEEM), where a retarded electron beam is imaged by electron lenses. Illuminating and imaging beams are separated by a magnetic prism [2.58, 59].

Miniaturized SEM. A small SEM unit, 140 mm in length and 26 mm in diameter, using a field-emisison gun, working between 0.3–3 kV with probe sizes of 300–10 nm can be attached to an ultrahigh vacuum chamber for solid-state surface investigations [2.60].

By using a field-emission cathode with an electrostatic einzel- or immersion lens consisting of centred silicon slices isolated by quartz spacers with a bore diamater of a few μm, a miniaturized SEM smaller than 1 cm can be built [2.61–64]. Such a miniaturized SEM can be placed in molecular-beam-epitaxy or other processing chambers and these instruments are creating considerable interest for electron-beam lithography in the sub-100 nm region.

2.2.4 Geometric Optical Theory of Electron-Probe Formation

A small electron probe of diameter d_0 at the specimen is produced by successive demagnification of the crossover of diameter d_c by means of two or three lenses (Fig. 2.13). Provided that each intermediate crossover image is formed at a large distance L in front of the next lens, the subsequent demagnified image will be near the lens focus, at a distance $(1/f-1/L)^{-1} \simeq f$, and the total demagnification of a three-lens system results in a spot with the geometric diameter

$$d_0 = \frac{f_1 f_2 f_3}{L_1 L_2 L_3} d_c = m d_c \ , \tag{2.24}$$

where $m \ll 1$ is known as the demagnification.

For a thermionic cathode with $d_c \simeq$ 20–50 μm, the demagnification has to be $m \leq 1/5000$ to get $d_0 \leq 10$ nm, so that $f_i/L_i \simeq (1/5000)^{1/3} \simeq 1/17$. With $f_i = 5$ mm, the minimum length of the electron optical column is thus $3L_i \simeq 25$ cm. A Schottky or field-emission gun can already produce a virtual source of less than 10 nm, so that only one probe-forming (objective) lens is needed to demagnify the electron probe to $d_0 \simeq 1$ nm in a STEM or SEM equipped with such a gun.

A diaphragm of radius r in the objective lens (Figs. 2.7 and 13) defines the electron-probe aperture α_p, which is given in first-order approximation by

$$\alpha_p = r/f \ . \tag{2.25}$$

The real value of α_p may be somewhat different if there is a magnetic field behind this diaphragm. Diaphragms of 50–200 μm in diameter are used to produce apertures of the order of 10–20 mrad with $f \simeq$ 5–20 mm. Back-projection of this final diaphragm (aperture) results in virtual images of the aperture (FAI) near the front foci of the condenser lenses (Fig. 2.13). The electron beam can only pass the final diaphragm if it can pass through all the FAI. This will prove to be important when discussing electron-beam chopping in Sect. 2.3.3. Diaphragms are used in the condenser lenses (Fig. 2.13) only to intercept any scattered electrons which should not reach the specimen; these are known as spray diaphragms.

A very important feature of SEM operation is that the electron-probe current I_p, the probe aperture α_p and the geometric probe diameter d_0 cannot be changed independently. The electron current density j_p and the probe aperture α_p are related by the gun brightness β (2.6)

$$j_p = \pi \beta \alpha_p^2 \ . \tag{2.26}$$

Assuming for simplicity that the current density is uniform over a circle of diameter d_p, the total probe current is given by

$$I_p = \frac{\pi}{4} d_0^2 j_p \tag{2.27}$$

or

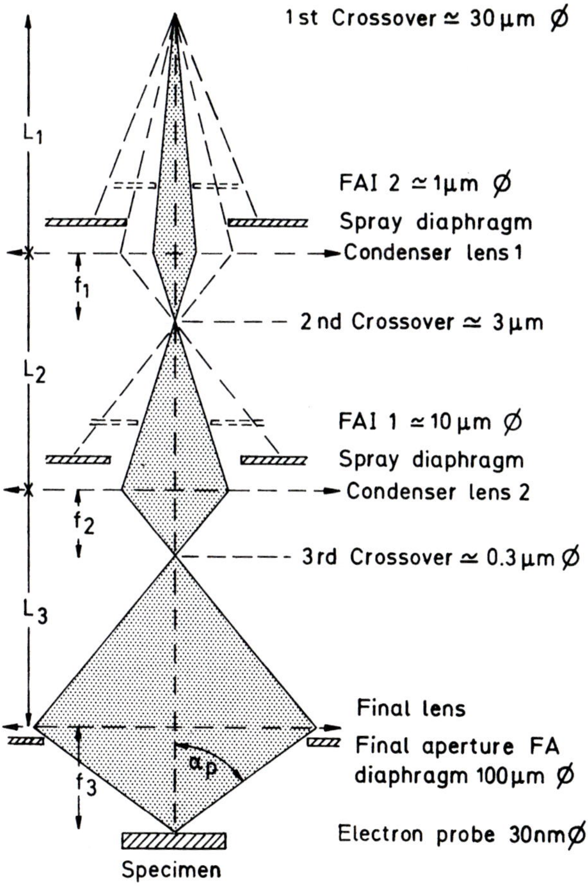

Fig. 2.13. Schematic ray in the electron-optical column of a SEM (FAI = final aperture image)

$$d_0 = \left(\frac{4I_\mathrm{p}}{\pi^2\beta}\right)^{1/2} \alpha_\mathrm{p}^{-1} = C_0\alpha_\mathrm{p}^{-1} \, . \tag{2.28}$$

This geometric probe diameter d_0 is broadened by the action of the lens aberrations of the objective lens. In order to estimate the influence of these aberrations, we assume that the geometric electron-probe profile of half-width d_0 and the error discs of widths d_s (2.16), d_c (2.18) and d_d (2.21) follow Gaussian distributions $\propto \exp(-r^2/a_i^2)$. This has the advantage that a blurring (convolution) of two Gaussians with widths a_1 and a_2 results in a blurred Gaussian of width $a^2 = a_1^2 + a_2^2$. We adopt this quadratic superposition of widths though in reality not all radial distributions of the error discs are Gaussian. This estimate of the effective probe diameter gives a good indication of the orders of magnitude involved and shows how different parameters influence the result. Adding the diameters of the aberration discs in quadra-

has to be calculated wave-optically (Sect. 2.2.5). For the field-emission gun, the formula corresponding to (2.32) [2.68] is found to be

$$I_{\mathrm{p,max}} = c\, d_{\mathrm{p,min}}^{2/3} \,. \tag{2.33}$$

The constant c depends strongly on the spherical aberration of the electrode system in the gun (triode or tetrode) and on the presence of a further magnetic condenser lens. The power 2/3 in (2.33) shows that $I_{\mathrm{p,max}}$ increases more slowly with increasing $d_{\mathrm{p,min}}$ than in the case of the thermionic cathode, for which the corresponding power is 8/3 according to (2.32) (Fig. 2.15). Field-emission guns are therefore better than thermionic guns if we want a small probe diameter or a high probe current for probe diameters $\leq 0.1\mu$m. But for probe diameters larger than 0.1 μm, a thermionic gun can provide a higher probe current, which is a valuable feature for x-ray and Auger electron microanalysis and in the specimen-current, cathodoluminescence, and acoustic thermal-wave microscopy, for example.

In practice, it is more important to know the dependence of probe diameter on the focal length or the working distance w, which can be continuously varied, and for a discrete set of final apertures. The aperture α_{p} decreases as f^{-1} according to (2.25). However, the spherical and chromatic aberration coefficients increase as f^3 according to (2.17) and as f, respectively (see also Fig. 2.10). Substitution in (2.29) shows that the last two terms, which contain C_{s} and C_{c} are approximately independent of f or w whereas the first term is proportional to f^2 for a constant probe current I_{p}.

2.2.5 Wave-Optical Theory of Electron-Probe Formation

The foregoing geometric-optical treatment of lens aberrations is adequate for the discussion in Sect. 2.2.4 of electron-probe formation when using thermionic cathodes. In this case we have completely incoherent illumination and wave-optical effects can be neglected. For the discussion of electron-probe formation when using Schottky or field-emission guns, it will be necessary to use a wave-optical treatment [2.66, 69–71] and hence to define the lens aberrations in wave optical terms.

The action of a lens as shown in Fig. 2.9a can be interpreted in wave optics as a phase shift, which deforms the plane wavefronts of a plane incident wave in front of the lens into a spherical converging wave behind the lens. For an ideal lens without aberration, the wavefronts are concentric spheres with the focal point as the centre. The electron trajectories in Fig. 2.9a are perpendicular to the wavefronts or surfaces of equal phase. Therefore, the stronger refraction of non-paraxial rays in the presence of spherical aberration is equivalent to a stronger curvature of the wavefronts (full curves in Fig. 2.9a). The deviation Δ from an ideal converging wave is equivalent to an additional phase shift $W(\theta) = 2\pi\Delta/\lambda$ which is called the wave aberration and depends only on the angle θ with the optic axis (Fig. 2.9a) [2.72] (for details see [1.1]):

$$W(\theta) = \frac{\pi}{2\lambda}(C_s\theta^4 - 2\Delta z\theta^2) . \tag{2.34}$$

The last term takes into account any defocus Δz.

Whereas all elementary waves from points R of the lens plane are summed with equal phases at the focal point (constructive interference) (Fig. 2.9d), the wave aberration can result in destructive interference if $W(\theta)$ becomes an odd multiple of π and hence $\Delta = \lambda/2$ or an odd multiple of this.

For an off-axis point $\boldsymbol{r}$ we have to consider the geometric-optical path difference, Δ_g in Fig. 2.9d, which becomes $\Delta_g = r\theta = rR/f$ if $\boldsymbol{r}$ and $\boldsymbol{R}$ are parallel or $\Delta_g = \boldsymbol{r}\cdot\boldsymbol{R}/f$ for arbitrary directions of these vectors. The resulting phase shift is

$$\phi = \frac{2\pi}{\lambda}\Delta_g = \frac{2\pi}{\lambda}\frac{\boldsymbol{r}\cdot\boldsymbol{R}}{f} . \tag{2.35}$$

This phase shift can be expressed in terms of a phase factor $\exp(\mathrm{i}\phi)$ together with a phase factor $\exp[-\mathrm{i}W(\theta)]$ coming from the wave aberration (2.34), which depends only on θ. We assume an incident plane wave of uniform amplitude ψ_0 and introduce polar coordinates R and χ in the lens plane. The sum over all wavelets from points at the lens plane then becomes

$$\psi(r) = \psi_0 \int_0^{R_0}\int_0^{2\pi} \exp[-\mathrm{i}W(\theta)]\exp\left(\frac{2\pi\mathrm{i}}{\lambda f}rR\cos\chi\right) R\,\mathrm{d}R\,\mathrm{d}\chi . \tag{2.36}$$

The azimuthal angle appears only inside the second phase factor and integration over χ introduces the Bessel function J_0

$$\psi(r) = 2\pi\psi_0 \int_0^{R_0} \exp[-\mathrm{i}W(R/f)]\,\mathrm{J}_0\left(\frac{2\pi}{\lambda f}rR\right) R\,\mathrm{d}R . \tag{2.37}$$

If $C_s = 0$ and $\Delta z = 0$ and hence $W = 0$, the relation $\int x\mathrm{J}_0(x) = x\mathrm{J}_1(x)$ enables us to write down the probability of finding an electron at a distance r from the optic axis ($\alpha_p = R_0/f$ is the electron-probe aperture)

$$|\psi(r)|^2 \propto \left|\frac{\mathrm{J}_1(x)}{x}\right|^2 \quad \text{with} \quad x = \frac{2\pi}{\lambda f}rR_0 = 2\pi\alpha_p r\,\lambda , \tag{2.38}$$

which is none other than the Airy distribution, the Fraunhofer diffraction pattern from a diaphragm of radius R_0 (Fig. 2.9d).

If $C_s \neq 0$ and hence $W \neq 0$, the phase factor can be split into real and imaginary parts by Euler's formula $\exp(\mathrm{i}\theta) = \cos\theta + \mathrm{i}\sin\theta$ and the probability density, which is proportional to the local current density $j(r)$, becomes

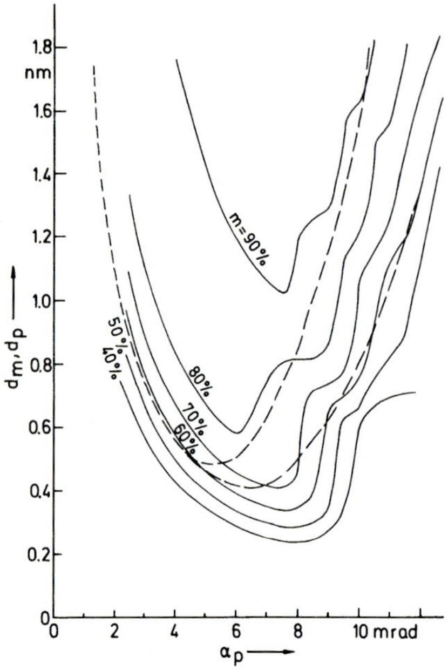

Fig. 2.16. Wave-optical calculation of electron-probe diameter d_m containing m% of the total electron-probe current versus electron-probe aperture α_{p} (E = 100 keV, λ = 3.7 pm, C_{s} = 3.1 mm). *Upper dashed line*: geometric-optical theory for d_{p} using formula (2.29); *lower dashed line*: substitution of 0.25 $C_{\mathrm{s}}\alpha^3$ for the spherical aberration term in (2.16) and (2.38) [2.70]

$$|\psi(r)|^2 \propto \left| \int_0^{R_0} \cos[W(R/f)] \mathrm{J}_0 \left(\frac{2\pi}{\lambda f} rR \right) R \,\mathrm{d}R \right|^2 + \left| \int_0^{R_0} \sin[W(R/f)] \mathrm{J}_0 \left(\frac{2\pi}{\lambda f} rR \right) R \,\mathrm{d}R \right|^2 \qquad (2.39)$$

Figure 2.16 contains calculated probe diameters d_m for $E = 100$ keV which contain different percentages m of the total current as a function of the probe aperture α_{p}, where the optimum defocus has been selected to get the smallest d_m for a fixed value of α_{p}.

Figure 2.16 also contains geometric calculations of d_{p} using the second and third terms of (2.29) (dashed lines), which represent the diffraction error and the spherical aberration. At low α_{p}, this calculation coincides with the wave-optical curves for d_m at $m = 60\%$, but the minimum of d_{p} appears at a lower value of α_{p}. A better fit of d_{p} to the $m = 60\%$ curve can be obtained by using $d_{\mathrm{s}} = 0.25 C_{\mathrm{s}}\alpha^3$ instead of (2.16) (lower dashed curve in Fig. 2.16).

The calculations presented in Fig. 2.16 assume temporal and spatial coherence. The former means that the electron beam is monochromatic. In reality, we have a polychromatic source as a result of the energy spread of the electron gun (Sect. 2.1.1) and $j(r, \Delta E)$ profiles have to be calculated and summed. If $N(\Delta E)\mathrm{d}(\Delta E)$ is the energy distribution of emitted electrons (approximately a Maxwellian distribution), the electron-probe profile becomes

$$j_{\text{poly}}(r) = \int_0^\infty j_{\text{mono}}(r, \Delta E) N(\Delta E) \mathrm{d}(\Delta E) . \tag{2.40}$$

Complete spatial coherence means that there are no effects due to finite source size and that the diameter d_0, (2.28), of the demagnified crossover is much smaller than d_m. When $d_0 \leq d_m$ (partial spatial coherence), the intensity distribution j_{poly} has to be convolved two-dimensionally with the current density $j_0(r)$ of the demagnified crossover, which can be approximated by a Gaussian (2.5),

$$j_{\text{p}}(\boldsymbol{r}) = \int \int j_{\text{poly}}(\boldsymbol{r} - \boldsymbol{r}') j_0(\boldsymbol{r}') \mathrm{d}^2 \boldsymbol{r}' = j_{\text{poly}}(\boldsymbol{r}) \otimes j_0(\boldsymbol{r}) \tag{2.41}$$

where $\otimes$ denotes a convolution.

2.2.6 Stabilisation of the Electron-Probe Current

We have to distinguish between the cathode emission current I_{c}, the electron-probe current I_{p} and the specimen current I_{s}. The probe current must be free of rapid fluctuations, which would modulate the image intensity during a single frame, and free of long-term drift: x-ray microanalysis, for example, may take a few minutes or even hours when a concentration profile or an elemental distribution map is being recorded.

The emission current of a thermionic cathode depends very sensitively on the temperature of the cathode tip. The cathode is heated by an ac current supply and the electric power oscillates at twice the mains frequency. Most TEM power supplies work at 50 Hz, which results in a 100 Hz modulation of cathode temperature and emission, but this modulation does not disturb the image formation in TEM because all points are recorded simultaneously. Owing to the thermal inertia, the temperature becomes constant for frequencies above a few kilohertz. Therefore, by driving the insulated transformer that produces the cathode current (Fig. 2.2a) from a primary source at a few tens of kHz, short-term fluctuations can be avoided in a SEM. Long-term drift of thermionic cathodes occurs because of differences in the thermal expansion, resulting in a shortening of the distance between cathode tip and Wehnelt cup, as a result of the thinning of the tungsten wire by evaporation or caused by changes of the work function. In the case of LaB_6 cathodes, the lobes can change their emission. Field-emission guns not only show long-term drift but also rapid fluctuations caused by ion bombardment and gas adsorption.

Even for a constant emission current I_{c}, the probe current I_{p} can be changed by deterioration of the gun alignment typically caused by small shifts of the tip position [2.73, 74] or by charging effects in the column.

The specimen current I_{s} varies as the specimen is scanned and can be used as a signal for image recording (Sect. 6.2.6). Even for a stationary electron probe, however, the specimen current can be altered by contamination and/or charging of the specimen.

This means that stabilisation of the cathode emission current is not sufficient and that of the specimen current undesirable. Stabilisation of the probe current will be necessary but this can be measured exactly only by means of a Faraday cage (Sect. 5.1). Some suitable signal has therefore to be found that is proportional to the probe current and that can serve as a feedback signal. The probe current can be varied by changing the heating power of the cathode, the bias of the Wehnelt cup or the lens excitation. Such a method should not alter the focus at the specimen.

Several stabilising methods have been reported. In one, the hot gun filament is connected into a branch of a Wheatstone bridge and its resistance and hence the filament temperature are kept constant [2.75]. In another, the electron current absorbed by a fixed diaphragm very close to the diaphragm of the final lens is used as a feedback signal to regulate the power supply of the gun filament [2.7]. An electronically variable resistor, represented by a transistor, may be added to the resistor R_W that creates the Wehnelt bias as a voltage drop and thus keep the cathode emission current constant [2.76]. A diaphragm of four segments between anode and first condenser lens may be used to keep the cathode emission current constant by using the sum signals of the four quadrants [2.77]. The two difference signals of opposite quadrants are used to drive motors which operate the adjustment controls of the gun alignment so that opposite quadrants get equal currents. A similar technique using a full annular collector below the anode has been used to stabilize the emission from a field-emission gun [2.78]. The current from an insulated final-lens diaphragm can be used as a feedback signal to vary the first condenser lens current [2.73]. Separated stabilized power supplies for filament heating, Wehnelt bias and high voltage allow automatic compensation of variations and drift of the emission current by a computer-controlled feedback loop [2.79].

2.3 Electron Beam Deflection Modes

2.3.1 Electron Beam Deflection by Transverse Fields

Electron-beam deflection is needed for scanning and rocking the electron probe on the specimen and for electron-beam blanking. This can be achieved by applying transverse electrostatic and magnetic fields. Electrons move along parabolic trajectories in the uniform field $|\boldsymbol{E}| = u/d$ of a parallel-plate capacitor (Fig. 2.17) with a plate distance d and an applied bias $\pm u/2$ and around a circle of radius $r = mv/eB$ in a uniform magnetic field $\boldsymbol{B}$. For small deflection angles $\epsilon = x/L$, for which $\sin\epsilon \simeq \tan\epsilon \simeq \epsilon$, the momentum approximation can be used and the pivot point of deflection then lies at the centre of the deflection unit. Here it is assumed that the electrons pass through the transverse field of length h with a constant component $p_z = mv$ of momentum. The Lorentz force $\boldsymbol{F} = \mathrm{d}\boldsymbol{p}/\mathrm{d}t$, (2.10), transfers a momentum

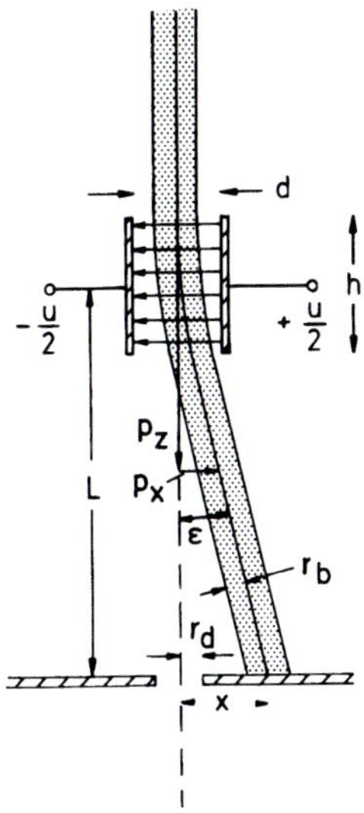

Fig. 2.17. Principle of beam blanking: the electrons are deflected across a diaphragm by means of the transverse electric field of a parallel-plate capacitor

p_x to the electrons during their time-of-flight $T = h/v$ through the deflection field:

$$p_x = \int_0^T F_x \mathrm{d}t = e\int_0^T (E_x + vB_y)\mathrm{d}t = \frac{e}{v}\int_0^h (E_x + vB_y)\mathrm{d}z = eh(E_x/v + B_y)\ . \tag{2.42}$$

The angle of deflection is given by $\epsilon = |p_x|/p_z$ (Fig. 2.17) and using (2.2–2.4) we obtain

$$\epsilon = \frac{eh}{mv^2}E_x = \frac{h}{2d}\frac{u}{U}\frac{1 + E/E_0}{1 + E/2E_0} \tag{2.43}$$

and

$$\epsilon = \frac{eh}{mv}B_y = \frac{ehB_y}{[2m_0E(1 + E/2E_0)]^{1/2}} \tag{2.44}$$

for transverse electric and magnetic fields, respectively (U: acceleration voltage between anode and cathode).

For example, in order to deflect 20 keV electrons through $\epsilon = 5° \simeq 0.1$ rad in a distance $h = 1$ cm, a field $|\boldsymbol{E}| = u/d = 20$ kV cm^{-1} would be required, which could be provided by applying a voltage $u = \pm\ 2$ kV across the two plates of the capacitor, a distance $d = 2$ mm apart. A magnetic field $B = 10^{-2}$ T would be needed for the same conditions. The magnetic field produced by an electro-magnet with a slit width d is given by $B = \mu_0 NI/d$ ($\mu_0 = 4\pi \times 10^{-7}$ H m^{-1}, N: number of windings, I: coil current) so that $NI = 20$ A or $N = 200$ turns and $I = 0.1$ A, for example.

This shows that electrostatic fields need a high tension and magnetic fields a low current. A scan generator using magnetic fields is therefore simpler to design and the scan coils for cathode-ray tubes also need currents of this order of magnitude.

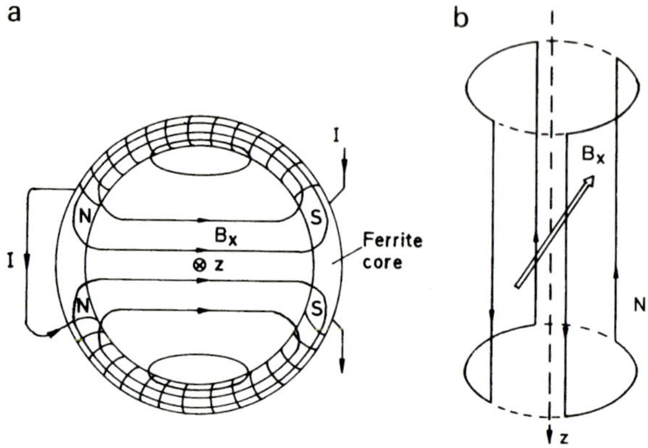

Fig. 2.18. Practical design of scan coils: (**a**) ring core and (**b**) saddle coils

The scan frequency of a saw-tooth excitation is limited by the eddy currents in the magnetic core. Ferrite cores or air coils have therefore be used for scanning and rocking the electron probe whereas electrostatic deflection fields are predominately used for electron-beam blanking up to gigahertz frequencies, since only small deflection angles are needed.

Figure 2.18 shows two possible ways of constructing coil systems for scanning in the x and y directions. In Fig. 2.18a, the coil is wound around an iron or ferrite core. The magnetic field distribution is shown for the y direction. The x coil pair is rotated by 90° and can be wound on the same core. The density of the windings varies sinusoidally in azimuth to produce a uniform field at the core centre. Another possibility is shown in Fig. 2.18b, where pairs of iron-free coils (saddle coils) are employed.

2.3.2 Scanning, Magnification and Rocking

The specimen has to be scanned by the electron probe in a raster and the electron beam must pass through the aperture-limiting diaphragm in the final lens without deflection (Fig. 2.6b). This means that the pivot point about which the beam rocks has to lie in the plane of this diaphragm. For this, two pairs of deflection coils S_1 and S_2 are needed for the x and y deflections. The deflection angles ϵ_1 and $\epsilon_2 = 2\epsilon_1$ determine the width b of the scanned area and the magnification of the image will be

$$M = \frac{\text{Width}\, B \,\text{of the CRT}}{\text{Width}\, b \,\text{of the scanned area}} . \tag{2.45}$$

The magnification can easily be varied by altering the scan-coil currents. Low magnification requires large deflection angles and lens aberrations may therefore cause larger distortions of the raster, whereas higher magnifications need deflection angles $\epsilon < 0.1$ mrad and the distortions are negligible. The main image distortion will then be the permanent distortion of the raster on the recording CRT.

For low magnifications and rough specimens, the image must be regarded as a central projection with the projection centre Z situated a few millimetres in front of the diaphragm owing to the curvature of electron trajectories behind the diaphragm; the exact position of Z depends on the lens excitation. The distance z_p of the projection centre Z from the specimen will be a distance k_z larger than the working distance w, defined as the distance between specimen and lower polepiece (Fig. 2.7)

$$z_\mathrm{p} = w + k_z \ . \tag{2.46}$$

The magnitude of z_p can easily be determined by choosing a calibrated cylindrical well of known depth and diameter as a test specimen. The top is imaged with a larger diameter than the bottom. This shows that it is not possible to characterize an image by one value of magnification because the magnification for rough or inclined specimens depends on their depth. A correct value of magnification or a micrometre-bar can be specified only for nearly plane specimens. For this reason, the magnification of a SEM micrograph is often given as the horizontal-field-width of a specimen area of mean depth. This has the advantage that this quantity does not change when recording or printing at different optical magnifications. The magnification can be checked by imaging commercial replicas of optical gratings or calibrated etched structures on silicon. Such a standard has to be positioned at the same working distance as the specimen without changing the lens (focusing) [2.80] and the scan coil current (deflection angle) because the width b of the scanned area depends on both parameters.

The fact that the scan coils are placed in front of the probe-forming final lens implies that the image rotation of this lens produces a rotation of the raster, which can be of the order of $\phi = 45°–90°$ and depends strongly on the lens excitation or the working distance w. From (2.11) and (2.14), we see that for a weak lens ($k^2 \ll 1$), the image rotation φ is proportional to $f^{-1/2}$. This raster rotation disturbs the angular relation between specimen directions, the take-off direction of a detector and the edges of the CRT. Compensation of raster rotation can be achieved electronically by a sine-cosine potentiometer or by a digital raster rotation system [2.81], whereby the deflection currents I_x and I_y are modified to

$$\begin{aligned} I'_x &= I_x \cos\varphi - I_y \sin\varphi \\ I'_y &= I_x \sin\varphi + I_y \cos\varphi \ . \end{aligned} \tag{2.47}$$

Keeping the electron probe at the specimen in a fixed position and varying the angle of incidence in a raster within the range $\gamma_x, \gamma_y = \pm\gamma$ is called rocking. This technique is of interest for the production of electron-channelling patterns (see Sect. 9.2.2 where methods of rocking the beam are described in detail). Another interesting application of beam rocking is stereoscopy and stereogrammetry (Sect. 6.4.5).

2.3.3 Electron Beam Blanking and Chopping

Blanking means interrupting the electron beam briefly and chopping, which is periodic blanking, means the production of short beam pulses of duration $t_p < T$ with a repetition frequency $f = 1/T$. This technique has several uses: separation of a weak signal generated by the electron beam, cathodoluminescence for example, from a stationary signal, which might be light emitted from a thermionic cathode [2.82] or the dc electroluminescence of a semiconductor diode [2.83], by using a lock-in amplifier; or measurement of the signal decay after a short pulse of irradiation, fluorescence or electron-beam-induced current decay curves, for example. The most important application is stroboscopy for examining high-frequency voltages on semiconductor devices by following the voltage contrast (Sect. 8.2). The chopping frequency is then the same as the driving frequency at the specimen. The electron beam hits the specimen at regular intervals at the same phase and a stationary pattern can be recorded during one frame period. A typical application is the stationary imaging of a high-frequency voltage pattern on an integrated circuit. By shifting the phase of the electron beam pulses relative to that of the driving voltage, the pattern can be studied at another phase. With the electron beam at rest, the local waveform of the signal can be recorded by averaging over a larger number of pulses and slowly shifting the phase. Electron beam blanking can also be employed in electron beam lithography (Sect. 8.8). The following methods can be used for blanking and chopping.

Blanking of the Beam by the Wehnelt Bias. During the pulse time, positive pulses are applied to the negatively biased Wehnelt electrode [2.84]. Frequencies of 100 MHz and pulse widths of 5 ns at 30 keV have been obtained [2.84–86]. Blanking pulses of 30 MHz at E = 2 keV and of 5 MHz at E = 20 keV can be directly coupled to the Wehnelt cup through a 500 pF capacitor [2.87]. The capacitance of the Wehnelt electrode sets an upper limit on the frequency. Furthermore, the energy spread increases with increasing frequency and is an order of magnitude larger at 30 MHz than at 1 MHz [2.88], which can result in a loss of resolution due to an increase of the chromatic error.

Blanking by Deflection Across a Diaphragm. The beam is deflected by transverse magnetic or electrostatic fields generated by magnetic coils or parallel-plate capacitors in front of or behind the first condenser lens. If r_d denotes the radius of the diaphragm and L the distance behind the deflection unit and the diaphragm, the beam of radius r_b will be absorbed if the deflection angle $|\epsilon| > (r_d + r_b)/L$ (Fig. 2.17). Substituting the deflection angle ϵ from (2.43) and solving for u yields the minimum voltage needed to blank the beam by a capacitor

$$u_{\min} = \pm \frac{2Ud}{hL}(r_d + r_b) . \tag{2.48}$$

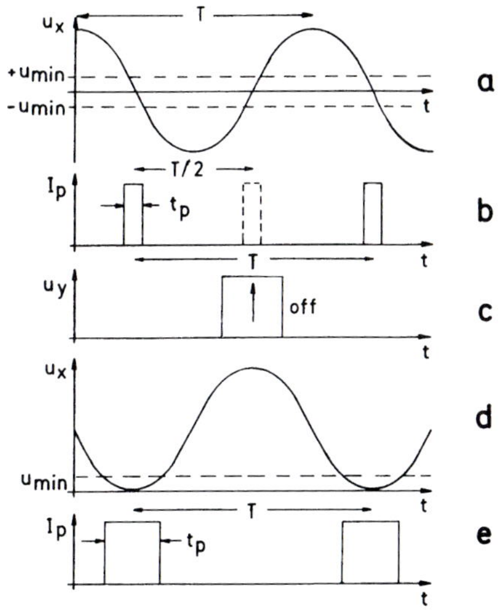

Fig. 2.19. Beam blanking by a sinusoidal deflection voltage (**a**) which causes (**b**) two beam pulses of width t_p per period T. A single deflection pulse per period in the y direction (**c**) can blank each second pulse in (b). One pulse per period (**e**) can also be obtained by using a constant offset (**d**)

The beam will go through the diaphragm if the voltage across the capacitor is in the range $-u_\mathrm{min} < u < +u_\mathrm{min}$. The pulse width t_p is determined by the slope of $u(t)$ in this range (Fig. 2.19a). This method generates two pulses per period T (Fig. 2.19b). This can be reduced to one pulse per period by deflecting the beam in the y direction with an additional capacitor during the fly-back in the negative x direction (Figs. 2.19b,c) or by adding a constant offset bias (Fig. 2.19d,e).

Using magnetic deflection coils, the maximum repetition frequency is of the order of 100 MHz with a pulse width of 5 ns at 30 keV [2.89]. The high-frequency limit is set by the self-inductance and capacitance of the coils. Electrostatic deflection by a parallel-plate capacitor is therefore widely used [2.88–90]. The high-frequency limit is reached when the time-of-flight $t_\mathrm{f} = h/v$ of the electrons through the capacitor of length h becomes of the order of the repetition time $T = 1/f$. When the deflection field alters during the time-of-flight, the deflection angle decreases as

$$\epsilon_\mathrm{eff} = \frac{h\,u}{2\,d\,U}\,\frac{\sin(\pi t_\mathrm{f}/T)}{\pi t_\mathrm{f}/T}\ . \tag{2.49}$$

Values of $L = 10$ mm and $E = 10$ keV or $v = 0.6\times10^8$ m s^{-1} result in $\epsilon_\mathrm{eff} = 0$ at $f = 6$ GHz. One way of increasing the frequency limit is to reduce the length h of the capacitor. Frequencies of 10 GHz are possible with $t_\mathrm{p} =$ 10 ps at 30 keV and $h = 5$ mm [2.91].

A few GHz have been realized by scanning the demagnified electron probe in the focal plane of a lens on a circle with a frequency of 2 GHz across a diaphragm with 64 radial slits of width 15 μm [2.92].

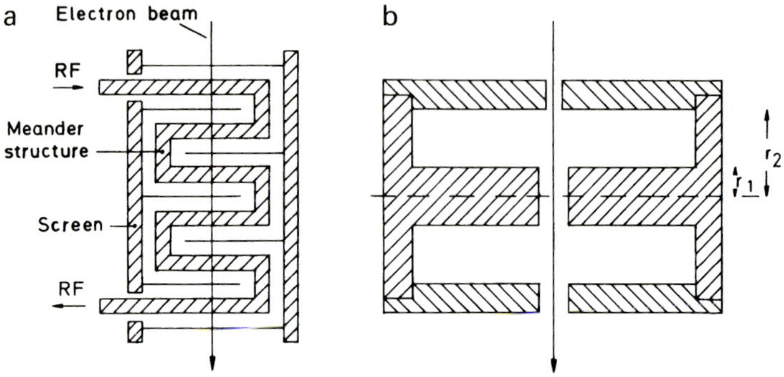

Fig. 2.20. Beam blanking in the GHz range by (**a**) a travelling-wave structure and (**b**) a re-entrant cavity

Travelling Wave Structures and Reentrant Cavities. Other solutions to the problem of chopping at very high frequencies have been found using travelling-wave structures (Fig. 2.20a) or re-entrant cavities (Fig. 2.20b). In the travelling-wave system, the deflection occurs along a meander-line structure facing an earthed plate, so that the electron deflection voltage travels down at the same velocity as the electron beam [2.93, 94]. At high frequencies, re-entrant cavities tuned to the working frequency with an electric field transverse to the electron beam can also be used [2.96]. The pulse width can be decreased by means of a buncher, which consists of a longitudinal re-entrant cavity. The first electrons to reach the buncher are accelerated by a weaker longitudinal field than those that reach it at later times. Pulse widths of the order of 2 ps could be decreased to 0.2 ps at a 1 GHz repetition rate in this way [2.97]. Both types of high-frequency deflection units have the disadvantage that they can only be used for one single frequency in the microwave region, whereas the electrostatic deflection method by a capacitor is applicable to all frequencies from static deflection to the high frequency-limit.

Deflection Across the Final Aperture Image (FAI). The simple diagram of Fig. 2.17 is not sufficient for optimizing a beam blanking or chopping system. The action of a blanking system in front of the first condenser lens is shown in more detail in Fig. 2.21. It was shown in Sect. 2.2.4 and Fig. 2.13 that a final aperture image 2 is formed in front of the lens by back-projection of the final aperture in the probe-forming lens. Only electrons passing the FAI 2 can pass through the final diaphragm. A real diaphragm, as shown in Fig. 2.17, is therefore not needed and we can use the FAI 2 to blank the beam if the deflection angle $\epsilon > \epsilon_{\rm min}$ (Fig. 2.21a). During the period $t_{\rm p}$ of the electron pulse ($\epsilon < \epsilon_{\rm min}$), the apparent image of the crossover moves inside the field-of-view diameter $d_{\rm f}$ in the crossover 1 plane. As shown in Fig. 2.21a, this field-of-view diameter is determined by the diameter of diaphragm 1 and the position of FAI 2. Because the final electron probe is a demagnified image of crossover 1, the apparent movement of this crossover with increasing

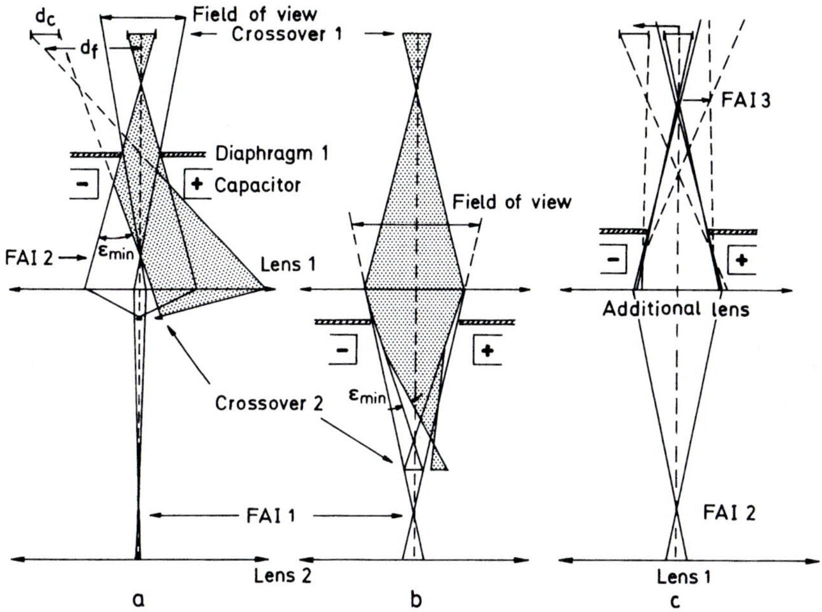

Fig. 2.21. Electron beam chopping with a deflection system (**a**) between crossover 1 and condenser lens 1 and (**b**) behind this lens. (**c**) Additional lens of long focal length above condenser lens 1. FAI = final aperture image (Fig. 2.13) acting as a virtual chopping diaphragm

deflection voltage causes a shift of the electron probe on the specimen which results in an increase of the effective probe size parallel to the direction of deflection by a degradation ratio given by

$$D = (d_\mathrm{f} + d_\mathrm{c})/d_\mathrm{c} \,. \tag{2.50}$$

The apparent movement of the crossover is coupled with an apparent movement of the diaphragm in front of the capacitor so that the field-of-view (determined by this diaphragm) also moves during deflection. The beam in Fig. 2.21a will be blanked at a deflection angle ϵ_min when the extremal electron rays shown there are outside the field-of-view. The degradation can only be decreased to a limited degree by use of a very small alignable diaphragm.

If the deflection unit is installed between condenser lenses 1 and 2 (Fig. 2.21b), the degradation is reduced to a minimum when the diameter of crossover 2 has the same size as the field-of-view diameter at the plane of this crossover. This needs lower deflection angles but a longer focal distance of lens 1, which leads to a loss of spatial resolution [2.98].

A further decrease of the necessary deflection angle ($\epsilon \simeq 0.2$ mrad) and of the degradation ratio (2.50) to below 2 can be achieved by inserting an additional lens of long focal length between the anode and lens 1 [2.91, 94]. Then, FAI 3 is in front of the capacitor (Fig. 2.21c). When the apparent image of the crossover is deflected to the left, the apparent image of FAI 3 and hence the field-of-view move to the right. For example, a deflection voltage of only $u = 50$ mV is sufficient to blank a 10 keV beam when using a 20 μm field-of-view diaphragm in front of the capacitor with $d = 30$ μm and $h = 5$ mm [2.91].

2.4 Focusing

2.4.1 Focusing and Correction of Astigmatism

The image can be focused by varying the current of the last probe-forming lens. Focusing can be controlled by examining small specimen details at high magnification. Focusing is equivalent to the reproduction of as many specimen details as possible and can therefore be effectively performed by observing the Y modulation of a linescan, which can be recorded on a separate cathode-ray tube or observed on the video display (Fig. 2.22). Such a record also has the advantage of indicating the upper and lower signal levels, permitting optimum use to be made of the grey levels available in the recording system. In focus, the intensity step at edges shows the largest slope and the oscillogram shows reproducible fine details (Fig. 2.22a,b), but this method becomes difficult to use in the presence of noise (Fig. 2.22c).

The quality of focusing can also be described by introducing the notion of spatial frequency q. If a specimen has a single periodicity of period Λ

$$S(x) = S_0 + S_1 \sin(2\pi x/\Lambda + \phi_0) \tag{2.51}$$

the spatial frequency is defined as $q = 1/\Lambda$ with the dimension of a reciprocal length (periods per unit length). More generally, the spatial frequencies q are the components of the Fourier spectrum of a non-periodic signal

$$\begin{aligned} S(x) &= \int_{-\infty}^{+\infty} F(q) \exp(2\pi \mathrm{i} q x) \mathrm{d}q = \mathcal{F}^{-1}\{F(q)\} \\ \text{with} \quad F(q) &= \int_{-\infty}^{+\infty} S(x) \exp(-2\pi \mathrm{i} q x) \mathrm{d}x = \mathcal{F}\{S(x)\} , \end{aligned} \tag{2.52}$$

where $\mathcal{F}$ is a symbol for a Fourier transform and $\mathcal{F}^{-1}$ for an inverse Fourier transform.

If v denotes the velocity of the scanning beam at the specimen, a spatial frequency q results in a temporal frequency $f = vq$ in units of s^{-1} or Hz. If L denotes the width on the CRT, M the magnification, T the frame scan time and n the number of lines per frame, the velocity becomes

$$v = Ln/MT . \tag{2.53}$$

The maximum spatial frequency q_{max} is the reciprocal of the minimum detectable object periodicity Λ_{min}, which is related to twice the diameter d_{b}

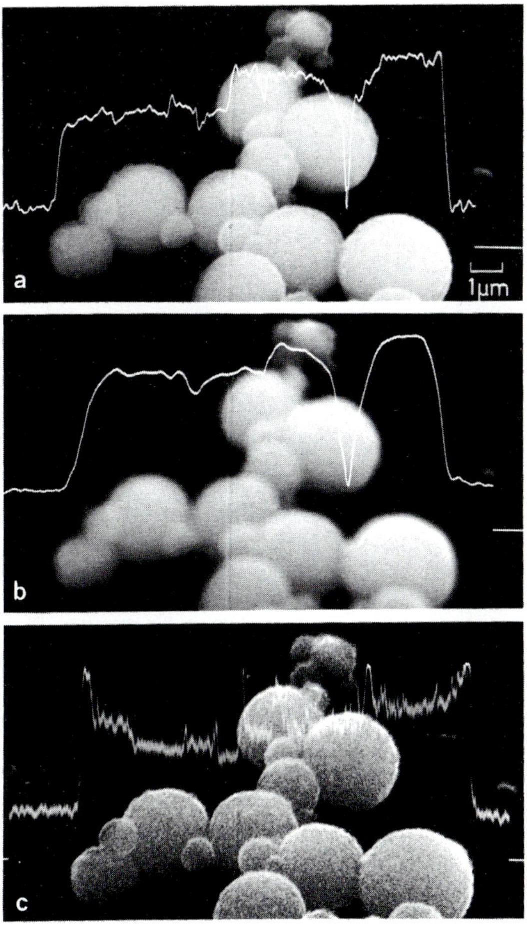

Fig. 2.22. SE micrographs from glass spheres in fly dust with superposed line-scans (**a**) in focus, (**b**) defocused and (**c**) with noise caused by a low electron-probe current

of the CRT beam (Sect. 2.4.4) or limited by the electron-probe diameter d_p. The maximum temporal frequency then becomes

$$f_{\max} = vq_{\max} = \frac{Ln}{Td_\mathrm{b}} \quad \text{or} \quad f_{\max} = \frac{Ln}{MTd_\mathrm{p}} \,. \tag{2.54}$$

With $d_\mathrm{p} = 10$ nm, $L = 10$ cm, $M = 10000$, $n = 1000$ lines per frame and a frame time $T = 1$ s, this gives $f_{\max} = 1$ MHz.

In the frequency spectrum, a focused image should therefore show as high spatial or temporal frequencies as possible. High-pass filtering of the video signal can hence be used for automatic focusing (Sect. 2.4.3). The use of a two-dimensional Fourier transform either in a light-optical diffraction experiment or by digital computation is discussed in Sects. 2.4.4 and 6.4.4.

Astigmatism (Sect. 2.2.2) results in an anisotropic blurring of the image. In a through-focus series, edges parallel to the line foci $\mathrm{F_s}$ and $\mathrm{F_m}$ (Fig. 2.9c)

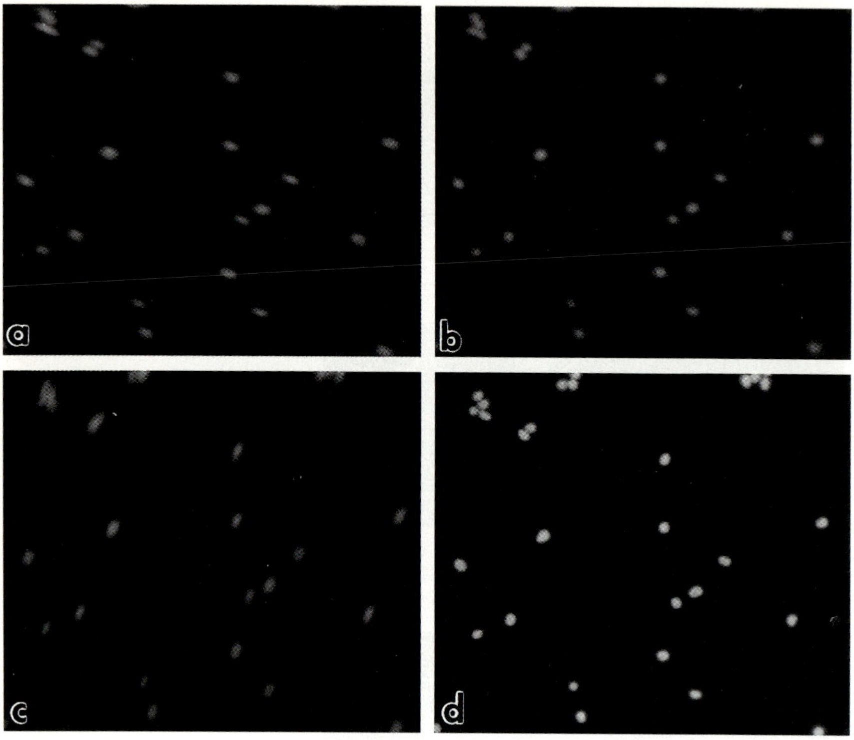

Fig. 2.23. Influence of astigmatism on focusing. The edges of the particles can be focused only in two perpendicular directions at the foci F_s (**a**) and F_m (**c**) and the image is blurred by astigmatism between the foci (**b**); sharp image (**d**) of 40 nm gold colloid after compensation of astigmatism

appear sharp, those normal to the foci diffuse. Images of small particles are blurred even at the line foci (Fig. 2.23a,c) and are not optimally focused between the two foci (Fig. 2.23b). The direction of blurring rotates through 90° when changing the focus from F_s to F_m.

Thus, two images at F_s and F_m indicate the direction of astigmatism when the stigmator coils are not excited. To compensate the astigmatism, the strength of the stigmator is first set at maximum and the astigmatism is rotated to be normal to its original direction. The strength of the stigmator can then be decreased until the astigmatism is as small as possible. Small further changes of strength and direction of the stigmator can optimally compensate the astigmatism (Fig. 2.23d). The original astigmatism has to be reasonably low to be capable of being compensated because otherwise stronger high-order astigmatisms remain after compensation of the first-order astigmatism. Charging of dirty diaphragms is a common source of astigmatism. This can be reduced by using thin-film diaphragms [2.99–101].

The digital detection and automatic compensation of astigmatism will be discussed in Sect. 2.4.3.

2.4.2 Depth of Focus

The depth of focus is affected by the electron-probe aperture α_p. The resolution on the CRT screen will be $d_b \simeq 0.1$ mm, which corresponds to one line-spacing for 1000 lines on a 10×10 cm screen. The corresponding resolution on the specimen will be $\delta = d_b/M$ and specimen details at a depth

$$T = \frac{\delta}{\tan\alpha_p} \simeq \frac{\delta}{\alpha_p} = \frac{d_b}{\alpha_p M} \tag{2.55}$$

will appear sharp (Fig. 2.24a).

It is informative to introduce a reduced depth of focus t to take account of the different magnifications on the CRT tube, the photographic film or the paper copy. We call B the field width of the image. The width of the raster on the specimen will be $b = B/M$ and the reduced depth of focus is defined as $t = T/b = MT/B$. Equation (2.55) then can be written as

$$t = \frac{d_b}{\alpha_p B} = \frac{10^{-3}}{\alpha_p} . \tag{2.56}$$

For high resolution, α_p has to be of the order of 10 mrad (Sect. 2.2.4) and we expect $t = 0.1$ for all magnifications, which means that the depth of focus extends over one tenth of the field width of the image (Fig. 2.24b). At low magnification, α_p can be reduced below 1 mrad by using a smaller diaphragm in the probe-forming lens and/or by increasing the working distance. Then,

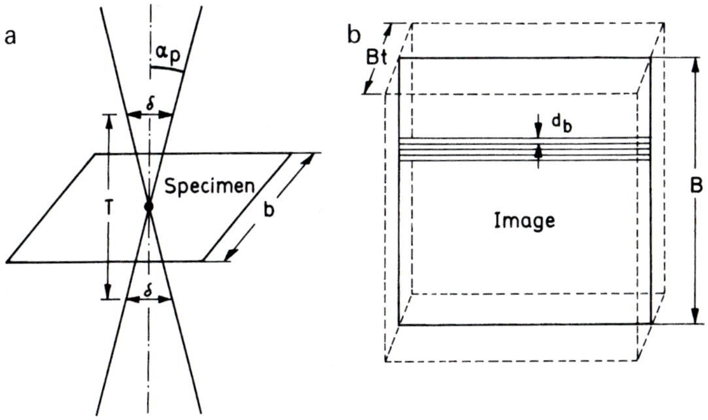

Fig. 2.24. Definition of geometric parameters of (**a**) the specimen and (**b**) the image used to describe the depth of focus T: δ = resolution, α_p = electron-probe aperture, t = reduced depth of focus, d_b = separation of lines on the CRT, B = width of the CRT screen, b = width of the raster on the specimen

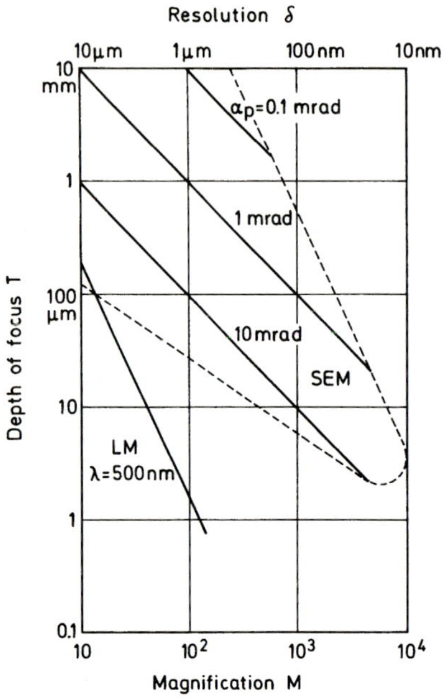

Fig. 2.25. Dependence of the depth of focus T on magnification M and resolution δ for different electron-probe apertures α_p. *Dotted line*: resolution limit due to Fig. 2.14. For comparison: depth of focus for a light microscope (LM)

t can become unity and the image will be sharp inside a cube of side equal to the field width.

The dependence of T on the resolution δ is plotted in Fig. 2.25 for different apertures α_p. The dotted line indicates the limit where the electron-probe diameter d_p from (2.29) becomes larger than $\delta = d_b/M$. For comparison, we look at the corresponding relation for a light microscope, for which the resolution will be

$$\delta = \lambda/n \sin\alpha \,, \tag{2.57}$$

where n denotes the refractive index of the medium between specimen and objective lens. This relation not only contains the resolution limit for the maximum possible value of $n \sin\alpha$ but also tells us how far α can be decreased for a preset value of δ to get the maximum depth of focus. The substitution of $\sin\alpha$ from (2.57) into (2.55) results in

$$T = \delta \sqrt{\delta^2/\lambda^2 - 1} \,. \tag{2.58}$$

This relation between T and δ is also plotted in Fig. 2.25 and shows the great advantage of the scanning electron microscope so far as the depth of focus is concerned even at low magnifications.

2.4.3 Dynamic and Automatic Focusing

Though the depth of focus is relatively large, inclined flat specimens cannot be focused over the whole specimen area. In this case, dynamic focusing can

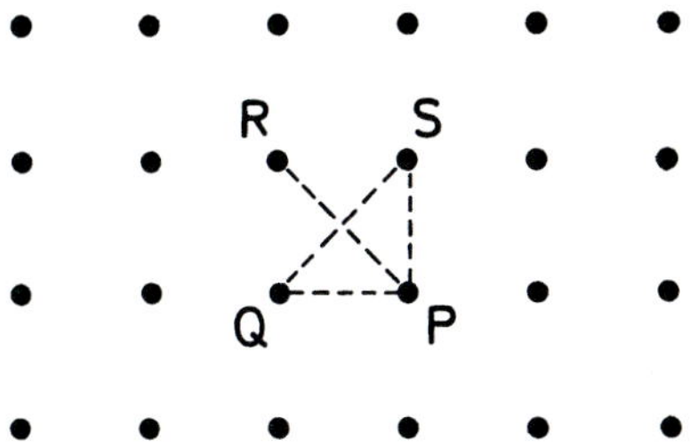

Fig. 2.26. Digital raster for automatic focusing by calculating the image intensity difference between P and Q and P and S for each raster point P. The additional difference between P and R is used for automatic correction of astigmatism

be employed [2.102], the focal length being changed during the scan in the y direction. This can be achieved by a linear change of the current of the probe-forming lens as a function of the current through the scan coils. However, this changes the magnification between the top and the bottom of the micrograph.

Since much time is occupied in focusing an image, it is of interest to seek automatic focusing methods. We have seen in Sect. 2.4.1 that the Fourier spectrum of the spatial frequencies q or the temporal frequencies f extends furthest in the q or f directions at focus and defocusing results in a decrease of the high-frequency part of the spectrum if this is not caused by noise. An analogue automatic focusing method has been proposed [2.103, 104], which uses the maximum of the high-frequency signal to control the lens current. The video signal behind the detector and preamplifier goes through a low-pass filter (f_{max}= 1–1000 Hz, preferably 200 Hz); this part of the signal is then subtracted from the total signal and the lens current is changed to maximize the difference signal. An accuracy in focus of ± 2 μm is obtainable using a 200 μm diaphragm. As Fig. 2.24 shows, this value lies inside the depth of focus. The method can also be used for dynamic focusing.

A digital automatic focusing method has been proposed [2.105, 106] that uses a digital scan (Sect. 5.5.3) of a 64×64 point frame, for example. A two-dimensional intensity gradient

$$G = \Delta I_1 + \Delta I_3; \quad \Delta I_1 = |I_\mathrm{P} - I_\mathrm{Q}|; \quad \Delta I_3 = |I_\mathrm{P} - I_\mathrm{S}| \tag{2.59}$$

is calculated for the point P and its neighbours Q and S (Fig. 2.26) in the same and the preceding line, respectively. The sum of the G values for all points of the frame is stored for a series of stepwise increasing lens currents. The optimum lens current will be that with the highest value of G. The procedure is repeated with progressively smaller current steps centred on the previously calculated best value.

For the measurement and correction of astigmatism it is necessary to detect differences in directionality of specimen details. In the presence of astigmatism, the specimen details will be blurred below and above the focus in perpendicular directions. Additional gradients $\Delta I_2 = |I_\mathrm{P} - I_\mathrm{R}|$ and $\Delta I_4 = |I_\mathrm{Q} - I_\mathrm{S}|$ are formed (Fig. 2.26) and together with ΔI_1 and ΔI_3, (2.59), provide four spatial directional derivatives which form a four-component vector $\boldsymbol{V}$. The vector components are summed for all points of the frame: ΔI_2 and ΔI_4

are divided by $\sqrt{2}$ to compensate for the wider point spacing. Two vectors $\boldsymbol{V}_1$ and $\boldsymbol{V}_2$, one below and one above the focus are normalized, so that the totals of both sets are equal. This ensures that the change in directionality is not masked if one of the scans is closer to the focus than the other. The directionality change is defined by

$$S = \sum_{i=1}^{4} |V_{1i} - V_{2i}| \tag{2.60}$$

and should be smallest when the astigmatism is compensated. The components of the vectors can also be used to calculate the strength and the direction of the remaining astigmatism to generate values for an automatic astigmatism control [2.105].

2.4.4 Resolution and Electron-Probe Size

Probe size and resolution are interrelated in SEM. The resolution can be worse but not better than the probe size. In principle, digital deblurring methods (Sect. 6.4.4) can provide a resolution better by a factor of two or three than the probe size. When using the SE mode, the high resolution information is the result of the excitation of SE by the incident electron probe within an exit depth t_{SE}, of the order of a few nanometres. If the electron probe size is made smaller than t_{SE} by using a field-emission gun, the resolution can be limited to an information volume of the order of t_{SE}^3. We have also to consider the delocalization of SE generation. Surface plasmons can be excited in small particles even if the electrons pass outside the particle at some distance x [2.107]. The excitation probability decreases exponentially as $\exp(-x/x_{\text{o}})$, where x_{o} is of the order of one nanometre. Part of the SE emission of solids is excited by the decay of plasmons [2.108]. Furthermore, each inelastic scattering process resulting in the excitation of SE will be delocalized. The resolution of the SE mode will hence be limited to the order of 1–2 nm.

When the BSE and x-ray modes are used, the resolution will be limited by the diameter of the diffusion cloud or the generation volume of x-rays; this is of the order of the electron range, typically a few tenths of a micrometre depending on the material and the electron energy (Sect. 3.4.2). However, smaller details can be resolved, depending on the specimen geometry, small particles of high atomic number on a substrate or cracks for example; the resolution then depends on the contrast relative to the signal background caused by the extended diffusion. The contribution of diffusely scattered BSE will be lower for thin electron-transparent film specimens. The resolution and contrast of BSE images can also be increased by energy filtering, especially by using low-loss electrons (LLE in Fig. 1.5). In LVSEM, the electron range and the exit depth of SE become of the same order of magnitude.

This means that a measurement of electron-probe size is not necessarily identical with a measurement of resolution. The latter is also related to the

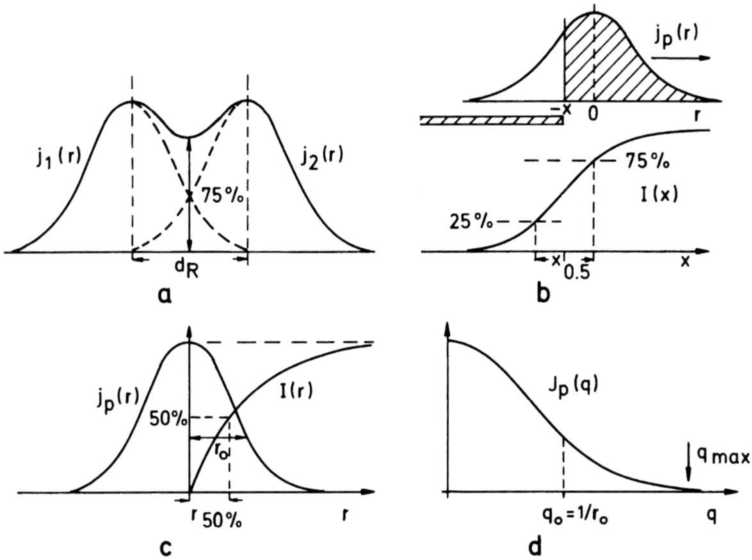

Fig. 2.27. Definition of resolution by (**a**) Rayleigh criterion, (**b**) edge resolution, (**c**) radial intensity distribution and (**d**) maximum spatial frequency $q_{\max} = 1/\Lambda_{\min}$

need for sufficient contrast, which in turn depends on the specimen structure and the imaging mode.

The geometric theory of electron-probe formation in Sect. 2.2.4 assumed Gaussian profiles of the geometric probe and the aberration discs. The square of the total width of the electron probe was then obtained by summing the squares of the individual contributions (2.29). A more general definition of resolution and/or electron-probe size should be free of assumptions about the electron-probe profile. The following definitions can be used:

a) Rayleigh Criterion. Two points a distance d_R apart will be resolved when their current density distributions $j(r)$ overlap at half their separation with an intensity drop of $\leq$75% of the maximum intensities (Fig. 2.27a).
b) Edge Resolution. When an electron probe scans across and perpendicular to a sharp edge, the transmitted intensity becomes a smoothed step function (Fig. 2.27b):

$$I(x) = \int_{-\infty}^{x} \left[\int_{-\infty}^{+\infty} j(\sqrt{x^2 + y^2}) \mathrm{d}y \right] \mathrm{d}x; \quad -\infty < x < +\infty \,. \tag{2.61}$$

The edge resolution $x_{0.5}$ can be defined as the distance between points corresponding to 25% and 75% of the total step height, for example.

c) Radial Intensity Distribution. The radial intensity is defined by

$$I(r) = \int_0^r j(r) 2\pi r \mathrm{d}r / I_\mathrm{p} \,. \tag{2.62}$$

The size of the electron probe can then be characterized by the diameter of the circle that contains $m = 50\%$, 75% or any other fraction of the total probe current I_p (Fig. 2.27c). This method has been used to obtain the results of Fig. 2.16.

d) Maximum Spatial Frequency. The resolution is defined to be the maximum spatial frequency $q_{\mathrm{max}} = 1/\Lambda_{\mathrm{min}}$ at which the contrast-transfer drops to such a small value that this minimum periodicity Λ_{min} cannot be detected in the micrograph (Fig. 2.27d).

The Rayleigh criterion has limited practical value because point-like specimen structures are needed that are smaller in size than the electron-probe diameter and show a distribution of separations inside a range of the order of the probe size. Resolution limits of 10–15 nm when using a probe diameter of 5–7.5 nm [2.109] and of 5 nm when using a 3 nm probe provided by an $\mathrm{LaB_6}$ cathode [2.110] have been demonstrated. The distance between two practically resolved points is thus of the order of twice the electron-probe diameter. This has been confirmed in experiments with larger probe diameters [2.111]. Such experiments also demonstrate that both resolution and adequate contrast are necessary to resolve a structure. The resolution also becomes worse when transparent specimens are prepared on a solid substrate of high Z. The SE signal generated directly by the electron probe decreases in relative amplitude owing to the contribution of the backscattered electrons to the SE yield.

Resolution Testing. The image is often focused by looking at the sharpness of edges; by recording a linescan across an edge, profiles like that of Fig. 2.27b can be obtained from which an edge resolution $x_{0.5}$ can be derived. The best results are obtained in the transmission mode with a scintillator or semiconductor detector placed below a sharp non-transparent edge. For high resolution, a sharp edge can be realized by evaporating indium crystals on a carbon supporting film [2.66, 112] or by preparing a sharp silicon edge [2.113].

A related problem is the imaging of surface steps of height h, which will also depend on the contrast of the steps. Artificial elliptically-shaped surface steps can be prepared by oblique evaporation of germanium films behind polystyrene spheres [2.114]. When thin gold layers are evaporated by the same method, the island structure results in decreasing gold particle sizes at the elliptical shadow edge. This is often used as a high-resolution test [2.115]. If the layers have been prepared on thin supporting films, the SE or BSE image can be compared with a TEM micrograph of the same area.

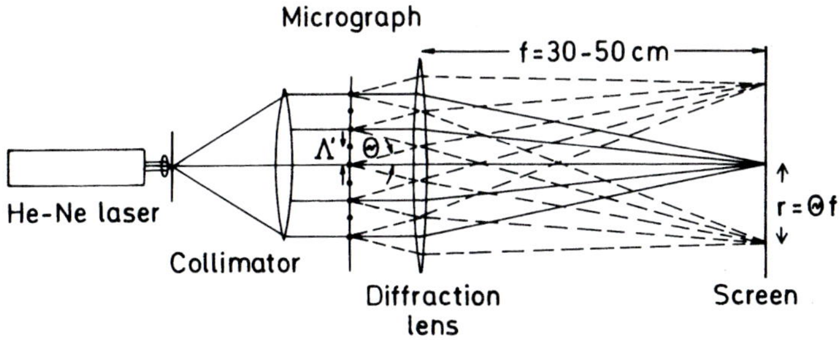

Fig. 2.28. Ray path of a light-optical Fraunhofer diffraction unit for light-optical Fourier transform of a micrograph ($\Lambda' = M\Lambda$ spacing of periodic structure on the micrograph, $\theta = \lambda/\Lambda'$ diffraction angle of first-order maxima)

There is no way of measuring the radial distribution directly. Larger cross-sections, the caustic cross-section when defocusing, for example, can be imaged by examining their contamination spots [2.116].

The quality of a micrograph can be assessed by looking at the smallest resolvable details and using the Rayleigh criterion, for example. When noise is present, it can be difficult to distinguish small resolved details from noise fluctuations. This problem can be avoided by taking two successive micrographs of the scanned area: only those structures that appear on both micrographs will be considered as resolved. This subjective technique of visual comparison can be made more objective by applying the light-optical diffraction technique (Fig. 2.28) [2.117], which is also used in TEM for the control of high resolution. A periodicity $\Lambda' = M\Lambda$ in the micrograph can be regarded as a diffraction grating and will thus generate first-order diffraction maxima at diffraction angles $\theta = \pm\lambda/\Lambda'$. A lens of long focal length f = 30–50 cm creates a Fraunhofer diffraction pattern by concentrating all rays of a parallel beam of direction θ at a single point of the focal plane at a distance $r = \theta f = \lambda f/\Lambda'$ from the axis. We call $q' = 1/\Lambda'$ the spatial frequency, which will be proportional to r. The diffraction pattern shows the distribution of spatial frequencies of the micrograph. This intensity distribution may contain quite sharp diffraction maxima if pronounced periodicities exist in the micrograph but often, the distribution will be diffuse extending to larger spatial frequencies as a result of noise. When we put the second micrograph behind the first one, shifted laterally by a distance d, all resolved structures will occur twice, distance d apart. The diffraction pattern of a double-point source consists of Young's interference fringes, with a fringe spacing $x_{\mathrm{f}} = \lambda f/d$, which lie perpendicular to the direction of the shift (Fig. 2.29). We can then estimate the radial distance (spatial frequency) at which we can still see such an overlapping Young fringe pattern, which means that the corresponding spatial frequencies appear twice in both micrograph. When applying this method to SEM micrographs, we have to consider that the latter also contain a grainy structure coming from the CRT screen. The two micrographs

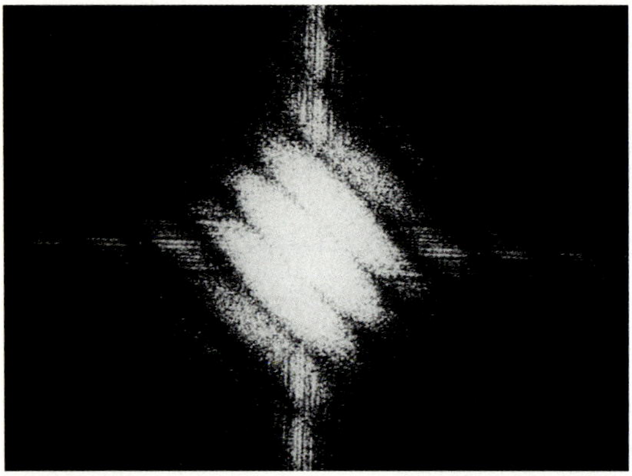

Fig. 2.29. Young's interference fringes obtained by digital Fourier transform of two superposed micrographs shifted in the x and y directions by 10 pixels

should therefore be recorded by shifting the image on the CRT screen between the two exposures by a distance of the order of one centimetre so that the granularity results in very small Young's fringe spacings which cannot be resolved.

This optical analogue technique, applied to developed emulsions, can also be performed numerically by calculating the on-line two-dimensional Fourier transform of the digitally stored image [2.118]. The Fraunhofer diffraction pattern that is observed at the focal plane in a light-optical diffraction experiment (Fig. 2.28) is none other than the squared Fourier transform of the light amplitude transmitted through the micrograph. In a first-order approximation, we can assume that an ideal resolved specimen structure $j_s(x, y)$ is convolved (blurred) with the normalized current-density profile $j_p(r)$ of the electron probe

$$j(x, y) = j_s(x, y) \otimes j_p(r) . \tag{2.63}$$

The convolution theorem of Fourier transforms says that the Fourier transform of two convolved functions is the product of their Fourier transforms

$$J(q_x, q_y) = J_s(q_x, q_y) \cdot J_p(q) . \tag{2.64}$$

The Fourier transform $J_p(q)$ of the radial density distribution $j_p(r)$ therefore acts as a contrast-transfer envelope (Fig. 2.27d). A Gaussian profile (2.5) of j_p with width r_0 results in a Gaussian profile $J_p(q)$ with a width $q_0 = 1/r_0$. This reciprocal relation between r_0 and q_0 implies that increasing the electron-probe diameter results in a decrease in q_0 or q_{max}.

3. Electron Scattering and Diffusion

Elastic and inelastic scattering processes result in zig-zag trajectories of electrons in a solid until the electrons come to rest by gradual deceleration or leave the specimen as backscattered electrons. Elastic large-angle scattering differs considerably from that characterized by the widely used Rutherford cross-sections and Mott cross-sections have to be used for more accurate calculations. The ionization cross-section of inner shells is important for calculating the number of characteristic x-ray quanta generated. The influence of inelastic scattering on deceleration can be treated by Bethe's continuous-slowing-down approximation without knowing the inelastic scattering processes in detail. The angular, spatial and energy distributions after passage through thin films or surface layers can be treated by multiple-scattering theories. The total electron diffusion is a very complex process. The properties of practical interest are the dependence of transmission on specimen thickness, the electron range and also the depth and spatial distributions of dissipated energy since this can generate electron-hole pairs in semiconductors, phonons or heat and can cause radiation damage and charging by the electron beam. The so-called diffusion models are very crude and detailed calculations using the transport equation or the Monte Carlo method can only be made numerically on a computer.

3.1 Elastic Scattering

3.1.1 Elastic Differential Cross-Section

The elastic scattering of electrons at the nuclei of the specimen is one of the most important interactions that influence electron diffusion and backscattering. It is evident that, owing to the wave character of electrons, scattering can be described exactly only by quantum mechanics. Nevertheless, a scattering model based on classical mechanics is presented in Fig. 3.1a to demonstrate the meaning of the differential cross-section $\mathrm{d}\sigma/\mathrm{d}\Omega$ and to give readers unfamiliar with quantum mechanics an idea of what elastic scattering means.

The electron is treated as a small particle of mass $m = 9.109 \times 10^{-31}$ kg and electric charge $Q = -e = -1.602 \times 10^{-19}$ C. It is attracted to the nucleus of charge $+Ze$ (Z: atomic number) by the Coulomb force

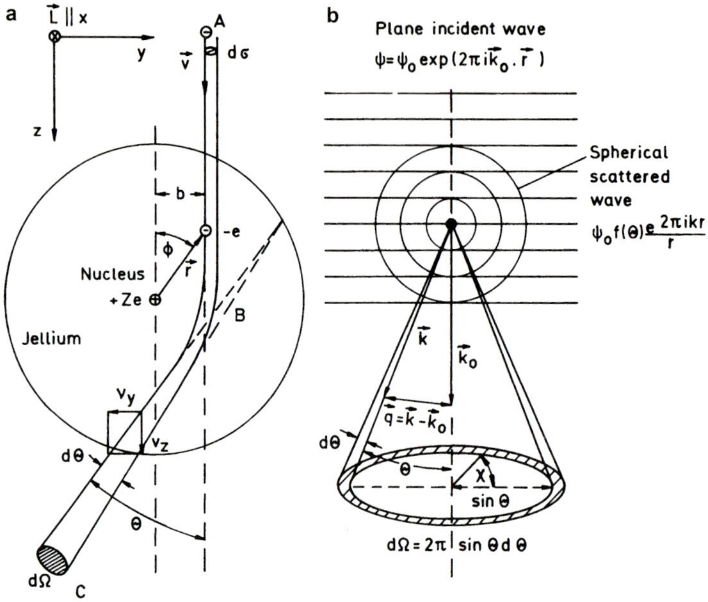

Fig. 3.1. (**a**) Classical model of elastic electron scattering at a nucleus (b = impact parameter). Electrons that hit an area $\mathrm{d}\sigma$ are scattered through an angle θ into a solid angle $\mathrm{d}\Omega$ ($\mathrm{d}\sigma/\mathrm{d}\Omega$ = differential cross-section). (**b**) Wave-mechanical model of elastic scattering (asymptotic solution). A plane incident wave with wave vector $\boldsymbol{k}_0$ generates a spherical wave of scattered amplitude $f(\theta)$. The scattering direction is described by θ or $\boldsymbol{q} = \boldsymbol{k} - \boldsymbol{k}_0$, where $|\boldsymbol{k}| = |\boldsymbol{k}_0| = 1/\lambda$

$$\boldsymbol{F} = -\frac{e^2 Z}{4\pi\epsilon_0 r^2}\boldsymbol{u}_r \tag{3.1}$$

where $\boldsymbol{u}_r$ denotes a radial unit vector. We assume that the nucleus remains at rest since its mass is substantially greater and that there is no interaction with the orbiting electrons of the atomic jellium. The latter interaction falls into the category of inelastic scattering (Sect. 3.2). For elastic scattering, the atomic electrons only produce a screening of the nuclear charge, causing the Coulomb force (3.1) to decrease more strongly than r^{-2} and to become zero outside the (neutral) atom.

The central, attractive Coulomb force proportional to r^{-2} acts on the electrons like the gravitational force of the sun on a comet, which results in hyperbolic trajectories. The scattering angle θ decreases with increasing impact parameter b (Fig. 3.1a). Electrons from a parallel beam that pass an area $\mathrm{d}\sigma$ are scattered into a cone of solid angle $\mathrm{d}\Omega$. The ratio $\mathrm{d}\sigma/\mathrm{d}\Omega$ is called the differential cross-section for scattering through an angle θ.

In quantum mechanics, the atom causes a phase shift of the plane, incident wave and the wavefront is strongly distorted behind the atom. At a large distance, the resulting 'asymptotic wavefield' can be described by a superposition of the plane wavefield and a spherical wave with a scattering amplitude $f(\theta)$ depending on the scattering angle θ (Fig. 3.1b).

$$\psi(r) = \psi_0 \left[\exp(2\pi \mathrm{i} \boldsymbol{k}_0 \cdot \boldsymbol{r}) + f(\theta)\frac{\exp(2\pi \mathrm{i} kr)}{r}\right] \tag{3.2}$$

where $|\boldsymbol{k}| = 1/\lambda$ denotes the wave number, and $\boldsymbol{k}$ is a vector parallel to the wave propagation. The differential cross-section $\mathrm{d}\sigma/\mathrm{d}\Omega$ and the scattering amplitude $f(\theta)$ are related by

$$\mathrm{d}\sigma/\mathrm{d}\Omega = |f(\theta)|^2 \tag{3.3}$$

and the problem is to find an expression for $f(\theta)$.

For most TEM applications, it is only necessary to know the scattering cross-sections into smaller angles. Methods of calculating small-angle cross-sections are discussed in detail in the TEM book [3.1]. For SEM it is essential to know the scattering through large angles between 0 and π (180°). An important weakness of the theoretical treatments of electron scattering developed during the last decades is the assumption that the electron large-angle scattering can be described as Rutherford scattering. This treatment neglects the existence of electron spin and of spin-orbit coupling during scattering. The exact solutions are Mott cross-sections, which can differ considerably from Rutherford cross-sections (Sect. 3.1.6).

3.1.2 Classical Mechanics of Electron-Nucleus Scattering

The first theoretical treatment of the scattering of a charged particle by a nucleus was presented by Rutherford in 1911 to explain the scattering of alpha particles at thin metal foils. The experiments suggested in the model in which the great majority of the mass of the atom is concentrated within the nucleus, of diameter less than 10^{-2} pm, and the nucleus has a charge $+Ze$.

In the following classical treatment of Rutherford scattering of an electron at a nucleus (see [3.2], for example), we neglect the influence of screening of the Coulomb force by the atomic jellium which is treated quantum-mechanically in Sect. 3.1.5, and we assume that the nucleus stays at rest. The incident electron of velocity v would pass the nucleus at a distance b – the impact parameter – if there were no interaction (Fig. 3.1a). The interaction can be described by the attractive Coulomb force (3.1) and Newton's equation becomes

$$\frac{\mathrm{d}\boldsymbol{p}}{\mathrm{d}t} = m\ddot{\boldsymbol{r}} = -\frac{e^2 Z}{4\pi\epsilon_0}\frac{\boldsymbol{r}}{r^3}\ . \tag{3.4}$$

Conside now the angular momentum $\boldsymbol{L} = \boldsymbol{r} \times \boldsymbol{p}$; its time derivative

$$\frac{\mathrm{d}\boldsymbol{L}}{\mathrm{d}t} = \dot{\boldsymbol{r}} \times \boldsymbol{p} + \boldsymbol{r} \times \dot{\boldsymbol{p}} = 0 \tag{3.5}$$

vanishes because $\dot{\boldsymbol{r}}$ and $\boldsymbol{p} = m\dot{\boldsymbol{r}}$ are parallel as are $\boldsymbol{r}$ and $\dot{\boldsymbol{p}}$ as shown by (3.4). The angular momentum is therefore constant in time, which means that the electron will move in a $y - z$ plane normal to $\boldsymbol{L}$ = const. This is a general result for central forces. In this plane, we introduce polar coordinates r and φ (Fig. 3.1a). The magnitude of $\boldsymbol{L}$,

$$|\boldsymbol{L}| = L = mvb \tag{3.6}$$

is obtained by considering the electron at a point A far from the atom. At any other position, the angular momentum will be

$$mr^2 \mathrm{d}\varphi/\mathrm{d}t = mvb = \text{const} . \tag{3.7}$$

The y component of (3.4) becomes

$$m\frac{\mathrm{d}v_y}{\mathrm{d}t} = F_y = F \sin\varphi = \frac{e^2 Z}{4\pi\epsilon_0 r^2} \sin\varphi . \tag{3.8}$$

Elimination of r^2 by using (3.7) results in

$$\frac{\mathrm{d}v_y}{\mathrm{d}t} = \frac{e^2 Z}{4\pi\epsilon_0} \frac{1}{mvb} \sin\varphi \frac{\mathrm{d}\varphi}{\mathrm{d}t} . \tag{3.9}$$

We integrate (3.9) from A to C in Fig. 3.1a to find the scattering angle θ of the electron. At point A, the value of v_y is zero and $\varphi \to 0$. At point C, we have $v_y = v \sin\theta$ and $\varphi \to \pi + \theta$. At C, the velocity is again v because the electron is accelerated from A to B and decelerated from B to C. This is a consequence of conservation of energy. Integration of (3.9) gives

$$\int_0^{v \sin\theta} \mathrm{d}v_y = \frac{e^2 Z}{4\pi\epsilon_0} \frac{1}{mvb} \int_0^{\pi+\theta} \sin\varphi \mathrm{d}\varphi \tag{3.10}$$

or

$$v \sin\theta = \frac{e^2 Z}{4\pi\epsilon_0} \frac{1}{mvb} (1 + \cos\theta) \tag{3.11}$$

and with $\cot(\theta/2) = (1 + \cos\theta)/\sin\theta$, we obtain

$$\cot\frac{\theta}{2} = \frac{4\pi\epsilon_0}{e^2 Z} mv^2 b \quad \text{or} \quad b = \frac{e^2 Z}{4\pi\epsilon_0} \frac{1}{mv^2} \cot\frac{\theta}{2} , \tag{3.12}$$

which gives the scattering angle θ in terms of the impact parameter b. The cross-section per atom for scattering through all azimuths into a cone between θ and $\theta + \mathrm{d}\theta$ corresponding to a solid angle $\mathrm{d}\Omega = 2\pi \sin\theta \mathrm{d}\theta$ (Fig. 3.1a) becomes

$$\mathrm{d}\sigma = 2\pi b \mathrm{d}b = \frac{e^4 Z^2}{4(4\pi\epsilon_o)^2 m^2 v^4} \frac{2\pi \sin\theta \mathrm{d}\theta}{\sin^4(\theta/2)} \tag{3.13}$$

where b and $\mathrm{d}b$ are substituted from (3.12), or

$$\frac{\mathrm{d}\sigma}{\mathrm{d}\Omega} = \frac{e^4 Z^2}{4(4\pi\epsilon_0)^2 m^2 v^4} \frac{1}{\sin^4(\theta/2)} = \frac{e^4 Z^2}{(4\pi\epsilon_0)^2 E^2} \left(\frac{E+E_0}{E+2E_0}\right)^2 \frac{1}{(1-\cos\theta)^2} . \tag{3.14}$$

This is the differential Rutherford cross-section for scattering by an unscreened nucleus. The large decrease of $\mathrm{d}\sigma/\mathrm{d}\Omega$ with increasing θ is shown as dashed lines in the polar diagram of Fig. 3.2 where a logarithmic scale is used. The Rutherford cross-section (3.14) has a singularity at $\theta \to 0$. Small scattering angles correspond to large values of b, and the singularity is a consequence of the long-range Coulomb field of the unscreened nucleus and disappears when screening is considered in Sect. 3.1.5.

For later applications in Sects. 3.4.3 and 4.1.2, we need the partial cross-section for scattering through angles $\theta \geq \alpha$, i.e.,

$$\sigma(\alpha) = \int_\alpha^\pi \frac{\mathrm{d}\sigma}{\mathrm{d}\Omega} 2\pi \sin\theta \mathrm{d}\theta . \tag{3.15}$$

Using the relation $\sin\theta = 2\sin(\theta/2)\cos(\theta/2)$ and substituting $u = \sin(\theta/2)$ and $\mathrm{d}u = \cos(\theta/2)\mathrm{d}\theta/2$, we obtain

$$\sigma(\alpha) = \frac{\pi e^4 Z^2}{(4\pi\epsilon_0)^2 m^2 v^4} \cot^2(\alpha/2) . \tag{3.16}$$

3.1.3 The Relativistic Schrödinger and Pauli–Dirac Equations

The total electron energy is the sum of the kinetic energy $E = p^2/2m$ and the potential energy $V = -e\Phi$, and the conservation of energy can be written

$$H = (p^2/2m) + V . \tag{3.17}$$

For example, at the cathode of an electron gun, we have $p = 0$ and a potential $\Phi = -U$ and at the anode $\Phi = 0$ and $E = eU$.

The Schrödinger equation of quantum mechanics for a free electron ($\Phi = 0$) is obtained by replacing the Hamiltonian H and the momentum $\boldsymbol{p}$ by the operators

$$H = \mathrm{i}\hbar\partial/\partial t ; \quad \boldsymbol{p} = -\mathrm{i}\hbar\nabla \tag{3.18}$$

and applying these operators to a wavefunction $\Psi(\boldsymbol{r}, t)$

$$(\hbar^2/2m)\nabla^2\Psi + \mathrm{i}\hbar\partial\Psi/\partial t = 0 . \tag{3.19}$$

If we seek a separated solution of the form

$$\Psi(\boldsymbol{r}, t) = \psi(\boldsymbol{r}) \exp[-\mathrm{i}(E - V)t/\hbar] \tag{3.20}$$

we obtain the time-independent Schrödinger equation for $\psi(\boldsymbol{r})$:

$$\nabla^2\psi + (2m/\hbar^2)(E - V)\psi = 0 , \tag{3.21}$$

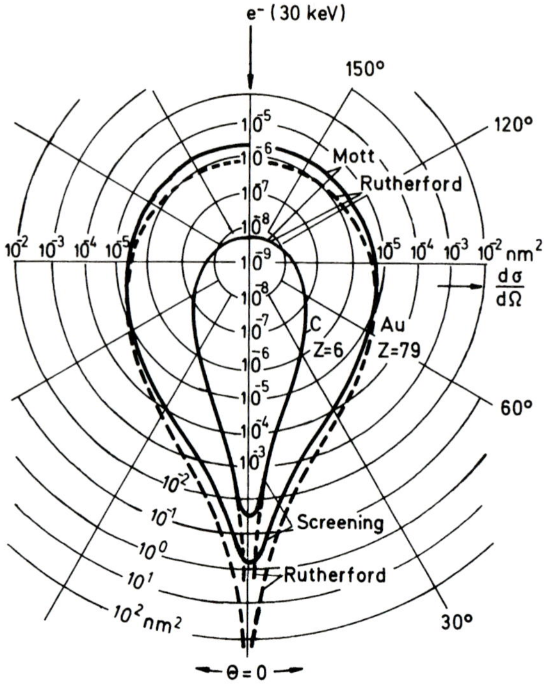

Fig. 3.2. Logarithmic polar diagram of the differential cross-sections $d\sigma_{el}/d\Omega$ for single carbon and gold atoms. The Rutherford formula (– – –) fails at low scattering angles θ because of screening effects and at large scattering angles because of the neglect of relativistic effects, which are considered by the Mott cross-section (——)

where $\nabla^2 = \partial^2/\partial x^2 + \partial^2/\partial y^2 + \partial^2/\partial z^2$ is the Laplace operator.

After substituting the relativistic mass (2.2) for m, this equation can also be used for high energies. A relativistically exact formula must however be covariant under a Lorentz transformation. Instead of (3.17), therefore we set out from the relativistic form of the law of energy conservation

$$(H - V)^2 = (\boldsymbol{p} - e\boldsymbol{A})^2 c^2 + E_0^2 \,, \tag{3.22}$$

where $\boldsymbol{p} - e\boldsymbol{A}$ denotes the canonical momentum with $\boldsymbol{A}$ as the magnetic vector potential ($\boldsymbol{B} = \nabla \times \boldsymbol{A}$) and $E_0 = 511$ keV is the electron rest energy.

When the operators (3.18) for $\boldsymbol{p}$ and H are substituted in (3.22), the Klein–Gordon equation is obtained. Dirac had the idea of splitting this equation into a product of two linear factors

$$\left[H - V - c\sum_i \alpha_i (p_i - eA_i) - \beta E_0\right] \times \left[H - V + c\sum_i \alpha_i (p_i - eA_i) + \beta E_0\right] \psi = 0 \,, \tag{3.23}$$

where $i = 1, 2, 3$ signify the x, y, z components of $\boldsymbol{p}$ and $\boldsymbol{A}$, respectively. It can easily be seen by multiplying these factors that the α_i and β have to fulfil the anticommutation conditions

$$\alpha_i\alpha_j + \alpha_j\alpha_i = 2\delta_{ij} \; ; \quad \alpha_i\beta + \beta\alpha_i = 0 \quad \text{and} \quad \beta^2 = 1 \; , \tag{3.24}$$

which can be satisfied not with ordinary numbers but with the 4×4 matrices (Clifford algebra):

$$\alpha_1 = \begin{bmatrix} 0 & 0 & 0 & 1 \\ 0 & 0 & 1 & 0 \\ 0 & 1 & 0 & 0 \\ 1 & 0 & 0 & 0 \end{bmatrix} ; \qquad \alpha_2 = \begin{bmatrix} 0 & 0 & 0 & -\mathrm{i} \\ 0 & 0 & \mathrm{i} & 0 \\ 0 & -\mathrm{i} & 0 & 0 \\ \mathrm{i} & 0 & 0 & 0 \end{bmatrix} ;$$

$$\alpha_3 = \begin{bmatrix} 0 & 0 & 1 & 0 \\ 0 & 0 & 0 & -1 \\ 1 & 0 & 0 & 0 \\ 0 & -1 & 0 & 0 \end{bmatrix} ; \qquad \beta = \begin{bmatrix} 1 & 0 & 0 & 0 \\ 0 & 1 & 0 & 0 \\ 0 & 0 & -1 & 0 \\ 0 & 0 & 0 & -1 \end{bmatrix} . \tag{3.25}$$

If we can solve the equation

$$\left[H - V - c\sum_i \alpha_i(p_i - eA_i) - \beta E_0\right]\psi = 0 \tag{3.26}$$

or the corresponding equation from the second factor in (3.23) with the positive signs of the last terms, then (3.23) is also solved. The linearized equation (3.26) has the advantage that it is of first order in $\partial/\partial t$ just like the Schrödinger equation (3.19), and is also linear in p. The solution ψ has to be a four-component vector. Equation (3.26) represents a system of four simultaneous first-order differential equations. Solving this system gives two linearly independent solutions $\psi_{1,2}(\boldsymbol{r}, t)$. These represent electrons with spin parallel or antiparallel to the direction of propagation. In the literature, it is often argued that these electron spin-eigenfunctions are a consequence of the relativistic Dirac equation (3.26). However, it can be shown that spin-eigenfunctions also result if the non-relativistic Schrödinger equation is linearized [3.3, 4]. The consequences on the elastic scattering cross-section and resulting strong deviations from the Rutherford formula (3.14) are discussed in Sect. 3.1.6.

3.1.4 Quantum Mechanics of Scattering

We begin this review of the basic concepts of the quantum mechanics of electron scattering by discussing the scattering at a hydrogen atom ($Z = 1$) (see [3.5], for example). If $\boldsymbol{r}_1, \boldsymbol{r}_2$ are the coordinates of the incident and atomic electrons, respectively, and $r_{12} = |\boldsymbol{r}_1 - \boldsymbol{r}_2|$ the distance between them (Fig. 3.3), the Coulomb potential energy V becomes

$$V = -\frac{e^2}{4\pi\epsilon_0}\left(\frac{1}{r_1} + \frac{1}{r_2} - \frac{1}{r_{12}}\right) . \tag{3.27}$$

where the first two terms are caused by the attractive force of the nucleus and the last term by the repulsive force between the two electrons. Substitution into the Schrödinger equation (3.21) results in

$$\left[\nabla_1^2 + \nabla_2^2 + \frac{2m}{\hbar^2}\left(E_{\text{tot}} + \frac{e^2}{4\pi\epsilon_0 r_1} + \frac{e^2}{4\pi\epsilon_0 r_2} - \frac{e^2}{4\pi\epsilon_0 r_{12}}\right)\right]\psi = 0 , \tag{3.28}$$

where $E_{\text{tot}} = E_0' + E$ is the sum of the energy E_0' of the ground state of the atomic electron and $E = eU$, the kinetic energy of the incident electron. The total wavefunction ψ will be a product of the wavefunction $\psi_0(\boldsymbol{r}_2)$ of the atomic ground state and of the incident plane wave $\exp(2\pi i \boldsymbol{k}_0 \cdot \boldsymbol{r}_1)$, which depends only on the coordinate $\boldsymbol{r}_1$ of the incident electron, and after scattering we can write for the asymptotic solution

$$\psi = \exp(2\pi i \boldsymbol{k}_0 \cdot \boldsymbol{r}_1)\psi_0(\boldsymbol{r}_2) + \Phi(\boldsymbol{r}_1, \boldsymbol{r}_2) . \tag{3.29}$$

The wave function $\psi_0(\boldsymbol{r}_2)$ and the wave functions $\psi_n(\boldsymbol{r}_2)$ of the atomic excited states obey the Schrödinger equation of the isolated hydrogen atom

$$\left[\nabla_2^2 + \frac{2m}{\hbar^2}\left(E_n' + \frac{e^2}{4\pi\epsilon_0 r_2}\right)\right]\psi_n = 0 , \tag{3.30}$$

where E_n' denotes the energy of the nth excited state. The ψ_n are orthonormal, which means that

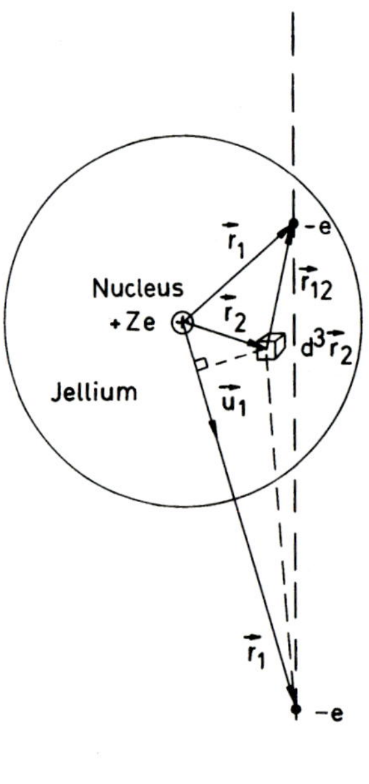

Fig. 3.3. Coordinates for formulating the wave-mechanical theory of scattering, $\boldsymbol{r}_1$ = coordinate of the beam electron, $\boldsymbol{r}_2$ = coordinate of the volume element $d^3\boldsymbol{r}_2$ of the atomic electrons

$$\int \psi_n \psi_m^* \mathrm{d}^3 \boldsymbol{r}_2 = \begin{cases} 1 & \text{for} \quad m = n \\ 0 & \text{for} \quad m \neq n \end{cases} \quad . \tag{3.31}$$

The unknown function $\Phi(\boldsymbol{r}_1, \boldsymbol{r}_2)$ in (3.29) can be expanded as a series of atomic wave functions

$$\Phi(\boldsymbol{r}_1, \boldsymbol{r}_2) = \sum_n F_n(\boldsymbol{r}_1) \psi_n(\boldsymbol{r}_2) \; . \tag{3.32}$$

Substitution of (3.32) into (3.29) and then into (3.28), making use of (3.30), gives

$$\begin{aligned} \sum_n \; & (\nabla_1^2 + 4\pi^2 k_n^2)\, F_n(\boldsymbol{r}_1)\, \psi_n(\boldsymbol{r}_2) \\ = \; & \frac{2me^2}{4\pi\epsilon_0 \hbar^2} \left[\left(\frac{1}{r_{12}} - \frac{1}{r_1} \right) \exp(2\pi \mathrm{i} \boldsymbol{k}_0 \cdot \boldsymbol{r}_1)\, \psi_0(\boldsymbol{r}_2) \right. \\ + \; & \left. \sum_n \left(\frac{1}{r_{12}} - \frac{1}{r_1} \right) F_n(\boldsymbol{r}_1)\, \psi_n(\boldsymbol{r}_2) \right] , \end{aligned} \tag{3.33}$$

with

$$k_n^2 = \frac{2m}{\hbar^2}(E - E_n' + E_0') = \frac{2m}{\hbar^2}(E - \Delta E) \; ; \tag{3.34}$$

$\Delta E = E_n' - E_0'$ denotes the energy loss of the incident electron. Multiplying both sides of (3.33) by $\psi(\boldsymbol{r}_2)$ and integrating over $\boldsymbol{r}_2$, using the orthogonality relations (3.31), gives

$$(\nabla_1^2 + 4\pi^2 k_m^2) F_m = U_{0m} \exp(2\pi \mathrm{i} \boldsymbol{k}_0 \cdot \boldsymbol{r}_1) + \sum_n U_{nm}(\boldsymbol{r}_1) F_m(\boldsymbol{r}_1) \; , \tag{3.35}$$

where

$$\begin{aligned} U_{nm} &= \frac{2m}{\hbar^2} V_{nm} = \frac{2m}{\hbar^2} \langle \psi_n | V | \psi_m^* \rangle = \frac{2m}{\hbar^2} \int \psi_n(\boldsymbol{r}_2) V(r_1) \psi_m^*(\boldsymbol{r}_2) \mathrm{d}^3 \boldsymbol{r}_2 \\ &= \frac{2m}{\hbar^2} \frac{e^2}{4\pi\epsilon_0} \int \psi_n(\boldsymbol{r}_2) \left(\frac{1}{r_{12}} - \frac{1}{r_1} \right) \psi_m^*(\boldsymbol{r}_2) \mathrm{d}^3 \boldsymbol{r}_2 \; . \end{aligned} \tag{3.36}$$

So far, no approximation has been made. We now assume that the first term on the right-hand side of (3.35), which contains the amplitude of the incident wave, is much larger than the other terms on the right-hand side and we obtain the approximation

$$(\nabla_1^2 + 4\pi^2 k_m^2) F_m = U_{0m} \exp(2\pi \mathrm{i} \boldsymbol{k}_0 \cdot \boldsymbol{r}_1) \; . \tag{3.37}$$

This very general treatment of the electron–atom interaction will be further discussed in Sect. 3.1.5 for elastic scattering ($m = 0$ only) and in Sect. 3.2.3 for inelastic scattering ($m \geq 1$) and will be extended to atoms with $Z \geq 1$.

3.1.5 Rutherford Scattering at a Screened Nucleus

Whereas Rutherford scattering at an unscreened nucleus was treated in Sect. 3.1.2 by classical mechanics, the influence of the atomic jellium, which screens the Coulomb potential of the nucleus, can only be treated by quantum mechanics. We use the method introduced in Sect. 3.1.4. For elastic scattering, no transition into higher atomic states occurs and the quantum number m remains zero in (3.37):

$$(\nabla_1^2 + 4\pi^2 k_0^2)F_0 = U_{00}(r_1)\exp(2\pi \mathrm{i}\boldsymbol{k}_0 \cdot \boldsymbol{r}_1) \ . \tag{3.38}$$

This equation can be solved in terms of a Green's function $G(r)$, which is a solution of

$$(\nabla^2 + 4\pi^2 k_0^2)\, G(r) = 4\pi\delta(r) \ . \tag{3.39}$$

This represents elastic scattering at a point scatterer represented by a δ-function. A suitable Green's function is

$$G(r) = -\exp(2\pi \mathrm{i}kr)/r \ , \tag{3.40}$$

which represents a spherical wave. The solution of (3.38) becomes

$$F_0 = \frac{1}{4\pi}\int G(|\boldsymbol{r}_1 - \boldsymbol{r}_2|)U_{00}(\boldsymbol{r}_2)\exp(2\pi \mathrm{i}\boldsymbol{k}_0 \cdot \boldsymbol{r}_2)\mathrm{d}^3\boldsymbol{r}_2 \tag{3.41}$$

as we may confirm by substitution of (3.41) in the left-hand side of (3.38). We find

$$\begin{aligned} &\frac{1}{4\pi}\int (\nabla_1^2 + 4\pi^2 k_0^2)\, G(|\boldsymbol{r}_1 - \boldsymbol{r}_2|)U_{00}(\boldsymbol{r}_2)\exp(2\pi \mathrm{i}\boldsymbol{k}_0 \cdot \boldsymbol{r}_2)\mathrm{d}^3\boldsymbol{r}_2 \\ = &\int \delta(\boldsymbol{r}_1 - \boldsymbol{r}_2)U_{00}(\boldsymbol{r}_2)\exp(2\pi \mathrm{i}\boldsymbol{k}_0 \cdot \boldsymbol{r}_2)\mathrm{d}^3\boldsymbol{r}_2 = U_{00}(\boldsymbol{r}_1)\exp(2\pi \mathrm{i}\boldsymbol{k}_0 \cdot \boldsymbol{r}_1) \ . \end{aligned} \tag{3.42}$$

which is identical with the right-hand side of (3.38). The asymptotic solution at a large distance r_1 of the incident electron from the nucleus is found by writing (Fig. 2.4):

$$|\boldsymbol{r}_1 - \boldsymbol{r}_2| \simeq r_1 - \boldsymbol{r}_2 \cdot \boldsymbol{u}_1 \tag{3.43}$$

and

$$\frac{\exp(2\pi \mathrm{i}k|\boldsymbol{r}_1 - \boldsymbol{r}_2|)}{|\boldsymbol{r}_1 - \boldsymbol{r}_2|} \simeq \frac{\exp(2\pi \mathrm{i}kr_1)}{r_1}\exp(-2\pi \mathrm{i}\boldsymbol{k} \cdot \boldsymbol{r}_2) \ , \tag{3.44}$$

where $\boldsymbol{u}_1$ is a unit vector. The last exponential term arises because $k\boldsymbol{u}_1 = \boldsymbol{k}$ is a wave vector in the scattering direction, whereas $\boldsymbol{k}_0$ denotes the wave vector of the incident plane electron wave. We introduce a scattering vector (Fig. 3.1b)

$$\boldsymbol{q} = \boldsymbol{k} - \boldsymbol{k}_0 \quad \text{with} \quad |\boldsymbol{q}| = |\boldsymbol{k} - \boldsymbol{k}_0| = 2\sin(\theta/2)/\lambda \ . \tag{3.45}$$

Substituting (3.40) and (3.44) in (3.41) and considering (3.45), we obtain

$$F_0 = -\frac{1}{4\pi}\frac{\exp(2\pi \mathrm{i}kr_1)}{r_1}\int U_{00}(\boldsymbol{r}_2)\exp(-2\pi \mathrm{i}\boldsymbol{q} \cdot \boldsymbol{r}_2)\mathrm{d}^3\boldsymbol{r}_2 \ . \tag{3.46}$$

Using (3.36), comparison with (3.2) yields the Born approximation for the scattering amplitude

$$f(\theta) = -\frac{m}{2\pi\hbar^2} \int V_{00}(r_2) \exp(-2\pi \mathrm{i} \boldsymbol{q} \cdot \boldsymbol{r}_2) \mathrm{d}^3\boldsymbol{r}_2 \,. \tag{3.47}$$

The potential $V_{00}(r_2) = V_{00}(r_1)$ is simply the screened Coulomb potential of the nucleus, as can be seen by rewriting (3.36) for $n, m = 0$ and representing an atom with Z electrons by means of the wavefunctions ψ_{0i} of the ground states

$$V_{00}(r_1) = \frac{e^2}{4\pi\epsilon_0} \sum_{i=1}^{Z} \int |\psi_{0i}(\boldsymbol{r}_2)|^2 \left(\frac{1}{r_{12}} - \frac{1}{r_1} \right) \mathrm{d}^3\boldsymbol{r}_2 \,. \tag{3.48}$$

The integral is split into a difference of two terms with the factors $1/r_{12}$ and $1/r_1$ inside the bracket. The last term in $1/r_1$, corresponds to the Coulomb potential of the nucleus allowing for the normalization of the atomic wavefunctions (3.31); the first term denotes the probability of finding an electron at the position $\boldsymbol{r}_2$:

$$V_{00}(r_1) = \frac{e^2 Z}{4\pi\epsilon_0 r_1} + \frac{e^2}{4\pi\epsilon_0} \int \frac{\rho(\boldsymbol{r}_2)}{|\boldsymbol{r}_1 - \boldsymbol{r}_2|} \mathrm{d}^3\boldsymbol{r}_2 \,, \tag{3.49}$$

where

$$\rho(\boldsymbol{r}_2) = \sum_{i=1}^{Z} \psi_{0i}\psi_{0i}^* \,. \tag{3.50}$$

The integral in (3.49) can be simplified if $\rho(r_2)$ is assumed to be rotationally symmetric. The Coulomb potential can then be calculated as if the charge inside the sphere of radius r_1 were concentrated at the centre

$$V_{00}(r_1) = -\frac{e^2 Z}{4\pi\epsilon_0 r_1} + \frac{e^2}{4\pi\epsilon_0 r_1} \int_0^{r_1} \rho(r_2) 4\pi r_2^2 \mathrm{d}r_2 = -\frac{e^2 Z_{eff}}{4\pi\epsilon_0 r_1} \,. \tag{3.51}$$

This screening effect can be approximated by a series of exponentials

$$V(r) = -\frac{e^2 Z}{4\pi\epsilon_0 r} \sum_j b_j \exp(-a_j r) \,, \tag{3.52}$$

where the coefficients a_j and b_j are calculated using Hartree–Fock–Slater functions ψ_{0i} or the Thomas–Fermi model for higher Z atoms [3.6].

We use here the approximate Wentzel model with one exponential term only

$$V(r) = -\frac{e^2 Z}{4\pi\epsilon_0 r} \mathrm{e}^{-r/R} \,. \tag{3.53}$$

where $R = a_{\mathrm{H}} Z^{-1/3}$ ($a_{\mathrm{H}} = h^2\epsilon_0/\pi m_0 e^2 = 0.0569$ nm, the Bohr radius) denotes the screening radius of the atom. This simplified model will be sufficient to bring out the most important properties of the solution of the Schrödinger equation.

Substitution of $V(r)$ from the Wentzel model (3.53) in (3.47) [3.7] with the polar coordinates φ and χ and $\boldsymbol{q}\cdot\boldsymbol{r} = qr\cos\varphi$ and $\mathrm{d}^3\boldsymbol{r} = r^2\sin\varphi\mathrm{d}\varphi\mathrm{d}\chi\mathrm{d}r$ gives

$$
\begin{aligned}
f(\theta) &= \frac{m}{2\pi\hbar^2}\int_0^\infty \frac{e^2 Z}{4\pi\epsilon_0 r} r^2 \mathrm{e}^{-r/R}\mathrm{d}r \underbrace{\int_0^\pi \exp(-2\pi \mathrm{i} qr\cos\varphi)\sin\varphi\mathrm{d}\varphi}_{\sin(2\pi qr)/\pi qr} \underbrace{\int_0^{2\pi}\mathrm{d}\chi}_{2\pi} \\
&= \frac{m}{\pi\hbar^2}\frac{e^2 Z}{4\pi\epsilon_0 q} \underbrace{\int_0^\infty \mathrm{e}^{-r/R}\sin(2\pi qr)\mathrm{d}r}_{2\pi q R^2/(1+4\pi^2 q^2 R^2)} \ . \qquad (3.54)
\end{aligned}
$$

With the de Broglie wavelength $\lambda = h/mv$, with (3.45) and with the introduction of a characteristic angle θ_0:

$$\sin(\theta_0/2) \simeq \theta_0/2 = \lambda/4\pi R \ , \qquad (3.55)$$

(3.54) may be written

$$f(\theta) = \frac{e^2 Z}{2(4\pi\epsilon_0)mv^2}\frac{1}{\sin^2(\theta/2)+\sin^2(\theta_0/2)} \ . \qquad (3.56)$$

The resulting scattering amplitude $f(\theta)$ is independent of the azimuth χ owing to the central symmetry of the potential $V(r)$. The differential cross-section becomes

$$
\begin{aligned}
\frac{\mathrm{d}\sigma_{\mathrm{el}}}{\mathrm{d}\Omega} &= |f(\theta)|^2 = \frac{e^4 Z^2}{4(4\pi\epsilon_0)^2 m^2 v^4}\frac{1}{[\sin^2(\theta/2)+\sin^2(\theta_0/2)]^2} \\
&\simeq \frac{e^4 Z^2}{4\pi^2\epsilon_0^2 m^2 v^4}\frac{1}{(\theta^2+\theta_0^2)^2} \quad \text{for} \quad \sin\frac{\theta}{2} \simeq \frac{\theta}{2} \qquad (3.57)
\end{aligned}
$$

and for $\theta \ll \theta_0$ is proportional to $\sin^{-4}(\theta/2)$ and identical with (3.14). In the absence of screening, the screening angle θ_0 becomes zero and $\mathrm{d}\sigma/\mathrm{d}\Omega$ has a singularity at $\theta = 0$. Screening results in a finite value at $\theta = 0$ (Figs. 3.2 and 3.4). A more accurate solution at small θ is obtained by using the approximation (3.52) for $V(r)$ and also by introducing a modification due to the dense packing of atoms in a solid because $V(r)$ is normally the screened potential of a free neutral atom. This is done by overlapping the potentials of neighbouring atoms:

$$
\begin{aligned}
V_{\mathrm{eff}}(r) &= V(r) + V(2a-r) - 2V(a) \quad &\text{for} \quad r \le a \\
&= 0 \quad &\text{for} \quad r > a \ ,
\end{aligned} \qquad (3.58)
$$

where $2a$ is the distance between neighbouring atoms or the diameter of the sphere with the same volume as the Wigner–Seitz cell of the lattice [3.8, 9]. This is known as the *muffin-tin model*. Furthermore, the WKB (Wentzel, Kramer, Brillouin) method (or Molière approximation) and the partial-wave

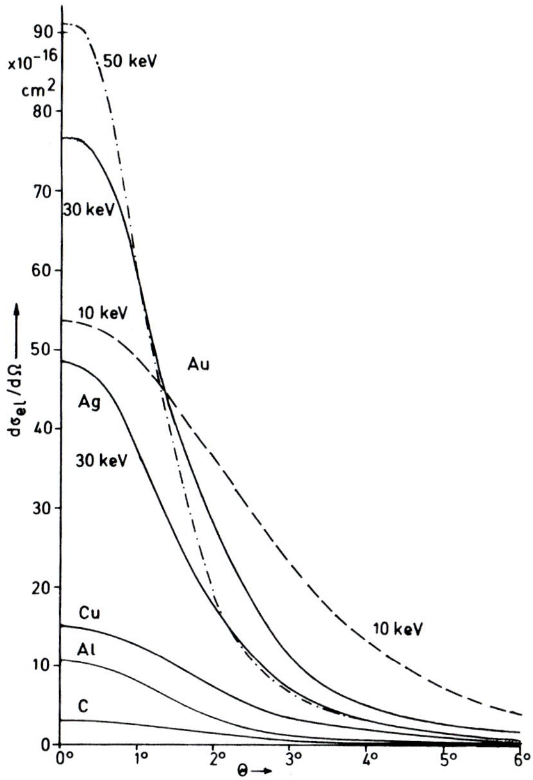

Fig. 3.4. Calculated differential elastic cross-sections $\mathrm{d}\sigma_{\mathrm{el}}/\mathrm{d}\Omega$ with $E = 30$ keV electrons for C, Al, Cu and Ag atoms and with 10 and 50 keV electrons as well for Au, dashed and dash-dotted curves, respectively

method (Sect. 3.1.6) give more accurate values at low scattering angles (see [3.1] for details). Figure 3.4 shows values of $\mathrm{d}\sigma/\mathrm{d}\Omega$ [3.9] at $E = 30$ keV for different elements and also at 10 and 50 keV for Au atoms.

3.1.6 Exact Mott Cross-Sections

Exact cross-sections for elastic large-angle scattering may be obtained by substituting a screened Coulomb potential in the relativistic Schrödinger or Pauli–Dirac equations (3.26). There is not much difference between the results given by these two equations though the latter is more accurate. It will also be convenient to consider the dense packing of atoms in a solid by a muffin-tin model (3.58) and the electron exchange effect by a modified potential [3.10]. For the Pauli–Dirac equation, the scattering problem has to be solved separately for the two spin directions (see [3.11–13] for details). Assuming that the spin is in the $+z$ spin direction, the asymptotic solution as $r \to \infty$ becomes

$$\begin{aligned}\psi_1 &= \psi_0 \left[\exp(2\pi i k z) + f_1(\theta,\chi)\frac{\exp(2\pi i k r)}{r}\right] \\ \psi_2 &= \psi_0\, g_1(\theta,\chi)\frac{\exp(2\pi i k r)}{r}\,.\end{aligned} \tag{3.59}$$

The spin-flip amplitude $g_1(\theta,\chi)$ takes into account the fact that the second spin direction may also be present after the scattering process and the scattering amplitude can therefore also depend on the azimuth χ. An exact solution can be expressed as an infinite series of Legendre polynomials:

$$\begin{aligned}f_1(\theta) &= \frac{1}{2ik}\sum_{l=0}^{\infty}\{(l+1)[\exp(2i\eta_l)-1]+l[\exp(2i\eta_{-l-1})-1]\}\,\mathrm{P}_l(\cos\theta) \\ g_1(\theta,\chi) &= \frac{1}{2ik}\sum_{l=0}^{\infty}\{-\exp(2i\eta_l)+\exp(2i\eta_{-l-1}\}\,\mathrm{P}_l^1(\cos\theta)\mathrm{e}^{i\chi} = g(\theta)\mathrm{e}^{i\chi}\,.\end{aligned} \tag{3.60}$$

For the $-z$ spin direction, the series becomes

$$f_2 = f_1 = f(\theta) \quad \text{and} \quad g_2 = -g_1\mathrm{e}^{-2i\chi} = -g(\theta)\mathrm{e}^{-i\chi}\,. \tag{3.61}$$

If the incident beam consists of a superposition of both spin directions:

$$A\begin{pmatrix}1\\0\end{pmatrix}\exp(2\pi i k z) + B\begin{pmatrix}0\\1\end{pmatrix}\exp(2\pi i k z)\,, \tag{3.62}$$

the differential cross-section becomes

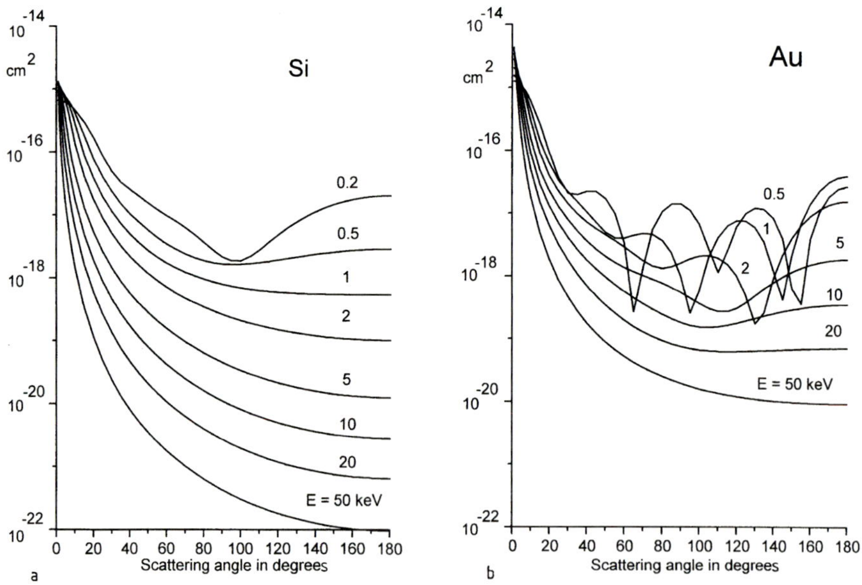

Fig. 3.5. Mott cross-sections $\mathrm{d}\sigma_\mathrm{M}/\mathrm{d}\Omega$ at different electron energies E for (**a**) Si and (**b**) Au

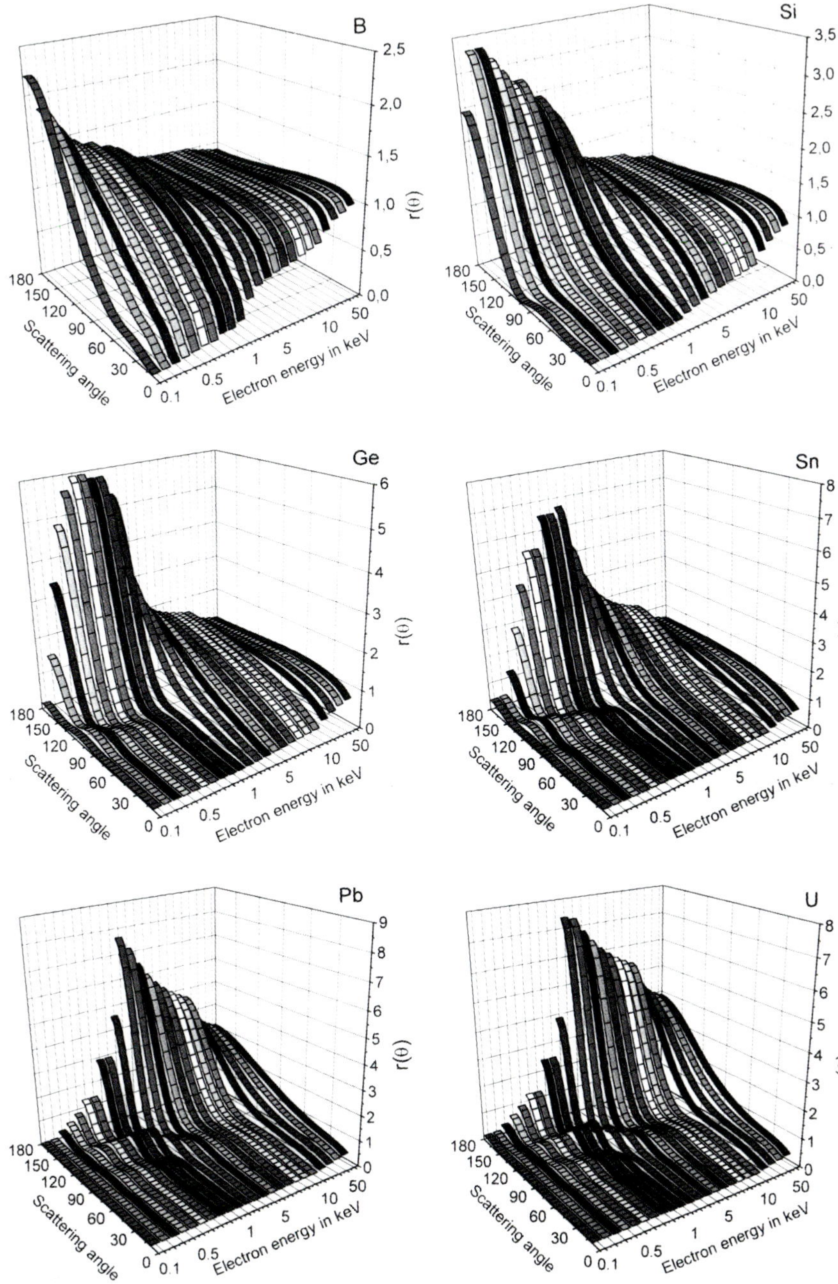

Fig. 3.6. The ratio $r(\theta) = (\mathrm{d}\sigma_{\mathrm{M}}/\mathrm{d}\Omega)/(\mathrm{d}\sigma_{\mathrm{R}}/\mathrm{d}\Omega)$ of Mott and unscreened Rutherford cross-sections for various elements and electron energies. At $\theta = 0$, the ratio r goes to zero because of the singularity of the unscreened Rutherford cross-section $\mathrm{d}\sigma_{\mathrm{R}}/\mathrm{d}\Omega$

$$\frac{\mathrm{d}\sigma_\mathrm{M}(\theta,\chi)}{\mathrm{d}\Omega} = |f|^2 + |g|^2 + \frac{-AB^*\mathrm{e}^{\mathrm{i}\chi} + A^*B\mathrm{e}^{-\mathrm{i}\chi}}{|A|^2 + |B|^2}(fg^* - f^*g)\,. \tag{3.63}$$

For a totally transversally polarized electron beam ($A = B = 1$), we find

$$\frac{\mathrm{d}\sigma_\mathrm{M}(\theta,\chi)}{\mathrm{d}\Omega} = (|f|^2 + |g|^2)[1 - S(\theta)\sin\chi] \tag{3.64}$$

where

$$S(\theta) = \mathrm{i}\,\frac{fg^* - f^*g}{|f|^2 + |g|^2} \tag{3.65}$$

is the so-called Sherman function, which is real because the numerator is the difference of two complex conjugate numbers. For example, the large-angle scattering at a thin gold film shows an azimuthal χ-dependence of the scattering intensity at scattering angles $\theta = 120°$. Left- and right-hand semiconductor detectors at $\chi = \pi/2$ and $3\pi/2$ measure different counts N_l and N_r:

$$\frac{N_l - N_r}{N_l + N_r} = PS(120°) \tag{3.66}$$

where P is the degree of polarization. Such a *Mott detector* can be used to investigate the surface magnetization of ferromagnetics (Sect. 8.1.4).

When the electron beam is unpolarized, (3.63) reduces to the Mott cross-section

$$\mathrm{d}\sigma_\mathrm{M}/\mathrm{d}\Omega = |f|^2 + |g|^2\,, \tag{3.67}$$

which is independent on χ. In contrast to Rutherford scattering no analytical expression for $\mathrm{d}\sigma_\mathrm{M}/\mathrm{d}\Omega$ can be formed. Different numerical methods have been developed to calculate the phase shifts η_l and η_{-l-1} in (3.60) by the partial-wave method [3.10, 14, 15]. Calculated Mott cross-sections are tabulated in [3.16–18]. Figure 3.5 shows calculated Mott cross-sections for Si and Au in a semi-logarithmic plot. Figure 3.2 shows the influence of Mott scattering at large scattering angles in a logarithmic polar diagram.

The difference between Rutherford and Mott cross-sections can be described by the ratio of Mott to unscreened Rutherford differential cross-sections:

$$r(\theta) = \frac{\mathrm{d}\sigma_\mathrm{M}/\mathrm{d}\Omega}{\mathrm{d}\sigma_\mathrm{R}/\mathrm{d}\Omega}\,. \tag{3.68}$$

When $r(\theta) = 1$, the Rutherford and Mott cross-sections agree. Figure 3.6 shows this ratio for electron energies between 0.1 and 50 keV. All plots decrease to zero at small θ because of the screening effect discussed in Sect. 3.1.5, which results in a finite value of $\mathrm{d}\sigma_\mathrm{el}/\mathrm{d}\Omega$ at $\theta = 0$ (Fig. 3.4), whereas the unscreened Rutherford cross-section in the denominator of (3.68) has a singularity at $\theta = 0$ because of the term $\sin^4(\theta/2)$ in the denominator of (3.14). For low atomic numbers and large electron energies and scattering angles, $r(\theta) = 1$ remains close to unity, whereas at low energies and high atomic numbers, strong deviations occur. In particular, a maximum in $r(\theta)$ at $\theta = 180°$ moves to higher energies with increasing Z (note the changes in scale for $r(\theta)$ in Fig. 3.6 with increasing Z). Oscillations with deep minima can be attributed to the angular momentum (3.6) of the incident electrons on their trajectories around the nuclei.

The results of Figs. 3.5 and 3.6 show that there can be strong deviations from the Rutherford cross-section, especially for high atomic numbers, and these affect the electron backscattering and diffusion (Sect. 3.4.6).

For Monte-Carlo simulations based on Mott cross-sections, universal formula for the total elastic cross-section σ_{el} as a function of $E/Z^{4/3}$ are available [3.19, 20]. These can be used for the calculation of the elastic mean-free-path $\Lambda_{\text{el}} = 1/N\sigma_{\text{el}}$. Analytical fits to integrated partial cross-sections [3.21] can be used to simulate random scattering angles. However, the memory capacity of modern PCs is so large that such fits are no longer necessary. This is advantageous since they do not always describe accurately the less frequent large-angle scattering processes. Integration over $\pi/2 \leq \theta \leq \pi$ results in cross-sections σ_{b} for backscattering that can be used to calculate probabilities of elastic backscattering [3.22].

3.2 Inelastic Scattering

3.2.1 Electron Excitation Processes and Energy Losses

During the elastic scattering of electrons (Sect. 3.1), the total momentum and total kinetic energy of the collision partners are conserved. The energy loss of the incident electron and the kinetic energy transferred to the nucleus can be neglected for the electron energies used in SEM since the electron mass is so much smaller than that of the nucleus. Even when 30 keV electrons are scattered through an angle of 180°, the energy transferred to a Cu nucleus is only of the order of one electronvolt and such scattering processes have a much lower probability than inelastic processes with energy losses larger than 5 eV. Only for electrons in the MeV region can the energy transferred become larger than the displacement energy necessary to dislodge an atom from its lattice site to an interstitial position, which is of the order of 10–30 eV.

Inelastic scattering is an interaction between the incident electron and the atomic jellium. Atomic electrons can only be excited with discrete energy differences between the ground state and excited states. In solids, the ground states of the inner shells are little influenced by the dense-packing of atoms but those of the valence electrons form broad bands (valence and conduction bands), which are filled with electrons up to the Fermi level. The three most important excitation processes are:

1. Plasmon excitation and inter- and intraband transitions [3.23, 24]. A plasmon is a collective longitudinal charge-density wave of the valence or conduction electrons, which can be excited by the incident electrons. Such an induced oscillation is concentrated near plasmon frequencies ω_{pl} and the energy can be transferred only as plasmons of energy $\hbar\omega_{\mathrm{pl}} \simeq 5$–30 eV. These plasmon losses and inter- and intraband transitions correspond to energy losses W in the range 0–50 eV and scattering angles below 10 mrad.
2. Electron–electron binary collisions can be treated as quasi-free when the ionization (binding) energy of the atomic electron is much smaller than the transferred energy. This interaction is also called Compton scattering and results in energy losses with a maximum of E/2 and scattering angles between 0° and 90°.
3. Ionization of inner shells with a final state above the Fermi level or inside the continuum, which results in an ionization of the atom [3.25, 26] Values of the ionization energy of the K shell range from 110 eV for Be to 80 keV for Au.

These interactions result in energy losses W, which can be recorded as the electron energy loss spectra (EELS) of 80–1000 keV electrons transmitted through thin films and scattered through a cone of a few milliradians [3.25–28]. Figures 3.7a and c show the low-loss region with the broad plasmon loss (P) of carbon and the sharper surface (S) and volume (V) plasmon losses of aluminium. Ionization of the K shell results in a saw-tooth-shaped ionization edge with a large jump ratio at $W = I_{\mathrm{K}}$ – the corresponding ionization energy – and a long tail for $W = E_{\mathrm{K}} + E_{\mathrm{kin}}$ where the latter is the kinetic energy of electrons ejected beyond the Fermi level (see examples in Figs. 3.7b and e for the carbon and aluminium K edges). The L shell ionization of aluminium (Fig. 3.7d) creates delayed maxima of the L_{23} and L_1 sub-edges. M edges in particular show a strongly delayed maximum as shown for the Ag M edge in Fig. 3.7f. All edges show a fine structure containing information about the electronic states and nearest-neighbour coordination.

These examples of energy-loss spectra demonstrate the complicated mechanism of successive energy losses along the electron trajectories in a solid. These spectra cannot normally be observed in a SEM, because of the low mean-free-path lengths (Sect. 3.3.1) between inelastic scattering processes for 5–50 keV electrons and consequently the demand for very thin films.

Plasmon losses of relatively fast electrons ($E > 1$ keV) reflected at bulk specimens can be observed with clean surfaces in ultra-high vacuum [3.23, 29, 30]. This method has not been applied in a SEM. The main disadvantage is that the energy resolution of the spectrometer must be very high; in the latter, the electrons are first decelerated in a retarding field to energies below 100 eV and are then analysed by an electrostatic sector field, which can only work with small apertures of the order of a few milliradians. Another very effective method for the investigation of surfaces is energy-loss spectroscopy

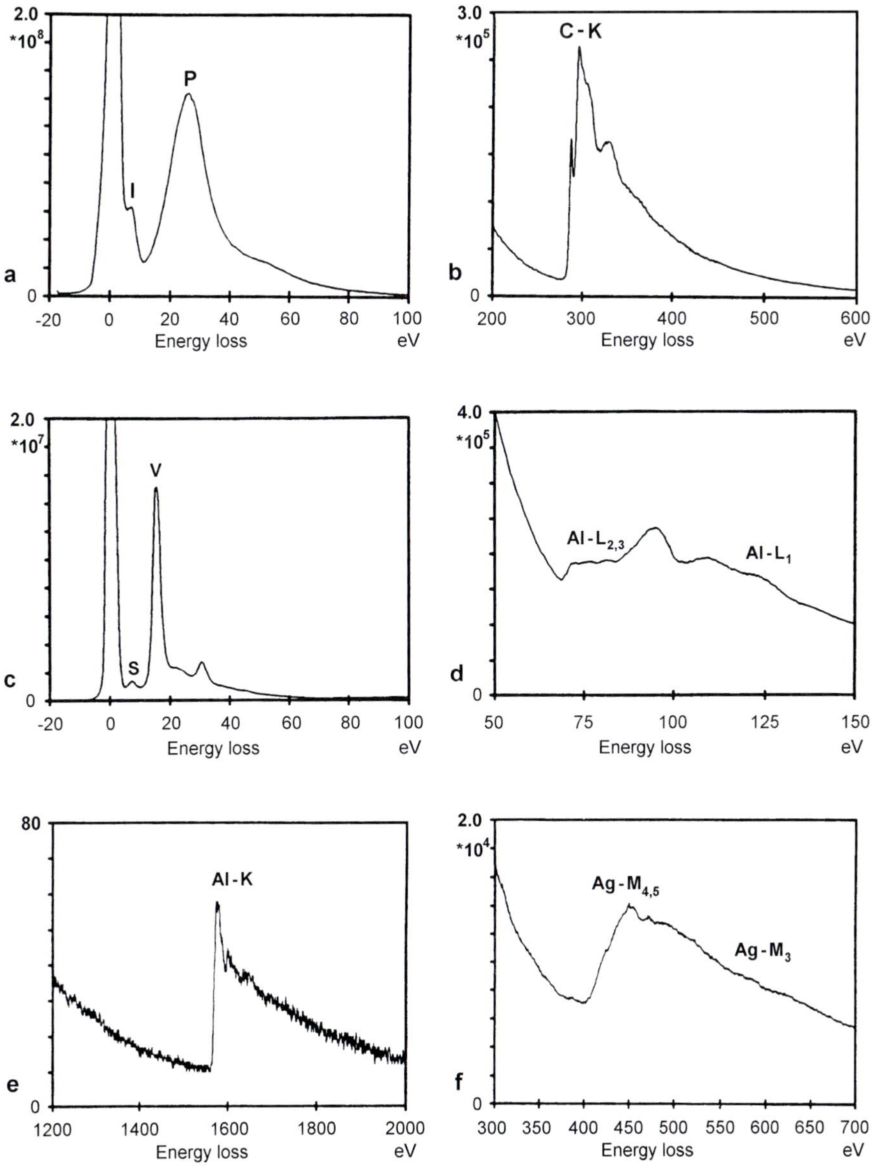

Fig. 3.7. Examples of electron energy-loss spectra (EELS) with the plasmon-loss region of (**a**) carbon and (**c**) aluminium, the K ionization edges of (**b**) carbon and (**e**) aluminium, (**d**) the L edge of Al and (**f**) the M edge of Ag [3.28]

with low-energy electrons ($E < 1$ keV) where the primary electrons can be monochromatized to a few meV and can also be polarized [3.31]. However, these methods have not been used in conjunction with high spatial resolution.

3.2.2 Electron–Electron Collisions

In this section we discuss the hypothetical case in which an incident high-energy electron interacts with an electron at rest, which may be considered free of binding forces. The resulting cross-section $d\sigma/dW$ for an energy transfer between W and $W + dW$, obtained using a collision model that relies on classical mechanics, can be employed to describe the electron–electron collision in a first-order approximation.

In Fig. 3.8a, the incident electron 1 passes the free electron 2 with a velocity v and an impact parameter b. We use the momentum approximation also employed in Sect. 2.3.1 to analyse the deflection of high-energy electrons in a transverse field. In our case, we assume that the free electron does not change its position during the collision time, but acquires momentum only. The repulsive force components will be (Fig. 3.8a):

$$F_z = \frac{e^2}{4\pi\epsilon_0 r^2}\cos\varphi; \quad F_y = \frac{e^2}{4\pi\epsilon_0 r^2}\sin\varphi \tag{3.69}$$

and the components of momentum transfer become

$$\begin{aligned} p_z &= \int_{-\infty}^{+\infty} F_z \mathrm{d}t = \frac{e^2}{4\pi\epsilon_0}\int_0^\pi \frac{1}{r^2}\cos\varphi\frac{\mathrm{d}t}{\mathrm{d}\varphi}\mathrm{d}\varphi = \frac{e^2}{4\pi\epsilon_0 bv}\int_0^\pi \cos\varphi \mathrm{d}\varphi = 0\,, \\ p_y &= \int_{-\infty}^{+\infty} F_y \mathrm{d}t = \frac{e^2}{4\pi\epsilon_0 bv}\int_0^\pi \sin\varphi \mathrm{d}\varphi = \frac{e^2}{2\pi\epsilon_0 bv} \end{aligned} \tag{3.70}$$

where we have used the relations $r\sin\varphi = b$; $\cot\varphi = -v/tb$; $\mathrm{d}t/\mathrm{d}\varphi = b/v\sin^2\varphi$. The energy transfer W in terms of the impact parameter b becomes

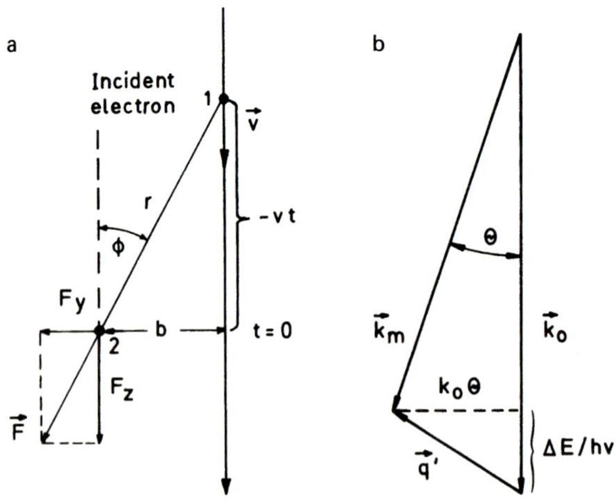

Fig. 3.8. (**a**) Coordinates for the classical model of inelastic electron–electron scattering, (**b**) relation between $\boldsymbol{k}_0$ and $\boldsymbol{k}_m$ for the wave-mechanical model

$$W = \frac{p_y^2}{2m_0} = \frac{2e^4}{(4\pi\epsilon_0)^2 b^2 m_0 v^2} = \frac{e^4}{(4\pi\epsilon_0)^2 E b^2} \tag{3.71}$$

where $E = m_0 v^2/2$ is the non-relativistic kinetic energy of the incident electron. Solving for b^2 and differentiating, we obtain

$$2b\mathrm{d}b = -\frac{e^4}{(4\pi\epsilon_0)^2} \frac{\mathrm{d}W}{EW^2} \, . \tag{3.72}$$

Electrons that hit an annular ring with radii b and $b + \mathrm{d}b$ and an area $\mathrm{d}\sigma = 2\pi b\mathrm{d}b$ will transfer an energy between W and $W + \mathrm{d}W$. Substitution of $\mathrm{d}\sigma$ in (3.72) results in

$$\frac{\mathrm{d}\sigma}{\mathrm{d}W} = \frac{\pi e^4}{(4\pi\epsilon_0)^2} \frac{1}{EW^2} \, . \tag{3.73}$$

which means that the cross-section $\mathrm{d}\sigma/\mathrm{d}W$ for an energy loss between W and $W + \mathrm{d}W$ is inversely proportional to E and W^2: small energy losses W occur with a larger probability.

Since the collision partners have the same mass, the angle between the momenta $\boldsymbol{p}_2'$ of the target electron and $\boldsymbol{p}_1'$ of the incident electron after the collision will be 90° for all scattering angles θ. With $p'^2_2 = 2mW$ and $p'^2_1 = 2m(E - W)$ this results in

$$\sin^2\theta = W/E \, . \tag{3.74}$$

As a result of the larger probability of small W, small scattering angles dominate.

Quantum mechanically, we cannot distinguish between electrons 1 and 2 with energies $E - W$ and W, respectively, and the more energetic electron after scattering will be called the incident electron. Therefore, we have to add in (3.73) the cross-section for the case in which the incident electron loses the energy $E - W$ [first two terms of the bracket in (3.75)]. A relativistic theory that includes the spin interaction [3.32] leads to the Møller formula

$$\frac{\mathrm{d}\sigma}{\mathrm{d}W} = \frac{\pi e^4}{(4\pi\epsilon_0)^2 E} \times \left(\frac{1}{W^2} + \frac{1}{(E-W)^2} - \frac{1}{W(E-W)} \frac{E_0(E_0+2E)}{(E_0+E)^2} + \frac{1}{(E_0+E)^2} \right) . \tag{3.75}$$

Equation (3.75) approaches (3.73) for $W \ll E \ll E_0 = m_0 c^2 = 511$ keV.

A semi-empirical approach proposed by Gryziński [3.33] considers a binding energy E_{nl} of the electron. It is assumed that the incident electron only collides with this electron and that the interaction of the nucleus with the atomic electron can be neglected during collision. This assumption is justified when the transferred energy W is much larger than E_{nl}. The atomic electron is not at rest before collision but has an energy distribution with a mean kinetic energy $m\overline{v^2}/2$ which is of the order of the binding energy E_{nl}. This results in

$$\frac{d\sigma}{dW} = \frac{\pi e^4}{(4\pi\epsilon_0)^2}\frac{E_{nl}}{EW^3}\left(\frac{E}{E+E_{nl}}\right)^{3/2}\left(1-\frac{W}{E}\right)^{E_{nl}/(E_{nl}+W)}$$
$$\left\{\frac{W}{E_{nl}}\left(1-\frac{E_{nl}}{E}\right)+\frac{4}{3}\ln\left[2.7+\left(\frac{E-W}{E_{nl}}\right)^{1/2}\right]\right\} . \quad (3.76)$$

For $E_{nl} \to 0$ this expression tends to (3.73).

3.2.3 Quantum Mechanics of Inner-Shell Ionization

We start from (3.37) but now in contrast to elastic scattering (Sect. 3.1.5), an atomic electron is excited by electron impact to a higher state $m > 0$. For the inelastic scattering amplitude analoguous to (3.47), we obtain

$$f_{0m}(\theta) = -\frac{m}{2\pi\hbar^2}\int V_{0m}(r_2)\exp(-2\pi i\boldsymbol{q}'\cdot\boldsymbol{r}_2)d^3\boldsymbol{r}_2 , \quad (3.77)$$

where $\boldsymbol{q}' = \boldsymbol{k}_m - \boldsymbol{k}_0$ (Fig. 3.8b) differs from the elastic $\boldsymbol{q} = \boldsymbol{k} - \boldsymbol{k}_0$ in (3.45) because the magnitudes of $\boldsymbol{k}_m$ and $\boldsymbol{k}_0$ are different after inelastic scattering owing to the energy loss

$$W = E'_m - E'_0 = \frac{h^2}{2m}(k_m^2 - k_0^2) = \frac{h^2}{2m}(k_m + k_0)(k_m - k_0) \simeq hv(k_m - k_0) . \quad (3.78)$$

From Fig. 3.8b, we see that

$$\begin{aligned} q'^2 &= k_0^2 + k_m^2 - 2k_0k_m\cos\theta \simeq k_0^2 + k_m^2 - 2k_0k_m\left(1-\frac{\theta^2}{2}\right) \\ &\simeq (k_m - k_0)^2 + (k_0\theta)^2 \simeq \left(\frac{W}{hv}\right)^2 + (k_0\theta)^2 . \end{aligned} \quad (3.79)$$

Substitution of f_{0m} from (3.36) gives

$$f_{0m}(\theta) = -\frac{m}{2\pi\hbar^2}\int \exp(-2\pi i\boldsymbol{q}'\cdot\boldsymbol{r}_2)[\int \psi_0(\boldsymbol{r}'_2)V(r_2)\psi_m^\star(\boldsymbol{r}'_2)d^3\boldsymbol{r}'_2]d^3\boldsymbol{r}_2 . \quad (3.80)$$

Recalling (3.49), we write $V(r_2)$ in the form

$$V(r_2) = -\frac{e^2 Z}{4\pi\epsilon_0 r_2} + \sum_{i=1}^{Z}\frac{e^2}{4\pi\epsilon_0|\boldsymbol{r}_2 - \boldsymbol{r}'_{2i}|} . \quad (3.81)$$

Equation (3.70) then contains integrals of the type

$$\begin{aligned} \int\frac{\exp(-2\pi i\boldsymbol{q}'\cdot\boldsymbol{r}_2)}{|\boldsymbol{r}_2 - \boldsymbol{r}'_2|}d^3\boldsymbol{r}_2 &= \frac{1}{\pi q'^2}\int\delta(|\boldsymbol{r}_2 - \boldsymbol{r}'_2|)\exp(-2\pi i\boldsymbol{q}'\cdot\boldsymbol{r}_2)d^3\boldsymbol{r}_2 \\ &= \frac{1}{\pi q'^2}\exp(-2\pi i\boldsymbol{q}'\cdot\boldsymbol{r}'_2) \end{aligned} \quad (3.82)$$

and (3.80) becomes

$$f_{0m}(\theta) = \frac{m}{2\pi\hbar^2}\frac{e^2}{4\pi^2\epsilon_0 q'^2} \times \Big[Z \int \psi_0(\boldsymbol{r}_2')\psi_m^*(\boldsymbol{r}_2')\mathrm{d}^3\boldsymbol{r}_2'$$
$$-\sum_i \int \psi_{0i}(\boldsymbol{r}_2')\exp(-2\pi\mathrm{i}\boldsymbol{q}'\cdot\boldsymbol{r}_2')\psi_{mi}^*(\boldsymbol{r}_2')\mathrm{d}^3\boldsymbol{r}_2'\Big] . \tag{3.83}$$

The first term in the square bracket is zero due to the orthogonality relations (3.31). The inelastic cross-section $\mathrm{d}\sigma/\mathrm{d}\Omega$ is proportional to the absolute square of (3.83) and, because it has the meaning of a flux density, we have to multiply by the ratio of velocities $v_m/v_0 = k_m/k_0$

$$\frac{\mathrm{d}\sigma_{0m}}{\mathrm{d}\Omega} = \left(\frac{me^2}{2\pi\epsilon_0 h^2}\right)^2 \frac{1}{q'^4}\frac{k_m}{k_0}\left|\sum_i \int \psi_{0i}(\boldsymbol{r}_2')\exp(-2\pi\mathrm{i}\boldsymbol{q}'\cdot\boldsymbol{r}_2')\psi_{mi}^*(\boldsymbol{r}_2')\mathrm{d}^3\boldsymbol{r}_2'\right|^2 . \tag{3.84}$$

For the calculation of the total cross-section of an excitation, the integration over the element of solid angle $\mathrm{d}\Omega = 2\pi\sin\theta\,\mathrm{d}\theta$ will be replaced by integration over q'. Differentiation of (3.79) gives

$$q'\mathrm{d}q' = k_0 k_m \sin\theta\mathrm{d}\theta = \frac{k_0 k_m}{2\pi}\mathrm{d}\Omega . \tag{3.85}$$

Then, with $k_0 = mv/h$ and excitation of only one of the Z atomic electrons, we find

$$\sigma_{0m} = \int_{q'} \frac{\mathrm{d}\sigma_{0m}}{\mathrm{d}\Omega}\mathrm{d}q' = 2\pi\left(\frac{e^2}{2\pi\epsilon_0 hv}\right)^2 \int \left|\int \psi_0 \exp(-2\pi\mathrm{i}\boldsymbol{q}'\cdot\boldsymbol{r}')\psi_m^*\mathrm{d}^3\boldsymbol{r}\right|^2 \frac{\mathrm{d}q'}{q'^3} . \tag{3.86}$$

The exponential term can be expanded in a series: $\exp(-2\pi\mathrm{i}\boldsymbol{q}'\cdot\boldsymbol{r}) = 1 - 2\pi q'x + \ldots$ Inelastic scattering is concentrated around small values of θ and it is sufficient to consider only the first-order term of the series. The inner integral in (3.86) then becomes the dipole moment $(d_x)_{0m}$ for the transition from $n = 0$ to m where x is parallel $\boldsymbol{q}'$:

$$|2\pi\mathrm{i}q' \int \psi_0 ex\psi_m^*\mathrm{d}^3\boldsymbol{r}|^2 = 4\pi^2 q'^2|(d_x)_{0m}|^2 . \tag{3.87}$$

The dipole moments are related to the oscillator strengths

$$\tilde{f}_{0m} = \frac{2m}{e^2\hbar^2}(E_m' - E_0')|(d_x)_{0m}|^2 \tag{3.88}$$

which satisfy the Bethe sum rule

$$\sum_m \tilde{f}_{0m} = Z \tag{3.89}$$

and (3.86) finally becomes

$$\begin{aligned}\sigma_{0m} &= \frac{2\pi e^4}{(4\pi\epsilon_o)^2 E}\frac{\tilde{f}_{0m}}{(E_m' - E_0')}\int_{q_{\min}}^{q_0}\frac{\mathrm{d}q'}{q'} = \frac{2\pi e^4 \tilde{f}_{0m}}{(4\pi\epsilon_0)^2 E(E_m' - E_0')}\ln\frac{\sqrt{2mE_0'}v}{E_m' - E_0'} \\ &= \frac{\pi e^4 \tilde{f}_{0m}}{(4\pi\epsilon_0)^2 E(E_m' - E_0')}\ln\frac{2mv^2 E_0'}{(E_m' - E_0')^2} .\end{aligned} \tag{3.90}$$

where $q_{\min} = (E'_m - E'_0)/hv = W/hv$, as shown by (3.79) for $\theta = 0$, and $q_0 = \sqrt{2mE'_0}/h$ is equal to the reciprocal de Broglie wavelength of the atomic electron on its orbital of energy E'_0.

This formula will now be employed to discuss the inner-shell ionization cross-section. Recalling that $2mv^2 = 4E$ and that for an inner-shell ionization energy $E_{0m} = E'_m - E'_0 \simeq -E'_0$, (3.90) gives

$$\sigma_{\mathrm{K}} = \frac{\pi e^4 z_{\mathrm{K}} b_{\mathrm{K}}}{(4\pi\epsilon_0)^2 E E_{\mathrm{K}}} \ln\frac{4E}{B_{\mathrm{K}}} \tag{3.91}$$

for the ionization of the K shell, for example, where $\tilde{f}_{0m} = b_{\mathrm{K}} z_{\mathrm{K}}$ and $z_{\mathrm{K}} = 2$ is the number of electrons in the K shell and $B_{\mathrm{K}} \propto E_{\mathrm{K}}$.

A value of $b_{\mathrm{K}} = 0.35$ and $B_{\mathrm{K}} \simeq 1.65 E_{\mathrm{K}}$ for $E \gg E_{\mathrm{K}}$ has been proposed by *Mott* and *Massey* [3.11] to describe the ioniation cross-section σ_{K} of the K shell. Instead of E, the overvoltage ratio $u = E/E_{\mathrm{K}}$ is often used and (3.91) becomes

$$\sigma_{\mathrm{K}}(u) = \frac{\pi e^4 z_{\mathrm{K}} b_{\mathrm{K}}}{(4\pi\epsilon_0)^2 E_{\mathrm{K}}^2} g_{\mathrm{K}}(u) = 6.15 \times 10^{-20} \frac{z_{\mathrm{K}}}{E_{\mathrm{K}}^2} g_{\mathrm{K}}(u) \tag{3.92}$$

where σ_{K} and E_{K} are measured in units of cm^2 and keV, respectively, and

$$g_{\mathrm{K}}(u) = \frac{1}{u}\left[0.35 \ln(2.42u)\right] . \tag{3.93}$$

In order to ensure that $\sigma_{\mathrm{K}} \to 0$ when $u \to 1$, *Worthington* and *Tomlin* [3.34] modified the relation (3.92) of Mott and Massey by writing

$$g_{\mathrm{K}}(u) = \frac{1}{u}\left(0.35 \ln\frac{4u}{1.65 + 2.35\exp(1-u)}\right) . \tag{3.94}$$

If the constants in $g_{\mathrm{K}}(u)$ are only a function of the x-ray series, the product $\sigma_{\mathrm{K}} E_{\mathrm{K}}^2$ becomes only a function of the overvoltage ratio u, which is demonstrated in Fig. 3.9 for different elements. The agreement is remarkable when we consider the large range of ionization energies E_{K} for the different elements.

Integration of the Gryziński formula (3.76) results in

$$g_{\mathrm{K}}(u) = \frac{1}{u}\left(\frac{u-1}{u+1}\right)^{3/2}\left\{1 + \frac{2}{3}\left(1 - \frac{1}{2u}\right)\ln[2.7 + (u-1)^{1/2}]\right\} \tag{3.95}$$

which fits the experimental results for large u very well whereas for $u < 3$ all calculated cross-sections differ from the experimental ones (Fig. 3.9). A review of theoretical and experimental values concerning the inner-shell ionization cross-sections has been published by *Powell* [3.35, 36].

Equation (3.92) can be simplified [3.37] to

$$\sigma_{\mathrm{K}} = \frac{\pi e^4 z_{\mathrm{K}} b_{\mathrm{K}}}{(4\pi\epsilon_0)^2 E_{\mathrm{K}}^2} \frac{\ln u}{u} \tag{3.96}$$

and is used in this form for the calculation of the x-ray quantum emission, for example (Sect. 10.1.2).

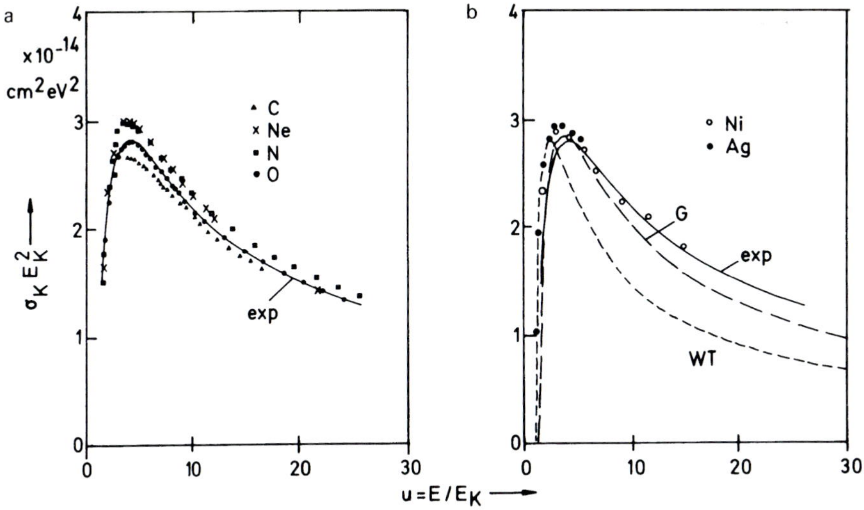

Fig. 3.9. Demonstration that the curve relating $\sigma_K E_K^2$ to the overvoltage ratio $u = E/E_K$ (σ_K = ionization cross-section and E_K = ionization energy of the K shell) is independent of atomic number. Experiments for (**a**) C, Ne, N and O atoms and (**b**) Ni and Ag atoms. Comparison with formulae of *Worthington* and *Tomlin* (WT) (3.94) and *Gryziński* (G) (3.95) (see [3.35] for references)

3.2.4 Angular Distribution of Total Inelastic Scattering

Equation (3.84) describes the differential cross-section of inelastic scattering for the excitation $0 \to m$ of one excited state. The inelastic angular distribution for a single atom can be obtained by summing over all $m \neq 0$ in (3.84). The calculation has been discussed in detail in [3.1] and results in [3.7]

$$\frac{d\sigma_{\text{inel}}}{d\Omega} = \sum_{m\neq 0} \frac{d\sigma_{0m}}{d\Omega} = \frac{4e^4 Z}{(4\pi\epsilon_0)^2 m_0^2 v^4} \frac{1 - \left(\dfrac{1}{1 + (\theta^2 + \theta_E^2)/\theta_0^2}\right)^2}{(\theta^2 + \theta_E^2)^2} \tag{3.97}$$

with $\theta_E = J/4E$ where J denotes the mean ionization potential (3.131).

This formula for the inelastic differential cross-section may be compared with its elastic counterpart (3.57). The characteristic angle θ_0 (3.55) is of the order of a few tens of mrad and is responsible for the decrease of $d\sigma_{el}/d\Omega$, whereas the angle θ_E in (3.97) is of the order of tenths of a mrad and is responsible for the decrease of $d\sigma_{inel}/d\Omega$. This shows that inelastic scattering is concentrated within much smaller angles than elastic scattering.

For very large scattering angles $\theta \gg \theta_0 \gg \theta_E$, we find that

$$\frac{d\sigma_{\text{inel}}/d\Omega}{d\sigma_{\text{el}}/d\Omega} = \frac{1}{Z} . \tag{3.98}$$

In view of this relation, the probability for scattering through large angles $\theta > 10°$ described for elastic scattering by (3.57) is sometimes modified by replacing the factor Z^2 by $Z(Z+1)$ in order to consider both elastic and inelastic large-angle scattering in Monte Carlo simulations, for example.

The ratio ν (Table 3.1) of total inelastic to total elastic cross-section decreases with increasing atomic number Z [3.7, 38, 39]:

$$\nu = \frac{\sigma_{\mathrm{inel}}}{\sigma_{\mathrm{el}}} \simeq 20/Z\ . \tag{3.99}$$

3.3 Multiple Scattering Effects

3.3.1 Mean-Free-Path Length

We have discussed single-scattering processes in Sects. 3.1 and 3.2. The experimental conditions for single scattering cannot be realised except when scattering occurs at a dilute gas target. Even in a very thin solid film, we encounter plural scattering. This needs special mathematical procedures to calculate the angular and spatial distributions and the energy-loss spectrum of electrons after passing through a specimen layer. The range of validity of the calculations also defines the range of film thicknesses or depths of a solid that can be treated by such a plural scattering theory.

An important quantity for describing plural and multiple scattering is the mean-free-path length Λ. A total cross-section can be defined by

$$\sigma_{\mathrm{t}} = \sigma_{\mathrm{el}} + \sigma_{\mathrm{inel}} = 2\pi \int_0^{\pi} \left(\frac{\mathrm{d}\sigma_{\mathrm{el}}}{\mathrm{d}\Omega} + \frac{\mathrm{d}\sigma_{\mathrm{inel}}}{\mathrm{d}\Omega} \right) \sin\theta \mathrm{d}\theta \tag{3.100}$$

in units of cm^2 and we can represent each atom by a disc of area σ_{t}. There will be a scattering process if this area is struck by an electron. In a film of thickness t and density ρ or mass-thickness $x = \rho t$, we have

$$n = \frac{N_{\mathrm{A}}\rho}{A} t = Nt \tag{3.101}$$

atoms per unit area, where N denotes the number of atoms per unit volume. The incident electron current is I_0. In a depth z, only a fraction $I < I_0$ will be unscattered and in an additional layer of thickness $\mathrm{d}z$, the decrease $\mathrm{d}I$ between z and $z + \mathrm{d}z$ will be

$$\mathrm{d}I = -IN\sigma_{\mathrm{t}}\mathrm{d}z \tag{3.102}$$

The remaining unscattered intensity is then

$$I = I_0 \exp(-N\sigma_{\mathrm{t}}t) = I_0 \exp(-t/\Lambda_{\mathrm{t}}) = I_0 \mathrm{e}^{-p} \tag{3.103}$$

in which we have introduced the mean-free-path length $\Lambda_{\mathrm{t}} = 1/N\sigma_{\mathrm{t}}$ and the mean number of collisions $p = t/\Lambda_{\mathrm{t}}$. Table 3.1 contains values of σ_{el}, Λ_{el} and Λ_{t} using the listed values of $\nu = \sigma_{\mathrm{inel}}/\sigma_{\mathrm{el}}$.

Table 3.1. Total elastic cross-sections σ_{el} in units 10^{-16} cm^2, elastic mean-free-path length $\Lambda_{el} = 1/N\sigma_{el}$ in nanometres ($N = N_A\rho/A$: Number of atoms per unit volume), total mean-free-path length $\Lambda_t = \Lambda_{el}/(\nu+1)$ with $\nu = \sigma_{inel}/\sigma_{el}$ and the electron range R in μm for different elements and electron energies E = 1–50 keV

E [keV]		1	5	10	20	30	50	
C Z=6	σ_{el}	0.65	0.11	0.055	0.027	0.018	0.012	$\times 10^{-16}$ cm^2
$\rho = 2$ g/cm^3	Λ_{el}	1.5	9	18	37	55	83	nm
$\nu \simeq 3$	Λ_t	0.4	2.3	4.5	9	14	20	nm
	R	0.033	0.49	1.55	4.9	9.7	22.6	μm
Al Z=13	σ_{el}	1.26	0.31	0.16	0.08	0.053	0.034	$\times 10^{-16}$ cm^2
$\rho = 2.7$ g/cm^3	Λ_{el}	1.3	5	10	21	31	49	nm
$\nu \simeq 1.5$	Λ_t	0.5	2	4	8	12	20	nm
	R	0.025	0.36	1.14	3.6	7.1	16.7	μm
Cu Z=29	σ_{el}	1.84	0.64	0.37	0.21	0.15	0.11	$\times 10^{-16}$ cm^2
$\rho = 8.9$ g/cm^3	Λ_{el}	0.64	1.8	3.2	5.6	7.8	10.7	nm
$\nu \simeq 0.6$	Λ_t	0.4	1.1	2.0	3.5	4.9	6.70	nm
	R	0.007	0.11	0.35	1.10	2.26	5.1	μm
Ag Z=47	σ_{el}	3.09	1.15	0.71	0.43	0.32	0.22	$\times 10^{-16}$ cm^2
$\rho = 10.5$ g/cm^3	Λ_{el}	0.5	1.5	2.4	4.0	5.3	7.7	nm
$\nu \simeq 0.4$	Λ_t	0.4	1.0	1.7	2.8	3.8	5.5	nm
	R	0.006	0.09	0.29	0.93	1.8	4.3	μm
Au Z=79	σ_{el}	3.93	1.60	1.05	0.67	0.52	0.37	$\times 10^{-16}$ cm^2
$\rho = 19.3$ g/cm^3	Λ_{el}	0.43	1.0	1.6	2.5	3.3	4.6	nm
$\nu \simeq 0.2$	Λ_t	0.36	0.9	1.3	2.1	2.7	3.8	nm
	R	0.003	0.05	0.17	0.51	1.0	2.3	μm

Figures 3.10a,b show for Al and Cu a double-logarithmic plot of the inverse mean-free-path lengths $1/\Lambda = N\sigma$ for the elastic Mott scattering and the inelastic excitation of conduction electrons, plasmons and the different subshells as a function of electron energy [3.40]. $1/\Lambda_t$ is obtained by summing the cross-sections of the different scattering processes

The probability for n-fold scattering (n = 0,1,2,...) in a film of thickness t with a mean number p of collisions is given by the Poisson coefficients

$$\Pi_n(p) = \frac{p^n}{n!}e^{-p}; \quad \sum_{n=0}^{\infty} \Pi_n(p) = 1 . \tag{3.104}$$

Plural scattering theories will be valid for $p < 25$. For thicknesses or depths of a solid for which $p > 25 \pm 5$ we enter the range of multiple scattering and then have to use models of electron diffusion (Sect. 3.4). In the light-emission image of an electron diffusion cloud shown in Fig. 3.23, the region of plural scattering is only the small paintbrush-shaped primary beam with a depth of the order of one tenth of the total range R.

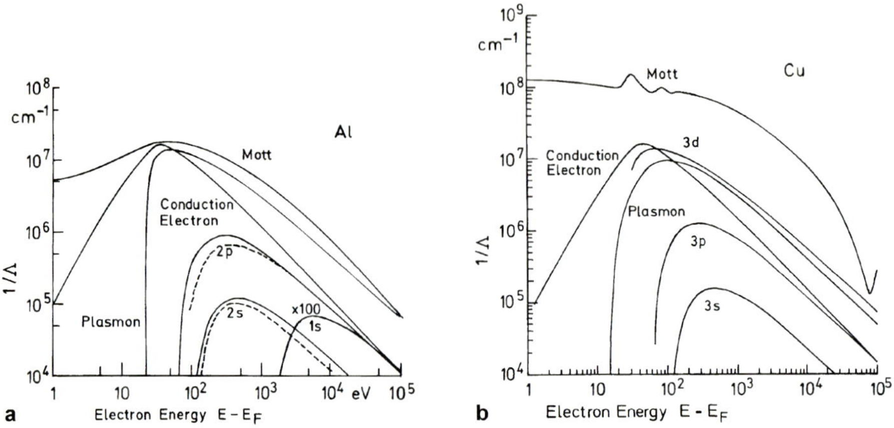

Fig. 3.10. Inverse mean-free-path lengths $1/\Lambda = N\sigma$ of Mott elastic scattering and the inelastic excitation of conduction electrons, plasmons and different subshells for (**a**) Al and (**b**) Au [3.40]

Though the most widely used specimens in SEM are solids and not thin films, the results of plural and multiple scattering can be used for discussion of the decrease of resolution in backscattered electron micrographs or in x-ray microanalysis of specimen details at increasing depths below the surface. Again, the angular distribution of plural scattering can be used to evaluate a mean scattering angle for Monte Carlo simulations, for example.

3.3.2 Angular Distribution of Transmitted Electrons

When we assume single scattering only, the intensity $I_1(\theta)\,\mathrm{d}\Omega$ scattered into a solid angle $\mathrm{d}\Omega$ can be described by

$$I_1(\theta)\mathrm{d}\Omega = I_0 n \frac{\mathrm{d}\sigma}{\mathrm{d}\Omega}\mathrm{d}\Omega = I_0 N\sigma t\left(\frac{1}{\sigma}\frac{\mathrm{d}\sigma}{\mathrm{d}\Omega}\right)\mathrm{d}\Omega = I_0\frac{t}{\Lambda}S_1(\theta)\mathrm{d}\Omega\,, \tag{3.105}$$

where I_0 is the intensity of the incident beam with a negligible aperture and $S_1(\theta) = (1/\sigma)\mathrm{d}\sigma/\mathrm{d}\Omega$ is a normalized single-scattering function such that $\int S_1(\theta)2\pi\sin\theta\mathrm{d}\theta = 1$.

A scattering function for the case in which all electrons are scattered twice is obtained by a self-convolution of S_1: $S_2(\theta) = S_1(\theta)\otimes S_1(\theta)$ where the symbol $\otimes$ stands for a two-dimensional convolution

$$f(\theta)\otimes g(\theta) = \int\int f(\boldsymbol{\theta}')g(\boldsymbol{\theta}-\boldsymbol{\theta}')\mathrm{d}^2\boldsymbol{\theta}' \tag{3.106}$$

of two functions $f(\theta)$ and $g(\theta)$. A scattering function for m-fold scattering is consequently defined by

$$S_m(\theta) = S_{m-1}(\theta)\otimes S_1(\theta)\,. \tag{3.107}$$

Using the Poisson coefficients (3.104), the angular distribution after passing through a film of thickness t with a mean number $p = t/\Lambda$ of collisions will be

$$I(\theta)\mathrm{d}\Omega = I_0\mathrm{d}\Omega \sum_{m=0}^{\infty} \Pi_m(p)S_m(\theta) = I_0\mathrm{d}\Omega \sum_{m=0}^{\infty} \frac{p^m}{m!}\mathrm{e}^{-p}S_m(\theta) \,. \qquad (3.108)$$

When $S_0(\theta)$ is taken to be a δ-function, the first term of the sum describes the exponential decrease (3.103) of the unscattered beam and the single-scattering approximation (3.105) results for $m = 1$ and $\Pi_m \ll \Pi_1$ for $m > 1$. With the typical approximations of plural scattering theory, namely:

1) Negligible energy loss or no change of $\mathrm{d}\sigma/\mathrm{d}\Omega$ inside the film, and
2) neglect of the increased path-length of scattered electrons,

(3.108) will be exact and the main problem will be to find simpler formulae for numerical calculation.

To this end, *Goudsmit* and *Saunderson* [3.41] expanded $S_1(\theta)$ in a series of Legendre polynomials

$$S_1(\theta)\mathrm{d}\Omega = \frac{\mathrm{d}\Omega}{4\pi} \sum_{n=0}^{\infty} (2n+1)a_n P_n(\cos\theta) \qquad (3.109)$$

where

$$a_n = \int_0^{\pi} S_1(\theta)P_n(\cos\theta)2\pi\sin\theta\mathrm{d}\theta \,. \qquad (3.110)$$

The incident beam, represented by a δ-function, can also be written as such a series by setting all $a_n = 1$. The m-fold convolution of $S_1(\theta)$ results in

$$S_m(\theta)\mathrm{d}\Omega = \frac{\mathrm{d}\Omega}{4\pi} \sum_{n=0}^{\infty} (2n+1)a_n^m P_n(\cos\theta) \qquad (3.111)$$

and substitution in (3.108) gives

$$\begin{aligned} \frac{1}{I_0}I(\theta)\mathrm{d}\Omega &= \frac{\mathrm{d}\Omega}{4\pi} \sum_{n=0}^{\infty} (2n+1) \left(\sum_{m=0}^{\infty} \frac{p^m\mathrm{e}^{-p}}{m!} a_n^m \right) P_n(\cos\theta) \\ &= \frac{\mathrm{d}\Omega}{4\pi} \sum_{n=0}^{\infty} (2n+1) \exp[-p(1-a_n)]P_n(\cos\theta) \,. \end{aligned} \qquad (3.112)$$

Using the definition of Λ given in (3.103) and of a_n in (3.110), the last exponential can be written

$$\exp[-p(1-a_n)] = \exp\left[-N \int_0^s \left\{ \int_0^{\pi} \frac{\mathrm{d}\sigma}{\mathrm{d}\Omega}[1 - P_n(\cos\theta)]2\pi\sin\theta\mathrm{d}\theta \right\} \mathrm{d}s \right] \,, \qquad (3.113)$$

where the film thickness t has been replaced by the path length s inside the film, which makes the theory applicable for $s > t$ as well.

A small-angle approximation suggested by *Lewis* [3.42] results when only the terms with $n = 0$, 1 of the series (3.112) are retained. With $P_0 = 1$

and $P_1 = \cos\theta$, the mean value of the cosine of the scattering angle can be calculated

$$\overline{\cos\theta} = \frac{1}{I_0}\int_0^{\pi} \cos\theta\, I(\theta)\mathrm{d}\Omega = \exp[-\chi(s)] \;, \tag{3.114}$$

where

$$\chi(s) = N\int_0^s \left[\int_0^{\pi} \frac{\mathrm{d}\sigma}{\mathrm{d}\Omega}(1-\cos\theta)2\pi\sin\theta\mathrm{d}\theta\right] \mathrm{d}s \;. \tag{3.115}$$

For small scattering angles, the Taylor series $\cos\theta = 1 - \theta^2/2 + ...$ and $\exp(x) = 1 + x + ...$ can be used, whereupon equation (3.114) becomes

$$\overline{\theta^2} = \frac{s}{\Lambda}\overline{\theta_1^2} = p\,\overline{\theta_1^2} \tag{3.116}$$

where $\overline{\theta_1^2}$ is the mean-square angle of the single-scattering distribution $S_1(\theta)$ in (3.105). This means that the mean square of the scattering angles increases as the path length.

The treatments of plural and multiple scattering proposed by *Molière* [3.43], and *Snyder* and *Scott* [3.44] used the small-angle approximation ($\sin\theta \simeq \theta$) from the outset. This means that the scattering angles θ lie in a $\boldsymbol{\theta}$ plane with polar coordinates θ and χ. The interconnections between the different scattering theories have been reviewed by *Bethe* [3.45]. The treatment of Molière using a transport equation will be discussed in Sect. 3.4.5.

For $p \geq 25 \pm 5$, the angular distribution can be analysed by error theory [3.46] on the assumption that successive scattering processes are independent. When large-angle scattering and energy losses are neglected, the angular distribution of transmitted electrons is Gaussian in form:

$$\frac{1}{I_0}I(\theta)\mathrm{d}\Omega = \frac{\mathrm{d}\Omega}{\pi\overline{\theta^2}}\exp(-\theta^2/\overline{\theta^2}) \;. \tag{3.117}$$

where $\overline{\theta^2}$ can be approximated by [3.47]

$$\overline{\theta^2} = 1.2\times 10^7\frac{Z^{3/2}}{AE}x \tag{3.118}$$

with $\overline{\theta^2}$ in rad^2, E in eV and x in g/cm^2.

The plot of $\log_{10}N(\theta)$ versus θ^2 in Fig. 3.11 [3.47] shows that mass-thicknesses $x = \rho t \leq 80\mu$g/cm^2 of copper at $E = 20$ keV can be treated by such a multiple scattering theory because the curves reduce to straight lines. This criterion for the validity of the Bothe law (3.117) can be used to establish the lower limit of multiple scattering. From Table 3.1, the mass-thickness $x = 80\mu$g/cm^2 corresponds to a multiple $p = 26$ of the mean-free-path length $\Lambda_{\mathrm{t}} = 3.1\mu$g/cm^2.

The root-mean square angle $\theta_{\mathrm{rms}} = (\overline{\theta^2})^{1/2}$, which increases as $x^{1/2}$ (3.118), saturates for very thick copper foils to a value $\theta_{\mathrm{rms}} \simeq 50°$. This saturation thickness can be defined as the limit between multiple scattering and diffusion. An increase of foil thickness beyond this value does not result

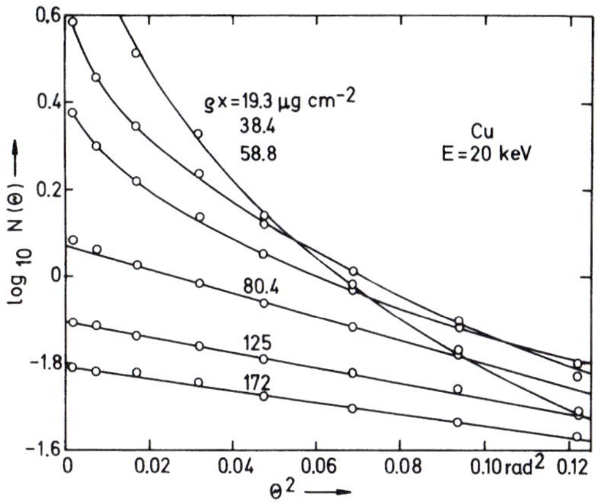

Fig. 3.11. The angular distribution $N(\theta)$ of 20 keV electrons scattered in Cu foils for various mass-thicknesses $x = \rho t$. The curves show $\log_{10} N(\theta)$ versus θ^2, demonstrating that $N(\theta) \propto \exp(-\theta^2/\overline{\theta^2})$ for large x [3.47]

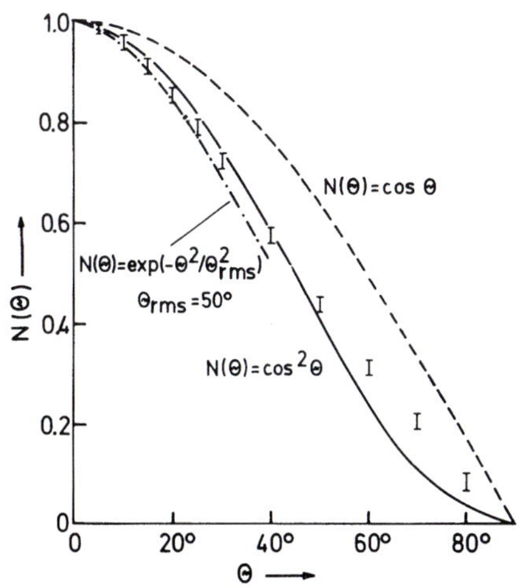

Fig. 3.12. Expressions for the angular distribution $N(\theta)$ in the region of diffuse multiple scattering: $N(\theta) \propto \cos^2\theta$ (*continuous line*), $N(\theta) = \exp(-\theta^2/\theta^2_{\mathrm{rms}})$ with $\theta_{\mathrm{rms}} = 50°$ (*dash-dotted line*) and $N(\theta) = \cos\theta$ (*dashed line*) [3.47]

in a further broadening of the angular distribution but only in a decrease of the total transmitted intensity. The angular distribution can be approximated by a $\cos^2\theta$ law (Fig. 3.12) [3.47] rather than the $\cos\theta$ law which is found for the angular distribution of backscattered electrons (Figs. 3.12 and 4.12a, respectively). At low scattering angles θ, the angular distribution can be approximated by a Gaussian distribution (3.117) with $\theta_{\mathrm{rms}} \simeq 50°$.

3.3.3 Spatial Beam Broadening

The angular distribution of scattering also results in beam broadening normal to the electron beam, which increases with increasing film thickness or depth below the surface as demonstrated in Fig. 3.13 by Monte Carlo simulations. This broadening limits the resolution in scanning transmission electron microscopy. Only structures on the top of a thick specimen are scanned by

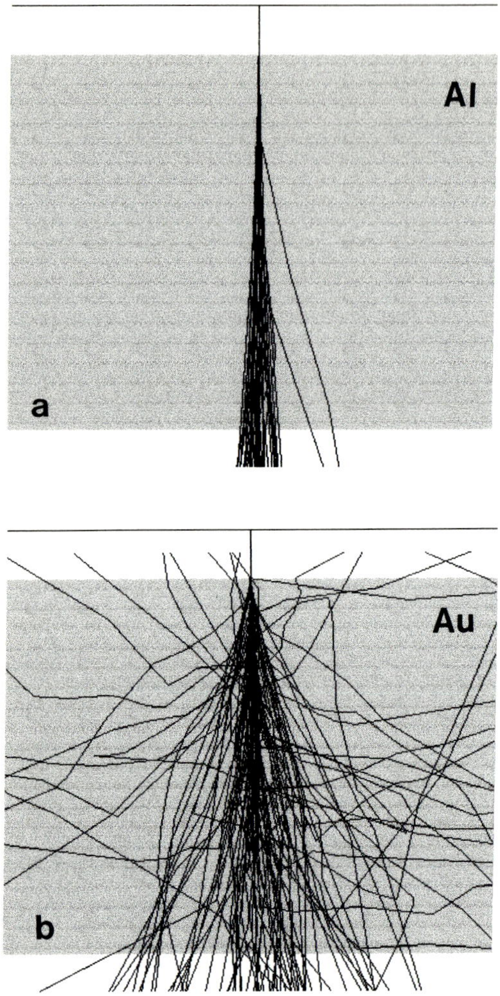

Fig. 3.13. Monte Carlo simulation of 100 20-keV electron trajectories in 60 nm foils of (**a**) Al and (**b**) Au to demonstrate the spatial broadening

the primary unbroadened electron probe with a resolution of the order of the probe diameter whereas structures at the bottom show a worse resolution owing to the probe broadening by multiple scattering [3.48]. This top-bottom effect has been discussed in detail in [3.1]. However, the same broadening of the primary electron probe with increasing depth is also present in bulk specimens. It results in an increasing blur of structures at increasing depths when these are imaged by backscattered electrons or by the EBIC mode or are analysed by x-ray excitation. In a photograph of the luminescence of a diffusion cloud (Fig. 3.23), this region is the first paintbrush-like volume of high light intensity near the electron impact.

For the calculation of the beam broadening, we make use of the approximation whereby the single-scattering function $S_1(\theta)$ in (3.105) is represented by a Gaussian function (3.117): $S_1(\theta) \simeq \exp(-\theta^2/\overline{\theta_1^2})$ and the increase of $\overline{\theta^2}$ is assumed to be proportional to the foil thickness (3.118). We can evaluate this broadening of the angular distribution, using the fact that a convolution of two Gaussians results in a Gaussian with the sum of the mean-square angles.

Scattering in a layer between z and $z + \mathrm{d}z$ results in $\overline{\theta^2} + \mathrm{d}\overline{\theta^2}$, where $\mathrm{d}\overline{\theta^2} = \overline{\theta_1^2}\mathrm{d}z/\Lambda$. Integration over z gives $\overline{\theta^2} = \overline{\theta_1^2}z/\Lambda = p\overline{\theta_1^2}$, which is identical with (3.116).

We now apply this method to the spatial broadening. We assume a spatial broadening $I(r) \propto \exp(-r^2/\overline{r^2})$ at a depth z and $\overline{r^2}$ is changed to $\overline{r^2} + \mathrm{d}\overline{r^2}$ in the layer $\mathrm{d}z$. A scattering angle θ causes a displacement $r = \theta z$ so that $\mathrm{d}\overline{r^2} = \overline{\theta^2}z\mathrm{d}z$. Substitution of $\overline{\theta^2} = \overline{\theta_1^2}z/\Lambda$ and integration from $z = 0$ to t (foil thickness) yields

$$\overline{r^2} = \frac{2\overline{\theta_1^2}}{3\Lambda}t^3 \quad \text{or} \quad \overline{x^2} = \frac{\overline{\theta_1^2}}{3\Lambda}t^3 \,. \tag{3.119}$$

If the two-dimensional distribution $I(r) \propto \exp(-r^2/\overline{r^2})$ at a depth $z = t$ with $r^2 = x^2 + y^2$ is projected onto the x-axis, we get a one-dimensional distribution $I(x) \propto \exp(-x^2/\overline{x^2})$. This means that the root-mean-square value r_{rms} or x_{rms} of the spatial broadening increases as $t^{3/2}$. This proportionality also resulted from a simpler model [3.49] in which the scattering was concentrated at the foil centre. Equation (3.119) is exact and the only approximation is the fit of the single-scattering function $S_1(\theta)$ to a Gaussian.

When we believe that elastic scattering will result in a larger contribution than inelastic scattering, which is concentrated in smaller scattering angles, then we replace $\overline{\theta_1^2}$ in (3.119) by the square of the characteristic angle θ_0 (3.55) and use $\Lambda_{\mathrm{el}} = 1/N\sigma_{\mathrm{el}}$ from Table 3.1 or from integration of $\mathrm{d}\sigma_{\mathrm{el}}/\mathrm{d}\Omega$ (3.57) between the limits $\theta = 0$ and π. This gives

$$x_{\mathrm{rms}} = 1.05 \times 10^5 \left(\frac{\rho}{A}\right)^{1/2} \frac{Z}{E} t^{3/2} \tag{3.120}$$

with x_{rms} and t in cm, ρ in g cm^{-3} and E in eV. Substitution of $t = 0.2$ µm and $E = 100$ keV gives $x_{\mathrm{rms}} = 10$ nm for Cu and 23 nm for Au, which is in reasonable agreement with the Monte Carlo simulations in Fig. 3.13.

This formula for electron-probe broadening will be valid in the same range of thicknesses as the plural scattering theory for the angular distribution, which means for $p = t/\lambda \leq 25 \pm 5$ because for larger p the electron energy loss and the increase of path-lengths inside the foil have to be taken into account. Like the plural scattering theory of *Molière* [3.43] (Sect. 3.4.5), the spatial broadening can also be treated by a transport equation (3.180), which has the advantage that a single-scattering function $S_1(\theta)$ that need not be a Gaussian can be used [3.50–53]; alternatively, the beam broadening may be calculated by a Monte Carlo simulation [3.54, 55].

3.3.4 Continuous-Slowing-Down Approximation

To describe the energy loss along an electron trajectory, we assume that the energy transfer W in a Coulomb interaction between incident and atomic electrons is much smaller than the energy E of the incident electrons. The sequence of collisions can then be regarded as a continuous slowing down and to each path element of the electron trajectory can be attributed a mean energy loss $-\mathrm{d}E_\mathrm{m}$ which results in the so-called stopping power or Bethe-loss formula [3.56]

$$S = \left|\frac{\mathrm{d}E_\mathrm{m}}{\mathrm{d}s}\right| = NZ\int_{W_{\min}}^{W_{\max}} \frac{\mathrm{d}\sigma}{\mathrm{d}W} W\mathrm{d}W = \frac{2\pi e^4 N_\mathrm{A}\rho Z}{(4\pi\epsilon_0)^2 AE}\int_{b_{\min}}^{b_{\max}} \frac{\mathrm{d}b}{b}, \tag{3.121}$$

where $N = N_\mathrm{A}\rho/A$ is the number of atoms and NZ the number of atomic electrons per unit volume. In the first integral $\int(\mathrm{d}\sigma/\mathrm{d}W)W\mathrm{d}W = \int W\mathrm{d}\sigma$, we have substituted $\mathrm{d}\sigma = 2\pi b\mathrm{d}b$ (3.13) and taken W from (3.71) to express the right-hand side of (3.121) in terms of the impact parameter b. The introduction of a lower and upper limit in the integral is necessary for convergence.

These limits $b_{\min}$ and $b_{\max}$ can be estimated by the following idea: $b_{\max}$ corresponds to the smallest energy transfer and the time of collision is of the order of $b_{\max}/v$. In a classical model of atomic orbitals, electrons are bound with an energy of the order of $\hbar\omega$ and the period of their motion on the orbital is of the order of $1/\omega$. For an interaction, the collision time should be smaller than this atomic period which results in

$$\frac{b_{\max}^{(i)}}{v} = \frac{1}{\omega_i}. \tag{3.122}$$

In a quantum-mechanical treatment [3.56], the single frequency ω is replaced by an assembly of oscillators with frequencies ω_i and oscillator strengths $\tilde{f}_i$ (3.88), the latter obeying the sum rule $\Sigma_i \tilde{f}_i = Z$ (3.89). The lower limit $b_{\min}$ is obtained from the condition that impact parameters smaller than the de Broglie wavelength λ (2.22), will be meaningless. This results in

$$b_{\min} = h/mv. \tag{3.123}$$

Substitution of (3.122) and (3.123) into (3.121) gives

$$\begin{aligned} S = \left|\frac{\mathrm{d}E_\mathrm{m}}{\mathrm{d}s}\right| &= \frac{4\pi e^4 N_\mathrm{A}\rho}{(4\pi\epsilon_0)^2 A m_0 v^2}\sum_i \tilde{f}_i \int_{b_{\min}}^{b_{\max}^{(i)}} \frac{\mathrm{d}b}{b} \\ &= \frac{4\pi e^4 N_\mathrm{A}\rho}{(4\pi\epsilon_0)^2 A m_0 v^2}\sum_i \tilde{f}_i \ln\frac{mv^2}{h\omega_i}. \end{aligned} \tag{3.124}$$

The lack of exact information about the values of $\tilde{f}_i$ and ω_i for most substances makes it necessary to define a mean ionization energy J of an atom such that

$$\Pi_i(h\omega_i)^{\tilde{f}_i} = J^Z \tag{3.125}$$

and with $\mathrm{d}x = \rho \mathrm{d}s$ as a mass-thickness element (3.124) becomes

$$\left|\frac{\mathrm{d}E_\mathrm{m}}{\mathrm{d}x}\right| = \frac{4\pi e^4 N_\mathrm{A} Z}{(4\pi\epsilon_0)^2 A m_0 v^2} \ln\frac{m_0 v^2}{J} \ . \tag{3.126}$$

Bethe [3.57] has also established a solution, using the relativistic Møller formula (3.75) for the electron–electron cross-section:

$$\left|\frac{\mathrm{d}E_\mathrm{m}}{\mathrm{d}x}\right| = \frac{2\pi e^4 N_\mathrm{A} Z}{(4\pi\epsilon_0)^2 A m_0 v^2} \left\{ \ln\frac{\epsilon^2(\epsilon+2)}{2(J/E_0)^2} + \left[1 + \frac{\epsilon^2}{8} - (2\epsilon+1)\ln 2\right] / (\epsilon+1)^2 \right\} . \tag{3.127}$$

where $\epsilon = E/E_0$ and $E_0 = m_0 c^2 = 511$ keV.

For practical purposes, the following formula is sufficient for all (apart from extreme relativistic) energies [3.58]

$$\left|\frac{\mathrm{d}E_\mathrm{m}}{\mathrm{d}x}\right| = 1.53 \times 10^5 \frac{1}{\beta^2}\frac{Z}{A} \left[\ln\left(\frac{\epsilon^2(\epsilon+2)}{(J/511000)^2}\right) - \beta^2\right] \tag{3.128}$$

where $\beta = v/c$. Measurement of E and J in units of eV results in units of eV g^{-1} cm^2 for $|\mathrm{d}E/\mathrm{d}x|$.

For non-relativistic energies $J \ll E \ll m_0 c^2$ the formula

$$S = \left|\frac{\mathrm{d}E_\mathrm{m}}{\mathrm{d}x}\right| = \frac{2\pi e^4 N_\mathrm{A} Z}{(4\pi\epsilon_0)^2 A}\frac{1}{E} \ln\left(b\frac{E}{J}\right) \tag{3.129}$$

is in use but different values of the constant b have been adopted:

$b = 1$ classical theory [3.59]
$b = 2$ semi-classical theory [3.2]
$b = (\mathrm{e}/2)^{1/2} = 1.166$ quantum-mechanical theory [3.60].

These different approaches are listed to demonstrate why formulae are encountered in the literature that differ in the values of these factors. The choice of the factor has little influence on the final value, however, because it figures in the argument of the logarithmic term. Therefore, for non-relativistic energies, a practical formula will be

$$S = \left|\frac{\mathrm{d}E_\mathrm{m}}{\mathrm{d}x}\right| = 7.8 \times 10^{10} \frac{Z}{A}\frac{1}{E} \ln\left(1.166\frac{E}{J}\right) \tag{3.130}$$

with E and J in units eV and S in eV g^{-1} cm^2.

The mean ionization potential J increases with increasing Z and can be described by the semi-empirical formula [3.61]

$$J = 9.76Z + 58.8Z^{-0.19} \quad \text{and} \quad J = 11.5Z \quad \text{for} \quad Z \leq 6 \tag{3.131}$$

where J is in units of eV.

It can be assumed that in a composite target the stopping effects of the atomic electrons of different elements combine additively (Bragg's rule). We recall that the number of atoms in 1 g of a pure-element substance is equal to N_A/A. In a multi-element target the number of atoms of an element 'a' in 1 g of substance is $m_a N_A / \sum m_i A_i$ where the m_i are the atomic fractions. Mass fraction c_a and atomic fraction m_i are related by

$$c_a = \frac{m_a A_a}{\sum m_i A_i} \quad . \tag{3.132}$$

where $M = \sum\limits_i m_i A_i$ is the molecular weight.

The Bethe formula of type (3.129) for a composite target can therefore be written in terms of the atomic fractions

$$S = \frac{2\pi e^4 N_A}{(4\pi\epsilon_0)^2 E} \frac{\sum m_i Z_i \ln(bE/J_i)}{\sum m_i A_i} \tag{3.133}$$

or of the mass fractions

$$S = \frac{2\pi e^4 N_A}{(4\pi\epsilon_0)^2 E} \sum_i c_i \frac{Z_i}{A_i} \ln(bE/J_i) \ . \tag{3.134}$$

The decrease of the mean electron energy $E_m(x)$ with increasing path length x can be obtained by numerical integration of (3.129), for example, and permits us to calculate the Bethe range R_B for which $E_m \to 0$ (Sect. 3.4.1 and Fig. 3.24). If we neglect the energy dependence of the logarithmic term in (3.129), which varies slowly with E, we find

$$-\frac{dE_m}{dx} = \frac{2\pi e^4 N_A Z l}{(4\pi\epsilon_0)^2 A} \frac{1}{E_m} = \frac{c}{E_m} \to -E_m dE_m = c dx \to E^2 - E_m^2 = cx \tag{3.135}$$

where l denotes the constant mean value of the logarithmic term. This type of formula is known as the Thomson–Whiddington law [3.62], which was introduced as an empirical law with Terrill's empirical constant $c_T = 4 \times 10^{11}$ eV^2 cm^2 g^{-1} [3.63]. However, we see that such a law will be an acceptable approximation only for thin films, so long as the logarithmic term can be neglected. Furthermore, (3.135) implies that the Thomson–Whiddington range is given by $R_{TW} = E^2/c$ when $E_m \to 0$. The proportionality of the range to the square of E does not agree with experiment (Sect. 3.4.1) and the Thomson–Whiddington law is often modified by fitting the constant c_T so that R_{TW} approaches the practical range. Another problem arises because $(-dE_m/dx)$ in (3.135) has a singularity as $E_m \to 0$, which overestimates the energy loss per unit path length at the end of the electron trajectories. These limitations of this law have to be kept in mind when employing this formula because of its simplicity.

The Bethe formula for the mean energy E_m can be compared with calculated mean energies from experimental energy spectra $N(E)\,dE$ of electrons transmitted through foils of mass-thickness $x = \rho t$ (Fig. 3.15):

$$E_{\mathrm{m}} = \int_0^E N(E) E \mathrm{d}E / \int_0^E N(E) \mathrm{d}E \,. \tag{3.136}$$

Such values of E_{m} are plotted in Fig. 3.16 and agree with the integrated Bethe formula $E_{\mathrm{m}}(x)$ [3.64]. For the determination of E_{m} from (3.136), it is necessary to measure the low-energy tail of $N(E)\,\mathrm{d}E$ accurately, which was achieved in the experiments of Figs. 3.15 and 3.16 by means of an electrostatic cylindrical spectrometer in which the electrodes consisted of wires to minimize the backscattering of electrons at the electrodes. The stronger decrease of the experimental values of E_{m} in Fig. 3.16b for Au can be attributed to an increase of the mean path-length inside the foil, which is expected to be greater in Au because of the higher probability of large-angle scattering.

3.3.5 Modification of the Stopping Power at Low Electron Energies

When the electron energy is decreased, atomic shells with ionization energies higher than the actual energy E do not contribute to the energy loss. This can be corrected either by using (3.124) instead of (3.129) or by the following approximations.

The mean ionization energy J (3.131) is replaced by an energy-dependent value [3.65]

$$J' = \frac{J}{1 + kJ/E} \tag{3.137}$$

with k values varying between 0.77 (C) and 0.85 (Au). When using this correction, the maximum of the stopping power S resulting from the Bethe formula (3.129) is shifted to lower energies but still goes to zero for $E \to 0$ (Fig. 3.14). In another proposal [3.66], Z in (3.129) is replaced by N_{eff} and J by J_{eff}, where these quantities are determined from the energy-loss spectrum (EELS) of thin films recorded at high electron energies of 100-200 keV [3.67, 68]. For energies below 5 keV the plasmon excitation has a strong influence on the stopping power calculated with values of J_{eff} and N_{eff} and the Bragg rule (3.134) can fail [3.68] because the plasmon spectrum of a compound does not follow the Bragg rule.

When the double-differential inelastic cross-section $\mathrm{d}^2\sigma/\mathrm{d}W\mathrm{d}\Omega$ is known, the stopping power can be obtained from [3.25]

$$\frac{\mathrm{d}E_{\mathrm{m}}}{\mathrm{d}s} = N \int\int \frac{\mathrm{d}^2\sigma}{\mathrm{d}W\mathrm{d}\Omega} W \mathrm{d}W \mathrm{d}\Omega \,. \tag{3.138}$$

Because these corrections need a detailed knowledge of the energy-loss spectrum, which is not available for all compounds, a correction proposed by *Rao-Sahib* and *Wittry* [3.69] is in common use in Monte Carlo simulations (Sect. 3.5.6). Below $E/J = 6.3$, the reciprocal stopping power is approximated by a parabolic curve

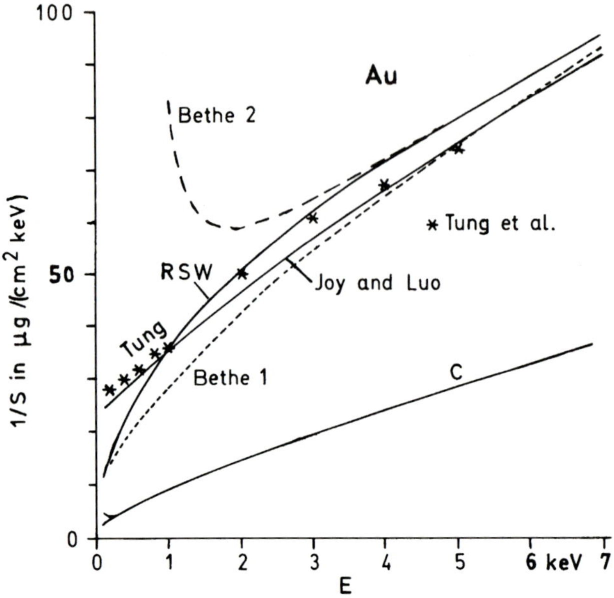

Fig. 3.14. Inverse Bethe stopping power $1/S$ of gold and carbon given by (3.124) (Bethe 1) and (3.129) (Bethe 2), the correction (3.137) due to *Joy* and *Luo* , the Rao-Sahib and Wittry approach (3.139) and calculations by *Tung et al.* [3.70]

$$1/S \propto E^{0.5} \,. \tag{3.139}$$

Figure 3.14 compares $1/S(E)$ for gold calculated from (3.129) and (3.124), the Rao-Sahib and Wittry approach (3.139), and values of the stopping power calculated by *Tung et al.* [3.70] (see also [3.71]). Since the value of J is much lower for carbon, the $1/S(E)$ dependence of the inverse Bethe stopping power calculated from (3.124) then holds down to 1 keV (lower curve in Fig. 3.14).

3.3.6 Distribution of Energy Losses

The electrons are decelerated statistically by different individual energy losses, resulting in an energy-loss spectrum. The deviation from the mean energy E_{m} discussed in Sect. 3.3.4 is known as straggling.

Landau [3.72] developed the following theory for calculating the straggling distribution. Let $f(W,z)\mathrm{d}W$ be the normalized probability that an electron that penetrates a foil of thickness z has lost an energy between W and $W+\mathrm{d}W$. The probability per unit path length for a further energy loss in the interval K, $K+\mathrm{d}K$ is denoted by $\Phi(K)\mathrm{d}K = N(\mathrm{d}\sigma/\mathrm{d}K)\mathrm{d}K$ where N denotes the number of atoms per unit volume and $\mathrm{d}\sigma/\mathrm{d}K$ the differential cross-section. When the electron proceeds from a depth z to $z+\mathrm{d}z$, there are two contributions to the change of $f(W,z)$. An electron arriving at z with an energy loss $W-K$ and losing energy K will increase the probability for the

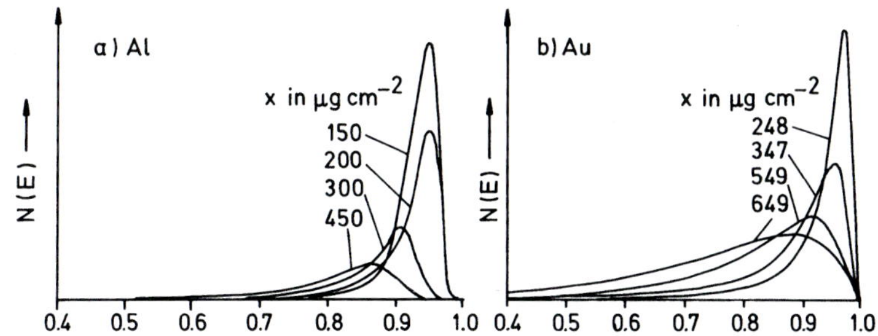

Fig. 3.15. Energy distributions $N(E)$ of electrons transmitted through (**a**) Al ($E =$ 20.2 keV) and (**b**) Au ($E=$ 20.5 keV) films of increasing mass-thickness $x = \rho t$[3.64]

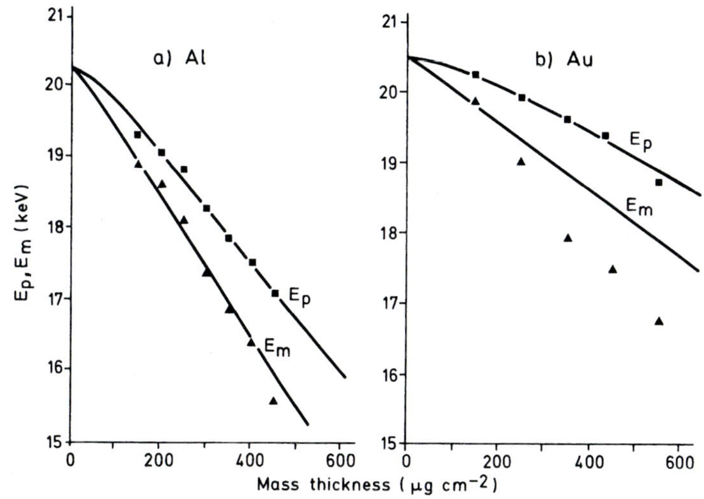

Fig. 3.16. Most probable energy E_{p} and mean energy E_{m} of electrons transmitted through (**a**) Al (20.2 keV) and (**b**) Au (20.5 keV) versus foil thickness $x = \rho t$. Experimental values of E_{p} and E_{m} and calculated full curves for E_{p}. (Landau) and E_{m} (Bethe) [3.64]

energy loss W at $z + \mathrm{d}z$, and the probability of this process is the product of the two probabilities: $\Phi(K)f(W-K)\mathrm{d}K$. An electron arriving at z with an energy loss W may lose an energy K thus decreasing the probability for the energy loss W at $z + \mathrm{d}z$ by $\Phi(K)f(W)\mathrm{d}K$. Combining these two effects, we obtain

$$\frac{\partial f}{\partial z} = \int_0^\infty \Phi(K)[f(W-K) - f(W)]\mathrm{d}K\,. \tag{3.140}$$

This integral equation for $f(W)$ can be solved by introducing the Fourier integral

$$f(W) = \int_{-\infty}^{+\infty} F(s)\exp(2\pi \mathrm{i} s W)\mathrm{d}s \tag{3.141}$$

in (3.140):

$$\int_{-\infty}^{+\infty} \frac{\mathrm{d}F(s)}{\mathrm{d}z} \exp(2\pi \mathrm{i} sW)\mathrm{d}s \tag{3.142}$$

$$= -\int_{-\infty}^{+\infty} F(s) \left\{ \int_0^{\infty} \Phi(K)[1 - \exp(-2\pi \mathrm{i} sK)]\mathrm{d}K \right\} \exp(2\pi \mathrm{i} sW)\mathrm{d}s \; .$$

Comparison of the integrands yields a differential equation for $F(s)$ with the solution

$$F(s) = \exp[-zg(s)] \quad \text{with} \quad g(s) = \int_0^{\infty} \Phi(K)[1 - \exp(-2\pi \mathrm{i} sK)]\mathrm{d}K \tag{3.143}$$

and substitution in (3.141) gives

$$f(W) = \int_{-\infty}^{+\infty} \exp[2\pi \mathrm{i} sW - zg(s)]\mathrm{d}s \; . \tag{3.144}$$

This formula obtained from a kind of transport equation (3.140) can also be established by describing the energy-loss spectrum after passing a layer z in terms of Poisson coefficients (3.104), in a similar fashion to the calculation of the angular distribution in (3.108):

$$f(W)\mathrm{d}W = \sum_{m=0}^{\infty} \Pi_m(p) f_m(W)\mathrm{d}W \; , \tag{3.145}$$

where $f_1(W) = (\mathrm{d}\sigma/\mathrm{d}W)/\sigma$ and $f_m(W) = f_{m-1}(W) \otimes f_1(W)$. These functions have the following Fourier transforms, the latter obtained with the aid of the convolution theorem

$$F_1(s) = \int_{-\infty}^{+\infty} f_1(W) \exp(-2\pi \mathrm{i} sW)\mathrm{d}W \quad \text{and} \quad f_m(s) = F_1^m(s) \; . \tag{3.146}$$

Substitution in (3.145) gives, with $p = t/\Lambda$

$$f(W) = \sum_{m=0}^{\infty} \frac{p^m \mathrm{e}^{-p}}{m!} F_1^m(s) \exp(2\pi \mathrm{i} sW) = \int_{-\infty}^{+\infty} \exp[-p(1 - F_1)] \exp(2\pi \mathrm{i} sW)\mathrm{d}s \tag{3.147}$$

which is identical with (3.144) when we recall that $\Lambda = 1/N\sigma$.

At this point the theory is exact provided that the effective path length in the foil is not considerably larger than the foil thickness t. The energy distribution $f(W)$ dW can be calculated by (3.144) or (3.147) when we know $f_1(W)$ from deconvoluted energy-loss spectra of thin films; the latter contain the plasmon and interband transition peaks at low-energy losses and the sawtooth-like steps caused by inner-shell ionization at high energy losses (Fig. 3.7).

The reasoning becomes approximate when we make assumptions about $\mathrm{d}\sigma/\mathrm{d}W$ based on elementary theories that do not consider the individual

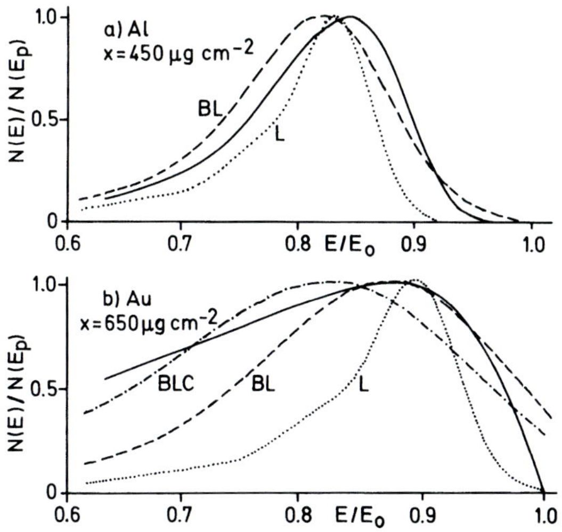

Fig. 3.17. Comparison of theoretical and experimental energy loss spectra (—) of thick (**a**) Al and (**b**) Au films (E = 20.2 and 20.5 keV, respectively) of mass-thickness x=ρt. (· · ·) Landau (L), (– – –) Blunck and Leisegang (BL), (– · –) BL corrected for increase of path lengths (BLC)[3.64]

atomic and band structure of a solid. The original Landau theory results in the following expression for the most probable energy

$$E_{\mathrm{p}} = E - \Delta E_{\mathrm{L}} = E - \xi\left(\ln\frac{\xi}{Q_{\mathrm{min}}} - \beta^2 + 0.198\right) \tag{3.148}$$

with

$$\xi = \frac{2\pi e^4}{(4\pi\epsilon_0)^2 m v^2}\frac{N_{\mathrm{A}} Z}{A}\rho t \quad \text{and} \tag{3.149}$$

$$Q_{\mathrm{min}} = \frac{J^2(1-\beta^2)}{2mv^2}\,, \tag{3.150}$$

which agrees with experiment (Fig. 3.14). However, the full width at half maximum

$$\Delta E_{1/2} = 4.04\,\xi \tag{3.151}$$

predicted by the Landau distribution is smaller than the experimental value.

A correction proposed by *Blunck* and *Leisegang* [3.73] gives better agreement and for large thicknesses of high Z material the increase of the path lengths has to be taken into account (Fig. 3.17) [3.74, 75]. However, the Landau theory becomes exact when theoretical or experimental values of $\mathrm{d}\sigma/\mathrm{d}W$ are used and the calculated half-widths $\Delta E_{1/2}$ also agree better with experiments [3.76]. A consideration of the double-differential cross-section $\mathrm{d}\sigma/\mathrm{d}W\mathrm{d}\Omega$ can also describe accurately the influence of multiple scattering on electron energy-loss spectra [3.77].

3.4 Electron Diffusion

3.4.1 Transmission and Electron Range

The transmission $T(x)$ is the fraction of the incident electrons penetrating foils of mass-thickness $x = \rho t$ regardless of their exit angles ($\theta = 0° - 90°$) and energies. The secondary electrons emerging from the exit side should be eliminated by a retarding field, as for the measurement of the backscattering coefficient (Sect. 4.1.1). Figure 3.18 shows the decrease of $T(x)$ with increasing mass-thickness x for different electron energies E in the range 10–100 keV [3.78]. The curves for different energies can be scaled to a unique transmission curve (Fig. 3.19) by plotting T versus $x/x_{0.5}$ where $x_{0.5}$ denotes the mass-thickness corresponding to a transmission of 50%. Characteristic differences among these normalized transmission curves are observed for low and high Z material. For low Z (Al in Fig. 3.19a), the slope of $T(x)$ first increases with increasing x, the curve then becomes linear and terminates in a relatively short tail, whereas for high Z (Au in Fig. 3.19b), the transmission curve starts linearly and has a long tail. These different shapes of the transmission curves have also been observed by other authors [3.79, 80], who have proposed the following formula to describe the experimental results:

$$T(x/x_0) = \exp[-(x/x_0)^p] , \tag{3.152}$$

where $p \simeq 2$ for Al and Cu and $\simeq 1$ for high Z elements (Bi, Au).

The differences in the transmission curves complicate the definition of an electron range. For this reason, different definitions of electron range exist and energy-range formulae should specify how the range has been extrapolated from experiments. The most important definitions are as follows:

1) Bethe range R_B. This quantity is the mean path length of electron trajectories. Numerical integration of dE_m/dx (3.129) results in the energy–path length relations $E_m(x)$ that are plotted in Fig. 3.24. Extrapolation to $E_m = 0$ results in the Bethe range R_B.

2) Maximum electron range R_{max}. This quantity (Fig. 3.19) cannot be measured easily because the points of measurement must be close and the foil thickness must be varied in very small steps at the tail of $T(x)$, or for a constant mass-thickness the electron energy has to be varied continuously to determine that energy for which $T = 0$. The latter condition also depends on the sensitivity of the measurement of the transmitted electron current.

3) Extrapolated ranges R_E and R_x. The linear part of $T(x)$ at medium mass-thicknesses (Figs. 3.18 and 3.19) can be extrapolated and the intersection with the axis is then the extrapolated range R_E. Similarly, this procedure can be applied in a plot $T(E)$ for a constant foil thickness x (Fig. 3.20) [3.81], resulting in R_x. Normally R_x and R_E are different.

4) Extrapolated ranges R_p and R_m. These ranges are obtained by extrapolating the decrease of the most probable energy $E_p(x)$ or of the mean energy $E_m(x)$ (Fig. 3.16) to zero values of E_p and E_m, respectively.

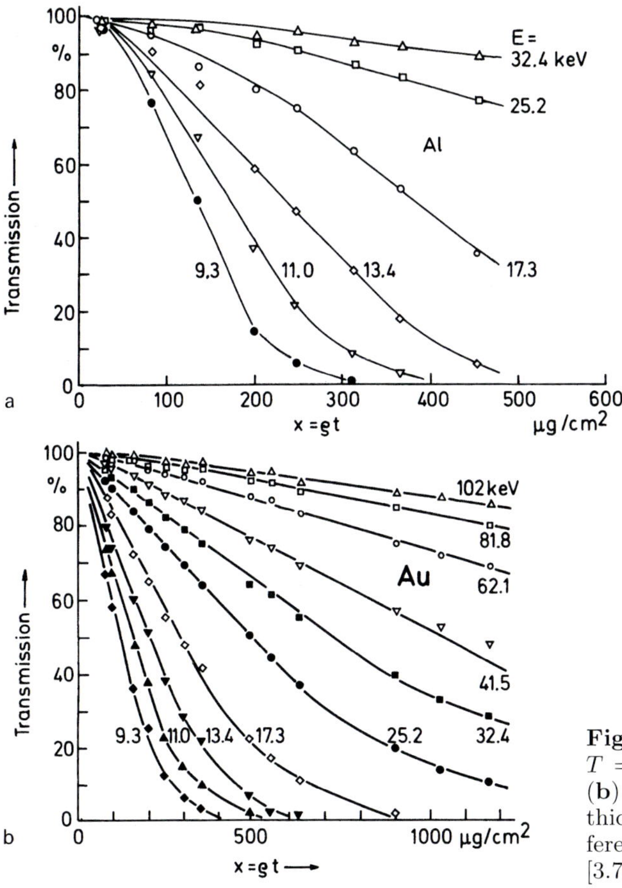

Fig. 3.18. Transmission $T = I/I_0$ for (**a**) Al and (**b**) Au foils versus mass-thickness $x = \rho t$ for different electron energies E [3.78]

For low Z material ($\overline{Z} \leq 20$) we have $R_{\max} \simeq R_B$ and $R_{\mathrm{x}} \simeq 0.75 R_B$, whereas for high Z material ($\overline{Z} \geq 50$) $R_{\max} < R_B$ and $R_{\mathrm{x}} \simeq 0.4 R_B$. The larger difference between $R_{\max}$ and R_{x} (or R_{E}) is a consequence of the long tail of $T(x)$ for high Z (Fig. 3.19b). This tail and the increasing discrepancy between R_{B} and the practical range with increasing Z can be understood from the fact that the former is a mean path length for complete deceleration and that the probability for large-angle scattering increases with increasing Z due to the term Z^2/A resulting from the product of the Rutherford cross-section (3.14) and the number of atoms $N_{\mathrm{A}}\rho/A$ per unit volume. In low Z material, a large fraction of the electrons is scattered through small angles only and electron diffusion results only in a brush-like broadening of the lateral distribution. In high Z material, the electron trajectories are more strongly coiled up, as is demonstrated by the simulated electron trajectories in Fig. 3.21.

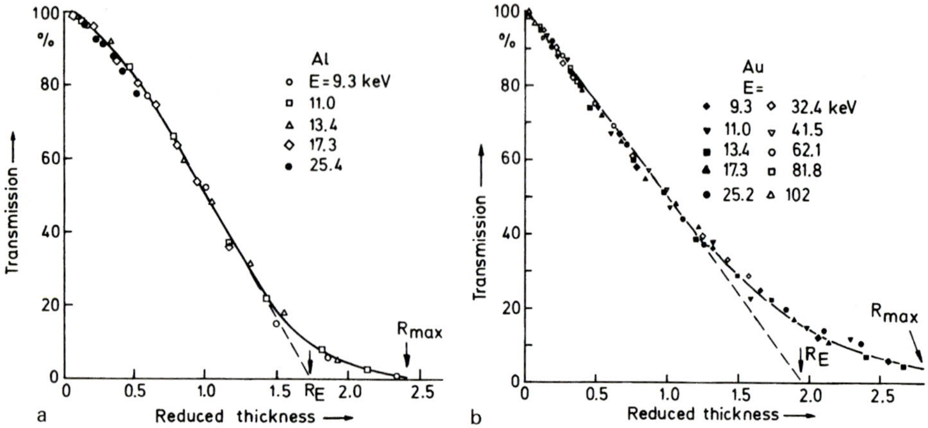

Fig. 3.19. Transmission $T = I/I_0$ for (**a**) Al and (**b**) Au foils versus reduced mass-thickness $x/x_{0.5}$ ($x_{0.5}$ = mass-thickness for $T = 50\%$, R_E = extrapolated range, R_{max} = maximum range) [3.78]

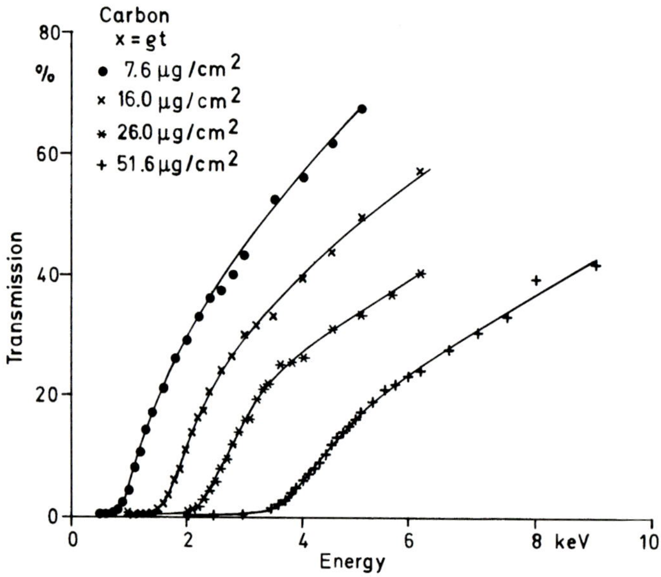

Fig. 3.20. Transmission $T(E)$ of evaporated carbon films of different mass-thickness $x = \rho t$ at low electron energies [3.81]

For the estimation of electron ranges, range–energy relations that fit the experimental values over a large range of energies are required. A power law

$$R = aE^n \tag{3.153}$$

can be used, where the exponent can vary between 1.3 and 1.7. This exponent can be obtained from the slope of $R(E)$ in a double-logarithmic plot (Fig.

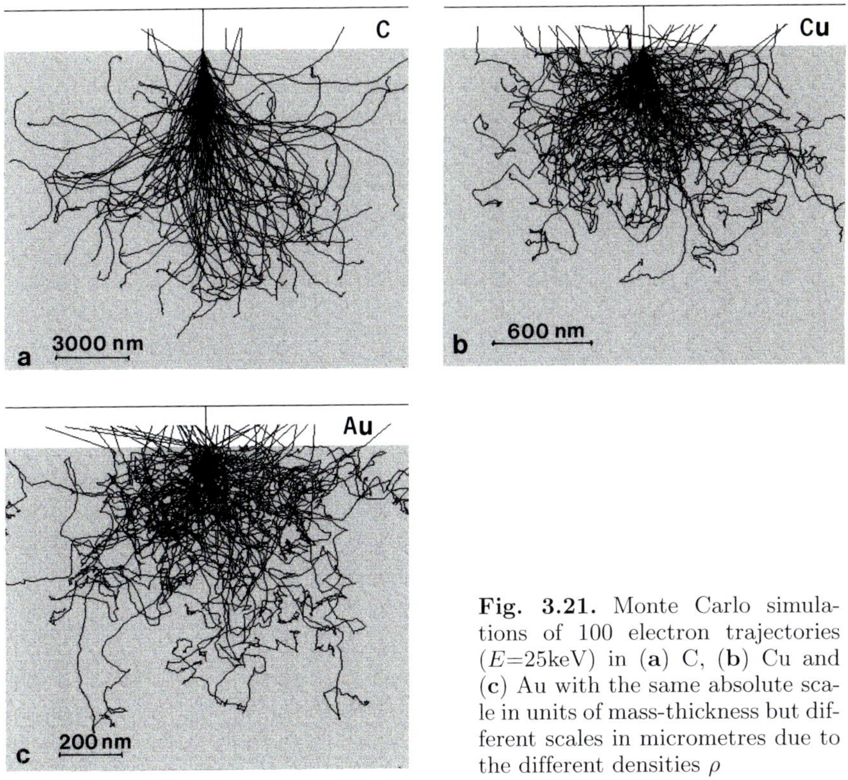

Fig. 3.21. Monte Carlo simulations of 100 electron trajectories (E=25keV) in (**a**) C, (**b**) Cu and (**c**) Au with the same absolute scale in units of mass-thickness but different scales in micrometres due to the different densities ρ

3.22). Table 3.2 contains some of the electron–range formulae [3.81–95] that have been proposed and their energy intervals of validity.

The values of R do not depend significantly on the atomic number when measured as mass-thicknesses in units of g cm^{-2} (see also Fig. 4.4). Some of the formulae can therefore be used for an estimation of the range for all elements. The range in units of centimetres is obtained by dividing by the density ρ in units of g cm^{-3} (1 μg cm^{-2} of density $\rho = 1$ g cm^{-3} corresponds to a geometric thickness of 10 nm).

3.4.2 Spatial and Depth Distribution of Energy Dissipation

The loss of energy along the electron trajectories results from ionization processes. Excited electrons can only escape as secondary electrons from a thin surface layer (Sect. 4.2). They are, however, generated throughout the diffusion cloud and their energy is finally converted into heat (phonons). Moreover, electrons excited with a larger excess energy and Auger electrons are decelerated by inelastic scattering within a short distance. A knowledge of the spatial distribution (local density) $g(\boldsymbol{r})$ or of the depth distribution $\Phi(z)$

Table 3.2. Range–energy relations

Formula E[keV], R[μg/cm^2]	Energy [keV]	Substance	Ref.
$R_x = \frac{20}{3}E^{5/3}$	10–200	Al	[3.93]
$R_E = 10.0E^{1.43}$	10–100	Au	[3.78]
$R_x = 7.4E^{1.5}$	5–11	Al	[3.87]
$R_x = 9.0E^{1.5}$	5–15	Cu, Ag, Au	[3.87]
$R_x = 11.5E^{1.35}$	2–40	Al	[3.92]
$R_x = 17.3E^{1.67}$	2–40	Al	[3.94]
$R_x = 4.0E^{1.5}$	50–25	Al, Si	[3.95]
$R_x = 10.3E^{1.34}$	2–5	Al	[3.87]
$R_x = 10.6E^{1.37}$	2–5	Cu	[3.87]
$R_x = 10.5E^{1.38}$	0.5–20	Al	[3.84]
$R_x = 14.0E^{1.31}$	0.5–20	Au	[3.83]
$R_x = 9.0E^{1.35}$	0.5–10	C	[3.81]
$R_x = 9.9E^{1.37}$	0.5–10	Al	[3.81]
$R_x = 11.8E^{1.44}$	0.5–10	Au	[3.81]

of the dissipated energy will be important for the discussion of heat generation, cathodoluminescence, excitation of electron-hole pairs in semiconductors, radiation damage and the exposure of photo-resists in electron-beam lithography.

Some of these interactions can be used to measure $g(z,r)$ or $\Phi(z)$. When an electron beam is allowed to leave the vacuum of the column through a differentially pumped system of diaphragms and enters a gas target at a reduced pressure of 50–500 mbar, the electron range and the diameter of the diffusion cloud can be increased to 10–20 cm. Figure 3.23 shows the luminescence excited by 100 keV electrons in N_2 and Ar at atmospheric pressure (1 bar). This experiment demonstrates the difference in the shape of the diffusion cloud for different values of Z. The more forwardly directed scattering in N_2 (low Z) results in a pear-shaped diffusion cloud, while that for the more diffuse scattering in Ar (higher Z) is apple-shaped. The light emission is proportional to $g(z,x)$ and can be represented as isodensities [3.96–98]. Similar experiments have been performed in fluorescent crystals by observing the cathodoluminescence with a light microscope [3.99]. Another possibility for imaging isodensities of equal $g(z,r)$ is the exposure of a photo-resist with an electron probe. The irradiated part becomes soluble (Sect. 7.3.6) and the shape of the diffusion cloud can be examined in cross-section with a SEM [3.100]. In low-pressure gas targets it is possible to measure the ionization density inside the diffusion cloud, 10–20 cm in diameter, by means of a semiconductor detector [3.101].

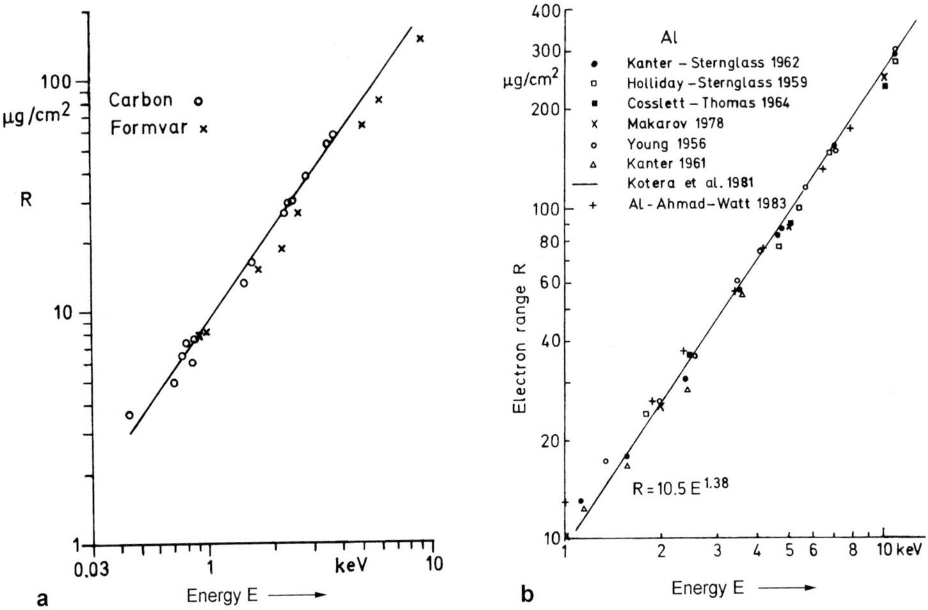

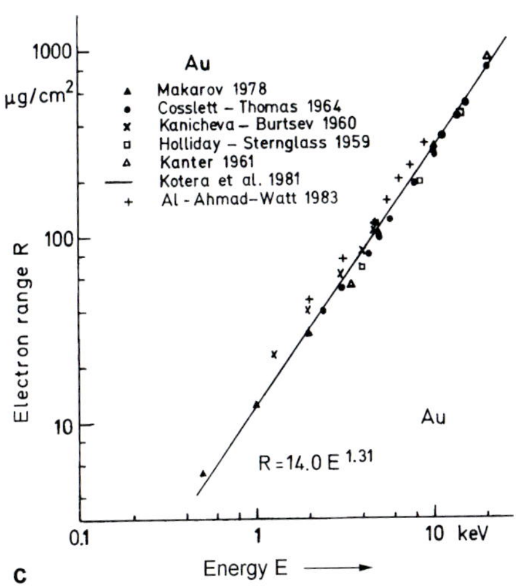

Fig. 3.22. Double-logarithmic plots of the electron range $R(E)$ vs. electron energy for (**a**) carbon [3.82], (**b**) Al and (**c**) Au [3.83–92]

The shape of the diffusion cloud does not vary significantly with energy. Only for relativistic energies ($E > 100$ keV) does the shape of the diffusion cloud start to become more forwardly directed.

The dissipation of electron energy per element of path length is well described by the Bethe stopping power (Sect. 3.3.4). Integration of $S = |dE_m/ds|$ yields E_m, which is plotted in Fig. 3.24 for carbon, copper and gold. Extrapolation to $E_m = 0$ results in the Bethe range R_B (Sect. 3.4.1), which

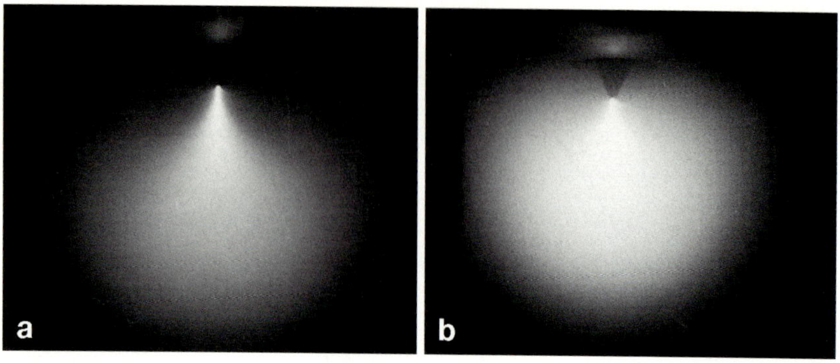

Fig. 3.23. Light emission caused by a 100 keV electron beam in (**a**) N_2, and (**b**) argon at atmospheric pressure

is twice as large for gold as for carbon, whereas the ranges practically observed in transmission experiments are approximately material-independent when measured in units of mass-thickness (Table 3.2). In high Z material (Au) no electrons can penetrate through the material without frequent large-angle scattering and no electron can move straight through the Bethe range, whereas in low Z material large-angle scattering is less frequent and many electrons travel through a thickness of the order of R_B though with frequent small-angle scattering processes; only a few electrons will be backscattered as indicated by the Z dependence of the backscattering coefficient in Fig. 4.7 and by the Monte Carlo simulations of electron trajectories in Fig. 3.21.

The depth distribution $\Phi(z)$ will not be proportional to the slope of $E_m(s)$, which is the stopping power S. The electrons lose energy continuously along their trajectories but the dependence on depth below the surface is not simple. In consequence, the depth distribution of energy dissipation can only be calculated numerically by using a transport equation or Monte Carlo simulations [3.102]. Figure 3.25 shows Monte Carlo simulations of $\Phi(z)$ for foils of increasing mass-thickness x. In thin foils, $\Phi(z)$ will be directly proportional to S. The energy dissipation increases with increasing x due to the deceleration coupled with the proportionality to $E^{-1}\ln(E/J)$ of the Bethe law (3.129) and due to the increase of path length caused by multiple scattering. Furthermore, the backscattering from deeper layers causes an increase of $\Phi(z)$ at low depths which is more pronounced in high Z material. This results in a maximum of $\Phi(z)$ not at the end of the range but nearer to the surface. Such $\Phi(z)$ curves for C and Au at normal incidence are also included in Fig. 3.24 for comparison with the Bethe ranges R_B.

Similar curves will be of interest for x-ray emission (Sect. 9.1.3). However, the curves for x-rays $\Phi_x(z)$ are different because we have to take into account the energy dependence of the cross-section of inner-shell ionization and the fact that electrons with energies lower than the ionization energy cannot

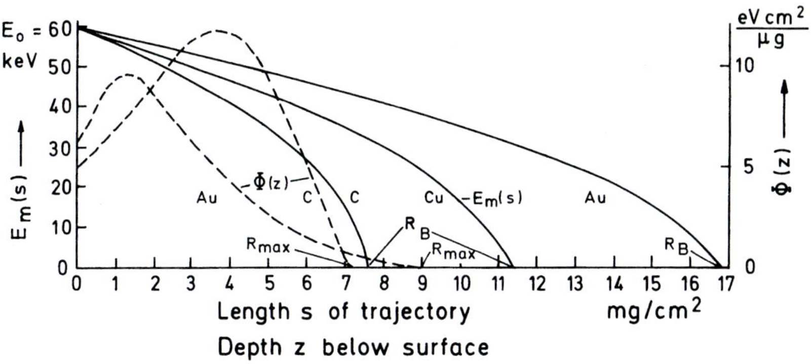

Fig. 3.24. Mean energy E_{m} versus path length s of the electron trajectories in the continuous-slowing-down approximation obtained by integrating the Bethe stopping power $S = -\,\mathrm{d}E_{\mathrm{m}}/\mathrm{d}s$ for C, Cu and Au and an initial energy $E = 60$ keV (——). Depth distribution $\Phi(z)$ of ionization (– – –) for C and Au obtained from Monte Carlo simulations [3.102] (R_{B} = Bethe range, $R_{\max}$ = maximum range)

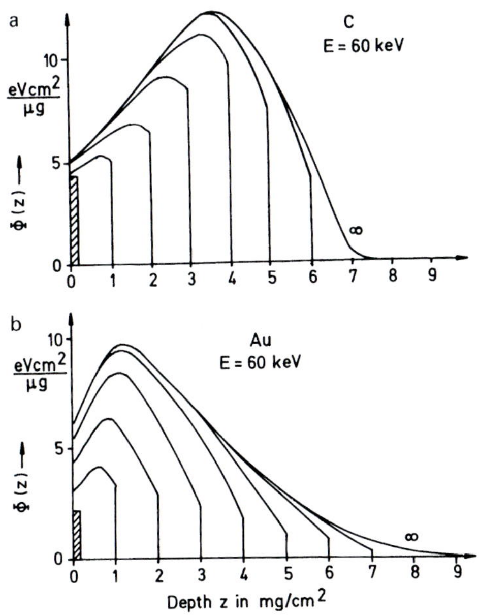

Fig. 3.25. Depth distribution of dissipated energy $\Phi(z)$ in foils of increasing mass-thickness $x = \rho t$ for (**a**) carbon and (**b**) Au ($E = 60$ keV) obtained from Monte Carlo simulations [3.102]

ionise the inner shell. These $\Phi_{\mathrm{x}}(z)$ curves are therefore shifted nearer to the surface.

Because no analytical formula can be established for $\Phi(z)$, this function is often approximated by a power or exponential series in z (see also Sect. 9.1.3 for the case of x-rays). The depth distribution is of considerable interest for the discussion of the EBIC mode (Sect. 7.1.2) and the series

$$\Phi(z') = 0.60 + 6.21z' - 12.4z'^2 + 5.69z'^3 \tag{3.154}$$

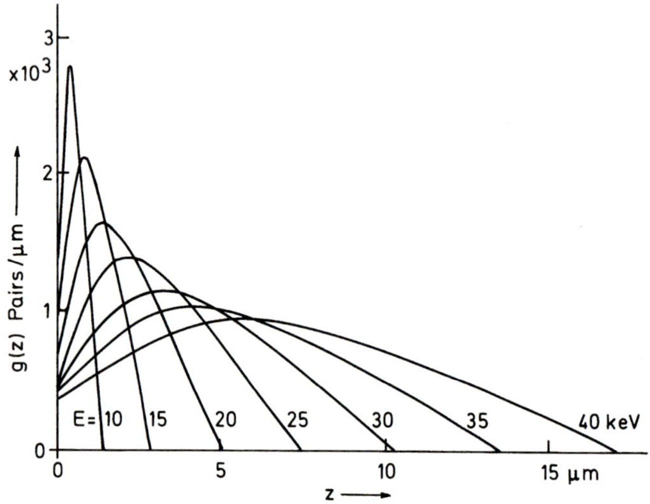

Fig. 3.26. Depth distribution $g(z)$ of electron-hole pairs generated in silicon for different electron energies E [3.95]

has been proposed for silicon [3.95] where $z' = z/R$ and the corresponding formula for R is listed in Table 3.2. Figure 3.26 shows how the number of electron–hole pairs with mean excitation energy $E_i = 3.65$ eV varies with depth and electron energy. Another approximation [3.103] is a shifted Gaussian

$$\Phi(z) \propto \exp[-a(z - z_m)^2] \tag{3.155}$$

with the maximum of $\Phi(z)$ at $z = z_m$.

The spatial distribution $g(z, r)$ for normal incidence cannot be satisfactorily described by an approximate formula. The generation of electron–hole pairs in silicon has been presented as isodensities [3.104]. However, it will often be sufficient to approximate the energy dissipation by a homogeneous density inside a sphere, the centre of which lies a certain distance below the surface [3.105] (see also diffusion models in Sect. 3.4.3).

3.4.3 Diffusion Models

The following diffusion models will be discussed:

1) Everhart's single-scattering model,
2) Archard's point-source diffusion model,
3) Thümmel's continuous diffusion model,
4) Combinations of single-scattering and diffusion models.

Diffusion models have mainly been developed to obtain an analytical formula for the backscattering coefficient (Sect. 4.1). Any agreement between the calculated and measured backscattering coefficients does not mean that other quantities evaluated with the same model are also correct!

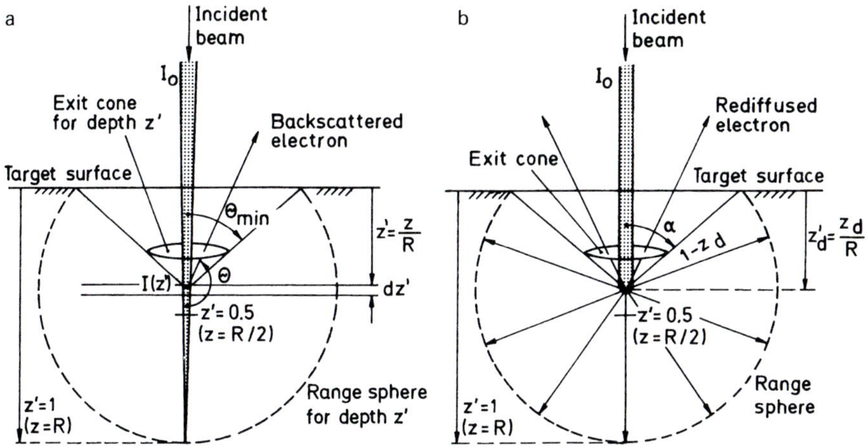

Fig. 3.27. (**a**) Everhart's single-scattering model, (**b**) Archard's point-source model ($z' = z/R$ = reduced depth coordinate, R = electron range)

a) Everhart's Single-Scattering Model. This model is based on the following assumptions (Fig. 3.27a) [3.106]:

1) The electron energy decreases with increasing depth z below the surface according to the Thomson–Whiddington law $v^4 - v_0^4 = c_T \rho z$ (3.135), which results in the range

$$R = v_0^4 / c_T \,, \tag{3.156}$$

where $v_0 = (2E/m)^{1/2}$ is the primary electron velocity.

2) The intensity $I(z)$ of the unscattered primary beam of intensity I_0 decreases with increasing z. The decrease in a layer between z and $z + \mathrm{d}z$ is given by

$$\mathrm{d}I(z) = -\sigma(\pi/2) \frac{N_A \rho}{A} I(z) \mathrm{d}z \,, \tag{3.157}$$

where $\sigma(\pi/2)$ is the cross-section for scattering through angles $\theta \geq 90°$ defined in (3.16). This means that all electrons scattered through angles $\theta \leq 90°$ are assumed to be still concentrated in the primary beam. Substituting for v^4 from (3.135) in $\sigma(\pi/2)$ and introducing a reduced depth $z' = z/R$ with the Thomson–Whiddington range R (3.156), we find

$$\mathrm{d}I(z') = -\frac{\pi e^4 N_A Z^2}{(4\pi\epsilon_0)^2 m^2 c_T A} I(z') \frac{\mathrm{d}z}{1 - z'} = -a \frac{I(z')\mathrm{d}z'}{1 - z'} \,, \tag{3.158}$$

where $a = 0.024 Z^2/A \simeq 0.012 Z$. The solution of (3.158) is

$$I(z') = I_0 (1 - z')^a \,. \tag{3.159}$$

3) Electrons are backscattered out of this beam only by single-scattering through angles $\theta > \pi/2$ and only electrons with scattering angles such that

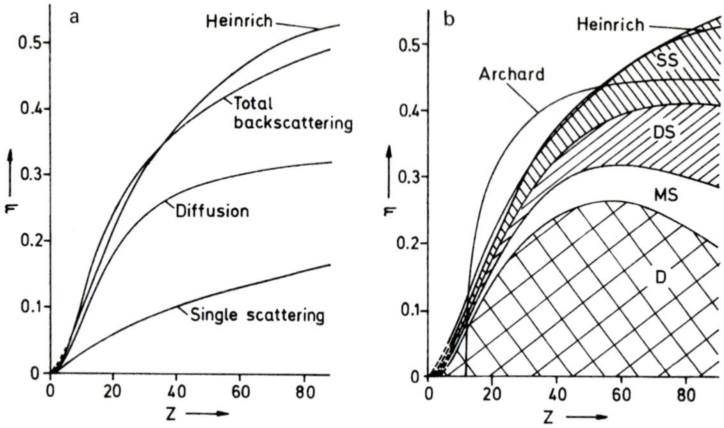

Fig. 3.28. (**a**) Calculation of the total backscattering coefficient η as the sum of contributions from Everhart's single-scattering model [3.106] and diffusion described by a diffusion model of *Niedrig* [3.111, 112]. (**b**) Contribution of single (SS), double (DS) and multiple (MS) scattering and diffusion (D) from a diffusion model of *Thümmel* [3.110] and comparison with Archard's model [3.107] and the measurements of *Heinrich*

$\pi - \theta_{\min} < \theta < \pi$ can leave the specimen, where $\cos(\pi - \theta_{\min}) = z'/(1-z')$ (Fig. 3.27a).

The total backscattering coefficient η_E is obtained by integrating from $z' = 0$ to $z' = 0.5$ ($z = R/2$):

$$\eta_E = \int_0^{0.5} \frac{N_A \rho}{A} \sigma(\theta_{\min}) I(z') dz'/I_0 \,. \tag{3.160}$$

Substitution of the quantities defined above gives

$$\eta_E = \frac{a - 1 + 0.5^a}{a + 1} \,. \tag{3.161}$$

This single-scattering contribution is plotted versus Z in Fig. 3.28a. There is a serious discrepancy between the values given by (3.161) and the experimentally observed values of η, which are of the order of a factor 3 larger than η_E. Everhart introduced a larger constant $a' = 0.0457Z$ instead of a to make the curve fit the experimental results but this choice is, however, arbitrary.

b) Archard's Point-Source Diffusion Model. Here it is assumed that the state of complete, isotropic diffusion is reached after multiple scattering at a diffusion depth z_d (Fig. 3.27b) [3.107]. Such a depth of complete diffusion was originally introduced [3.46, 108] as the depth at which, by plural scattering, $\overline{\cos\theta} = 1/e$. Archard used this model by drawing a sphere of radius 1–z_d with its centre at a reduced depth $z'_d = z_d/R$. The backscattering coefficient then becomes $\eta = \Omega/4\pi$ where Ω is the solid angle of the cone with

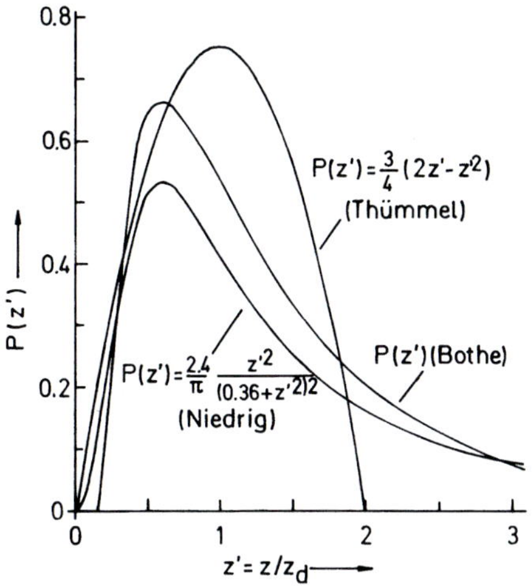

Fig. 3.29. Depth distribution $P(z')$ of isotropic point sources from the calculation of *Bothe* [3.46] and analytical fits of *Thümmel* [3.110] and *Niedrig* [3.112] ($z' = z/R$)

a semi-apex angle α and $\cos\alpha = z'_\mathrm{d}/(1 - z'_\mathrm{d})$ (Fig. 3.27b). With $z'_\mathrm{d} \simeq 40/7Z$ this results in

$$\eta_\mathrm{A} = \frac{1}{2}\frac{1-2z'_\mathrm{d}}{1-z'_\mathrm{d}} = \frac{7Z-80}{14Z-80} . \tag{3.162}$$

This formula only works for $Z > 11$ and even then does not agree well with experiment (Fig. 3.28b). In a modified version of this model [3.109], the diffusion depth was taken to be the depth of largest energy dissipation, which is somewhat smaller than z_d.

c) Thümmel's Continuous Diffusion Model. A better approximation [3.110] than the Archard model is a continuous depth distribution $P(z')$ of isotropic point sources (Fig. 3.29) [3.46], which may be written as an infinite series. Such a distribution can be approximated by simpler formulae (Fig. 3.29) [3.110–112]. The distribution $P(z')$ is normalized: $\int P(z')\mathrm{d}z' = 1$ and the contribution from a layer between z' and $z' + \mathrm{d}z'$ to the backscattering becomes, with the Archard formula (3.162),

$$\mathrm{d}\eta_\mathrm{D} = \int_0^{0.5} P(z')\frac{1}{2}\frac{1-2z'}{1-z'}\mathrm{d}z' . \tag{3.163}$$

Werner et al. [3.113] also considered electron beam spreading by small-angle multiple scattering.

d) Combination of Models. When the single-scattering and diffusion models are combined, both scattering processes result in a decrease of the primary beam. The relative importance of the two contributions can be obtained by fitting the curves to experiments [3.114] or evaluated from the condition that negligible single scattering ($a \to 0$) will result in the pure diffusion model [3.111, 112]. Figure 3.28b shows how the dependence of the backscattering coefficient on Z can be approximated by a single-scattering contribution (SS), a double-scattering contribution (DS) [3.115] and a contribution from diffusion (D) [3.110]. Figure 3.28a contains calculations of η for single scattering and diffusion using the model of *Niedrig* [3.111, 112].

This presentation of the diffusion models can give only a superficial idea of how such models work; for details the reader is referred to reviews [3.110, 112]. It can be seen that the simplifications of the complex diffusion process are very crude. The rather good agreement of the total backscattering coefficient, even at non-normal incidence, and of the backscattering from thin films does not mean that other quantities that can be evaluated from diffusion models also agree with experiment. For example, a large discrepancy is found in the shape of the angular distribution of backscattered electrons [3.111].

3.4.4 General Transport Equation

The ideal information about electron diffusion would be the phase-space density $n(r, \boldsymbol{v}, t)$. At a time t, $n\mathrm{d}^3\boldsymbol{r}\mathrm{d}^3\boldsymbol{v}$ electrons occupy the volume $\mathrm{d}^3\boldsymbol{r}$ around $\boldsymbol{r}$ with velocities $\boldsymbol{v}$ in $\mathrm{d}^3\boldsymbol{v}$. The absolute values of $|\boldsymbol{v}|$ contain the energy distribution and the directions of $\boldsymbol{v}$ contain the angular distribution. This means that $\mathrm{d}^3\boldsymbol{v}$ can equally well be written $\mathrm{d}\Omega\mathrm{d}E$. The phase-space flux density is defined as

$$J(\boldsymbol{r}, \boldsymbol{v}, t)\mathrm{d}S\mathrm{d}^3\boldsymbol{v} = \boldsymbol{v}n(\boldsymbol{r}, \boldsymbol{v}, t)\mathrm{d}S\mathrm{d}^3\boldsymbol{v} \tag{3.164}$$

and represents the number of electrons passing per unit time through a surface element $\mathrm{d}S$ at a time t with velocities $\boldsymbol{v}$ in $\mathrm{d}^3\boldsymbol{v}$. Inside a small volume V of surface S around a point $\boldsymbol{r}$, the number N of electrons in V per phase-space element $\mathrm{d}^3\boldsymbol{v}$ changes in time by

$$\frac{\mathrm{d}N}{\mathrm{d}t} = \frac{\partial}{\partial t}\int_V n(\boldsymbol{r}, \boldsymbol{v}, t)\mathrm{d}^3\boldsymbol{r} \quad \text{or} \quad \frac{\mathrm{d}N}{\mathrm{d}t} = v\frac{\partial}{\partial s}\int_V n(\boldsymbol{r}, \boldsymbol{v}, t)\mathrm{d}^3\boldsymbol{r} \,. \tag{3.165}$$

where the time element $\mathrm{d}t = \mathrm{d}s/v$ is related to the path length element $\mathrm{d}s$. The following terms contribute to this change:

1) Generation of electrons in a source:

$$\frac{\mathrm{d}N_1}{\mathrm{d}t} = \int_V s(\boldsymbol{r}, \boldsymbol{v}, t)\mathrm{d}^3\boldsymbol{r} \tag{3.166}$$

with a source function $s(\boldsymbol{r}, \boldsymbol{v}, t)$ that can be a δ-pulse at $\boldsymbol{r} = 0$ at time $t = 0$ with an initial velocity $\boldsymbol{v} = \boldsymbol{v}_0$ of the primary beam, for example, but will be zero inside the specimen.

2) Flux of electrons through S:

$$\frac{\mathrm{d}N_2}{\mathrm{d}t} = -\int_S J(\boldsymbol{r}, \boldsymbol{v}, t)\mathrm{d}S = -\int_V \boldsymbol{v} \cdot \nabla n \,\mathrm{d}^3\boldsymbol{r} \,. \tag{3.167}$$

3) Loss of electrons in $\mathrm{d}^3\boldsymbol{v}$ by scattering from the directions $\boldsymbol{v}$ into other directions $\boldsymbol{v}'$:

$$\begin{aligned}\frac{\mathrm{d}N_3}{\mathrm{d}t} &= -\int_\Omega\int_V v\,n(\boldsymbol{r},\boldsymbol{v},t)N_V\frac{\mathrm{d}\sigma(\boldsymbol{v}\to\boldsymbol{v}')}{\mathrm{d}\Omega}\mathrm{d}^3\boldsymbol{v}'\mathrm{d}^3\boldsymbol{r}\\ &= -\frac{1}{\Lambda}\int_V v\,n(\boldsymbol{r},\boldsymbol{v},t)\mathrm{d}^3\boldsymbol{r}\,. \end{aligned} \tag{3.168}$$

where $\mathrm{d}\sigma(\boldsymbol{v}\to\boldsymbol{v}')/\mathrm{d}\Omega$ is the scattering cross-section (elastic + inelastic) from $\boldsymbol{v}$ to $\boldsymbol{v}'$ and N_V the number of atoms per unit volume. The second line (3.168) results from the relation

$$N_V\int\frac{\mathrm{d}(\boldsymbol{v}\to\boldsymbol{v}')}{\mathrm{d}\Omega}\mathrm{d}^3\boldsymbol{v}' = N_V\sigma(E) = 1/\Lambda(E) \tag{3.169}$$

where Λ is the mean-free-path length introduced in Sect. 3.3.1.

4) Gain of electrons in $\mathrm{d}^3\boldsymbol{v}$ around $\boldsymbol{v}$ by scattering from other directions $\boldsymbol{v}'$ into $\boldsymbol{v}$:

$$\frac{\mathrm{d}N_4}{\mathrm{d}t} = +\int_\Omega\int_V v'n(\boldsymbol{r},\boldsymbol{v}',t)N_V\frac{\mathrm{d}\sigma(\boldsymbol{v}'\to\boldsymbol{v})}{\mathrm{d}\Omega}\mathrm{d}^3\boldsymbol{v}'\mathrm{d}^3\boldsymbol{r}\,. \tag{3.170}$$

From the total rate $\mathrm{d}N/\mathrm{d}t = \sum_i \mathrm{d}N_i/\mathrm{d}t$, we find that for the case of stationarity in time ($\mathrm{d}N/\mathrm{d}t = 0$), we have

$$\begin{aligned}&v_x\frac{\partial n}{\partial x} + v_y\frac{\partial n}{\partial y} + v_z\frac{\partial n}{\partial z} = \\ &s(\boldsymbol{r},\boldsymbol{v}) + \tfrac{1}{\Lambda}\int[v'n(\boldsymbol{r},\boldsymbol{v}')S(\boldsymbol{v}'\to\boldsymbol{v}) - vn(\boldsymbol{r},\boldsymbol{v})S(\boldsymbol{v}\to\boldsymbol{v}')]\mathrm{d}^3v'\end{aligned} \tag{3.171}$$

or with (3.168),

$$\boldsymbol{v}\cdot\nabla n = s(\boldsymbol{r},\boldsymbol{v}) - \frac{v}{\Lambda}n + \frac{1}{\Lambda}\int v'n'S(\boldsymbol{v}'\to\boldsymbol{v})\mathrm{d}^3v'\,. \tag{3.172}$$

Here,

$$S(\boldsymbol{v}'\to\boldsymbol{v}') = \frac{\mathrm{d}\sigma(|\boldsymbol{\theta}-\boldsymbol{\theta}'|)}{\mathrm{d}\Omega}\Bigg/\int\frac{\mathrm{d}\sigma}{\mathrm{d}\Omega}\mathrm{d}\Omega = \Lambda N_\mathrm{V}\frac{\mathrm{d}\sigma(|\boldsymbol{\theta}-\boldsymbol{\theta}'|)}{\mathrm{d}\Omega} \tag{3.173}$$

is a normalized single-scattering function analoguous to S_1 introduced in (3.105) and S/Λ is the probability of scattering per unit path length. If we prefer to introduce the path length s, we can use (3.165):

$$\frac{\partial n}{\partial s} + \boldsymbol{v}\cdot\nabla n = s(\boldsymbol{r},\boldsymbol{v}) - \frac{v}{\Lambda}n + \frac{1}{\Lambda}\int v'n'S(\boldsymbol{v}'\to\boldsymbol{v})\mathrm{d}^3v'\,. \tag{3.174}$$

Equations (3.171, 3.172 and 3.174) contain six variables x, y, z, v_x, v_y, v_z and in (3.174) the path length s as well. If we assume an energy–path length relation $E(s)$ (continuous-slowing-down approximation), we again have six variables x, y, z, θ, χ, s. These equations are far more general than is needed and are too complex to be solved. The number of variables can be reduced by assuming a semi-infinite target and neglecting the dependence of n on x

and y. This corresponds to the case in which the specimen is irradiated with a broad uniform electron beam, and we have to consider only the dependence of n on depth z below the surface. The variables are now z, θ, χ, E (or s). Equation (3.171) reduces to

$$v\cos\theta\frac{\mathrm{d}n(z,\theta,\chi,E)}{\mathrm{d}z}=\frac{1}{\Lambda}\int\int[v'n'S(|\boldsymbol{\theta}-\boldsymbol{\theta}'|,E')-vnS(|\boldsymbol{\theta}'-\boldsymbol{\theta}|,E)]\mathrm{d}\Omega'\mathrm{d}E'\,. \tag{3.175}$$

It proves convenient to rewrite (3.175) in terms of the flux density J according to (3.164), giving

$$\cos\theta\frac{\mathrm{d}J(z,\theta,\chi,E)}{\mathrm{d}z}=\frac{1}{\Lambda}\int\int[J'S(|\boldsymbol{\theta}-\boldsymbol{\theta}'|,E')-JS(|\boldsymbol{\theta}'-\boldsymbol{\theta}|,E)]\mathrm{d}\Omega'\mathrm{d}E'\,. \tag{3.176}$$

3.4.5 Special Solutions of the Transport Equation

Let us apply (3.172) to the simpler case of small-angle scattering, where large-angle scattering processes and any energy losses are neglected; this means that v, E and $\mathrm{d}\sigma/\mathrm{d}\Omega$ are constant throughout a foil of thickness t, which we recognize as the conditions for plural scattering in Sect. 3.3.2. We can write $v = v(\theta_x, \theta_y, 1)$, where $\theta = (\theta_x^2 + \theta_y^2)^{1/2}$ is the scattering angle, to a good approximation. Equation (3.172) then reduces to

$$\theta_x\frac{\partial n}{\partial x}+\theta_y\frac{\partial n}{\partial y}+\frac{\partial n}{\partial z}=-\frac{n}{\Lambda}+\frac{1}{\Lambda}\int nS(|\boldsymbol{\theta}-\boldsymbol{\theta}'|)\mathrm{d}^2\boldsymbol{\theta}' \tag{3.177}$$

for $n(x, y, z, \boldsymbol{\theta})$).

To solve this equation, we take the four-fold Fourier transform of n with respect to the variables $\boldsymbol{r}$ (x, y plane) and $\boldsymbol{\theta}$ (θ_x, θ_y plane)

$$\tilde{n}(R,z,\Theta)=\int\int n(r,z,\theta)\exp[-2\pi\mathrm{i}(\boldsymbol{r}\cdot\boldsymbol{R}+\boldsymbol{\theta}\cdot\boldsymbol{\Theta})]\mathrm{d}^2\boldsymbol{r}\mathrm{d}^2\boldsymbol{\theta} \tag{3.178}$$

and the two-dimensional Fourier transform of $S(\theta)$:

$$\tilde{S}(\Theta)=\int S(\theta)\exp(-2\pi\mathrm{i}\boldsymbol{\theta}\cdot\boldsymbol{\Theta})\mathrm{d}^2\boldsymbol{\theta}\,. \tag{3.179}$$

The last integral in (3.177) is a convolution of $n(\theta)$ and $S(\theta)$, the Fourier transform of which becomes $\tilde{n}\cdot\tilde{S}$ thanks to the convolution theorem of Fourier transforms. Taking the Fourier transform of both sides of (3.177), we obtain the simpler form

$$X\frac{\partial\tilde{n}}{\partial\theta_x}+Y\frac{\partial\tilde{n}}{\partial\theta_y}+\frac{\partial\tilde{n}}{\partial z}=-\frac{\tilde{n}}{\Lambda}(1-\tilde{S})\,. \tag{3.180}$$

If we are interested only in the angular distribution $n(z, \theta)$ of electrons that have transversed a foil of thickness $z = t$, we can omit the first two terms on the left-hand sides of (3.177) and (3.180). If we retain these two terms, we get the transport equation, which also describes the lateral displacement of the electron beam and the beam broadening, discussed in Sect. 3.3.3 [3.50, 52, 53].

For the former case, we have $\mathrm{d}\tilde{n}/\tilde{n} = -(1-\tilde{S})\mathrm{d}z/\Lambda$ with the solution

$$\tilde{n} = \exp\left[-\frac{1}{\Lambda}\int_0^t (1-\tilde{S})\mathrm{d}z\right] \tag{3.181}$$

and n is then obtained from the inverse Fourier transform

$$n(t,\theta) = \int \exp\left[-\frac{1}{\Lambda}\int_0^t (1-\tilde{S})\mathrm{d}z\right]\exp(2\pi\mathrm{i}\boldsymbol{\theta}\cdot\boldsymbol{\Theta})\mathrm{d}^2\Theta\,. \tag{3.182}$$

We assume rotational symmetry, which means $\boldsymbol{\theta}\cdot\boldsymbol{\Theta} = \theta\Theta\cos\chi$ and $\mathrm{d}^2\Theta = \Theta\mathrm{d}\chi\mathrm{d}\Theta$. An integral of the form

$$\int_0^{2\pi} \exp(\mathrm{i}x\cos\chi) = 2\pi\mathrm{J}_0(x)$$

introduces a Bessel function and (3.182) becomes

$$n(z,\theta) = 2\pi\int_0^\infty \exp\left\{-\frac{t}{\Lambda}\left[1-2\pi\int_0^\infty S(\theta)\mathrm{J}_0(2\pi\theta\Theta)\theta\mathrm{d}\theta\right]\right\}\mathrm{J}_0(2\pi\theta\Theta)\Theta\mathrm{d}\Theta\,. \tag{3.183}$$

This is the plural scattering integral of *Molière* [3.43], which is exact for any differential cross-section $\mathrm{d}\sigma/\mathrm{d}\Omega$ or $S(\theta)$, provided the scattering angles are small. A quite similar method has been applied in Sect. 3.3.5 to the problem of energy losses. There we showed that the solution (3.144) of the corresponding transport equation (3.140) is identical with the multiple scattering formula (3.145) using Poisson coefficients. We have the same relations between the solution (3.183) of the transport equation (3.180) and the solution (3.108) with Poisson coefficients.

Another method of solving the transport equation, which includes large-angle scattering, is to expand $n(\theta)$ and $S(\theta)$ as a series of Legendre polynomials, as described by *Goudsmit* and *Saunderson* [3.41] (Sect. 3.2.2). Equations (3.112) and (3.113) are valid for any scattering angles and differential cross-sections. This solution also satisfies the transport equation. A disadvantage is that a large number of terms of the series, of the order of 50–100, is needed to consider both large- and small-angle scattering and it becomes difficult to introduce the boundary conditions. A special momentum method has been proposed by *Spencer* [3.116], which assumes a unidirectional source of electrons in an infinite medium and allows the depth distribution of energy dissipation (Sect. 3.4.2) to be calculated.

Two other numerical methods for solving the transport equations (3.175) and (3.176) have been developed, which can be applied directly to problems that are of interest in SEM. These are the matrix method of *Fathers* and *Rez* [3.117] and the transformation of the integro-differential equation (3.175) into an integral equation by *Hoffmann* and *Schmoranzer* [3.118].

3.4.6 Monte Carlo Simulations

In Monte Carlo simulations [3.102, 119–129], a large number of electron trajectories are simulated on a computer using random numbers to guide the choice of the most important parameters of the scattering processes. Depending on the final quantity that is to be calculated, which may be the backscattering coefficient, the depth distribution of dissipated energy, the local x-ray emission or the energy distributions of backscattered or transmitted electrons, 10^3–10^6 trajectories will be necessary to achieve a sufficient accuracy. Due to the increasing speed of personal computers, Monte Carlo simulations are becoming of considerable interest and the agreement with experiment is in most cases better than calculations with the approximations described in Sects. 3.5.3 and 3.5.4, especially because there are no limits on the choice of boundaries.

Although no real electron will follow a calculated trajectory (Fig. 3.21) and all trajectories will be different because of the large number of scattering processes, the statistical mean value of a quantity obtained from a large number of trajectories will agree with measurements. An advantage of Monte Carlo simulation is that quantities that are not attainable directly by experiment can also be calculated (Figs. 3.13, 25, and 4.20, 28, for example). In order to calculate statistical values of a scattering parameter x, such as the free-path length s between two collisions, the scattering angle θ or its azimuth χ, we introduce $p(x)\mathrm{d}x$, the probability that an event occurs inside the interval $x, x+\mathrm{d}x$; $p(x)$ will be defined over some interval x_1, x_2. Then a normalized probability function

$$P(x) = \int_{x_1}^{x} p(x)\mathrm{d}x \Big/ \int_{x_1}^{x_2} p(x)\mathrm{d}x \tag{3.184}$$

can be introduced with values between zero and one and each value of $P(x)$ will correspond to a value of x and conversely.

We illustrate this procedure by considering the free-path length s with $p(s) = \exp(-s/\Lambda)$ where $\Lambda = 1/N(\sigma_{\mathrm{el}} + \sigma_{\mathrm{inel}})$ (3.103) denotes the mean-free-path length. We find

$$P(s) = \int_{0}^{x} \exp(-s/\Lambda)\mathrm{d}s \Big/ \int_{0}^{\infty} \exp(-s/\Lambda)\,\mathrm{d}s = 1 - \exp(-s/\Lambda)\,. \tag{3.185}$$

After choosing a random number R in the interval (0,1) and setting $P(s) = R$, solution of (3.185) for s gives

$$s = -\Lambda \ln(1-R) = -\Lambda \ln R' \tag{3.186}$$

because if R is a random number, so is $R' = 1 - R$. By generating a great many random numbers R, the corresponding s values are distributed with the correct probability $p(s)\Delta s$ inside the interval $s, s+\Delta s$.

It is now possible to decide whether the scattering process is elastic or inelastic. If $R < q = \sigma_{\mathrm{el}}/(\sigma_{\mathrm{el}} + \sigma_{\mathrm{inel}})$ then the process is elastic, otherwise it is inelastic.

For calculating the scattering angle θ, we remember that

$$p(\theta)\mathrm{d}\theta = (\mathrm{d}\sigma/\mathrm{d}\Omega)2\pi \sin\theta\mathrm{d}\theta \tag{3.187}$$

will be the probability for scattering through an interval $\theta, \theta + \mathrm{d}\theta$. It will be useful to start the interval of θ not at zero but at some minimum scattering angle θ_{min} of the order of $5° - 10°$. Otherwise, the number of scattering processes or the corresponding mean-free-path lengths Λ become too large or too small, respectively. Scattering through $\theta < \theta_{\mathrm{min}}$ can be considered by means of a multiple scattering theory, which calculates a mean scattering angle by use of (3.114), for example. Equation (3.187) gives

$$P(\theta) = \int\limits_{\theta_{\mathrm{min}}}^{\theta} \frac{\mathrm{d}\sigma}{\mathrm{d}\Omega} 2\pi \sin\theta\mathrm{d}\theta \bigg/ \int\limits_{\theta_{\mathrm{min}}}^{\pi} \frac{\mathrm{d}\sigma}{\mathrm{d}\Omega} 2\pi \sin\theta\mathrm{d}\theta \ . \tag{3.188}$$

We have to substitute $\mathrm{d}\sigma_{\mathrm{el}}/\mathrm{d}\Omega$ or $\mathrm{d}\sigma_{\mathrm{inel}}/\mathrm{d}\Omega$ according as the scattering is elastic or inelastic ($R < q$ or $R > q$).

The Rutherford formula (3.14) can be substituted for $\mathrm{d}\sigma_{\mathrm{el}}/\mathrm{d}\Omega$. However, it was demonstrated in Sect. 3.1.6 that Mott scattering cross-sections can differ significantly from the Rutherford cross-sections, and a discrete table $\theta(R)$ has to be computed from which the value of θ corresponding to a random number R can be interpolated. The correspondence between azimuth angle and a random number is very simple

$$\chi = 2\pi R \, . \tag{3.189}$$

A plot of the random numbers R in an interval $R, R + \Delta R$ should be approximately constant for all intervals between $R = 0$ and 1. The random numbers are calculated by a computer subroutine, which generates so-called pseudo-random numbers: there is a periodicity in these but only after a very large number of calculations. It is possible, therefore, to start each program with the same initial random number and the same sequence of random numbers, which has the advantage that the calculation is reproducible but the disadvantage that the computation is not completely random. Alternatively, it is possible to start each new calculation with the last random number of the preceding run.

Each step consists in calculating the new position $x_{n+1}, y_{n+1}, z_{n+1}$ of the electron and its new angles θ_{n+1} relative to the z axis and its azimuth χ_{n+1} relative to the x axis by using the calculated values of s_{n+1}, θ and χ of the next scattering process (Fig. 3.30), which can be obtained from (3.186), (3.188) and (3.189), respectively:

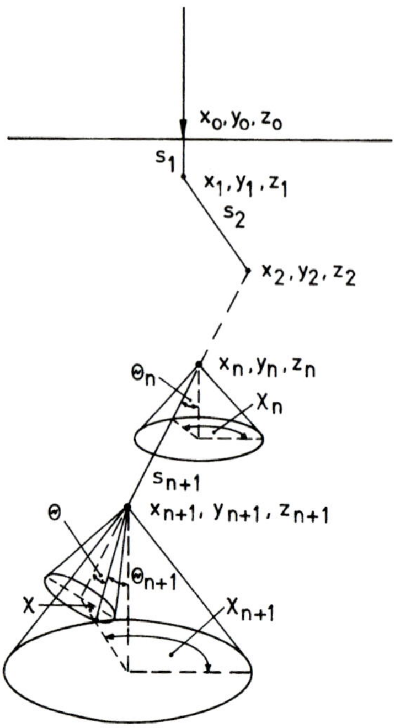

Fig. 3.30. Sequence of scattering processes in a Monte Carlo simulation with s_i = free path lengths, x_i, y_i, z_i. = coordinates of electron at the ith collision, θ_i and χ_i = scattering and azimuth angles after the ith collision

$$
\begin{aligned}
x_{n+1} &= x_n + s_{n+1}\sin\theta_n\cos\chi_n \\
y_{n+1} &= y_n + s_{n+1}\sin\theta_n\sin\chi_n \\
z_{n+1} &= z_n + s_{n+1}\cos\theta_n \\
\cos\theta_{n+1} &= \cos\theta_n\cos\theta - \sin\theta\cos\chi\sin\theta_n \\
\sin\theta_{n+1} &= (1-\cos^2\theta_{n+1})^{1/2} \\
{}^{\cos}_{\sin}\chi_{n+1} &= [{}^{\cos}_{\sin}\theta_n(\sin\theta\cos\chi\cos\theta_n + \cos\theta\sin\theta_n){}^{-\sin}_{+\cos}\sin\theta\sin\chi]/\sin\theta_{n+1}.
\end{aligned}
\tag{3.190}
$$

The energy loss along the trajectories can be represented by the Bethe continous-slowing-down approximation (Sect. 3.3.4) and is then proportional to the path length $\Delta(\rho s)$ in units of g/cm^2:

$$
\Delta E = \left[\left(\frac{\mathrm{d}E_\mathrm{m}}{\mathrm{d}x}\right)_\mathrm{Bethe} - \left(\frac{\mathrm{d}E_\mathrm{m}}{\mathrm{d}x}\right)_\mathrm{single}\right]\Delta(\rho s)\,. \tag{3.191}
$$

If single large-angle, inelastic scattering events are considered by means of the Gryziński formula (3.76), for example, the contributions of these to the energy loss have to be subtracted from the Bethe stopping power (first term in (3.191). *Shimizu* et al. [3.124] used three loss mechanisms and completely neglected any continuous loss:

1) Inner-shell excitation with the cross-sections of *Gryziński* [3.33],
2) plasmon losses with the mean-free-path length of *Quinn* [3.130], and

3) excitation of conduction electrons with the cross-section (4.28) of *Streitwolf* [3.131].

This scheme also makes it possible to include secondary electron production in cascade [3.132]. The best results have been obtained for aluminium. For other elements the elementary excitation processes are less well known and for copper, the number of conduction electrons had to be adjusted so that the total stopping power approached the Bethe value.

In the calculation, each scattering process is followed by an 'IF' condition, which examines the quantity in question. In the simple case of the backscattering coefficient, the number of backscattered electrons is increased by unity if $z \leq 0$ and a new electron trajectory is then started. In addition, the BSE can be sorted according to their energies or exit angles. Alternatively, the Bethe loss may be used as a guide to sum up depth distributions of energy dissipation, for example. However, these 'IF' conditions after each scattering process consume a considerable amount of computation time and it is not possible to ask for too many quantities in one Monte Carlo run. In contrast to the diffusion models and transport equations, the Monte Carlo method can use any desired boundary such as a sphere on a substrate, for example, and it is possible to configure specimen geometries stored in a matrix [3.133–135].

3.5 Specimen Heating and Damage

3.5.1 Specimen Heating

The power $P = UI_\mathrm{p}$ of the electron probe reaches $P = 20\ \mu\mathrm{W}$ with a probe current $I_\mathrm{p} = 10^{-9}$ A and an acceleration voltage $U = 20$ kV, for example. The fraction that does not leave the specimen as BSE or SE and which is of the order of $f = 40$–80% will be converted into heat in the specimen. Most of the excitation processes will also finally result in the production of phonons. High-current electron beams can be used for melting and welding, but in the SEM the power P is so small that melting processes and thermal decomposition can be observed only in materials of low melting or decomposition temperature and of low thermal conductivity, such as polymers or organic substances.

The increase of temperature of a plane specimen surface can be estimated by assuming that the power $P = fUI_\mathrm{p}$ is homogeneously dissipated inside a hemisphere of radius $R/2$ (R: electron range). The electron probe is assumed to be at rest and the probe diameter $d_\mathrm{p} \ll R$. Then we have a radially symmetric problem and the equation for thermal conduction $P = -\lambda \nabla T$ becomes

$$P\frac{r^3}{(R/2)^3} = -\lambda \underbrace{2\pi r^2}_{S} \frac{\mathrm{d}T}{\mathrm{d}r} \quad \text{for} \quad 0 < r < R/2 \tag{3.192}$$

$$P = -\lambda 2\pi r^2 \frac{\mathrm{d}T}{\mathrm{d}r} \quad \text{for} \quad R/2 < r < \infty\,, \tag{3.193}$$

where $S = 2\pi r^2$ is the surface area of a hemisphere of radius r and λ denotes the thermal conductivity in units of J s^{-1} m^{-1} K^{-1}. Integration of (3.193) gives the temperature difference

$$\Delta T_1 = T(0) - T(R/2) = P/2\pi\lambda R \tag{3.194}$$

between the centre ($r = 0$) and $r = R/2$, and integration of (3.193) gives

$$\Delta T_2 = T(R/2) - T(\infty) = 2P/2\pi\lambda R\,. \tag{3.195}$$

The total increase of temperature for compact matter becomes

$$\Delta T_\mathrm{c} = \Delta T_1 + \Delta T_2 = 3P/2\pi\lambda R\,. \tag{3.196}$$

A similar formula was derived by *Dudek* [3.136] using a more detailed calculation.

Consider, for example, 20 keV electrons, which have a range $R \simeq 1\ \mu$m in copper. The maximum probe current for a probe diameter of the order of 0.1 μm is $I_\mathrm{p} \simeq 10^{-9}$ A. Using the value 420 J s^{-1} m^{-1} K^{-1} for the thermal conductivity and $f = 70\%$ [3.137], (3.196) gives $\Delta T = 0.02°$C. This value will be even lower when the beam is scanned over a larger area; see [3.138, 139] for the calculation of temperature near a moving probe.

However, polymers have a thermal conductivity $\lambda \simeq 4$ J s^{-1} m^{-1} K^{-1} and can be melted or deteriorated when large probe currents are used. For thin fibres especially, the temperature can increase considerably. We consider a fibre of length L and diameter D with a cross-section $S = \pi D^2/4$, which is irradiated at one end whereas the other is connected to a heat sink of constant temperature. The equation of thermal conductivity becomes

$$P = -\lambda \frac{\pi D^2}{4} \frac{\mathrm{d}T}{\mathrm{d}x} \tag{3.197}$$

and the temperature difference between the two ends of the fibre is

$$\Delta T_\mathrm{f} = \frac{4PL}{\pi \lambda D^2}\,. \tag{3.198}$$

Taking the ratio of (3.196) and (3.198), we have

$$\frac{\Delta T_\mathrm{f}}{\Delta T_\mathrm{c}} = \frac{8}{3}\frac{L}{D}\frac{R}{D} \tag{3.199}$$

which shows that there will be a large temperature rise for high values of L/D.

A natural question is whether the specimen temperature of organic material can be decreased by coating it with metal. We assume that the heat power P can be conducted through a freely-supported metal layer of thickness t. The equation of thermal conductivity becomes

$$P = -\lambda 2\pi r t\, \mathrm{d}T/\mathrm{d}r\,. \tag{3.200}$$

Solving for $\mathrm{d}T/\mathrm{d}r$ and integrating from $r = R/2$ to R_a, the distance to a heat sink of constant temperature, we find

$$\Delta T_\mathrm{m} = T(R/2) - T(R_\mathrm{a}) = \frac{P}{2\pi\lambda t}\ln\left(\frac{R_\mathrm{a}}{R/2}\right) . \tag{3.201}$$

For $R = 1\ \mu$m, $R_\mathrm{a} = 1$ cm and $t = 10$ nm, ΔT_m will be a factor 400 larger than ΔT_c for the same bulk material. Recalling that the thermal conductivities of organic materials are two orders of magnitude lower than those of metals, we see that there will be a small gain in the thermal conductivity when a metal film is superposed on a plane substrate. However, the thermal conductivity of a thin evaporated film can be lower than that of the bulk material.

If a metal coating on a fibre is to have a better thermal conduction than the fibre, the product λS for the coating must be larger than that for the fibre

$$\lambda_\mathrm{m}\pi D t \gg \lambda_\mathrm{f}\frac{\pi D^2}{4} \quad \rightarrow \quad t \gg \frac{\lambda_\mathrm{f} D}{4\lambda_\mathrm{m}} \simeq 2.5\times 10^{-3} D . \tag{3.202}$$

This means that fibres of $D \leq 10\mu$m can be heat-protected more effectively by a 10 nm metal coating.

The dependence of specimen heating on electron energy shows the following tendency. We have $P \propto E$ and the range increases as $R \propto E^{5/3}$ with increasing E. For equal probe currents I_p, we expect from (3.196) a decrease $\Delta T_\mathrm{c} \propto E^{-2/3}$ with increasing energy. For a stationary probe, this would result in an increase of specimen heating in LVSEM. However owing to the increase of the secondary electron yield $\delta \propto E^{-0.8}$ for $E \geq 2-3$ keV, we can work with lower probe currents. In the case of a scanning probe, the probe scans faster over the now reduced area of radius R. This estimate can explain why waxes of low melting points ($40° - 60°$C) can be observed at 2–5 keV but melt at high energies.

These are some examples with simple geometries that give an idea of how the increase of temperature can be estimated. Organic material also suffers from radiation damage by electron bombardment (Sect. 3.5.3), which can result in considerable loss of mass. The formation of gaseous products inside a layer of the thickness of the electron range ($\simeq 10\mu$m in organic material of density $\rho = 1$ g/cm^{-3} at $E = 20$ keV) results in surface damage, which can look like thermally induced damage but will in fact be radiation damage or both in most cases.

We must not forget that thermal damage can also occur during the metal coating by sputtering or high-vacuum evaporation, to avoid charging at high electron energies. This has been demonstrated by comparison of coated and uncoated low-density polystyrene and cellulose acetate foams at high and low voltages [3.140].

The application of periodic heating by electron-beam chopping creates damped thermal waves; these excite ultrasound waves, which can be picked up by a transducer. This is the basis of the acoustic SEM mode (Sect. 8.7).

3.5.2 Charging of Insulating Specimens

Dependence of Surface Potential on Energy. Irrespective of the distribution of accumulated space charge and its drift inside the specimen, it is the surface potential that influences the trajectories of secondary, backscattered and primary electrons. Below the surface, the signs of the space charge and of the electric field strength can change [3.141, 142], which can influence the distribution of ions in x-ray microanalysis (Sect. 3.5.3). Whether the accumulation of electrons in insulators results in a reduction of the electron range, decrease of backscattering and lower x-ray emission due to retardation inside the specimen is still in doubt. Monte Carlo simulations of this retardation [3.143] do not consider the electron-beam-induced conductivity inside the diffusion cloud and the drift of charge carriers to deeper non-irradiated zones, observed in high polymers [3.144]. Most of the influence of charging on x-ray emission can be attributed to a reduction of the landing energy E_L by a high negative surface potential or to the change of ion migration in the internal electric field. The latter needs much lower field strengths than retardation.

Whether the surface potential of an insulating specimen will be positive or negative depends on the total electron emission yield $\sigma = \eta + \delta$ (Figs. 3.31a and 4.18b). The dependence of σ on the primary electron energy E shows two crossover energies, $E_1 = eU_1$ and $E_2 = eU_2$, at which $\sigma = 1$ and no charging occurs, but E_1 is lower than 100–200 eV and is of interest in SEM only for the detection of polarized electrons (Sect. 8.1.4). Here therefore we discuss only the influence of E_2, which falls on the high-energy tail of the $\sigma(E)$ curve.

If $E_1 < E < E_2$ and hence $\sigma > 1$, the number of electrons emitted as backscattered and secondary electrons per unit time is greater than the number of primary electrons hitting the specimen. This excess emission results in a positive charging of the specimen surface. At a potential $+U_s$ of the surface, all secondary electrons with exit energies smaller than eU_s return to the specimen and equilibrium will be reached when the effective SE emission coefficient is reduced to

$$\delta_{\text{eff}} = \int_{U_s}^{\infty} \frac{dN_{\text{SE}}}{dE_{\text{SE}}} dE_{\text{SE}} = 1 - \eta \tag{3.203}$$

or $\sigma_{\text{eff}} = \eta + \delta_{\text{eff}} = 1$ where $dN_{\text{SE}}/dE_{\text{SE}}$ denotes the energy distribution of the SE (Fig. 4.18a). This positive charging for $E_1 < E < E_2$, will only be of the order of a few volts therefore. The decrease of the effective SE emission can be observed as a lowered image intensity, similar to the voltage contrast of a positively biased conductive surface (Sect. 8.2). However, if the electric

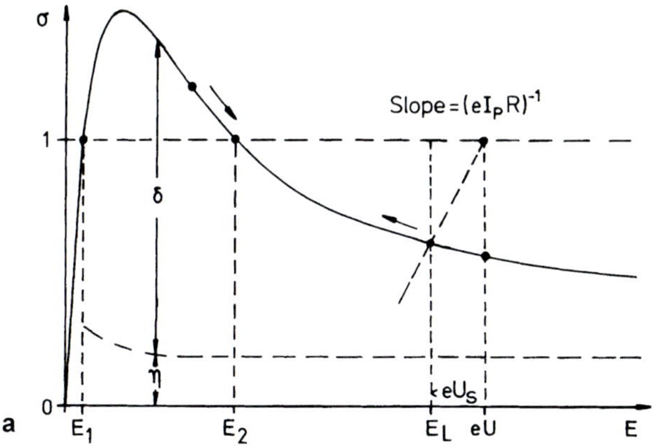

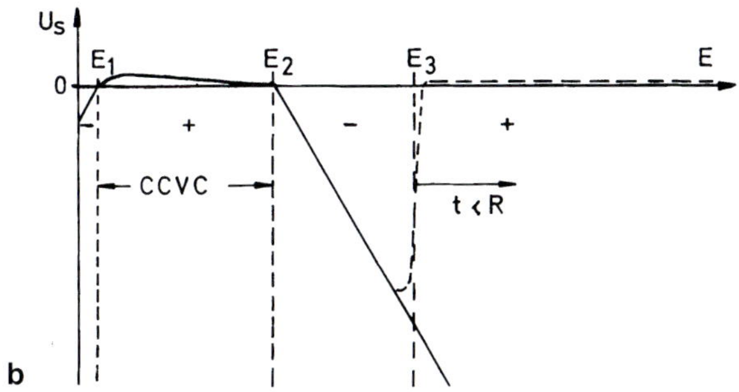

Fig. 3.31. (**a**) Dependence of the total yield $\sigma = \eta + \delta$ on electron energy with two critical energies E_1 and E_2 where $\sigma = 1$; positive charging occurs for $\sigma > 1$ and negative charging for $\sigma < 1$. At an energy $E = eU > E_2$, the potential U_s can be constructed as the point of intersection of $\sigma(E)$ with a straight line of slope $1/eI_pR$ (I_p = electron-probe current, R = leak resistance). (**b**) Surface potential U_s. Dashed line: charging of an insulating layer of thickness t on a conductive substrate with strongly reduced charging when the electron range $R \geq t$ for $E \geq E_3$

field of an Everhart–Thornley detector acts on the specimen, more SE are collected; the surface potential U_s increases until the field strength at the specimen surface is compensated. This increase of U_s can be confirmed with a Kelvin probe [3.145].

If $E > E_2$ and hence $\sigma < 1$, more electrons fall on the specimen than can leave it as secondary or backscattered electrons and a negative space charge is produced in a region of the order of the electron range. In consequence, the surface potential becomes negative and acts as a retarding field for the primary electrons of energy E, the landing energy E_L of which will be reduced. This negative charging reaches an equilibrium when the primary electron energy is retarded to $E_L = E_2$ where $\sigma = 1$ or $U_s = U_2 - U$. The negative charging can therefore reach a few kilovolts in contrast to the positive charging. The negative charging of the surface can be reduced if the specimen shows a surface or volume conductivity with a resistance R to earth.

The increase of the surface potential U_s corresponds to a charge accumulation Q that can be treated as the charging of a capacitance C with a leak resistance R to ground. The current equilibrium results in

$$\dot{Q} = I_p(1 - \sigma_{\text{eff}}) - U_s/R \quad \text{or} \quad \dot{U}_s + U_s/RC = I_p(1 - \sigma_{\text{eff}})/C \tag{3.204}$$

since $Q = CU_s$.

For negative charging, the stationary potential U_s is formed as a particular integral of (3.204):

$$[1 - \sigma(E_L)]I_p = U_s R \quad \text{or} \quad \sigma(E_L) = 1 - \frac{E_L - E}{eI_p R} . \tag{3.205}$$

The solution can be found graphically (Fig. 3.31a) as the point of intersection of the $\sigma(E)$ curve with a straight line through the point $\sigma = 1$, $E = eU$ of slope $(eI_pR)^{-1}$. For an ideal insulator, $R = \infty$, the intersection will be at $E = E_2$ (maximum value of $U_s = U_2 - U$) and for a metal, $R \to 0$, no charging will be observed.

The charging can change in the presence of intermittent irradiation during scanning; the stationary state conditions discussed above are not fulfilled and we have to take into account charging and discharging during the scan period [3.146]. When switching on and off the electron probe current I_p for a pixel time t_p, the solution of the differential equation (3.204) results in an increase and an exponential decrease (discharging) of U_s, respectively:

$$\begin{aligned} \text{Charging} &: \quad U_s = U_{s0}[1 - \exp(-t/\tau_0)] \\ \text{Discharging} &: \quad U_s = U_{s0}\exp(-t/\tau_0) \end{aligned} \tag{3.206}$$

where the time constant $\tau_0 = RC = \rho\epsilon$ (ρ: resistivity, $\epsilon = \epsilon_r\epsilon_0$: permittivity of the specimen) is approximately independent on the specimen geometry and $U_{s0} = I_p(1 - \sigma_{\text{eff}})$.

Charging artifacts can thus look quite different when changing from a fast scan for visual observation to a slow scan for image recording. This can be demonstrated by (3.206) when a) using a slow scan with a large frame time T and a pixel time $t_p = T/N$ (N: number of pixels and b) a fast scan with frame time T/n and pixel time t_p/n ($n \gg 1$). When the electron probe has scanned a pixel after one or n scans, respectively, the potential will have increased to

$$\begin{aligned} &\text{a) Slow scan:} \quad U_s = U_{s0}[1 - \exp(-t_p/\tau_0)] \\ &\text{b) Fast scan:} \quad U_s = U_{s0}\sum_n [1 - \exp(-t_p/n\tau_0)]\exp[-T(n-1)/n\tau_0] . \end{aligned} \tag{3.207}$$

If the time constant $\tau_0 > T$, no significant difference in U_s is expected, but U_s will be lower in a fast scan when $t_p < \tau_0 < T$.

When an insulating specimen is irradiated by a stationary electron probe (spot) below E_2 the core also shows positive charging but is surrounded by a region of negative charging caused by SE that have been retarded by the positive core charging [3.147, 148]. This can be seen for a short time as a doughnut-shaped voltage contrast when switching to the scanning mode. With increasing irradiation the whole spot becomes negatively charged [3.149].

Insulating Layers on Conductive Substrates. For insulating layers of thickness t on conductive substrates (e.g. resin or passivation layers, see also Sect. 8.3.2) we will have the same situation as for bulk insulating specimens so long as the electron range $R(E) < t$, including the strong negative charging if $E \geq E_2$. When $R(E) > t$, free movable charge carriers (electrons or holes) are generated through the whole layer and a potential difference between the surface and the substrate will result in an electron-beam-induced current (EBIC). The surface potential becomes equal or near to that of the substrate beyond an incident electron energy E_3 that fulfils the condition $R(E_3) = t$ [3.150, 151]. For an electron energy $E \geq E_3$, the strong negative surface potential U_s suddenly drops to a small positive value (Fig. 3.31b). The rapid increase of the discharging EBIC has also been demonstrated by measuring the gain $G = I_{EBIC}/I_p$ on insulating Se, As_2S_3, Sb_2S_3, tantalum and silicon oxide, and polymer films (Sect. 7.1.8). A photoresist on silicon of 0.8 μm shows $E_3 = 8.5$ keV, for example [3.152]. When $E_3 < E_2$, no negative charging will be observed beyond E_2. For $E_1 < E < E_2$ and $R(E) < t$, the capacitive-coupling contrast (CCVC, Sect. 8.2.5) can be observed. When the electron probe scans a pixel again after discharging, the re-built charging can be recorded by the opposite charge induced on a wire loop placed in front of the specimen (Rau-detector, Sect. 8.2.6).

These charging effects can also be demonstrated in the SE image of polystyrene spheres with a diameter $d = 1$ μm. At 6 keV ($E > E_3$), the sphere appears bright without charging (Fig. 3.32a), at 2 and 1 keV with $E_2 < E < E_3$ the top of the sphere becomes negatively charged with typical inhomogeneous charging artifacts (Fig. 3.32b,c), and at $E = 0.6$ keV with $E < E_2$ a positive charging is observed that decreases the image intensity at the centre of the sphere (Fig. 3.32d).

In the case of free-supported insulating polymer films, the total yield σ also becomes larger than unity when the range R becomes larger than the film thickness t with increasing E because the transmitted electrons are not stopped and generate SE at the exit side as well [3.150].

Methods of Measuring E_2. At normal incidence ($\phi = 0$), E_2 is of the order of 1–4 keV; it shifts to higher values when the specimen is tilted [3.153] owing to the increase of η and δ with increasing tilt angle (Figs. 4.7, 4.20). A knowledge of E_2 is of interest for the discussion of charging effects and in LVSEM it enables us to establish an optimal energy $E < E_2$ at which there is

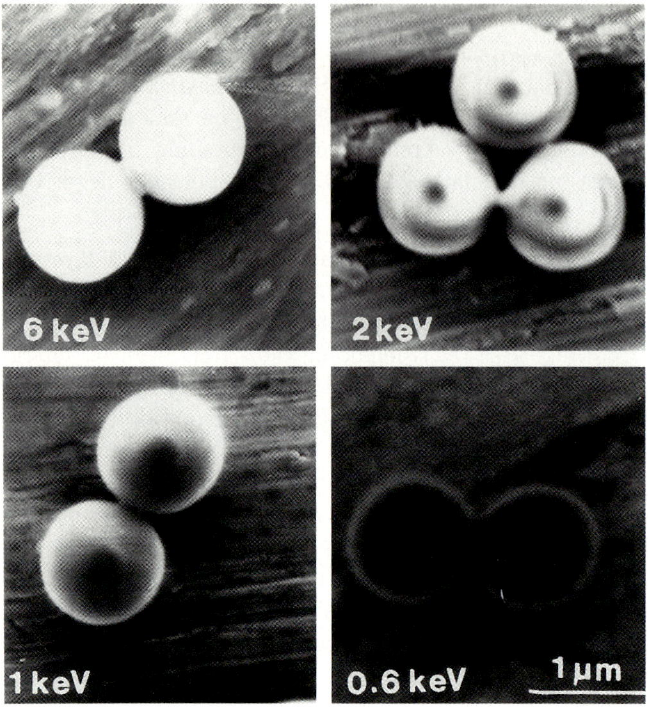

Fig. 3.32. Charging of polystyrene spheres with $E = 6$ keV $> E_3$, 2 and 1 keV ($E_2 < E < E_3$) and 0.6 keV ($E < E_2$)

no negative charging, though this is not guaranteed for all types of specimen, especially those with a rough surface.

The following methods can be used to measure the crossover energy E_2:

1. Measurement of $\sigma(E)$ or calculation of σ by Monte Carlo simulation [3.154] and determination of E_2 from the intersection at $\sigma = 1$.
2. Measurement of the floating potential U_s of an insulated metal specimen [3.155]; this can also be measured by a MOSFET [3.156].
3. Measurement of the specimen current to earth and adjustment to zero by varying E [3.157, 158].
4. Deflection of an additional 400–1000 eV electron beam travelling parallel to the surface at a distance of $\simeq$1 mm [3.159]. Conversely, the specimen may be irradiated with an additional source and the deflection of primary electrons measured [3.160].
5. Irradiation at normal incidence and detection of charging by beam deflection after tilting the target through 90° [3.161].
6. Change in SE image intensity of a preirradiated area when the magnification is decreased and adjustment to energy E_2 at which the image shows neither darkening nor whitening [3.162].

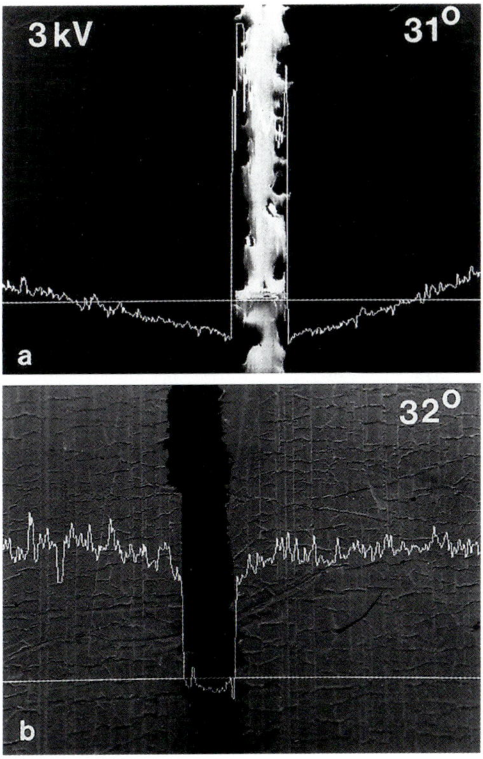

Fig. 3.33. Secondary image of a stripe of uncoated teflon between evaporated gold films irradiated with 3 keV electrons at tilt angles (**a**) $\phi = 31°$ and (**b**) $32°$

7. Coating of insulators with a conducting metal film except for a parallel stripe and determination of the energy E_2 at which the insulator is not negatively charged (without bright negative charging artifacts) [3.150, 153].
8. Measurement of the electric field in front of the specimen by means of a vibrating Kelvin probe [3.145, 163].
9. Shift of the SE energy spectrum [3.164].

Methods 1–3 can be used for conductive specimens only. The specimen-current method 3 is the easiest one put into practice. When all the SE leave the specimen and no BSE or SE scattered or emitted at parts of the specimen chamber hit the specimen again, the specimen current becomes

$$I_s = I_p[1 - (\eta + \delta)] \tag{3.208}$$

This current can be measured by connecting an electrometer between specimen and ground. At the crossover energy E_2, the specimen current changes sign (Fig. 6.18).

Methods 5–7 can be used for insulating specimens only. Method 7 is the easiest and most effective one. The insulator, e.g. MgO or teflon, is coated with a thin gold film except for parallel stripes 50–200 μm in width. The

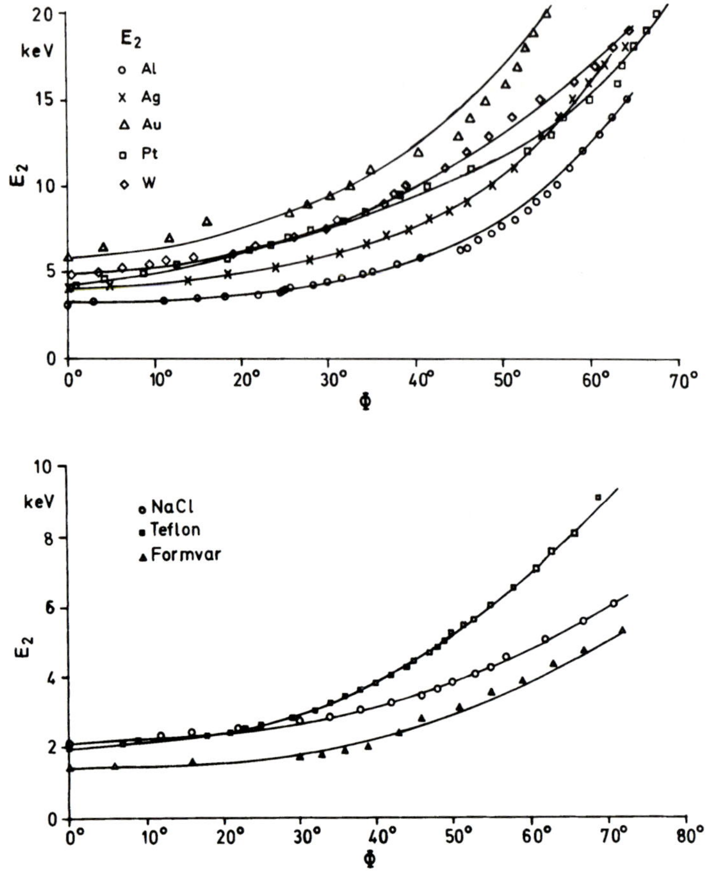

Fig. 3.34. Examples of the dependence of the crossover energy E_2 on the tilt angle ϕ of the specimen for different materials

method can also be used to measure the dependence of E_2 on specimen tilt angle ϕ. When the specimen is tilted, the image intensity of the stripes changes dramatically when the insulator is negatively charged. This is demonstrated in Fig. 3.33 for teflon and $E = 3$ keV and a change of tilt angle from 31° to 32°; η and δ increase with increasing ϕ (Sects. 4.1.3 and 4.2.1). Figure 3.34 shows the tilt dependence of E_2 for some metals and insulators. Empirically, the relation $E_2(\phi) = E_2(0) + \sec^2\phi$ was proposed [3.154, 164], but this relation has no physical background and is not capable of describing the results in Fig. 3.34.

Method 7 can also be employed for conductive specimens when the metal of interest is insulated from ground (floated) and mounted behind a slit between two metal plates connected to ground [3.150]. This method seems to

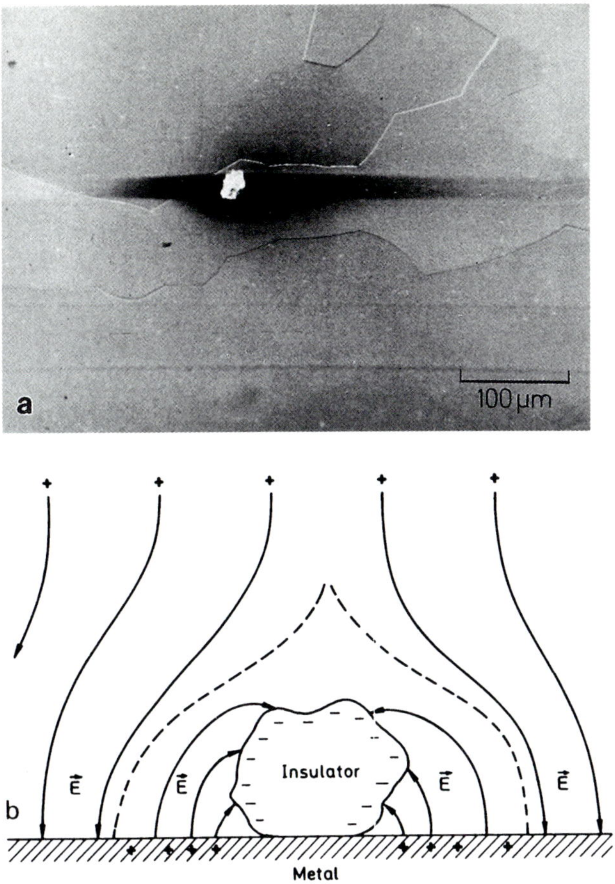

Fig. 3.35. (**a**) SE micrograph of a dust particle (bright), which is charged negatively, on an aluminium substrate causing (**b**) an electric field, which connects induced positive charges on the metal substrate to the negative charges on the particle, resulting in dark voltage contrast beside the particle within the range indicated by the dashed line

be the only one that allows E_2 to be measured for both conductive and insulating specimens with a minimum of effort.

Further measurements of E_2 are reported for vanadate-phosphorus glasses of different composition [3.165] and for silicon rubber and ethylene propylene diene monomer [3.162].

Further Examples of Charging Artifacts. Figure 3.35 shows a typical example of how the SE image can be affected by the negative charging of a dust particle on a conductive aluminium surface. The particle appears bright due to the increased emission of SE at the rough surface. The size of the particle is such that more electrons are absorbed than emitted, which results in a negative charging. This negative charge causes a distortion of the electric

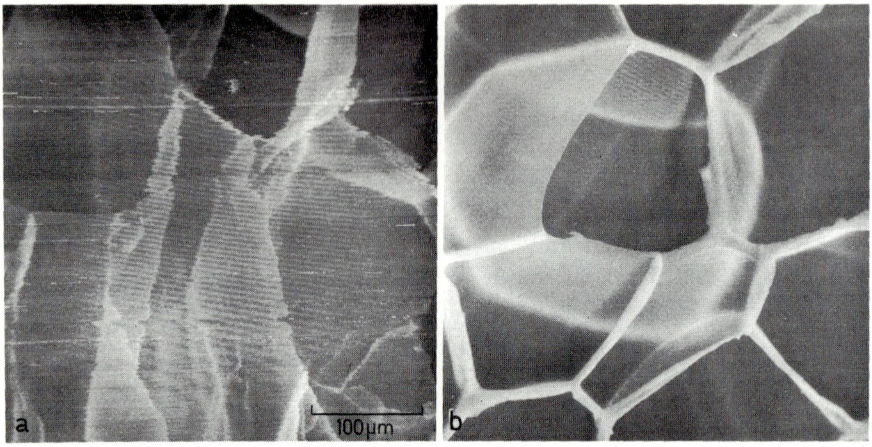

Fig. 3.36. (**a**) Styropore evaporated with an Au-C layer only and (**b**) pre-treated with OsO_4 vapour and coated with a C-Au-C layer [3.173]

field, which is schematically shown in Fig. 3.35b. We note that the electrons move against $\boldsymbol{E}$ since they are negatively charged. A region therefore exists on the conductive surface with induced positive surface charge from which those electric field lines that end on the particle start and no SE can leave the substrate from this zone. This results in the dark areas around the particle in Fig. 3.35a caused by voltage contrast (Sect. 8.2.1). This darkening of the metal substrate starts when the electron probe, which moves from top to bottom and from the left-hand to the right-hand side of the figure, first strikes the particle. The dark area is extended more to the right-hand side than to the left-hand side because the charging of the particle is decreased by conduction as the electron probe moves along each line and always approaches from the left-hand side. This decrease of negative charging can also be seen as a dark area below the particle when the electron probe moves to the bottom and cannot restore the negative charge of the particle. Though this is a typical example, charging can result in quite different image artifacts depending on the specimen geometry, conductivity and electron emission.

If the surface potential is lower than a few hundred volts, only the SE (and not the BSE and PE) are affected. However, when the negative surface potential for $E > E_2$ reaches values of 10–90% of the acceleration voltage, BSE as well as PE trajectories are influenced by the highly inhomogeneous fields in front of the specimen. This leads to strong image distortions when the deflected primary beam lands on specimen details quite far from those it would have encountered in the absence of charging. This is demonstrated in Fig. 3.36a with styropore imperfectly coated with a Au-C conductive layer. In this case, the charging effects disappear when the surface conductivity is raised by treatment in OsO_4 vapour (Fig. 3.36b).

Such deflections can also be used to give a quantitative measure of the charging [3.153, 166]. A tilted specimen at a negative potential U_s deflects primary electrons of acceleration voltage U if $U_\mathrm{s} \geq U \cos^2 \phi$. The signal depends on where the deflected electrons land, either on wrong surface positions resulting in image distortions or on the specimen stage and chamber walls where they can excite SE. A very bright signal will be observed when the primaries are directly deflected into the detector. A distorted projection of the collector grid, for example, can be superposed on the image [3.146, 167]. These strong charging artifacts are often not stationary. Statistical oscillations can be observed when the high potential is discharged by electric breakdown, which may also generate light quanta and low energy electrons resulting in random bright spots or lines.

When a glass sphere or polymer foil is scanned with high-energy electrons (10–20 keV) and then observed at lower energies, the electrons are deflected after being retarded and hit points of the specimen chamber where they excite SE. This allows observation of the interior of the specimen chamber from a fish-eye perspective [3.167] and control of local detection efficiences of detectors when these are scanned by the reflected primary beam [3.168, 169]. This can also be realized with a biased metal ball [3.170].

Even the low positive charging for electron energies $E_1 < E < E_2$ can influence the trajectories of primary electrons in electron-beam lithography and metrology (Sect. 8.8).

Prevention of Charging. These principles of the physics of specimen charging artifacts also suggest ways of preventing charging. The most common prodecure is sputter-coating of the specimen with a conductive layer [3.171, 172]. Sputter-coating provides a more uniform coating film on rough surfaces than high-vacuum evaporation, because of the scattering of atoms in the residual argon atmosphere, of the order of a few millibars. A disadvantage of this method is that the fine structure of the coating film caused by the grain size is visible at high resolution and in addition, specimen details can be buried by the coating film to a depth of the order of a few nanometres. Some organic specimens can become conductive after treatment in OsO_4 vapour [3.173–175].

There are specimens, ferroelectrics for example (Sect. 8.2.2), that cannot be coated without losing the information provided by voltage contrast from the ferroelectric domains. The observation of insulating ferroelectrics and other substances is possible if the electron energy is reduced, so that $E < E_2$. Another possibility is to work at $E > E_2$ and switch off the collector bias while switching on an additional electron beam of 100–1000 eV during the flyback period of the scanning beam from the end to the start of a line [3.176] or during each second line [3.177], so that the time average of σ becomes larger than unity.

Charging can also be prevented by surface conductivity, which is often orders of magnitude higher than the volume conductivity. For example, a charging of only –200 V was observed at the centre of a slit of width d = 100 μm in an evaporated metal layer on teflon [3.153] which corresponds to a field gradient of 4 keV/mm. This gradient remains constant as d is increased and the charging at the centre increases as d. It is therefore possible in some cases to prevent charging by covering the specimen with a hole in a metal foil [3.178] or by painting the surface with a silver conductive paste apart from a small area [3.179].

A further possibility is to furnish positive charges in the form of ionized atoms. A negatively charged surface detail has to collect enough ions to compensate the negative charge. This can be achieved by means of a thermionic Li ion source [3.180], an environmental chamber [3.181] or a gas stream in which the ions are produced by the primary and backscattered electrons.

3.5.3 Radiation Damage of Organic and Inorganic Specimens

The increase of specimen temperature can result in irreversible thermally induced damage but most of the damage processes are a consequence of ionization.

Damage of Organic Specimens. The radiation damage of organic specimens has been extensively investigated for thin films in TEM ([3.1, 182–186]. Bond rupture after electron excitation destroys –CH, –COOH, –CNH2 and other bonds. The formation of carbon double bonds results in cross-linking and the final product will be a carbonaceous polymerized substance. A large fraction of the non-carbon atoms leave the specimen as volatile fragments. When the electron intensity of irradiation is high, this can lead to the formation of bubbles, which may alter the surface structure [3.187].

If there is no increase of temperature, the damage will be proportional to the incident charge density $q = jt$ in C/cm^2 where j is the current density in A/cm^2 and t the irradiation time. The interpretation of the radiation damage of thin films as used in TEM is easier because the ionization density will be uniform through the film. In bulk material, the ionization density is a function of depth (Figs. 3.24–3.26) and a layer below the surface at the maximum of the ionization density will be damaged faster than those at other depths inside the electron range R. The damage rate becomes very inhomogeneous. When a large loss of mass occurs, the electrons can penetrate deeper into the material and damage a zone that may be thicker than the electron range of the undamaged material. This has been demonstrated by interferometric measurements of the height of surface steps at scanned and damaged areas [3.187] and by profilometry [3.188].

The results obtained in TEM concerning the loss of mass and of crystallinity are summarized in the following list of charge densities $q = jt$ needed to destroy the compounds by 60 keV electrons:

Amino acids	$jt \simeq$	10^{-3}	$\mathrm{C/cm^2}$
Higher hydrocarbons		10^{-2}	$\mathrm{C/cm^2}$
Aromatic compounds		10^{-1}	$\mathrm{C/cm^2}$
Phthalocyanines		$1-10$	$\mathrm{C/cm^2}$

According to the Bethe stopping power (3.130), the damage will be proportional to $\ln(E/J)/E$, which is proportional to $E^{-0.8}$ in the range 10–100 keV. This means that the critical charge density is lower by a factor 0.24 and 0.42 for 10 and 20 keV electrons, respectively.

Scanning with a probe current $I_\mathrm{p} = 10^{-11}$ A in a frame time $T = 100$ s at a magnification $M = 10\,000$ corresponds to a charge density

$$q = jt = \frac{I_\mathrm{p} M^2}{B^2} \simeq 10^{-3}\mathrm{C/cm^2} \tag{3.209}$$

where $B = 10$ cm is the width of the CRT. This means that amino acids are already strongly damaged by one high-resolution scan.

However, the cathodoluminescence and photoconductivity of organic substances already begins to decrease at lower charge densities, of the order of $10^{-5} - 10^{-4}$ C/cm^2 [3.182, 189]. This limits not only the lifetime of a plastic scintillator but also of the cathodoluminescence of organic substances (see also Sect. 7.2.5). For example, the light intensity of a plastic scintillator NE102A decreases to 50% of its initial value after irradiation with 20 keV electrons and a charge density $q = 2 \times 10^{-4}$ C/cm^2 [3.190].

For the investigation of freeze-fractured and freeze-etched biological specimens by low-temperature LVSEM at 150 K [3.190], the results of experiments in a 100 keV TEM need to be considered. Contamination changes over to etching of carbon and organic material when the specimen is cooled below 200 K [3.208]. This is attributed to a radiation-induced chemical reaction with adsorbed residual gas molecules (H_2O, O_2, CO, H_2). These effects observed at 100 keV should also occur at 1 keV and need further consideration in low-temperature SEM.

Frozen-hydrated specimens obtained by cryo-fractioning can be observed at high magnification only by one slow scan and the specimen is usually damaged by about 3000 $\mathrm{e^-/nm^2}$. A double layer of 2 nm of platinum and 5–10 nm carbon can protect the surface, which can then be observed by the electron backscattering from the platinum layer. Structural information comparable with that provided by freeze-fractured replicas in TEM can be obtained [3.191].

Damage of Inorganic Specimens. Most inorganic substances are much more resistant to radiation damage than organic specimens. The following damage processes can occur [3.192, 193]:

1. Knock-on process. The momentum transferred to a nucleus by elastic large-angle scattering can knock an atom from its lattice site to an interstitial

site. However, this process needs electron energies larger than 100 keV and is not relevant in SEM.

2. Damage by ionization (radiolysis). The energy released in an electron excitation can change the electronic and atomic configuration. Ionization processes in metals and semiconductors are normally reversible. Excited electrons reach the ground state in a cascade of radiative and non-radiative transitions. Typical radiolysis effects occur in ionic crystals such as alkali halids, for example. At first, colour (F) centres are formed, which can coalesce to form M centres and finally larger defect clusters and metal colloid particles. The damage process depends not only on the charge density but also on the current density because of the complicated kinetics of cluster formation. This type of damage can also lead to a decrease of cathodoluminescence, though the formation of F centres can also result in an initial increase of CL. The necessary charge densities for this damage to occur are of the order of a few tens of C/cm^2 [3.189]. A degradation in the intensity of electron channelling patterns can also be observed (Sect. 9.2.4). Electron-beam-induced defects at the Si-SiO_2 interface of integrated circuits can affect the parameters of the device (Sect. 7.1.7). Irradiation of ice also results in radiolysis, which has to be considered when investigating frozen-hydrated biological specimens.

3. Field-induced drift of ions. Insulating compounds accumulate charges, which generate strong electric fields in the irradiated area. These can cause a drift of ions of high mobility such as Na^+ and K^+. Changes of Na and K concentrations have been observed during x-ray microanalysis of glasses [3.194-196], for example. However, the interpretation of such damage processes can be complex because an increase of specimen temperature has also to be taken into account. Generally, the analysis of radiation-sensitive substances should be verified by varying the probe current and/or the counting time.

Another damage effect, which is caused by negative O^- ions formed at the cathode and accelerated by the anode, has also to be taken into account. The damaged specimen area is of circular shape and represents the shadow of the final lens diaphragm. The damage can be observed as a decrease of the cathodoluminescence of CdS [3.197] or by examining the EBIC signal of silicon [3.198].

3.5.4 Contamination

The presence of hydrocarbon molecules on the specimen surface is virtually unavoidable. These are commonly formed in the vacuum of the microscope as a result of the partial pressure of hydrocarbons or silicon oils from the diffusion pump and the grease of vacuum seals and fingermarks (see [3.199] for a review of the contamination problem in TEM). However, hydrocarbons are already deposited on the specimen in the open air so that an ultra-high vacuum alone does not solve the problem if the specimen itself has not been prepared by cleaving, evaporation or ion-beam etching inside the ultra-high vacuum specimen chamber.

During electron irradiation, these hydrocarbon molecules are cracked by the same mechanisms as those discussed for organic material in Sect. 3.5.3. The result is polymerization by cross-linking, and a carbon-rich contamination film grows over the scanned area or a contamination needle grows up when the electron probe is stationary. At a thickness of the order of 10 nm, the SE emission will be that of carbon and the irradiated areas appear darker in the SE image, though a brighter area can also be observed in some circumstances [3.200]. Obviously the SE yield depends on the thickness of the contamination layer and can be influenced by changes in the vacuum conditions. The contamination layer may also bury underlying structures and this decreases the resolution.

The hydrocarbon molecules show a high surface mobility and if they are cracked and cross-linked, further molecules diffuse to the irradiated area by a random-walk process. At low current densities, the thickness t_c of the contamination film will be proportional to the dissipated energy; this is proportional to the stopping power S (Sect. 3.3.4) and the accumulated charge density $q = jt$, where j denotes the current density and t the irradiation time, or $q = I_p T/a$, where T denotes the frame time and a the scanned area:.

$$t_c \propto q\,S \propto q \ln(J/E)/E \propto \frac{I_p T}{a} E^{-0.8} \,. \tag{3.210}$$

The energy dependence of the stopping power $S \propto E^{-0.8}$ means that the contamination rate is higher in LVSEM [3.201, 202].

At high current densities, in particular in conditions of spot irradiation, the growth rate saturates when more molecules are polymerized than can reach the area by diffusion. The mass of contaminated material then only grows in proportion to the irradiation time t and is independent of j. A theory of this effect has been developed for the case of TEM [3.203]. When a rectangular area is scanned, a stronger contamination may appear on the side region of the irradiated area [3.201, 202] that can be explained by this theory. Molecules arriving by surface diffusion at the irradiated area are immediately cracked and cannot reach the central area. There is no reason to interpret these contamination rings as charging effects, as in [3.204].

Contamination can be decreased by taking the following precautions:

1) Use of a diffusion pump oil of low vapour pressure. However, most oil contamination is caused by back-streaming of oil from the rotary pump. A fully fluorinated oil can be used and prolonged evacuation by the rotary pump should be avoided, the changeover to the high-vacuum pump being made at 0.1–0.2 mbar. Viton rings and apiezon grease should be used instead of silicon grease for vacuum seals. If plastic material is used for insulation, polytetrafluor-ethylene (PTFE) or polyimides should be preferred [3.205]. The best way of obtaining an oil-free vacuum is to use a turbomolecular

pump. Another precaution is to flood the microscope column and the specimen chamber with dry nitrogen [3.206].

2) The partial pressure of hydrocarbons and other gases can be decreased in the specimen chamber by means of cool-traps. Analysis of the residual gas by a mass spectrometer provides an effective check of the efficacy of these precautions [3.207]. When the specimen itself is cooled down to between −80°C and −150°C, an etching effect can dominate [3.208, 219] by interaction with reactive ions, commonly oxygen, which are formed by ionization after adsorption at the surface, and it is also possible for carbonaceous material to be etched [3.208, 209]. However, the main effect of specimen cooling is the strong decrease of the mobility of hydrocarbon molecules and cooling below −100°C is the most effective way of preventing contamination [3.202]. A cool trap at a lower temperature can avoid the condensation of oil vapour.

Specimen heating to 200°C decreases the probability of adsorption of the residual gas but gases already adsorbed need higher temperatures of the order of 400–500°C to be desorbed [3.210, 212].

3) Because most of the contamination is caused by the diffusion of adsorbed hydrocarbon molecules to the irradiated area, the specimen itself should be clean before being introduced into the vacuum. An effective method is to wash the specimen in methylene when this is possible [3.213]. After one day in air, a new layer of hydrocarbons has already formed, which increases the contamination rate again. Other methods of cleaning are those that damage the adsorbed hydrocarbon molecules by ion-beam irradiation or during sputter coating of conductive layers. Electron preirradiation of a larger area also cracks the adsorbed hydrocarbon molecules and a smaller area can be observed contamination-free for a longer period. Sources of contaminating material at the walls of the specimen chamber can be reduced by a gas discharge [3.214]. Another possibility is to spray the specimen with N_2 or Ar through a capillary during the electron irradiation [3.206].

4. Emission of Backscattered and Secondary Electrons

Backscattered electrons (BSE) and secondary electrons (SE) are the most important signals for image recording. A knowledge of the dependence of the backscattering coefficient η and the secondary electron yield δ on surface tilt, material and electron energy and their angular and energy distributions is essential for the interpretation of image contrast. The spatial exit distributions and information depths of these electrons govern the resolution if the latter is not limited by the electron-probe size. The shot noise introduced by the incident electron current will be increased during SE and BSE emission.

4.1 Backscattered Electrons

4.1.1 Measurement of the Backscattering Coefficient and the Secondary Electron Yield

By convention, electrons with exit energies $E_{\mathrm{B}} > 50$ eV are called backscattered electrons (BSE) and those with $E_{\mathrm{SE}} < 50$ eV secondary electrons (SE). Full information about backscattering would require a knowledge of the probability $\mathrm{d}^2\eta(E, \phi, \zeta, \chi, E_{\mathrm{B}})/\mathrm{d}E_{\mathrm{B}}\mathrm{d}\Omega$ of finding an electron inside the energy interval $E_{\mathrm{B}}, E_{\mathrm{B}} + \mathrm{d}E_{\mathrm{B}}$ with $50\ \mathrm{eV} < E_{\mathrm{B}} \leq E = \mathrm{eU}$ and inside a solid angle element $\mathrm{d}\Omega$ with an exit angle ζ relative to the surface normal and an azimuth χ (ϕ: tilt angle between surface normal and incident electron beam). The complexity of this probability function obliges us to accept less information and to look at either the angular distribution regardless of the exit energy or at the energy distribution regardless of the exit angle:

$$\frac{\mathrm{d}\eta}{\mathrm{d}\Omega} = \int_{50\mathrm{eV}}^{E} \frac{\mathrm{d}^2\eta}{\mathrm{d}E_{\mathrm{B}}\mathrm{d}\Omega}\mathrm{d}E_{\mathrm{B}} \tag{4.1}$$

and

$$\frac{\mathrm{d}\eta}{\mathrm{d}E_{\mathrm{B}}} = \int \frac{\mathrm{d}^2\eta}{\mathrm{d}E_{\mathrm{B}}\mathrm{d}\Omega}\mathrm{d}\Omega\,, \tag{4.2}$$

respectively. A further integration over each distribution yields the (total) backscattering coefficient

$$\eta = \int \frac{d\eta}{d\Omega} d\Omega = \int_{50\mathrm{eV}}^{E} \frac{d\eta}{dE_B} dE_B \tag{4.3}$$

which is the probability that an electron will escape through the surface as a backscattered electron.

The backscattering coefficient η can be measured by using a device that collects all electrons scattered into the whole hemisphere above the specimen with exit energies $E_B \geq 50$ eV; if the resulting current is I_{BSE}, then

$$\eta = I_{BSE}/I_p \tag{4.4}$$

where I_p denotes the incident electron-probe current.

By analogy with (4.3) and (4.4), a secondary electron yield may be defined by

$$\delta = \int_{0}^{50\mathrm{eV}} \frac{d\eta}{d\Omega} dE_B = I_{SE}/I_p \, . \tag{4.5}$$

It is not useful to distinguish between BSE and SE below $E = 1$ keV and a total yield

$$\sigma = \eta + \delta = (I_{BSE} + I_{SE})/I_p \tag{4.6}$$

can then be introduced. This can also be useful at higher energies because $\sigma > 1$ means that more electrons leave the specimen than are absorbed and an insulator becomes positively charged whereas $\sigma < 1$ results in negative charging (Sect. 3.5.2).

A SEM is not designed to measure η and δ precisely because of the differences in the collection efficiencies of the detector systems normally used. A knowledge of the exact values of η and δ does not mean that the observed signals are proportional to these quantities. Nevertheless, these values will be important for the discussion of signal amplitude and its dependence on different parameters and for quantitative SEM.

An accurate measurement of the backscattering coefficient requires the following [4.1]:

1. A retarding field for separating the SE with energies $E_{SE} \leq 50$ eV from the backscattered electrons.
2. A collector electrode of low η to suppress double and multiple backscattering.
3. A grid in front of this collector electrode biased at −50 V relative to the collector to prevent SE generated by the BSE from flying back to the specimen.
4. The dimensions of the specimen should be much smaller than the diameter of the collector electrode, so that electrons backscattered at the collector are unlikely to hit the specimen again but can fly to the opposite side of the collector electrode (merely placing a large specimen stub below the polepiece does not fulfil this condition).

5. A small hole for the incident electron beam and a negatively biased diaphragm in front of the device to exclude electrons that stream down the column by multiple backscattering.

Additional grids and electrodes have been proposed [4.2–4] to overcome some of the difficulties in obtaining exact values of η. Figure 4.1 shows a device for measuring η and δ inside a SEM [4.5], which covers points 1–5. The BSE current is measured by applying a bias U_s of +50 V at the specimen and the sum of the SE and BSE current is obtained by applying a negative bias of –50 V. The total incident current I_p is measured by connecting the specimen and collector electrodes. With a lid on the spherical collector electrode, it is possible to observe the specimen in the SE mode.

The angular distribution $d\eta/d\Omega$, (4.1), can be measured by using a small Faraday cage of solid angle $\Delta\Omega = A/r^2$, where A denotes the detector area and r the specimen–detector distance. The cage is moved round a circle to explore different take-off directions. If detectors other than a Faraday cage are used, the integrand in (4.1) has to be multiplied by the detection efficiency $\epsilon(E_B)$, which means that the weighting of the BSE varies with E_B. In a semiconductor or scintillation detector, the number of electron-hole pairs or light quanta is proportional to $E_B - E_{th}$, with a threshold energy $E_{th} \simeq 1$ keV arising from the dead layer. The recorded signal therefore depends not only on $d\eta/d\Omega$ but also on the energy spectrum $d\eta/dE_B$. Energy distributions (spectra) can be measured by electron spectrometers (Sect. 5.6.3).

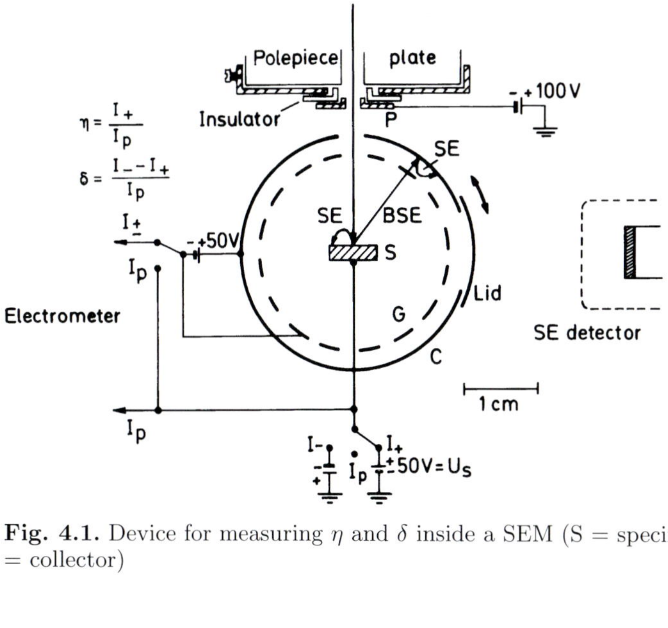

Fig. 4.1. Device for measuring η and δ inside a SEM (S = specimen, G = grid, C = collector)

4.1.2 Backscattering from Films and Particles

If it is assumed that only single scattering occurs, the backscattering coefficient of thin films of thickness t and density ρ will be proportional to the number of atoms per unit area $Nt = N_A\rho t/A$, where N is the number of electrons per unit volume, $N_A = 6.023 \times 10^{23}$ Avogadro's number and A the atomic weight, and proportional to the total backscattering through angles $\theta \geq \pi/2 = 90°$:

$$\eta = \frac{N_A \rho t}{A} \int_{\pi/2}^{\pi} \frac{d\sigma}{d\Omega} 2\pi \sin\theta d\theta = \frac{N_A \rho t}{A} \sigma(\pi/2) . \tag{4.7}$$

Using relation (3.16) for $\sigma(\pi/2)$ results in the non-relativistic formula for $E \ll E_0 = 511$ keV [4.6]:

$$\eta(t) = \frac{\pi e^4 Z^2 N_A}{4(4\pi\epsilon_0)^2 A E^2} \rho t = c(E) N Z^2 t \tag{4.8}$$

with the backscattering constant

$$c_{\text{theor}}(E) = \frac{\pi e^4}{4(4\pi\epsilon_0)^2} \frac{1}{E^2} = \frac{\lambda^4}{16\pi^3 a_H^2} \tag{4.9}$$

where λ denotes the de Broglie wavelength (2.22) and $a_H = 0.0529$ nm the Bohr radius of the hydrogen atom.

Experiments [4.1, 6–11] confirm the linear increase of η with increasing mass–thickness $x = \rho t$ in units of μg cm^{-2} (Fig. 4.2) and the proportionality to Z^2/A and E^{-2} rather well. The thickness limits t_{max} for the linearity of η vs. t are identical with the half-thicknesses $t_{1/2}$, where $\eta(t_{1/2})$ is half of the saturation value η_∞ [4.11] and are plotted in Fig. 4.3. Extrapolation of the linear part of $\eta(t)$ to η_∞ gives an intercept at a thickness T, which can be interpreted as the total depth from which electrons can be backscattered (Fig. 4.4). This Figure also contains the extrapolated electron ranges R from transmission experiments (Sect. 3.4.1). The comparison shows that T is of the order of half the electron range R [4.1, 12, 13]. Like the electron range R, the exit depth T is approximately independent of the material when measured as mass–thickness in units of μg cm^{-2}. The experimental results of Fig. 4.4 can be fitted for $E = 10$–100 keV to the formulae [4.1]

$$T = 2.8 E^{1.54} \quad \text{and} \quad R = 10 E^{1.43} \tag{4.10}$$

where T and R are in μg cm^{-2} when E is in units of keV.

However, the experimental backscattering constant $c(E)$ in (4.8) taken from the slope of the $\eta(t)$ curves is found to be a factor 2–3 larger than that predicted by the theoretical formula (4.9), for which we assumed single Rutherford scattering (Fig. 4.5). This discrepancy can be explained by

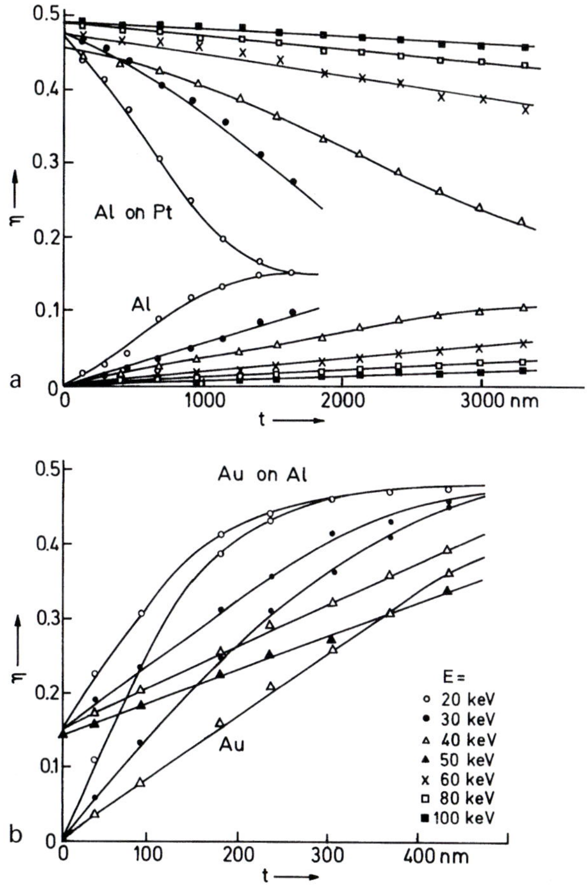

Fig. 4.2. Dependence of the backscattering coefficient η on film thickness of self-supported films of (**a**) Al and (**b**) Au and coating films of (a) Al on Pt and (b) Au on Al [4.16]

1. consideration of multiple scattering even in thin films,
2. use of Mott cross-sections (Sect. 3.1.6) instead of the Rutherford approximation, and
3. consideration of the production of fast secondary electrons by electron–electron scattering (Sect. 3.2.2).

Monte Carlo simulations [4.14], which do consider points 1–3, can fully explain the shape and slope of the $\eta(t)$ curves (Fig. 4.5).

For a self-supporting film of two (or more) elements X and Y with the chemical composition X_pY_q, the product NZ^2 in (4.8) has to be replaced by

$$NZ^2 \to N_\mathrm{m} Z_\mathrm{eff}^2 = N_\mathrm{m}(pZ_\mathrm{X}^2 + qZ_\mathrm{Y}^2) \tag{4.11}$$

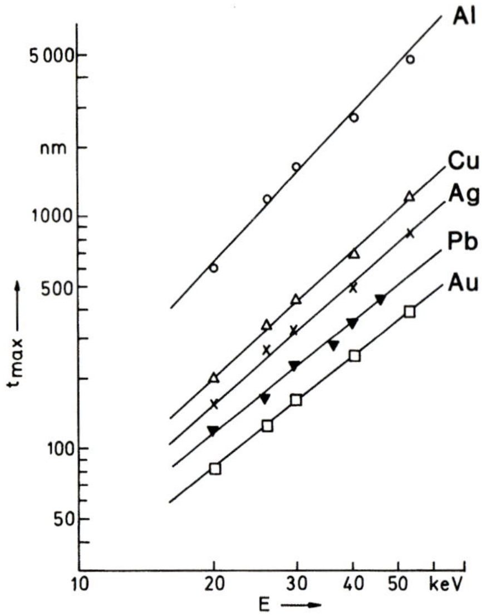

Fig. 4.3. Thickness limits $t_{\max}$ for linearity between backscattering coefficient η and film thickness t versus electron energy [4.6]

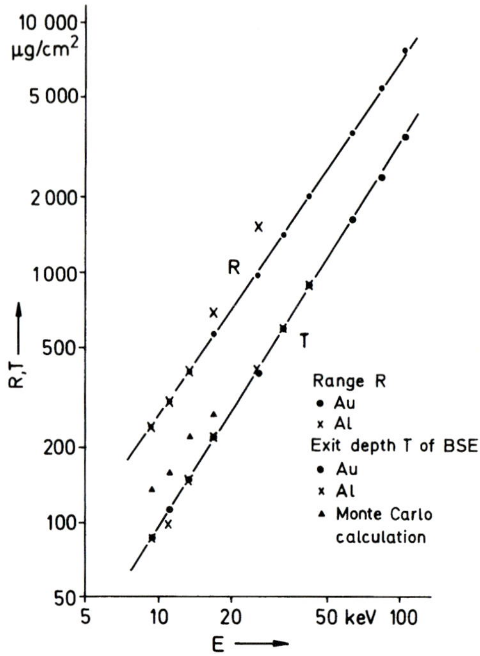

Fig. 4.4. Total depth T of BSE emission and electron range R in Al and Au versus electron energy (double-logarithmic plot) [4.1]

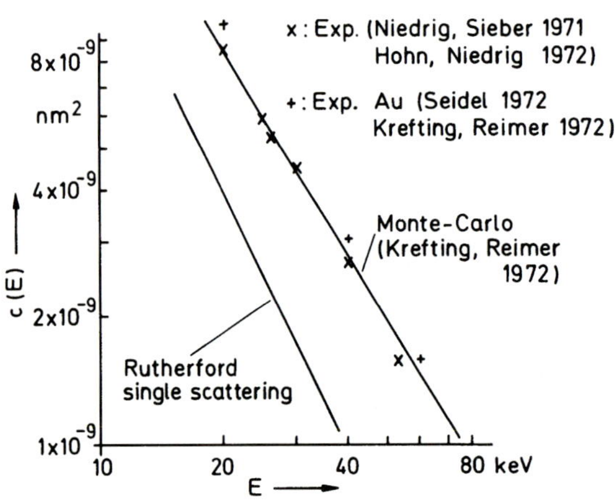

Fig. 4.5. Experimental and theoretical values of the constant $c(E)$ describing the increase of η with increasing thickness by $\eta = c(E)NZ^2t$ [4.11]

where N_m is the number of molecules of composition X_pY_q per unit volume. The validity of this formula has been tested for ZnS and PbF_2 films, for example [4.9].

The linear increase of η with increasing thickness can be used to measure the latter and record thickness profiles [4.10, 15]. If a Faraday cage, which collects all BSE with a solid angle of collection near 2π, is not used, the collection angle should be made large when semiconductor or scintillation detectors are placed below the polepiece because the increase of backscattering with increasing film thickness can change with the take-off angle (see below).

Coating films of material F and thickness t_F on a solid substrate S show different $\eta_{tot}(t_F)$ curves depending on whether $\eta_F < \eta_S$ (Fig. 4.2a) or $\eta_F > \eta_S$ (Fig. 4.2b). The backscattering coefficient shows an approximately linear decrease or increase, respectively, with increasing t_F, and varies between the saturation value $\eta_{S\infty}$ of the clean substrate and the saturation value $\eta_{F\infty}$ of the coating film [4.16, 17]. Coating films of increasing thickness can therefore also be used to determine the exit depth T of BSE [4.12, 13]. Analytical studies of $\eta_{tot}(t_F)$ are discussed in [4.16, 18–20]:

$$\eta_{tot} = \eta_{S\infty} + \eta_F(t_F)(1 - \eta_{S\infty}/\eta_{F\infty}) , \tag{4.12}$$

for example, where $\eta_F(t_F)$ is the backscattering coefficient of a self-supporting film of thickness t_F. Theoretical calculations are based on the single-scattering theory of Everhart (Sect. 3.4.3) [4.21], a combined single-scattering and diffusion model for $t_F > R/2$ [4.22], a simple layer-to-layer scattering theory [4.23] or Monte Carlo simulations [4.24]. The minimum detectable coating thickness decreases with decreasing electron energy. However, the different dependence of η on E in the LVSEM region (Fig. 4.9) has to be taken into account. Experiments agree with Monte Carlo simulations [4.25]. A dependence on surface tilt angle is reported in [4.26] and on the BSE take-off angle in [4.27].

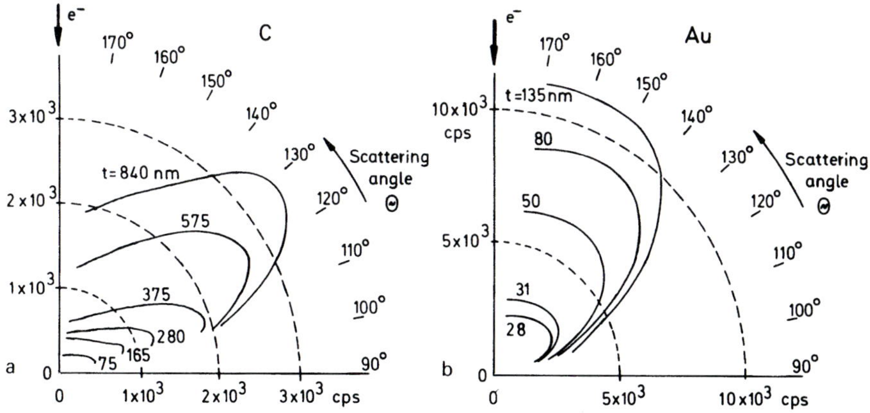

Fig. 4.6. Angular distribution of BSE plotted as a polar diagram for (**a**) amorphous carbon and (**b**) polycrystalline freely supported films of thickness t [4.29]

The angular distribution $d\eta/d\Omega$ of electrons backscattered from thin self-supporting films shows a flat distribution (Fig. 4.6) in accordance with the $\mathrm{cosec}^4(\theta/2)$ term of the Rutherford cross-section and has been recorded using a movable Faraday cage [4.28] or a semiconductor [4.29]. With increasing film thickness, an angle of maximum emission can be observed (Fig. 4.6). These characteristic angular distributions are also predicted by transport theory [4.30] and Monte Carlo simulations [4.14] and agree with experiment when Mott rather than Rutherford cross-sections are used. In contrast to the total backscattering coefficient of thin films discussed above, the dependence of the angular distribution on film thickness shows a non-linear increase of $d\eta/d\Omega$ when recorded with a detector of small solid angle at a fixed take-off direction [4.29].

The total increase of the BSE signal S_{BSE} of small particles on supporting films, membrane filters or bulk substrates of low Z increases as the particle mass m [4.19]. The signal can be obtained by summing the pixel intensities inside an area that includes the particle; This signal must be set against the signal S_{S} from a substrate inside an equal area:

$$m \propto (S_{\mathrm{BSE}} - S_{\mathrm{S}})/S_{\mathrm{S}} \,. \tag{4.13}$$

This allows a mass determination by calibration over several orders of magnitude in the sub-picogram region. The minimum detectable mass will be limited by the existing noise and the maximum mass by the exit depth of BSE. The BSE signal should show a maximum at the thickest particle dimension and should not show an increased signal at edges.

The backscattering of single-crystal films also depends on orientation due to channelling effects, which will be discussed in Sect. 9.1.4.

4.1.3 Backscattering from Solid Specimens

This Section summarizes some important experimental results on backscattering at solid, plane surfaces. There exists no exact theory of backscattering owing to the complexity of electron diffusion as discussed in Sect. 3.4.

Backscattering Coefficient for Energies Beyond 5 keV. Variation of the atomic number Z and of the tilt angle ϕ ($\phi = 0$: normal incidence) results in the curves of Fig. 4.7 [4.1, 31]. Values of η for normal incidence are listed in Table 4.1, see also Fig. 4.8. An empirical formula [4.32]

$$\eta(Z,\phi) = (1+\cos\phi)^{-9/\sqrt{Z}} \tag{4.14}$$

fits the experiments rather well. Another formula [4.33, 34]

$$\eta(Z,\phi) = \eta_0 \exp[\gamma l(1-\cos\phi)] \tag{4.15}$$

originally developed for SE emission [4.35] can give a reasonable fit to the experimental values of Fig. 4.7, where $\eta_0 = \eta(Z,0)$, γ is an absorption coefficient and l is a diffusion range. A plot of ln $[\eta(Z,\phi)/\eta_0]$ versus $1-\cos\phi$

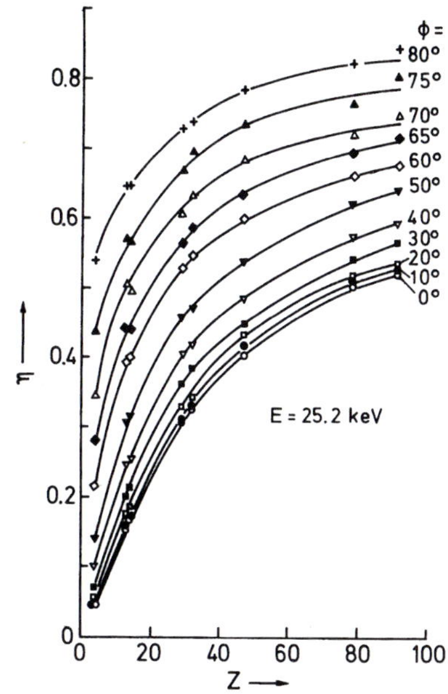

Fig. 4.7. Increase of the backscattering coefficient η of 25.2 keV electrons with increasing atomic number Z for different tilt angles ϕ ($\phi = 0$: normal incidence [4.1]

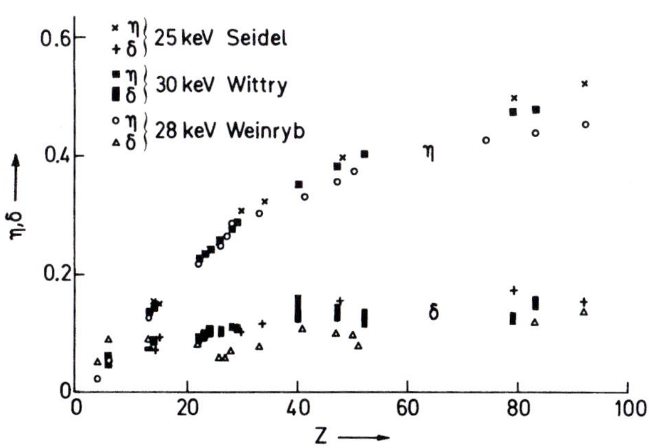

Fig. 4.8. Increase of the backscattering coefficient η and the secondary electron yield δ of 25–30 keV electrons with increasing Z

results in a straight line from which the relation $\gamma l = -\ln \eta_0 - 0.119$ was derived. Substitution in (4.15) results in

$$\eta(Z, \phi) = B \left(\frac{\eta_0}{B} \right)^{\cos \phi} \quad \text{with} \quad B = 0.89 \, . \tag{4.16}$$

Measurements at normal incidence and $E = 20$ keV can also be fitted to a polynomial in Z [4.18]

$$\eta = -0.0254 + 0.016Z - 1.86 \times 10^{-4} Z^2 + 8.3 \times 10^{-7} Z^3 \, . \tag{4.17}$$

Table 4.1. Backscattering coefficients η in % as a function of atomic number and electron energy E for normal incidence ($\phi = 0$)[4.1]

E [keV]	C	Al	Si	Cr	Cu	Ge	Ag	Sb	Au	Bi
0.5	14.5	23.0	26.9	27.0	21.0	21.0	20.0	18.3	18.0	17.4
0.7	12.9	20.4	26.4	24.0	24.0	23.0	25.0	22.0	22.7	22.0
1.0	11.7	19.5	23.7	24.0	27.0	29.0	30.0	26.3	28.0	28.0
1.4	10.9	18.3	23.5	23.0	27.0	33.0	32.0	28.4	34.0	32.0
2.0	9.5	17.0	22.8	24.0	30.0	33.0	35.0	34.0	37.5	36.0
3.5	8.7	15.4	20.2	23.7	31.2	32.8	37.1	37.0	43.0	42.0
5.0	7.5	15.0	20.2	23.0	32.6	33.2	37.6	40.0	46.0	47.0

E [keV]	Be	Al	Si	Cu	Ge	Ag	Au	U
9.3	5	17.5	18.3	31.3	33.6	40.7	47.8	49.8
11.0	5	17.1	17.8	31.0	33.6	41.1	48.0	50.2
13.4	5	16.4	18.4	31.4	32.4	40.2	49.1	51.3
17.3	5	15.9	17.8	31.0	32.7	40.6	49.2	51.6
25.2	4.5	15.1	16.5	30.7	32.7	40.3	50.1	52.7
41.5	4	14.5	15.5	30.1	31.7	40.5	50.7	52.4
62.1	4	13.7	14.9	29.9	31.9	39.9	51.3	54.1
81.8	4	13.5	14.5	29.4	31.5	40.2	51.0	54.2
102	3.5	13.3	14.5	29.1	31.3	39.9	51.3	56.2

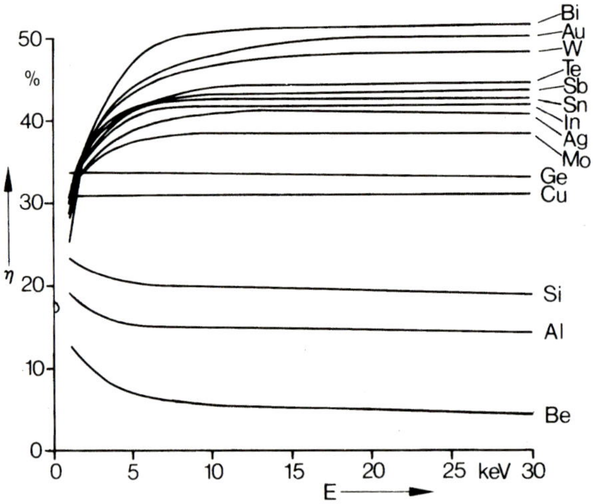

Fig. 4.9. Backscattering coefficient η as a function of electron energy in the range 1–30 keV [4.5]

Deviations from the smooth and monotonic increase of η with increasing Z in the range $Z = 22$–29 have been interpreted in terms of variations of the ratio A/Z [4.3]. However, the accuracy in measuring η is insufficient to detect

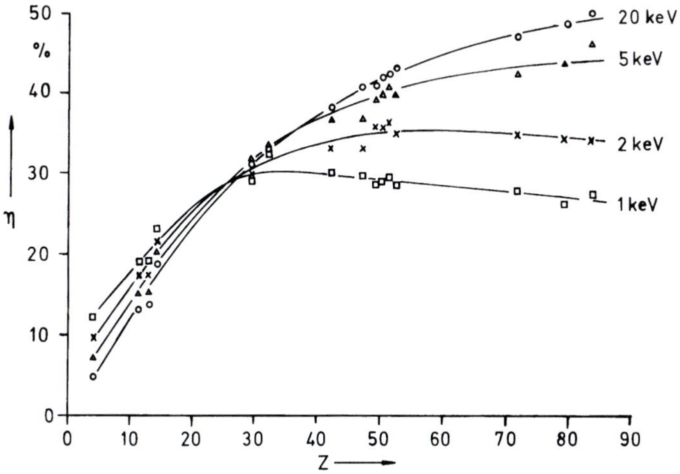

Fig. 4.10. Backscattering coefficient η as a function of atomic number Z for different electron energies in the range 0.5–5 keV [3.210]

such second-order influences unambiguously and the monotonic increase will be a sufficiently good approximation.

The measured backscattering coefficient of multicomponent targets can be fitted to the formula

$$\eta = \sum_i c_i \eta_i \tag{4.18}$$

where the η_i are the backscattering coefficients of the pure elements and the c_i their mass fractions [4.3, 37, 38].

Backscattering Coefficient for Energies Below 5 keV. The backscattering coefficient is approximately independent of the primary electron energy E in the range 5–100 keV. Below 5 keV in the LVSEM mode, η decreases for high Z and increases for low Z with decreasing E (Fig. 4.9). There is no longer a monotonic increase with increasing Z and there is even a maximum at medium Z (Fig. 4.10). With some differences due to different specimens and contamination layers, this tendency has been confirmed by several authors [4.5, 33]. These deviations from the approximately constant values of η at high energies can be explained by the much stronger deviations of elastic Mott cross-sections from Rutherford cross-sections. When using a Monte Carlo simulation with the Rutherford cross-section, the backscattering coefficient even increases for gold with decreasing E [4.36], whereas simulations with Mott cross-sections agree with the observed decrease of η for Au in Fig. 4.9 and also show the increase for low Z material.

The increase of η with increasing tilt angle ϕ (Fig. 4.7) is approximately independent of energy for $E \geq 5$ keV, whereas for $E < 5$ keV a lower increase of $\eta(\phi)/\eta(0)$ is observed for increasing ϕ as shown in (Fig. 4.11) for $E = 1$ and 20 keV [4.39, 40].

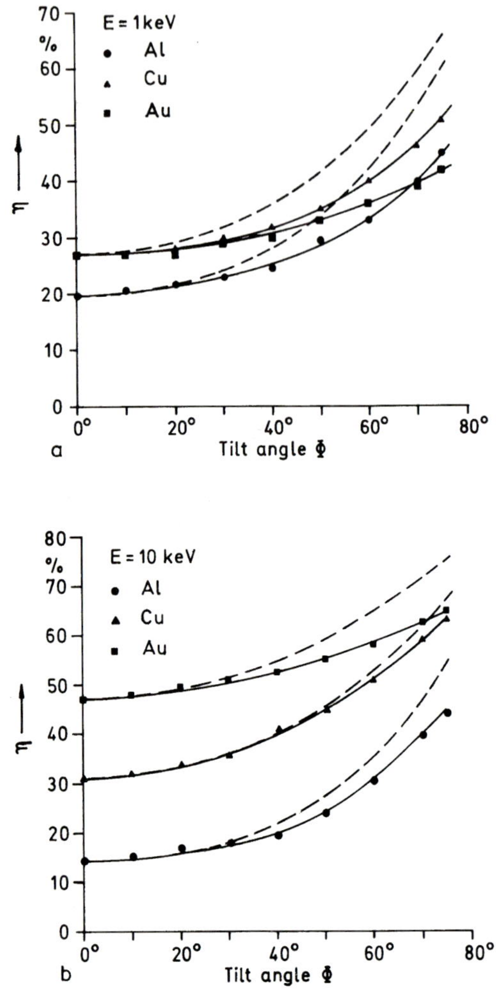

Fig. 4.11. Dependence of the backscattering coefficient η on the surface tilt angle ϕ at electron energies of (**a**) 1 keV and (**b**) 20 keV. For comparison, the dashed curves are calculated by (4.16)

4.1.4 Angular Distribution of Backscattering

Measurements of the angular distribution $\mathrm{d}\eta/\mathrm{d}\Omega$ are plotted as polar diagrams in Fig. 4.12 [4.1, 31, 41]. At normal incidence (Fig. 4.12a), $\mathrm{d}\eta/\mathrm{d}\Omega$ can be approximated by Lambert's cosine law

$$\frac{\mathrm{d}\eta}{\mathrm{d}\Omega} = \frac{\eta}{\pi} \cos\zeta \tag{4.19}$$

where ζ is the angle between the surface normal and the direction of emission. This results in a circle on a polar diagram, though the exact distributions for low and high Z materials are significantly more flat and elongated, respectively. For oblique incidence ($\phi = 60°$ and 80° in Figs. 4.12b,c) a reflection-like

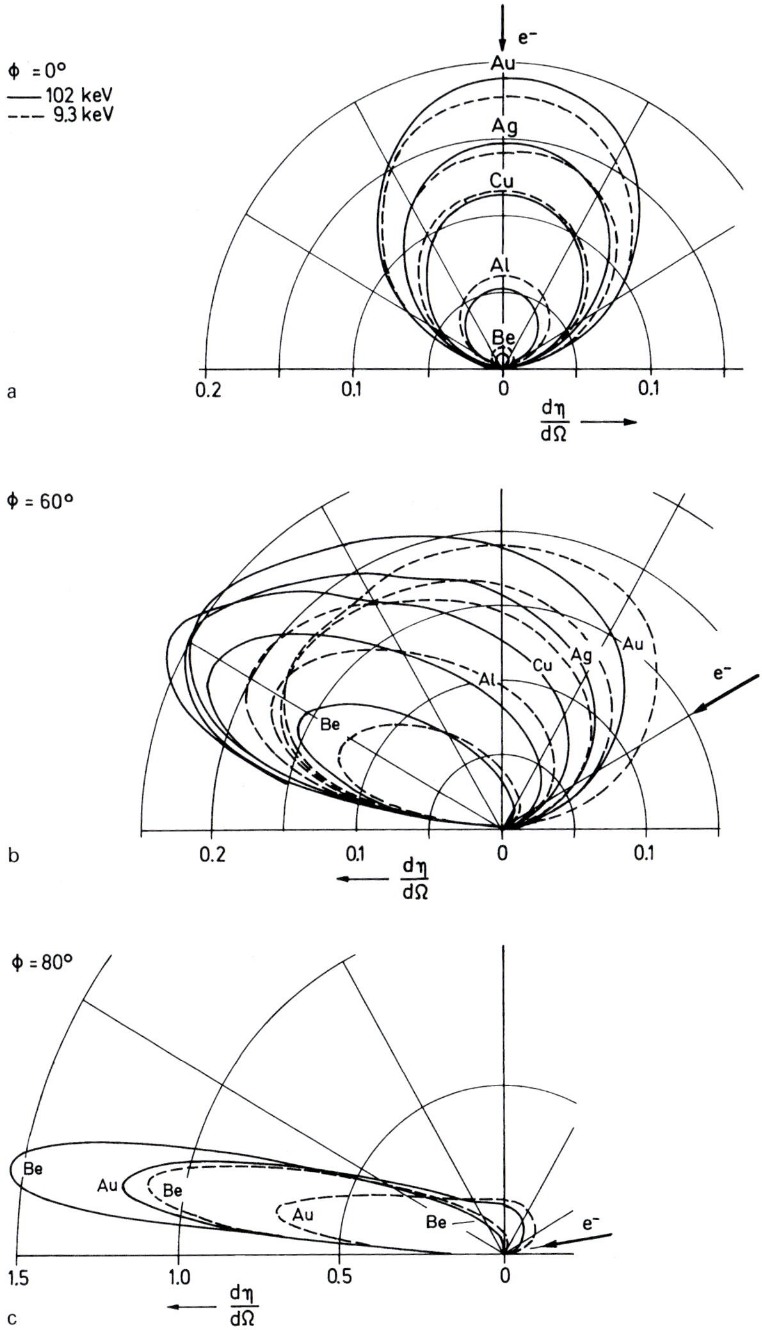

Fig. 4.12. Polar diagrams $d\eta/d\Omega$ for increasing tilt angles: (**a**) $\phi = 0$ (normal incidence), (**b**) $\phi = 60°$ and (**c**) $\phi = 80°$ (—— $E = 100$ keV, - - - $E = 9.3$ keV) [4.1]

emission maximum can be observed and the angular distribution consists of two parts. One part is generated by electrons that have been diffusely scattered in the material and traverse the whole diffusion cloud, resulting in Lambert's distribution. For take-off directions opposite to the direction of electron incidence, the polar diagrams in Fig. 4.12c contain such a fraction, which still obeys Lambert's law (see also Fig. 6.14). The other reflection-like part is generated within a smaller exit depth and these electrons can escape after a few scattering processes through small angles. However, the term 'reflection-like' has to be used with care. In dependence on the azimuth χ, the maximum extends to a much larger azimuth than expected from a reflection [4.42]. This has also been confirmed by Monte Carlo simulations [4.43].

Though the total backscattering coefficient $\eta = 0.53$ of Be at $\phi = 80°$ is smaller than the corresponding value $\eta = 0.81$ of Au (Fig. 4.7), the angular distribution $d\eta/d\Omega$ in Fig. 4.12c shows larger values for Be in the direction of maximum emission while the polar diagram of Au shows a large diffusely scattered part. Inside the tilt range $\phi = 0° - 60°$ the diffusely scattered part remains approximately constant. This means that the signal given by a scintillator or semiconductor detector situated below the polepiece is approximately independent of the tilt angle ϕ, which is an advantage for the observation of material contrast (Sect. 6.2.1). Only at large ϕ, when a larger fraction is scattered into the emission maximum, will the signals given by these detectors decrease with increasing ϕ.

At energies below 5 keV, Monte Carlo simulations [4.36,43] still result in a $\cos\zeta$ law for Al at normal incidence ($\phi = 0$), whereas the angular characteristics of 1–5 keV electrons backscattered by Au increase more strongly than $\cos\zeta$ with decreasing ζ ($\zeta = 0$: normal emission). This is a consequence of the increased Mott cross-sections near scattering angles $\theta = 180°$ (Fig. 3.6). Quantitatively, this has been confirmed by measuring the BSE current falling on annular segments for different ranges of the emission angle ζ [4.43]. For strongly tilted specimens, the angular distribution shows minima of backscattering at angles where the Mott cross-sections show minima and an even stronger BSE emission can be observed opposite to the direction of incidence and corresponding to backscattering near $\theta = 180°$ [4.43–46].

4.1.5 Energy Distribution of Backscattered Electrons and Low-loss Electrons

The spectrum $d\eta/dE_B$ consists of an elastically scattered peak and a broad maximum (Figs. 4.13 and 4.14), which extends from the primary energy E to very low energies where it is overlapped by the Auger electron and SE energy spectra (Fig. 1.5). The broad maximum shifts to lower energies with decreasing Z. An energy resolution $\Delta E/E \simeq 1$–2% is not sufficient to resolve the elastic peak at electron energies $E > 10$ keV. Several measurements of energy spectra for $E > 10$ keV have been reported that either go to zero at

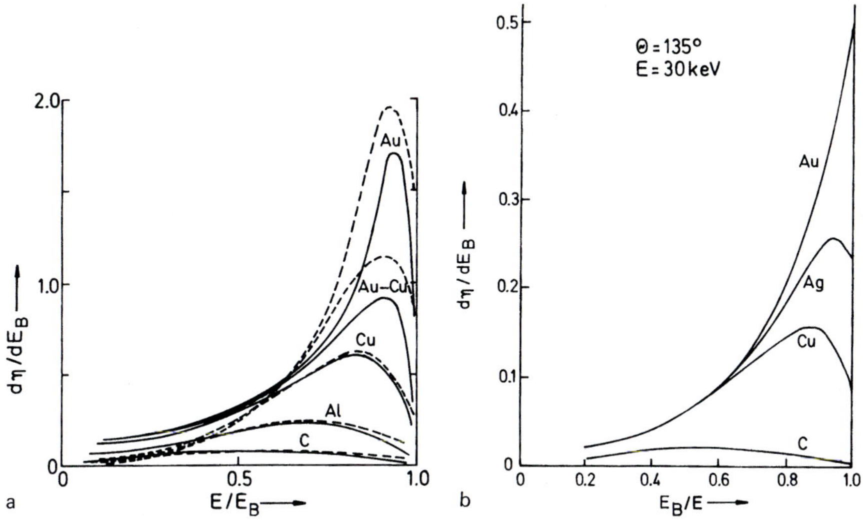

Fig. 4.13. Energy spectra $d\eta/dE_B$ of backscattered electrons measured (**a**) with the retarding field method for C, Al, Cu, Au and Au-Cu (50:50) (E = 20 keV) [4.52] and simulated by the Monte Carlo method (– – –) in (a), (**b**) measured at exit angle $\theta = 135°$ at E = 30 keV [4.51]

$E_B/E = 1$ [4.47, 48] or end at a finite value (Fig. 4.13) [4.49–52]. The latter is due to the fact that the authors cut off their recorded spectra at $E_B/E = 1$ and did not show the tails arising from the finite resolution $\Delta E/E$, which means that the measured spectra have not been deconvolved, using the transmission curve of the spectrometer for monoenergetic electrons. The reported energy spectra also differ inside the range $E_B/E = 0.1–0.4$ as a result of systematic errors such as the generation of SE at grids and electrodes or the reflection of BSE at the massive electrodes of an electrostatic prism spectrometer, for example.

The BSE energy spectrum from layered structures can be used to analyse the thickness of layers [4.53, 54]. When an Al film of thickness t on a Au substrate is examined, BSE of high energy near the incident energy will be backscattered in the Al film with low intensity. Primary electrons reaching the Au substrate with a mean energy $E_m(t)$ (Fig. 3.24) can be backscattered with a larger probability. The backscattered electrons lose energy on the return path t and, at an energy of approximately $E_m(2t)$, the BSE spectrum intensity shows a steep increase extending in a maximum to lower BSE energies. This process is analogous to the Rutherford backscattering spectroscopy of ions with the difference that ions shows less straggling and the mean energy $E_m(s)$ is a unique function of the straight path length s.

The existence of an elastically scattered no-loss peak for high electron energies can be shown by using a high-resolution spectrometer [4.55–57]. Figures 4.15 and 4.16 show measured [4.57] and calculated [4.30] energy spectra

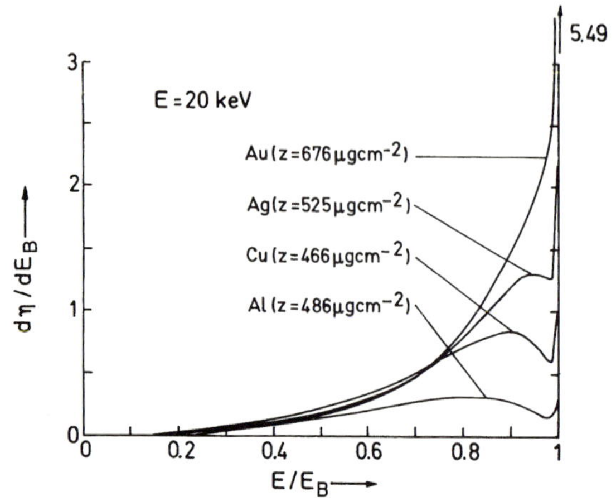

Fig. 4.14. Energy spectra $d\eta/dE_B$ of 20 keV electrons calculated by the transport-equation method [4.30]. The indicated mass–thicknesses x are larger than the exit depths of the BSE

$d^2\eta/dE_B d\Omega$ at different take-off angles θ relative to the direction of electron incidence. The elastic peak can be better resolved for energies $E \leq 5$ keV

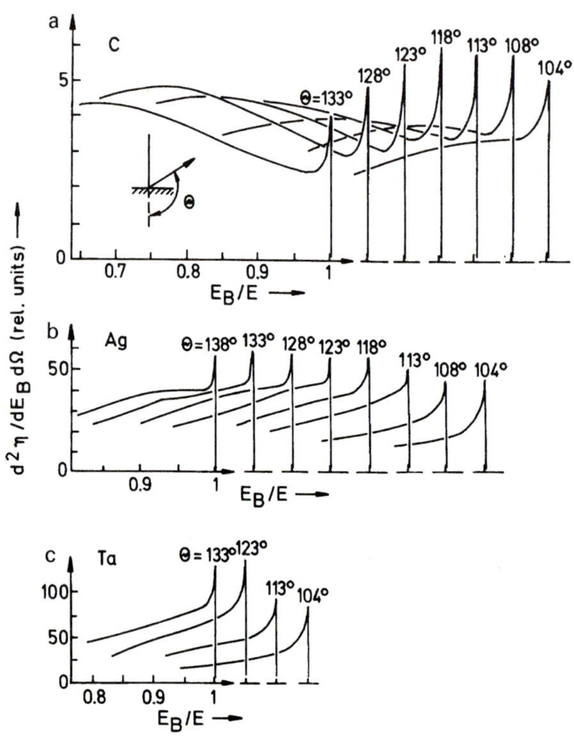

Fig. 4.15. Measured energy spectra $d^2\eta/dE_B d\Omega$ of 25 keV electrons backscattered at (**a**) C, (**b**) Ag and (**c**) Ta for different take-off angles θ relative to the direction of incidence (see inset in (a)). The spectra are shifted on the energy scale to avoid overlap [4.30]

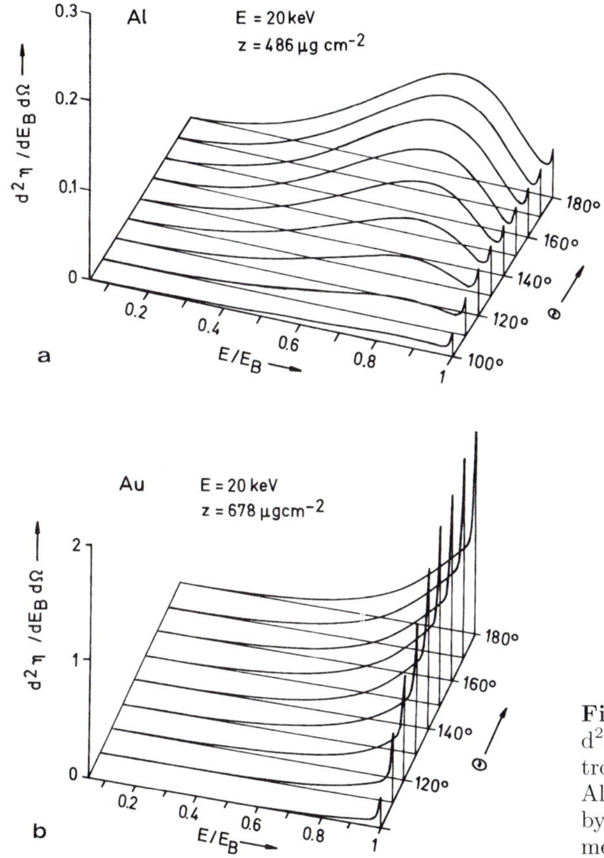

Fig. 4.16. Energy spectra $d^2\eta/dE_B d\Omega$ of 20 keV electrons backscattered at (**a**) Al and (**b**) Au calculated by the transport-equation method [4.30]

because the resolution $\Delta E/E$ of an electrostatic spectrometer is independent of energy. Another reason why the elastic peak can be observed better for low energies is that it increases as E^{-1} with decreasing energy as discussed below. It has been observed that the magnitude of the elastic peak is very sensitive to contamination layers [4.57], which may also be the reason why this peak was not found in earlier experiments whereas it is no problem to observe it in Auger electron microprobe instruments, which operate with an ultra-high vacuum.

The elastic peak formed by no-loss electrons and the fraction of low-loss electrons inside an energy window $E - \Delta E \leq E_B \leq E$ with $\Delta E = 10$–200 eV can both be attributed mainly to single large-angle scattering processes. This low-loss fraction of the BSE spectrum therefore arises within a very thin surface layer only a few nanometres thick in contrast to the signal of all the other BSE, which comes from a depth T of the order of half the electron range (Fig. 4.4). The intensity and the angular distribution of these low-loss electrons can be estimated by assuming Rutherford scattering and the Bethe

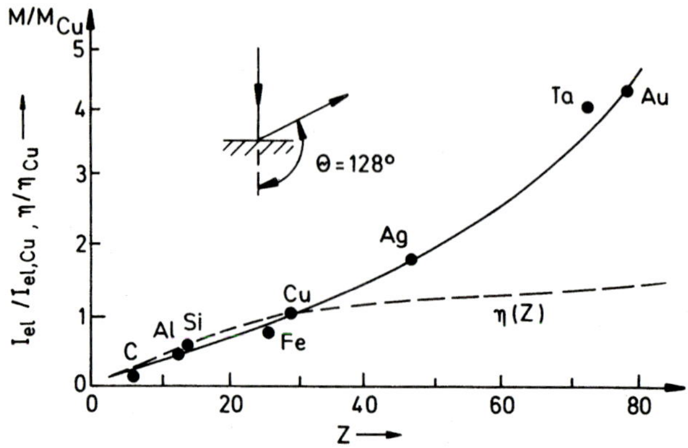

Fig. 4.17. Increase of the intensity of the elastic peak in the energy spectrum of 25 keV electrons with increasing atomic number at an angle $\theta = 128°$ and the increase of the backscattering coefficient η (– – –) for comparison. The quantities are matched to unity for copper [4.57]

stopping power [4.58]. The no-loss electrons will be scattered from a layer of the order of the mean-free-path length Λ_{pl} for plasmon losses in thickness, which increases as the electron energy E. The layer thickness from which low-loss electrons can be observed can be estimated from the Bethe stopping power $S = |\Delta E_{\mathrm{m}}/\Delta x| \propto Z/AE$, (3.130), resulting in $\Delta x \propto AE\Delta E_{\mathrm{m}}/Z$. Both Λ_{pl} and Δx are therefore approximately proportional to E. Because there are $N_{\mathrm{A}}\Delta x/A$ atoms per unit area inside a layer of mass-thickness Δx or Λ_{pl}, multiplication by the Rutherford cross-section $\sigma_{\mathrm{R}} \propto Z^2/E^2$ (3.14), results in a low-loss contribution $\eta_{\mathrm{LLE}}/\eta_0 \propto Z/E$. The dependence of the elastic peak on Z is shown in Fig. 4.17. For comparison this intensity and the total backscattering coefficient η are scaled to unity for copper.

4.2 Secondary Electrons

4.2.1 Dependence of Secondary Electron Yield on Beam and Specimen Parameters

Many measurements of the secondary-electron yield δ (Sect. 4.1.1) have been made for primary electron energies below 2 keV [4.59–65] because of interest in the SE yield for the development of dynodes in electron multipliers.

At these low primary electron energies, SE and BSE are often not distinguished and the total yield $\sigma = \eta + \delta$ has been measured, whereas at high electron energies all electrons that leave the specimen with an exit energy $E_{\mathrm{SE}} \leq 50$ eV are called SE by convention.

The spectrum of SE energies $\mathrm{d}\delta/\mathrm{d}E_{\mathrm{SE}}$ (Figs. 1.5 and 4.18a) reaches a maximum with a most probable energy E_{p} and a full width ΔE_{W} at half

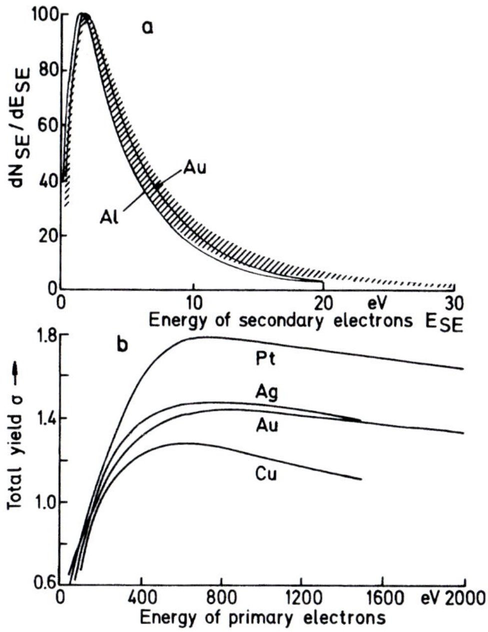

Fig. 4.18. (**a**) Energy spectra of secondary electrons emitted from different metals, normalized to unity at their most probable energies. Shaded area: experimental results for different metals, —: calculations for Al and Au by (4.30). (**b**) Total yield $\sigma = \eta + \delta$ as a function of primary electron energy E [4.61]

maximum. Both quantities depend on the material. The full width ΔE_{W} is smaller for insulators (KCl, for example [4.66]) than for pure metals and depends strongly on very thin surface layers. Experiments on clean metal surfaces [4.67] result in $E_{\mathrm{p}} = 1$–5 eV and $\Delta E_{\mathrm{W}} = 3$–$15$ eV whereas older measurements [4.61] gave values in the range $E_{\mathrm{p}} = 1.3$–2.5 eV and $\Delta E_{\mathrm{W}} = 4$–$7$ eV, the shaded region in Fig. 4.18a. The full curves for Al and Au are the calculated energy spectra discussed in Sect. 4.2.2. Superposed on the high-energy tail of the SE spectrum, structures can be observed that can be attributed to excitations of single electrons by decaying plasmons, for example [4.68–70] (Sect. 4.2.2). However, all other inelastic scattering processes also contribute to the SE emission. Coincidence measurements of scattering events in electron energy-loss spectra (EELS) and of SE emission in thin films at 100 keV show that the probability of SE emission increases with increasing energy loss [4.71–73], which confirms the simple model that the SE yield is proportional to the Bethe stopping power (4.22).

The dependence of δ on the primary electron energy E shows a maximum $\delta_{\mathrm{m}} = 0.35$–$1.6$ for metals and 1–10 for insulators at energies of $E_{\mathrm{m}} = 100$–800 eV and 300–2000 eV, respectively (Fig. 4.18b and 3.31). The shapes of the $\delta(E)$ curves are very similar for different materials when $\delta/\delta_{\mathrm{m}}$ is plotted against E/E_{m}. A plot of δ_{m} and E_{m} against Z shows the influence of the periodic system [4.63, 74, 75] whereas the ratio $\delta_{\mathrm{m}}/E_{\mathrm{m}} \simeq 2\ \mathrm{keV}^{-1}$ is approximately constant (see Tables of δ_{m} and E_{m} in [4.74], for example). This has been attributed to the Z dependence of the mean ionization energy J of the

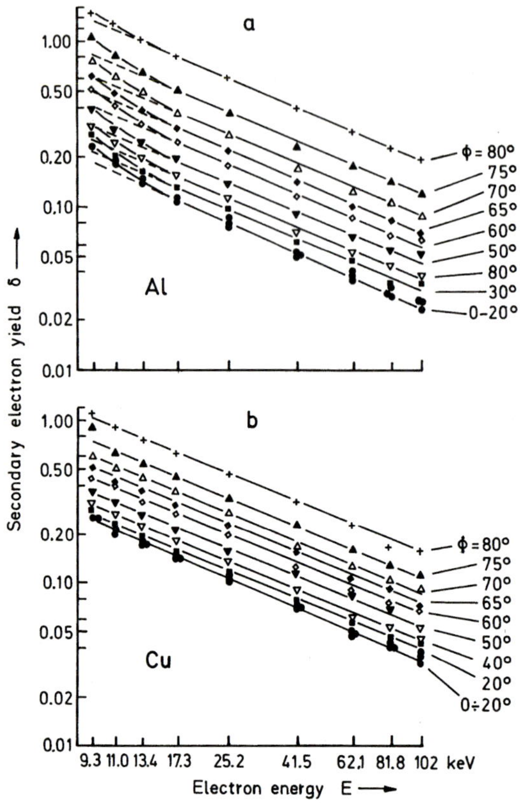

Fig. 4.19. Decrease of the secondary electron yield δ with increasing electron energy E for (**a**) Al and (**b**) Cu and different tilt angles $\phi = 0$: normal incidence) in a double-logarithmic plot which confirms the relation $\delta \propto E^{-0.8}$ [4.1]

surface atoms (δ_m and $E_m \propto J^{4/5}$) [4.74] or of the atomic radii [4.63], atoms with a large diameter showing small values of δ_m.

However, all measurements of SE emission have to be used with caution because the SE emission depends on three mechanisms: the generation of SE, the scattering and diffusion of SE and the penetration through the surface barrier. Whereas thermionic and field emission depend strongly on the work function ϕ_w, the influence of ϕ_w on δ is not so strong because the penetration through the surface barrier of height ϕ_w is only one stage in the mechanism of SE emission (Sect. 4.2.2). For example, reduction of the work function from 4.79 to 2.3 eV by deposition of Na on Ge increases δ from 1.2 to 3.6 [4.76]; see also [4.67] where Na on Pt is studied.

Measurements of δ at primary energies in the range 1–100 keV [4.1] show a decrease with increasing E, which can be represented by $\delta \propto E^{-0.8}$. This is demonstrated by straight lines of slope –0.8 in a double-logarithmic plot (Fig. 4.19) and can be confirmed for all specimen tilt angles ϕ (see also discussion of Eqn. (4.22) below). Figure 4.8 shows a plot of η and δ versus Z for 25–30 keV electrons, which shows that the total backscattered electron signal of metals becomes larger than the SE signal and that the scatter of the measured δ

values is much larger than that of η for BSE because the SE yield depends on more parameters (Sect. 4.2.2). It will be shown in Sect. 4.2.3 that the small increase of δ with increasing Z can be mainly attributed to the generation of SE by the BSE on their trajectories through the surface layer.

The SE can leave the specimen only from a small exit depth t_{SE}, which is of the order of 0.5–1.5 nm for metals and increases to the order of 10 nm for carbon and 10–20 nm for insulators [4.63]. The stopping power increases with increasing SE energy beyond the Fermi level and the total mean path lengths in aluminium are Λ = 5 nm and 7 nm for SE of energies 9 and 20 eV, respectively, and for higher energies little extra travel distance is gained. The SE follow a random walk and are scattered through large angles with a mean free path λ_p which results in an escape depth of the order of $(\lambda_p\Lambda)^{1/2}/3$ [4.73]. Values between 1 and 30 nm are reported for metal oxides and alkali halides [4.77]. Measurements on potassium films showed that this exit depth also depends on the SE exit energy [4.78]. The probability of escape from a depth z below the surface can be described in a first-order approximation by an exponential law

$$p(z) = p(0)\exp(-t/t_{SE}) . \tag{4.20}$$

The large value of t_{SE} for some metal oxides and alkali halides may be the reason for their large values of δ.

The SE yield δ increases with increasing specimen tilt angle ϕ, a very important effect for imaging of the surface topography (Sect. 6.1.2). This can be seen in Fig. 4.19, and Fig. 4.20 shows measurements at different primary electron energies, which have been normalized to unity for normal incidence (ϕ =0). The results (small dots) do not vary systematically with energy (10–100 keV) and show an increase with increasing ϕ which is more rapid than $\sec\phi$ for low Z (Be in Fig. 4.21a) and slower than $\sec\phi$ for higher Z (U in Fig. 4.20c). The proportionality to $\sec\phi$ is a good approximation for medium atomic numbers. These differences will be discussed further in Sect. 4.2.3. Here, we merely attribute this increase of δ with increasing ϕ to the increase of the path length of the primary electron inside the exit depth, which is $t_{SE}\sec\phi$. A similar dependence of δ on ϕ and Z has been observed for a primary energy of 3 keV [4.79]. These authors fitted their results to $\delta \propto (\sec\phi)^n$ with $n \simeq 1.3$ for light and $n \simeq 0.8$ for heavy elements.

Below 5 keV (LVSEM) the electron range becomes comparable with the exit depth of SE and the assumption of a path length $t_{SE}\sec\phi$ of primary electrons and BSE inside the exist layer breaks down. The increase of δ with increasing ϕ becomes lower than $\sec\phi$ (Fig. 4.21) [4.40], which affects the topographic contrast (Sect. 6.1.2)

The angular distribution $d\delta/d\Omega$ of SE has been observed in all experiments [4.80] to be proportional to $\cos\zeta$, where ζ denotes the emission angle relative to the surface normal, and thus to conform to Lambert's law. This is found irrespective of primary electron energy, surface tilt and material. Together with the proportionality to $\sec\phi$, we can describe the dependence

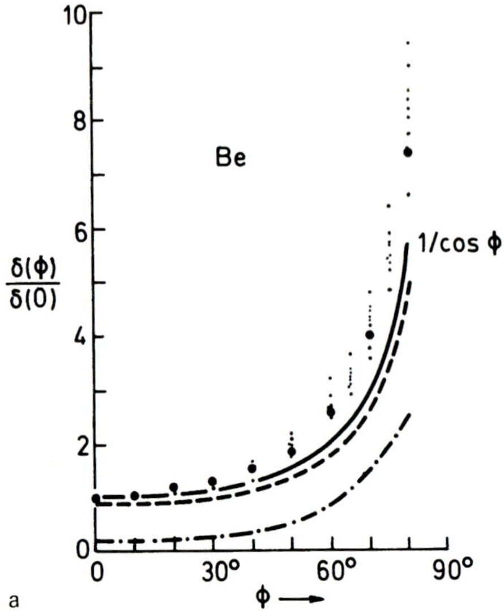

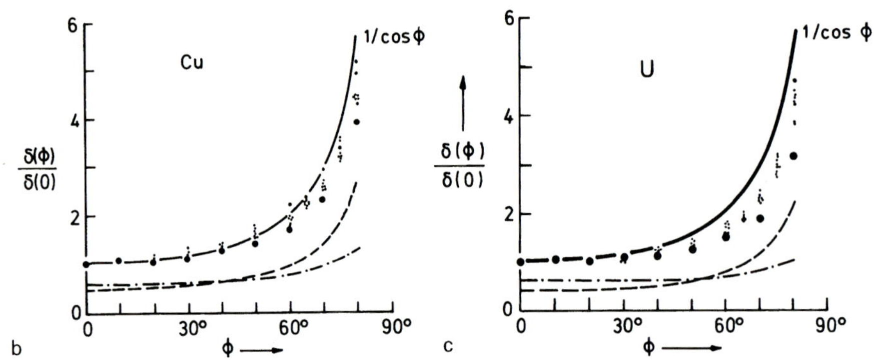

Fig. 4.20. Dependence of the relative SE yield (normalized to unity at $\phi = 0$) on the tilt angle ϕ of the specimen for (**a**) Be, (**b**) Cu and (**c**) U targets. *Small points:* measured values in the range $E = 10$–100 keV, *large dots:* Monte Carlo simulations, *full lines:* $\propto \sec\phi$ for comparison. *Dashed lines:* contribution of primary electrons and *dash-dotted lines:* contribution of the BSE to the SE yield obtained from Monte Carlo simulations [4.1]

of the SE yield on ϕ and and ζ by

$$\frac{d\delta}{d\Omega} = \frac{\delta(0,Z)}{\pi} \sec\phi \cos\zeta \,, \tag{4.21}$$

(optionally adding the exponent n to $\sec\phi$, for greater accuracy), where $\delta(0, Z)$ denotes the SE yield for normal incidence. Integration of $d\delta/d\Omega$ over $d\Omega = 2\pi \sin\zeta d\zeta$ will give the total SE yield $\delta(0, Z)\sec\phi$.

The angular distribution of SE from single crystals may be anisotropic [4.81] and angular-resolved SE spectroscopy is a powerful method for the investigation of bulk and surface states in solid state physics [4.82–84]. This technique works with primary energies of a few hundreds of eV and beam currents of $\simeq 5 \times 10^{-8}$ A, for example. However, no application in SEM with a better spatial resolution has been reported. When the primary electron beam is rocked on a single crystal, the total secondary electron yield varies as a result of the electron channelling effect (Sect. 9.1.4).

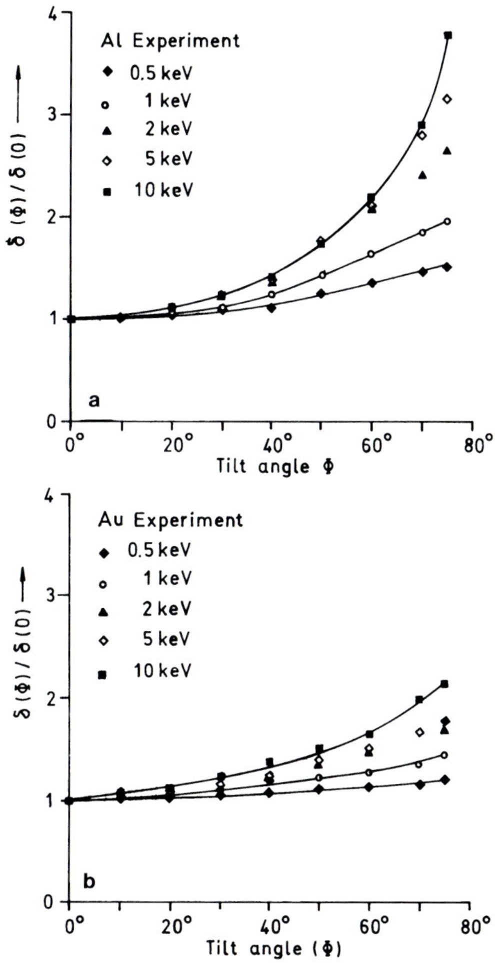

Fig. 4.21. Relative secondary electron yield $\delta(\phi)/\delta(0)$ for increasing tilt angle ϕ and electron energies E = 0.5–10 keV for (**a**) Al and (**b**) Au

Normally, a saturation value of the SE emission current is observed when the collector is positively biased by a few volts. Some insulators in the form of compact evaporated layers and especially porous layers evaporated at a pressure of a few 100 Pa show an increase of the SE current with increasing bias. Porous films of BaF_2, KCl and MgO can reach values δ = 30–70 at E = 10 keV and a collector bias of a few hundreds of volts [4.60, 63] (see also SE field-emission in Sect. 8.3.2).

The secondary-electron yield can be changed during electron irradiation by the growth of a contamination layer, which normally results in a decrease of δ but can also result in an increase after exposure to air [4.85]. Irreversible radiation damage of ionic crystals can also change their SE emission. For example, a decrease from an initial value δ = 4.2 to 2.0 has been found for Na_3AlF_6 after an excitation with 300 eV electrons and an electron exposure of 5×10^{-2} C cm^{-2}. Beyond this minimum, δ increases again with increasing electron exposure. The decrease can be attributed to a loss of fluorine and the increase to an oxidation of the metallic sodium to Na_2O by CO_2, O_2 or H_2O molecules of the residual gas [4.86].

4.2.2 Physics of Secondary-Electron Emission

First, we discuss a simplified theory of SE emission for electron energies $E \geq$ 5 keV capable of explaining most of the experimental findings described in the last section. Further ideas will then be presented, which provide a more detailed description of the elementary processes.

The simplest theory of SE emission assumes that the generation of SE per path length of primary electron is proportional to the Bethe stopping power $S = |\mathrm{d}E_\mathrm{m}/\mathrm{d}s| \propto E^{-1}\ln(E/J)$ (3.129) and is constant within the exit depth t_SE, which is much smaller than the electron range R. The SE generated at a depth z below the surface can escape with the probability given by (4.20). At a specimen tilt ϕ, the path length element between z and $z+\mathrm{d}z$ becomes $\mathrm{d}s = \mathrm{d}z \sec\phi$ and the SE emission caused by primary electrons becomes

$$\delta = \frac{0.5}{\epsilon}\left|\frac{\mathrm{d}E_\mathrm{m}}{\mathrm{d}s}\right| \sec\phi \int_0^\infty \exp(-z/t_\mathrm{SE})\mathrm{d}z \propto E^{-0.8} \sec\phi\, t_\mathrm{SE} \tag{4.22}$$

where ϵ denotes the mean energy loss required to produce one SE. This formula adequately describes the dependence of δ on the tilt angle ϕ and the primary electron energy E (Figs. 4.19 and 4.20). If we plot E^{-1} ln (E/J) versus E in a double-logarithmic diagram in the range E = 1–100 keV, the resulting curve can be approximated by a straight line of slope –0.8, which corresponds to the observed slope in Fig. 4.17. The two parameters ϵ and t_SE can be fitted to experimental values when this model is to be used in Monte Carlo simulations, for example [4.87, 88]. This simple model will also be used in Sect. 4.2.3 to consider the contribution of backscattered electrons to the SE yield. A more detailed simulation includes the generation of SE by fast secondary electrons [4.89].

We now discuss a theory of SE excitation [4.90] that results in a reasonable analytical formula for the energy distribution of the SE. We consider an excited electron with an energy E' above the bottom of the conduction band. Only electrons with exit energies $E_\mathrm{SE} = E' - W > 0$ can overcome the work function ϕ_w (Fig. 4.22a). This process has to be treated quantum-mechanically as the scattering of an electron wave at a potential barrier, which results in refraction and total internal reflection (Fig. 4.22b). The angle of incidence α' and the angle of refraction α are related by

$$\cos^2 \alpha'_\mathrm{c} = \frac{E_\mathrm{SE}\cos^2\alpha + W}{E_\mathrm{SE} + W} \; . \tag{4.23}$$

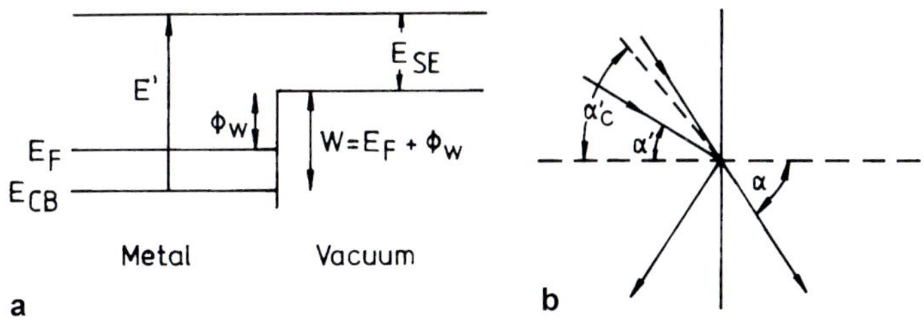

Fig. 4.22. (**a**) Energy barrier for SE at the metal–vacuum interface (E_F = Fermi energy, E_{CB} = bottom of conduction band, ϕ_w = work function). (**b**) Refraction and total internal reflection of SE at the interface

Setting $\alpha = 90°$, we obtain the critical angle α'_c for total internal reflection

$$\cos^2 \alpha'_c = W/(E_{SE} + W) = W/E' \; . \tag{4.24}$$

The probability of crossing the surface barrier can be assumed to be unity for $\alpha' < \alpha_c$ and zero for $\alpha' > \alpha_c$. Electrons excited at a depth z with an angle α' will reach the surface without collision with a probability

$$P(E', \alpha', z) = \exp[-z/\Lambda(E') \cos \alpha'] \tag{4.25}$$

where $\Lambda(E')$ is the mean-free-path length for scattering of an electron of energy E' by the electron gas of the solid. The number of excited electrons in the interval $E', E' + \mathrm{d}E'$ is denoted by $S(E', z)$ and is measured in units of $\mathrm{cm}^{-1}\,\mathrm{eV}^{-1}$. Unlike the more detailed theory discussed below, we here assume that $S(E', z)$ is isotropic in α'. The emitted energy spectrum then becomes

$$\frac{\mathrm{d}N_{SE}}{\mathrm{d}z} = \frac{1}{2} \int_0^{\alpha'_c} \int_0^{\infty} S(E', z) \exp[-z/\Lambda(E') \cos \alpha'] \sin \alpha' \mathrm{d}\alpha' \mathrm{d}z \; . \tag{4.26}$$

If we assume that $S(E', z)$ is independent of z, which means that the primary electrons travel a distance $\Lambda(E')$ along a straight trajectory with no appreciable energy loss, the integration over z and α' in (4.26) gives

$$\frac{\mathrm{d}N_{SE}}{\mathrm{d}E'} = \frac{S(E')\Lambda(E')}{2} \int_0^{\alpha'_c} \sin \alpha' \cos \alpha' \mathrm{d}\alpha' = \frac{S(E')\Lambda(E')}{4}(1 - W/E') \; . \tag{4.27}$$

The calculations of *Streitwolf* [4.91] lead to

$$S(E') = \frac{e^4 k_F^3}{3(4\pi\epsilon_0)^3} \frac{1}{E(E' - E_F)^2} \tag{4.28}$$

where k_F, denotes the wavevector of electrons at the Fermi level E_F, and those of *Quinn* [4.92] to

$$\Lambda(E') = \frac{1.47(E_F\beta)^{3/2}}{(m^*/m)^{1/2}[\tan^{-1}\beta^{-1/2} + \beta^{1/2}/(1+\beta)]} \frac{E'}{(E' - E_F)^2} \tag{4.29}$$

with $\beta = (4/9\pi)^{1/3} r_s/\pi a_H$ and where Λ is measured in nm and energies in eV, m^*/m is the ratio of the effective and free electron mass and r_s the radius of the volume element per electron with a_H as the Bohr radius. Substitution of (4.28) and (4.29) into (4.27) gives the following expression

$$\frac{\mathrm{d}N_{\mathrm{SE}}}{\mathrm{d}E_{\mathrm{SE}}} \propto \frac{E'(1 - W/E')}{E(E' - E_{\mathrm{F}})^4} = \frac{1}{E} \frac{E_{\mathrm{SE}}}{(E_{\mathrm{SE}} + \phi_{\mathrm{w}})^4} \,. \tag{4.30}$$

Energy spectra of SE calculated with this formula are plotted for Al and Au in Fig. 4.18a normalized to unity at the most probable energy. Measured energy spectra of metals spread over the shaded area. The most probable energy of (4.30) can be derived by setting the derivative with respect to E_{SE} equal to zero, giving

$$E_{\mathrm{p,SE}} = \phi_{\mathrm{w}}/3 \,. \tag{4.31}$$

A more detailed theory of SE emission can be developed by considering the following three steps:

1. Generation of SE by excitation of core and conduction electrons and by the decay of plasmons.
2. Diffusion of hot electrons by elastic and inelastic collisions.
3. Escape of the SE through the surface barrier.

Such calculations have been made for aluminium [4.93–95] because numerous reliable data are available, the valence electrons behave like nearly free electrons and there is only one plasmon loss at $\Delta E = 15$ eV.

Without going into details, we discuss the consequences of the most important mechanisms that come into play in the three steps for 2 keV primary electrons [4.95].

The atomic configuration of the Al core is given by $1s^2 2s^2 2p^6$. The contribution of the K shell with a binding energy $\simeq 1550$ eV can be neglected due to the low ionization probability and we take into account only the excitation of the L shell electrons with binding energies of 106 and 61 eV relative to the bottom of the conduction band. The excitation functions $4\pi S_{\mathrm{c}}(E, E')$ of the $2s$ and $2p$ core electrons are plotted in Fig. 4.23 for $E = 2$ keV. Only electrons with $E' \geq 16$ eV (vacuum level) can later overcome the work function and be emitted as SE into the vacuum. Integration of $S_{\mathrm{c}}(E, E')$ over E' gives the mean-free-path length $\Lambda_{\mathrm{c}}(E')$ for core excitation:

$$\frac{1}{\Lambda_{\mathrm{c}}(E')} = 4\pi \int\limits_{E' \geq E_{\mathrm{F}}}^{\infty} S_{\mathrm{c}}(E, E') \mathrm{d}E' \,. \tag{4.32}$$

Values of $\Lambda(E') = 23.5$, 28 and 32 nm are found for $E = 1$, 1.5 and 2 keV, respectively. The excitation of the secondary electrons is not isotropic but is described by the angular characteristics plotted in Fig. 4.24a as a polar diagram ($\boldsymbol{k}_0$: direction of the primary electron).

A dynamical screening theory for the electron–electron scattering of the conduction electrons results in the excitation function $4\pi S_{\mathrm{c}}(E, E')$ of Fig.

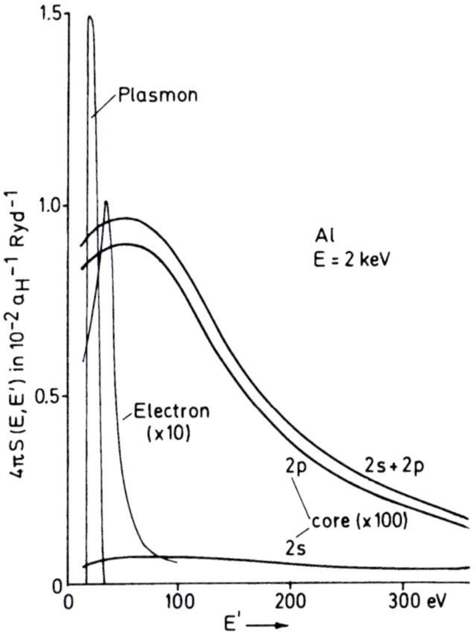

Fig. 4.23. Excitation function $S(E', E)$ as a function of SE energy E' for core electrons ($2s$ and $2p$ electrons), for electron–electron scattering and for plasmon decay in aluminium [4.95], (Ryd denotes the Rydberg constant)

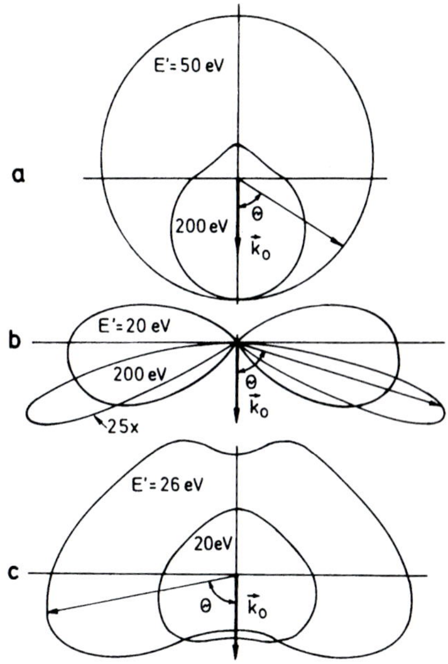

Fig. 4.24. Angular dependence of the excitation function of (**a**) core electrons for $E' = 50$ and 200 eV, (**b**) of electron–electron scattering for $E' = 20$ and 200 eV and (**c**) plasmon decay for $E' = 20$ and 26 eV ($E = 2$ keV, $\boldsymbol{k}_0$ = wave vector of incident electron) [4.95]

4.23; examples of the angular characteristics are shown in Fig. 4.24. Excited plasmons can decay and transfer their energy to conduction electrons near the Fermi level, resulting in the contribution $4\pi S_{\mathrm{pl}}(E, E')$ of Fig. 4.23 with the angular characteristics of Fig. 4.24c. This mechanism gives the largest contribution to the excitation of SE in Al. However, the contribution from electron–electron and core excitations (Figs. 4.23, 24a,b), which are a factor 10 and 100 smaller in magnitude than the plasmon contribution, will not be negligible due to the extension of S_{c} and S_{e} to a larger range of E' values.

These excited hot electrons with energies above the Fermi level are elastically and inelastically scattered before they can reach the surface. Figure 4.25 shows the calculated mean-free-path lengths for these processes. The inelastic mean-free-path length Λ_{inel} is calculated by the dielectric theory and the elastic value Λ_{el} by a random-phase-shift calculation using a muffin-tin model and considering exchange and correlation effects. Since the elastic free-path length is only of the order of 0.3 nm, the scattering of the hot electrons has to be treated by solving a transport equation. The transmission of electrons

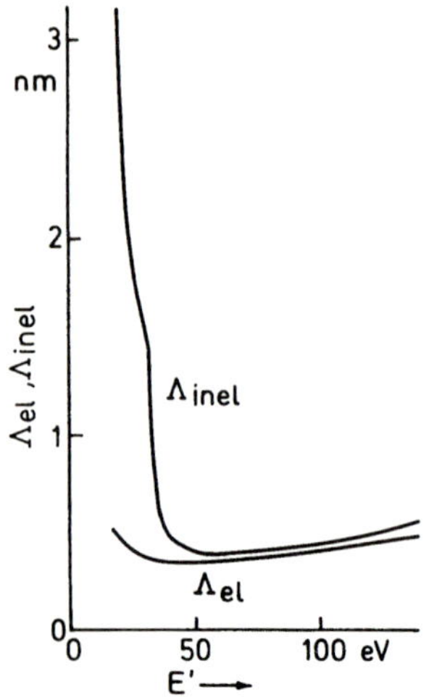

Fig. 4.25. Energy dependence of the mean-free-path lengths for elastic and inelastic scattering in aluminium [4.95]

with exit energies $E_{\mathrm{SE}} = E' - W \geq 0$ can be treated as discussed above by using the critical angle α' of total internal reflection (4.24).

Figure 4.26a shows the emission spectrum (4) of SE finally obtained from the three different contributions (1–3) and Fig. 4.26b the angular distributions, which follow Lambert's $\cos \zeta$ law exactly though the primary excitation is anisotropic (Figs. 4.23a–c). In consequence, the SE yield will not depend on the direction of $\boldsymbol{k}_0$, nor does it differ at the top and bottom of a thin foil, where $\boldsymbol{k}_0$ is directed into and out of the foil, respectively [4.97]. The main reason for this result is the short mean-free-path length Λ_{el} in Fig. 4.25, which results in a total randomization of the directions of the SE inside the specimen.

4.2.3 Contribution of Backscattered Electrons to the Secondary Electron Yield

We discussed in the last section the basic processes that result in SE emission by primary electrons (PE). Owing to the low exit depth t_{SE}, the emission of this group SE1 that can be excited by the PE is concentrated near the electron probe and contributes to the high resolution of the SE mode. However, backscattered electrons also excite secondaries (SE2) on their path through the surface layer of thickness t_{SE} (Fig. 4.27). These BSE of energy E_{B} show a reduced mean energy (Figs. 4.13 and 4.14) and pass the surface layer with exit angles ζ_{B}, relative to the surface normal (Fig. 4.12). The two effects result in a mean number of SE per BSE that is greater by a factor $\beta > 1$ than the number of SE per PE. In a SEM, a further group SE3 is produced by BSE at certain parts of the specimen chamber. Here, we only discuss the number of SE1 and SE2. We use the simple model first discussed in Sect. 4.2.2, which predicts a SE production proportional to $E^{-0.8}$ and to $\sec \zeta_{\mathrm{B}}$.

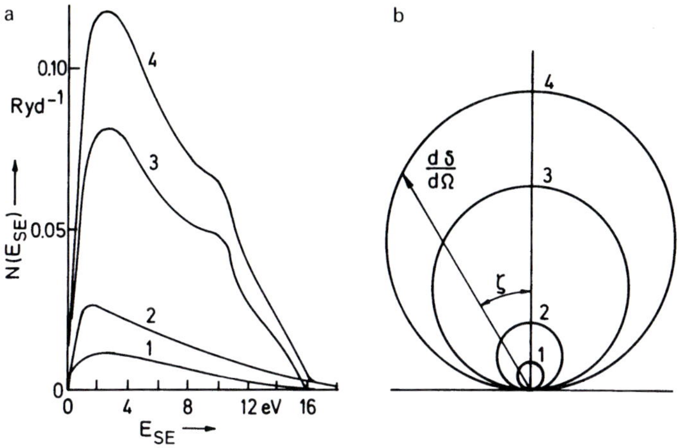

Fig. 4.26. (**a**) Calculated energy distribution of SE with contributions from (1) core electrons, (2) electron–electron scattering, (3) plasmon decay; (4) shows the total distribution. (**b**) Angular distribution of SE for $E' = 2$ eV with the same contributions as in (**a**) [4.95]

The total SE yield, therefore, consists of SE1 excited by the PE and of SE2 excited by the BSE:

$$\delta = \delta_{\mathrm{PE}} + \delta_{\mathrm{BSE}} = \delta_0[\sec\phi + \beta(\phi)\eta] \tag{4.33}$$

where δ_0 denotes the SE1 contribution at normal incidence ($\phi = 0$). The SE2 contribution is proportional to the fraction η of backscattered electrons multiplied by the factor β given by

$$\beta = \frac{1}{\eta E^{-0.8}} \int_0^E \int_\Omega \frac{\mathrm{d}^2\eta}{\mathrm{d}E_\mathrm{B}\mathrm{d}\Omega} E_\mathrm{B}^{-0.8} \sec\zeta_\mathrm{B} \mathrm{d}E_\mathrm{B}\mathrm{d}\Omega\,. \tag{4.34}$$

This factor $\beta = 2$–3 tells us how many more SE2 are excited per BSE than SE1 per primary electron. The sizes of groups 1 and 2 predicted by Monte Carlo simulations are plotted in Fig. 4.20. Group 2 causes the deviations from the $\sec\phi$ law in Fig. 4.20.

At energies below 3 keV, the factor β becomes meaningless since the electron range becomes of the same order as the SE exit depth of $\simeq 3\,t_{\mathrm{SE}}$. A large fraction of electrons is then scattered inside this near-surface zone and SE are generated all along the trajectories inside this zone, so that we cannot know whether the SE are generated by scattered primaries or backscattered electrons.

The angular and energy distributions of the transmitted electrons can also cause a higher SE yield at the exit side of a transparent foil [4.97–100]. All the experiments described in [4.97] can be explained in terms of the proportionality of the SE yield to $E^{-0.8}$ and $\sec\zeta$, where ζ is the angle between the transmitted electrons and the film normal. There is no reason

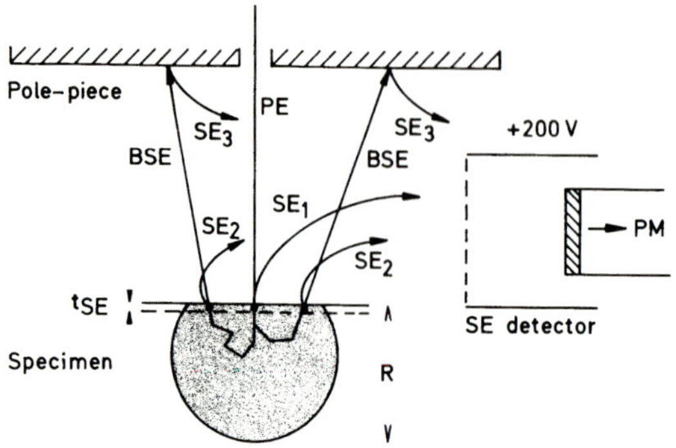

Fig. 4.27. Contributions SE1, SE2 and SE3 to the SE signal excited by the primary electrons, the BSE at the surface and the BSE at the specimen chamber, respectively

to assume that the forward-backward ratio is different from unity [4.101] because the mean momentum of SE is more nearly parallel to the incident electron beam direction $\boldsymbol{k}_0$. We saw at the end of Sect. 4.2.2 that the diffusion of SE excited inside the specimen randomizes such SE momenta.

For high electron energies, the spatial exit distributions of SE1 and SE2 have been imaged directly by means of an immersion lens [4.96]. The exit distributions of BSE and SE given by Monte Carlo simulations are shown in Fig. 4.28 for $E = 1$ and 10 keV at normal incidence (see also [4.102]). Mirror symmetric continuations of the two-dimensional plots of exit probability in the x-y plane should be imagined behind the figure plane. The BSE exit distribution for 10 keV (a) is very broad, of the order of the electron range. The SE distribution (b) shows a strong central peak of SE1, the diameter of the electron probe and the diameter of the excitation volume of the order of t_{SE} contribute to the FWHM. For high resolution we have to take into account the delocalization of SE generation, of the order of 1–2 nm. The SE2 are distributed over a much broader area where the BSE pass the surface and their density in the x-y plane is so low that (b) only shows the central peak. However, Fig. 4.28 also contains integrated distributions which can be obtained by shifting an edge from the left- to the right-hand side:

$$\delta(x) = \int_{-\infty}^{x} \left[\int_{-\infty}^{+\infty} \delta(x, y)\mathrm{d}y \right] \mathrm{d}x \tag{4.35}$$

with a corresponding formula for $\eta(x)$. Now the contribution of SE1 in (b) results in a strong central step with extended fringes caused by SE2. The corresponding distribution (a) of BSE does not of course show this step. When $E = 1$ keV (c,d), a comparison of the scaling shows the strong decrease of the exit diameter due to the decrease of the range R. Because range R and $3\,t_{\mathrm{SE}}$ become comparable, we can no longer distinguish between SE1 and SE2 as discussed below and (d) shows no step caused by SE1 in the integral distribution. The width of the integral distribution between 25% and 75% of the normalized integral distribution can be used as a resolution measure

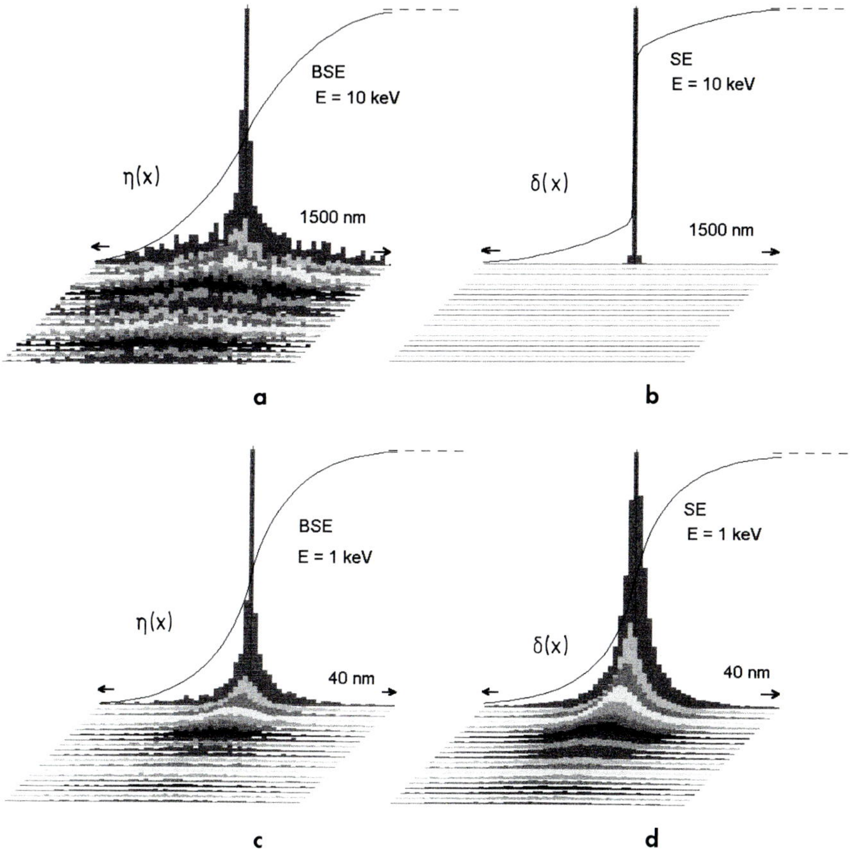

Fig. 4.28. Spatial exit and integrated distributions (4.35) for (**a**) BSE and (**b**) SE from silicon at E = 1 keV and (**c**) and (**d**) BSE and SE at 10 keV, respectively (calculated by the Monte Carlo program MOCASIM [3.135])

and is even smaller for BSE (c) than for SE (d); the latter because some SE excited by BSE at a large distance from the impact can travel a further distance of the order of t_{SE} away from the impact before leaving the surface. At energies beyond 5 keV only the SE1 excited in a volume of the order of t_{SE}^3 contribute to high resolution and the SE2 and BSE contain information from a larger volume of the order of R^3; conversely, there is no significant difference in the information volumes of SE and BSE in LVSEM.

4.2.4 Statistics of BSE and SE Emission

Not only does the primary electron beam exhibit shot noise, which means that the number of PE hitting the specimen during a given pixel time is statistically distributed, but the statistics also increases the noise in SE and

BSE emission. We therefore discuss some of the basic laws of statistics [4.103, 104], which will be reconsidered in Sect. 5.1.5 in connection with the cascading of different detector systems.

We assume that one particle (electron or photon) can randomly generate $i = 0, 1, 2, \ldots$ particles in the subsequent step of an information channel with a probability $P_x(i)$ where $\sum_i P_x(i) = 1$. The mean yield of this stage will be

$$x = \sum_i iP_x(i) \tag{4.36}$$

and the variance of x, which is the square of the standard deviation σ, is

$$\begin{aligned} \operatorname{var}(x) &= \sigma^2 = \sum_i (i-x)^2 P_x(i) = \sum_i i^2 P_x(i) - 2x \underbrace{\sum_i iP_x(i)}_{=x} + x^2 \underbrace{\sum_i P_x(i)}_{=1} \\ &= \sum_i i^2 P_x(i) - x^2 \ . \end{aligned} \tag{4.37}$$

The relative variance is defined by

$$v(x) = \operatorname{var}(x)/x^2 \ . \tag{4.38}$$

Whereas $P_x(i)$ has been introduced as the probability that one incident particle generates i particles, the probability $P_x(n,i)$ is defined to be the probability that n incident particles will result in a total number of particles i:

$$P_x(n,i) = \sum_j P_x(j) P_x(n-1, i-j) \ . \tag{4.39}$$

$P_x(i)$ is called a binomial distribution if only the values $i = 0,1$ are possible. This means that the particles are either absorbed ($i = 0$) or transmitted, backscattered or ejected ($i = 1$), and

$$P_x(0) = 1 - x; \quad P_x(1) = x \tag{4.40}$$

with the mean yield x. The variances become

$$\operatorname{var}(x) = x(1-x); \quad v(x) = (1-x)/x \ . \tag{4.41}$$

For a binomial distribution the relation (4.39) gives

$$P_x(n,i) = \frac{n!}{i!(n-i)!} x^i (1-x)^{n-i} \tag{4.42}$$

with mean yield $y = nx$ and the variances

$$\operatorname{var}(nx) = nx(1-x); \quad v(nx) = (1-x)/nx \ . \tag{4.43}$$

For example, $P_x(2,0) = (1-x)^2$; $P_x(2,1) = 2x(1-x)$; $P_x(2,2) = x^2$ etc.

The binomial distribution degenerates to a Poisson distribution for $y = xn$ if $x \ll 1$ and n is very large:

$$P_y(i) = \frac{y^i}{i!} e^{-y}; \quad \mathrm{var}(y) = y; \quad v(y) = 1/y \ . \tag{4.44}$$

The following equations are relevant to the statistics of cascade processes. If one particle generates i particles in the first stage with a probability $P_{x1}(i)$ and if these i particles enter the second stage and generate j particles with a probability $P_{x2}(j)$ per incident particle, then the probability $P_y(k)$ of generating k particles in the second stage per incident particle in the first stage is given by

$$P_y(k) = \sum_i P_{x1}(i) P_{x2}(i,k) \tag{4.45}$$

with mean yield y and variances:

$$y = x_1 x_2; \ \mathrm{var}(y) = \mathrm{var}(x_1) x_2^2 + x_1 \mathrm{var}(x_2); \ v(y) = v(x_1) + v(x_2)/x_1 \ . \tag{4.46}$$

For m statistical processes, the cascade has the mean yield and variances:

$$y = x_1 x_2 ... x_m; \quad v(y) = v(x_1) + \frac{v(x_2)}{x_1} + ... + \frac{v(x_m)}{x_1 x_2 ... x_{m-1}} \ . \tag{4.47}$$

A cascade of two binomial distributions again results in a binomial distribution with $y = x_1 x_2$ and $\mathrm{var}(y) = y(1-y)$. If a Poisson distribution with mean value x, is followed by a binomial distribution, we get a Poisson distribution with $y = x_1 x_2$ and $\mathrm{var}(y) = y$. However, two Poisson distributions in cascade do not result in a Poisson distribution.

We first apply these general laws to the statistics of the primary electrons. The mean number of electrons per pixel becomes

$$\overline{n}_\mathrm{p} = I_\mathrm{p} \tau / e \ . \tag{4.48}$$

The shot noise may be analysed by the following argument. The time τ for one pixel can be divided into a large number n of time intervals, so that the probability x of observing one electron in one of these time intervals is much less than unity and the probability of observing more than one electron per time interval is negligible. We then expect the mean value of the number of electrons y in the time interval to follow a Poisson distribution (4.44), $y = nx = \overline{n}_\mathrm{p}$ with variance $\mathrm{var}(n_\mathrm{p}) = \overline{n}_\mathrm{p}$. The electron current consists of pulses of charge e and can be Fourier analysed, from which we obtain the rms current of the ac or noise component [4.104].

$$I_{n,\mathrm{rms}} = (\overline{I_n^2})^{1/2} = (2e\Delta f I_\mathrm{p})^{1/2} \tag{4.49}$$

where Δf is the bandwidth of the detection system. Furthermore, equation (4.48) gives the noise amplitude

$$I_{n,\mathrm{rms}} = (e/\tau)[\mathrm{var}(n_\mathrm{p})]^{1/2} = e\overline{n}_\mathrm{p}^{1/2}/\tau = (eI_\mathrm{p}/\tau)^{1/2} \tag{4.50}$$

of the primary current and these equations become identical for $\tau = 1/(2\Delta f)$. The signal-to-noise ratio of the primary electrons becomes

Table 4.2. SE yield δ, noise factor b and increase of noise by a factor $b\,\delta$

	Al 10 keV	20 keV	Au 10 keV	20 keV
δ	0.152	0.088	0.355	0.205
$1/\delta$	6.6	11.4	3.8	4.9
b	8.4	13.5	4.3	6.6
$b\,\delta$	1.3	1.2	1.5	1.5

$$(S/N)_{\mathrm{PE}} = I_{\mathrm{p}}/I_{n,\mathrm{rms}} = \overline{n}_{\mathrm{p}}/[\mathrm{var}(n_{\mathrm{p}})]^{1/2} = \overline{n}_{\mathrm{p}}^{1/2} = (I_{\mathrm{p}}/2e\Delta f)^{1/2}\,. \tag{4.51}$$

The first step of the electron cascade is the electron–specimen interaction. In the case of backscattered electrons, we have a binomial distribution (4.40, 4.41) because the only possibilities are $i = 0$ (primary electron absorbed) and $i = 1$ (electron backscattered) and $x = \eta$ is the backscattering coefficient. The cascade of the Poisson distribution of the PE and the binomial distribution of the BSE becomes a Poisson distribution. Application of (4.44) or (4.46) and (4.41) yields the variance of the BSE emission.

$$\mathrm{var}(\overline{n}_{\mathrm{p}}\eta) = \eta^2\mathrm{var}(n_{\mathrm{p}}) + \overline{n}_{\mathrm{p}}\mathrm{var}(\eta) = \eta^2\overline{n}_{\mathrm{p}} + \overline{n}_{\mathrm{p}}\eta(1-\eta) = \overline{n}_{\mathrm{p}}\eta \tag{4.52}$$

or

$$(S/N)_{\mathrm{BSE}} = \overline{n}_{\mathrm{p}}\eta/[\mathrm{var}(\overline{n}_{\mathrm{p}}\eta)]^{1/2} = (\overline{n}_{\mathrm{p}}\eta)^{1/2} = (I_{\mathrm{p}}\eta/2e\Delta f)^{1/2}\,. \tag{4.53}$$

The distribution of SE emitted is neither binomial nor Poisson because one PE can excite zero, one or more SE with decreasing probability (for example, $i = 2$ when SE are excited by the PE and by its BSE trajectory through the surface exit depth). Equation (4.46) becomes [4.105–107]

$$\mathrm{var}(\overline{n}_{\mathrm{p}})\delta^2[1 + (\mathrm{var}(\delta)/\delta^2)] = \overline{n}_{\mathrm{p}}\delta^2(1+b) \tag{4.54}$$

or

$$(S/N)_{\mathrm{SE}} = \overline{n}_{\mathrm{p}}\delta/[\mathrm{var}(\overline{n}_{\mathrm{p}}\delta)]^{1/2} = [\overline{n}_{\mathrm{p}}/(1+b)]^{1/2} = [I_{\mathrm{p}}/2e\Delta f(1+b)]^{1/2}\,. \tag{4.55}$$

In the case of a Poisson distribution, we would expect $b = 1/\delta$. However, owing to the deviation from this statistics, b becomes larger than this value by a factor 1.2–1.5 for 10–20 keV electrons, see Table 4.2 [4.107].

Let us now use (4.54) to estimate the minimum electron-probe current I_{p} necessary to detect a signal with a signal-to-background ratio $\Delta I_{\mathrm{SE}}/I_{\mathrm{SE}}$ ($I_{\mathrm{SE}} = I_{\mathrm{p}}\delta$). If we assume that all SE are detected without any additional detector noise, a question which will be discussed in Sect. 5.1.5, the noise amplitude ΔI_{n} obtained from (4.50) and (4.54) is

$$\Delta I_{\mathrm{n}} = \frac{e}{\tau}[\mathrm{var}(\overline{n}_{\mathrm{p}}\delta)]^{1/2} = \frac{e}{\tau}\delta\,[\overline{n}_{\mathrm{p}}(1+b)]^{1/2}\,. \tag{4.56}$$

The signal-to-noise ratio becomes

$$\frac{S}{N} = \frac{\Delta I_{\mathrm{SE}}}{\Delta I_{\mathrm{n}}} = \left(\frac{\tau I_{\mathrm{p}}}{e(1+b)}\right)^{1/2} \frac{\Delta I_{\mathrm{SE}}}{I_{\mathrm{SE}}} > k \tag{4.57}$$

and k should be of the order of 3–5 for detectability. For the probe current therefore, we find

$$I_{\mathrm{p}} > \left(\frac{k}{\Delta I_{\mathrm{SE}}/I_{\mathrm{SE}}}\right)^2 \frac{e(1+b)}{\tau} \,. \tag{4.58}$$

Numerical Example: For $e = 1.6 \times 10^{-19}$ C, $b = 10$, $\Delta I_{\mathrm{SE}}/I_{\mathrm{SE}} = 5\%$, $k = 3$ and for $\tau = 4 \times 10^{-4}$ s (500×500 pixels in a frame time of 100 s) the minimum probe current is $I_{\mathrm{p}} = 1.6 \times 10^{-11}$ A. It has been shown in Sect. 2.2.4 that such a probe current is too high to reach the smallest possible probe size.

5. Electron Detectors and Spectrometers

The most effective detection systems for secondary electrons (SE), which has a low noise level and a large bandwidth, is the Everhart–Thornley detector. Electrons are collected by a positively biased grid in front of a scintillator biased at +10 kV. The light emission is recorded by a photomultiplier tube. Scintillation detectors can also be used for backscattered electrons (BSE) when the solid angle of collection is increased. Other alternatives for BSE are semiconductor detectors, microchannel plates, or the conversion of BSE to SE. The recording of SE in the in-lens mode or in LVSEM needs special detector designs for rotationally symmetric collection fields. For the separation of different types of contrast and for more quantitative signals, improvements can be obtained by employing multiple detector systems.

Energy filtering of SE is important for the measurement of surface potentials and of BSE for increasing the resolution by studying low-loss electrons or for enhancing those contrast effects that are produced in a thin surface layer or by applying an energy window to the BSE energy spectrum. BSE spectrometers can be used for studying multilayer systems.

5.1 Measurement of Small Currents

The simplest signal for imaging is the specimen current (see applications in Sect. 6.2.6). It is also useful to measure the electron-probe current quantitatively, for the discussion of statistical noise (Sect. 5.5) and the control of electron-probe current stability, for example. The specimen current that flows from an insulated specimen to earth is not exactly equal to the probe current because parts of the electron beam leave the specimen as SE and BSE. If the sum of the SE and BSE yields is larger than the incident current, $\sigma = \eta + \delta > 1$, a negative specimen current can even be recorded (Fig. 6.18). However, the true electron-probe current can be measured by mounting an aperture diaphragm with a small bore inside a hole in the specimen stub. Then BSE and SE generated inside the hole cannot escape through the small bore and the diaphragm acts as a miniature Faraday cage. A further precaution to reduce BSE and SE emission at the bottom is to apply a carbon coating.

Electron-probe and specimen currents are of the order of $I_\mathrm{p} = 10^{-13} - 10^{-10}$ A for high resolution and 10^{-10}–10^{-7} A for x-ray microanalysis. Such small electron currents can be measured by commercial electrometers. These consist of a FET (field-effect transistor), which has to work with a high input impedance $R > 1$ MΩ, so that the voltage drop $U = RI_\mathrm{p}$ is of the order of a few millivolts. Many SEMs are equipped with an electrometer to record the specimen current as an image signal and this can also be used for quantitative measurement of the specimen current if it is suitably installed. The sum C of the capacitance between the specimen stub and the chamber, that of the coaxial cable and the input capacitance of the FET gives a time constant $\tau = RC$, so that fast recording requires a low RC product. To keep the capacitance small, the FET should be mounted as close as possible to the specimen stub [5.1, 2]. With $C \simeq 10$ pF we find

$$\begin{aligned}&\tau = 0.1\ \mathrm{s}\quad \text{with } R = 10\ \mathrm{G}\Omega \text{ for } I_\mathrm{p} = 10^{-12}\ \mathrm{A} \text{ and } U = 10\ \mathrm{mV} \text{ and}\\ &\tau = 10\ \mu\mathrm{s}\ \text{ with } R = 10\ \mathrm{M}\Omega \text{ for } I_\mathrm{p} = 10^{-9}\ \mathrm{A}\ \text{ and } U = 1\ \mathrm{mV}\quad .\end{aligned}$$

In the latter conditions, a slow-scan specimen-current image can be recorded. However, the bandwidth is limited to 100 kHz for a specimen current of $10^{-8} - 10^{-6}$ A and 15 kHz for $10^{-11} - 10^{-9}$ A can be realized [5.3].

5.2 Scintillation Detectors and Multipliers

5.2.1 Scintillation Materials

Scintillators convert electrons to photons by cathodoluminescence (Sect. 7.2). An average of 10–15 photons are generated by a 10 keV electron. Most of the electron energy absorbed during electron excitations is converted to heat by radiationless transitions.

Luminescence centres are excited in scintillator materials. The decay of excitation is exponential, $n = n_0 \exp(-t/\tau)$, with a decay time τ that must be less than 20 ns when scanning at TV frequencies. ZnS and CdS scintillators, which are used for fluorescent screens, have a high light efficiency but their decay times are of the order of milliseconds. They cannot be used for recording images, for which decay times less than 1 μs for 1 s frame time and 10^6 pixels per frame are essential. A great many scintillating materials with decay times less than 10 ns are in use for γ-spectroscopy in nuclear physics. However, some show an afterglow with a very long decay time. This does not affect the analysis of pulse-heights by a multichannel analyser but, in the analogue recording mode necessary for recording an image of low noise amplitude, the afterglow sums up to a varying background signal.

The maximum light emission should lie in the range $\lambda = 300$–600 nm, which is the range of the normal S20 photocathode used in photomultipliers. The efficiency of luminescence should not be decreased by radiation damage.

Organic scintillators are sensitive to radiation damage (Sect. 3.5.3) whereas the lifetime of inorganic scintillators is mainly limited by ion damage and contamination. A scintillator material for SEM should be easily machinable so that scintillators of the required shape can be formed and damaged surface layers repolished. The material should be heat-resistant if bake-out in a UHV chamber is likely to be necessary.

Several scintillation materials are in use in SEM. Plastic scintillators (NE 102A of Nuclear Enterprises, for example) have a decay time of the order of 2 ns. They can be machined easily. Radiation damage decreases the lifetime to the order of one month of daily use. However, these detectors can easily be regenerated by repolishing and recoating with a conductive layer, because only a thin film of the order of the electron range $R = 10\ \mu$m is damaged at $E = 10$ keV. The scintillation material can also be dissolved in ultra-pure toluene or ethyl acetate so that the end of a perspex light-pipe can be coated with a thin layer of this material [5.4, 5].

Lithium activated glass [5.6] and CaF_2:Eu [5.7] have been proposed as inorganic scintillators. The most frequently used material is P-47, which is employed for the blue component of TV colour tubes and consists of cerium-doped yttrium silicate ($\tau = 70$ ns, $\lambda_{max} = 550$ nm). Several methods of preparing P-47 powder layers have been described [5.8–11]. The optimum thickness of the powder layer will be of the order of the electron range. The number of light quanta entering the light-pipe can be doubled by coating the sedimented powder layer with an aluminium film floated from a glass slide on a water surface [5.11, 12].

Autrata et al. [5.13–15] succeeded in growing large single crystals of cerium-activated yttrium aluminium garnet YAG:Ce^{3+} and yttrium aluminium perovskite (YAP:Ce^{3+}) with the composition $YAlO_3$ ($\tau = 30$ ns, $\lambda_{max} = 370$ nm). These are at present the best available scintillator materials because single-crystal scintillators are more uniform and light-transparent. The light emission shows a multiexponential decay with a weak component of large τ [5.16]. At present, thin-layer powder and single-crystal scintillators are mainly used.

5.2.2 Scintillator–Photomultiplier Combination

An excellent and widely used detector for secondary electrons is the scintillator–photomultiplier combination (Fig. 5.1) first proposed by *Everhart* and *Thornley* [5.17] for SEM and often known as the Everhart–Thornley detector. The secondary electrons (SE) generated by the primary electrons (PE) are collected by a grid biased between +100 and +200 V or by an equipotential surface formed by an earthed cup and the bias of 10 kV at the scintillator behind it. However, the SE collected are not only those generated at the specimen surface but also those created by backscattered electrons (BSE) at the polepiece and other parts of the specimen chamber (Sect. 5.4.1). The SE are not collected when the bias of the collector grid is more negative than the

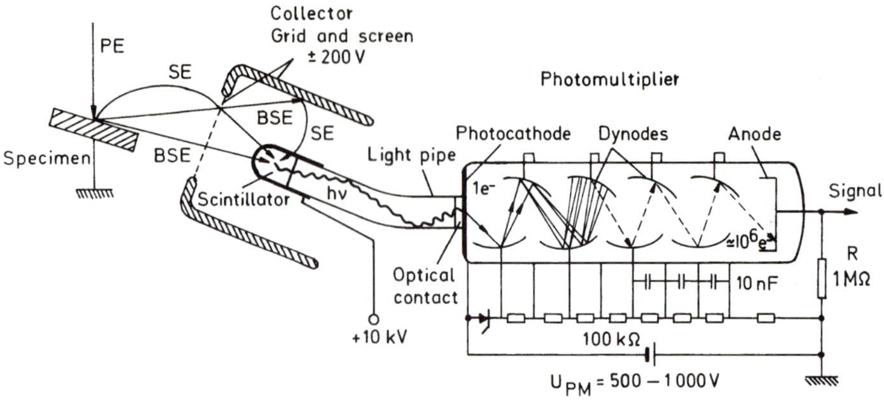

Fig. 5.1. Scintillator–photomultiplier combination (Everhart–Thornley detector) for recording secondary electrons

specimen. In this case only BSE can reach the scintillator on approximately straight trajectories because of their higher exit energies. This detection mode is however very inefficient for BSE because the solid angle of collection is too small. Better detectors for BSE using scintillator–photomultiplier combinations are discussed in Sect. 5.4.2.

The SE that pass through the collector grid are accelerated to the scintillator with a bias of 10 kV at the conductive coating, which can be an evaporated film of aluminium of the order of a few tens of nanometres thick or a conductive indium–tin oxide layer [5.18]. The accelerated electrons can penetrate this metal coating but the latter is thick enough to reflect and absorb light quanta which come from the thermionic cathode or from cathodoluminescence of the specimen. The current that falls on the scintillator with an energy $E_{sc} \simeq 10$ keV will be given by

$$I_{SE} = I_p \delta f_{SE} \tag{5.1}$$

where I_p denotes the electron-probe current, δ the secondary electron yield and f_{SE} the collection efficiency of the detector for SE (see also Sect. 5.4.1).

The SE accelerated to $E_{sc} = 10$ keV can produce a large number of electron–hole pairs inside the scintillator. A fraction η_{sc} will be backscattered. The mean energy needed to produce an electron–hole pair is of the order of $E_P \simeq 3$ eV, so that of the order of $E_{sc}/E_P \simeq 3000$ pairs are produced per collected SE. The recombination of electron–hole pairs in luminescence centres generates light quanta but a large fraction recombine without radiating and only a fraction $q_{sc} = 1$–3% of the ratio E_{sc}/E_P will be converted into light quanta, on average

$$\overline{m} = (1 - \eta_{sc}) q_{sc} E_{sc}/E_P \tag{5.2}$$

in number. The light quanta are guided by reflection at the metal coating and by total reflection in the light-pipe to the photomultiplier. Only a fraction

r of the light quanta will enter and pass through the light-pipe. The latter is made of perspex or quartz glass and should not be metal-coated because multiple reflection at a metal surface causes a stronger absorption whereas the total internal reflection at the light-pipe–vacuum boundary becomes 100% for angles of incidence smaller than $\theta_t = \arcsin(1/n_r)$, where n_r is the light optical refractive index of the light-pipe material; for $n_r = 1.5$, $\theta_t = 42°$ for example (see also Sect. 7.2.2). The shape of the light-pipe can be optimized by a Monte Carlo simulation of the light reflection [5.19].

A photoelectron will be produced at the photocathode of the photomultiplier tube (PM) with a quantum efficiency $q = 5$–20%. The optical contacts at the interfaces between scintillator, light-pipe and photomultiplier should have a low reflectance by using a cement with a medium refractive index [5.14]. The photoelectrons are accelerated to the first dynode at a bias of +100 V relative to the photocathode. In the design shown in Fig. 5.1, this voltage is kept constant by a Zener diode, whereas the biases of the other dynodes are generated by the voltage drops of the photomultiplier voltage $U_{\mathrm{PM}} = 500$–1000 V along the resistor chain. The photoelectrons are collected by the first dynode with a probability c close to unity. The dynodes consist of a material of high secondary electron yield $\delta_{\mathrm{d}} = 2$–10. The successive stages of acceleration and generation of secondary electrons at $n = 8$–10 dynodes result in a gain $G_{\mathrm{PM}} = \delta_{\mathrm{d}}^n$ of the multiplier. By altering U_{PM}, the gain G_{PM} can be varied over several orders of magnitude (Fig. 5.17). The pulse of G_{PM} electrons per generated photoelectron causes a voltage pulse at the resistor $R = 100$ kΩ–1 MΩ in the connection line to the final collector electrode which can be further amplified in the subsequent pre-amplifier. The pulse is large enough for single photoelectrons to be counted [5.20] and, for very low probe currents especially, images show a better signal-to-noise ratio because the noise due to smaller afterpulses (see below) can be suppressed by pulse-height discrimination. A gain of 2–3 in signal-to-noise ratio can be obtained by this method [5.21]. However, the system is normally not used in the pulse-counting mode; instead an analogue video signal of magnitude

$$U_{\mathrm{V}} = I_{\mathrm{p}} \delta f_{\mathrm{SE}}\, \overline{m}\, r\, q\, c\, G_{\mathrm{PM}} R = I_{\mathrm{p}} \delta f_{\mathrm{SE}} G_{\mathrm{D}} R \tag{5.3}$$

is recorded, where G_{D} represents the gain of the whole detector system. The mean number of photoelectrons per incident SE will be mainly influenced by the factor $\overline{m}q$. With the typical values given above, of the order of 1–10 photoelectrons will be generated per SE that arrives at the scintillator. Though this detector system is inefficient for the conversion of electron–hole pairs into light quanta and of light quanta into photoelectrons, it has the lowest noise, an aspect which will be discussed in detail in Sect. 5.5.1.

If the decay time of the scintillation material is less than a few nanoseconds, the time constant will be limited by the drift-time of the electrons through the photomultiplier, which is of the order of 10 ns. This detector system therefore forms a wideband amplifier, which converts all spatial frequencies into a video signal even when working at TV scan rates.

The signal of a photomultiplier is not completely free of a memory effect, reflecting earlier irradiation. As well as the main pulses generated by photons at the photocathode, after-pulses with delay times between 20ns and 10 ms may occur [5.22–25]. Ions of the residual gas, generated by the main pulse, can be accelerated to the photocathode and generate several SE. The delay is less than 40 ns and they represent of the order of 10^{-4} to 10^{-2} of the number of main pulses. The main pulses can also generate photons by ionization of the atoms of the residual gas and by cathodoluminescence at the dynodes or at the glass walls; these can excite photoelectrons at the photocathode with a delay time less than 10 ms. Surface effects at the dynodes and the photocathode can also liberate single electrons. These effects may create about 0.2 after-pulses per primary photoelectron.

5.2.3 Electron Channel Multipliers

Channeltron. An electron channel multiplier (channeltron) is a special type of electron multiplier tube, which is not sealed in vacuum like a photomultiplier; instead, the secondary electrons are excited at the input by direct ion or electron impact. These multipliers take the form of tubes about 1 mm in diameter. A bias of about 1 kV causes a continuous voltage drop along the highly resistive inner wall of the tube and SE can be generated and accelerated, thus forming a cascade. The entrance of the channeltron can be biased at +200 V to collect the SE and the distance of the detector from the specimen can be reduced to a few millimetres. To avoid amplifying the exit current at the high floating bias of 1 kV, the current at the entrance to and flowing through the channeltron can be amplified at earth potential [5.26].

At a first glance, this type of multiplier seems to be attractive, because the scintillator and its 10 kV bias, the light-pipe and the photocathode are not necessary. However, channeltrons have the following disadvantages that discourage their application in SEM. The SE yield at the entrance decreases with increasing electron energy, and backscattered electrons can only be detected with a signal contribution inversely proportional to their energy. The frequent exposure to air and the presence of hydrocarbons in the residual gas cause considerable contamination, especially at the end of the tube. We must not forget that the detector was originally conceived for single-particle counting, not to produce the strong analogue signal that is necessary for image recording. We shall see in Sect. 5.5.1 that the signal-to-noise ratio of a scintillator-photomultiplier combination is nearly ideal and indeed no significant improvement of signal-to-noise ratio has been found for channeltrons [5.27].

Microchannel Plate. Another type of channel multiplier is the microchannel plate (MCP) [5.28, 29], which consists of a slice from a bundle of tightly packed, fused tubes of lead-doped glass with an inner diameter of 10–20 μm and a resistance of 10^8–10^9 Ω over their length. An annular disc-shaped channel plate with a thickness of only 3–4 mm can be mounted below the polepiece, and a working distance of $\simeq$10 mm can be used. Biasing of the front plate at +100 V collects SE and BSE, at −20 V, BSE only. The incident electrons produce SE at the inner tube walls; these SE are accelerated by the continuous voltage drop along the tube with a bias of 1 kV or 3 kV for a three-plate unit. The gain is better than 10^4 and may even reach 10^8 in a three-plate unit and electron probe currents of $I_p = 10^{-13}$–10^{-10} A can thus be employed. At this signal level no degradation in performance has been observed with a clean vacuum. The MCP starts to become non-linear above 1 nA and saturates at 10–100 nA. The anode plate and a preamplifier at a potential of 1–3 kV are electrically insulated by an optical decoupler or by transmitting a modulated 30 MHz signal. This amplification system limits the bandwidth to about 2 MHz.

Because the SE signal shows a maximum $\delta > 1$ at 300–800 eV, MCPs are very efficient for the detection of low-energy BSE. For this reason, MCPs are of interest for IC metrology (Sect. 8.8) with low currents in the LVSEM range of 0.5–5 keV where semiconductor detectors decrease in sensitivity.

Another application of the MCP is as a position-sensitive detector, especially when followed by a fluorescent screen and a CCD array. This allows us to observe and measure the deflection of the primary beam by magnetic stray fields [5.30], for example, or to observe a LEED pattern in ultrahigh vacuum.

5.3 Semiconductor Detectors

5.3.1 Surface Barrier and p-n Junction Diodes

In a semiconductor, an electron energy E produces a mean number $\overline{n} = E/\overline{E}_i$ of electron–hole pairs, where the mean energy per excitation, $\overline{E}_i = 3.6$ eV in silicon. These charge carriers can be separated before recombination and, as a consequence, an external charge-collection current I_{cc} is generated. Two types of detectors are in use, based on the surface-barrier or Schottky diode (Fig. 5.2a) and the p-n junction diode (Fig. 5.2b). In both cases a depletion layer of width w is formed, which is free of majority carriers and contains a space charge density $\rho = eN_D^+$ or $\rho = -eN_A^-$ in n and p type material with concentrations of N_D^+ donor and N_A^- acceptor atoms per unit volume, respectively. The Fermi levels on each side are at the same potential. This results in a diffusion potential or built-in voltage, $U_d < \Delta E$ (band gap), which guarantees a zero net current across the junction. The diffusion current that results from thermal diffusion of minority carriers (holes in n-type and electrons in p-type material) across the junction is decreased by the opposition of the electric field gradient in the junction; it thus becomes equal to the reverse current caused by charge collection of thermally activated electron–hole pairs inside the junction.

The width of the junction increases with increasing reverse, external bias U_e (Fig. 5.2a). The Poisson equation

$$\nabla^2 \Phi = -\rho/\epsilon\epsilon_0 \tag{5.4}$$

can be used to calculate the potential distribution for a Schottky diode with $\rho = eN_D^+$, N_D^+ being the concentration of uncompensated donor atoms, for example. We need only consider the one-dimensional Poisson equation

$$\frac{\partial^2 \Phi}{\partial z^2} = -eN_D^+/\epsilon\epsilon_0 \, . \tag{5.5}$$

A first integration results in the electric field strength

$$E_z = -\partial\Phi/\partial z = eN_D^+(z-w)/\epsilon\epsilon_0 \quad \text{for} \quad z \le w;\ E_z = 0 \quad \text{for} \quad z > w \tag{5.6}$$

and a second integration, between $z = 0$ and w, in the potential difference

$$U_d - U_e = eN_D^+ w^2/2\epsilon\epsilon_0 \quad \text{for} \quad U_e < U_d \tag{5.7}$$

which can be solved for w

$$w = \left[\frac{2\epsilon\epsilon_0}{eN_B}(U_d - U_e)\right]^{1/2} \tag{5.8}$$

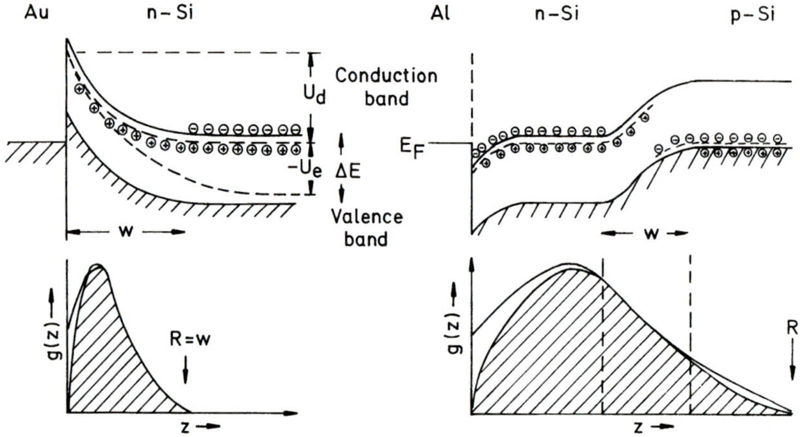

Fig. 5.2. Potential variation for (**a**) a surface barrier (Schottky) diode and (**b**) a p-n junction with an ohmic contact; the fraction of the depth distribution $g(z)$ of the electron–hole pairs generated that contributes to the charge-collection current I_{cc} (shaded area) as a result of charge collection efficiency $\epsilon_c \simeq 1$ in the depletion layer, of width w, and $\propto \exp(-x/L)$ outside the depletion layer and an increased surface recombination rate is also shown

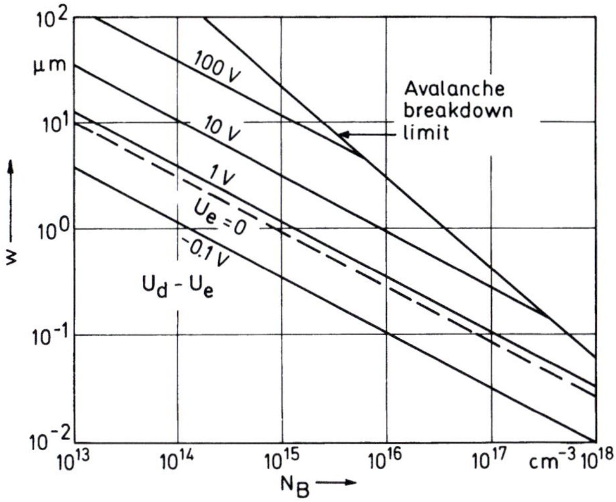

Fig. 5.3. Dependence of the depletion layer width w on an external, reverse bias $-U_\mathrm{e}$ and on the concentration N_B of uncompensated fixed charges of donor and acceptor atoms (U_d = diffusion voltage)

where $N_\mathrm{B} = N_\mathrm{D}^+$ for a Schottky and $1/N_\mathrm{B} = 1/N_\mathrm{D}^+ + 1/N_\mathrm{A}^-$ for a p-n junction. This dependence of w on U_e and N_B is plotted in Fig. 5.3. Avalanche breakdown limits the increase of width with increasing U_e to the point at which the field strength $|\boldsymbol{E}| = 2(U_\mathrm{d} - U_\mathrm{e})/w$ inside the junction exceeds a value of the order of 10^4 V cm^{-1}. With $U_\mathrm{e} = -10$ V and dopant concentrations N_B less than 10^{14}, it is possible to produce widths larger than 10 μm, which is the range of 30 keV electrons.

Both the Schottky and the p-n junction diodes form a non-ohmic contact with the rectifying current-voltage characteristics

$$I = I_\mathrm{s}\left[\exp(eU_\mathrm{e}/kT) - 1\right] - I_\mathrm{cc} \tag{5.9}$$

plotted in Fig. 5.4. The dark-current I_s for reverse biasing ($U_\mathrm{e} < 0$) results from the generation of charge carriers by thermal activation. The charge collection current I_cc results in a shift of the current–voltage characteristics.

The magnitude of the charge collection current becomes

$$I_\mathrm{cc} = I_\mathrm{p}(1 - \eta_\mathrm{c})[(E - E_\mathrm{th})/\overline{E}_i]\epsilon_\mathrm{c} = I_\mathrm{p} G \epsilon_\mathrm{c} \tag{5.10}$$

where η_c takes into account the loss of electron–hole pair generation by electron backscattering, which is only of the order of 10% for silicon. I_p denotes the incident electron current and G the gain. The threshold energy E_th of the order of a few keV, is determined by the absorption and energy loss of electrons inside the front metal coating. The charge-collection efficiency ϵ_c will be discussed below. Experimental values of the gain G for different semiconductor detectors are shown in Fig. 5.5. As in the scintillation detector, the signal contribution is proportional to the BSE energy, so that high-energy BSE contribute to the signal with a larger gain. This will be advantageous for the recording of electron channelling patterns, for example.

The Schottky diode has the advantage that the depth distribution $g(z)$ of electron–hole formation (Fig. 3.26) can lie completely inside the depletion layer when the electron range $R < w$ (Fig. 5.2a). The charge collection efficiency ϵ_c is near unity. A p-n junction, which is formed by doping n-type silicon with acceptor atoms by diffusion or ion-implantation, lies below the surface. For good charge collection, the depletion layer should lie at the maximum of the depth distribution $g(z)$, which will not be possible for all electron energies. Minority carriers excited at a distance $x = |z - z_d|$ from the junction will reach the depletion layer with a probability $\exp(-x/L)$ where L is the diffusion length of the minority carriers, which is normally larger than w. Both Schottky and p-n junction diodes show an effect due to the increased recombination rate at the surface. The charge-collection efficiency ϵ_c is the shaded fraction of the area under $g(z)$ in Fig. 5.2 and will be discussed in detail in Sect. 7.1.2.

Commercially available semiconductor detectors are either annular or semi-annular devices of the surface-barrier type [5.31], which only need a non-ohmic contact layer of gold on a high-resistivity n-type silicon wafer, for example, or of the p-n junction type (Fig. 5.6) produced by ion implanta-

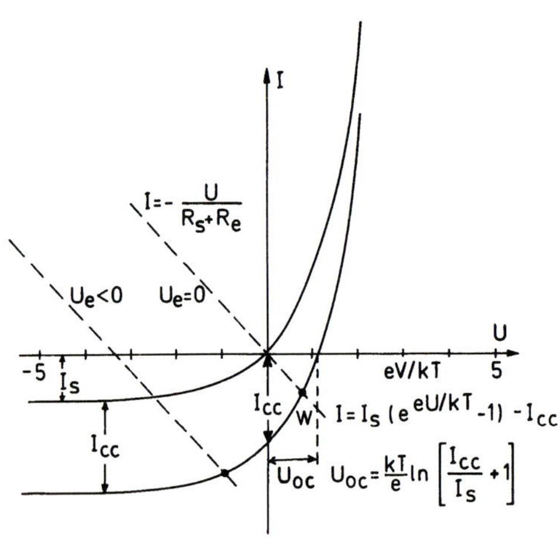

Fig. 5.4. Current–voltage characteristics of a semiconductor detector without (I_{cc}=0) and with electron irradiation ($I_{cc} \neq 0$). I_{cc} = charge-collection current (short-circuit), U_{oc} = open-circuit photovoltage, W = working point with an external resistor R_e, R_s = internal resistance of the semiconductor

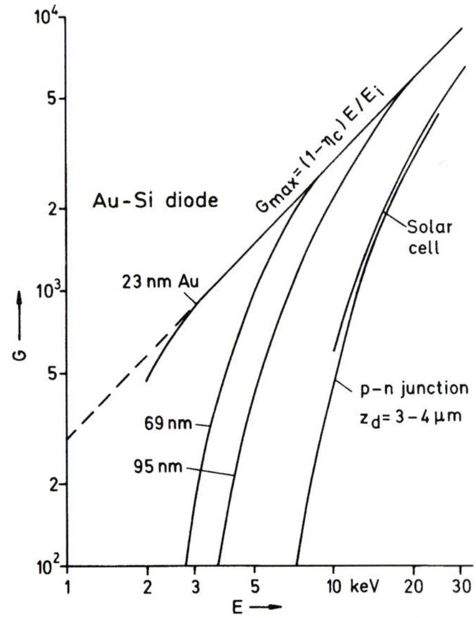

Fig. 5.5. Measured gain G versus electron energy E of a Au-Si Schottky diode with different thicknesses of the Au contact layer [5.32], of a p-n junction 3–4 μm below the surface [5.33] and of a solar cell [5.34]

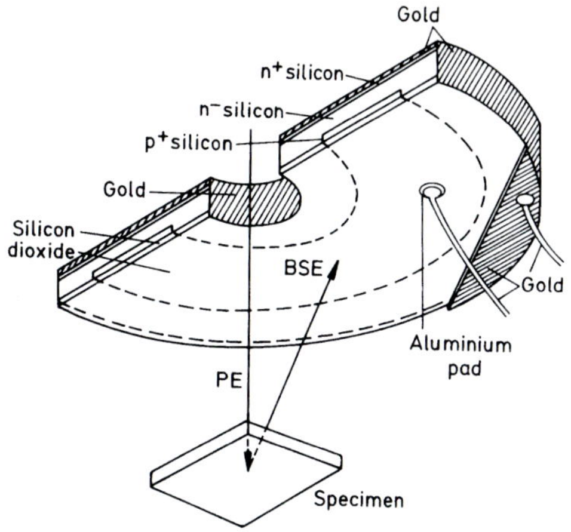

Fig. 5.6. Cross-section through an annular p-n junction diode prepared by ion implantation and having a SiO_2 protection layer [5.35]

tion [5.35]. The Al contact layer on n-type silicon forms an ohmic contact. Such annular or semi-annular detectors can be mounted below the polepiece to record BSE with a large solid angle of collection. Surface barrier diodes are more sensitive to mechanical damage, which can cause short-circuiting of the surface barrier; they can also be destroyed by very high electron-beam currents, whereas the p-n junction, which effects the charge collection, lies a few micrometres below the surface and is better protected against damage. Damage in the Al contact layer does not destroy the ohmic contact to the semiconductor.

Small silicon solar cells, available in sizes $2\times5\times1$ up to $10\times10\times1$ mm^3 with shallow p-n junctions, also form very useful and inexpensive detectors once the lacquer outer layer has been removed. Being more versatile, they can be placed in positions inaccessible to the relatively compact and encapsulated annular detectors [5.1, 34, 36].

5.3.2 The Electronic Circuit of Semiconductor Detectors

Though the gain of the semiconductor detector is of the order of a few thousands, special care must be taken over the preamplifier. The charge-collection current I_{cc} or electron-beam-induced current (EBIC) is a short-circuit current. In the case of an open circuit, the diffusion voltage U_d decreases, which results in a photovoltage or an electron-beam-induced voltage (EBIV) $U_{oc} < U_d$ that saturates at high radiation levels. In reality, a mixture of both effects will be observed, depending on the sum of the internal bulk resistance R_s of the semiconductor and the external resistance R_e. The working point W on the $I-U$ characteristics of Fig. 5.4 is the point of intersection of the $I-U$ curve with a straight line through the origin of slope $-(R_s+R_e)^{-1}$ or a shifted line when a reverse negative bias U_c is applied. The preamplifier therefore, has to be a low-current, low-impedance amplifier.

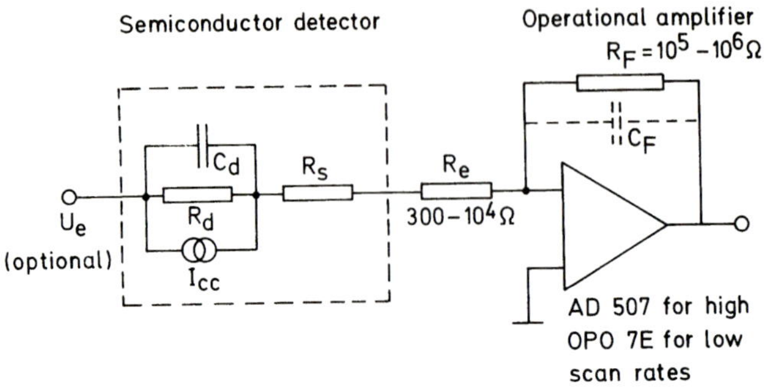

Fig. 5.7. Electronic circuit of a semiconductor detector system. I_{cc} = charge collection current, R_{d}, C_{d} = resistance and capacitance of the depletion layer, R_{s} = resistance of the semiconductor, R_{e} = input resistance and R_{F} = feedback resistor of the preamplifier, U_{e} = reverse bias of the diode (optional)

In order to establish the high-frequency limit of the detector, we have to consider that the equivalent circuit consists of a parallel arrangement of the current source I_{cc} and the resistance R_{d} and capacitance C_{d} of the depletion layer in series with the bulk resistance R_{s} of the semiconductor (Fig. 5.7). The latter can often be neglected or combined with the external resistance R_{e}. The capacitance is given in terms of the width w and area A of the depletion layer by the formula for a parallel-plate capacitor

$$C_{\mathrm{d}} = \frac{\epsilon\epsilon_0 A}{w} \ . \tag{5.11}$$

Numerical Example: For $\epsilon_0 = 8.8544\times10^{-12}$ F m^{-1}, $\epsilon = 11.7$ for Si, A = 1 cm^2, w = 20 μm we find $C_{\mathrm{d}} \simeq 0.5$ nF.

Because I_{cc} charges C_{d}, which is then discharged across R_{e}, the time constant $\tau = R_{\mathrm{e}}C_{\mathrm{d}}$ limits the maximum frequency, which for $R_{\mathrm{e}} = 10$ kΩ and $C_{\mathrm{d}} = 0.5$ nF becomes 200 kHz. The amplification of the operational amplifier depends on the ratio $R_{\mathrm{F}}/R_{\mathrm{e}}$. For too large a ratio $R_{\mathrm{F}}/R_{\mathrm{e}}$ the amplifier can become unstable and begin to oscillate; this also depends on the capacitative impedance of the detector. An additional C_{F} in the feedback loop of the amplifier can in some cases decreases the tendency to oscillate.

For the recording of material contrast arising from the dependence of the backscattering coefficient on the mean atomic number (Sect. 6.2.1), which is the main application of semiconductors in the BSE mode, electron-probe currents of $10^{-10} - 10^{-9}$ A can be used without loss of resolution and in this case the detection and amplifier system remains perfectly stable with the values mentioned above. The main source of noise is the shot-noise of the BSE, and the high electron-probe current guarantees a high signal-to-noise ratio so that even small differences in the backscattering coefficient can be detected.

For frame times of 1 s or TV scan rates, it is necessary to reduce R_{e} and C_{d}. R_{e} can be decreased to a few hundreds of ohms if I_{cc} is correspondingly large, which means that the primary electron energy and the BSE current must both be high; this can be achieved by increasing I_{p} and/or the solid angle of collection. The capacitance C_{d} of the depletion layer can be decreased by reducing the detector area A and increasing the width w (5.8). If, however, A is decreased, so too is the solid angle of collection if the distance between specimen and detector is not correspondingly reduced. The width w can be increased by reverse biasing of the diode. By paying attention to all these points, it becomes possible to work at TV scan rates [5.31, 36, 37].

5.4 Secondary and Backscattered Electron Detectors

5.4.1 Secondary Electron Detectors

Scintillator–photomultiplier combinations are the most effective detectors for SE, though a semiconductor detector for SE can also be designed by accelerating the SE to energies higher than 10 keV [5.38]. The conventional arrangement of an Everhart–Thornley detector mounted on one side of the specimen chamber has already been described in Sect. 5.2.2.

It is an advantage of the Everhart–Thornley detector that the low-energy SE with exit energies $E_{\mathrm{SE}} \leq 50$ eV can be collected with a high efficiency by a collector grid biased between +100 and +300 V. The collection efficiency is nevertheless less than 100%. Figure 5.8 shows measured equipotentials between the collector and the specimen. The lines of force $\boldsymbol{F}$ of this electrostatic field are at each point parallel to those of the electric field $\boldsymbol{E}$ which is the gradient of the potential Φ ($\boldsymbol{F} = -e\boldsymbol{E} = +e\nabla\Phi$). The conductive surface of the specimen is at a uniform potential ($\Phi_{\mathrm{s}} = 0$, for example) and the lines of force are thus normal to the surface. The SE are accelerated in this field and their kinetic energy will be the sum of their exit energy E_{SE} and the local difference in potential energy $V = e\Phi$. The conservation of energy $E + V =$ const results in

$$E_{\mathrm{kin}} = E_{\mathrm{SE}} + e(\Phi - \Phi_{\mathrm{s}}). \tag{5.12}$$

With increasing kinetic energy, the radii of curvature of the electron trajectories produced by the component of $\boldsymbol{E}$ normal to the trajectories increase as E_{kin} and the SE trajectories do not follow the lines of force. The collector grid in front of the scintillator (Fig. 5.1) can be omitted when the 10 kV bias of the scintillator is shielded by a cylindrical earthed electrode, so that an equipotential of a few hundreds of volts lies near the end of this electrode.

With the specimen stub close to the polepiece and zero tilt, an increasing number of SE hit the polepiece when the distance between the electron impact point on the specimen and the detector is increased and when the momentum component of the SE away from the detector increases (Fig. 5.8). This can be observed as an intensity gradient across the specimen stub at low magnification. Furthermore, SE that pass through the gap between specimen

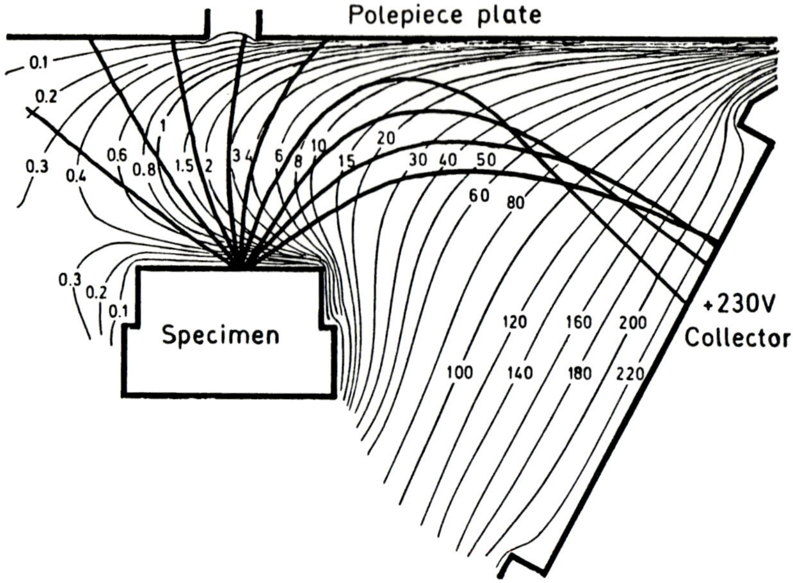

Fig. 5.8. Equipotentials and trajectories of 3 eV SE between the specimen and the positively biased grid of an Everhart–Thornley detector

and polepiece can follow trajectories that miss the collector. The fraction of SE collected depends on the design of the specimen chamber and detector, on the working distance and on the tilt of the specimen stage.

Figure 5.9 shows measurements of the video signal as a function of the specimen tilt ϕ [5.39]. With a positively biased collector grid (+200 V) the main contribution to the signal comes from the SE collected but some BSE, which strike the scintillator directly though with a small solid angle of collection, are also present. A negatively biased grid (–50 V) suppresses the SE contribution and only the BSE contribution remains; this is plotted in Fig. 5.9 at a five-fold larger scale and is subtracted from the SE+BSE curve to get the pure contribution of SE to the video signal. This curve is quite different from the $\delta(\phi)$ curve plotted in Fig. 4.20. The BSE signal shows a maximum due to the increase of the differential backscattering coefficient $d\eta/d\Omega$ with increasing ϕ; beyond this point, it decreases as a result of the reflection-like angular characteristics of the BSE at large ϕ (Fig. 4.12). The remaining SE signal in Fig. 6.2 also contains a large number of SE that have been generated by the BSE at the polepiece and other parts of the specimen chamber (group SE3 in Fig. 4.27). Because of multiple backscattering, the BSE fill the whole specimen chamber like bouncing balls. This multiple backscattering can be suppressed by placing a thin carbon foil below the polepiece, which has the further advantage of suppressing the generation of Fe Kα x-ray quanta for x-ray microanalysis. The SE contribution from the polepiece can also be suppressed by inserting a flat grid in front of a positively biased metal plate

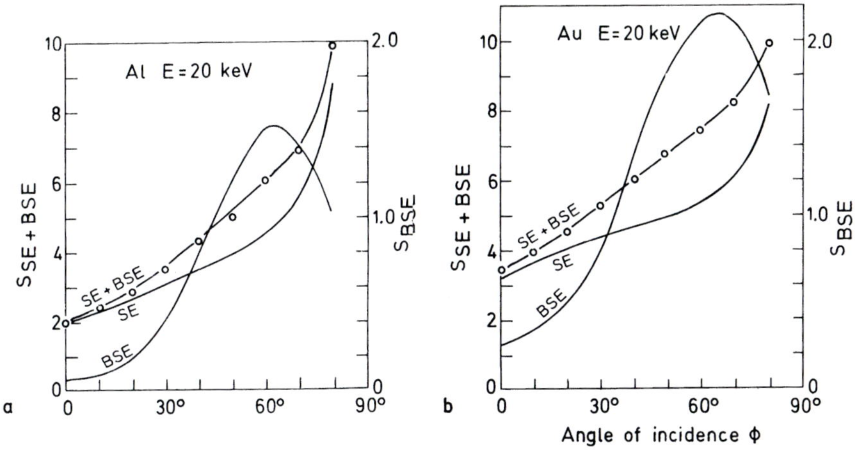

Fig. 5.9. Signal S_{SE+BSE} recorded with a positively biased collector grid (+200 V) in an Everhart–Thornley detector and S_{BSE} (5-fold scale) with a negatively biased grid (–50 V) from (**a**) Al and (**b**) Au plane specimens as a function of tilt angle ϕ (E = 20 keV) [5.39]

(Sect. 5.4.3). This additional external SE contribution generated by the BSE enhances the contribution of BSE to the SE signal that is caused by BSE on their trajectories through the surface layer (group SE2 in Fig. 4.27). The fraction of this group of SE3 contributing to the total SE signal depends on the geometry of polepiece, specimen and detector; typically the fraction is of the order of 30% [5.39] but can increase to 50% [5.40] and even 60–70%.

The recorded signal S_{SE} therefore consists of four parts (Fig. 4.27):

1) The signal S_{SE1} excited directly by the primary electrons of the electron probe with a yield which can be described by $\delta_0 \sec\phi$ where ϕ is the local tilt angle.

2) The signal S_{SE2} excited by the BSE at the specimen surface with a yield $\delta_0\beta\eta$,

3) The signal S_{SE3} excited externally by the BSE with a yield δ_{ext} and

4) The signal of BSE that strike the scintillator directly with a small solid angle $\Delta\Omega$ of collection.

The first two contributions form the SE-yield (4.30) and the total SE signal detected becomes

$$S_{SE} = f_1(\delta_0 \sec\phi + \delta_0\beta\eta) + f_2\delta_{ext} + f_3(\mathrm{d}\eta/\mathrm{d}\Omega)\Delta\Omega \,. \qquad (5.13)$$

where f_1, f_2 and f_3 denote the corresponding collection and detection efficiences. Only some 20–50% of the SE signal can be attributed to the primary beam (S_{SE1}) therefore and it is only these SE that convey high-resolution information of the order of the electron-probe diameter. The other part (S_{SE2+3}) is modulated in intensity by deeper specimen structures, which differ

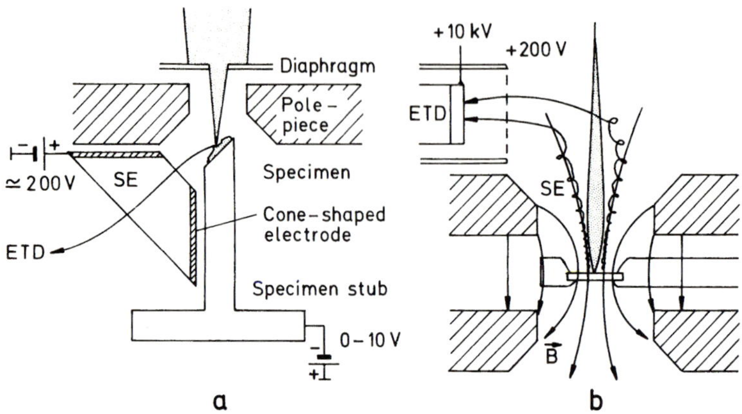

Fig. 5.10. (**a**) Extraction electrode for SE with the specimen at nearly zero working distance [5.42]. (**b**) extraction of SE in the in-lens mode with the specimen immersed at the centre of a condenser–objective lens [5.43]

in their backscattering properties, and by the surface topography, which also influences the escape probability of the BSE. This part contains, therefore, information from a hemispherical volume with a radius of the order of $R/2$ (R: electron range). As a consequence, the typical types of contrast caused by SE differ strongly for different operating conditions and instruments and every SE image contains a more or less significant BSE contribution [5.41].

Special problems of SE collection arise when the working distance is small or when the specimen is immersed in the magnetic field of the probe-forming lens, to decrease the electron-probe size by reduction of the spherical aberration coefficient to a few millimetres. A cone-shaped extraction electrode biased at +200 V can be used [5.42] to extract SE at a working distance nearly equal to zero (Fig. 5.10a). Strongly excited objective lenses in TEM can work in the in-lens mode (Sect. 2.2.3). The specimen lies inside the lens field and is introduced on a side-entry stage through the polepiece gap. Thanks to the small spherical aberration, electron-probe diameters of the order of 1–2 nm are attainable when a Schottky or field-emission cathode is used. The SE emitted move on screw trajectories of small radius around the magnetic field lines and can be collected by placing a positively biased grid in front of a scintillator outside the lens where the magnetic field is sufficiently weak (Fig. 5.10b) [5.43]. The specimen area that can be observed falls to a few millimetres in diameter. Outside this area, the magnetic field lines and the SE trajectories end on the polepiece. A similar situation arises when a single-polepiece lens is used with the specimen in front of the polepiece

The detector configuration of Figs. 5.10a,b can also be used with asymmetric collection fields. For LVSEM with $E \leq 1$ keV it will be necessary to use detector systems that are free of primary beam deflection and do not cause additional astigmatism. This can be achived in various ways.

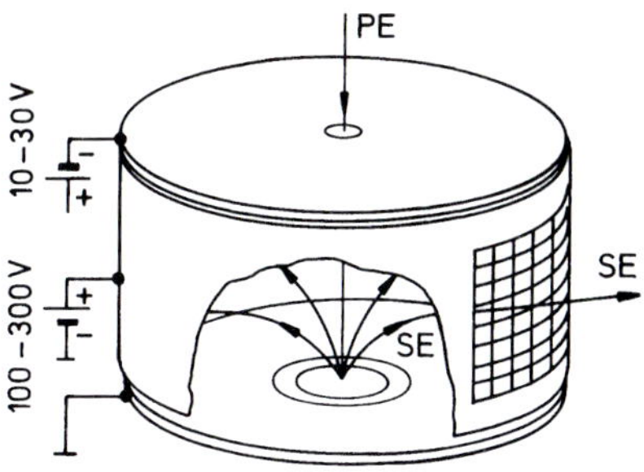

Fig. 5.11. Banbury–Nixon cage for the directional selection of SE consisting of a positively biased cylindrical electrode and a negatively biased top electrode

1. Banbury–Nixon Cage. A device proposed by *Banbury* and *Nixon* [5.44] separates the SE extraction field from the collection field. It consists of a cylindrical electrode with a window covered by a grid and directed towards the SE detector, a plane top electrode with a bore for the primary electrons and a grounded plate, which can be the specimen stub (Fig. 5.11). When the cylindrical electrode is biased more positively than the top electrode, which can also be negatively biased, SE trajectories with different exit momenta will spray out like a fountain and the fraction of the angular characteristics that goes through the grid-coated window can be influenced by the biases of the electrodes. Though originally designed for higher electron energies, this detector fulfils the condition of low beam deflection.

2. Wien Filter. A superposition of electric and and magnetic quadrupoles (Fig. 5.12) decreases the magnetic and electric field on the axis to zero and behaves like a Wien filter [5.45-47]. (In a Wien filter with crossed $\boldsymbol{E}$ and $\boldsymbol{B}$ fields, the Lorentz forces $e\boldsymbol{E}$ and $e\boldsymbol{v} \times \boldsymbol{B}$ cancel each other.) Two Everhart–Thornley detectors placed opposite each other form the positively biased electrodes of an electric quadrupole. This allows us to record two SE signals A and B simultaneously (Sect. 5.4.3).

3. Electrostatic Detector–Objective Lens. In an electrostatic objective lens system (Fig. 5.13) [5.48, 49] SE are accelerated on-axis by the positively biased electrodes. Examples of calculated SE trajectories show that they can be collected by an annular or two semi-annular scintillation detectors. This system has the advantage that the SE trajectories do not follow screw trajectories as they do in magnetic fields. This lens can also be combined with a multipole corrector for chromatic and spherical aberrations (Sect. 2.2.2).

4. Rotationally Symmetric In-lens Detector. Instead of the Everhart–Thornley detector mounted laterally in Fig. 5.10b, an annular YAG crystal disc is mounted on-axis in front of the field-free region of the probe-forming lens [5.50]. The electron passes through an earthed metal tube. The front face of the YAG crystal is metal-coated and biased to a few kV. An earthed or negatively biased grid in front of the detector shields this acceleration field from the column. BSE and SE3 cannot reach the detector and a nearly 100% collection efficiency for SE1 and SE2 can be achieved [5.51].

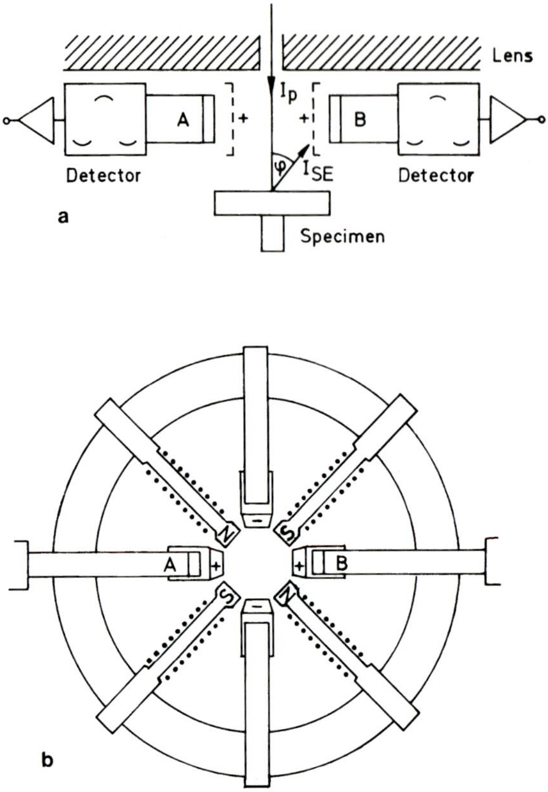

Fig. 5.12. Electric/magnetic quadrupole combination acting like a Wien filter for near-axis electrons with two opposite Everhart–Thornley detectors A and B in (**a**) side and (**b**) top view mounted below the polepiece [5.45]

5.4.2 Backscattered Electron Detectors

The use of semiconductor detectors and microchannel plates has already been discussed in Sects. 5.3 and 5.2.3.

BSE Detection Using an Everhart–Thornley Detector. There are more BSE than SE for primary electron energies above 5 keV (Fig. 4.8). However, since the trajectories of the BSE are straight, it is necessary to use a large solid angle of collection $\Delta\Omega$. For the Everhart–Thornley detector (Sect. 5.2.2), $\Delta\Omega$ is of the order of only a few 10^{-2} sr. The number of BSE recorded is very low and the image becomes very noisy. Nowerdays, therefore, Everhart–Thornley detectors are not used with a negatively biased grid to retard the SE; a comparison of the SE and BSE images in Sects. 6.1. and 6.2 recorded in this way provides a good demonstration of their differences.

BSE-to-SE Conversion. Another alternative is to use the conventional Everhart–Thornley detector in the SE mode with a positively biased collector

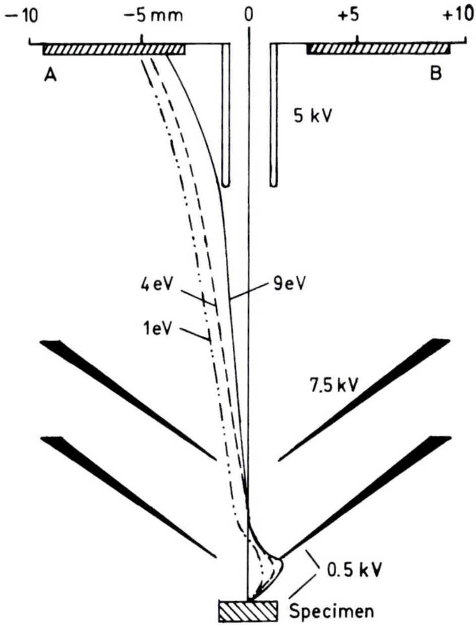

Fig. 5.13. Electrostatic detector–objective lens with in-lens SE detectors A and B and calculated SE trajectories of 1, 4, and 9 eV with a starting angle of 60°; landing energy of the primary electrons is 500 eV [5.45]

grid and collect the SE that are excited by the BSE at the polepiece, SE3 in Fig. 4.27. Each SE signal contains a large fraction of these converted BSE, which can be observed separately by screening the SE excited at the specimen with a retarding field created by a negatively biased grid or ring electrode placed around the specimen (Fig. 5.15). The generation of SE by BSE at the polepiece is sufficient to produce a usable signal, which contains only BSE information [5.52–54]. However, the SE production at the metal polepiece favours BSE with low exit energies E_{B} because the high-energy tail of the $\delta(E)$ curves decreases as $E_{\mathrm{B}}^{-0.8}$ with increasing E_{B} (Fig. 4.19). The BSE-to-SE conversion can be increased by using a copper plate coated with MgO smoke particles [5.53, 55]. The BSE-to-SE signal becomes much larger than the SE signal from the specimen and indeed an earthed grid has to be placed in front of the positively biased converter plate (Fig. 5.15) to suppress these converted electrons when looking at the SE signal from the specimen. This mode has the advantage that the SE3 group excited at the polepiece is also suppressed. The BSE-to-SE conversion mode can be switched on by negatively biasing the converter plate. A BSE-to-SE converter plate with a front grid can also be positioned at other take-off directions and can be divided into separate segments [5.53].

Owing to the charging of the surface of the MgO crystals, the SE yield of such a BSE-to-SE converter becomes independent of the BSE energy so long as $\sigma = \eta + \delta$ is larger than unity for $E_{\mathrm{B}} \leq$ 20–30 keV. This independence of the BSE energy means that the BSE are recorded regardless of their energy and the image becomes complementary to a specimen-current image (Sect. 6.2.6). This shows that the dependence of detection efficiency on BSE energy can be varied from proportionality to E_B (scintillator) to independence of E_B (MgO converter) and finally proportionality to $E_\mathrm{B}^{-0.8}$ (metal converter).

If an electron-transparent foil is placed in front of the Everhart–Thornley detector, it can act as an absorbing filter for low-energy BSE and the only BSE recorded are those that can penetrate the foil and excite SE, which are collected by the Everhart–Thornley detector (ETD) [5.56]. The signal will be lower than in the BSE-to-SE conversion mode discussed above because of absorption and multiple scattering in the foil but contrast effects, which can be attributed to high-energy BSE, can be recorded with a higher contrast, channelling contrast (Sect. 6.2.2) and magnetic contrast type 2 (Sect. 8.1.2), for example.

BSE Scintillation Detectors. Backscattered electrons have sufficient energy E_B to produce a large number of light quanta $\propto (E_\mathrm{B} - E_\mathrm{th})$ in a scintillator without additional acceleration. The threshold energy E_th of 1–2 keV is caused by absorption in the conductive and light-absorbing evaporated metal layer. This favours the detection of high-energy BSE, which is no disadvantage because the depth of escape and information of low-energy BSE is larger. The solid angle $\Delta\Omega$ can be increased by using a scintillator of large area mounted on a light-pipe close to the specimen. Figure 5.14 shows various practical arrangements. Detectors at (a) high and (b) low take-off angles [5.57–60] mainly record material and topographic contrast, respectively (Sect. 6.2). A turret mechanism can be used to change from one position to the other without disturbing the vacuum [5.58]. A flat scintillator sheet below the polepiece (c) [5.57, 61, 62] or YAG and YAP single-crystal scintillators [5.15] with a central bore for the primary beam can record BSE at high take-off angles over a

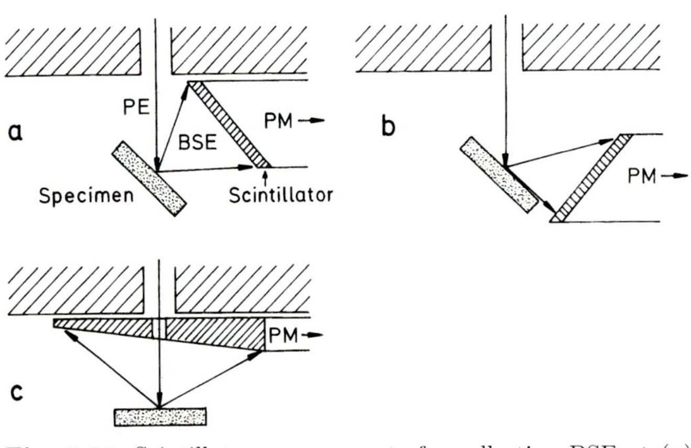

Fig. 5.14. Scintillator arrangements for collecting BSE at (**a**) high and (**b**) low take-off angles, (**c**) as a scintillator sheet below the polepiece

very large solid angle. Further arrangements of BSE detectors are discussed in Sect. 5.4.3.

Although the threshold energy E_{th} can be made smaller than 1 keV by applying a special coating of the scintillator [5.63, 64], the signal is not strong enough for LVSEM and therefore the BSE have to be post-accelerated by a bias of 4–5 kV at the scintillator surface, which is shielded from the specimen by a grid [5.64, 65], or by a grid held at 3 kV in direct contact to the scintillator [5.66].

5.4.3 Multiple Detector Systems

Commercial SEMs are normally equipped with an Everhart–Thornley detector and a top BSE detector in the form of a scintillator, semiconductor detector or microchannel plate. To make the best use of the electron–specimen interactions and to obtain more quantitative information, it will be necessary to improve the detector strategy [5.67, 68] as demonstrated by the following examples.

Two-Detector System. When the electric field at the specimen generated by the collector bias of the SE detector is reduced either by placing a biased ring around the specimen or by inserting a hemispherical grid of a few thin wires, the SE initially follow trajectories corresponding to their exit momenta. By using two oppositely placed Everhart–Thornley detectors, it will be possible to separate the electrons according to their exit momenta when they reach the collection fields (Fig. 5.15) [5.55, 67, 68]. A pair of ETDs placed opposite each other and coupled to light-pipes can be pivoted so that the line joining them is parallel to either the x or the y direction [5.69].

The detector system shown in Fig. 5.15 also contains a BSE/SE converter plate below the polepiece. A positive bias U_{G} at the ring or grid electrode and a small positive bias U_{C} at the converter plate suppresses the contribution of SE3 excited at the polepiece and the signal only contains information

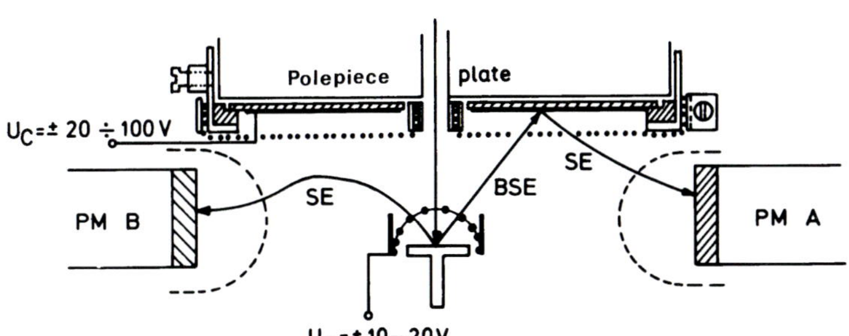

Fig. 5.15. Two-detector system for SE consisting of two Everhart–Thornley detectors facing each other and a grid or ring electrode to suppress the collection field at the specimen. The polepiece plate contains a BSE-to-SE converter plate behind an earthed grid

from the SE1 and SE2. When both U_G and U_C are negative, the direct SE contribution is suppressed and only a converted BSE signal is recorded. Both modes select SE or BSE emitted to the right and left by the detectors A and B, respectively. The separation of the topographic and material contrast by forming the signals A–B and A+B, respectively, and the quantitative use of these signal intensities will be shown in Sect. 6.1.4 and the enhancement of magnetic contrast type-1 by the SE A–B image in Sect. 8.1.1.

Multiple-Detector Systems for BSE. Multiple-detector systems for BSE are reviewed in [5.73]. The first multiple detector system proposed for BSE [5.70] used two semi-annular semiconductor detectors A and B below the polepiece (Fig. 5.16a).

Figure 5.16b shows the arrangement of a side A and top scintillation detector B for the recording of topographic and material contrast, respectively. The side detector A can have either a semi-cylindrical front face [5.73] (Fig. 5.16c) to increase the solid angle of collection or a closed cylindrical face (Fig. 5.16d) [5.72]. The former design increases the topographic contrast. The latter design does not collect all light quanta from the right-hand side. This is no disadvantage because the resulting contrast corresponds to a portrait illumination.

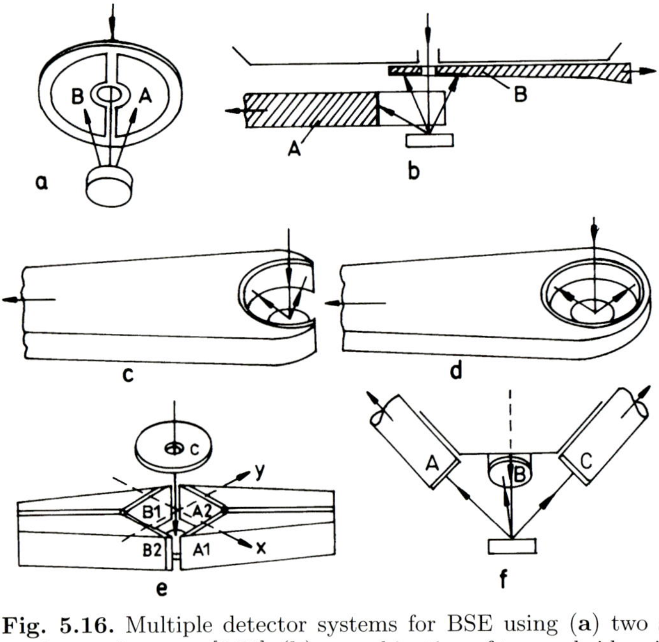

Fig. 5.16. Multiple detector systems for BSE using (**a**) two semi-annular semiconductor detectors [5.70], (**b**) a combination of top and side scintillation detectors with (**c**) a semi-cylindrical [5.73] or (**d**) a full-cylindrical side detector [5.72], a four-detector system with (**e**) side detectors [5.76] and (**f**) scintillation detectors at 60° [5.78]

In order to exploit direction-sensitive detectors, four-detector systems with a take-off angle of the order of 45° relative to the electron beam have also been proposed, using semiconductor detectors [5.74, 75], four scintillator sheets mounted like a pyramid [5.76] or as four rectangles [5.77] on two light-pipes with a shutter system (Fig. 5.16e) or four scintillators coupled by fibre optics to two photomultipliers (Fig. 5.16f) [5.78]. To change the take-off direction during observation, a small scintillator mounted on a curved light-pipe can be turned around the specimen about an axis normal to the electron beam [5.79]. The effectiveness of multi-detector systems can be controlled by adopting a fish-eye perspective using charged balls as a target (Sect. 3.5.2).

5.5 Detector Noise

5.5.1 Noise of a Scintillator–Photomultiplier Combination

We discussed in Sect. 4.2.4 the shot noise of the incident electrons and the increase of noise associated with SE and BSE emission. To estimate the noise introduced by the Everhart–Thornley detector [5.12, 22, 80, 81], we assume that a mean number $\overline{n}$ of electrons hits the scintillator (input signal) and a mean number $\overline{N} = \overline{n}G_{\mathrm{D}}$ is recorded at the output, where G_{D} is the total detector gain introduced in (5.3). From (4.46) the relative variance is thus

$$v(\overline{N}) = v(\overline{n}) + v(G_{\mathrm{D}})/\overline{n}\,. \tag{5.14}$$

The output current will be $I_{\mathrm{out}} = e\overline{N}/\tau$ and, using (4.51), its rms noise amplitude $\Delta I_{\mathrm{n}} = e[\mathrm{var}(\overline{N})]^{1/2}/\tau$. With (4.38), we obtain the output signal-to-noise ratio

$$\left(\frac{S}{N}\right)_{\mathrm{out}} = \frac{\overline{N}}{[\mathrm{var}(\overline{N})]^{1/2}} = \left(\frac{S}{N}\right)_{\mathrm{in}} \left(\frac{1}{1+v(G_{\mathrm{D}})}\right)^{1/2} \tag{5.15}$$

in which we have assumed a Poisson distribution for $\overline{n}$ with $(S/N)_{\mathrm{in}} = 1/v(\overline{n}) = \overline{n}$ (4.44).

The quality of the detector can be described either by a factor

$$r_{\mathrm{n}} = \frac{(S/N)_{\mathrm{in}}}{(S/N)_{\mathrm{out}}} = [1 + v(G_{\mathrm{D}})]^{1/2} > 1\,, \tag{5.16}$$

which is a measure of the increase of the rms noise amplitude by the detector [5.81], or by the detection quantum efficiency [5.82]

$$\mathrm{DQE} = \frac{(S/N)^2_{\mathrm{out}}}{(S/N)^2_{\mathrm{in}}} = \frac{1}{r_{\mathrm{n}}^2} = \frac{1}{1+v(G_{\mathrm{D}})} < 1\,. \tag{5.17}$$

A low-noise detector system should satisfy the condition $v(G_{\mathrm{D}}) \ll 1$. From (5.3), we see that $G_{\mathrm{D}} = \overline{m}rqcG_{\mathrm{PM}}$ is the result of cascading the collection and detection processes, for which the relative variance can be calculated by (4.47). The quantities r_n, q and c and hence T obey binomial distributions, with the result that $v(T) = (1-T)/T$ (4.41). Collecting up we find

$$v(G_{\rm D}) = v(\overline{m}) + \frac{1-T}{\overline{m}T} + \frac{v(G_{\rm PM})}{\overline{m}T} \,. \tag{5.18}$$

If we assume a Poisson distribution for the generation of $\overline{m}$ quanta, we have $v(\overline{m}) = 1/\overline{m}$ from (4.44), and (5.18) simplies to

$$v(G_{\rm D}) = \frac{1}{\overline{m}T}[1 + v(G_{\rm PM})] \,. \tag{5.19}$$

We introduce a mean secondary electron yield $\delta_{\rm d}$ of the n dynodes of the photomultiplier whereupon (4.47) becomes a geometric progression

$$v(G_{\rm PM}) = v(\delta_{\rm d}) \sum_{i=0}^{n} 1/\delta_{\rm d}^{i} = v(\delta_{\rm d}) \frac{\delta_{\rm d}^{n} - 1}{(\delta_{\rm d} - 1)\delta_{\rm d}^{n-1}} \simeq \frac{1}{\delta_{\rm d} - 1} \,. \tag{5.20}$$

In order to estimate $v(\delta_{\rm d})$, we have assumed Poisson statistics with $v(\delta_{\rm d}) = 1/\delta_{\rm d}$, which leads to the final expression on the right-hand side of (5.20). Normally, $\delta_{\rm d}$ lies between 2 and 5 and $v(G_{\rm PM})$ decreases with increasing $\delta_{\rm d}$ or photomultiplier voltage $U_{\rm PM}$ (dashed line in Fig. 5.17). Measured values of $v(G_{\rm PM})$ obtained from the pulse-height spectrum in the single-photon-counting mode [5.81] give larger values for $v(G_{\rm PM})$ (full curves). This shows that the contribution of $v(G_{\rm PM})$ to the increase of the rms noise amplitude of the detector system can be kept low for high $U_{\rm PM}$ but increases rapidly for small $U_{\rm PM}$.

The output noise amplitude can be measured by means of special operational amplifiers, which furnish the rms signal directly. When measuring the incident electron-probe current $I_{\rm p}$ with a Faraday cage, its shot-noise amplitude can be calculated from (4.49) with the same bandwidth Δf as that used in the output circuit. This allows the total increase $r_{\rm n}$ of the primary shot-noise, including the noise of SE and BSE emission discussed in Sect. 4.2.4, to be determined. Figure 5.18 shows the dependence of $r_{\rm n}$ on the primary electron energy E for the following modes when using a P-47 powder layer as the scintillator of an Everhart–Thornley detector:

1. Normal SE mode with a positively biased collector grid (+250 V). The increase of $r_{\rm n}$ with increasing E results from the decrease of δ and increase of $v(\delta)$.
2. Pure SE mode with suppression of the SE contribution from the BSE at the polepiece by placing a positively biased plate behind an earthed grid. $r_{\rm n}$ increases as a result of the reduction in the mean number of SE per PE.
3. BSE-to-SE conversion mode where the SE from the specimen are suppressed by a retarding field and only the SE generated by the BSE at the polepiece are collected.

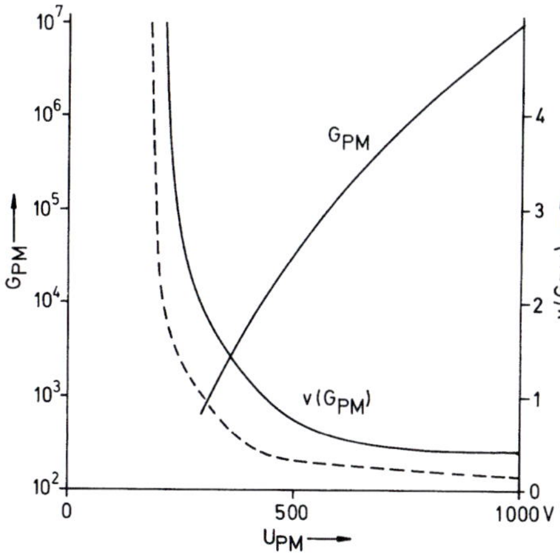

Fig. 5.17

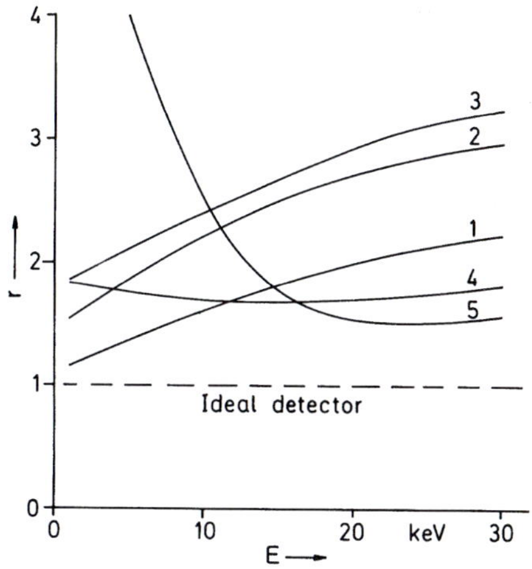

Fig. 5.18

Fig. 5.17. Example of the increase of photomultiplier gain G_{PM} and decrease of the relative variance $v(G_{\mathrm{PM}})$ with increasing photomultiplier voltage U_{PM}, *dashed line:* $v(G_{\mathrm{PM}}) = 1/(\delta_{\mathrm{d}} - 1)$ [5.81]

Fig. 5.18. Ratio r_{n} of the rms noise amplitude at the detector output to the rms shot-noise amplitude of the incident electron-probe current for (1) the conventional SE mode, (2) suppression of the SE3 contribution from the polepiece, (3) the SE3 contribution obtained by BSE-to-SE conversion from the polepiece only, (4) BSE-to-SE conversion at a MgO-coated plate and (5) a scintillator detector for BSE below the polepiece (Fig. 5.14c) [5.81]

4. BSE-to-SE conversion mode with a MgO-coated plate at the polepiece to increase the BSE/SE conversion. The value of r_{n} remains approximately constant because of the order of one SE is generated per BSE irrespective of E, due to charging effects.
5. Use of a cone-shaped scintillation detector above the specimen, which records nearly all BSE in a P-47 powder layer of 7 mg/cm^2. The value of r_{n} decreases with decreasing E owing to the decrease of luminescence. This demonstrates the advantage of mode 4 for the detection of BSE at low E.
6. Recording of BSE with the conventional Everhart–Thornley detector as in mode 1 but with a negatively biased collector grid. The r_{n} values are of the order of 10–14 for a specimen tilt of 30° (not shown in Fig. 5.18).

These curves demonstrate that it is possible to work in most of the detection modes with an increase r_{n} of the rms noise amplitude less than 2 or in other words a DQE better than 25%.

The r_{n} values of an Everhart–Thornley detector using a plastic scintillator (NE 102A) and an optimum P-47 powder layer of 7 mg/cm^2 have been measured to be 2.5 and 1.5, respectively [5.81]. A somewhat better performance can be expected when using the single-crystal scintillators discussed in Sect. 5.1.1.

5.5.2 Noise of Semiconductor Detectors

The following sources of noise have to be considered for semiconductor detectors:

1. The shot noise of the incident BSE current,
2. the statistics of the electron gain G,
3. shot noise caused by the reverse-leakage current when the detector is reverse-biased, and
4. noise of the preamplifier.

The first contribution has already been discussed in Sect. 4.2.4. There, we showed that the combination of the Poisson distribution of the incident electron-probe current and the binomial distribution of BSE emission again results in a Poisson distribution with $\mathrm{var}(n_{\mathrm{BSE}}) = n_{\mathrm{BSE}}$. The average number of electron-hole pairs is $N = n_{\mathrm{BSE}}G$. Estimation of the factor r_{n} in (5.16), which describes the increase in the rms noise amplitude caused by the detector, requires a knowledge of $v(G)$ for semiconductor detectors. We assume that $G = E/\overline{E}_i$ and neglect the decrease caused by backscattering, by absorption in the metal film and by incomplete collection as considered in (5.10). The variance of the number G of electron–hole pairs would be $\mathrm{var}(G) = G$ if it followed a Poisson distribution. However, the continuous slowing-down is not a statistically independent process. In the extreme case of constant $\overline{E}_i$ the number G would be the same for all electrons of energy E resulting in zero variance. The statistics of E_i results in a reduction of the variance by the Fano factor F; for silicon, $F = 0.12$ (see also Sect. 10.2.2):

$$\begin{aligned} \mathrm{var}(G) &= FG; \quad v(G) = F/G \ll 1 \\ \mathrm{var}(n_{\mathrm{BSE}}G) &= G^2 n_{\mathrm{BSE}}[1 + v(G)] \,. \end{aligned} \tag{5.21}$$

This means that the second source of noise is negligible compared to the shot-noise of the first source. The rms noise amplitude therefore becomes

$$\Delta I_{n1} = (2\Delta f e I_{\mathrm{BSE}})^{1/2} G \,. \tag{5.22}$$

The shot noise of the detector reverse-leakage current I_{s} in Fig. 5.3 is now

$$\Delta I_{\mathrm{n}3} = (2\Delta f e I_{\mathrm{s}})^{1/2} \,. \tag{5.23}$$

This term can be neglected if $I_{\mathrm{s}} \ll I_{\mathrm{BSE}}G^2$.

The most significant noise contribution of the amplifier is generated in the feedback resistor R_{F} (Fig. 5.7):

$$\Delta I_{\mathrm{n}4} = (4kT\Delta f/R_{\mathrm{F}})^{1/2} \tag{5.24}$$

where $k = 1.3805 \times 10^{-23}$ J K^{-1} denotes Boltzmann's constant and T the temperature of the resistor. For $R_F = 100$ kΩ, $\Delta f = 600$ kHz and room temperature, this results in a value of about 3×10^{-10} A. With these values an amplifier with a suitable gain of the order of $10^6 - 10^8$ V A^{-1} can be designed [5.83].

5.6 Electron Spectrometers and Filters

5.6.1 Types of Electron Spectrometer

We first describe the different types of spectrometers in general and then draw attention to designs that are particularly suitable for SE and BSE; for reviews on electron spectrometers, see [5.84, 85]. All energy spectrometers make use of the Lorentz force (2.9), which also acts on electrons in electron lenses (Sect. 2.2) and electron-beam-deflection systems (Sect. 2.3). The deflection of an electron beam that passes through a transverse electric or magnetic field is inversely proportional to the electron energy $E = mv^2/2$ or the momentum $p = mv$. An electron spectrometer should show high values of collection efficiency, of transmittance and of energy resolution $E/\Delta E$. It is no problem to construct electron spectrometers of high resolution $E/\Delta E > 10^4$ but the entrance aperture is low, of the order of only a few milliradians. Such spectrometers are therefore of interest mainly in TEM and STEM, where a large fraction of the transmitted electrons with energy losses is concentrated in a cone of such a small aperture.

The magnetic sector field in Fig. 5.19a is shown merely to demonstrate how focusing and dispersion $\Delta y/\Delta E$ can be obtained with a transverse magnetic field between plane polepieces. By this we mean that electrons of the same energy E emerging from a specimen point are focused in the plane of the exit slit of the spectrometer and that electrons of energy $E - \Delta E$ are focused at another position, distance Δy away. This focusing and in consequence the energy resolution are limited by the aberrations of the deflection system. The dimensions of such a magnetic spectrometer are too large and the entrance aperture α too small for it to be employed in SEM.

Electrostatic spectrometers can be classified into deflection spectrometers with transverse fields and retarding-field spectrometers. A radial electric field can be created by means of cylindrically or spherically curved electrodes (Fig. 5.19b). When the electrodes are massive, electrons with energies very different from the pass energy of the spectrometer can hit the electrodes and be scattered to the exit slit. The electrodes should therefore be made transparent by simulating their profile with a series of parallel wires, so that these electrons can escape from the spectrometer [5.86]. Other types of focusing electrostatic fields are the plane mirror (Fig. 5.19c) and the cylindrical mirror analyser (CMA) (Fig. 5.19d). These spectrometers record $E \cdot n(E)$

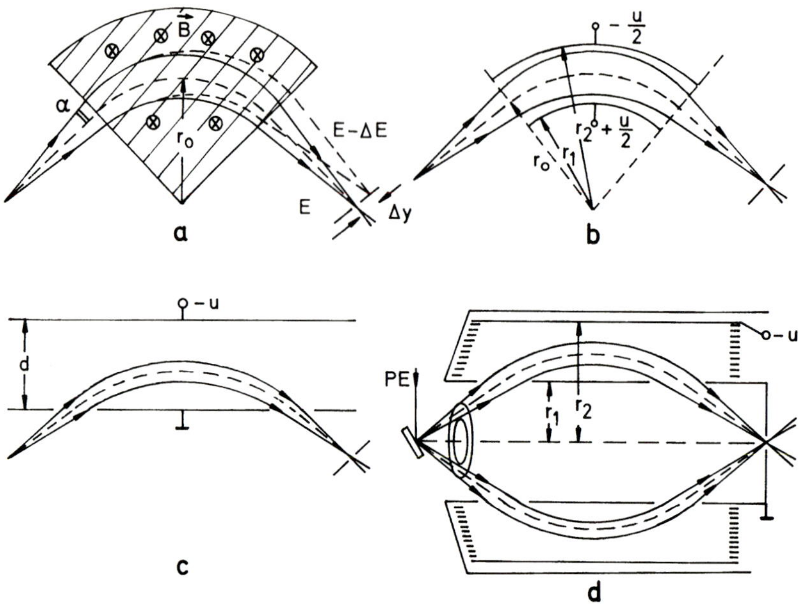

Fig. 5.19. Electron spectrometers of the deflection type: (**a**) magnetic sector field, (**b**) electrostatic radial field between cylindrical or spherical capacitor plates, (**c**) plane-mirror and (**d**) cylindrical-mirror analyser (CMA)

spectra because $\Delta E/E$ is constant, they do not give the energy spectrum $n(E)$ directly.

The principle of the retarding-field spectrometer is illustrated in Fig. 5.20a. Electrons of energy $E = eU = m(v_{0x}^2 + v_{0y}^2)/2$, which leave the source plane with an angle α, are decelerated and deflected in the homogeneous field $|\boldsymbol{E}| = U_G/d$ between this plane and the retarding grid biased at $-U_G$. The equations of motion are now

$$\begin{aligned} m\ddot{x} &= -eU_G/d & m\ddot{y} &= 0 \\ mv_x &= mv_{0x} - (eU_G/d)t & v_y &= v_{0y} \\ x &= v_{0x}t - (e/m)(U_G/2d)t^2 & y &= v_{0y}t\,. \end{aligned} \tag{5.25}$$

Elimination of the time t leads to the parabolic trajectory

$$(x - x_0) = (y - y_0)^2;\ x_0 = Ed\cos^2\alpha/eU_G;\ y_0 = 2(Ed/eU_G)\sin\alpha\cos\alpha \tag{5.26}$$

where x_0, y_0 are the coordinates of the point of reflection. The condition that the electrons pass through the grid, $x_0 \geq d$, may be written

$$E\cos^2\alpha \geq eU_G\,. \tag{5.27}$$

For normal incidence ($\alpha = 0$), this becomes $E \geq eU_G$, which can also be derived from the conservation of energy. For $\alpha > 0$, the electrons need a

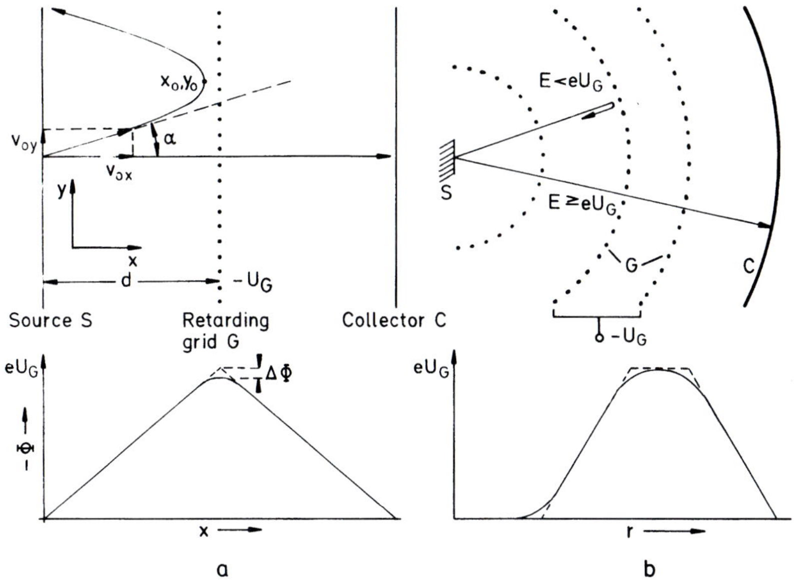

Fig. 5.20. (**a**) Electron trajectories and potential $\Phi(x)$ with voltage drop $\Delta\Phi$ at the centres of the retarding-grid meshes for one plane retarding grid at $\Phi = -U_G$ between two earthed plates. (**b**) Trajectories and potential $\Phi(r)$ for a combination of three spherical grids, the last two at the retarding potential $\Phi = -U_G$

larger kinetic energy to overcome the retarding field and pass through the grid. Equation (5.27) can also be interpreted as stating that this retarding field between the plane electrodes selects the electrons according to their z component of momentum $p_z = mv\cos\alpha$.

When a real grid is used as a retarding electrode, the potential energy will not be $\Phi = eU_G$ across the whole grid plane but instead there will be potential drops with a maximum value of $\Phi - \Delta\Phi$ at the centre of each mesh (Figs. 5.20 and 5.21). In the ideal case of a circular opening of radius r in an infinite conducting plane, this potential drop depends on the potential gradients (electric field strengths) E_1 and E_2 on either side of the electrode [5.87].

$$\Delta\Phi = (E_2 - E_1)r/\pi \; . \tag{5.28}$$

Numerical Example: For an $r = 0.5$ mm diaphragm at a potential $\Phi = 100$ V between two earthed plates each at a distance of 5 mm, we find $E_2 = -E_1 = 200$ V cm^{-1} and $\Delta\Phi \simeq 7$ V.

This means that the bias of the grid has to be increased to more negative values to retard electrons that arrive at the centre of the meshes, and the inhomogeneous field in the mesh can strongly influence the trajectories of retarded electrons of low kinetic energy. Figure 5.21 shows that the field in the mesh centre acts as an electrostatic lens [5.88]. From (5.28), we see that the potential drop can be decreased by having low potential gradients

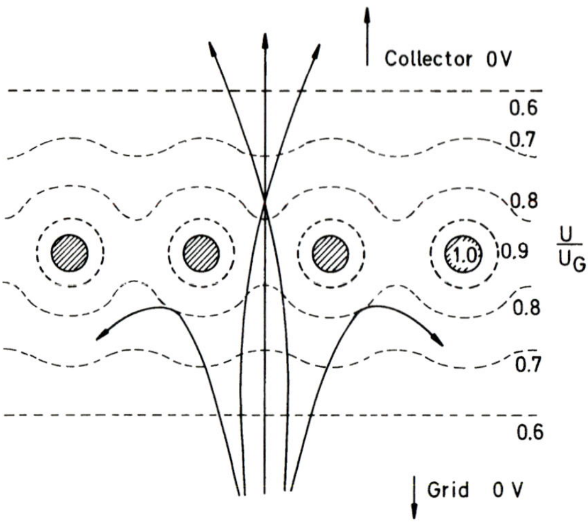

Fig. 5.21. Equipotentials Φ = const of a retarding between plane electrodes at earth potential and examples of electron trajectories with kinetic energy just sufficient to overcome the potential drop at the mesh centres [5.88]

on both sides (large distance between the meshes and the electrodes) or by using a double grid at the same bias. In order to satisfy condition (5.27) in the presence of the potential drop (5.28), at least three spherical grids must be used (Fig. 5.20b); the first, at the same bias as the specimen, allows the electrons to move on straight trajectories without deflection and to enter the retarding field between the first and second grid normally ($\alpha = 0$). The presence of two grids at the retarding bias $-U_G$ ensures that the potential drop $\Delta\Phi$ does not drastically reduce the energy resolution $E/\Delta E \simeq U_G/\Delta\Phi$.

In the case of a spherical retarding field, the fraction

$$N(U_G) = \int_{eU_G}^{\infty} n(E)\mathrm{d}E \Bigg/ \int_{0}^{\infty} n(E)\mathrm{d}E \tag{5.29}$$

of the energy distribution $n(E)$ is transmitted and the retarding field curve plotted in Fig. 8.8a for SE, for example, is found; the energy distribution is obtained by differentiation of this retarding curve (Fig. 8.8b). The differentiation can also be performed on-line by modulating the retarding grid bias $-U_G$ with a small ac amplitude and measuring the resulting ac current of the transmitted electrons with the aid of a lock-in amplifier.

By taking the second derivative of the retarding curve or the first derivative of $E \cdot n(E)$ when using an energy-selective spectrometer such as a CMA (Fig. 5.19d), a curve is obtained (Fig. 10.12b) on which small peaks corresponding to Auger electrons (AE), for example, are enhanced against a high background of BSE.

5.6.2 Secondary Electron Spectrometers

Spectrometers for SE with exit energies $E \leq 50$ eV can be used to measure the local surface potential (Sect. 8.2.2). An investigation of the SE energy spectrum, and of the fine structure caused by plasmon decay (Sect. 4.2.2) for example, will only be possible in an ultra-high vacuum microscope.

The spectrometers in use contain electron-optical elements for pre-acceleration (extraction), energy analysis and deflection. Pre-acceleration or extraction by means of an extraction electrode biased positively to a few hundreds of volts is optional but becomes indispensable when the influence of neighbouring surface potentials in the output record must be avoided, an essential condition for studying quantitative voltage contrast (Sect. 8.2.2). The extraction field at the specimen surface should be of the order of a few hundreds of V/mm. Deflection will be necessary when the SE behind the energy-analysing unit cannot be directly collected by the positive bias of the collector grid of the Everhart–Thornley detector. In some spectrometers, analysis and deflection are combined in a single unit.

These principles will be illustrated by considering four practical spectrometer designs; see review [5.88] for further details and references. Fig-

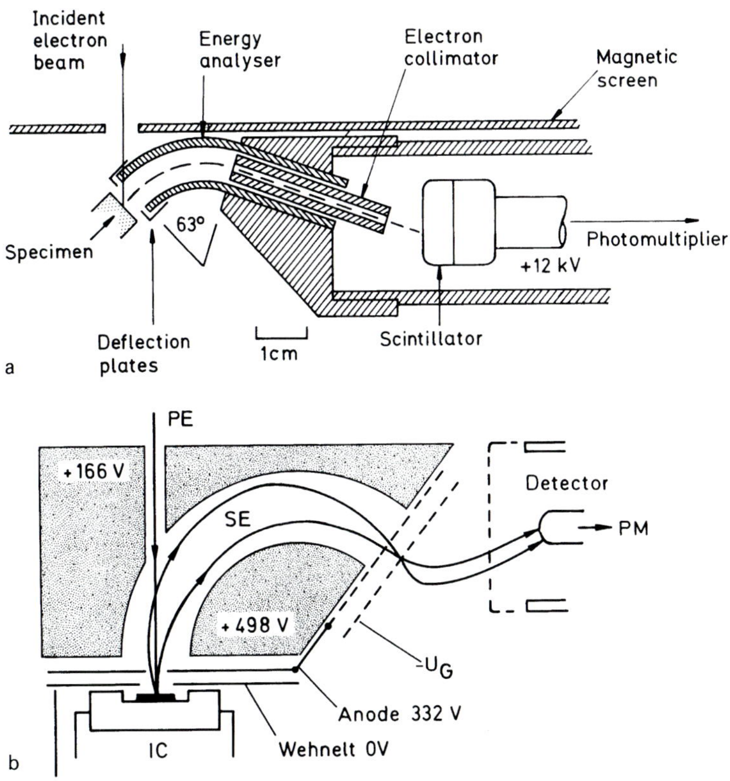

Fig. 5.22. SE spectrometers using radial electrostatic deflection fields: (**a**) cylindrical electrodes and a slit between plane plates, no pre-acceleration [5.89], (**b**) pre-acceleration by an extraction field, deflection by a spherical-plate capacitor and analysis by a retarding grid [5.90]

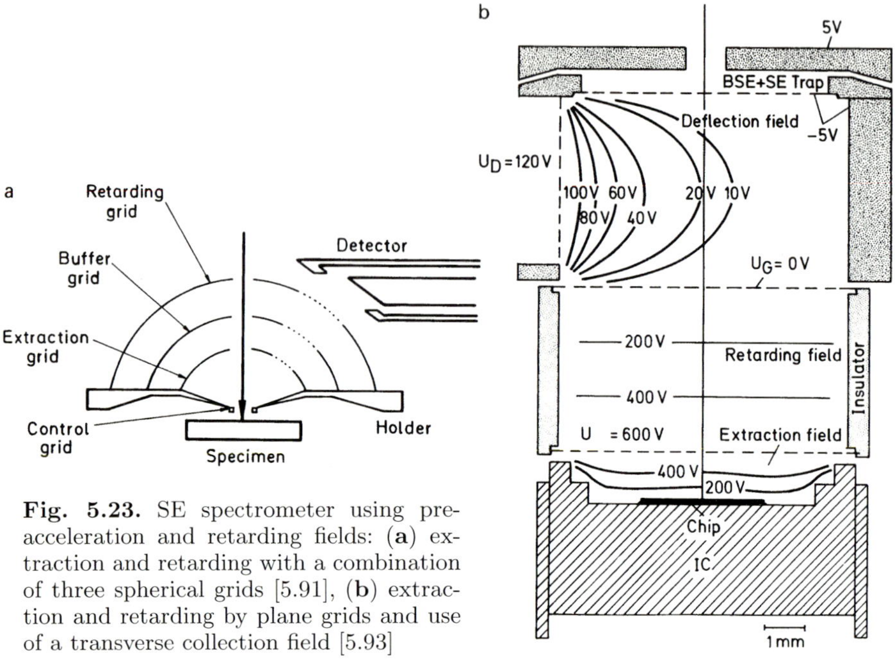

Fig. 5.23. SE spectrometer using pre-acceleration and retarding fields: (**a**) extraction and retarding with a combination of three spherical grids [5.91], (**b**) extraction and retarding by plane grids and use of a transverse collection field [5.93]

ures 5.22a,b represent spectrometers that use a radial electrostatic deflection field without and with pre-acceleration, respectively. In the former [5.89], the cylindrical plate capacitor works as an energy-analysing and deflection unit. The energy selection is effected by the narrow slit between the plates and is combined with angular selection. In the latter [5.90], the energy selection is performed by a retarding grid behind the spherical plate capacitor, which works more as a deflection unit through which electrons with energies of the order of 330 eV, which is the kinetic energy of the pre-acceleration, can pass.

Examples of spherical and planar extraction and retarding fields are shown in Figs. 5.23a,b. The control grid in Fig. 5.23a is at half the potential (1 kV) of the extraction grid (2 kV) and renders the extraction field near the specimen more closely spherical [5.91]. The buffer grid is at a potential of 240 V. A hemispherical retarding field analyser can also be used for voltage measurement with a microchannel plate detector and a high extraction field [5.92]. The device of Fig. 5.23b first accelerates and then retards the SE, which are deflected by a positively biased grid on one side and in front of a scintillator–photomultiplier combination [5.93]. The upper negatively biased grid in front of the top electrode suppresses SE excited by BSE which could otherwise pass the retarding grid.

Because of radiation damage effects caused by high-energy electrons in integrated circuits (Sect. 7.1.6), it is clearly desirable to measure the potentials in a LVSEM. The high extraction and deflection fields of the spectrometer

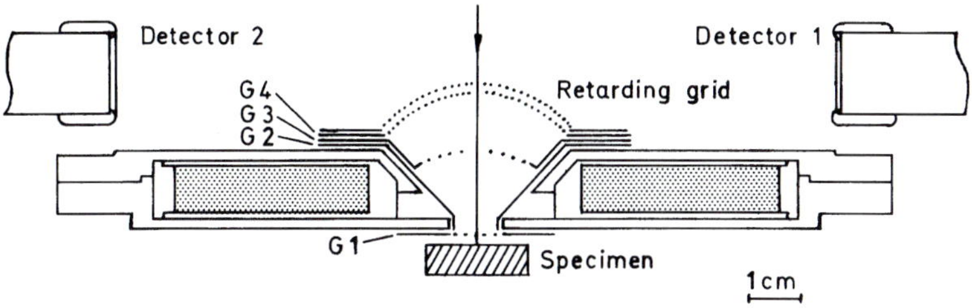

Fig. 5.24. Example of a through-lens SE spectrometer for quantitative voltage contrast [5.94]

shown in Fig. 5.23b perturb the LVSEM mode. Through-lens spectrometers of rotationally symmetry have therefore been developed that work with a low extraction field and a superposed adiabatically varying magnetic field [5.94, 96, 97]. The SE trajectories spiral then around the magnetic field lines with a radius $r = mv_r/eB$ where $v_r = v \sin\alpha$ is the radial component of velocity. Because of the conservation of angular momentum

$$L = rmv_r = m^2 v_r^2 / eB = \text{const} \tag{5.30}$$

the initial angle α_1 relative to the field line decreases to a lower angle α_2 when the magnetic induction B_1 decreases to B_2:

$$\sin^2\alpha_1 / \sin^2\alpha_2 = B_1/B_2 \tag{5.31}$$

and the SE trajectories become more parallel to the field lines [5.98]; $\alpha_1 = 80°$ at $B_1 = 100$ mT results in $\alpha_2 = 5.6°$ in a final field $B_2 = 1$ mT. Using a retarding grid normal to the field lines will decrease the influence of neighbouring potentials, resulting in smaller apparent voltage shifts and more exact waveforms in the stroboscopic voltage-contrast mode (Sect. 8.2.3).

Figure 5.24 shows an example of a through-lens SE spectrometer [5.94, 95]. The first concentric grid G1 and grid G4 in front of the last retarding grid are biased at 40 V, and grids G2 and G3 at 70 V.

Such retarding field spectrometers record the low-energy part of the retarding curve. The contribution of BSE can be suppressed if the construction of the spectrometer is sufficiently energy-selective to prevent BSE from hitting the scintillator directly and designed so that SE excited by the BSE cannot be collected. This needs careful shielding. A retarding-field spectrometer with spherical grids as shown in Fig. 5.20b with a massive outer collection electrode [5.99] does not fulfil this condition, and the BSE form a background in the retarding-field spectrum of the SE.

5.6.3 Backscattered and Auger Electron Spectrometers

The recording of Auger electrons (AE) (Sect. 10.1.5) in the range 50–5000 eV with line widths of the order of one electronvolt needs a spectrometer of high

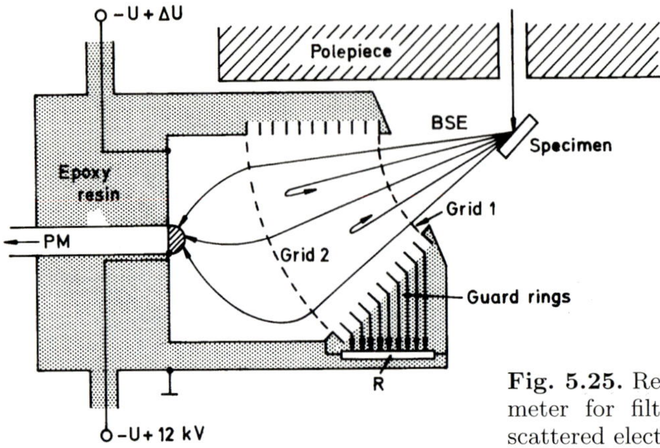

Fig. 5.25. Retarding-field spectrometer for filtering low-loss backscattered electrons [5.104]

energy resolution $E/\Delta E > 10^3$ and high collection efficiency. The most widely used spectrometer for Auger electron spectroscopy is the cylindrical mirror analyser (CMA) as shown in Fig. 5.19d. Because the Auger electrons can only leave the specimen without energy loss from a small exit depth of a few monolayers, the primary electron energy is lowered to a few keV to increase the probability of generating AE in this surface layer. In a particularly good design, the electron gun lies inside the inner cylinder of the CMA.

BSE energy spectra are of interest not only to calculate detector efficiencies. The shift of the most probable energy of pure elements depends on the atomic number (Fig. 4.14), though information about the mean value of Z can be obtained more easily from the total backscattering coefficient. However, energy spectra show a structure and become of interest for the depth analysis of layered structures (Sect. 4.1.5). A high-pass filter allows us to form an image with the low-loss electrons (LLE) only (Sects. 4.1.4 and 6.2.5). These almost elastically scattered electrons, with only plasmon or interband losses, depending on the resolution, are backscattered in a very thin surface layer. Imaging with LLE can therefore avoid the large information depth, of the order of half the electron range, which is inevitable when imaging with all BSE regardless of their energy. Shifting a selected energy window to lower energies (band-pass filter) can increase the brightness of structures at the maximum depths the BSE of selected energy can reach before being backscattered.

In a SEM it is advantageous to reduce the size of BSE spectrometers. Avalanche photodiodes in the pulse-counting mode can be operated with an energy resolution of only 1–2 keV [5.100]. Cooled p-i-n diodes can reach a resolution of 0.7 keV [5.101]. Energy spectra of BSE as shown in Fig. 4.15 have been recorded with large retarding-field spectrometers. In a SEM, such a filter can only be used for $E \leq 5$ keV. The problems that arise with this type of spectrometer have been discussed in Sect. 4.1.1. The resolution of a spherical retarding-field analyser will be of the order of only $E/\Delta E < 10^2$, which is not sufficient to resolve the elastic peak (Figs. 4.15 and 4.16). A retarding-field spectrometer of low solid angle can be used for recording BSE spectra at different take-off angles with a resolution of 150 eV for 8.5 keV electrons [5.102].

Figure 5.25 shows a retarding-field spectrometer, which consists of spherical grids in front of a scintillator–photomultiplier combination [5.103, 104]. The filter allows only BSE with no loss or energy losses $\leq$10–100 eV to pass the retarding grids. Applications of this filter will be discussed in Sect. 6.2.5. When using a STEM with the specimen immersed in the strong rotationally symmetric field of a condenser–objective lens, the magnetic field can act as a spectrometer for LLE [5.105].

A well-designed toroidal electrostatic spectrometer, which collects a hollow cone of emitted BSE with angles of 20° to the axis and a resolution $E/\Delta E = 200$, can be used both for BSE spectroscopy and for imaging with a selected BSE energy window [5.106, 107].

6. Image Contrast and Signal Processing

The most important topographic contrast mode with secondary electrons is a consequence of the dependence of the SE yield on the local tilt of the specimen surface. A fraction of the SE signal is excited by the primary electron probe and carries high-resolution information since the exit depth of the SE is small. Another fraction of poorer resolution is excited by the BSE. If, instead of using the conventional SE detector, the SE are sorted according to their exit momenta, a more quantitative interpretation of the topography may be possible.

The dependence of the backscattered electron signal on the specimen tilt, on the mean atomic number and on crystal orientation results in topographic contrast at low take-off angles and in material contrast and channelling contrast at high take-off angles. Multiple detector systems can separate and improve the different types of contrast.

Scanning electron microscopes are equipped with a standard set of analogue signal processing modes with which the range of signal intensities can be optimally displayed on the cathode-ray tube. Modern instruments offer digital image acquisition and the recorded images can be transferred to a powerful computer for digital image processing.

6.1 Secondary Electron Imaging Mode

6.1.1 Types of Image Contrast

Depending on the specimen structure we can distinguish the following types of contrast, often superimposed in SE and BSE images.

1. Topographic contrast can be generated by SE (Sect. 6.1.2) or BSE (Sect. 6.2.1). In both cases the contrast depends on the selected angular range of the electrons collected. In BSE images with a low take-off angle relative to the surface stronger shadowing can be observed.
2. Material contrast in BSE images (Sect. 6.2.1) corresponds to an increase in intensity with increasing mean atomic number owing to the increase of the backscattering coefficient. This contrast is optimally recorded by a BSE top detector at high take-off angles. SE images also contain this

type of contrast in the SE2 contribution (Sect. 6.1.3). Different materials can also be distinguished by SE by virtue of differences in the SE yield, which arise from differences in the work function and surface layers even for the same material.

3. Crystal orientation or channelling contrast (Sect. 6.2.2) caused by the orientation anisotropy of backscattered electrons can generate images in which grains of different orientations in polycrystalline material have different grey levels; these depend very sensitively on specimen tilt. SE images also contain this contrast though mostly superimposed by a stronger topographic contrast.
4. Magnetic contrast (Sect. 8.1) in SE images (type-1) can represent external magnetic stray fields and, in BSE images (type-2), internal directions of magnetization.
5. Voltage or potential contrast (Sect. 8.2) results in dark positively biased and bright negatively biased areas by retarding or repelling SE, respectively, though the external electric field also influences the contrast. For a quantitative measurement of surface potentials it is necessary to use a strong extraction field and a SE spectrometer.
6. Charging contrast should usually be avoided (Sect. 3.6.2) but can be useful in some cases (Sect. 8.2.4). Charging is also used in the capacitive coupling voltage contrast (Sect. 8.2.5), or the variation of induced charges in an external metal electrode (Sect. 8.2.6) can be used as a signal.

This listing shows that some types of contrast are peculiar to SE or to BSE while others can be observed by both. For the interpretation of images we need such a classification but we have to consider that the contrast always changes when the electron energy, the specimen tilt or the collected angular range of emitted electrons is altered. The contrast may depend on the detector configuration used and on any energy selection of the electrons emitted. A well-designed detector strategy (Sects. 6.1.4 and 6.2.3) can enhance special types of contrast superimposed on conventional images.

6.1.2 Topographic Contrast in the SE Mode

Types of Topographic Contrast. The topographic contrast in SE images is caused by several contrast mechanisms, which will be summarized with the aid of the schematic surface profile at the top of Fig. 6.1:

1. *Surface-tilt contrast* (Fig. 6.1a) results from the dependence of the secondary electron yield $\delta(\phi)$ on the tilt angle ϕ of the local surface normal relative to the incident beam.
2. *Shadowing contrast* is superimposed on the surface tilt contrast (Fig. 6.1b) when an Everhart–Thornley detector ETD on the right-hand side does not collect all SE; surface normals directed away from the detector then appear darker; the topographic contrast also depends on the azimuth angle χ of the surface normal.

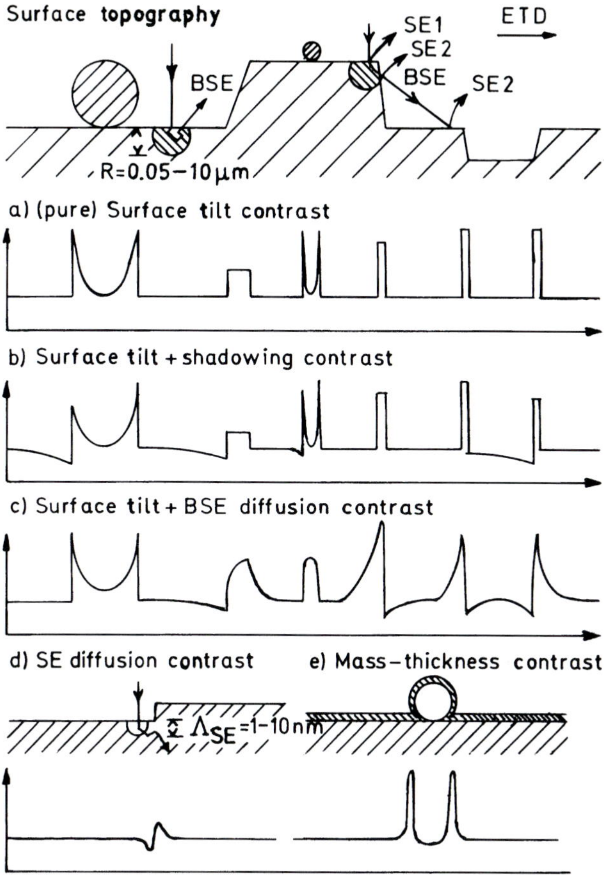

Fig. 6.1. Contributions to topographic contrast demonstrated schematically by surface contours (top) and linescans of SE signals; (**a**) surface tilt contrast, (**b**) shadowing contrast, (**c**) BSE diffusion contrast, (**d**) SE diffusion contrast and (**e**) mass-thickness contrast

3. *Diffusion contrast* is a consequence of electron diffusion with a diameter of the order of the electron range $R \simeq 0.05$–$10\ \mu$m depending on electron energy (Fig. 6.1c). Near edges, more SE2 are excited by BSE leaving the side wall, for example, and BSE can also excite SE2 when they strike the specimen again (mutual illumination, see Fig. 6.6).
4. *SE diffusion contrast* (Fig. 6.1d) arises on a much lower scale when the distance from an edge is of the order of the exit depth $t_{SE} \simeq 0.5$–20 nm of the secondary electrons.
5. *Mass-thickness contrast* can be obtained when the number of SE1 produced in a coating film is approximately proportional to the mass–thickness the primary electrons have to penetrate (Fig. 6.1e).

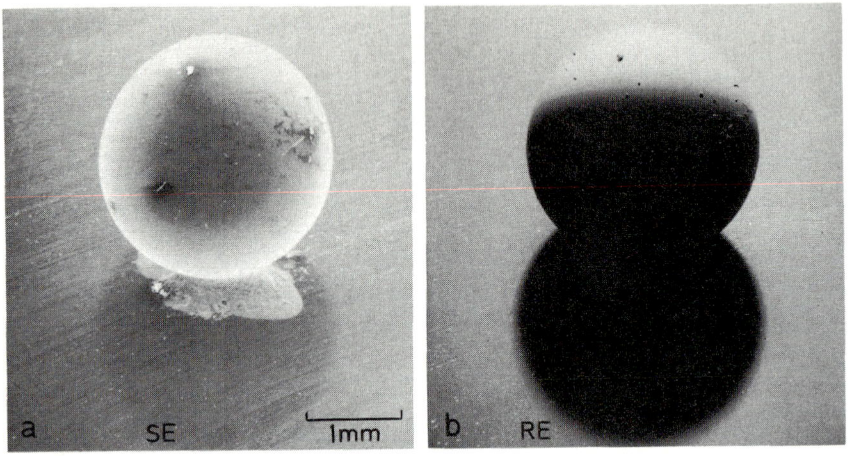

Fig. 6.2. Imaging of a 3 mm steel ball in the (**a**) SE and (**b**) BSE mode with biases of +200 V and –50 V at the collector grid of an Everhart–Thornley detector, respectively, to demonstrate the topographic contrast and the stronger shadow casting in the BSE mode

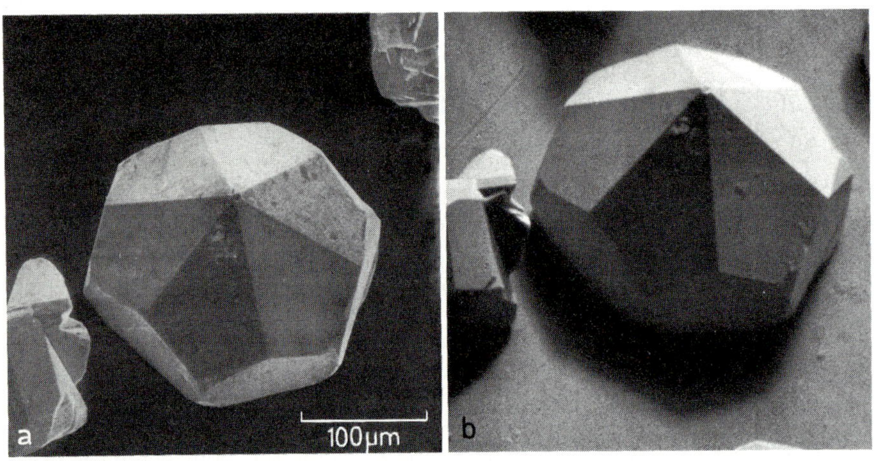

Fig. 6.3. Illustration of surface tilt and shadow contrast with micrographs of $Ge_{38}P_8I_8$ crystals in the (**a**) SE and (**b**) BSE mode

Surface-Tilt and Shadowing Contrast. Surface-tilt and shadowing contrast in the SE mode are illustrated in Fig. 6.2a, which shows a steel ball and in Fig. 6.3a where a crystal with facets is imaged. The Everhart–Thornley detector is situated at the top of the images. The sphere is imaged with surface-tilt contrast depending mainly on $\delta \propto \sec\phi$. On the side opposite to the detector, the increase of the SE signal due to high tilt angles ϕ is lower because more SE have momenta in this direction and do not reach the detec-

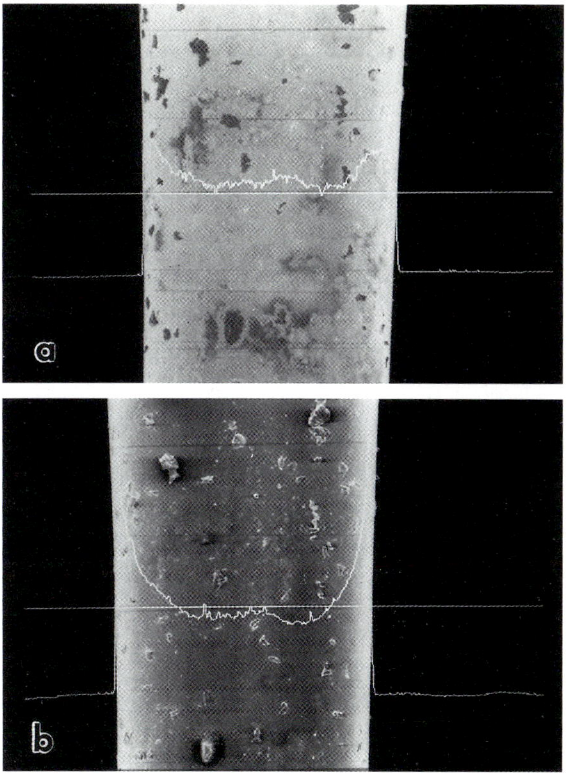

Fig. 6.4. SE imaging of a glass fibre coated with a 0.1 μm gold layer with superposed central linescans using (**a**) 1 keV and (**b**) 10 keV electrons. The Everhart–Thornley detector is situated at the top

tor (Fig. 5.8, see also the schematical lowered signal intensity of the large ball in Fig. 6.1b with the detector at the right-hand side). This type of shadowing contrast results in non-concentric isodensities (Fig. 6.13a) which should become concentric when the contrast depends only on ϕ and not on the azimuth χ of the surface normal. This dependence of the signal on azimuth for constant ϕ can be reduced when the SE are collected by an ETD inside the last lens in the in-lens mode (Fig. 5.10b) or by one of the other rotationally symmetric SE detectors described in Sect. 5.4.1.

In LVSEM the increase of δ with increasing ϕ becomes less than $\propto \sec\phi$ for $E \leq 5$ keV (Fig. 4.21). The ratio of the SE video signal at the periphery and to that at the centre of balls or fibres is therefore smaller at 1 keV owing to the weaker increase of $\delta(\phi)$ with increasing tilt angle ϕ (Fig. 6.4), especially for high Z [6.1] (see also superimposed linescans in Fig. 6.4).

When the shadowing contrast is less pronounced – and especially when it is independent of the azimuth of the surface normal when using rotationally symmetric SE detectors – it becomes difficult to decide whether surface structures are humps or depressions of complementary size. We show in Sect. 6.1.4 that the presence of additional electrodes between the specimen and the collector can decrease the electric field $\boldsymbol{E}$ at the specimen, which results in a better angular selection of the SE according to their exit momenta and in a more pronounced shadow effect. Another way of seeing the real surface topography is to record and observe stereo-pairs (Sect 6.4.5). An objective way of controlling and documenting the influence of the detector–specimen arrangement on the surface tilt contrast is to record isodensities over a small sphere, which represents all tilt and azimuth angles [6.2] (see Fig. 6.13, for example).

The shadow contrast blocks the imaging of structures inside deep holes and surface cracks. Secondary electrons can be extracted either by placing a positively biased grid in front of the specimen [6.3] or by superimposing a magnetic field in which the SE move on screw trajectories around the magnetic field lines until they reach the collecting field [6.4–6].

Diffusion Contrast. BSE diffusion contrast in the SE mode is caused by an increase of the ratio SE2/SE1 of SE emission when more electrons scattered in the forward or lateral directions leave the surface; the SE2 contribution is greater than for a flat surface with the same local tilt angle at the point of electron impact. A typical effect of diffusion contrast is the enhanced SE emission at small particles and at edges. When, as in Figs. 6.1c and 6.2a, the radius of a sphere is larger than the electron range R, each surface element scanned looks like a flat surface, and the SE signal shows the surface-tilt contrast with enhanced emission at the periphery of spherical or cylindrical specimen structures (Figs. 6.2a and 6.4). When the radius becomes comparable with or smaller than the electron range, diffusely scattered electrons generate more SE2 over the whole surface and the SE emission becomes greatest at the centre of the sphere. Figure 6.5 shows contrast reversal of such small balls when the tilt angle ϕ of the substrate is increased [6.7]. The signal from the polystyrene balls remains constant whereas the signal from the substrate increases approximately as sec ϕ.

The increase of the SE and BSE signal at edges is shown schematically in Fig. 6.1c. Diffusely scattered electrons, which could not leave a flat specimen, can leave the specimen at the edge and increase the number of BSE and SE, where the term BSE is used for all fast electrons that leave the specimen. The Monte Carlo simulations of Fig. 6.6 show how BSE emerging at the edge can hit the specimen again (mutual illumination) where they may be absorbed, backscattered or excite further SE far from the point of impact.

The increase of SE generation creates a bright zone near the edge, the width of which increases with increasing electron energy, proportional to the electron range. As shown in Sect. 3.4.1, the range increases as $E^{5/3}$ and from 10 to 30 keV the width of the bright zone increases by the ratio $(10{:}30)^{5/3} =$ 1:6. This is demonstrated in Fig. 6.7, where parts of the crystals also become transparent at 30 keV so that SE are excited at the opposite side of the crystal as well. This latter effect can also result in a brighter image intensity at the intersection of two electron-transparent fibres, for example [6.8].

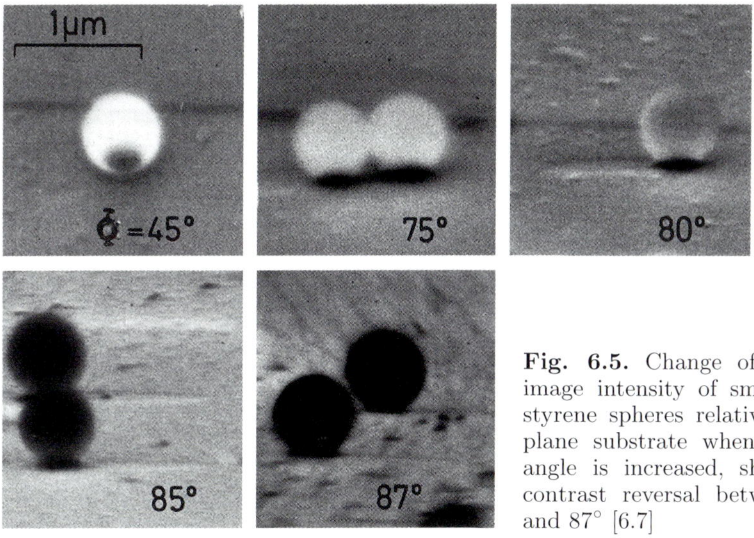

Fig. 6.5. Change of the SE image intensity of small polystyrene spheres relative to the plane substrate when the tilt angle is increased, showing a contrast reversal between 85° and 87° [6.7]

Fig. 6.6. 5 keV electron trajectories from a Monte Carlo simulation near a 50 nm surface edge of gold show BSE leaving the steep step surface and the mutual illumination when the BSE strike the specimen again

The dramatic reduction of diffusion contrast in LVSEM is demonstrated by the series of micrographs of an etched Y-shaped bar on a silicon substrate shown in Fig. 6.8. At 30 keV the exit area of BSE exciting SE2 is larger than the width of the bar. The zones of SE2 diffusion contrast overlap and the centre of the bar appears brighter than the surrounding Si surface. Reducing the electron energy decreases the exit area, and the bright zones of SE2 diffusion contrast become separated and can hardly be resolved at 1 keV.

SE diffusion contrast becomes apparent when the electron-probe diameter is of the order of the exit depth t_{SE} of electrons. Particles with dimensions

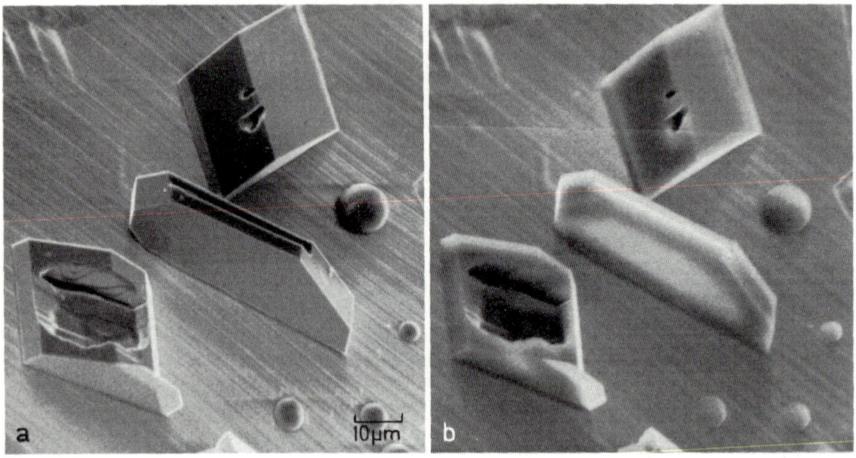

Fig. 6.7. SE micrographs of tridymite crystals and hyalite spheres on a gold substrate at (**a**) E = 10 keV and (**b**) E = 30 keV demonstrating the enhancement of the edge effects and of electron transmission at higher energy (courtesy R. Blaschke)

below one nanometre will show an increased SE signal. The topographic image can be enhanced by means of a detector at grazing incidence [6.8]. Monoatomic surface steps can be imaged at high tilt angles with $5°$–$10°$ grazing angles [6.9]. Steps seen upwards from the direction of incidence appear black and those directed downwards, bright. The same contrast can be observed for larger step heights; the contrast is then more likely to be due to BSE diffusion contrast [6.7].

SEM micrographs have much the same appearance as if the specimen were illuminated with light from the detector and observed visually in the direction of the electron beam. However, the strong increase of the signal for large ϕ and the diffusion contrast at edges have no counterparts in light optics and can result in a totally wrong impression of the surface topography (Fig. 6.8a,b). The difference image A–B recorded with two Everhart–Thornley detectors A and B opposite each other (Fig. 6.11c) corresponds more closely to light-optical illumination from the right [6.10].

6.1.3 Material Contrast in the SE Mode

Material contrast in the SE mode is caused by local variations of the secondary electron yield that cannot be attributed to differences in surface tilt. As shown in Fig. 4.8, the SE yield δ excited by high-energy primary electrons increases monotonically with increasing atomic number though the increase is less rapid than that of the backscattering coefficient η and the measured values of δ show a larger scatter around a mean curve. The main contribution to this increase of $\delta(Z)$ can be attributed to SE2 excited by the BSE on their trajectories through the surface layer – second term in (4.21) and (5.13) –

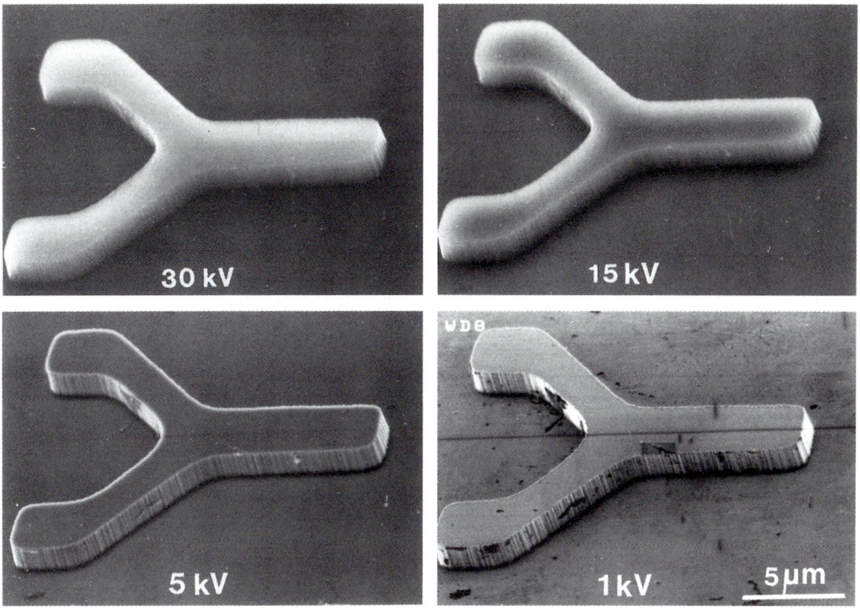

Fig. 6.8. SE image of a Y-shaped bar on silicon imaged with a primary electron energy of (**a**) 30 keV, (**b**) 15 keV, (**c**) 5 keV and (**d**) 1 keV (surface tilt $\phi = 45°$)

and at the polepiece plate – third term in (5.13). At electron energies above 5 keV, therefore, the SE image shows material contrast similar to that of the BSE image (Sect. 6.2.1). Even if a heterogeneous material like basalt (Fig. 6.9a) is covered with an evaporated conductive gold film, the SE signal is not uniform but increases with increasing mean atomic number or backscattering of the basalt components. The SE are all excited in the gold layer but areas where the underlying material has a large backscattering coefficient, like magnetite (Fe_3O_4) for example, appear brightest and those for which the backscattering is very low, like the alkali-earth–alkali-aluminosilicate glass matrix ($\simeq$ 80% Si, 10% Al), appear darkest. This contrast is much more pronounced in the corresponding BSE image (Fig. 6.9b). Typically, the SE image in Fig. 6.9a also reveals the topography caused by the polishing process. This topographic contrast is caused mainly by the diffusion process; some of these structures have a small influence on backscattering and they can be identified with weak contrast in the BSE image of Fig. 6.9b only by comparison with the SE image. Thus, each SE micrograph contains information not only from an exit depth of the order of a few nanometres but also from the larger information depth of the BSE, of the order of half the electron range, which increases with increasing energy. In Sect. 6.1.4, methods will be described for suppressing this matrix effect due to the BSE by forming the difference signal of two SE detectors placed opposite each other or by using the difference between simultaneously recorded SE and BSE signals.

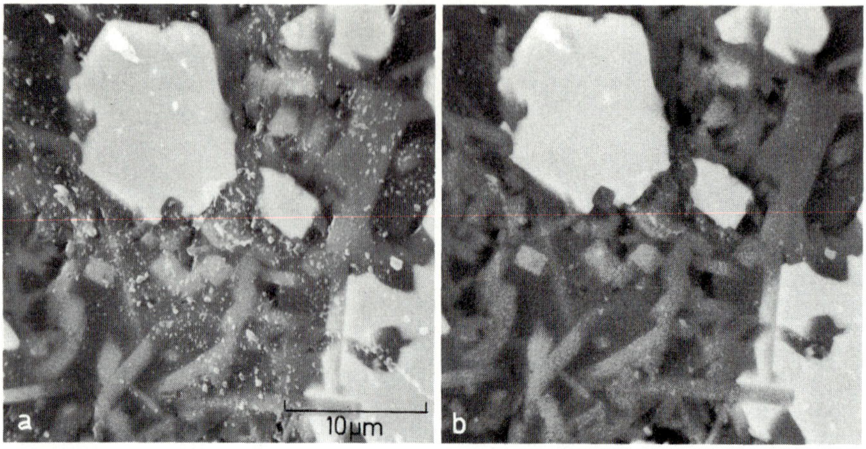

Fig. 6.9. (**a**) SE and (**b**) BSE micrographs of basalt coated with a conductive gold layer showing material contrast: bright (Fe_3O_4, magnetite), grey ($MgFeSi_2O_6$, pyroxene) and the dark matrix (alkali-earth–alkali-aluminosilicate glass, $\simeq$80% Si, 10% Al) (courtesy R. Blaschke)

Material differences in δ due to differences in the work function ϕ_w [6.13] are much more pronounced near the maximum of the $\delta(E)$ curve (Fig. 4.18b) at primary energies of a few hundreds of eV. This type of SE material contrast should, therefore, be of greater interest when an ultra-high vacuum chamber is used and further contrast effects can be created by energy filtering of the SE because their energy spectrum depends on material, crystal orientation and surface layers [6.11, 12].

After reducing the influence of the collection field of an ETD, biasing of the specimen in the range ± 20 V will change the contrast because only the more energetic SE are emitted at a positive bias. In a specimen with MgO smoke cubes with neighbouring small Cu particles, the MgO cubes dominate in brightness at negative biases; at positive biases, however, MgO and Cu are nearly equal in brightness at positive biases and more surface details can be detected [6.14]. The reason is that the higher SE yield of insulators is concentrated in the SE emission spectrum at lower SE energies than that of metals. The biasing technique is also capable of detecting Cs on Si with a 0.5% monolayer sensitivity [6.15].

With angular-resolved SE detection a contrast difference between 1×2 and 2×1 reconstructed regions on Si(100) surfaces has been observed [6.16]; real time observation of reconstruction transitions on GaAs are possible by 5°–10° grating incidence [6.17]. Doping of semiconductors influences the bending of the energy bands at the surface, resulting in differences of the work function, which can result in contrast between p- and n-type regions [6.18–21].

6.1.4 Detector Strategies for the SE Mode

We have seen that, in the SE mode, the reproducibility of SE collection with a well-defined angular selection is poor and that the SE contrast is a mixture of several different contributions. Most SEM micrographs are taken in the conventional SE mode and most problems can obviously be solved without separating and identifying the different types of contrasts. Nevertheless, a better detector strategy becomes necessary if the best use of the electron–specimen interactions is to be made and more quantitative information is required [6.22, 23]. In this section we discuss detector strategies for topographic, material and crystal orientation contrast. Special detector strategies for magnetic contrast and voltage contrast will be discussed in Sects. 8.1 and 8.2, respectively.

Suppression of the Externally Generated SE. The unwanted external generation of the signal SE3 by BSE – third term in (5.13) – can be suppressed by introducing in front of the polepiece plate a potential barrier consisting of a positively biased conductive plate behind a grid (Fig. 5.15) [6.23, 24]. The grid should be earthed so as not to disturb the electric field in the specimen chamber. The conductive plate, electrically insulated from the polepiece, can be formed by etching a copper-coated resin plate as used for the etching of electronic circuit boards. The surface of this plate should consist of a low atomic number material to absorb BSE and excite a large number of SE (MgO, for example), the latter for the BSE-to-SE conversion mode (Sect. 5.1.3) when the plate is negatively biased. A carbon-coated plate with no front grid and no biasing [6.25] will not be capable of suppressing this group of SE3 totally.

A further contribution to SE3 is furnished by BSE that leave the specimen at a low take-off angle and hit other parts of the specimen chamber or the specimen support. This contribution can be reduced by surrounding the specimen with a carbon-coated ring a few millimetres higher than the specimen plane, so that the straight BSE trajectories hit either this ring or the plate–grid combination below the polepiece. This also suppresses the contribution of BSE that directly hit the scintillator of the Everhart–Thornley detector – last term in (5.13). This ring electrode can be biased to decrease the electric field at the specimen caused by the positively biased collector grid of the SE detector (see below).

Additional Electrodes for Angular Selection. It was shown in Sect. 5.4.1 that the arbitrary angular selection of the SE exit momenta is caused by the inhomogeneous electric field between the specimen and the positively biased grid of the Everhart–Thornley detector (Fig. 5.8). This problem can be solved by separating the extraction field from the collecting field, as in the Banbury–Nixon cage (Fig. 5.11), for example. This cage can increase the topographic contrast by creating a more pronounced shadow contrast. It can be employed to increase the type-1 magnetic contrast (Sect. 8.1.1) and to linearize the voltage contrast (Sect. 8.2.1). A further advantage is that the video

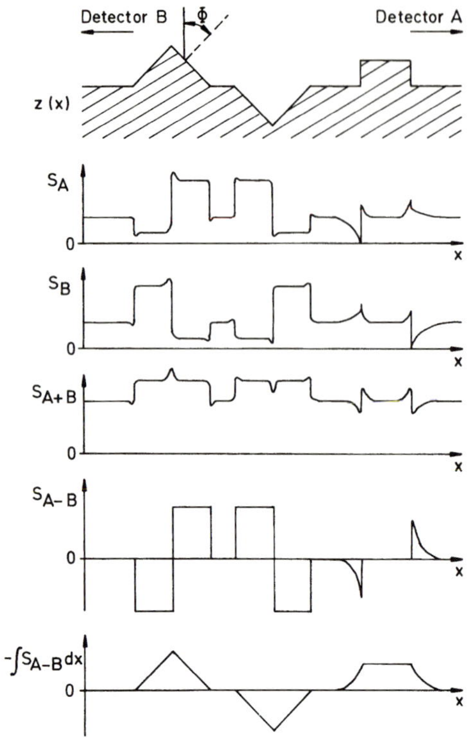

Fig. 6.10. Signal variation of different surface structures in the S_{A}, S_{B}, $S_{\mathrm{A+B}}$ and $S_{\mathrm{A-B}}$ SE signals and $-\int S_{\mathrm{A-B}}\mathrm{d}x$ for the two-detector system of Fig. 5.15

signal from central parts of the specimen stub becomes more uniform at low magnification because the electric field at the specimen becomes more uniform and the field inside the cage is rotationally symmetric. A disadvantage is that the external generation of SE by BSE at the electrode can contribute to the signal. The electrodes should be carbon-coated to attenuate multiple scattering of BSE inside the cage. The closed geometry of the cage hinders the use of other imaging and analysing modes.

Two-Detector System. When the electric field at the specimen generated by the collector bias of the SE detector is reduced, either by placing a biased ring around the specimen or by inserting a hemispherical grid of a few thin wires, the SE initially follow trajectories corresponding to their exit momenta. By using two oppositely placed Everhart–Thornley detectors, it will be possible to separate the electrons according to their exit momenta when they reach the collection fields (Fig. 5.15) [6.22, 23, 26]. This technique can also be applied to LVSEM [6.27] when the field-free space is extended so that the low-energy primary electrons are not deflected; alternatively, a special arrangement of electrostatic and magnetic quadrupole fields forming a Wien filter (Fig. 5.12) can be used, as can the electrostatic detector–objective lens in Fig. 5.13. However, any magnetic lens field between specimen and SE detector causes the SE to spiral around the field lines and selection in azimuth will be lost.

As shown schematically in Fig. 6.10, the sum A+B of the two video signals will result in contrast caused mainly by surface tilt and diffusion (and material) contrast; shadow contrast, however, will be suppressed. In the difference signal A–B, the diffusion contrast is suppressed, the material and channelling contrast are cancelled and the surface tilt and shadow contrast as well as the type-1 magnetic contrast (Sect. 8.1.1) are enhanced. The effect on the topographic contrast is demonstrated in Fig. 6.11 with the example of surface steps in the form of conductive pads on an integrated circuit. In the A and B images (Fig. 6.11a, b) and the A+B signal, all steps appear bright irrespective of their orientation relative to the detectors. This can be an advantage if this signal is used for automatic pattern recognition, for example. The contrast would be the same if the pads were grooves and, in general, surface hillocks and depressions cannot be distinguished in the A, B or A+B modes, as is also the case with a single, conventional Everhart–Thornley detector, which collects nearly all the SE. On the contrary, the A–B SE mode in Fig. 6.11c clearly shows bright and dark steps, depending on their orientation relative to the detectors, and hillocks and indentations can be distinguished immediately by virtue of the pronounced shadow effect. Such images give the impression that a light is being shone on the specimen from one of the detectors [6.10]. When the negative amplitudes of an A–B signal are suppressed, real shadows can be observed. The addition of a constant background signal as in Fig.6.11 can also fill in the shadows but has the disadvantage that the range of contrast is decreased for specimen structures with positive amplitude of the A–B signal because only about 16–20 grey levels are distinguishable by the human eye on a TV screen or print.

The two-detector system for SE not only allows us to separate different types of contrast but can also be used to obtain more quantitative information. We assume that the SE yield is proportional to sec ϕ and the angular characteristic of the emission is described by Lambert's law $\propto \cos\zeta$, ζ denoting the angle between take-off direction and surface normal. The surface tilt is characterized by the tilt angle ϕ and its azimuth χ relative to the x axis, which is parallel to the line joining the two detectors A and B. The signals detected by A and B can then be calculated by integrating over all SE with exit momenta towards the right- and left-hand sides, respectively [6.2, 6.28] (δ_0: SE yield at normal incidence):

$$S_{\mathrm{A}} \propto \delta_0 0.5\,\sec\phi(1+\sin\phi\cos\chi) \tag{6.1}$$

$$S_{\mathrm{B}} \propto \delta_0 0.5\,\sec\phi(1-\sin\phi\cos\chi) \tag{6.2}$$

$$S_{\mathrm{A+B}} \propto \delta_0 \sec\phi \tag{6.3}$$

$$S_{\mathrm{A-B}} \propto \delta_0 \tan\phi\cos\chi \propto \partial z/\partial x\,. \tag{6.4}$$

If we describe the topography of the surface by $z(x,y)$ where z is parallel to the electron beam, then the difference signal $S_{\mathrm{A-B}}$ will be proportional to the

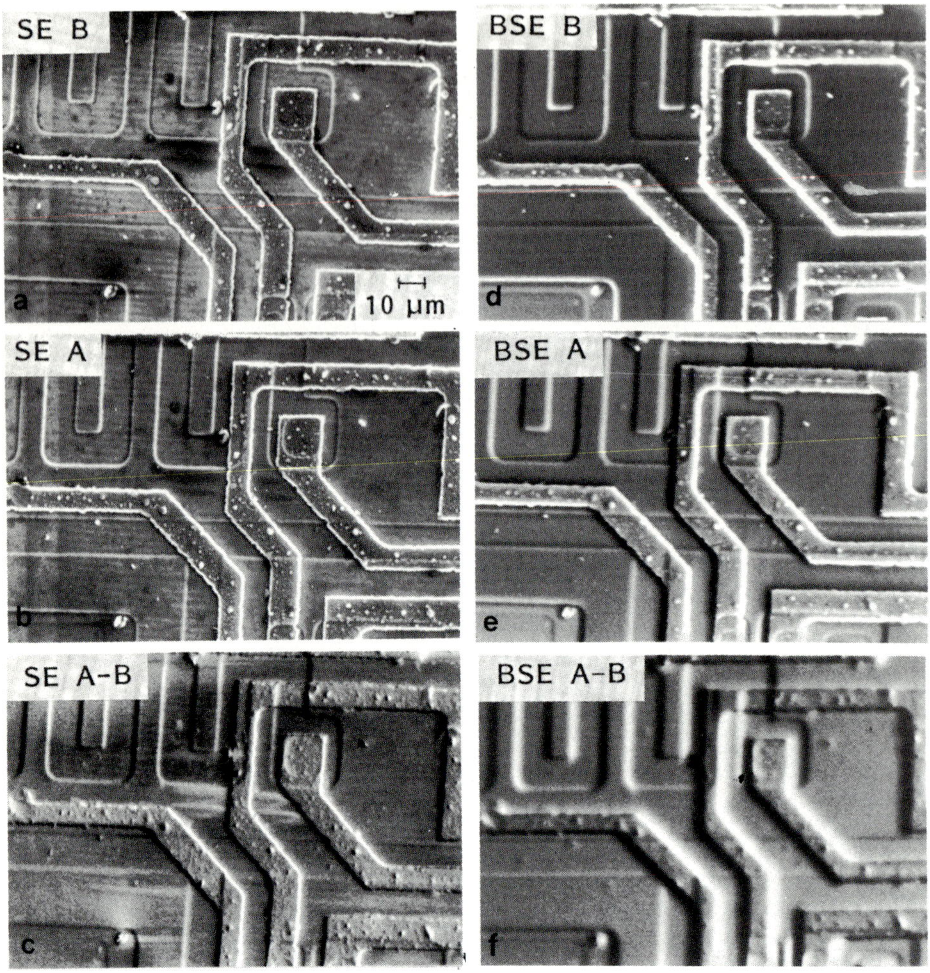

Fig. 6.11. SE and BSE micrographs of an integrated circuit recorded with (**a**) S_{B}, (**b**) S_{A} and (**c**) $S_{\mathrm{A-B}}$ SE signals and with (**d-f**) the corresponding BSE-to-SE converted signals obtained with the detector system of Fig. 5.15

partial derivative $\partial z/\partial x$ and the surface profile can be calculated by digital or on-line analogue integration of $S_{\mathrm{A-B}}$ and displayed in Y modulation on the CRT (Figs. 6.10 and 6.12) [6.22, 23, 28, 29]. Difficulties arise when we try to produce a two-dimensional map, because the value of $\partial z/\partial x$ is unknown at the start of each line. In Fig. 6.12, the analogue amplifier is set to zero at the start of each line. The derivative in the y direction can be obtained by taking a second $S_{\mathrm{A-B}}$ micrograph with the specimen stub rotated through 90° so that the y axis becomes parallel to x. The profile $z(0, y)$ of the starting points can then be calculated digitally and stored for use as initial points for

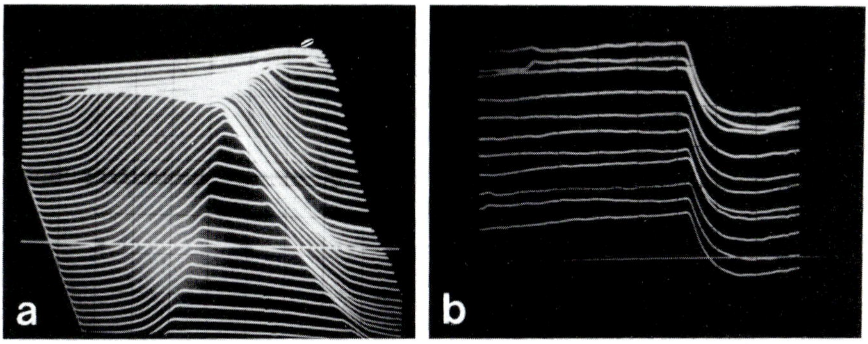

Fig. 6.12. Two-dimensional reconstruction of the surface profile by analogue integration of the $S_{\mathrm{A-B}}$ SE signal from the two detectors A and B in Fig. 5.15 for (**a**) a pyramid, 1 mm edge length and (**b**) a surface step, 2 μm in height

the calculation of $z(x,y)$ from the first micrograph. Shadow effects at steps, for example, create false step profiles but the height of the steps is correct (Figs. 6.10 bottom and 6.12b). If the specimen consists of different materials (variable δ_0),the signal $S_{\mathrm{A-B}}/S_{\mathrm{A+B}} \propto \sin\phi\cos\chi$ can be used. Such a signal can also be obtained from two oppositely placed BSE detectors (Sect. 6.2.4) but digital image processing is needed to reconstruct $z(x,y)$.

An objective method of comparing different detection and signal processing modes is to record the isodensities of a sphere, a 1 mm steel ball, for example, which contains all possible tilt and azimuth angles [6.2]. Figure 6.13a illustrates the asymmetric collection of a conventional Everhart–Thornley detector. The isodensities in Fig. 6.13c demonstrate the proportionality to $\tan\phi\cos\chi$ in (6.4). If the specimen is rotated through 90°, so that the original y axis is parallel to x, the local tilt angle ϕ and its azimuth χ can be calculated. The BSE isodensities in Fig. 6.13b and d will be discussed in Sect. 6.2.5.

Difference Between SE and BSE Signals. Another way of suppressing the BSE contribution to a SE image is to record the signal SE–k×BSE, where the SE and BSE signals are recorded simultaneously by means of a conventional SE detector and a scintillator or semiconductor detector above the specimen, respectively. The detector arrangement for the BSE should be symmetrical, using four scintillators [6.30] or an annular semiconductor detector, for example. The difference signal SE–k×BSE contains mainly surface information from the exit depth of the SE. The cancellation of BSE contributions in the SE image is, however, limited by the angular selection of BSE scattered into the upper hemisphere. The formation of A–B SE or SE–k×BSE micrographs can be performed on-line by an operational amplifier or by digital image processing.

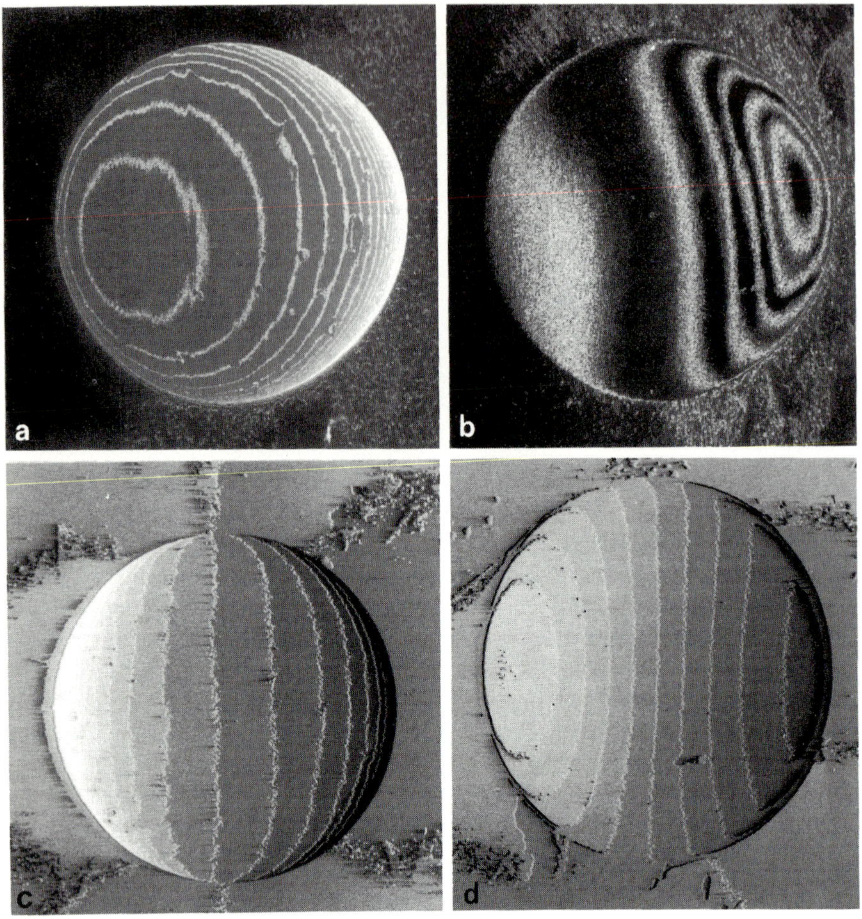

Fig. 6.13. Isodensities of SE and BSE signals from a 1 mm steel ball (**a**) in the conventional SE mode with the Everhart-Thornley detector (ETD) at the right, (**b**) in the BSE mode in which the collector grid of the ETD is negatively biased, (**c**) for the difference signal $S_{\mathrm{A-B}}$ SE $\propto \tan\phi \cos\chi \propto \partial z/\partial x$ and (**d**) for the $S_{\mathrm{A-B}}$ BSE signal $\propto \sin\phi \cos\chi$

6.2 Backscattered Electron Imaging Modes

6.2.1 Topographic and Material Contrast in the BSE Mode

The BSE images in Figs. 6.2b and 6.3b were obtained with a negatively biased collector grid, which corresponds to a BSE detector of small solid angle at low take-off. Unlike that of SE images, the contrast is now stronger for areas with a surface normal opposite to the detector and sharp shadows are seen because of the straight trajectories of the fast BSE. The shadow can be observed not only on the ball and the crystal but also behind them. The just detectable shadow in the SE image of Fig. 6.2a is produced by the same fraction of BSE in Fig. 6.2b, which also contributes of the order of 10–20% to the SE signal. At low magnification, therefore, BSE micrographs can give a better impression of the surface topography than SE micrographs, thanks to the more pronounced and sharper shadow effects obtained when a detector that only collects BSE emitted to one side of the specimen is used (Figs. 6.2b, 6.3b and 6.11d,e).

Furthermore, the surface tilt contrast is created only by BSE that travel on straight trajectories towards the detector. This contrast depends on the magnitude of the angular characteristics (Fig. 4.12) in the take-off direction. Such characteristics have been measured by placing a scintillation detector of small $\Delta\Omega$ in various take-off directions and plotting the recorded signals as superposed polar diagrams for increasing tilt angles ϕ as in Fig. 6.14. These curves show that an annular detector below the polepiece registers no significant signal variation for surface tilts below 40°–50°. At larger tilt angles, the signal decreases even though the total backscattering coefficient increases with increasing ϕ (Fig. 4.7) but into the forward direction. With a detector on one side, an Everhart–Thornley detector with a negatively biased collector grid, for example, the signal maximum occurs at medium tilt angles (Fig. 5.9). Thus, topographic and material contrast can be separated by changing the take-off angle of the BSE. If the detector is located in a more backward direction, which means at take-off angles larger than 135° relative to the direction of the incident electron beam (Figs. 5.14c), then the signal will be affected less by surface tilt and mainly by the dependence of η on the atomic number Z (Fig. 4.8). As a result, it is principally material contrast (Fig. 6.9), also known as Z or compositional contrast, that will be observed, especially for plane specimens with not too large local tilt angles. The material contrast on the basalt surface of Fig. 6.9b can be estimated by applying the sum rule (4.18); mean values $\overline{\eta} = 0.23$, 0.16 and 0.155 are found for magnetite, pyroxene and the matrix, respectively. An annular detector (Fig. 5.6 or 5.14c) allows us to suppress any shadow effect. This will be the optimum detection mode for observing material contrast and can also be used for quantitative measurement of the mean atomic number $\overline{Z}$ and the thickness of coating films (Sect. 6.2.4) or for the imaging of heavy-metal stains below the surface of biological tissues [6.31–33]. The variation of electron energy in the range 3–20 keV furnishes information from successively deeper layers [6.34]. Low energy images display the surface topography and material contrast from the uppermost layer. With increasing energy, the topographic contrast decreases and the contributions of BSE from deeper layer dominate the contrast. At low electron energies ($E \leq 5$ keV) η does not increase monotonically with increasing Z (Sect. 4.1.3). Nevertheless, differences in $\overline{Z}$ can be recognized for plane specimens, as has been demonstrated for unstained and uncoated composite polymers [6.35], for example.

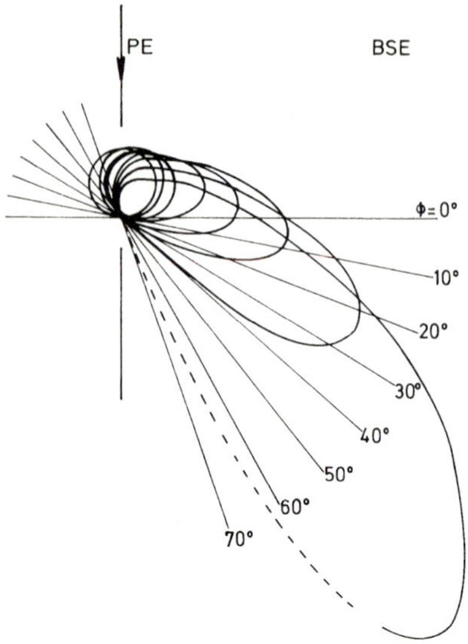

Fig. 6.14. Angular characteristic of BSE recorded with a scintillation detector of small solid angle at different take-off directions and tilt angles ϕ of the surface indicated by straight lines

When small particles of high Z on a low Z substrate are observed by BSE material contrast, the particle-to-substrate contrast increases with decreasing electron energy and passes through a maximum at a few keV depending on the particle size [6.36]. Such a maximum of contrast is also seen for thin films on a substrate. The measured contrast of colloidal Au particles used for labelling in immuno-cytochemistry and of Au films on a silicon substrate agrees with Monte Carlo simulations [6.37].

Detectors at low take-off angles show mainly surface tilt contrast, which may also be enhanced by shadow contrast. When the specimen stub is tilted, a detector at the high take-off position (Fig. 5.14a) shows predominantly material contrast and at the low take-off position (Fig. 5.14b) topographic contrast. This can also be seen by moving a small semiconductor or scintillation detector around the specimen [6.38–40]. The angular characteristics $d\eta/d\Omega$ (Fig. 4.12) for large tilt angles show that there are take-off angles for which the values of $d\eta/d\Omega$ for low and high Z material are equal; at take-off angles at the maximum of the angular characteristics (Fig. 4.12c), contrast reversal of material contrast can even be observed [6.38] so that Al becomes brighter than Au, for example. However, this effect is seen only for very flat specimens because otherwise, the topographic contrast will be much stronger. Separation of topographic and material contrast by means of a two-detector system will be discussed in Sect. 6.2.4.

In Sect. 6.1.2 we considered the diffusion contrast that arises from an increase of the SE emission at edges and from structures not plane within the exit area of the BSE. Such diffusion contrast is also to be expected in the BSE signal. However, this type of contrast depends on the take-off direction of the detector. If we consider a surface step (Fig. 6.1c), the number of BSE emitted to the right (detector A) will be increased within a distance of the order of half the electron range from the edge. This effect can be seen in Fig. 6.11e as bright zones on the right sides of the IC conductive pads (detector A) and vice versa on their left sides (detector B), see Fig. 6.11d. Whereas the SE images (a) and (b) show the diffusion contrast on both sides, the BSE images (d) and (e) show little or no decrease in intensity for the steps opposite the detectors but behind the steps is a dark shadow region, which we have seen to be a typical consequence of the straight BSE trajectories.

6.2.2 Channelling or Crystal Orientation Contrast

The backscattering coefficient of single crystals is strongly anisotropic as a function of orientation and depends sensitively on the direction of electron incidence relative to the crystal lattice planes owing to channelling effects (Sect. 9.1.4). Continuous variation of the tilt of the crystal generates a rocking curve (Fig. 9.16) and two-dimensional rocking of the electron beam produces an electron channelling pattern (ECP) (Fig. 9.23). When a large-area single crystal is scanned at low magnification $M = 20$–50, the angle of incidence varies during the scan. This is the idea behind the standard method of producing an ECP, in which an area of a few millimetres is scanned. In a polycrystalline material with smaller crystals and at higher magnification, the angle of incidence relative to the lattice planes does not vary significantly as the beam scans across each small crystal and contrast can only be observed between crystals of different orientations (Fig. 6.15) [6.41, 42]. This crystal-orientation or channelling contrast can be distinguished from the material contrast, for example, by tilting the specimen through a few degrees, which changes significantly the contrast differences between grains of different orientations (Figs. 6.15a,b) whereas no change of the contrast caused by the Z dependence of the backscattering coefficient is to be expected. Crystals appear very bright when the angle of incidence happens to lie near a low-indexed pole for which the backscattering is higher.

The following points should be borne in mind in order to obtain high crystal orientation contrast. High angular resolution of an ECP needs an electron-probe aperture smaller than 0.1 mrad (Sect. 9.2.1). For 1 mrad, fine-structured lines from high-order Laue zones (Fig. 9.23) will not be resolved but this is unimportant for channelling contrast. The rocking curve and ECP start to be blurred for apertures larger than 1 mrad. For good channelling contrast, the electron-probe aperture should lie in the range 1–10 mrad.

If crystals of polycrystalline material or single crystals are distorted by plastic deformation, the channelling contrast can vary across a single crystal and bend contours can occasionally be observed as lines where the incident electron beam is nearly parallel to a set of lattice planes [6.43, 44]. It will be shown in Sect. 6.2.5 that it is even possible to image the strain field of a single dislocation by this channelling contrast when using energy filtering

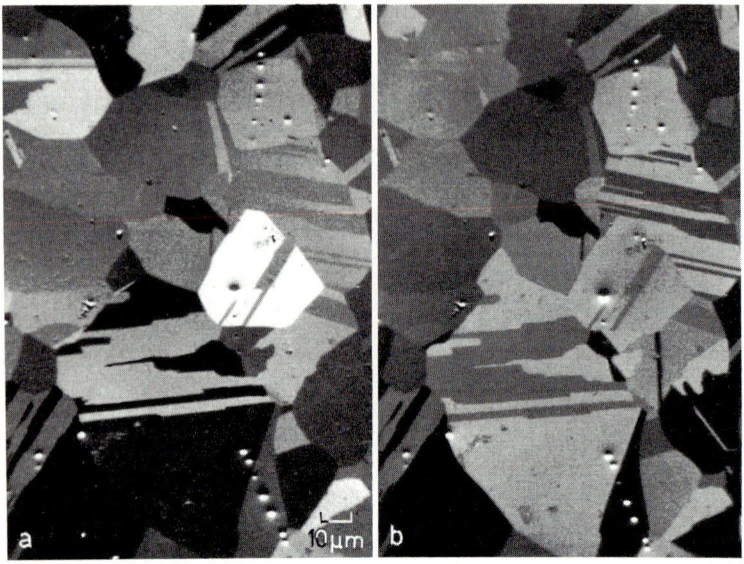

Fig. 6.15. Crystal orientation contrast in a BSE micrograph of an electrolytically polished, polycrystalline copper specimen. Note the change of contrast of single grains when the specimen is tilted through 1° between (**a**) and (**b**) [6.35]

of the BSE. The channelling contrast that is caused by the primary Bloch-wave field is generated only in a very thin surface layer of the order of a few nanometres thick since the primary Bloch waves are strongly absorbed. The information depth of channelling contrast is therefore very low and the contrast becomes very sensitive to the surface conditions of the crystal, a Beilby layer after mechanical polishing or a contamination or oxide film, for example. The best method of preparation is either electrolytic polishing or ion-beam etching (Sect. 9.2.3).

The greatest change $\Delta\eta/\eta$ of the backscattering coefficient is only a few percent for E = 10–20 keV and decreases with increasing energy (Fig. 9.17). As these variations in contrast are so small, the surface has to be flat so that the channelling contrast is not dwarfed by stronger topographic contrast, and a black level correction (Sect. 6.4.2) has to be applied to transform the small variations in η into larger relative variations of the CRT intensity. The signal-to-noise ratio has to be adequate and larger electron-probe currents have to be used to get a noise-free image. This aspect is also discussed in Sect. 9.2.1 for ECP.

An increase of primary electron energy decreases the channelling contrast but increases the information depth by augmenting the absorption length of the primary Bloch-wave field; the channelling contrast thus becomes less sensitive to the surface layers. A decrease of the primary electron energy increases the channelling contrast but also increases the multiple scattering inside any surface layer, thereby widening the effective beam aperture and blurring the channelling contrast. There is thus an optimum electron energy, which depends on the thickness of the surface layer [6.45].

The main contribution to channelling contrast and ECP comes from direct backscattering from a surface layer a few nanometres thick, and the energy lost by these electrons is very small. This backscattered fraction and hence the orientation contrast increases with increasing tilt angle $\phi = 50^\circ - 80^\circ$ and using a detector at low take-off angles [6.46]. Another group of electrons, which are to some extent forward scattered out of the primary Bloch-wave field and have a variable probability of escaping as BSE, will suffer larger energy losses. A scintillator or semiconductor detector will therefore provide better contrast because the number of photons or charge carriers is proportional to the BSE energy. In other modes, the specimen-current mode (Sect. 6.2.6), for example, the contrast will be weaker because both low and high energy BSE contribute one elementary charge to the specimen current.

The secondary electron yield δ also exhibits an orientation anisotropy (Sect. 9.1.5) and corresponding contrast can be observed in the SE mode though overlaid by a stronger, pronounced topographic contrast.

6.2.3 Contrast of BSE Using Multiple Detector Systems

The first multiple detector system proposed for BSE (Fig. 5.16a) used two semi-annular semiconductor detectors A and B below the polepiece. The detectors A and B indicate opposite surface tilt and shadow contrast. The latter can be seen in Fig. 6.11d,e at the conducting pads of the integrated circuit. These micrographs were obtained by using BSE-to-SE conversion at a converter plate situated below the polepiece and collection of the converted SE by a two-detector system (Fig. 5.15). This type of detector arrangement was introduced to increase the material contrast and decrease the topographic contrast with the A+B signal and to decrease material and increase topographic contrast with the A–B signal. An example of a eutectic Al–Ag alloy is shown in Fig. 6.16. Figure 6.17 shows schematically that the result is not exactly as predicted when surface tilt, shadow and diffusion contrast are taken into account. Surface tilt contrast is indeed cancelled in the A+B and enhanced in the A–B mode (left side of Fig. 6.17). However, the shadow contrast at steps (right side) is not completely cancelled in the A+B mode. The A–B mode shows for the right step a bright and for the left a dark intensity, which corresponds to illumination with light from the right. However, this results in a false topographic appearance. The conducting pads in Fig. 6.11f, for example, look quite different and more trapeziform when compared with the A–B SE mode in Fig. 6.11c. Another image artifact of the A–B BSE mode can be observed at boundaries between high- and low-Z materials without a surface step. When the electron beam approaches the boundary in Fig. 6.17 from the left and reaches a distance smaller than the diameter of the electron diffusion cloud, of the order of 0.1–3 μm, electrons that are scattered to the right can penetrate the weakly absorbing low-Z material and reach the detector A. The consequence is a surviving bright signal in the A–B mode, which looks like a surface step higher on the side of the high-Z material. Conversely, the contrast will be reversed if the low-Z material is on the left and the high-Z material on the right. Such intensity profiles at a boundary have been confirmed by Monte Carlo simulations [6.47]. Figure 6.16d shows this image artifact at the boundaries between the Ag and Al rich phases of the eutectic alloy. In this A–B BSE image recorded with 30 keV primary electrons, the material contrast, which appears in the A+B BSE image (Fig. 6.16c), has been cancelled but there appear to be steep steps at the boundaries though the specimen was only mechanically polished. A comparison with the A–B BSE image (Fig. 6.16b) recorded with 5 keV electrons does not show this image artifact because the electron range is smaller and hence the diffusion contrast reduced. The A+B BSE image in Fig. 6.16a demonstrates that, for the same specimen position, the decreased electron range results in quite different material contrast from that seen in Fig. 6.16c: it is not the lamellar structure at larger depths that is imaged but the surface distribution smeared by the polishing. We observe no difference between the A–B SE modes at 5 and 30 keV and the micrographs are very similar to Fig. 6.16b, but with a better topographic resolution and contrast. We conclude, therefore, that such a signal-mixing method should not be used for BSE for resolutions better than a few micrometres.

In spite of these artifacts of the A–B BSE image, a multiple detector system offers the possibility not only of recording a signal that is more or less proportional to the backscattering coefficient but also of making use of the exit characteristics of BSE. In order to exploit this, four-detector systems with a take-off angle of the order of 45° relative to the electron beam have also been proposed (Sect. 5.4.3), using semiconductor detectors, four scintillators coupled by fibre optics to two photomultipliers (Fig. 5.14f) or four scintillator sheets mounted on two light-pipes with a shutter system (Fig. 5.14e). The sum signal can be used as with an annular detector for imaging the material and channelling contrast and for quantitative measurement of the mean atomic number Z, while difference signals can be used to determine the surface orientation and to reconstruct the surface profile (Sect. 6.2.4).

6.2.4 Quantitative Information from the BSE Signal

As shown in Sect. 4.1.3, the backscattering coefficient increases monotonically with increasing atomic number Z [Fig. 4.8 and Eqs. (4.14–4.17), and a simple mixing rule (4.18) can be applied for compounds. A mean atomic number Z can be therefore related to a measured value of $\overline{\eta}$. We discussed in Sect. 4.1.1 a device for the accurate measurement of η inside a SEM. Measurement of η with a semiconductor detector or scintillator has the advantage that secondary electrons are not recorded because of their small energy. A

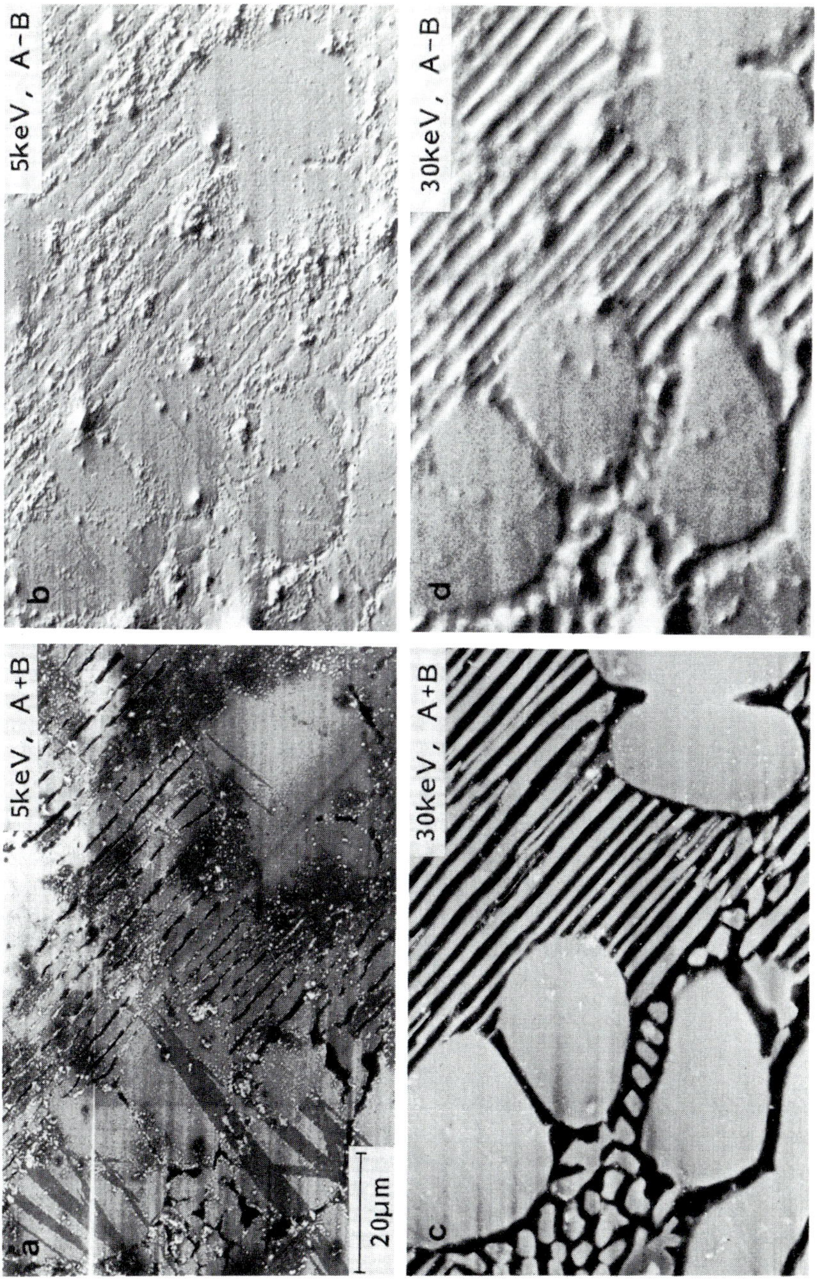

Fig. 6.16. BSE micrographs of the same area of a eutectic Al-Ag alloy, mechanically polished, in the A+B mode at (**a**) $E = 5$ keV and (**c**) 30 keV and in the A–B mode at (**b**) 5 keV and (**d**) 30 keV

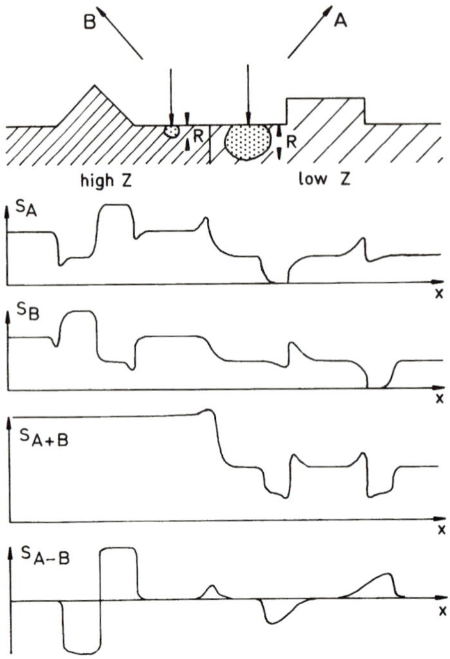

Fig. 6.17. Signal variations of different surface structures in the S_A, S_B, S_{A+B} and S_{A-B} BSE modes of a two-detector system with detectors A and B opposite one another below the polepiece

disadvantage is that a standard of known η_{st}, electropolished copper, for example, has to be used to correct the measured BSE signal by means of a calibration curve representing $S_{BSE}(Z)/S_{BSE}$(standard) [6.48, 49]; the latter is not identical with $\eta(Z)/\eta_{st}$ because the energy and angular selection of the detectors modify the increase of S_{BSE} with increasing Z. A quantification of calcium in bone mineral distribution is possible by relating the BSE signal to measured intensities of the Ca Kα x-ray line [6.50].

The sensitivity in the detection of small differences in $\overline{Z}$ also depends on the superimposed topographic and material contrast. The human eye is capable of comparing mean intensities of larger areas with a better resolution in intensity than a quantitative record by a linescan or digital computation of a mean value, which is affected by all the statistical variations of the image intensity. In favourable cases, it is possible to detect differences in Z of unity. The measurement of a mean value of $\overline{Z}$ can be used as an aid for identifying phases of known composition; elements of low atomic number, which are not detected by an energy-dispersive x-ray detector, also contribute to $\overline{\eta}$ by the sum rule (4.18). A mean value of $\overline{\eta}$ will also be useful for quantitative x-ray microanalysis because some correction programs (Sect. 10.3) make direct use of this quantity in the backscattering correction.

A further quantitative application is the measurement of the film thickness of self-supporting films, for which η increases linearly with thickness, or of coating films on a substrate (Fig. 4.2) [6.51, 52]. When particles on a low atomic number substrate are not too large, the total additional signal by backscattering from the particles in a two-dimensional scan can be integrated after subtraction of the background and used to measure the total particle mass if the composition is known and if the single large-angle scattering approach is valid [6.53].

Two oppositely placed BSE detectors A and B can be used to establish the orientation of a surface when the latter is plane over a radius of about $R/2$. Figure 6.14 shows that at take-off angles of the order of 45° relative to the electron beam and for tilt angles ϕ smaller than 40°–60°, the angular characteristic of the BSE signal can be approximated by Lambert's law $\propto \cos\zeta$, ζ being the angle between take-off direction and the surface normal; this law is represented by a circle in the polar diagram. Unlike the SE emission, where the SE yield increases as sec ϕ, the backward emission of BSE is nearly constant inside this range of tilt angles (Fig. 6.14). The signal recorded by two detectors A and B is therefore obtained from (6.4) by omitting sec ϕ:

$$\begin{aligned} S_{\mathrm{A}} &\propto 0.5\eta(1+\sin\phi\cos\chi) \\ S_{\mathrm{B}} &\propto 0.5\eta(1-\sin\phi\cos\chi) \\ S_{\mathrm{A+B}} &\propto \eta \\ S_{\mathrm{A-B}} &\propto \eta\sin\phi\cos\chi) \,. \end{aligned} \tag{6.5}$$

A second pair of detectors at an azimuthal angle $\chi_2 = 90°$ in a four-detector system (Sect. 6.2.4) records the signal

$$S'_{\mathrm{A-B}} \propto \eta\sin\phi\cos(90° - \chi) = \eta\sin\phi\sin\chi \,. \tag{6.6}$$

These equations can be solved for ϕ and χ by digital computation and from a knowledge of the local values of ϕ and χ, the surface profile can be reconstructed [6.54–58] as discussed also for the SE difference signal in Sect. 6.1.4.

6.2.5 Energy-Filtering of Backscattered Electrons

The information depth of a normal BSE image is large, of the order of half the electron range from which electrons can leave the specimen as BSE. Structures at progressively greater depths below the surface are increasingly blurred due to the broadening of the primary electron beam by multiple scattering and electron diffusion. BSE backscattered at greater depths travel along longer pathes and hence lose more energy. Energy-filtering of low-loss electrons (Sect. 5.6.3) is, therefore, a method of decreasing the information depth.

Before discussing the low-loss electron mode, we examine types of energy filtering that affect the contrast but are less effective than the introduction of an electron-energy spectrometer. The specimen current mode (Sect. 6.2.6) and the BSE-to-SE conversion at an MgO-coated converter plate (Sect. 5.4.2) generate signals that are proportional to the number of BSE emitted regardless of their energy. The conversion of BSE at a metal plate even produces the largest signal for low-energy BSE. Otherwise, the signal of semiconductor and scintillator detectors is proportional to the BSE energy. Indeed, the width of diffusion contrast at edges, for example, is reduced for the latter by a factor of the order of two. Another way of selecting high-energy BSE is to transmit them through foils that are just transparent to the high-energy BSE but absorb the low-energy part of the BSE spectrum. The transmitted BSE are converted to SE at the exit and these can be recorded by an Everhart–Thornley detector [6.59]. This is used to enhance magnetic type-2 contrast, for example.

The most effective way of reducing the information depth and increasing the resolution and contrast is to select low-loss electons (LLE) with a retarding-field spectrometer (Fig. 5.25) [6.60, 61]. It has been shown that the information depth can be reduced to that of secondary electrons [6.62, 63]. The signal only contains information from this strongly reduced information depth whereas the SE signal not only contains the high-resolution information from SE excited by PE (SE1) but also from SE excited by the BSE (SE2). The contrast of a LLE image can therefore be better than that of a SE image.

The channelling contrast (Sect. 6.2.3) is generated inside a thin surface layer of the order of the absorption length of the dynamical theory of electron diffraction (Sect. 9.1.4). Lattice defects like the strain field of dislocations or the phase shift of stacking faults influence the propagation of Bloch waves. This is an important technique in TEM for the investigation of lattice defects. Theoretical calculations also predict that BSE from thin foils will produce contrast [6.64] and this has been confirmed experimentally with thin foils in the scanning transmission electron microscope (STEM) mode of a TEM and in a field-emission STEM [6.65, 66].

For bulk specimens the channelling contrast is weaker owing to the strong background of multiply and diffusely scattered electrons. The total orientation anisotropy of the backscattering coefficient when the angle of electron incidence is changed is only of the order of a few percent (Fig. 9.17). The contrast anticipated is mainly produced by electrons that have been removed from the primary Bloch-wave field by single large-angle scattering, for which the energy losses will be low; energy filtering can therefore considerably increase the contrast, to the same values as in the STEM mode with thin foils. Using a field-emission gun and a retarding-field electron filter (Fig. 5.25), single defects in silicon and silicon layers on substrates can be resolved in the LLE mode [6.61, 67]. By rocking the electron beam, an electron channelling pattern (ECP, Fig. 9.16) can be recorded, from which the relative orientation of the electron beam and the specimen can be derived. The contrast of dislocations disappears when the angle of incidence corresponds to a Kikuchi band of a reciprocal lattice vector $\boldsymbol{g}$, so that $\boldsymbol{g} \cdot \boldsymbol{b} = 0$ where $\boldsymbol{b}$ is the Burgers vector of the dislocation. The same disappearance rule is used in TEM for determining the Burgers vector.

6.2.6 The Specimen-Current Mode

Part of the incident electron-probe current I_p passes through the specimen to earth, this is called the specimen current or absorbed current. All the SE and BSE that leave the specimen decrease this current. If $\sigma = \eta + \delta$ is the total yield of SE and BSE, the specimen current becomes

$$I_\mathrm{s} = I_\mathrm{p}[1 - (\eta + \delta)] \; . \tag{6.7}$$

This situation will be reached when all the SE and BSE are collected by a positively biased external electrode. Figure 6.18 shows that the measured values of I_s plotted against the specimen tilt angle ϕ agree rather well with the values calculated from (6.7) using the values of $\eta(\phi)$ and $\delta(\phi)$ of Figs. 4.7 and 4.19, respectively. The specimen current changes sign when $\sigma = \eta + \delta > 1$ for large tilt angles. This means that more SE and BSE leave the specimen than PE are absorbed. The tilt angle for zero current will also depend on the primary electron energy E because it is mainly δ that increases with decreasing E.

However, (6.7) will not be exactly correct and the situation depends on the specimen and specimen chamber geometry because multiply scattered BSE and SE produced externally by the BSE can find their way back to the specimen. Applying a negative bias to the specimen guarantees that all SE leave the specimen and that SE generated externally by the BSE are retarded. The multiply scattered BSE will be unaffected by this bias. If, for a signal $\propto (1 - \eta)$ complementary to a BSE signal, the SE contribution is suppressed by a positive bias of the specimen, the externally produced SE will be attracted by the specimen. For this case, it will be better to surround the specimen by a negatively biased grid, which retards both the internally and externally produced SE.

Such a specimen-current signal $\propto (1 - \eta)$ has the advantage that all BSE emitted into a full hemisphere of solid angle 2π contribute, a condition that can hardly be satisfied by a BSE detector. A consequence of this large solid angle of collection is that there are no shadow effects but the dependence of the backscattering coefficient on the local tilt of the surface normal produces a surface tilt component in the topographic contrast.

The specimen-current mode can be employed to provide quantitative material contrast, channelling contrast and type-2 magnetic contrast because these modes need a large electron-probe current to increase the signal-to-noise ratio [6.68]. The recording of low electron-probe currents has been discussed in Sect. 5.1.

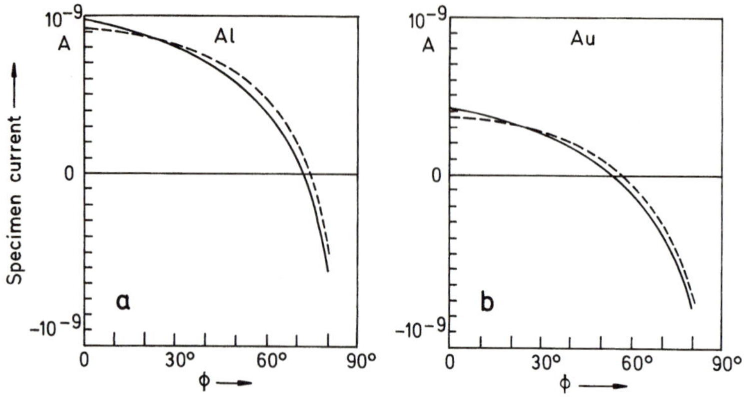

Fig. 6.18. Dependence of the specimen current on the surface tilt angle ϕ for (**a**) Al and (**b**) Au (——) experiment, (- - - -) calculated by (6.7)

6.3 Resolution and Contrast in the SE and BSE Modes

6.3.1 Secondary-Electron Mode

Resolution and contrast are related quantities. Very high resolution is worthless if the contrast and/or signal-to-background ratio are insufficient. The resolution of SE micrographs is limited by the exit depth t_{SE} of SE and the information volume of the order of t_{SE}^3. The expected distribution of the SE signal is blurred owing to the current density distribution in the electron probe of diameter d_p. When discussing such blurring effects, it must be remembered that the SE signal consists of a contribution SE1 with an exit diameter of the order of $(d_p^2 + t_{SE}^2)^{1/2}$ and of SE2 and SE3 with an exit diameter of the order of the electron range. Only SE1 contribute to high resolution. The SE3 should be reduced as far as possible as discussed in Sect. 5.4.1, because the SE1 contribute only a fraction of 10–50% of the total SE signal, depending on material and detector geometry. Also, delocalization of the excitation of SE has be taken into account (Sect. 2.4.4).

The exit depth of carbon and organic material is of the order of 10 nm and more, depending on density. Contrast and resolution can be increased by coating with a thin metal film. Unfortunately, sputter-coated films for preventing charging are often of the order of 10 nm in thickness and consist of coarse crystals; this grainy structure will be imaged at high resolution, especially when gold is used as the coating material [6.69, 70]. A metal coating intended to increase resolution and contrast should be uniform and only 1–2 nm thick [6.71, 72]. Evaporated films of Pt, W or Cr as used in TEM fulfil this condition best and can be prepared by planar magnetron sputtering [6.73], for example, or by evaporating 1.5 nm Pt(Ir)/C layers [6.74]. Shadow-casting by oblique evaporation is employed for the investigation of surface structures on replicas and inorganic and organic particles and macromolecules on substrate films in a TEM. This should be a useful technique for improving the contrast in SEM for conductive specimens also, as local variations of the thickness of the coating film will cause differences in mass-thickness contrast. A problem can, however, arise if the coating-film thickness that is appropriate for contrast enhancement provides insufficient conductivity to prevent charging. Cryo-fractured biological specimens have been coated unidirectionally at an angle of 45° with an evaporated 1.5–3 nm platinum-carbon film and covered by a 5–7 nm carbon film [6.75]

The resolution of biological macromolecules can also be increased by heavy-metal impregnation methods using phosphotungstic acid, uranyl acetate or osmium tetroxide mordanted by tannic acid [6.76]

6.3.2 Backscattered Electron Mode

The main aspects of the discussion of resolution and contrast with SE given in Sect. 6.3.1 can be transferred to BSE. When using a scintillator or semiconductor detector below the polepiece at normal incidence, we have to consider that the width of BSE emission (Fig. 4.28) has a diameter of $\simeq R$ and structures inside an information volume of radius $\simeq R/2$ can contribute to the BSE image. However, particular structures can be imaged with a better resolution [6.77] when their topographic or material contrast significantly exceeds the background signal of BSE from the whole information volume. The backscattering caused by the coarse crystals of a Au coating film, for example, allows 5–10 nm to be resolved in the BSE mode [6.31]. An extreme example is a groove or crack in a plane surface or a gap between particles that is much narrower than R but larger than the electron-probe diameter d_p. In such cases, diffused electrons have to penetrate a larger distance, which results in an appreciable decrease of the BSE signal. The material contrast of a 5 nm superlattice of $\mathrm{Al_{0.5}Ga_{0.5}As/GaAs}$ can be resolved [6.78]. In such cross-sections of layered structures, the contrast between low Z (dark) and high Z (bright) material depends on electron energy and the contrast decreases rapidly with tilting. At optimum energies depending on the width of the layers, the contrast can even be larger than expected from the backscattering coefficients of the materials. At the low Z layers, the scattering of electrons at the neighbouring high Z layers can result in decreased backscattering, whereas at high Z layers, electrons can more easily escape from the specimen as BSE through the low Z layers [6.79, 80].

We have seen that tilting the specimen and using a BSE detector at a low take-off angle increases the topographic contrast. Contrast and resolution can be increased by locating the detector at the reflection-like maximum of the BSE exit characteristic (Fig. 4.12), where most of the BSE have been backscattered at lower depths with lower energy losses, whereas the diffusely scattered BSE are also scattered to high take-off angles. The other extreme case is the use of low-loss electrons as discussed in Sect. 6.2.5, where the information depth can be reduced to the same order as for SE. Otherwise

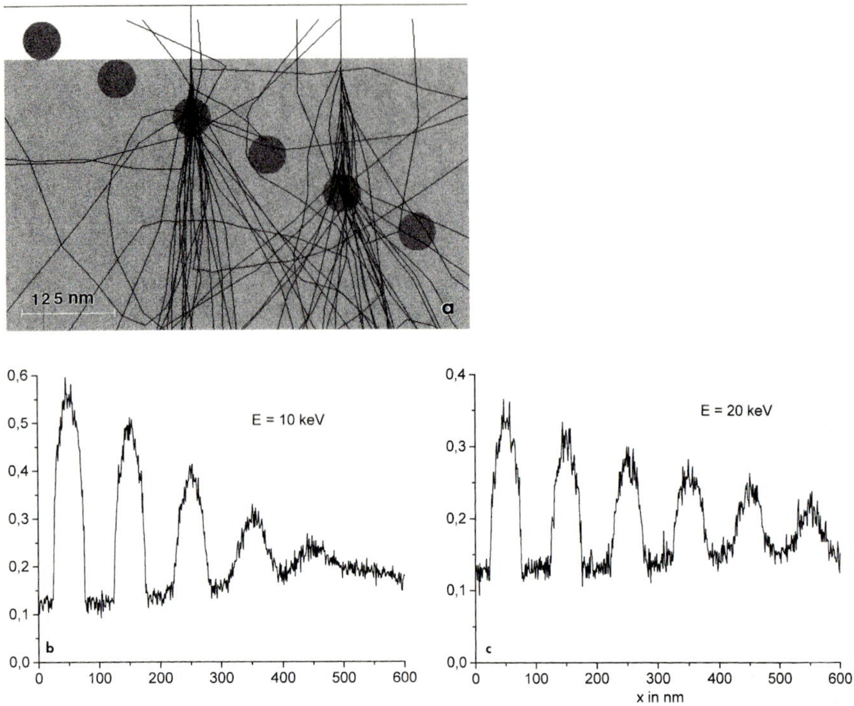

Fig. 6.19. Monte Carlo simulation of linescans across 50 nm cylinders of gold at different depths in an aluminium matrix. (**a**) Geometry with superimposed 20 keV trajectories. BSE linescans for (**b**) 10 keV and (**c**) 20 keV across this structure (calculated with the Monte Carlo program MOCASIM [3.135])

each BSE signal contains a certain fraction of LLE information, which is normally too small to be detected when all the BSE are recorded. Angular selection and energy filtering can appreciably increase this type of contrast and improve the resolution. It is much more important to know the physics of backscattering and the properties of the detection system than to discuss the contrast and resolution for special geometries.

BSE material contrast allows inhomogenities below the surface to be imaged within an information depth that increases with increasing electron energy. However, the beam broadening caused by multiple scattering increases with depth and limits the resolution. Figure 6.19a shows full cylinders of gold 50 nm in diameter in an aluminium matrix at different depths $d = -25 + 50n$ in nm (n=1–6). Superposed 20 keV trajectories obtained by Monte Carlo simulation show the increase of beam broadening with depth. The simulated linescans of the BSE signal across this structure, shown in Figs. 6.19b,c, for E= 10 and 20 keV, respectively, demonstrate the decreasing net intensity and the increasing blurring with increasing depth. Structures not resolved at 10 keV can be resolved at 20 keV.

6.4 Image Recording and Processing

6.4.1 Observation and Recording on Cathode-Ray Tubes

A SEM image is normally displayed on a cathode-ray tube (CRT). Most instruments have separate tubes, one with a green-yellow emission at the maximum of the human eye sensitivity and a long luminescence decay time for visual observation, the other with a blue emission and a short decay time for recording micrographs of area 5×5 cm^2 with a film camera. The frame time can be varied from TV frequency (50 frames per second) to 1–10 s for visual observation and can be extended up to a few minutes. Such long recording times are necessary to increase the signal-to-noise ratio. For resolving fast changes of the specimen structure, frame frequencies of 300 images per second have been achieved [6.81].

At the exit of the head amplifier, the video signal is normally proportional to the number of electrons or quanta recorded, which means that the detector and the head amplifier are linear. Even if the main amplifier, which modulates the bias of the Wehnelt electrode of the CRT, is linear, the CRT emission current versus Wehnelt bias becomes nonlinear owing to the triode characteristics of the electron gun; the light intensity of the CRT versus the video signal is often represented by the characteristic curve $\gamma = 1$ in Fig. 6.20 [6.82]. Further nonlinearities are introduced during the photographic recording of the CRT image and during photographic processing in the dark-room. By digital image acquisition (Sect. 6.4.3) these problems can be avoided. Grey levels can be resolved with 7–9 bits resulting in a large dynamic range and dark-room work is unneeded when the images are digitally processed and reproduced by a printer.

The number of lines per frame can be increased from a single line for generating a linescan as an oscilloscope trace, up to 1000–2000 lines per frame to record an image and it is assumed that a frame contains of the order of 10^5–10^6 image points (pixels), which is a useful number for estimating the noise (Sect. 4.2.4) and for digital recording.

The line structure of the image causes some typical artifacts, which result from the continuous recording in the x and the discontinuous recording in the y direction (or conversely if the scan direction is rotated by 90°). When scanning normal to a sharp edge, for example, excesive magnification of the image can suggest that the edge is limited by bows arising from the final diameter of the electron beam in the recording tube [6.83], or a small point can appear twice in adjoining lines [6.84]. Furthermore, the intensity of the CRT beam and its spot diameter cannot be varied independently. The diameter increases with increasing beam current (Fig. 6.21).

At the specimen too, the diameter of the electron probe should be of the same order as the line spacing. Information can be lost if the diameter is smaller (underscanning) and the image becomes blurred if the diameter is larger (overscanning) than the line spacing. Such artifacts can be avoided by

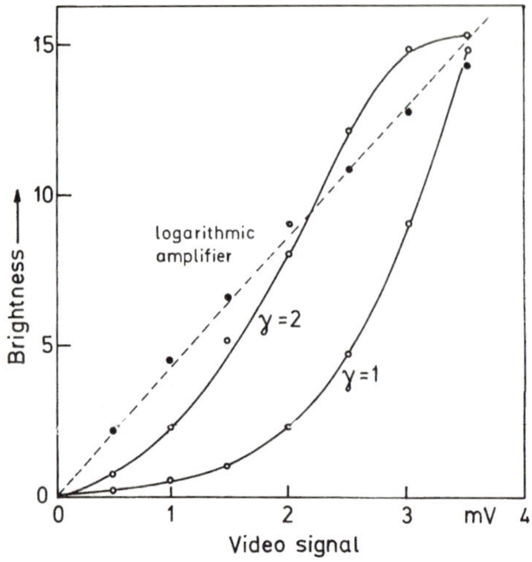

Fig. 6.20. Brightness of the cathode-ray tube versus the video signal for the normal mode ($\gamma = 1$). Partial linearization of the curve by going to $\gamma = 2$ and more complete linearization by using a logarithmic amplifier are also shown [6.82]

Fig. 6.21. Large optical magnification of the screen of a cathode-ray tube to demonstrate that increasing brightness is combined with an increase of spot diameter

ensuring that the recording-beam diameter d_c of the CRT and the lines on the specimen just overlap. To quote an extreme case, it will not be useful to work with a small electron probe at 1000-fold magnification. The maximum probe diameter that just shows no loss of resolution at a magnification M will be given by

$$d_{p,max} = d_c/M \tag{6.8}$$

which means that for $d_c = 0.1$ mm and $M = 100$, the probe diameter $d_{p,max}$ should be 1 μm, for example. The magnitude of the useful maximum magnification of a SEM can be estimated as the ratio of the recording-beam diameter d_c and the electron-probe diameter d_p

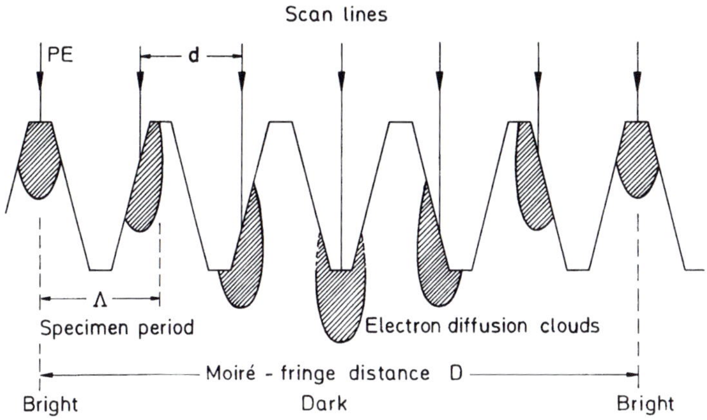

Fig. 6.22. Observation of moiré-fringe distance D when imaging a periodic specimen structure of period Λ with a separation d of the linescans perpendicular to the plane of the figure

$$M_{\max} = d_c/d_p \ . \tag{6.9}$$

For $d_c = 0.1$ mm and $d_p = 10$ nm, for example, this gives in $M = 10^4$. In the context of digital image recording, this problem of under- and overscanning is discussed in [6.85].

If a resolution of 0.2 nm is assumed with the naked eye, the maximum useful photographic magnification of a SEM micrograph will be fourfold only for a CRT spot size of 0.1 mm and a twofold optical demagnification of the 10×10 cm^2 screen onto the 5×5 cm^2 photographic film. This is quite different from a TEM, where micrographs can be recorded directly with a resolution of the order of 20 μm and can be magnified tenfold. In principle, a higher storage capacity is only possible if the electron-beam diameter of the recording system is decreased to $\simeq$10 μm, which needs a second SEM rather than a CRT [6.86] and direct exposure of the photographic emulsion inside the vacuum of this system. The useful optical magnification will be 20–fold and a 10×10 cm^2 exposed area can be magnified to 2×2 m^2 which corresponds to a hundredfold larger area at the same resolution.

Another imaging artifact that arises when the lines do not overlap at the specimen is the moiré effect [6.87]. In Fig. 6.22, an electron probe scans across a periodic structure of periodicity Λ parallel to the grooves and the electron-probe diameter d_p is smaller than the line spacing d. A minimum of the signal will be observed whenever the electron beam coincides with the bottom of the grooves, which results in an intensity modulation of large periodicity D

$$D = \Lambda/(\Lambda - d) \ . \tag{6.10}$$

D takes very large values when the difference between Λ and d is small. This effect is demonstrated in Fig. 6.23. The image is free of artifacts when

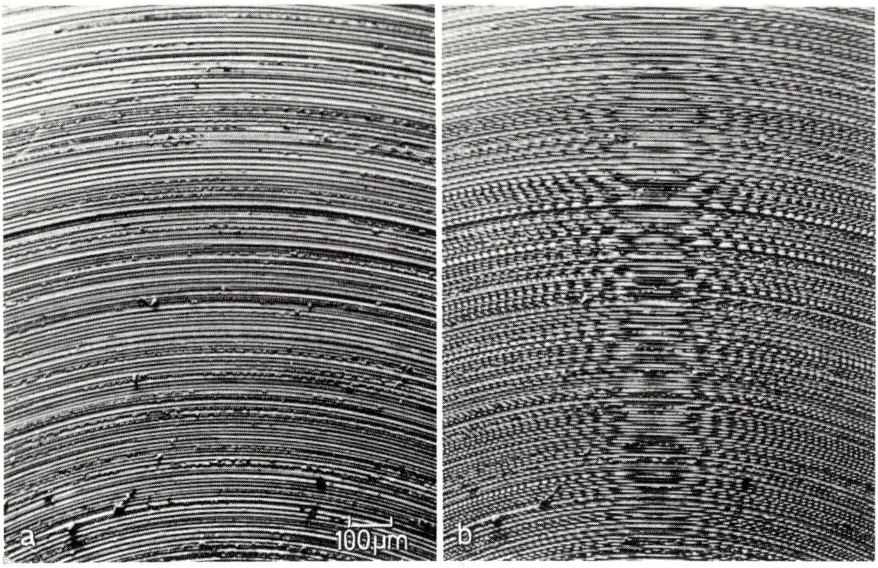

Fig. 6.23. Circular grooves in a copper specimen caused by polishing and observed (**a**) with the scan direction parallel to y and perpendicular to the grooves and (**b**) parallel to x and approximately parallel to the grooves, resulting in a moiré effect [6.87]

scanning in the y direction normal to the grooves (Fig. 6.23a) but shows a moiré pattern when scanning in the x direction approximately parallel to the grooves (Fig. 6.23b).

6.4.2 Analogue Signal-Processing Methods

Allthough analogue has been almost wholly supplanted by digital image processing, it is still worth examining many of the analogue techniques, some of which have been translated into digital processing programs.

Analogue signal processing can be applied between the head and main amplifiers. A γ-control, black level control (subtraction of a constant background U_0) and signal differentiation can modify the input video signal U_v to

$$U_{out} = c_1(U_v^{1/\gamma} - U_0) + c_2 \frac{dU_v}{dt} . \tag{6.11}$$

A γ-value larger than unity increases the contrast at low signal level and a γ-value smaller than unity increases the contrast at high signal levels (Fig. 6.24). The action on the light intensity of the CRT is illustrated for $\gamma = 2$ in Fig. 6.20. Figure 6.25 shows an example of how the image contrast changes from $\gamma = 1$ to $\gamma = 4$.

The black level control subtracts a constant signal U_0 thereby increasing the contrast of small signal amplitudes in the presence of an approximately constant signal. For example, signal variations $\Delta U_\mathrm{v}/U_\mathrm{v} = 0.1$–$1\%$ can be detected as 1–10% signals when 90% of U_v is subtracted and the difference is amplified up to the full scale that can be represented on the CRT screen. Normally, the human eye can at best distinguish intensity variations $\Delta I/I \geq 5\%$. The successful application of this method requires a low noise level and a uniform background intensity across the image frame.

The signal can be differentiated by coupling the signal to an operational amplifier by means of a capacitor. Each spatial or temporal frequency (Sect. 2.4.1) yields a signal

$$\frac{\mathrm{d}}{\mathrm{d}t} U_0 \sin \omega t = \omega U_0 \cos \omega t = \omega U_0 \sin(\omega t - \pi/2) \,, \tag{6.12}$$

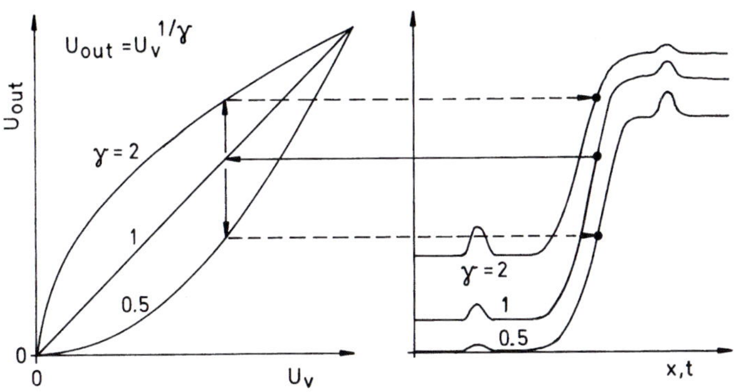

Fig. 6.24. *Left*: characteristics of an amplifier with $U_\mathrm{out} = U_\mathrm{v}^{1/\gamma}$. *Right*: Modification of small contrast differences at low and high signal levels after applying a γ-control

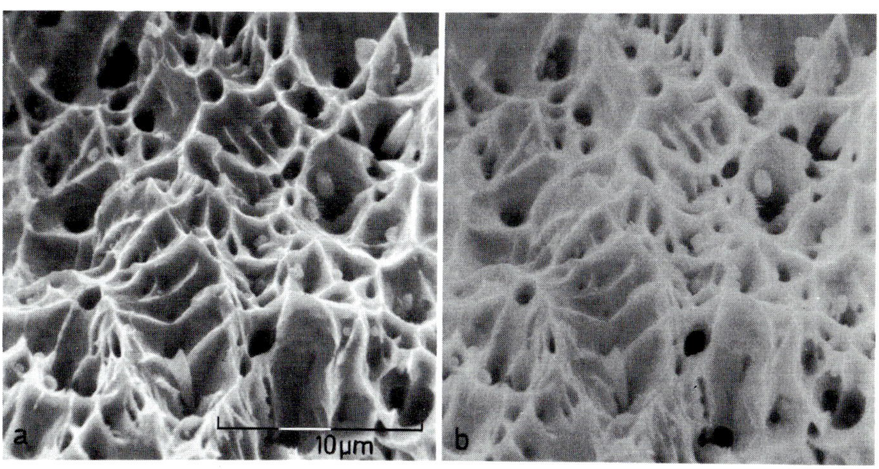

Fig. 6.25. Change of contrast when applying a γ control with (a) $\gamma < 1$ (normal mode) and (b) $\gamma = 4$ (fracture surface of copper)

which means that differentiation is equivalent to a frequency filter proportional to ω that suppresses slow and enhances fast signal variations and also introduces a phase shift of $\pi/2$. As an example, Fig. 9.23b shows the electron channelling pattern of a silicon single crystal obtained with the standard method and black level modification; in Fig. 9.23d the same diagram is displayed as $\mathrm{d}U_\mathrm{v}/\mathrm{d}t$. The increased intensity inside the central 111 pole is suppressed. Kikuchi lines and bands appear sharper and with higher contrast. In particular, the fine structure inside the 111 pole can be distinguished more clearly. For other specimens and imaging modes, this technique again has advantages whenever fast variations of small amplitude are to be imaged in the presence of slow variations of large amplitude. The sudden increase of the signal at an edge gives a positive value of the differentiated signal and conversely, a decrease results in a negative value. Edges parallel to the linescan will show no contrast. Differentiated signals therefore show typical asymmetries similar to shadow effects, which are more generally the consequence of the phase shift $\pi/2$ mentioned above. These asymmetries can be avoided by recording $|\mathrm{d}U_\mathrm{v}/\mathrm{d}t|$ or $\mathrm{d}^2U_\mathrm{v}/\mathrm{d}t^2$ or by superposition of an additional image with the linescan parallel to y [6.88]. The contrast of small intensity variations can also be enhanced by recording a linear combination (6.11) of the normal and the differentiated signal.

An integrating operational amplifier with a capacitance in the feedback loop can be used when the signal is proportional to the slope $\partial z/\partial x$ of the surface facet (z parallel to the specimen normal, x parallel to the scan direction, see Sect. 6.1.4 for details). Another application is the determination of the total mass of a particle in STEM when the signal is proportional to the local mass–thickness [6.89]. It has been proposed [6.82] that a logarithmic operational amplifier should be used to linearize the light intensity of the CRT screen (Fig. 6.20) or to create a signal proportional to the mass–thickness in STEM [6.89] (see also homomorphic filtering below).

Contrast reversal with $U_\mathrm{out} = c_1 - U_\mathrm{v}$ can be of interest for some applications. Figure 6.26a shows the palaeozoic replica of a coccolith in bitumen lias ϵ in the normal SE mode. The replica appears dark, as an indentation, and can be seen bright, in relief, by applying contrast reversal (Fig. 6.26b). In this particular case, we then get a better impression of the original form of the coccolith.

A signal can be displayed quantitatively not in the form of grey levels but as an oscillogram by employing the Y modulation technique, either by superposing one linescan on the normal micrograph (Figs. 2.22, 6.2) or by examining the specimen with a reduced number of linescans (Fig. 9.23c). For example, the intensity of a characteristic x-ray peak can be superimposed in Y modulation (Fig. 10.40h,i), thus presenting the image of the specimen and the elemental concentration along one line. Too large a number of linescans in Y modulation can result in a confusing image. With several linescans, Y

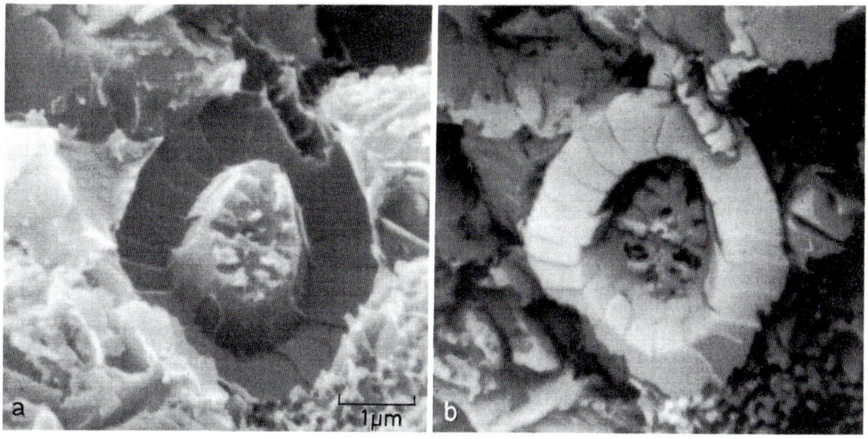

Fig. 6.26. Example of the use of contrast reversal: (**a**) the negative, palaezoic replica of a coccolith in bitumen lias ϵ and (**b**) a positive impression after contrast reversal

modulation offers the possibility of displaying a two-dimensional intensity map (Fig. 9.23b).

Isodensities can be plotted by first quantizing the signal (Fig. 6.27a) into a distinct number of grey levels (b), using threshold amplifiers and then differentiating the signal (c,d). By forming differences between the signals (a) and (b), contours can be mapped (e) [6.90, 91]; these enables us to present a larger range of grey levels per image. Normally, the human eye can distinguish on TV screens and paper copies only 16–20 grey levels (for examples of isodensities, see Fig. 6.13).

So-called homomorphic filtering [6.92] can be applied when the signal consists of a product $U_{\rm v}(t) = u(t) \cdot v(t)$ of a slowly varying function $u(t)$ and a rapidly varying function $v(t)$. A logarithmic amplifier forms the signal $\log U_{\rm v} = \log u + \log v$ and the two parts are then separated by high- and low-pass filters and amplified with different amplifications $1/N$ ($N = 1$–2) and $C = 1$–3, respectively. After passing the results through an exponential amplifier, a signal $U_{\rm out} = u^{1/N} \cdot v^C$ is obtained. This means that the signal components u and v are imaged with different values of γ, N or $1/C$, respectively.

Periodic signals can be selected by a phase-sensitive lock-in amplifier. The amplitude of the periodic signal can be 60 dB (a factor of 1000) smaller than the noise amplitude [6.93]. The operation of a lock-in amplifier can be characterized as follows. A reference signal of the same periodicity (frequency) as the input signal changes the polarity of the latter each half-period. A sine-wave signal in phase becomes a dc half-wave signal. The time average after passing through an integrating circuit becomes non-zero. This time average decreases as $\cos \varphi$ if there is a phase shift of φ between the input and the

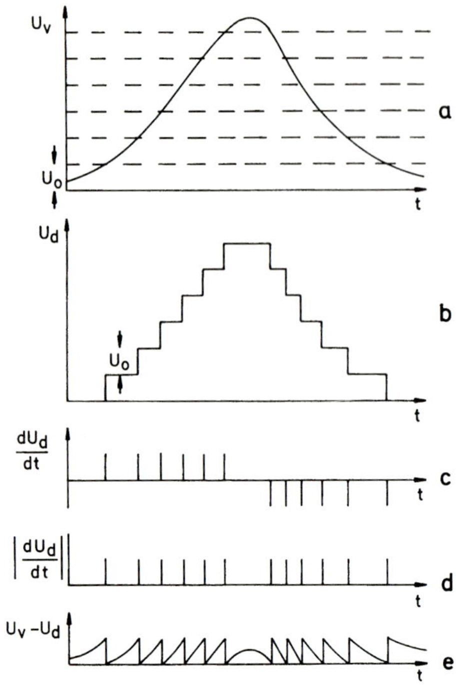

Fig. 6.27. (**a**) Original signal. (**b**) Digitized signal with a voltage step U_0. (**c**) Production of dark and bright isodensities by differentiation of (b). (**d**) Production of bright isodensities by using absolute values of (c). (**e**) Contour mapping by subtraction of (b) from (a)

reference signal. Signal contributions with the wrong frequency and statistical noise will be attenuated to very small values, which decrease with increasing averaging time per pixel.

6.4.3 Digital Image Acquisition

For all quantitative work that makes use of the recorded image intensity, it will be more convenient to acquire and store the image digitally [6.94]. Various image processing procedures can then be applied without repeatedly scanning the specimen, as would be necessary for the analogue signal processing methods described in Sect. 6.4.2.

It is difficult to adapt an analogue-to-digital converter (ADC) to the varying scan frequencies that are used for observing and recording an image. It is therefore much better to let the computer produce the signals for the x and y scan coils itself, so that the speed of scanning from point to point is dictated by the speed of the ADC and the data acquisition and storage rates. Alternatively, the digital image is stored temporarily in the SEM to produce a stationary image on the CRT and is transferred to the main computer on request. Slow-scan commercial acquisition cards are available for recording 4096×4096 pixels with 12–14 bit resolution.

Two main types of ADC are in use. The integrating-type (or Wilkinson ADC) is described in Sect. 10.2.2. Here, a capacitor is charged during the integrating cycle and the time of discharge by a constant current source is then a measure of the signal amplitude. This type is limited in frequency by the total cycling time, which is of the order of 1–50 ms. The approximating-type of ADC digitally increases a voltage, which is compared with the instantaneous signal value. This ADC needs cycling times of only 1–50 μs and is, therefore, in general faster than the integrating type. However, statistical noise of the input signal will be recorded when the integrating time constant is not of the order of the cycling time. The speed of ADCs depends also on the fabrication process. For recording at higher rates, there are other ADCs that can be used up to frequencies of 500 MHz.

The best way of decreasing the noise by averaging over one sampling interval per pixel is to use an integrating amplifier with a capacitor in the feedback loop and a sample-and-hold amplifier for the period during which the capacitor is discharging between the reading of two pixels. As in the case of photographic recording from a CRT, there is no point in averaging over a large number of scans because the scan can shift during the protracted recording time. It is better to use a single scan with a longer sampling time per pixel.

The standard frame size for digital image recording is 1024×1024 pixels. Although only of the order of $16 = 2^4$ grey levels can be distinguished on the TV screen, it is useful to store the signal with 256 levels corresponding to an 8-bit pixel depth and even pixel densities at 10 bit resolution are available. This requires 1024×1024×8 bit = 1 Mbyte of memory (1 byte = 8 bit, M = 1024×1024 in computing circles). Storage arrays of this size are available, which can be read at TV rate and displayed on a TV screen.

6.4.4 Digital Image Processing

Digital image processing can be performed on a small microcomputer when only the simpler procedures are needed. A larger computer system working with 16-bit processors and optionally a vector-array processor [6.95] must be used when faster operation is required and more complicated procedures are to be invoked. Image processing systems therefore vary in cost and flexibility. We shall discuss only a few important techniques to give some idea of the capabilities of image processing; for reviews about image processing see [6.96–103].

After the acquisition of the image, we have to distinguish between image enhancement, restoration and analysis [6.104].

Image enhancement procedures typically involve manipulation of the contrast and intensity levels and edge enhancement. Most of these procedures are familiar from the analogue signal-processing methods of Sect. 6.4.2 but can now be implemented two-dimensionally. It is an important advantage of digital image processing that it is no longer necessary to scan the specimen a large number of times to adjust accurately the black level and intensity so as to record the image within the intensity range of the CRT and the photographic emulsion. It will only be necessary to control the maximum signal level so that it does not exceed $2^8 = 256$ grey levels or more. The first step is to look at the histogram of the image, that is, the curve representing the frequency of occurrence of pixel intensities. Figure 6.28b is the histogram of Fig. 6.28a and shows that the intensities from the dark Al-rich regions of a eutectic alloy are separated from those from the bright Ag-rich regions. When there are no pixels at low intensity, a cursor can be set in the histogram to subtract this level (black level) and setting another cursor at the maximum intensity allows us to expand the intensity scale to fill the full range (scaling). Figure 6.29 shows how channelling and topographic contrast from a platinum surface can be accentuated by scaling the levels 190–230 to fill the full scale of 256 levels. Arbitrary γ-values can be used when γ-control or logarithmic scaling, for example, are implemented digitally. A further control procedure, which can be made quantitative, is the presentation of a linescan profile in Y modulation but now the line can start and end at arbitrary points selected by cursors. A Y-modulated image can easily be presented with a selected number of lines. If the image contains an intensity gradient across the image, a continuously varying black-level subtraction and/or a continuous scaling factor can be applied (shading). Displaying isodensities presents no problems because the image is already digitally stored and only the signal levels have to be selected.

A further typical analogue image enhancement procedure is differentiation of the signal. Here, we see a disadvantage of the analogue technique with which differentiation can be performed only along a scan line and not in other directions. This and related operations can be carried out digitally by introducing a 3×3 or 5×5 matrix, which contains weighting factors. The matrix is centred at a point of the image at which the computation is to be carried out and a new image intensity is calculated as a sum of the intensities of the central and the eight neighbouring points, multiplied by the appropiate weights. One- and two-dimensional differentiation can be effected by means of the following matrices $M(i,j)$, for example:

$$\frac{\partial}{\partial x} = \begin{pmatrix} 0 & 0 & 0 \\ -1 & 1 & 0 \\ 0 & 0 & 0 \end{pmatrix}; \quad \frac{\partial}{\partial y} = \begin{pmatrix} 0 & -1 & 0 \\ 0 & 1 & 0 \\ 0 & 0 & 0 \end{pmatrix};$$

$$\frac{\partial}{\partial x} + \frac{\partial}{\partial y} = \begin{pmatrix} 0 & -1 & 0 \\ -1 & 2 & 0 \\ 0 & 0 & 0 \end{pmatrix}; \quad \frac{\partial^2}{\partial x^2} + \frac{\partial^2}{\partial y^2} = \begin{pmatrix} 0 & -1 & 0 \\ -1 & 4 & -1 \\ 0 & -1 & 0 \end{pmatrix}.$$

Figure 6.28c shows the effect of applying $\partial/\partial y$ to Fig. 6.28a. These methods can generally be described as a convolution $\otimes$ of the image $f(n,m)$ with the $k \times k$ matrix $M(i,j)$ ($k = 3$ or 5, $\nu = (k-1)/2$):

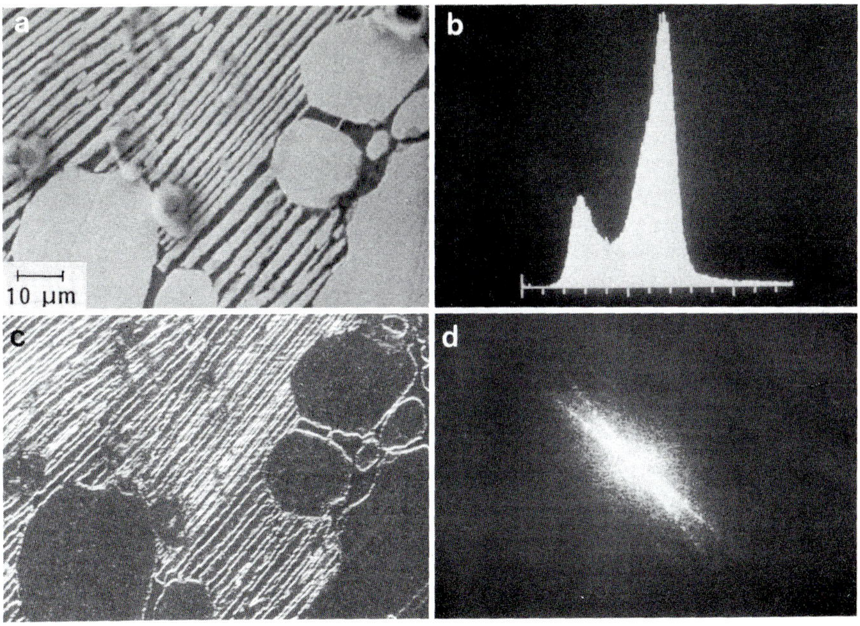

Fig. 6.28. Examples of digital image processing: (**a**) original SE micrograph of a eutectic Al-Ag alloy, (**b**) histogram of intensities, (**c**) derivative $\partial/\partial y$ perpendicular to the scan lines, (**d**) Fourier transform pattern $|F(u,v)|^2$

$$f'(n,m) = \sum_{i=-\nu}^{+\nu} \sum_{j=-\nu}^{+\nu} f(n+i, m+j) \cdot M(i,j) = f \otimes M \tag{6.13}$$

and correspond to the application of a filter in reciprocal or Fourier space.

The matrices

$$\begin{pmatrix} 0 & -1 & 0 \\ -1 & 5 & -1 \\ 0 & -1 & 0 \end{pmatrix} \quad \text{and} \quad \frac{1}{16}\begin{pmatrix} 1 & 2 & 1 \\ 2 & 4 & 2 \\ 1 & 2 & 1 \end{pmatrix}$$

represent superposition of the original image and its Laplacian $(\partial^2/\partial x^2 + \partial^2/\partial y^2)$, which enhances edges, for example, and averaging to suppress noise, respectively. A better way of removing noise is *complex hysteresis smoothing* [6.105].

Another possibility is to apply filtering processes in Fourier space, and Fourier transforms offer further possibilities for detecting image periodicities, for deblurring and for the implementation of routines requiring auto- and cross-correlation. For an analogue function $f(x,y)$ defined in the interval $0 \le x \le x_0$ and $0 \le y \le y_0$, the Fourier transform $\mathcal{F}$ is defined by

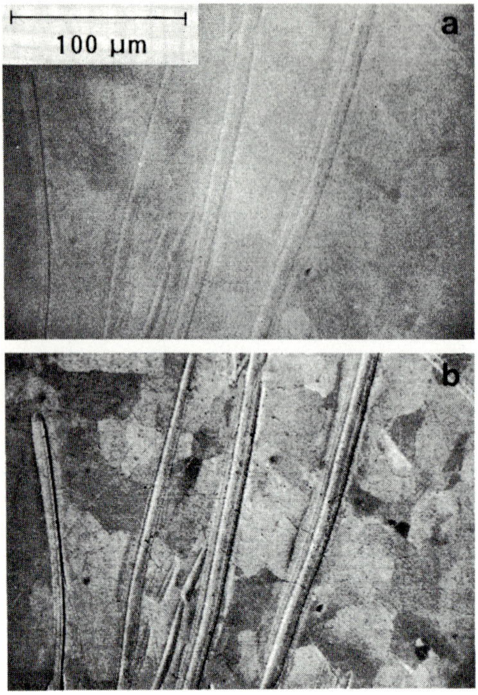

Fig. 6.29. Digital scaling of (**a**) the original SE micrograph of a platinum surface by (**b**) expanding the grey levels 190–230 to the full scale of 256 grey levels, showing the channelling contrast of the crystal grains [6.109]

$$\mathcal{F}\{f(x,y)\} = F(q_x,q_y) = \frac{1}{x_0 y_0}\int_0^{x_0}\int_0^{y_0} f(x,y)\exp[-2\pi\mathrm{i}(q_x x+q_y y)]\mathrm{d}x\mathrm{d}y \quad (6.14)$$

where q_x and q_y are spatial frequencies. The discrete Fourier transform $\mathcal{F}$ of $f(n,m)$ regarded as an $N \times N$ matrix can be written

$$\mathcal{F}\{f(n,m)\} = F(u,v) = \frac{1}{N^2}\sum_{n=1}^{N}\sum_{m=1}^{N} f(n,m)\exp[-2\pi\mathrm{i}(un+vm)/N] \,. \quad (6.15)$$

The exponential can be split into a cosine and sine term by Euler's formula, $\exp(\mathrm{i}x) = \cos x + \mathrm{i}\,\sin x$, which produces the real and imaginary parts of $F(u,v)$. Such a Fourier transform seems to need $2\,N^2$ multiplications of $f(n,m)$ with a trigonometric function. This enormous number of multiplications can be reduced to $1.5\,N\ln N$ by using the Fast Fourier Transform (FFT) algorithm [6.106].

The real function $|F(u,v)|^2 = F(u,v)\cdot F^\star(u,v)$ is identical with the light-optical Fraunhofer diffraction pattern, which can be used to detect and measure mean periodicities in the image [6.107–109]. Fourier transform of Fig. 6.28a. The lamellae of the Al-Ag eutectic alloy act as a diffraction grating. The two rows of diffuse diffraction spots correspond to the two main directions of the lamellae, which can be recognized in the micrograph of Fig. 6.28a.

We showed in Sect. 2.4.4 that the diffraction pattern of two superposed micrographs, recorded sequentially from the same specimen area and shifted by a small distance, contains periodic Young interference fringes from all structures that appear in both micrographs whereas statistical noise does not contribute to the fringes. Such a pattern as that shown in Fig. 2.29 was obtained by digital superposition of two images shifted by 10 pixels in the x and y directions and a digital FFT.

By suitably attenuating either low or high spatial frequencies u, v in $F(u,v)$, we can perform high-pass or low-pass frequency filtering, respectively; the filtered image is then obtained by an inverse Fourier transform. This method of Fourier filtering can also be used for image restoration. An image blurred (convolved) by the approximately Gaussian profile of an electron probe at high magnification or by the much broader exit distribution of backscattered electrons (Fig. 4.28) does not contain high spatial frequencies. The convolution theorem tells us that the Fourier transform of a convolution (2.41) of the specimen function $f(x,y)$ and the blurring function $b(x,y)$ is equal to the product of their Fourier transforms $F(u,v) \cdot B(u,v)$

$$\mathcal{F}\{f \otimes b\} = F \cdot B \tag{6.16}$$

and we see that B acts as an envelope function in Fourier space. For example, the Fourier transform of a Gaussian $b(x) \propto \exp[-(x/x_0)^2]$ is a Gaussian $B(u) \propto \exp[-(u/u_0)^2]$ with $u_0 = 1/x_0$. Multiplication of the Fourier transform (6.16) by $B^{-1}(u,v)$ corrects F for the attenuation caused by the multiplication with B. In the inverse Fourier transform $\mathcal{F}^{-1}$ of the corrected Fourier transform

$$\begin{aligned} F' &= (F \cdot B) \cdot B^{-1} \qquad \text{for} \quad u,v \le q_{\max} \quad \text{and} \\ F' &= F \cdot B \qquad\qquad\quad \text{for} \quad u,v > q_{\max} \,. \end{aligned} \tag{6.17}$$

the image function $b' = \mathcal{F}^{-1}\{F'\}$ is sharper and the image resolution can be increased by a factor 2–3. However, each sharp cut-off in Fourier space causes an oscillation in real space and conversely. Such a restoration procedure can only be employed when the contrast transfer is linear. This is the case for phase contrast in TEM and STEM but normally not for the contrast in SEM generated by SE and BSE emission, which depend non-linearly on surface tilt and azimuth, for example. Therefore, such a deconvolution can only be applied to the blurring caused by the profile of the electron probe [6.110] but this requires an exact knowledge of the profile.

The inverse Fourier transform of $|F|^2$ is the autocorrelation function of f, which has an autocorrelation peak at the centre and subsidiary maxima at directions and distances for which a superposition of two shifted micrographs would show a coincidence of image structures. In x-ray diffractography, this autocorrelation function is known as the Patterson function. For two similar micrographs $f(x,y)$ and $g(x,y)$, the inverse Fourier transform of $F(u,v) \cdot G^\star(u,v)$ is their cross-correlation function, which has a peak shifted by a vector $\boldsymbol{r}$ from the centre if the two micrographs are not aligned correctly and are shifted by $-\boldsymbol{r}$. The magnitude and the width of the cross-correlation peak are indicators of the presence or absence of common structures in the two images, a SE and a BSE micrograph, for example. Cross-correlation can also be used to check whether two sequentially recorded micrographs of the same area are aligned.

Image processing can also be applied to images recorded with a multiple-detector system, to reconstruct the surface profile from the dependence of the SE or BSE signals on surface tilt and azimuthal angles (Sects. 6.1.4 and 6.2.3) for example or to map the mean atomic number from the dependence of the backscattering coefficient on atomic number (Sect. 6.2.4).

A wide field of application of image analysis is stereology [6.111–120]. For example, the ratio of the areas of three phases (features) in polished sections is equal to the ratio of their volumes:

$$A_1 : A_2 : A_3 = V_1 : V_2 : V_3 \tag{6.18}$$

if there is no texture in the material. Sections through the material in perpendicular directions will reveal whether a texture is present since these ratios will then differ. The areas $A_i = a^2 N_i$ can easily be measured by counting the number of pixels N_i within A_i (a: pixel size) when the phases can be distinguished by the material contrast of BSE, for example, and when an unambiguous threshold signal can be set to generate binary on and off signals.

6.4.5 Stereogrammetry

For the observation of stereopairs by stereo binoculars and for quantitative stereogrammetry [6.121–123], two micrographs are recorded in turn with a defined specimen tilt $\pm\gamma = 5° - 10°$ between the exposures. Whereas the methods for the reconstruction of the surface profile by A–B SE images (Sect. 6.1.4) and A–B BSE images (Sect. 6.2.3) can be applied also to surfaces containing no sharp structures, photogrammetry needs sharp image details to measure the height difference h between two images points $i = 1$ and 2 from the parallax p. When the specimen is tilted between the left (L) and right (R) exposures about an axis normal to the electron beam the corresponding coordinates $X^{(i)}_{\mathrm{L,R}} = Mx^{(i)}_{\mathrm{L,R}}$ at the points (1) and (2) (Fig. 6.30) are

$$p = (X^{(2)}_{\mathrm{R}} - X^{(1)}_{\mathrm{R}}) - (X^{(2)}_{\mathrm{L}} - X^{(1)}_{\mathrm{L}}) \quad \text{and} \quad h = p/2M\sin\gamma; . \tag{6.19}$$

The formula for h is valid only for high magnification M (parallel projection). For low magnification, the fact that the SEM micrograph is a central projection must be taken into account (Sect. 2.3.2) and more complex formulae than (6.19) have to be used [6.123]. The manual determination of the parallax point by point is very time-consuming. In a pair of stereo images cross-correlation (Sect. 6.4.4) in small windows can be used to get the local

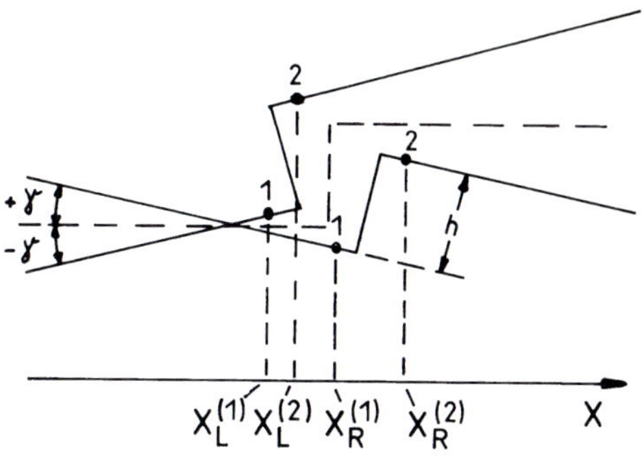

Fig. 6.30. Determination of the parallax p (6.19) from the image coordinates $X_{\mathrm{L,R}}^{1,2}$ of two points 1 and 2 for two specimen orientations tilted to the left (L) and right (R) by an angle $\pm\gamma$

parallaxe and scanning of the window across the image results in an automatic reconstruction of the topography by stereogrammetry [6.124, 125].

For micro-manipulation of a specimen during observation, it must be possible to observe stereo images at TV scan rates. This can be achieved by recording each line twice with rocking angles $\gamma = \pm(3° - 5°)$ and displaying the two micrographs either on two separate CRTs or superposed as red and green images on a colour CRT. For this post-lens deflection coils are employed or else the final-lens diaphragm is removed and the aperture is limited by means of a diaphragm behind the last condenser lens [6.126–129].

7. Electron-Beam-Induced Current and Cathodoluminescence

Electron beams generate electron–hole pairs or minority carriers in semiconductors within a small volume. They are therefore excellent tools for measuring semiconductor-device parameters such as the diffusion length, the surface recombination velocity, the relaxation time, and the position and width of depletion layers by recording the charge-collection current or electron-beam-induced current in a depletion layer. Schottky barriers as well as diffused and ion-implanted p-n junctions can be studied. By modulating the CRT with the charge-collection current, images of depletion layers and of crystal defects, which influence the recombination of minority carriers, can be displayed.

Cathodoluminescence (CL) is a further important source of information about semiconductors and allows the local composition, doping and temperature in devices to be measured by recording the spectrum of the cathodoluminescence emission and the decay times. Defects can also be imaged in the CL mode. For mineralogical and biological specimens, CL provides a method of distinguishing phases and elemental concentrations and of localizing specific fluorescent stains.

For recording light emission excited by cathodoluminescence, a light collection system of high efficiency must be employed, especially for spectroscopic resolution and at low light levels.

7.1 Electron-Beam-Induced Current (EBIC) Mode

7.1.1 EBIC Modes and Specimen Geometry

When electron–hole pairs are generated inside a depletion layer or when minority carriers diffuse to the layer, the electric field of the depletion layer separates the charge carriers and a charge-collection current I_{cc} or an electron-beam-induced current (EBIC) can be measured externally. This effect has been discussed in Sect. 5.2.2 in connection with semiconductor detectors. However, the most important application of EBIC is in semiconductor technology, where the EBIC signal can be used as a video signal for the imaging of p-n junctions and of crystal defects, since it is affected by differences in the

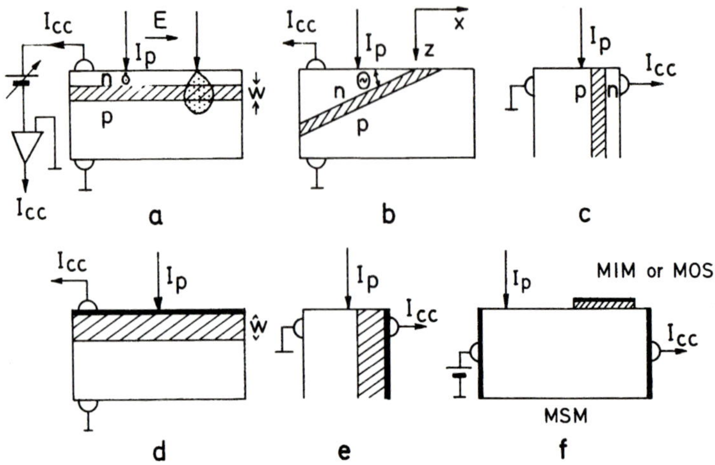

Fig. 7.1. Different geometries for observing EBIC with (**a**) the p-n junction parallel to the surface, (**b**) at a bevel angle θ to the surface and (**c**) perpendicular to the surface and with Schottky barriers (**d**) parallel and (**e**) perpendicular to the surface; (**f**) metal–semiconductor–metal (MSM) and MIM and MOS structures (I=insulator, O=oxide)

recombination rate, and of sites of avalanche breakdown, for example. Furthermore, EBIC can be used for quantitative measurement of the width w and depth z_d of depletion layers, of the diffusion length L_n and L_p of electrons and holes and of their lifetimes and of the surface recombination velocity S (see [7.1–5] for reviews). Another possibility is the use of beam pulsing and time-resolved EBIC signals (transient mode). Other useful modes are the voltage contrast of secondary electrons (Sect. 8.2) and the cathodoluminescence in semiconductors (Sect. 7.2).

Figure 7.1 shows several different geometries suitable for investigating and using EBIC. A p-n junction of width w at a depth z_d below the surface (Fig. 7.1a) can be produced by thermal diffusion or ion implantation. The figure shows schematically how an increase of electron energy increases the dimensions of the diffusion cloud and modifies the depth distribution of charge carriers generated. The possibility of varying the electron energy between one and 50 keV, corresponding to electron ranges of 0.03–20 μm in silicon, for example, is therefore an important advantage of EBIC, in contrast to OBIC (optical beam-induced current) where the depth of charge carrier generation is limited by the absorption of the exciting wavelength.

For the investigation of EBIC normal to the junction, a bevelled specimen geometry (Fig. 7.1b) can be prepared by lapping the specimen at an angle θ, or the device is cleaved, so that the junction lies perpendicular to the surface (Fig. 7.1c). The two latter geometries allow charge carriers to be generated at different distances from the depletion layer for investigating the influence of diffusion of minority carriers on the EBIC signal: the width

w and the depth z_d, quantities that can only be obtained approximately when using the geometry of Fig. 7.1a and varying the electron energy, can be measured directly. The geometries of Figs. 7.1a, c can also be employed with Schottky diodes (Figs. 7.1d, e). Such diodes can easily be prepared by evaporation of a non-ohmic metal contact layer, Al on p-type and AuPd on n-type silicon, for example, about 100 nm in thickness and a few hundred μm in diameter [7.6, 7]. Lattice defects in pure semiconductor material can be investigated within the depletion layer below the reversely biased Schottky contact and quantitative measurements of semiconductor parameters can be made by using the geometry of Fig. 7.1d, either by varying the width w and electron energy or by increasing the distance x from the contact edge. Separation of charge carriers also occurs at grain boundaries with ohmic contacts on both sides. The height of the potential barrier at the boundary can be determined from local $I-U$ characteristics at different electron-probe currents [7.8].

When two ohmic contacts and an external bias are applied to a semiconductor (MSM structure, Fig. 7.1f), local variations in conductivity can be recorded. Metal–insulator–metal (MIM) or MIS or MOS structures (I: insulator, O: oxide) are in use in all integrated circuits (IC). Leakage currents are excited by electron bombardment and irreversible defects caused by electron irradiation can influence the electronic properties of an IC (Sect. 7.1.7).

The EBIC flows through ohmic contacts and can be amplified by an operational amplifier with an input resistance that must be small, $R \simeq 1$ kΩ, because otherwise the time constant $\tau_{RC} = RC$ would become too large since the capacitance C of the depletion layer is large (Sect. 5.2.3). This means that an EBIC will be a short-circuit current. A reverse bias will not be necessary but does somewhat increase the EBIC signal and decrease the capacitance C because w becomes larger with increasing reverse bias (Fig. 5.3). The time needed for scanning a frame or a line has to be long enough to ensure that the recorded current is independent of the scan speed. The lifetime τ of minority carriers (Sect. 7.1.2) also has to be taken into account to approach steady-state conditions; this can vary from 10^{-10} to 10^{-6} s.

In the open-circuit situation, for which $I = 0$ in (5.9), the separation of charge carriers results in an electron-beam-induced voltage (EBIV) which appears as an open-circuit voltage U_{oc} (Fig. 5.4):

$$U_{oc} = \frac{kT}{e} \ln(1 + I_{cc}/I_s) \ . \tag{7.1}$$

This cannot, however, become larger than the diffusion voltage U_d, which is of the order of the semiconductor band gap. This allows us to obtain the diffusion voltage by measuring the saturation value of the EBIV signal at high electron-probe currents of the order of 10^{-8} A [7.9]

Table 7.1. Mean excitation energy $\overline{E_i}$ per electron–hole pair for different semiconductors

	Si	Ge	GaAs	InSb	CdS
$\overline{E_i}$ in eV	3.65	2.85	4.6	0.42	7.3

7.1.2 Charge Carrier Generation, Diffusion and Collection

The magnitude of the charge-collection current I_{cc} (5.10) has already been discussed in Sect. 5.2.2 for the case of semiconductor detectors. For an incident electron-probe current I_p and an electron energy $E = eU$, the power dissipated is given by $P = I_p U = I_p E/e$. The maximum number of electron–hole pairs excited per unit time is obtained by dividing by the mean excitation energy $\overline{E_i}$ per electron–hole pair (Table 7.1).

The value of $\overline{E_i}$ is larger than the energy gap ΔE between valence and conduction bands because electrons are also excited from lower states in the valence band to higher states in the conduction band. The depth distribution $\Phi(z')$ with $z' = z/R$ of the carriers in silicon is shown in Fig. 3.26 using an analytical approach (3.154). This is a normalized function of z': $\int_0^\infty \Phi(z')\mathrm{d}z' = 1$ and the depth distribution of charge carriers becomes, by analogy with (5.10):

$$g(z)\mathrm{d}z = \frac{E - E_{\text{th}}}{\overline{E_i}} \frac{I_p}{e}(1 - \eta_c)\Phi(z/R)\frac{\mathrm{d}z}{R} \tag{7.2}$$

where E_{th} denotes a threshold energy due to any metal or insulator films that have to be penetrated before the beam reaches the semiconductor. The loss of backscattered electrons is represented by η_c: for silicon $\eta_c = 0.08$.

For geometries that vary only in the z direction, such as those of Figs. 7.1a,d, this depth distribution will be sufficient. For other geometries, Figs. 7.1b,c and e, for example, the spatial distribution $g(r, z)$ is also required. This function can be obtained by measuring the light emission of a gas target [7.10, 11], by observing electron-beam-exposed resists [7.12, 13] or by Monte Carlo simulations [7.14]. Because no analytical expression for these results is known, the carrier density is often assumed to be uniform inside a sphere [7.15], which corresponds to Archard's model of electron diffusion (Sect. 3.4.3). A point-source model at the surface or at a depth $z_d \simeq R/3$ can be used when the distance from the source is one order of magnitude greater than the electron range R and the uniform sphere model can be employed for distances comparable with the range. In another proposal [7.16], the pear-shaped diffusion cloud is represented by

$$g(r, \theta) \propto \exp(-r^2/a^2)\cos\theta \tag{7.3}$$

with $a \simeq R/2$.

For $E = 20$ keV and $I_p = 10^{-9}$ A, 3×10^{13} s^{-1} electron–hole pairs are created per unit time in a volume that is approximately a sphere of diameter R. The generation rate n' of excess electrons in p-type material, for example, becomes

$$n' = \frac{I_p}{e} G \frac{6}{\pi R^3} \tag{7.4}$$

in which we have introduced the gain $G = (1 - \eta_c)E/\overline{E_i}$ as in (5.10). In the following, n' means the additional concentration of minority carriers excited by electrons. With the above numerical values and $R = 5$ μm this results in $n' = 5 \times 10^{22}$ s^{-1}cm^{-3}. However, this will not be the steady-state concentration because of the recombination and diffusion of minority carriers as discussed below.

The oppositely charged carriers generated can recombine, which normally occurs at traps with a trap cross-section σ_T and a trap density N_T. This results in a mean-free-path length $\Lambda_T = 1/N_T\sigma_T$ before trapping and in a mean-free-lifetime or relaxation time

$$\tau = 1/\sigma_T N_T v_{th} \tag{7.5}$$

which is different for holes and electrons. The mean thermal velocity v_{th} is given by

$$mv_{th}^2/2 = 3kT/2\ . \tag{7.6}$$

When we assume that the excess concentration n' of minority carriers is small, in p-type material for example, the number of available majority carriers (holes) is so large that the lifetime of a generated pair is determined by that of the majority carriers and the recombination rate becomes $r = n'/\tau$. When we assume that the diffusion length L (see below) is smaller than the electron range, so that all generated pairs recombine inside approximately the same volume with a relaxation time τ, the steady-state concentration of excess electrons can be estimated from

$$N' = n'\tau = \frac{I_p}{e} G \frac{6}{\pi R^3}\tau\ . \tag{7.7}$$

For $\tau = 1$ μs, for example, this gives $N' = 5 \times 10^{16}$ cm^{-3} with the value of n' estimated above. This figure gives an idea of the ability of an electron beam to produce such a large number of excess minority carriers per unit volume.

The thermal velocity results in a diffuse drift of charge carriers by random collisions with the lattice. An electric field $\boldsymbol{E}$ causes a uni-directional drift. During drifting, the majority carriers follow the minority carriers so that the charge neutrality is maintained. The two mechanisms result in a particle flux

$$\boldsymbol{J} = -D\nabla N' + N'\mu\boldsymbol{E} \tag{7.8}$$

where $D = \mu kT/e$ is the coefficient of thermal diffusion and μ the electron mobility. The change of excess concentration N' must therefore satisfy the following continuity equation

$$\frac{\partial N'}{\partial t} = g - \frac{N'}{\tau} - \nabla \cdot \boldsymbol{J} = g - \frac{N'}{\tau} + D\nabla^2 N' - \mu \nabla N' \cdot \boldsymbol{E} - \mu N' \nabla \cdot \boldsymbol{E} \,. \quad (7.9)$$

The rate of change $\partial N'/\partial t$ is zero for the stationary or steady state but not for the transient state, when we are working with a chopped electron beam or when the electron probe is scanned too rapidly. A corresponding equation describes the case of holes, the minority carriers in n-type material.

We now discuss some solutions of (7.9). In the absence of an electric field $\boldsymbol{E}$ and in regions far outside the generation volume, where $g = 0$, the steady-state continuity relation (7.9) simplifies to

$$D\nabla^2 N' - \frac{N'}{\tau} = 0 \quad (7.10)$$

with the following solution for the one-dimensional case corresponding to uniform radiation in the x-y plane

$$N' \propto \exp(-z/L), \quad \text{where} \quad L = (D\tau)^{1/2} \quad (7.11)$$

is the minority carrier diffusion length; this is an important quantity in semiconductor technology as it describes the mean distance which minority carriers can diffuse before recombination. In very pure Si and Ge, L can be of the order of 1 cm but it decreases rapidly with decreasing τ or increasing trap density N_{T}.

Far from a point source at $r = 0$, a solution will be

$$N'(r) \propto \exp(-r/L)/r \,. \quad (7.12)$$

Integration over the $x - y$ plane yields the depth distribution

$$N'(z) = \frac{GI_{\mathrm{p}}}{2\pi e D} \int_0^\infty \frac{\exp(-\sqrt{\zeta^2 + z^2}/L)}{\sqrt{\zeta^2 + z^2}} 2\pi\zeta \mathrm{d}\zeta = \frac{GI_{\mathrm{p}}L}{eD} \mathrm{e}^{-z/L} \,, \quad (7.13)$$

where $\zeta^2 = x^2 + y^2$.

Because the generation of carriers by the electron beam occurs near the surface, surface recombination has to be taken into account; this can be expressed in terms of the boundary condition

$$sN' = D \left| \frac{\partial N'}{\partial z} \right|_{z=0} \quad (7.14)$$

where s is the surface recombination velocity in units of cm s^{-1}.

A solution of the steady-state equation (7.10) with the boundary condition (7.14) has been found by *Van Roosbroeck* [7.18] for the one-dimensional case when a large area is irradiated with a uniform current density j. If we denote the source depth distribution of generated charge carriers per unit depth by $g(z_{\mathrm{s}})$, the contribution from a layer of thickness $\mathrm{d}z_{\mathrm{s}}$, at a depth z_{s} will be

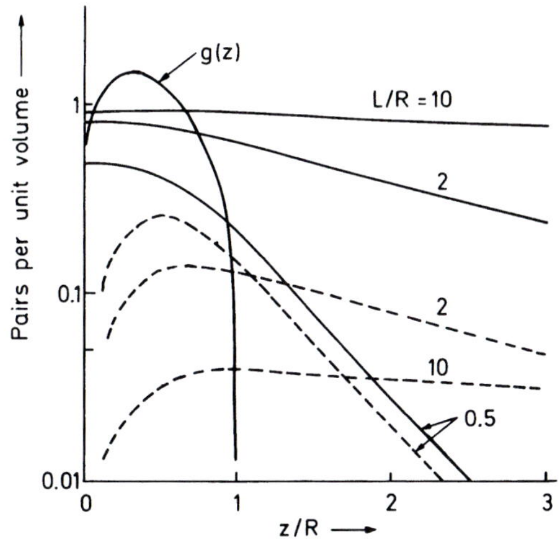

Fig. 7.2. Semi-logarithmic plot of the generation rate $g(z)$ of charge carriers versus the reduced depth z/R below the surface (R = electron range). The full lines represent the carrier concentration for different diffusion lengths L and zero surface recombination velocity ($S = 0$) and the dashed ones that for $S = \infty$ (perfect sink at the surface)

$$\mathrm{d}N'(z) = \frac{Gj}{e}\frac{\tau}{2L}\left\{\exp[-|z-z_\mathrm{s}|/L] - \frac{S-1}{S+1}\exp[-(z+z_\mathrm{s})/L]\right\} g(z_\mathrm{s})\mathrm{d}z_\mathrm{s} \tag{7.15}$$

where $S = s\tau/L$ is a reduced surface recombination velocity. For the source function $g(z_\mathrm{s})$, the polynomial fit to $\Phi(z/R)$ in (3.16) can be substituted. According to (7.14), $D|\partial N'/\partial z|_{z=0}$ will be the number of carriers that recombine at the surface per unit time. When we differentiate (7.15) with respect to z and set $z = 0$, we obtain the following fraction $\mathrm{d}n_\mathrm{r}(z_\mathrm{s})$ of surface-recombined minority carriers

$$\mathrm{d}n_\mathrm{r}(z_\mathrm{s}) = \frac{sN'}{N'_0} = \frac{D|\partial N'/\partial z|_{z=0}}{N'_0} = \frac{S}{S+1}\exp(-z_\mathrm{s}/L)g(z_\mathrm{s})\mathrm{d}z_\mathrm{s} \tag{7.16}$$

by dividing by the total excess concentration rate $N'_0 = G\tau j/e$. This means that surface recombination removes carriers at a rate proportional to $S/(S+1)$ and this rate decreases as $\exp(-z_\mathrm{s}/L)$ with increasing values of the ratio of source depth to diffusion length.

Figure 7.2 shows in a semi-logarithmic plot the depth distribution $g(z_\mathrm{s})$ of carrier production. The steady-state carrier concentration $N'(z)$ is calculated from (7.15) by integration over z_s within $0 \leq z_\mathrm{s} \leq R$, assuming different values of L/R (R: electron range); the extreme values for $S = 0$ (full curves) and $S = \infty$ (dashed curves) corresponding to the cases that the surface is a perfect reflector and a perfect sink, respectively, are plotted. These demonstrate the considerable influence of S. In the semi-logarithmic plot, the exponential decrease (7.11) takes over for $z/R > 1$, resulting in a straight line.

The observed charge-collection current I_cc is written as $I_\mathrm{p}G\epsilon_\mathrm{c}$ (5.10) where the charge-collection efficiency ϵ_c is determined by the approximately complete separation of the carriers inside the depletion layer and the minority carriers, which reach the layer by diffusion. We now consider how I_cc depends

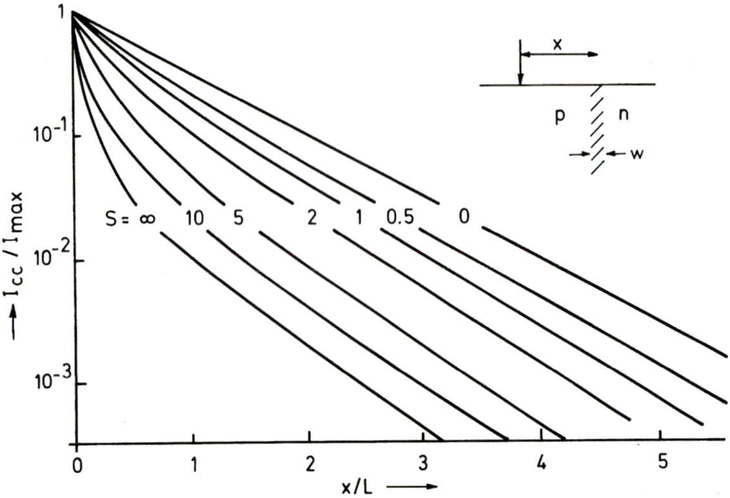

Fig. 7.3. Charge-collection current I_{cc} of a p-n junction perpendicular to the surface (Fig. 7.1c) and the electron probe at a distance x for a point source at the surface and increasing surface recombination velocity S [7.18]

on the location of the electron probe for some of the geometries of Fig. 7.1. If $w > R$ in the geometries of Figs. 7.1c, e (perpendicular junction), the width w of the depletion layer appears as a region of relatively constant $I_{cc} \simeq I_p G$. If surface recombination is neglected ($S = 0$), the minority carrier concentration can be described by (7.12) when the electron probe is at a distance $x \geq 3L$. The number of minority carriers that cross the junction and contribute to I_{cc} is given by the value of $-D\partial N'/\partial x$ at the junction. Integration of this quantity over the half-plane of the junction as for (7.13) results in

$$I_{cc} \propto \exp(-x/L) \; . \tag{7.17}$$

For a point source at the surface, the semi-logarithmic plot in Fig. 7.3 shows how I_{cc} decreases with increasing relative surface recombination velocity $S = s\tau/L$, because the electrons can recombine at the surface when diffusing to the depletion layer parallel to the surface.

Schottky diodes (Fig. 7.1d) have the advantage that the influence of surface recombination can be neglected because the surface is inside the depletion region of complete charge collection. If the electron range R is smaller than the width w ($w/R > 1$), the collection efficiency ϵ_c will be near unity. This will be the best situation for measuring G and $\overline{E_i}$ [7.19]. If R is larger than w ($w/R < 1$), only the carriers produced in the region $z < w$ are fully collected. Those produced at values of $z > w$ have to diffuse to the depletion layer. Figure 7.4 shows the dependence of the total collection efficiency ϵ_c on the ratio L/R [7.20].

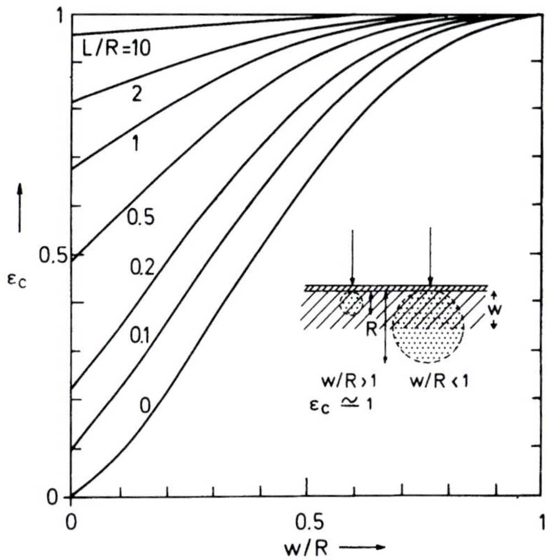

Fig. 7.4. Charge-collection efficiency ϵ_c of a Schottky diode of width w for the case $R > w$ $(w/R < 1)$ and for different relative diffusion lengths L/R [7.20]

7.1.3 Quantitative Measurement of Semiconductor Parameters

The dependence of the charge-collection current I_{cc} on L and S as discussed in the last section can be used for quantitative measurement of these important semiconductor parameters and their dependence on doping and temperature. Furthermore, local variations of these quantities can be used to image inhomogeneities (Sect. 7.1.5). The problem of getting quantitative results will be analysed for a depletion layer perpendicular to the surface (Fig. 7.1e) with I_{cc} given by (7.17) for $S = 0$ and the calculated curves for $S \neq 0$ in Fig. 7.3. Figure 7.5 shows the signal obtained when the beam is scanned across a p-n junction in GaP with the geometry of Fig. 7.1c [7.21]. At the maximum of I_{cc} at $x = 0$, the collection efficiency is close to unity. The semi-logarithmic plot becomes linear for large x with different slopes

$$\Delta(\log I_{cc})/\Delta x = -0.4343/L \qquad (7.18)$$

for the n- and p-type regions on either side of the junction. When using 10 keV electrons, which of course have a shorter range than 20 keV electrons, the electron beam produces the charge carriers nearer to the surface and the influence of surface recombination becomes stronger. The decay curve for increasing x becomes similar to that of Fig. 7.3 for large S. The diffusion length evaluated from the slope may not be the same for different energies. Figure 7.3 also shows that, for large S, the slope is less steep and the results obtained with 20 keV electrons will be more reliable because the charge carriers are generated at a greater depth and the influence of surface recombination, when the charge carriers diffuse parallel to the surface, is reduced.

It is easier to avoid this problem by adopting the bevelled specimen geometry of Fig. 7.1b where

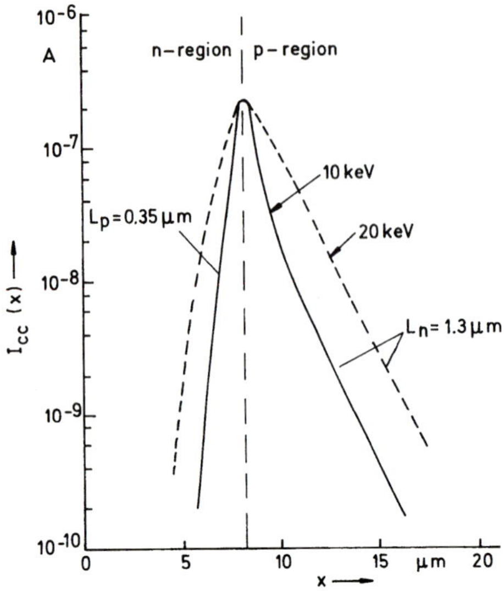

Fig. 7.5. Charge-collection current I_{cc} obtained when scanning across a perpendicular p-n junction in GaP excited by 10 and 20 keV electrons. The indicated values of the diffusion lengths L of minority carriers are obtained from the slope of the curve using (7.18) [7.21]

$$I_{cc} = I_{max} \exp(-x \sin\theta / L) \tag{7.19}$$

and the carriers have to diffuse further normal to the surface than parallel to it [7.22]. However, this requires that S be constant along x after the lapping process and the technique does not improve the limit of resolution for small L. By using a more quantitative theory including a finite spherical source size with uniform $g(r)$, for example, both L and S can be extracted at different electron energies [7.23–25].

When a linescan is recorded across the specimen at a distance x from the edge of a Schottky contact (Fig.7.1d), the EBIC current falls as

$$I_{cc} \propto x^{\alpha} \exp(-x/L) \tag{7.20}$$

with increasing x where $\alpha = -1/2$ for $S = 0$ and $-3/2$ for $S = \infty$ [7.17, 26, 27]. As another example we discuss a Schottky Au barrier on n-type GaAs using the charge-collection efficiency shown in Fig. 7.4 and working with a stationary electron probe [7.20, 28]. Increasing the electron energy E leads to an increase in the electron range R and measured values of the charge-collection efficiency $\epsilon_c(E)$ in Fig. 7.6 can be fitted to the calculated curves in Fig. 7.4 by transferring the ϵ_c values for corresponding w/R. The decrease at low E is caused by the absorption of electrons in the metal contact layer of the Schottky diode, represented by $(E - E_{th})$ in (7.2). Unlike the linescan method of Fig. 7.5 and that leading to (7.20), where L has to be uniform, this Schottky barrier technique allows L to be measured more locally.

For p-n junctions parallel to the surface (Fig. 7.1a), $\epsilon_c(E)$ [7.29] can again be investigated; the $L(z)$ profile then has to be taken into account using an iterative fitting program [7.30, 31]. It is more convenient to shift a wedge-shaped layer across the position of interest than to vary E; the decrease of

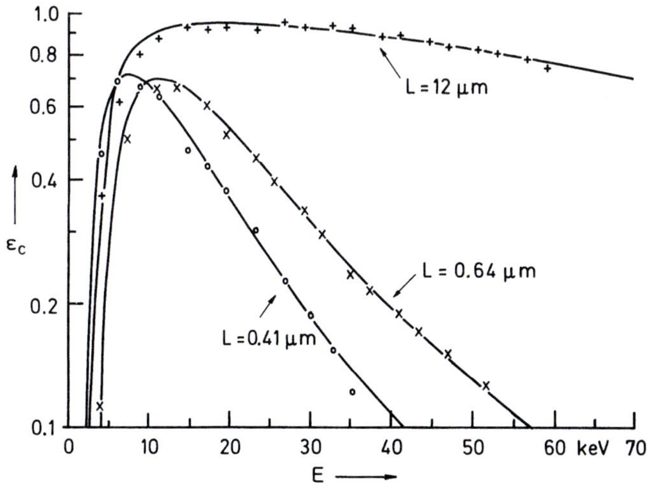

Fig. 7.6. Charge-collection efficiency ϵ_c as a function of electron energy E for Schottky Au barriers on n-type GaAs in the geometry of Fig. 7.1d (solid lines are computed curves using Fig. 7.4) [7.20]

mean transmitted electron energy with increasing layer thickness is then used instead of varying the primary energy [7.32].

$I_{cc}(E)$ curves can also be exploited to determine the depth of the depletion layer [7.33]. This method is non-destructive unlike to the technique for detecting the depth from the EBIC image of cleaved p-n junctions in the geometry of Fig. 7.1c. From (5.8), we see that the width w of the depletion layer depends on the reverse bias U and the dopant concentration, which can be determined from $w(U)$ [7.34, 35]. In semi-insulating semiconductors, it is important to use a correct value of U [7.36]. The accuracy in the measurement of the depth of parallel junctions can be increased when the signal contains only that fraction of I_{cc} that is produced by direct carrier generation inside the depletion layer and not the fraction contributed by diffusion. These fractions can be separated by electron-beam chopping and time-resolved EBIC, so that the time of measurement immediately after switching on the beam is short compared with the drift time of minority carriers to the junction [7.37].

The general continuity relation (7.9) simplifies in the transient case when the electron beam is suddenly switched off ($g = 0$) to

$$\partial N'/\partial t = -N'/\tau; \quad N' = N'_0 \exp(-t/\tau) . \tag{7.21}$$

The resulting exponential decay with time allows the relaxation time τ to be measured [7.8, 38–40]. From (7.11), the diffusion constant D can be calculated by simultaneous measurement of L and τ. It is obvious that such an experiment can also be performed by periodic electron-beam chopping, which has the advantage that the exponential decay signal can be averaged, by means of a sampling oscilloscope, for example. The geometry of a perpendicular depletion layer (Fig. 7.1c, e) can be used to measure both L and τ from semi-logarithmic plots of $I_{cc}(x)$ and $I_{cc}(t)$, respectively. The distance x should be at least two or three times greater than L because the finite excitation volume of carriers and the surface recombination also have to be taken into account, as discussed previously for the steady state.

A satisfactory signal-to-noise ratio can also be attained by using phase-sensitive signal detection with a lock-in amplifier. For periodic beam modulation of angular frequency ω, the excitation of amplitude and phase at a distance x becomes

$$I(x,\omega) = I(0,\omega)\exp[-x/L(\omega)]\exp[-\mathrm{i}\,x\gamma(\omega)/L_0]\ , \tag{7.22}$$

where L_0 is the dc diffusion length,

$$L(\omega) = L_0\{[1+(1+\omega^2\tau^2)^{1/2}]/2\}^{-1/2}\ , \tag{7.23}$$

is the ac diffusion length, and

$$\gamma(\omega) = \{[-1+(1+\omega^2\tau^2)^{1/2}]/2\}^{1/2} \tag{7.24}$$

is the phase factor. Determination of the phase and amplitude therefore enables us to obtain L_0 and τ [7.41–44].

7.1.4 EBIC Imaging of Depletion Layers

Owing to their importance for semiconductor technology, we have first discussed the quantitative methods. A two-dimensional scan with the EBIC signal fed to the CRT allows us to observe the position of p-n junctions below the surface [7.4, 45–47]. As in the case of the quantitative methods, variation of the electron energy and hence the electron range R provides information about the junction depth. As an example, Fig. 7.7 demonstrates the imaging of a p-n junction using the emitter–collector current of a planar MOS transistor for increasing values of E [7.48] (see SE image in Fig. 7.7a and the schematic cross-section). Electron energies below 5 keV do not produce an EBIC signal because the electrons cannot penetrate the oxide layers and the conductive pads. At 8 keV, electrons start to penetrate the oxide layer but not the additional conductive pads, and part of the perpendicular substrate–collector and collector–base junctions become visible in EBIC. At the contact window in the oxide layer of the base, the electrons have only to penetrate the contact pad and reach the bottom of the emitter zone. The surface-parallel part of the base–collector junction can be imaged as two bright areas at the centre. This contrast decreases with increasing energy because the ionization density at this shallow junction decreases with increasing energy (Fig. 3.26). The emitter–base junction is imaged at 8 keV as a dark line because the direction of the EBIC current is reversed. At 15 keV, electrons start to penetrate the conductive pads and the influence of the oxide and metal layers

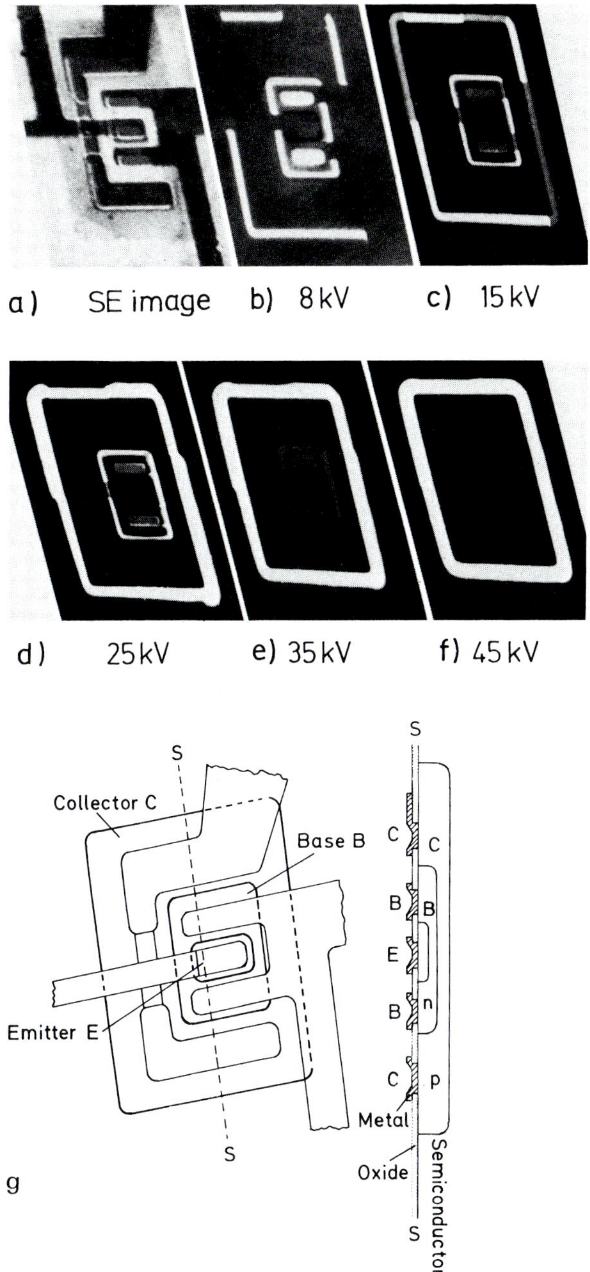

Fig. 7.7. Imaging of perpendicular p-n junctions in a MOS transistor: (**a**) SE image, (**b**–**f**) EBIC images with increasing electron energy, (**g**) arrangement of electrodes and cross-section through the MOS structure [7.48]

diminishes with increasing energy. The broadening of the perpendicular junction regions is caused by the broadening of the electron diffusion cloud with increasing energy. The parallel junction between substrate and collector shows no significant contrast even at 45 keV, indicating that this junction is deeper than 20 μm.

The depth profiles of p-n junctions and inversion layers can also be imaged on cross-sections obtained by crystal cleavage [7.49] and the width of the p-n junction can be determined if the electron range R, regarded as the diameter of the charge-carrier generation volume, is smaller than the width w [7.47].

Schottky barriers have the advantage that nearly complete charge collection can be attained in a depth $w = 1$–100 μm below the surface, depending on the applied reverse bias (Fig. 5.2). Such barriers can easily be prepared by evaporation of a non-ohmic metal contact, which allows them to be investigated in the form of as-grown crystals or chips at different steps of the fabrication process. Variations of the dopant concentration N_{B} are accompanied by variations in w, (5.26) and, therefore, in the collection efficiency ϵ_{c} if $R > w$. The contrast variation of the charge-collection current becomes $\Delta I_{\mathrm{cc}}/I_{\mathrm{cc}} \propto \Delta w/w \propto \Delta N_{\mathrm{B}}/N_{\mathrm{B}}$ [7.28, 50–52]. This type of contrast vanishes if $R \simeq w$ but can reappear as inverse contrast if $R < w$ [7.50]. At high doping levels, changes in the diffusion length also influence the contrast [7.52]. Lattice defects and grain boundaries near the surface can also be imaged by charge carrier separation in the zone where the Fermi level is bent near the surface. A signal can be recorded even with a side-faced ohmic contact at a distance up to 2 mm though with only one tenth of the amplitude that would be obtained with a normal depletion layer [7.53].

7.1.5 EBIC Imaging and Defect Analysis

Crystal defects are usually investigated by evaporating Schottky barriers. Extended crystal defects such as dislocations, stacking faults and precipitates usually increase the recombination rate of minority carriers in their vicinity. Defects that are electrically active can be imaged in the EBIC mode with dark contrast [7.50, 54, 55]. Depending on composition and pre-treatment, brightening may also be observed [7.57], as a bright zone around a dark kernel of dislocations due to a depleted zone (getter effect). The EBIC contrast of defects is commonly expressed quantitatively as $c = 1 - (I_{\mathrm{d}}/I_0)$, which is of the order of 1–10%, where I_{d} and I_0 are the signals at the defect and far from it, respectively. The resolution and contrast decrease with increasing depth and shallow defects are imaged with the best resolution. The resolution is of the order of 0.5–1.5 of the diffusion length L and depends on the depth of generation and the surface recombination velocity [7.56].

For the calculation of the EBIC contrast of dislocations, the dislocation line can be treated as a sink for minority carriers with a reduced relaxation time $\tau' < \tau$ inside a cylinder of radius r [7.58–60]. This decrease of τ can be attributed either to dangling bonds or the segregation of impurity atoms (Cottrell atmosphere). A recombination velocity

$$v_R = \pi r^2/\tau' = \pi r^2 N \sigma v_{th} \tag{7.25}$$

can be introduced, which is analogue of the surface recombination velocity S in (7.15). Measurements of v_R [7.61] show that the value for dissociated dislocations is twice that for undissociated ones [7.62, 63]. Measured values of $v_R/D = 0.03$ as the detection limit for an EBIC contrast of 0.05% with $\sigma = 10^{-14}$ cm^2 show that the dislocation line has to contain impurity atoms at least 5 nm [7.61] apart. Differences in lifetime τ near dislocations can be measured by time-resolved EBIC [7.64].

Qualitatively, the same defect contrast as discussed above for Schottky diodes can be observed with shallow p-n junction at a depth smaller than the electron range R [7.65]. Computer simulations of contrast agree well with experiment [7.58, 59]. Defects inside depletion layers are preferential sites of avalanche breakdown and when the reverse bias is increased, such sites can be imaged thanks to carrier multiplication in the microplasma, which can also be detected by the increase of cathodoluminescence [7.66, 67].

Most EBIC experiments have been done with the specimen at room temperature. The investigation of the dependence of contrast on temperature, typically between 80 and 300 K not only shows that the contrast can be improved but may also give more information about the mechanism of recombination near the defects. For example, screw parts of hexagonal dislocation loops in silicon show negligible contrast at room temperature compared to the 60° dislocation parts, whereas the contrast increases for both types to 5–6% at 80 K [7.68]. Experimental $c(T)$ and $c(I_p)$ results reported in [7.66, 69–86] show that, in principle, two different types of contrast behaviour can be found [7.83, 84]. One group shows a negative slope of the $c(I_p)$ and a positive slope for the $c(T)$ curves, which can be attributed to the interaction of beam-induced carriers with the potential barrier and the screening depletion cylinder around charged dislocations; deep-level dislocation-related states are filled with majority carriers [7.81]. $NiSi_2$ precipitates in n-Si show a depletion region , which acts as an inner Schottky barrier [7.85]. Another group with the opposite behaviour, positive slope of $c(I_p)$ and negative slope of $c(T)$, can be assumed to result from the activation of shallow centres near the band edges related to the defects. This behaviour is in agreement with the Schottky–Read–Hall recombination theory, which predicts a decrease of τ with increasing I_p at constant temperature or with decreasing temperature at constant I_p; c is roughly proportional to $1/\tau$. This behaviour is encountered for high intragrain dislocation densities in multi-crystalline Si with shallow centres 70 meV below the band edge [7.76] or 0.1 eV below the edge for stacking faults [7.6], for example.

Comparison of SEM EBIC images of thinned specimens with images of the same area in a scanning transmission electron microscope (STEM) or high-voltage TEM allow us to decide whether a defect is active or not because all defects can be imaged in the STEM mode by Bragg contrast with high resolution [7.87–90].

7.1.6 Scanning Deep-Level Transient Spectroscopy (SDLTS)

In SDLTS, carrier traps in reverse-biased Schottky diodes or p-n junctions are filled periodically at 20–100 Hz by electron excitation of charge carriers [7.91–96] whereas in conventional DLTS the traps are filled by lowering the reverse bias of a junction. In the beam-on period, the signal is cut off; in the beam-off period, in addition to a rapid transient related to carrier recombination, a slowly decaying transient current or change of transient capacitance is caused by emission of carriers from deep traps in the space charge region. The kinetics of emission is temperature-dependent and the variation of DLTS signal with temperature can pass through maxima corresponding to different types of traps.

The emission rate $n_{\mathrm{n,p}}$ for electrons and holes, respectively, is determined by the Shockley–Read equation

$$n_{\mathrm{n,p}} = \gamma T^2 \exp(-E_{\mathrm{t}}/kT) \tag{7.26}$$

where E_{t} denotes the activation energy and $\gamma T^2 = \sigma_{\mathrm{n,p}} \overline{v} N_{\mathrm{c,v}}/g$ in which $\sigma_{\mathrm{n,p}}$ is the capture cross-section for electrons or holes, $\overline{v}$ the averaged thermal carrier velocity, $N_{\mathrm{c,v}}$ the effective density of states in the conductive (c) or valence (v) band and g a degeneracy factor. This leads to a measurable current transient

$$I(t) = e n_{\mathrm{n,p}} V_{\mathrm{t}} N_{\mathrm{t}} \exp(-n_{\mathrm{n,p}} t) \tag{7.27}$$

(N_{t} = trap density; V_{t} = volume that contributes to the signal), or a capacitance transient can be used [7.92]

$$\Delta C(t) = \frac{e V_{\mathrm{t}} N_{\mathrm{t}}}{4(U + U_{\mathrm{d}})} \exp(-n_{\mathrm{n,p}} t) \tag{7.28}$$

(U = applied reverse bias; U_{d} = diffusion voltage).

Defects that trap and emit an electron and a hole simultaneously can only be detected by current-SDLTS. They are not detected in capacitance-SDLTS because the net charge balance of the emission is zero. For defects of other kinds, it is possible to distinguish between hole and electron traps.

In order to get a high signal, (7.27) shows that V_{t} has to be increased by using a defocused beam of 0.1 μm, for example. With an average current value of 10^{-10} A, resolutions of the order of 10 μm can be obtained for trap densities $N_{\mathrm{t}} = 10^{14} - 10^{15}\ \mathrm{cm}^{-3}$.

7.1.7 Radiation Damage Effects in Semiconductor Devices

Ionization effects in metals and semiconductors are normally reversible (Sect. 3.6.3). However, reversible and irreversible damage effects can occur as a result of the accumulation of space charge inside the insulating surface layers and the generation of surface recombination states near the interface between semiconductor and surface coatings. The insulating layers are typically oxide and nitride layers, as used in integrated circuits to insulate the conductive pads from the semiconductor.

We discuss the case of Al–SiO_2–Si. Electron–hole pairs are created in the SiO_2 layer with a mean energy dissipation of $\overline{E_i} \simeq 9$ eV per pair. Whereas electrons can move in SiO_2 with a mobility $\mu_n \simeq 20\ cm^2V^{-1}s^{-1}$, the holes are quite immobile [7.97]. The resulting positive space charge can be analysed by studying the capacitance–voltage characteristics of a metal–oxide–semiconductor (MOS) structure [7.98]. If a bias voltage V_g is applied to such a MOS capacitor, a fraction of the electrons created in the SiO_2 are collected, which leaves a positive space charge $\propto V_g$ corresponding to the immobile holes trapped in the SiO_2 layer [7.99]. This positive charge cancels the voltage across the oxide layer and finally no further charge-collection current is observed. This space charge can be discharged by applying a reverse bias $-V_g$ during electron irradiation and is, therefore, a reversible effect. Silicon nitride layers, on the other hand, contain a large concentration of traps for electrons and holes. The charge-collection current becomes very small and no space charge is built up [7.100].

In addition, electron irradiation can create surface defects at the SiO_2–Si interface. These so-called fast surface states increase the surface recombination velocity [7.101] and influence all device parameters that are sensitive to surface recombination. They can be attributed to broken and rearranged bonds at the interface. Annealing temperatures of 250°C are necessary to remove these fast states. Both the built-up space charge and the fast recombination states influence the device parameters [7.102–104]; the space charge can result in an inversion layer, which can in turn act as a short-circuit for the depletion layer [7.105].

By using primary energies below 5 keV (LVSEM), the electron range in passivation layers and conductive pads can be considerably decreased thus reducing the influence by space charge and the formation of recombination centres. However, radiation damage has still been detected at 2 keV and is attributed to the emission of weak x-ray quanta, which can penetrate more deeply into the material [7.106, 107]. It is therefore advisable to work at 1 keV for damage-free electron-beam testing.

The influence of negative ions from the SEM cathode on EBIC is discussed at the end of Sect. 3.5.3.

7.1.8 EBIC and EBIV Without Any Depletion Layer and in Superconductors

We now discuss the geometries of Fig. 7.1f without a depletion layer, corresponding to semiconductors or insulators with ohmic metal contacts. Figure 7.8a illustrates the geometry of a parallel MSM structure in more detail. Electron irradiation causes a small change $\Delta\sigma$ in conductivity within a volume represented approximately by a cube of side Δx; this is the volume of carrier generation plus the volume into which the carriers diffuse.

There are two possible ways of detecting the influence of electron irradiation. Either a constant current source is used and the change ΔV in the voltage drop is recorded [7.108–111] or the change ΔI in current is observed with a constant voltage source [7.112, 113]. These methods allow variations in doping to be recorded as two-dimensional maps of the resistivity. We discuss the former case (see also [7.110]). The current density inside a small bar of cross-section S_0 and resistance R_0 is

$$j = \sigma|\boldsymbol{E}| = U_s/R_0S_0 \,, \tag{7.29}$$

where U_s is the external applied voltage and $|\boldsymbol{E}| = U_s/l$ the internal electric field. Electron irradiation causes a change in both σ and $|\boldsymbol{E}|$:

$$j + \Delta j = (\sigma + \Delta\sigma)(|\boldsymbol{E}| - \Delta|\boldsymbol{E}|) \tag{7.30}$$

and for a constant current source ($\Delta j = 0$), we have

$$\Delta\sigma|\boldsymbol{E}| \simeq \sigma\Delta|\boldsymbol{E}| \,. \tag{7.31}$$

The additional voltage drop along the bar becomes

$$\Delta V_1 = \int_0^l \Delta|\boldsymbol{E}|\mathrm{d}x \simeq |\boldsymbol{E}|\frac{\Delta\sigma}{\sigma}\Delta x = \rho^2 j\Delta\sigma\Delta x \,, \tag{7.32}$$

where $\rho = 1/\sigma$.

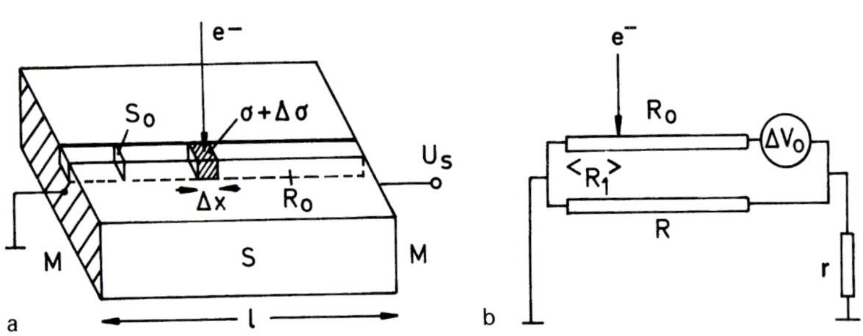

Fig. 7.8. (**a**) Geometry used for calculating the contribution of a bar of cross-section S_0 and length l and change of conductivity $\Delta\sigma$ inside a volume $(\Delta x)^3$ to the electron-beam-induced voltage ΔV_0 in a MSM structure for a constant current source. (**b**) Equivalent circuit for calculating the change of external voltage ΔV caused by the change ΔV_0 along the bar

Certain electromotive forces also have to be taken into account. Localized variation in the doping concentration causes a variation of the local resistivity and of the position of the Fermi level between the valence and conduction bands. In equilibrium, the Fermi level must remain constant and the bands are bent, which creates localized electric fields. This chemical voltage can be neglected because normally these fields are not high enough to influence the carrier diffusion. However, the so-called Dember voltage cannot be neglected. In the drift of charge carriers, the mobilities of electrons (μ_n) and holes (μ_p) will generally be different. One type of charge carrier will diffuse faster into the material away from the beam. This will not occur symmetrically on either side of the beam in the presence of non-uniform doping. This asymmetric separation of charge carriers sets up a Dember voltage [7.114]

$$\Delta V_2 = \frac{kT}{e}\frac{2}{1+\mu_n/\mu_p}\frac{\mathrm{d}\sigma}{\mathrm{d}x}\Delta\sigma\Delta x \,. \tag{7.33}$$

The total voltage at the end of the bar is therefore

$$\Delta V_0 = I_s R_1 + \Delta V_1 + \Delta V_2 \tag{7.34}$$

$$= I_s R_1 + \rho^2 j \Delta\sigma\Delta x + \frac{kT}{e}\frac{2}{1+\mu_n/\mu_p}\frac{\mathrm{d}\sigma}{\mathrm{d}x}\Delta\sigma\Delta x \,, \tag{7.35}$$

where the first term represents the voltage drop of the specimen current I_s across that part R_1 of R_0 (Fig. 7.8b) that lies on the flow to earth. The actual change of voltage can be read from the equivalent circuit of Fig. 7.8b where R is the resistance of the bulk crystal and r a shunt resistance:

$$\Delta V = \frac{r}{r+R_t}\frac{R_t}{R_p}\Delta V_0, \quad \text{where} \quad R_t = R_0 R/(R_0+R) \,. \tag{7.36}$$

The case of constant applied voltage has been treated for both linear and annular geometries and for low- and high-bias fields [7.112, 113] whereas in the case discussed above, only the low-field case has been considered. This technique can also be applied to disc-shaped TEM specimens thinned in the central region [7.115].

The electron-beam-induced conductivity in MIM and MIS layers perpendicular to the electron beam has been investigated for materials that offer a high gain such as amorphous As_2S_3, Sb_2S_3 or Se films [7.116, 117]. These structures look like solid-state ionization chambers. However, owing to the low mobilities of charge carriers and the high trap densities and recombination rates, they are far from the saturation range where all charge carriers are collected. The typical behaviour of the gain, the ratio of the electron-beam-induced current to the incident electron current is plotted in Fig. 7.9 against the electron energy E for a 2.9 μm Se film sandwiched between gold films and for the different values of the bias $U_s = 10$–20 V and both polarities of the bottom electrode. For low electron energies, the gain is very small; it increases rapidly to a maximum at electron energies for which the range R

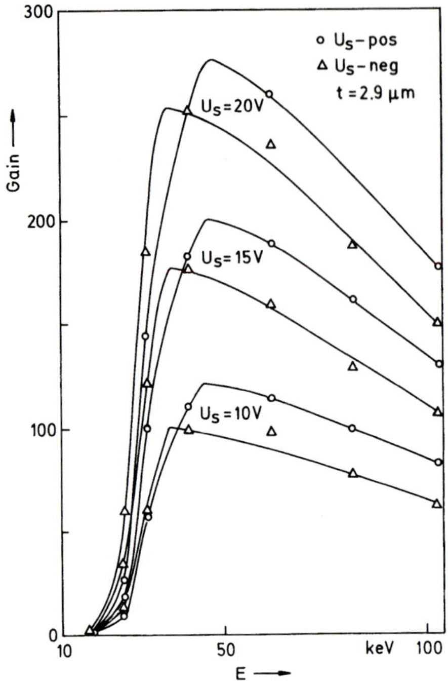

Fig. 7.9. Gain I_{cc}/I_p of a 2.9 μm Se layer sandwiched between Au films as a function of electron energy E for different values of the bias U_s of the bottom electrode; both polarities are shown [7.118]

is equal to the thickness t of the layer [7.118]. The subsequent decrease is due to the fall in the excitation rate with increasing E for $t < R$. The increase with U_s demonstrates that the conditions are far from the saturation regions. The dependence on polarity is a consequence of the inhomogeneous generation of charge carriers across the layer. Electrons are trapped and form a space charge, which influences the potential distribution inside the layer. Ti-Si-Ti [7.119] and metal–GaAs–metal structures [7.120] with gains of $\simeq$200 and 1000, respectively, for 1 keV electrons have been proposed for low-energy electron detection. In SiO_2 films, the electrons show a mobility $\mu_n \simeq 20$ $cm^2V^{-1}s^{-1}$ while holes are quite immobile. The gain is only of the order of 2–3.

In insulating MIS sandwich layers, conductive channels (current filaments) can be created by applying increasing voltages. These lead to regions of negative differential resistance in the I–U characteristics. Induced electric microfields at the surface of 40 nm Ag–80 nm SiO_2–36 nm Ge can be detected by the voltage contrast [7.121]. EBIC experiments on p-i-n structures allow reveal these filaments [7.122]. An increased EBIC signal at the filament boundaries has been interpreted as an induced change of conductivity in samples with shallow traps [7.123]. In p-n-p-n diodes the spatial filamentary as well as temporal current-density oscillations can be displayed and the analysis of EBIC signals can give information about both electric field and current-density distributions [7.124].

Low-temperature SEM allows current or voltage changes to be recorded in superconductors [7.125]. Irradiation of superconducting films by a chopped electron probe causes small increases of temperature and small decreases of the local critical current density. Current biasing of microbridges 10–100 μm wide and 20–200 μm long with the specimen near T_c causes a voltage drop $\Delta U(x, y)$, which can be recorded using the lock-in technique. This also allows us to image spatial inhomogenities in grain boundaries. In superconducting tunnel junctions, which normally consist of two superconducting films separated by a thin insulating barrier, the change of the tunnel junction current or voltage can be used as a signal when the electron probe scans across the junction. For example, trapped vortices or standing microwave patterns can be imaged in the Y-modulation mode.

7.2 Cathodoluminescence

7.2.1 Basic Mechanisms of Cathodoluminescence

Light emission can be stimulated by irradiation with ultraviolet light (photoluminescence) or by electron irradiation (cathodoluminescence, CL). In principle, there is no difference between the light spectra emitted and results obtained from photoluminescence experiments can also be used for the interpretation of CL experiments. However, the relative intensities of different luminescence bands may not be exactly the same because of differences in the excitation mechanism, the rate of excitation and the absorption of the excited radiation. CL is not only excited inside the diffusion cloud at a depth of the order of the electron range R but minority carriers can also recombine with or without radiating outside this region in materials with a diffusion length L larger than R (Sect. 7.2.3).

In this section, the fundamental physical processes leading to CL are described (see [7.2, 126–130] for reviews and [7.131] for bibliography). The experimental equipment needed to detect and investigate CL will be discussed in Sect. 7.2.2. The classification of CL emission into a few basic mechanisms is a consequence of the band structure of electronic states in a solid. In semiconductors and insulators, the filled valence band and the empty conduction band are separated by an energy gap $\Delta E = E_c - E_v$ (Fig. 7.10). Electrons excited by inelastic collisions from the valence band to unoccupied states above the bottom of the conduction band lose their excess energy by a cascade of non-radiative phonon and electron excitations (Fig. 7.10a) and are thermalized to states at the bottom of the conductive band in times of the order of 10^{-11} s. Most of the processes of recombination with holes in the valence band will be non-radiative and take place via multi-phonon processes. The following mechanisms can result in radiative transitions.

Intrinsic emission is due to direct combination of electron–hole pairs, the probability of which is strongly influenced by the shape of the $E(\boldsymbol{k})$ surface,

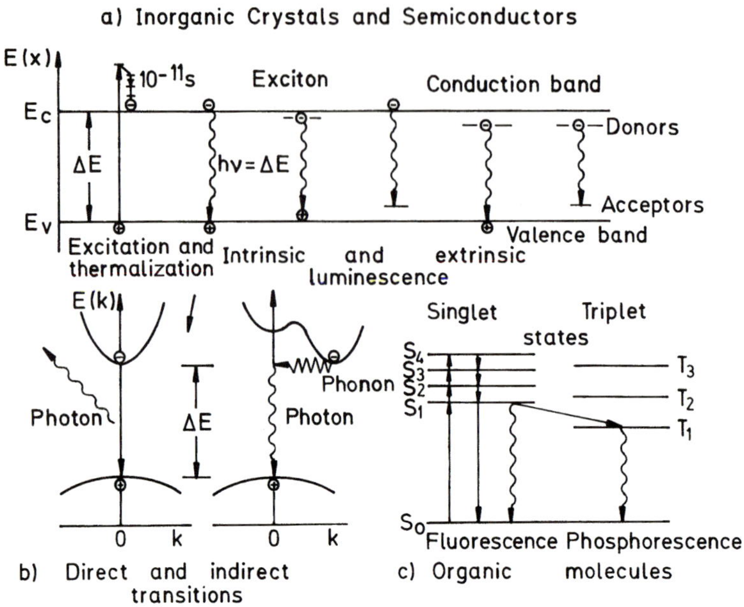

Fig. 7.10. Basic mechanisms of cathodoluminescence: (**a**) Thermalization and intrinsic and extrinsic transitions in inorganic scintillators, (**b**) direct and indirect transitions in semiconductors, (**c**) singlet–triplet arrangement of organic molecules

where $\boldsymbol{k}$ is the wave vector of the electron or hole. Free conduction electrons are concentrated near the minimum of the conduction band and free holes near the maximum of the valence band. In the case of a semiconductor with a direct band gap (Fig. 7.10b, left) such as GaAs and ZnS, for example, these maxima and minima lie at the same $\boldsymbol{k}$ value and the emitted light quanta have the energy $h\nu = \Delta E$; this is known as edge emission because there will also be a sharp edge-like increase of photoabsorption when the photon energy increases beyond this value. In materials with an indirect band gap (Fig. 7.10b, right) such as GaP and Ge, for example, a phonon also has to participate in the recombination to supply the difference $h(\boldsymbol{k}_{\mathrm{c}} - \boldsymbol{k}_{\mathrm{v}})$ in momentum because the minima and maxima of $E(\boldsymbol{k})$ are not directly above one another and because the momentum of a photon is negligible. In materials with a direct band gap, therefore, vertical transitions are more probable and can result in high CL efficiencies. This model can also explain the dependence of edge emission on temperature. At high temperature, the conduction electrons are thermally activated over a small range of energies beyond E_{c} and conversely, the holes extend over a small region below E_{v}, which results in a broadening of the emission line with increasing temperature. The temperature also influences the band gap ΔE by virtue of thermal expansion, which can cause a shift of the emission line and can be used to measure device temperature, for example [7.132]. In polymorphic materials, the energy gap and the position of

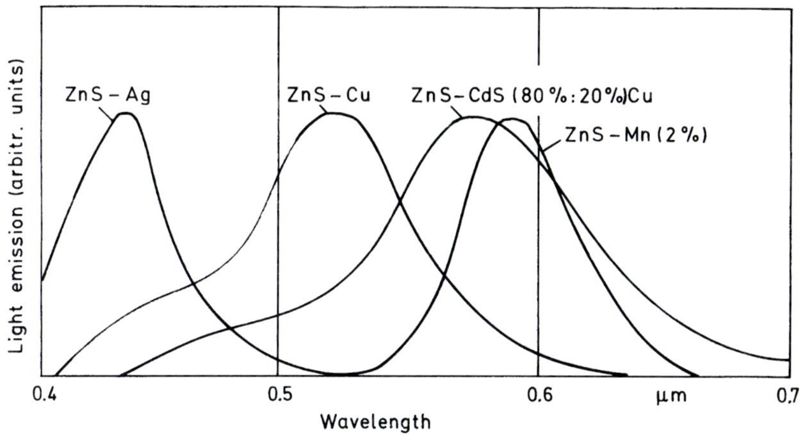

Fig. 7.11. Influence of different dopants on the CL emission spectrum of ZnS

the edge emission band depend on the crystal structure; ZnSe with cubic and wurtzite structures is an example of this [7.133]. Another example, in which information can be gained from the variation of ΔE is provided by miscible semiconductor compounds such as $GaAs_xAl_{1-x}$ or $GaAs_xP_{1-x}$, which can be used for composition measurement in heterostructures [7.8, 134] or in graded layers [7.135]. For Te-doped GaAs, for example, the peak intensity at the emission maximum shifts from 1.41 to 1.46 eV and the half-width increases from 0.04 to 0.15 eV when the dopant concentration is raised from 10^{17} to 10^{19} cm^{-3} [7.136].

At low temperature, electrons and holes can form intermediate bound states, known as excitons. The recombination of these excitons results in sharp lines with energies below the edge emission.

Extrinsic emission is the result of radiative recombination of trapped electrons and holes at donor and acceptor levels, respectively. The trapping increases the probability of recombination and the photon energy of the emitted radiation is lower than that of the edge emission. Figure 7.11 shows how different dopants in ZnS influence the emission spectrum. An increase of the dopant concentration of luminescence centres causes an increase of CL intensity until the distance between the dopant atoms is small and the wavefunction and lattice relaxation regions overlap. The curve representing luminescence versus dopant concentration may pass through a maximum as shown in Fig. 7.12 for GaAs doped with Te [7.137]. This figure also shows the concentration of donors or acceptors for which the wavefunction overlap and donor and acceptor levels become degenerate and form narrow impurity bands. At higher concentrations, donors and acceptors start to precipitate.

In the indirect band gap semiconductor GaP, radiative recombination can be enhanced by employing Zn–O donor–acceptor pairs, N isoelectric traps or N–N pairs, which give red, green and orange emission, respectively, in light-

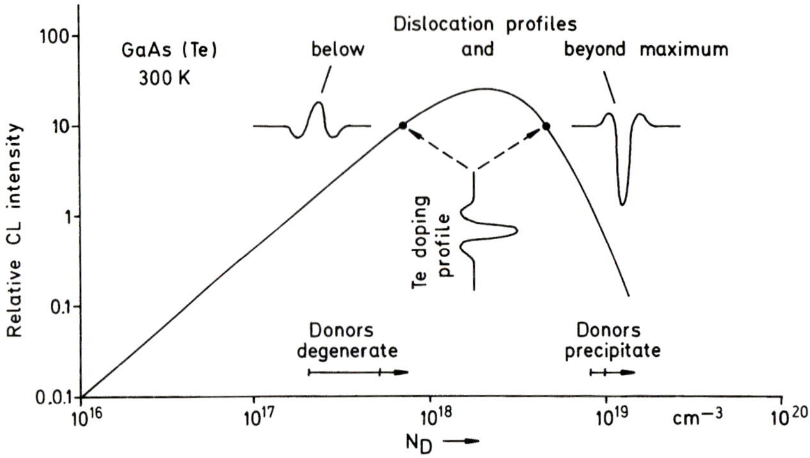

Fig. 7.12. Luminescence versus Te dopant concentration in GaAs and two profiles of the CL signal from the Cottrell atmosphere of Te atoms around a dislocation

emitting diodes. Here, the ratio of the peak height of the N–N line to that of other lines can also be used to determine the nitrogen concentration and its local variation [7.138].

These luminescence mechanisms, which are typical of semiconductors, cannot be expected to operate for metals which do, however, emit weak visible and ultraviolet light (Lilienfeld radiation). The origins of this radiation are threefold [7.139, 140]:

1. Transition radiation emitted when an electron of constant velocity penetrates an interface (vacuum–metal) between media with different dielectric constants,
2. Long-wavelength tail of the x-ray continuum, and
3. Radiative decay of excited plasmon oscillations of the electron gas.

7.2.2 Detection and Spectrometry of Light Quanta

Microscopy based on fluorescence excited by ultraviolet light is an important technique in mineralogy and biology. Excitation by electrons – cathodoluminescence (CL) (Sect. 7.2) – can also be combined with optical microscopy either by large-area bombardment [7.142] or inside a SEM [7.143]. Recording on colour film has the advantage that both the spatial distribution of CL and its colour information are recorded. However, this method is restricted to strongly cathodoluminescent materials and the resolution is limited by the relatively low aperture of the optical microscope. Better resolution and sensitivity can be attained in the scanning mode where the signal generated by the light quanta emitted is used to modulate the CRT intensity.

A photomultiplier is the ideal quantum detector. A GaAs photocathode has the advantage that the sensitivity is nearly constant over a wavelength range from 300 to 850 nm. Silicon cathodes extend this range to 1.1 μm but have a lower sensitivity. The only remaining problem is to collect and guide the emitted quanta with a high collection efficiency and to make an optical spectrometer of high transparency optionally available. Silicon p-i-n photodiodes have the advantage that they can be directly mounted in the specimen chamber and thus collect a large fraction of the emitted light quanta in the transmission mode, for example, but of the order of 1.5×10^8 photons per second are necessary to get a signal-to-noise ratio $S/N = 5$ at $\lambda = 500$ nm [7.144].

Strongly luminescent materials such as ZnS or CdS can be investigated by introducing a lens or a concave mirror [7.145] between specimen and light-pipe or by placing a glass-fibre light-pipe close to the specimen [7.146, 147]. However, because of the low CL intensity of most materials, a light collecting system should work with as large a solid angle of collection as possible. Reflections at mirrors and transparent interfaces cause losses of the order of 3–10%. The transparency of light-pipes is limited to angles of incidence of 20°–40° relative to the normal because otherwise, the light is not totally internally reflected in the light-pipe. It has been shown that an Al foil tube acting as a mirror to guide the light to the photomultiplier can be more efficient than a light-pipe [7.148]. Furthermore, the transmittance of the spectrometer and the quantum efficiency of the photomultiplier cathode decrease the sensitivity. All these parameters also depend on the optical wavelengths and should be measured for a real system to optimize the collection efficiency [7.149]. In view of all this, we conclude that only the following types of collection systems will be efficient:

1. A parabolic mirror (Fig. 7.13a) forming a parallel light beam [7.150, 151] or an elliptical mirror (Fig. 7.13b) [7.152, 153] can be used to focus the light on the entrance slit of a spectrometer [7.154]; they have the advantage that the solid angle of collection is of the order of π sr. There is enough space to detect secondary electrons with an Everhart–Thornley detector.
2. A mirror in the form of half an ellipsoid of rotation [7.155] with the electron impact point at one focus and the entrance to the light-pipe at the second focal point (Fig. 7.13c) has the largest possible angle of collection. However, the limitation of the acceptance angle of a light-pipe has to be taken into account. An optical microscope objective lens with its focus at the second focal point of the mirror can collect a solid angle of $\simeq 0.75\pi$ sr for an eccentricity $\epsilon = 0.56$ of the ellipsoid and refracts the light onto a light-pipe as a parallel beam [7.156]. A parallel light beam can also be formed by placing a parabolic mirror below the second focus [7.155].

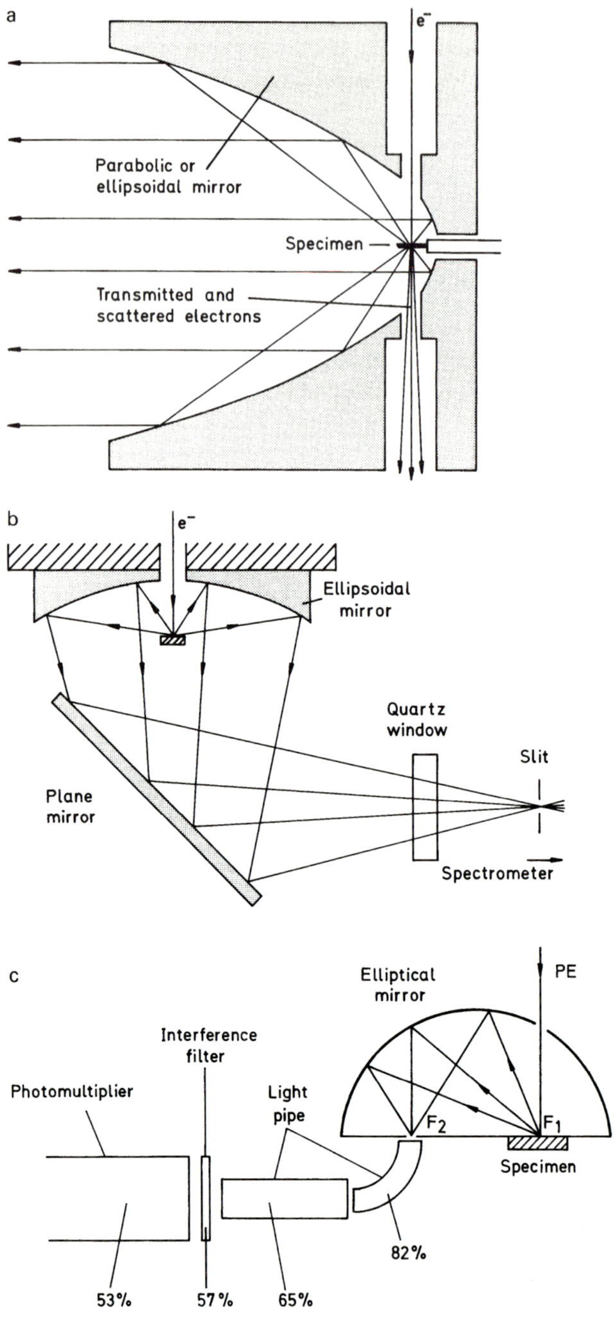

Fig. 7.13. Light-collection system for cathodoluminescence: (**a**) parabolic mirror with parallel light output [7.150], (**b**) ellipsoidal mirror with focusing on a spectrometer slit [7.153], (**c**) half a rotational-ellipsoidal mirror coupled to a light-pipe [7.161]

3. Optically transparent specimens can be investigated by placing a light-microscope objective lens below the specimen with a collection angle of 1.4 π sr [7.157] or by using fibre optics coupling [7.158].

Normally these CL detection systems are used in the non-dispersive (panchromatic) mode, which is not capable of separating phases having different emission spectra. Monochromatic images can be obtained by placing a colour or interference filter in front of the photomultiplier. The simplest way of getting colour images is to use red, green and blue colour filters in front of the photomultiplier and to record a coloured image on colour film by placing the same filter sequentially in front of the CRT [7.159]. Real colour images are generated by simultaneously recording red, green and blue components either by splitting the CL signal collected by an ellipsoidal mirror into three light guides [7.160–162] or by using a trisector ellipsoidal mirror [7.163] followed by colour or interference filters and photomultipliers. The best method is to store the images digitally. For many applications, the shifts in colour that are detectable by the human eye are sufficient; otherwise a monochromator with higher wavelength resolution has to be used. A digital record of a series of CL micrographs is useful when the intensity and spectral distribution of the CL changes as radiation damage advances. This can occur in both inorganic and organic material. The non-dispersive CL signal of NaCl increases, for example, by the formation of F centres but decreases at high accumulated charge densities due to the increase of defect concentration [7.188]. Such damage effects can be observed in many geological and mineralogical specimens. The higher beam sensitivity of organic specimens is discussed in Sects. 3.5.3 and 7.2.5.

Saparin and coworkers have shown in numerous examples from geology, mineralogy, semiconductor science and biology [7.165, 166] that the recording of real colour CL images has great advantages, in the detection of different phases and dopants, for example.

Backscattered electrons can produce spurious CL when they strike the collector or are backscattered to the specimen. Dust particles and pump oil contaminants often show intense CL. It must also be recalled that metals show a weak CL, the so-called Lilienfeld radiation (Sect. 7.2.1). The light from the thermionic cathode can also disturb weak CL signals. This contribution can be eliminated by chopping the electron beam (Sect. 2.3.3) and recording the CL emitted with a lock-in amplifier [7.164].

For recording CL spectra, conventional prism or grating spectrometers can be used or Fourier-transform spectroscopy by means of a Michelson interferometer can be applied, especially for the infrared. A device proposed by Oxford Instruments combines the possibility of specimen cooling in the range 5–300 K with spectrum recording between 200 nm and 2.4 μm. A sequential record of a spectrum normally takes a few minutes. This can be reduced by using an optical multichannel analyser, where the spectrum or part of it (depending on the dispersion and resolution) is focused on the photocathode of a SIT video camera (silicon intensifier target) [7.167]. The linescans of the spectrum are fed to a multichannel analyser where the signal can be averaged. Electron-beam chopping (Sect. 2.3.3) with repetition frequencies between 1 kHz and 1 MHz can be employed to study the kinetics of CL processes [7.168].

At very low CL levels or after passing through a spectrometer, it may be more convenient to use a single-photon counting mode than the analogue signal. At low light levels, the latter has to work with a high PM voltage, which results in more noise and fluctuations in amplification than at low PM voltages. Using a fast, high-gain amplifier behind the PM, the background can be reduced in a single-photon counting mode by a discriminator, and after-pulses (Sect. 5.1.2) can be eliminated by introducing an artificial dead-time of the order of 25 ns [7.169], which would not be possible in the analogue signal mode.

7.2.3 Quantitative Cathodoluminescence of Semiconductors

The cathodoluminescence signal J_{CL} has the advantage that it is excited in the bulk and does not need a depletion layer for charge collection. Information is available not only from the emission spectra but also from the dependence of J_{CL} on the dopant concentration (Fig. 7.12, for example), on the electron beam energy and on the electron-probe current and from lifetime measurements.

In a first-order approximation, J_{CL} should increase as the electron energy and electron-probe current. In many cases, these proportionalities are respected. In scintillator materials, for example, the linear relation between J_{CL} and I_{p} guarantees that the signal will be proportional to the number of SE and BSE. The generation depth of light quanta increases in proportion to the electron range $R \propto E^n$ with $n = 1.5$–1.7. In Sect. 3.4.2 references are given to experimental investigations of the distribution of CL in gas and solid targets. However, deviations from these proportionalities can be observed, due to

1. the presence of a dead layer of thickness d near the surface where the band structure is deformed, owing to the presence of a space-charge depletion layer caused by pinning the Fermi level by surface states (see left side of Fig. 5.2b), and where radiative recombination processes are decreased,
2. diffusion of minority carriers before radiative or non-radiative recombination for substances with a large diffusion length L, which also includes the surface recombination velocity S introduced in Sect. 7.1.2, and
3. absorption of the emitted radiation in materials with a large absorption coefficient.

The theoretical treatment of CL in semiconductors [7.141, 170] uses the same differential equation of diffusion (7.9) as that employed for the EBIC signal. We introduced in (7.5) the lifetime τ, which is inversely proportional to the capture cross-section σ_{T}. The cross-sections or reciprocal lifetimes have to be added for different recombination processes. When we distinguish between radiative and non-radiative recombinations, we find

$$\frac{1}{\tau} = \sum_i \frac{1}{\tau_{\mathrm{r},i}} + \sum_j \frac{1}{\tau_{\mathrm{nr},j}} = \frac{1}{\tau_\mathrm{r}} + \frac{1}{\tau_\mathrm{nr}} \tag{7.37}$$

where τ_r is the lifetime for radiative and τ_nr that for non-radiative recombinations. The radiative recombination efficiency becomes

$$\eta_\mathrm{r} = \frac{\sigma_\mathrm{r}}{\sigma_\mathrm{r} + \sigma_\mathrm{nr}} = \frac{\tau}{\tau_\mathrm{r}} = \frac{\tau_\mathrm{nr}}{\tau_\mathrm{r} + \tau_\mathrm{nr}} \ . \tag{7.38}$$

Again denoting the local concentration of excess electrons by the primary electrons by $N'(r)$, the overall recombination rate is $N'(r)/\tau$ and only the fraction $N'(r)\eta_\mathrm{r}/\tau = N'(r)/\tau_\mathrm{r}$ will recombine radiatively. In the case of semiconductors with a diffusion length $L > R$, we obtain $N'(r)$ from the same continuity relation (7.9) or (7.10). When we assume that a point source is situated at the surface, neglect surface recombination and consider light absorption, the radiation intensity becomes

$$J_\mathrm{CL} = \frac{\eta_\mathrm{T}}{\tau_\mathrm{r}} \int_0^\infty N'(z) \mathrm{e}^{-\alpha z} \mathrm{d}z = \frac{\eta_\mathrm{T} G I_\mathrm{p}}{e} \frac{\tau}{\tau_\mathrm{r}} \frac{1}{1+\alpha L} \tag{7.39}$$

using (7.13) and an exponential absorption term with the linear absorption coefficient α and the transmission η_T through the surface, see (7.45).

For finite source-size, which has to be taken into account when considering surface recombination and a dead layer and including the case $L > R$, the van Roosbroeck equation (7.15) can be used [7.141, 170]. We have calculated the fraction of electrons that recombine at the surface with the aid of (7.16). We assume that only non-radiative recombination occurs in the dead layer of thickness d. The CL efficiency is then given by

$$\epsilon_\mathrm{CL} = \left[\int_d^R g(z_\mathrm{s}) \mathrm{e}^{-\alpha z_\mathrm{s}} \mathrm{d}z_\mathrm{s} - \frac{S}{S+1} \int_d^R g(z_\mathrm{s}) \mathrm{e}^{-\alpha z_\mathrm{s}} \mathrm{e}^{-(z_\mathrm{s}-d)/L} \mathrm{d}z_\mathrm{s} \right] \Big/ \int_0^R g(z_\mathrm{s}) \mathrm{d}z_\mathrm{s} \ , \tag{7.40}$$

which is plotted in Fig. 7.14a for $d = 0$ and in 7.14b for $d/L = 0.1$ against the reduced range R/L, neglecting absorption ($\alpha = 0$). $\epsilon_\mathrm{CL} = 1$ means that J_CL will be proportional to the electron energy and Fig. 7.14 shows the departure from this linearity caused by surface recombination and a dead layer. The influence of surface recombination will decrease with increasing range or energy. For Fig. 7.14, the analytical fit (3.155) has been used for $g(z_\mathrm{s})$ or Monte Carlo simulations of $g(z_\mathrm{s})$ can be used [7.171]. If absorption has to be taken into account, the CL intensity can decrease further for electron ranges $R \gg 1/\alpha$. The dependence of J_CL on electron energy is therefore an analytical tool, providing information similar to that obtained from the corresponding EBIC experiments discussed in Sect. 7.1.3. Experimental $J_\mathrm{CL}(E)$ curves can be fitted to theoretical curves using Fig. 7.14 with a suitable set of L, S and

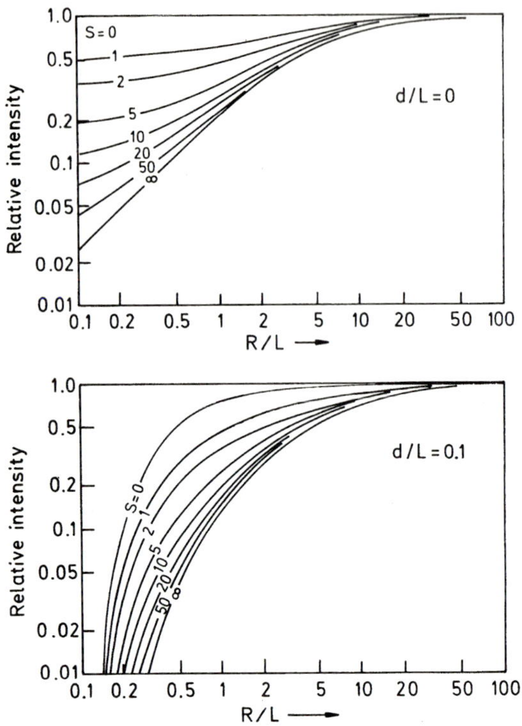

Fig. 7.14. Cathodoluminescence efficiency ϵ_{CL} versus reduced electron range R/L and influence of the reduced surface recombination velocity S (**a**) without and (**b**) with a dead layer $d/L = 0.1$ [7.170]

d values [7.170–172]. However, the accuracy is less than for the $I_{cc}(x)$ curves of EBIC.

The dependence of J_{CL} on the electron-probe current will be affected by the kinetics of the recombination processes, which is influenced by the local density $N'(r)$ and by the dependence of the lifetime on electron concentration. Increases faster than linear, $J_{CL} \propto I_p^n (1 \leq n \leq 2)$ [7.141, 172] and proportional to $I_p^{1/2}$ [7.173] have been observed. In direct band-gap semiconductors, the lifetime τ of minority carriers (see also Sect. 7.2.2) exhibits a temperature dependence $\tau \propto \exp(-b/kT)$ where the constant b depends on the material. The consequence is an increase of luminescence with decreasing temperature, which agrees with theoretical calculations based on the van Roosbroeck theory [7.174].

In the direct measurement of lifetime, smaller values of τ can be measured using CL than with EBIC because the practical frequency limitation due to the capacitance of the depletion layer does not exist. The lifetime τ is the time required for the intensity to drop to 1/e after blanking the electron beam (7.21). Response times of the order of nanoseconds can be detected with commercial photomultiplier tubes [7.175]. The electron beam is periodically chopped and a gated photon counting technique can be employed for recording fast CL decay [7.153].

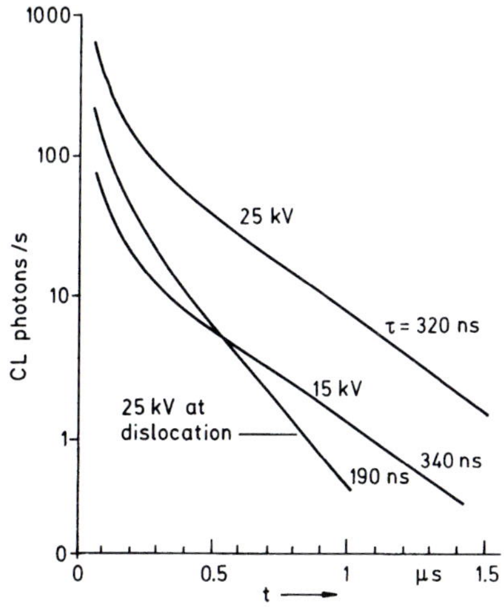

Fig. 7.15. Double-logarithmic plot of the cathodoluminescence decay curve excited by 15 and 25 keV electrons and by 25 keV electrons near a dislocation. The minority carrier lifetimes τ are extracted from the slope of the straight lines [7.153]

The example shown in Fig. 7.15 demonstrates the dependence on E and the decrease of τ near a dislocation in GaP. Though the decrease of E from 25 keV to 15 keV results in a strong decrease of J_{CL} by one order of magnitude owing to increased surface recombination, the lifetime derived from the slope of the semi-logarithmic plot is less affected. Another way of obtaining a simultaneous record combined with signal averaging over a large number of periodic scans, triggered by the electron beam blanking pulse, is to use a streak camera [7.176, 177]. For example, the increase from $\tau \simeq 3.8$ ns at $T = 80$ K to values of 9 ns at $T = 300$ K has been measured in this way for Se-doped GaAs ($N_{\mathrm{A}} = 1.6 \times 10^{-18}$ cm^{-3}).

Dislocations influence not only the lifetime τ but also the CL intensity, which shows a dot or dot-and-halo contrast around dislocations perpendicular to the surface [7.170, 177–179]. This can be attributed to segregation of impurity atoms (e.g. Te in GaAs) at the dislocation (Cottrell atmosphere) and the presence of a denuded region around the defect. In the $J_{\mathrm{CL}}(N_{\mathrm{D}})$ curve of Fig. 7.12, two concentrations near the maximum are indicated, which result in black-dot contrast with and without a halo. The decrease of J_{CL} beyond the maximum is most striking. Tellurium atoms have the largest misfit in the GaAs lattice and therefore tend to segregate at dislocations whereas no contrast is observed in Se or Si doped material. X-ray microanalysis, emission spectrum studies and lifetime measurements (Fig. 7.15) confirm that this type of contrast is caused by segregation. Nevertheless, dangling bonds at the core of the dislocation or at point defects surrounding dislocations can contribute to the defect contrast though their effect is much weaker than that caused by a Cottrell atmosphere. This has been demonstrated for defects introduced by plastic deformation [7.16, 180]. The question of whether CL is influenced by doping, quenching or plastic deformation is also important for the interpretation of CL in quartz and other minerals, for example [7.181, 183].

The magnitude of dislocation contrast depends on electron energy and the energy for which the contrast is greatest can be related to the depth of the defect [7.184]. A three-dimensional model calculation agrees with experiments [7.185].

7.2.4 Cathodoluminescence Imaging Mode

Detector systems for CL have been described in Sect. 5.4.5. Often, CL images are recorded non-dispersively without a colour filter or spectrometer and only the total CL intensity then modulates the CRT intensity, weighted, however by the spectral intensity of the photomultiplier. In Fig. 7.16, the SE (a,c) and non-dispersive CL images (b,d) of the red, green and blue luminescent dots of a colour TV screen are compared [7.186]. At low magnification, the dots appear dark, grey or bright, respectively. The blue component seen at a larger magnification in Fig. 7.16d indicates that some of the small grains appear dark and have obviously not received the correct doping during the fabrification of the luminescent powder. At the edges of the large triangular-shaped crystal, the CL is reduced; this can be attributed to the increased backscattering near edges. A bright luminescent area at the top face may not mean that there is a maximum of primary luminescence – it can also mean that the excited light can leave the crystal after multiple reflection (see below) through any of the surfaces with a higher probability. A light-optical observation of CL can therefore look quite different because the spatial exit distribution is observed for simultaneous excitation of the whole crystal.

Section 7.2.2 described theoretical approaches that have been employed for calculating the CL intensity. However, for the observation of CL, we have also to take into account the refraction and total internal reflection that occur when the light passes through the surface [7.163, 187]. We use Snell's law

$$n_1 \sin\theta_1 = n_2 \sin\theta_2 \tag{7.41}$$

where the indices 1 and 2 correspond to the materials on either side of the interface (1: material, 2: vacuum, for example) and the Fresnel formulae. For a polarisation normal and parallel to the plane of incidence, the reflected intensities are

$$R_\perp = \frac{\sin^2(\theta_1 - \theta_2)}{\sin^2(\theta_1 + \theta_2)}; \quad R_\parallel = \frac{\tan^2(\theta_1 - \theta_2)}{\tan^2(\theta_1 + \theta_2)} . \tag{7.42}$$

For the case $\theta_2 = 90°$, we see from (7.41) that the critical angle θ_t of total internal reflection is given by

$$\sin\theta_t = n_2/n_1 . \tag{7.43}$$

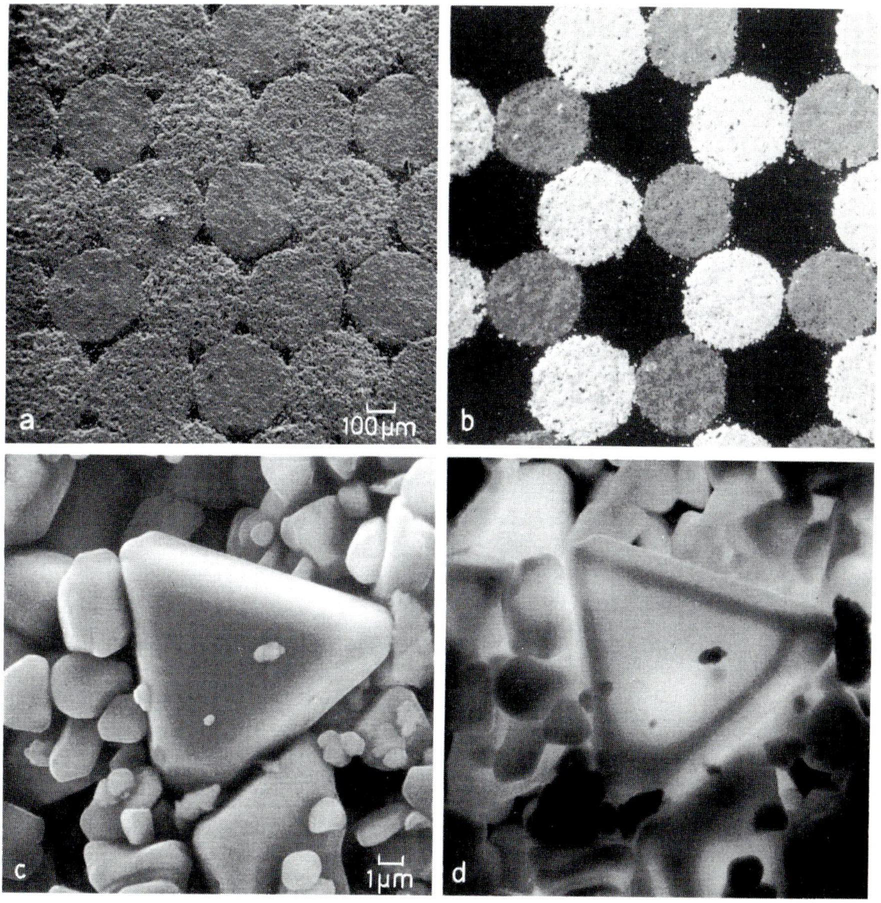

Fig. 7.16. Colour spots of a colour TV screen imaged with SE (**a,b**) and with a non-dispersive CL mode (**b,d**). (**a**) and (**b**) General view of red (*dark*), green (*grey*) and blue (*bright*) spots in the CL mode, (**c**) and (**d**) larger magnification of a blue spot [7.186]

For normal incidence, $\theta_1 = \theta_2 = 0$, the reflected intensity is found from (7.42) to be

$$R = \left(\frac{n-1}{n+1}\right)^2 . \tag{7.44}$$

For non-normal incidence und unpolarized light, the reflected intensity $R = (R_\perp + R_\parallel)/2$ does not change significantly with increasing θ_1 and only increases rapidly to unity at and near $\theta_1 = \theta_t$. We can therefore assume that (7.44) is valid to a first-order approximation over the whole interval $0 > \theta_1 > \theta_t$. The CL will be generated isotropically but only a small fraction inside a cone of semi-apex angle θ_t can cross the surface. Rays for which $90° > \theta_1 > \theta_t$ are totally internal reflected and those for which

$180° > \theta_1 > 90°$ are emitted into the material. Whether these rays are absorbed or can leave the specimen somewhere after multiple reflection depends on the absorption coefficient α. Thus, only a fraction η_T of the signal actually excited will leave the specimen:

$$\eta_T = (1-R)\frac{1}{4\pi}\int_0^{\theta_1} 2\pi \sin\theta_1 d\theta_1 = \left[1-\left(\frac{n-1}{n+1}\right)^2\right]\frac{n-\sqrt{n^2-1}}{2n} , \quad (7.45)$$

where $n_1 = n$, $n_2 = 1$ (vacuum). The refractive indices of semiconductors may be large, between 2 and 5.

Numerical Example: n	=	1.3	θ_t	=	50.3°	η_T	=	17%
		2			30°			6%
		5			11.5°			0.9%

This demonstrates that for higher values of n, the CL intensity can be drastically reduced by total internal reflection.

After refraction, an element of solid angle $d\Omega_1 = 2\pi \sin\theta_1 d\theta_1$ becomes $d\Omega_2 = 2\pi \sin\theta_2 d\theta_2$ so that for a uniform light source $dJ(\theta_1)/d\Omega = 1/4\pi$ we find

$$\frac{dJ(\theta_2)}{d\Omega} 2\pi \sin\theta_2 d\theta_2 = \frac{1-R}{4\pi} 2\pi \sin\theta_1 d\theta_1 \quad (7.46)$$

and

$$\frac{dJ(\theta_2)}{d\Omega} = \frac{1-R}{4\pi}\frac{\sin\theta_1}{\sin\theta_2}\frac{d\theta_1}{d\theta_2} = \frac{1-R}{4\pi}\frac{1}{n}\frac{1}{\sqrt{n^2-\sin^2\theta_2}}\cos\theta_2 . \quad (7.47)$$

Lambert's cosine law, for which $dJ(\theta_2)/d\Omega$ would be proportional to $\cos\theta_2$, is thus not obeyed.

Emission spectra, colour and decay times are also influenced by the specimen temperature. CL equipment for the SEM should, therefore, be capable of cooling the specimen to liquid nitrogen or helium temperature. For example, the CL intensity of quartz increases by a factor 50 when the specimen is cooled from room to liquid nitrogen temperature (see also [7.181]). However, the decay time also increases by an order of magnitude, so that only a very low scan frequency can be used.

The possibility of applying stroboscopy for the measurement of lifetimes or decay times has already been discussed in Sect. 7.2.2. Lifetimes measured at different positions of the electron beam can be displayed as a two-dimensional Y-modulation pattern [7.182], for example. CL materials that have approximately the same CL intensities and nearly equal spectral distributions but different lifetimes can also be distinguished in an intensity modulated image by using a delayed gate after switching off or on the electron beam pulse [7.189]. For example, owellite ($CaMoO_4$) and scheelite ($CaWO_4$) can be separated by their differences in lifetime in molybdoscheelite [$Ca(W,Mo)O_4$]. In all stroboscopic measurements and imaging, the electron beam has to be switched on or off with a rise-time smaller than the observed CL delay time and the pulse-length should be variable to produce intensity variations within the rise or decay time of the CL signal or to attain the steady-state intensity before switching off the beam. The pulse frequency also has to be adapted to the decay time.

Whereas the observation of CL by a light microscope will be limited in resolution by the light wavelength and the numerical aperture of the microscope, CL in SEM can in principle reach resolutions of the order of nanometres, which is the mean interaction distance required for a primary electron to generate an inelastic excitation of an atom or in a solid. However, the practical resolution will be limited by the diameter of the electron beam and of the electron diffusion cloud (electron range) and by the diffusion of minority carriers [7.190]. Often the electron-probe current has to be increased beyond the optimum value for obtaining minimum probe size (Sect. 2.2.3) to generate a sufficiently large CL signal. Strong absorption of CL and the absence of a dead layer can result in a smaller information depth, which can also increase the resolution. In any discussion of CL resolution, all these beam and specimen parameters have to be taken into account. The reduction of the information volume by decreasing the electron energy below 5 keV (LVSEM) can be used for the CL observation of GaAs at temperatures lower than 20 K [7.191], for example.

7.2.5 Cathodoluminescence of Organic Specimens

With a few exceptions, the CL of organic substances is excited inside a single molecule. Figure 7.10c shows the principle of the energy exchange scheme. Electrons from the ground state S_0 can only be excited to other singlet states S_n, $n > 1$. The electrons are de-excited to the state S_1 by radiationless transitions from which the de-excitation to S_0 can either be radiationless or radiative with decay times $< 10^{-7}$ s (fluorescence). The first de-excitation can also result in occupation of the triplet state T_1, which can also lead to radiative transition to S_0 with decay times $> 10^{-7}$ s (phosphorescence). The dependence of the luminescence spectra on the chemical structure has been investigated in [7.192–194].

Cathodoluminescence of organic specimens can be caused either by autoluminescent molecules, which are already present in the native state and which have not been damaged during the dehydration and fixation processes, or by selective staining with luminescent molecules (fluorochromes). Examples of the former are herbicides, leaves and roots with a high tannin and phenolic content, lichens, calcifications and nucleic acids. Typical fluorochromes are fluorescein, fluorescein isothiocyanate (FITC) and acridine orange. The applications of CL in biology and medicine are reviewed in [7.195, 196].

Normally, CL of organic substances is weak and one of the systems discussed in Sect. 7.2.2 should be used to increase the solid angle of collection and focus the light onto the detector or spectrometer. A severe limitation is the radiation damage of organic substances, which can destroy functional molecular groups, finally creating a cross-linked carbon-rich product (Sect. 3.5.3). Because a charge density $j \cdot t \simeq 10^{-4}$ C cm^{-2} can already cause complete fading of CL, and we estimated in (3.209) that a high-resolution scan needs $\simeq 10^{-3}$ C cm^{-2}, the question arises of whether bright areas in a CL image can indeed be attributed to the original fluorescent molecules. In the fringe of the diffusion cloud , the decreased ionisation probability can result in a tail of the fading curve and when the mass-loss of the material is high, the electron beam penetrates deeper into the material with increasing irradiation time. The best method is to record digitally the CL signal from a fresh specimen area in two subsequent scans and then display the change in CL by forming the difference image. If no further changes in contrast are observed after repeated scans, the CL may be excited by the cross-linked damage product or by backsacttered electrons striking a collecting mirror or other parts, which show spurious CL.

For comparison, radiation damage experiments in TEM based on the fading of electron diffraction intensities show that the crystallinity is better preserved at low temperatures by factor 2–10 depending on the molecular structure. However, after irradiating a specimen at low temperature and subsequently warming it up to room temperature, the damage is the same as that observed with the same charge density at room temperature. This means that the molecular damage is independent of temperature but that the damage products are frozen in and secondary damage processes are reduced, thereby preserving the crystal structure. Because CL is generated inside a single molecule, the damage rate should be independent of temperature. However, differences in CL are observed at low and room temperatures, which can be attributed to different recombination mechanisms. For example, cooling of erythrocytes from 300 K to 80 K and 4 K increases the CL intensity by a factor 8 [7.197] and this type of low-temperature CL decreases rapidly with increasing irradiation time.

8. Special Techniques in SEM

Magnetic stray fields in front of the specimen can be detected by the deflection of secondary electrons by the Lorentz force (magnetic contrast type–1) or by the deflection of primary electrons incident parallel to the surface. The Lorentz force due the internal magnetization affect the trajectories of backscattered electrons, resulting in type–2 magnetic contrast. The surface magnetization can be analysed by detecting the spin-polarization of SE.

Surface biasing and electrostatic fields in front of the specimen cause voltage contrast in the SE image. Quantitative measurement of surface potentials needs pre-acceleration and spectroscopy of the SE emitted.

The use of transmitted electrons, mirror-reflection of electrons at specimens biased at cathode potential, the use of increased partial pressure at the specimen in environmental SEM and the generation of thermal-acoustic waves furnish additional imaging and quantitative modes.

Electron-beam lithography and metrology is now widely used in the technology of semiconductor-device preparation. Multiple scattering of the electrons (proximity effect) limits the resolution of this technique.

8.1 Imaging and Measurement of Magnetic Fields

The four methods for observing magnetic stray fields or ferromagnetic domains are

1. Type–1 magnetic contrast created by the magnetic deflection of SE in external magnetic stray fields,
2. Type–2 magnetic contrast created by the Lorentz deflection of BSE by the internal magnetic induction $\boldsymbol{B}$,
3. measurement of magnetic stray fields by deflection of electron beams parallel to the surface and
4. the imaging of surface magnetization by the emission and detection of polarized electrons.

SEM can also improve the standard Bitter technique giving a resolution of $\simeq$80 nm when deposited ferromagnetic fine particles of $\simeq$20 nm are imaged in the SE mode [8.1].

8.1.1 Type–1 Magnetic Contrast

The SE, which have an exit angular distribution (4.18) and an exit energy spectrum with a most probable energy of a few electronvolts (Fig. 4.18a), are deflected in an external magnetic field. External fringing fields can be created by domains in ferromagnetics with uniaxial anisotropy such as cobalt, by the fields of recording heads and tapes or by the magnetic fields of thin-film conductors on integrated circuits or superconductors. The extent of the fringing field is of the order of the domain width, the recording-head gap, the wavelength on the magnetic tape or the distance between the thin-film conductors, respectively. This means that the fringing field is highly inhomogeneous and SE emitted normal to the surface do not pass through the same magnetic field as those with an exit direction inclined to the normal. In consequence, the contrast expected can only be estimated by considering SE emitted in the direction normal to the surface with a mean energy $\overline{E_{\text{SE}}} \simeq 3$ eV.

The specimen–detector direction is taken to be parallel to the x axis. The Lorentz force component in the x direction is $F_x = evB_y$ so that only the y component of $\boldsymbol{B}$ has any effect. The SE emitted acquire additional momentum in the x direction

$$p_x = \int_0^\infty F_x \mathrm{d}t = \frac{1}{v}\int_0^\infty F_x \mathrm{d}z = e\int_0^\infty B_y \mathrm{d}z \;, \tag{8.1}$$

where $\mathrm{d}t = \mathrm{d}z/v$ is the time taken to travel a distance $\mathrm{d}z$ through the magnetic field. If we assume that the angular deflection ϵ of the SE trajectories will be small, we can apply the momentum approximation (Sect. 2.3.1)

$$\epsilon = p_x/p = \frac{e}{mv}\int_0^\infty B_y \mathrm{d}z \;. \tag{8.2}$$

Numerical Example: A magnetic tape with a recorded periodicity of 0.1 mm and $B = 10^{-2}$ T and $\int B_y \mathrm{d}z \simeq 10^{-7}$ T m deflects SE of energy 4 eV by $\epsilon = 0.024$ rad $= 1.4°$.

The resulting deflection is thus proportional to the line integral over B_y in the z direction and in a first-order approximation is equivalent to a tilt ϵ of the angular characteristic (Fig. 8.1), which is in turn analogous to a tilt of the surface normal. Contrast can be observed, therefore, only by angular selection of the SE emitted, and any detector strategy that increases the sensitivity for topographic contrast also enhances type–1 magnetic contrast. The deflection of fast BSE can be neglected owing to the presence of the velocity v in the denominator of (8.2).

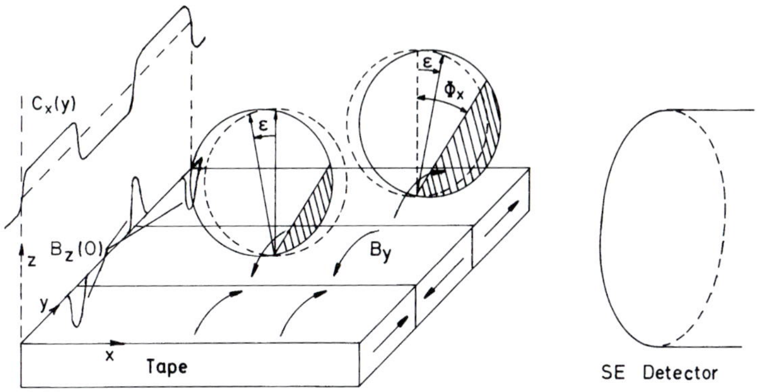

Fig. 8.1. Explanation of type–1 magnetic contrast. The original SE exit characteristic $\propto \cos\zeta$ (*dashed circles*) is modified by external stray field components $\pm B_y$, which appear to rock the exit characteristics through $\pm\epsilon$; ϕ_x denotes the limit in the collection angle of the SE collector

For the special case in which the detector selects only SE with exit momenta that lie below a plane tilted by ϕ_x and indicated by the shaded fraction of the exit characteristics in Fig. 8.1, this fraction of SE collected forms a signal

$$S_{\rm SE} = \frac{\delta_0}{2}(1 - \sin\phi_x) \ . \tag{8.3}$$

The action of magnetic field components B_y in opposite directions results in tilts $\pm\epsilon$ of the exit characteristics, so that the relative contrast between regions of antiparallel B_y becomes

$$C_x = \frac{\Delta S_{\rm SE}}{S_{\rm SE}} = \frac{1}{S_{\rm SE}}\left|\frac{\partial S_{\rm SE}}{\partial\phi_x}\right| 2\epsilon = \frac{\cos\phi_x}{1-\sin\phi_x}2\epsilon \ . \tag{8.4}$$

For $\phi_x = 0$, which is equivalent to a collection efficiency of 50%, (8.4) becomes

$$C_x = 2\epsilon = \frac{2e}{mv}\int_0^\infty B_y \mathrm{d}z = 3.4 \times 10^5 \int_0^\infty B_y \mathrm{d}z \ . \tag{8.5}$$

The numerical constant corresponds to $\overline{E_{\rm SE}} = mv^2/2 = 3$ eV and the integral is measured in units of T m = V s m^{-1}. If we assume that a relative difference $C = 1\%$ can be detected by applying a black-level shift, the minimum value of the line integral will be of the order of 3×10^{-7} T m. The fringe fields of cobalt cut perpendicular to the uniaxial direction of magnetization have domain widths of the order of 10 to 100 μm corresponding to $10^{-6}-10^{-5}$ T m and produce strong magnetic contrast, whereas the fringe fields of magnetic tapes or of a 50 μm wide thin-film conductor carrying a current of 250 mA are of the order of 10^{-7} T m.

When the line profile across a magnetic structure is to be evaluated, we consider a magnetic domain parallel to x with a side area $\Delta x \Delta y$. Gauss's law div $\boldsymbol{B} = 0$ gives the magnetic flux penetrating this area [8.2]

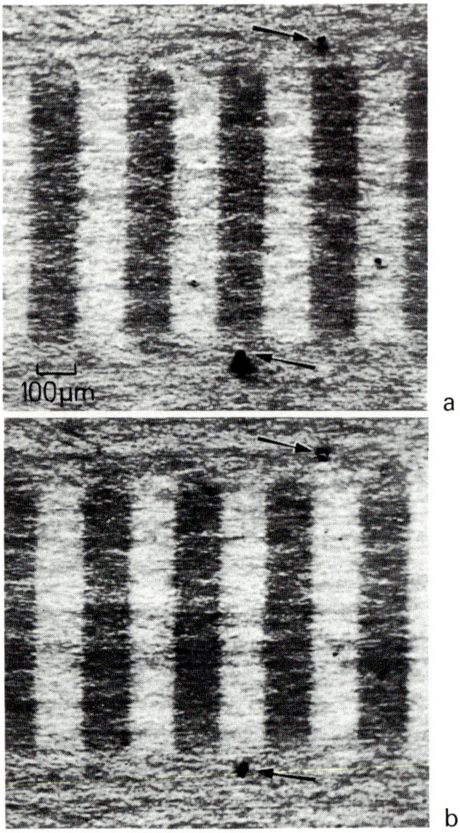

Fig. 8.2. Type–1 magnetic contrast from the stray field of a 400 Hz square wave recorded on magnetic tape. Contrast reversal obtained by rotating the specimen through 180°: (**a**) Detector at the top and (**b**) at the bottom (The arrows point to the same particle)

$$B_z(z=0)\Delta x \Delta y = \Delta x \left[\int B_y(y)\mathrm{d}z - \int B_y(y+\Delta y)\mathrm{d}z\right] , \tag{8.6}$$

or by using (8.5)

$$B_z(z=0) = \frac{\partial}{\partial y}\int\limits_0^\infty B_y \mathrm{d}z \propto \frac{\partial}{\partial y} C_x . \tag{8.7}$$

For an arbitrary direction of magnetization (8.7) becomes [8.2]

$$B_z(z=0) \propto \left(\frac{\partial}{\partial y} C_x - \frac{\partial}{\partial x} C_y\right) . \tag{8.8}$$

The signal may now be collected by a second detector in the y direction or by a segment of a quadrant detector.

Though B_z has no direct influence on the electron deflection, the normal component $B_z(z=0)$ at the surface can be obtained by partial differentiation of the signals C_x and C_y. The open domain structure in Fig. 8.1 results in the schematic linescan of C_x on the left. For Fig. 8.2, $\boldsymbol{B}$ is alternately parallel and antiparallel to the movement of the tape during recording. A normal component $B_z(z = 0) \neq 0$ only appears where C_x shows a jump. On the domain areas where C_x = const, $B_z(z = 0)$ will be zero. The distribution of B_x parallel to the tape will therefore be imaged correctly. This will also be the case for other line shapes of C_x (sine- or saw-tooth-shaped, for example).

It is no problem to observe the magnetic domain contrast of cobalt or other uniaxial ferromagnetic materials of high saturation magnetization in the normal SE mode [8.3, 4]. In order to observe the weaker contrast of magnetic recording tapes, for example, it becomes necessary to improve the angular selection of the SE. From (8.4), we see that the contrast can be increased by a factor two by working with $\phi_x \simeq 45°$ (Fig. 8.1), as demonstrated by the record of a 400 Hz square wave on magnetic tape in Fig. 8.2. The following methods have been proposed to perform an angular selection, which increases the contrast:

1) A negative tilt of the specimen opposite to the detector or insertion of a shield between specimen and detector [8.5].

2) Use of a metal disc behind the collector grid of the Everhart–Thornley detector with a bore of about 5 mm in diameter, which limits the solid angle of the SE collected [8.6, 7].

3) Use of a cage around the specimen with plane and cylindrical electrodes (Fig. 5.11), which are biased independently to select a limited solid angle [8.8, 9].

4) The two-detector system of Fig. 5.15.

The latter can be used to record type–1 contrast with the A–B SE signal, which also suppresses the material and channelling contrast [8.10], and the contrast becomes more uniform across a larger area. A two-detector system consisting of two metal collectors above the specimen was first used by Dorsey [8.3] and in another proposal, two oppositely situated spatial apertures and spectrometers were employed [8.11]. Contrast calculations show that low-pass energy filtering below the most probable SE energy can increase the type–1 magnetic contrast [8.12]

As a test of type–1 magnetic contrast, the contrast should be reversed when the specimen is rotated through 180° about its normal (Fig. 8.2). For strong magnetic fields and large values of the line integral, however, not only the deflection contrast but also an absorption contrast can be observed when the SE trajectories are bent back to the surface; this occurs for values of the line integral larger than 5×10^{-6} T m [8.13].

8.1.2 Type–2 Magnetic Contrast

Type–2 magnetic contrast [8.14] is caused by the action of the Lorentz force of the spontaneous induction $\boldsymbol{B}_s$ in ferromagnetic domains on the BSE trajectories inside the specimen. As demonstrated schematically in Fig. 8.3a with a much smaller bending radius of the electron trajectories between scattering

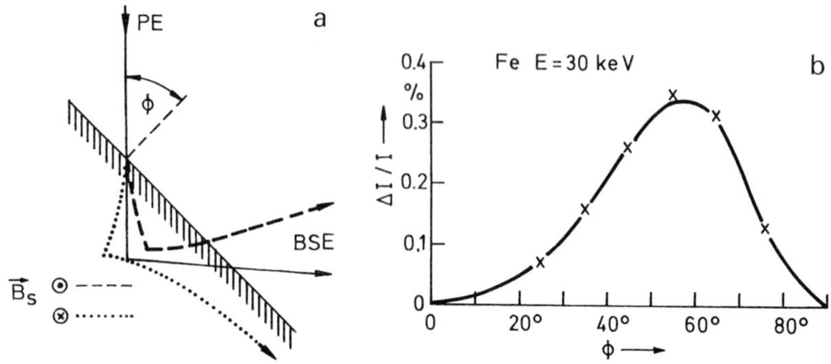

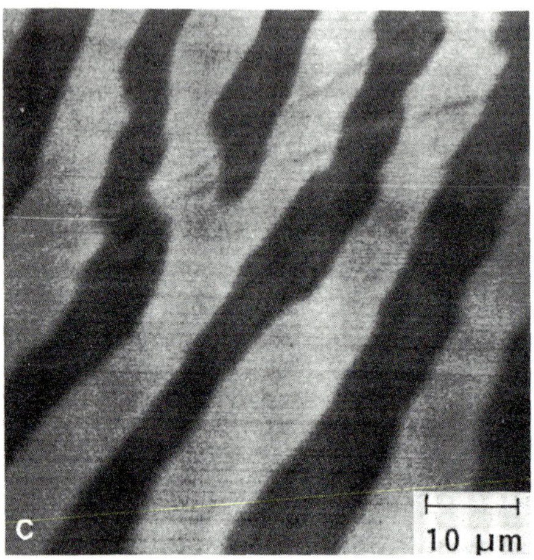

Fig. 8.3. (**a**) Explanation of type–2 magnetic contrast in terms of $\boldsymbol{B}_s$ in opposite directions. (**b**) Dependence of type–2 contrast $C = \Delta I/I$ on specimen tilt angle ϕ for opposite directions of $\boldsymbol{B}_s$ in iron at $E =$ 30 keV. (**c**) Imaging of magnetic domains in a Fe-Si transformer sheet at $E = 30$ keV and $\phi = 45°$, tilt axis horizontal (courtesy R. Herrmann)

events than in reality, the backscattering coefficient will be different for opposite directions of the induction $\boldsymbol{B}_s$, depending on whether the trajectories are bent away from or towards the surface. In reality, the induction $B_s = 2.1$ T for iron, for example, so that for 30 keV electrons the radius $r = mv/eB_s$ is 0.3 mm when $\boldsymbol{v}$ and $\boldsymbol{B}_s$ are perpendicular. The total electron range in iron at this energy is only $R = 2.4$ μm which explains why the observed contrast is so weak, only a few tenths of a percent.

The image intensity can be written as [8.14]

$$I = I_0 + Q(\phi, E)(\boldsymbol{u} \times \boldsymbol{n}) \cdot \boldsymbol{B}_s \;, \tag{8.9}$$

where I_0 denotes the background arising from the BSE collected by the detector system used, and $\boldsymbol{u}$ and $\boldsymbol{n}$ are unit vectors parallel to the direction of the electron beam and to the surface normal, respectively. The vector product $\boldsymbol{u} \times \boldsymbol{n}$ is a unit vector parallel to the specimen tilt axis. The factor Q depends on the tilt angle ϕ and on the incident electron energy. The contrast between two neighbouring domains becomes

$$C = \frac{I_2 - I_1}{I_0} = \frac{Q}{I_0}(\boldsymbol{u} \times \boldsymbol{n}) \cdot (\boldsymbol{B}_2 - \boldsymbol{B}_1) \,. \tag{8.10}$$

This formula expresses the following properties of type–2 contrast. The contrast will be proportional to the magnetic induction B_s. The factor Q and therefore C are proportional to $E^{3/2}$ [8.15, 16]; this dependence can be explained in terms of the following model. The type–2 contrast will be proportional to the mean-free-path length between two large-angle scattering processes and inversely proportional to the bending radius r. The mean-free-path length Λ is inversely proportional to the Rutherford cross-section (3.14) and hence varies as E^2 while the radius $r \propto v \propto E^{1/2}$. Combining these, we find $Q \propto \Lambda/r \propto E^{3/2}$.

The practical consequence is that type–2 magnetic contrast needs electron energies $E \geq 30$ keV to be observed with sufficient contrast, whereas low primary energies will be more favourable for type–1 SE contrast because the SE yield and therefore the signal-to-noise ratio increase with decreasing energy. High type–2 contrast has indeed been observed using 100–200 keV electrons [8.17].

The contrast falls to zero for normal incidence ($\phi = 0$) because $\boldsymbol{u} \times \boldsymbol{n}$ is then zero; it increases with increasing tilt angle ϕ but reaches a maximum at $\phi = 40° - 60°$ (Fig. 8.3b) [8.15, 16]. can be attributed to the decreasing path lengths of the BSE inside the ferromagnetic material with increasing tilt angle ϕ. For iron with $B_s = 2.1$ T, the maximum contrast with 30 keV electrons is only $C = 0.4\%$. The signal must remain very uniform over the scanned area if the image contrast is to be enhanced by black-level control. The dependence of C on ϕ shown in Fig. 8.3b is confirmed by Monte Carlo simulations considering the bending of trajectories for two opposite directions of $\boldsymbol{B}_s$ [8.15].

It can be assumed that the factor Q in (8.10) depends only on ϕ and E. If the specimen is rotated about its normal $\boldsymbol{n}$, the contrast becomes zero when $(\boldsymbol{u} \times \boldsymbol{n}) \cdot (\boldsymbol{B}_2 - \boldsymbol{B}_1) = 0$, which means that there is no change in the component of $\boldsymbol{B}$ parallel to the tilt axis across the domain wall. The contrast is greatest when the specimen is rotated through $\pm 90°$ from this position and the largest values are observed when $\boldsymbol{B}_1$ and $\boldsymbol{B}_2$ are parallel and antiparallel to the tilt axis, respectively. This allows the directions of $\boldsymbol{B}_1$ and $\boldsymbol{B}_2$ to be determined.

It can be shown by Monte Carlo simulations [8.15] that BSE that have lost less than 20% of their initial energy make the largest contribution to type–2 contrast. The contrast can therefore be improved by energy filtering (see below). This result also sheds light on the information depth and the resolution because the diameter of the diffusion cloud and the volume from which the high-energy BSE can escape increase with increasing energy. The information depth has been determined experimentally by evaporating copper films on iron and observing the decrease of contrast [8.17]. For a tilt angle $\phi = 45°$, the information depth is found to be 1, 4 and 15 μm for $E = 10$,

100 and 200 keV, respectively. The corresponding electron ranges in iron are 5, 17 and 55 μm and the BSE information depths can be assumed to be of the order of half these values (Sect. 4.1.2). The large range of information depths allows internal and surface domains to be observed.

The contrast depends not only on the tilt angle ϕ but also on the take-off angle ψ. For a specimen tilt $\phi = 45°$, the maximum signal was found at take-off angles $\psi = 60° - 70°$ [8.16–19] whereas the topographic contrast is highest for lower take-off angles; this difference can be used to enhance one type of contrast at the expense of the other by using the appropriate take-off angle.

The observation of type–2 magnetic contrast needs a relatively high electron-probe current to reveal contrast less than 0.5% with a good signal-to-noise ratio. It is therefore possible to record this contrast in the specimen-current mode (Sect. 6.2.8) [8.20], whereas type–1 contrast is produced outside the specimen and does not affect the specimen current. A Faraday cup at a low take-off angle has also been used [8.14]. An increase of contrast can be obtained by employing a scintillator or semiconductor detector because the signal is proportional to the BSE energy and most of the BSE contributing to type–2 contrast have lost less than 20% of their initial energy. This type of contrast can therefore be enhanced by placing an absorbing foil in front of the detector since this mainly absorbs the low-energy BSE [8.21].

So far we have been assuming that the magnetic contrast type–2 is caused by differences in absorption of BSE inside the specimen but differences in the exit angular distribution should also exert an influence. Two semi-annular BSE detectors A and B (Fig. 5.16a) and normal incidence of the electron beam give black-white contrast of domain boundaries with the A+B signal and of antiparallel domain areas with the A–B signal [8.22].

8.1.3 Deflection of Primary Electrons by Magnetic Fields

In order to produce a detectable deflection of fast primary electrons, $\int B_y \mathrm{d}z$ must be large. If 100 keV electrons penetrate a thin ferromagnetic foil 0.1 μm in thickness with a magnetic induction parallel to the foil, the deflection angle of the transmitted electrons is of the order of 0.1 mrad and this effect is used in the TEM technique of Lorentz microscopy [8.23, 24].

External magnetic stray fields can be measured quantitatively when the electron beam approaches a plane surface at a tilt $\phi = 90°$ and the magnetic structure consists of parallel domains, of records on tape or of the fringe field of a magnetic head with B_x and B_y components independent of the z direction, which is parallel to the electron beam. The beam, which would hit the plane of observation at x_0, y_0 in the absence of any magnetic field, will then be deflected (see Fig. 8.4 where L and l are defined) to the point

$$x_1 = x_0 + L\beta_x = x_0 - \frac{e}{mv} B_y l L \tag{8.11}$$

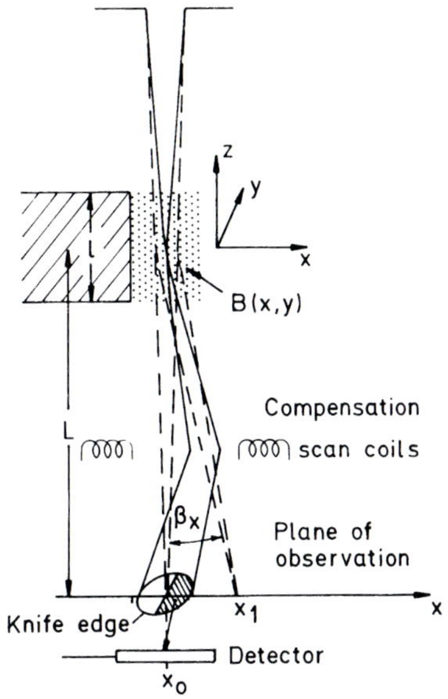

Fig. 8.4. Measurement of external magnetic field components $B_y(x, y)$ by the deflection of primary electrons through an angle β_x and detection by compensation coils and a knife-edge in front of a detector

$$y_1 = y_0 + L\beta_y = y_0 - \frac{e}{mv} B_x l L \,. \tag{8.12}$$

When the specimen is irradiated with a parallel beam in TEM, the shadow of the specimen is distorted by deflection. By placing a cross-grating or wires in front of the specimen, the magnitude and direction of the deflection can be deduced from the distortion of their shadow images [8.25, 26].

In the corresponding SEM technique, the cross-grating is placed at the plane of observation below the specimen, and deflection of the electron beam causes a distortion of this grating on the CRT [8.27–29] when the electron beam is focused on the plane of observation (dashed rays in Fig. 8.4). If the beam is focused at the specimen field (full lines), the circular shadow in the plane of observation can be partly obscured by a knife edge. A detector below this plane records a signal variation when the electron beam is deflected parallel to x by a B_y component [8.30]. This variation can be compensated via a feedback loop connected to additional scan coils below the specimen so that the recorded signal becomes constant. These additional scan coils also receive x and y currents opposite in polarity to the scan currents so that the centre of the circular area remains at x_0, y_0. The feedback current will then be proportional to B_y and when the scan is parallel to x, the profile $B_y(x)$ can be recorded. Rotation of the knife edge through 90° so that it becomes parallel to x likewise allows B_x to be recorded. The deflection of the primary beam can also be recorded by a CCD camera [8.31].

8.1.4 Imaging of Surface Magnetization by Polarized Electrons

Electron beams with a preferential direction of the electron spin are said to be polarized. We have already discussed the influence of electron spin on elastic large-angle (Mott) scattering in Sect. 3.1.6. The departure from the Rutherford cross-section can be interpreted as a consequence of the spin-orbit coupling. At low electron energies, electron exchange scattering also has to be taken into account [8.32]. The degree of polarization (Sherman function) has already been discussed in Sect. 3.1.6. Polarized electron beams of high brightness with polarizations of 30–80% may be produced either by field emission from a filament covered with a EuS film [8.33], which becomes ferromagnetic at 20 K, or by photoemission from GaAs [8.34] using a circularly polarized laser beam. However, only experiments with unpolarized incident electron beams and the emission of polarized secondary electrons from ferromagnetics have so far been reported in SEM. An explanation of this preferential emission of the majority spin direction is given in [8.35]; the effect can be caused by the spin dependence of the SE generation and of their elastic and inelastic mean-free-path lengths at the different steps of SE emission as discussed in Sect. 4.2.2.

The following methods can be used to detect any polarization:

1. The electrons are accelerated to 120 keV and scattered at a 100 μg/cm^2 Au foil through $\theta = 120°$ and azimuths $\chi = \pi/2$ and $3\pi/2$. Such a Mott detector [8.32, 36] is discussed in Sect. 3.1.6.
2. In a LEED detector [8.37–39] a 100 eV electron beam hits a W(100) single crystal and channeltron detectors are positioned at the opposite (2,0) and $(\overline{2},0)$ Bragg reflections of the low-energy electron diffraction (LEED) pattern (Sect. 9.1.6).
3. The absorbed-current detector [8.40–42] makes use of a small influence of spin polarization on the lower critical energy $E1$ (Fig. 3.31) where σ $=1$ and the specimen current to ground becomes zero. A gold surface is inclined at 25° to the electron beam. Incident electron beams of opposite spin cause a shift of $\Delta E1 = 1.8$ eV at $E1 = 150$ eV as a result of the anisotropic scattering; the value $(I_\uparrow - I_\downarrow)/I_0 = 0.006$ is found for the difference between opposite spin-directions where I_0 is the incident electron current.
4. A diffuse-scattering detector [8.43] makes use of spin-dependent scattering of 150 eV electrons from evaporated polycrystalline gold films resulting from spin-orbit interactions.

Secondary electrons emitted from a ferromagnetic material are spin polarized with the polarization vector antiparallel to the magnetization, which allows us to measure the direction of magnetization [8.36, 37, 44]. The maximum degree of polarization of 50% from a Fe(110) surface is observed for low SE energies of a few eV just at the most probable energy of the SE energy spectrum and decreases with increasing SE energy to 20–50% for 20–30 eV

[8.35]. Selection of low-energy SE by a spectrometer before the SE enter a spin analyser therefore increases the contrast. With decreasing primary electron energy the degree of polarization saturates at 1 keV. Because the SE yield shows a maximum at 500 eV, a primary energy of $\simeq$ 1 keV will be optimum. A spin rotator in front of the spin analyser allows us to detect all three components of the magnetization vector [8.45]

The escape depth of SE contributing to spin polarization is about 0.5 nm. Spin polarization is thus a sensitive method for studying surface magnetism (deviations of the magnetic structure in the uppermost layers) and very thin films. Owing to this low escape depth, contamination will strongly reduce the polarization and it is necessary to use ultra-high-vacuum and in-situ cleaning methods. To observe the bulk domain structure it is easier to use the type–2 magnetic contrast of BSE (Sect. 8.1.2).

For 50–150 eV electrons, the polarization of scattered electrons varies with electron energy, specimen tilt and take-off angle [8.46] and can be altered by exposure to residual gas [8.47]. These methods are in common use in surface physics and imaging methods are becoming of considerable interest, though the resolution will be limited by the low primary electron energy.

8.2 Voltage Contrast

8.2.1 Qualitative Voltage Contrast

Different potentials can be generated on specimen surfaces by charging effects at insulators (Sect. 3.5.2), or also by biasing of integrated circuits and on piezoelectric and ferroelectric substances, for example. The variation of the electrostatic field between the specimen and the Everhart–Thornley detector caused by these potentials influences the SE trajectories and the SE signal intensity. In principle, a positively biased area should appear darker when the SE are retarded back to the specimen, and a negatively biased area should appear brighter when all the SE are repulsed from it. Such changes of contrast are demonstrated for a MOS-FET in Fig. 8.5. We expect, therefore. to see dark and bright signal intensities superposed on the existing topographic and material contrast, which are, of course, present without biasing. The voltage contrast can therefore be separated from other types of contrast by digital image subtraction, for example. However, there is no unambiguous relation between voltage and image intensity because the contrast is also influenced by neighbouring potentials and by the overall deformation of the electrostatic field above the specimen.

To get an idea of how a particular SEM and detector arrangement influence the voltage contrast, model experiments should be done [8.48, 49]. This will be demonstrated by considering a specimen consisting of two areas A and B prepared by evaporating gold films on a 1 cm^2 glass slide; A and B are separated by a 50 μm gap and can be biased separately. In Fig. 8.6a, the

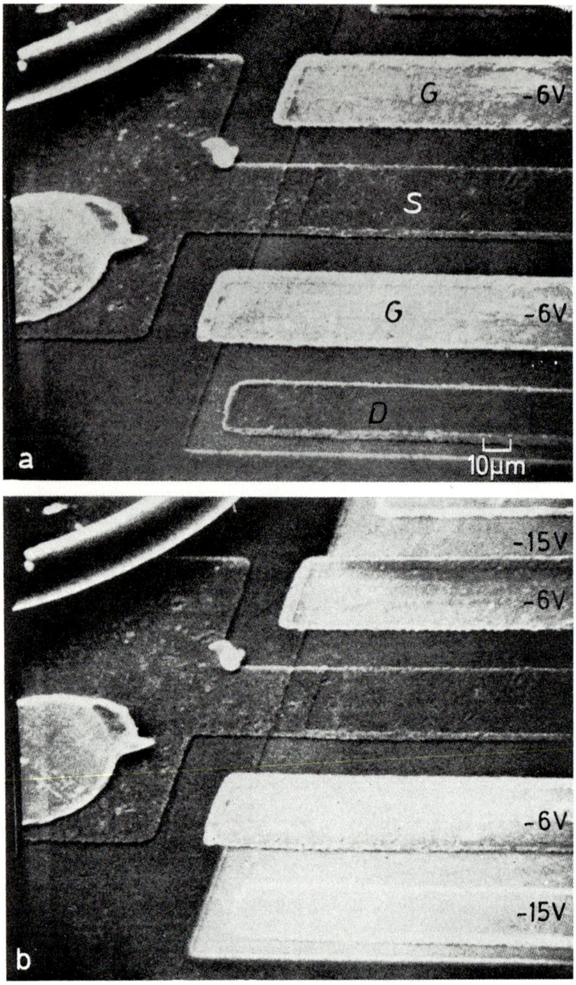

Fig. 8.5. Voltage contrast on a MOS-FET by biasing (**a**) the gate by –6 V and (**b**) additionally biasing the drain by –15 V (S = source, G = gate, D = drain)

electron beam scans across the shaded area on A and the SE signal is plotted versus U_A for different constant values of U_B. In Fig. 8.6b, the beam scans across area B and the signal is plotted versus U_B for constant values of U_A. In principle, we see a brighter signal at the left for a negative bias of the scanned area and a darker signal at the right for a positive bias. The relatively large remaining signal intensity for a positive bias can be attributed to SE excited by BSE at the polepiece and other parts of the specimen chamber. In (a) the signal at constant U_A increases when U_B becomes more negative. The resulting electrostatic field repels the SE from A to the detector at the top. However, the signal at constant U_B decreases in (b) for negative U_A because

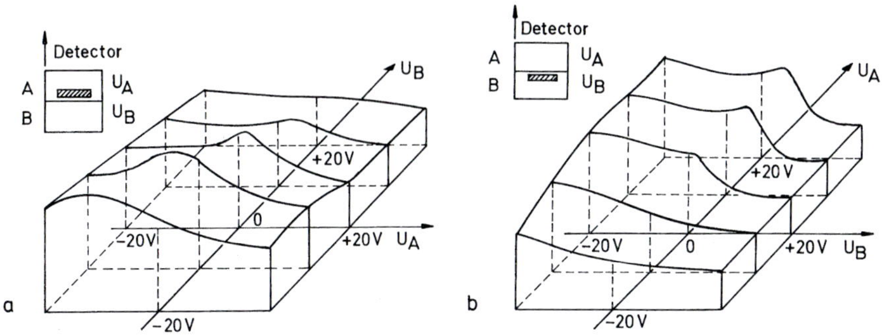

Fig. 8.6. Model experiment for studying the influence of neighbouring potentials on voltage contrast using 1 cm^2 of a glass slide covered by gold films A and B separated by a 50 μm gap with the electron beam (**a**) on area A and (**b**) on area B [8.49]

the negatively biased area A produces a potential barrier in front of the detector so that a fraction of the SE are repelled. Otherwise, a positive U_A increases the signal because SE trajectories that would normally end on the polepiece (Fig. 5.8) are more bent towards the detector. Another example of the influence of a positively charged particle on image intensity is discussed in Sect. 3.5.2.

These examples show that the SE trajectories are influenced by the microfield caused by neighbouring potentials and that the voltage contrast is also affected by the selection of SE in the macrofield between specimen and collector. The dependence of the signal on voltage becomes more linear with a slope of the order of 10% per 1 V difference when a Banbury–Nixon cage (Fig. 5.11) is used [8.50, 51] and depends less on neighbouring potentials. This improvement is, however, not sufficient for quantitative measurement, as discussed in Sect.8.2.3.

8.2.2 Voltage Contrast on Ferroelectrics and Piezoelectrics

Ferroelectric materials exhibit a spontaneous polarization and a domain structure similar to ferromagnetics. The resulting surface charge distribution can be investigated by exploring the voltage contrast. However, these materials are usually insulators, and charging due to electron irradiation becomes a serious problem. If a conductive coating is applied, the surface topography can be observed but not the surface charges. Only ferroelectrics such as $BaTiO_3$ can be investigated since they are so strongly polarized that the domain structure affects the surface topography [8.52].

Another way of observing insulating material without strong negative charging is to work below E_2 in Fig. 3.31 where $\sigma = \eta + \delta > 1$. In this way, it is possible to observe voltage contrast from domains of different polarity [8.53–57]. The lack of this type of contrast for BSE and its disappearance at

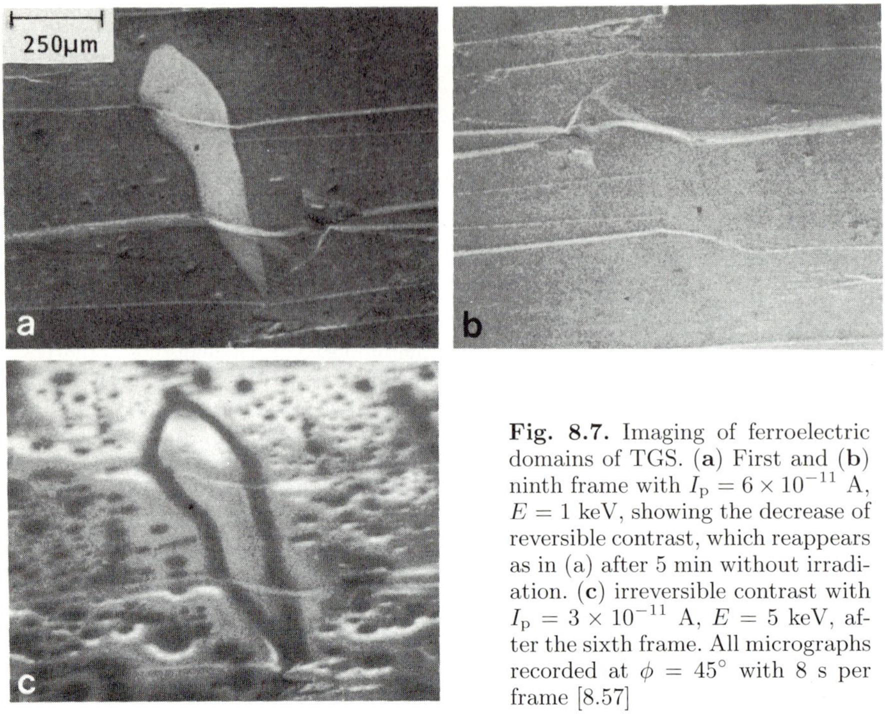

Fig. 8.7. Imaging of ferroelectric domains of TGS. (**a**) First and (**b**) ninth frame with $I_p = 6 \times 10^{-11}$ A, $E = 1$ keV, showing the decrease of reversible contrast, which reappears as in (a) after 5 min without irradiation. (**c**) irreversible contrast with $I_p = 3 \times 10^{-11}$ A, $E = 5$ keV, after the sixth frame. All micrographs recorded at $\phi = 45°$ with 8 s per frame [8.57]

temperatures above the Curie point as well as the correspondence with etching techniques confirm that the signal is caused by voltage contrast. However, the observed contrast is extremely sensitive to the irradiation condition and it becomes difficult to get reproducible results. Systematic experiments on TGS (triglycine sulphate) [8.57] showed that the domain contrast shown in Fig. 8.7a vanishes at energies for which $E\cos^2\phi \le 1.2 \pm 0.2$ keV (ϕ: tilt angle) and at charge densities of the order of $10^{-8} - 10^{-6}$ C cm^{-2} (Fig. 8.7b). The contrast does, however, reappear as in (a) after an interruption of the irradiation for a few minutes. In addition to this reversible effect, an irreversible contrast is observed for $E\cos^2\phi > 1.2$ keV (Fig. 8.7c) and no restoration of contrast then takes place. A model has been proposed, which takes into account the accumulation of charges in a thin surface layer, but more experiments are needed for a full understanding of these contrast phenomena.

The complex interaction of the electron beam with a ferroelectric specimen can be avoided by transferring the deflection technique of the primary beam described for magnetic fields in Sect. 8.1.3 to the electrostatic stray field in front of ferroelectrics [8.58].

Piezoelectrics, used for oscillators (quartz) and transducers ($LiNbO_3$), for example, also have to be investigated at low energies of the order of 2–3 keV. In principle, the energy should be as high as possible but below E_2 to increase the resolution. Standing-wave patterns on vibrating quartz oscillators can be observed, the oscillating parts (antinodes) appearing brighter than the nodes [8.59]. This voltage contrast effect can probably be attributed to a nonlinear relation between the SE signal and the surface potential, as discussed in Sect. 8.2.1. The variation of surface potential by excited surface acoustic waves can be imaged in the stroboscopic mode [8.60]. An alternative imaging method makes use of the superposition of the longe-range electric field from the exciting top electrode and the short-range field of the acoustic wave [8.61, 63]. These experiments show that this type of voltage contrast is capable of following high-frequency oscillations at 43.5 MHz and that the contrast of stationary patterns is not caused by different mean stationary charging at nodes and antinodes. The pyroelectric effect can be used to image the stripe domains of $LiNbO_3$ by cooling or heating the specimen ±5 °C [8.62]. Surface-acoustic waves invert the voltage contrast of the image by π on the domain walls, so that they can be imaged without thermoelectric treatment [8.63]

8.2.3 Quantitative Measurement of Surface Potential

Besides the micro- and macrofields described in Sect. 8.2.1, the signal also depends on the local yield. A quantitative method of measuring potentials must eliminate all these influences.

The 273 eV Auger electrons of carbon are less influenced by microfields owing to their larger exit energy. Biasing of the specimen results in a corresponding shift of the Auger electron peak, which can be measured with an Auger electron spectrometer (Sect. 5.3.3) [8.64]. However, measurement of the shift of the SE spectrum will be more effective because the SE yield is three to four orders of magnitude larger than the AE yield.

Retarding-field spectrometers for SE are described in Sect. 5.5.2 (Figs. 5.22–24). Ideal retarding curves of SE spectrometers for different specimen

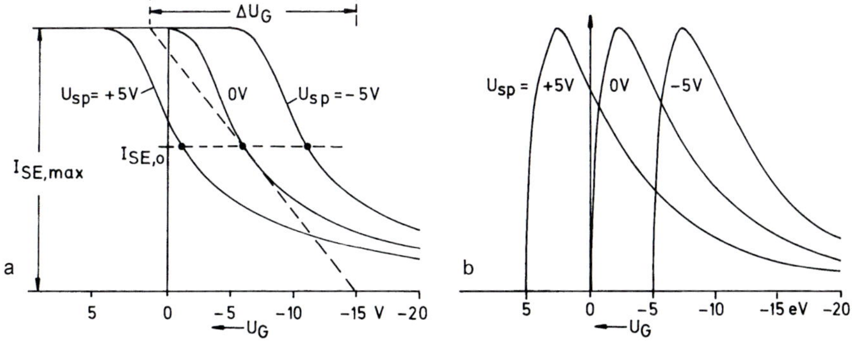

Fig. 8.8. (**a**) Shift of the retarding-field curve of the SE signal and (**b**) of the differentiated retarding-field curve obtained by applying a voltage U_{sp} to the specimen (U_G = bias of the retarding grid, $I_{SE,o}$ = detector current selected for quantitative measurement)

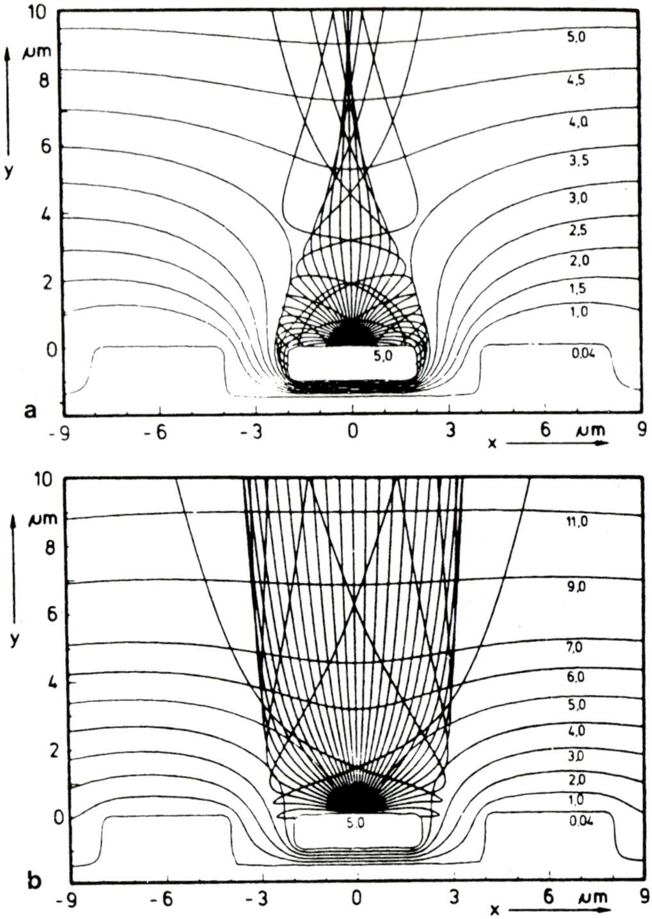

Fig. 8.9. Influence of the strength of the external extraction field, (**a**) 400 and (**b**) 1000 V mm^{-1} on the equipotentials and trajectories of SE emitted at angles α between $-85°$ and $+85°$ in steps of $5°$ from a central conduction pad biased at +5 V, the neighbouring pads being at zero bias [8.66]

biases U_{sp} are shown in Fig. 8.8a. The background of BSE, which will be observed for larger negative biases U_{G} of the retarding grid, can be suppressed by preselection of the BSE in a deflection field. Figure 8.8b shows the SE energy spectra obtained by differentiation of the retarding curve. Such spectra can be recorded directly by superposing an ac modulation on the retarding grid bias and amplifying the transmitted SE signal with a lock-in amplifier [8.65]. The shift of the retarding curves depends linearly on the specimen bias and can be measured by connecting a feedback loop from the exit of the preamplifier to the retarding-grid, the bias of which is thus allowed to float so that the exit signal is kept constant at a level $I_{\mathrm{SE,o}}$ (Fig. 8.8a). This feedback voltage is a linear measure of the local specimen bias. Such a linearizing feedback loop has a maximum operating frequency of about 300 kHz [8.66]. Another possibility is to measure the peak position of the differentiated retarding curve in Fig. 8.8b [8.65]. The feedback voltage can be used for two-dimensional mapping by Y modulation or by intensity modulation of the CRT beam [8.67].

Spectrometers without pre-acceleration of the SE, as in Fig. 5.14a or in a combination of a Banbury–Nixon cage (Fig. 5.11) and a retarding grid in front of the SE detector [8.68], have the disadvantage that the resulting shift and shape of the retarding curve depend on the microfield. The only way of avoiding this influence of the microfield is to use pre-acceleration with a high electric field of the order of 500–1000 V mm^{-1} at the specimen surface. The need for such a strong field is demonstrated in Fig. 8.9, which shows a + 5 V biased conduction line 5 μm in width and neighbouring lines at zero bias on an integrated circuit. The equipotentials in front of the device [8.66] indicate that there is still a potential barrier even with an extraction field strength of 400 V mm^{-1} (Fig. 8.9a). The calculated SE trajectories for $E_{\mathrm{SE}} = 1.5$ eV with emission angles varying from $\alpha = -85°$ to $+85°$ in steps of 5° show that those with large $|\alpha|$ return to the specimen. At an extraction field strength of 1000 V mm^{-1} (Fig. 8.9b), all the SE can be extracted.

Figures 5.22b and 5.23a,b show types of electron spectrometer in which pre-acceleration, beam deflection and retarding field are combined; in Fig. 5.22b the deflection is in front of and in Fig. 5.23b behind the retarding-field grid. These two types of spectrometer allow the specimen potential to be measured with a resolution of the order of one millivolt for a voltage range ±20 V (see [8.66, 69] for a comparison of the different spectrometers).

The resolution in the detection of small differences in surface biasing is limited by the noise of the signal current [8.70] at the operating level $I_{\mathrm{SE,o}}$ of the feedback loop (Fig. 8.8a), which is a fraction γ of $I_{\mathrm{SE,max}} = I_{\mathrm{p}}\delta T$, where I_{p} is the electron-probe current, δ the SE yield and T the transmission of the spectrometer. Due to (4.44) and (4.50), the noise amplitude of $I_{\mathrm{SE,o}}$ is given by

$$I_{\mathrm{n,rms}} = \delta T\gamma[2e\Delta f I_{\mathrm{p}}\gamma(1+b)]^{1/2} \,. \tag{8.13}$$

The slope of the retarding curve at the operating level is

$$a = \frac{\mathrm{d}I_{\mathrm{SE}}}{\mathrm{d}U_{\mathrm{G}}} = \frac{I_{\mathrm{SE,max}}}{\Delta U_{\mathrm{G}}} = \frac{I_{\mathrm{p}}\delta T}{\Delta U_{\mathrm{G}}} \tag{8.14}$$

where ΔU_{G} is indicated in Fig. 8.8a. For a voltage resolution ΔU, the variation ΔI_{SE} should be of the order of three times the noise amplitude

$$\Delta I_{\mathrm{SE}} = a\Delta U_{\mathrm{G}} \geq 3I_{\mathrm{n,rms}} \,. \tag{8.15}$$

Substitution of (8.13) and (8.14) into (8.15) leads to

$$\Delta U_{\mathrm{min}} = 2\Delta U_{\mathrm{G}}\gamma\left(\frac{2e\Delta f(1+b)}{I_{\mathrm{p}}}\right)^{1/2} \,. \tag{8.16}$$

Numerical Example [8.71]: Writing $\Delta U_G = 6$ V for Al, $T = 0.64$ (two retarding grids with $T = 0.8$ each), $\gamma = 0.6$, $b \simeq 1/\delta = 1/0.8$ for $E = 1$ keV, we find $U_{min} = 0.05$ mV for $I_p = 5 \times 10^{-8}$ A and $\Delta f = 2$ s^{-1}. A value of $\Delta U_{min} \simeq 0.5$ mV has been observed experimentally. At the frequency limit of the linearizing feedback loop, a value $\Delta U_{min} \simeq 40$ mV can be expected.

It must be borne in mind that voltages cannot be measured absolutely because of the material differences in the work function, which result in contact potential differences and are influenced by physisorbed and chemisorbed surface layers. These effects can result in a shift of the SE spectra of the order of a few 100 mV [8.73]. The surfaces of insulating SiO_2 layers on integrated circuits become charged when irradiated with electrons but the surface potential approaches that of the underlying conductive material (semiconductor or metal) when the electron beam has sufficient energy to penetrate the layer and produce charge carriers throughout the whole thickness of the insulating layer [8.67]. Topography too can result in an unpredictable shift [8.67].

8.2.4 Stroboscopic Measurement of Voltages

It is no problem to measure time-varying voltages in a linearizing feedback loop for frequencies below 100 kHz. An investigation of periodic ac voltages in the qualitative voltage contrast mode (Sect. 8.2.1) generates periodic variations of the collected SE signal that can be recognized on a larger scanned area as a dark–bright stripe pattern [8.75] up to frequencies smaller than 100 times the line frequency f_x. This technique is known as voltage-coding. If f_x and the investigated frequency are not synchronized, the observed stripe pattern will be inclined to the x axis. When observed at TV scan rates, the frame time $T = 1/25$ s and 1000 lines per frame result in a line frequency $f_x = 25$ kHz and frequencies of the order of megahertz can be detected by this technique. A shift in the phase will result in a shift of the stripe pattern.

Quantitative measurement of voltage with an electron spectrometer, as described in Sect. 8.2.3, is possible for frequencies higher than 100 kHz only by periodic pulsing of the electron beam at the same frequency, thereby maintaining a fixed phase position. Such electron-beam blanking methods (Sect. 2.3.3) can work up to GHz frequencies. By sweeping the phase continuously, the method of voltage coding can also be applied at high frequencies. When f_x and the frequency f_s of the specimen point are somewhat different, the signal contains mixed frequencies $|f_x - f_s|$ and $f_x + f_s$ (heterodyne principle) which allows f_s to be determined with an accuracy of 1 Hz [8.76]. The time-of-flight of the SE is of the order of 10 ns but this does not disturb the measurement when using a spectrometer and a linearizing feedback loop because the signal will first go through a low-pass filter and the electron beam hits the specimen at the same phase of the periodic voltage. The voltage waveform can be recorded by sweeping the phase [8.77–81].

The different possible techniques for local measurement and for recording a linescan or a two-dimensional distribution of voltage are reviewed in [8.71, 82, 83].

8.2.5 Capacitive-Coupling Voltage Contrast

We explained in Sect. 3.6.2 and Fig. 3.31 that a passivation layer can be charged positively when $E_1 < E < E_2$ and $R(E) < t$ where t is the passivation layer thickness. In equilibrium, this positive charging results in stationary voltage contrast and is independent of any dc bias of conductive pads below the layer. However, a capacitive-coupling voltage contrast (CCVC) is oberved for a short 'storage time' [8.74, 84–87]. Figure 8.10 demonstrates a) the sequence of positive and negative biasing as U_B is switched on and off and b) the resulting surface potential U_s. In switch 1 an additional positive charge on top of the layer capacitance is induced by a positive bias, resulting in an increased surface potential U_s and a darker voltage contrast. More SE will be retarded until the equilibrium value U_{s0} is reached after a storage time T_{s+}. Switch 2 (U_B off) induces a more negative surface charge, resulting in a bright voltage contrast that vanishes after about double the storage time T_{s-} because the absorbed current, responsible for reaching the equilibrium again, is lower for negative charging. Switching on/off negative potentials (switch 3 and 4 in Fig. 8.10) results in a reverse contrast. The following obervations agree with this model. The CCVC increases with increasing bias switch. The storage times T_{s+} and T_{s-} are proportional to the switched voltage difference and inversely proportional to the incident probe current and the passivation layer thickness t, the latter because the capacitance is given by $C = \epsilon\epsilon_0 A/t$. The storage times can therefore vary in the range 1–1000 s depending on these parameters. Capacitive-coupling error and capacitive-coupling crosstalk are

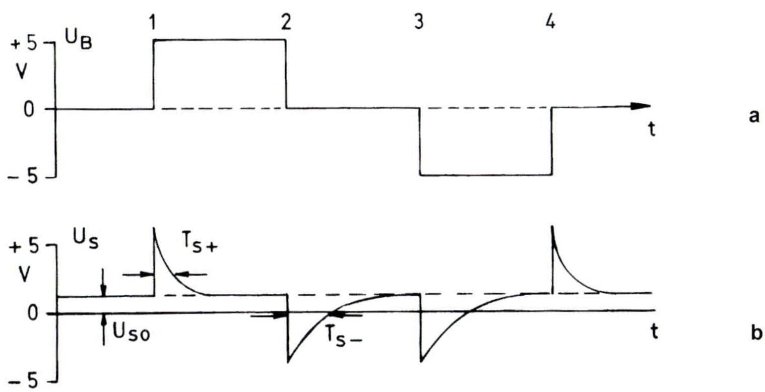

Fig. 8.10. Time sequence in the charge-coupling voltage contrast mode of (**a**) the bias applied to the substrate and (**b**) the surface potential U_s on top of the passivation layer

known to ceate problems with quantitative CCVC measurements; they are discussed in [8.88, 89].

8.2.6 Capacitively Induced Charging in an External Electrode

Changes in the surface potential during scanning can induce an opposite charge $Q = CU_s$ in a wire-loop in front of the specimen (Rau detector) [8.90] (e.g. a noble-metal wire of 0.1–0.3 mm diameter bent in a loop of $\simeq$2 mm diameter and at a distance of $\simeq$0.1 mm in front of the specimen). The capacitance C between loop and specimen is of the order of a few tenths of picofarad. The displacement currents caused by changes of the surface potential are of the order of picoamperes and can be used as an image-forming signal. This contactless method allows us to record changes in the surface potential caused by charging or an EBIV-like image in semiconductors. A weak BSE image formed by BSE striking the wire loop is superimposed on the signal. SE can be retarded or collected by applying a small negative or positive bias of the loop, respectively. Such a SE signal has the advantage that the signal is symmertic, as it is in an in-lens detector unlike the situtation with a side-mounted Everhart–Thornley detector.

By mechanically vibrating the loop, such a device can be made to operate as a Kelvin probe (Sect. 3.5.2). The recorded displacement current can be used to measure a stationary surface potential U_s.

8.3 Low-Energy Spectroscopies

8.3.1 Threshold Spectroscopies

When the energy of the primary electrons increases from below to beyond the ionization energy of an inner shell, the ionization starts when the primary energy exceeds the threshold energy, which is the energy difference between the Fermi energy and the core level. Electrons excited to higher unoccupied states beyond the Fermi level act as a probe for the density of states. The following types of threshold appearance potential spectroscopies (APS) can be applied [8.91]. When the elastically reflected and low-loss electrons (LLE in Fig. 1.5) are separated, by a retarding field analyser consisting of hemispherical grids, for example, a new channel of inelastic scattering opens beyond the threshold energy and the intensity of the LLE signal decreases (disappearance potential spectroscopy) or the target current increases (target current APS). When the vacancy in the inner shell is filled, either an Auger electron or an x-ray quantum is emitted. The energy-filtering of Auger electrons results in Auger APS and the recording of x-rays in the soft x-ray APS. The relative signal variations are of the order of 10^{-3} and it is necessary to apply a small oscillation of the slowly increasing primary energy and filter the signal by means of a lock-in amplifier; this produces the first derivative of the spectrum. A second derivative suppresses the background slope of the first derivative.

8.3.2 Electron Trap Spectroscopy and Microscopy

In thin insulating films of thickness $d < 150$ nm on a conductive substrate (e.g. SiO_2 on Si) and for $E < 1.5$ keV, the secondary-electron field emission (SEFE) by Fowler–Nordheim tunneling from the conductive substrate into the positively charged film maintains the total yield $\sigma > 1$ (Fig. 8.11). The extremely sensitivity of the Fowler–Nordheim current (8.19) to the electric field component normal to the substrate–film boundary

$$\boldsymbol{E}_z(d) = \frac{1}{\epsilon_0 \epsilon_r} \int_0^d \rho(z) \mathrm{d}z \tag{8.17}$$

caused by the charge density $\rho(z)$ allows charge changes of 10^9 electrons per cm^2 within the insulating layer to be detected and can be used for electron trap spectroscopy [8.92, 93]. Figure 8.12 shows the steady-state potential distribution of a 100 nm SiO_2 layer on Si.

The current density through a plane at depth z can be written as

$$j(z) = j_{\mathrm{PE}} + j_{\mathrm{SE}} + j_{\mathrm{h}} + j_{\mathrm{PF}} + j_{\mathrm{FN}} + j_{\mathrm{R}} \tag{8.18}$$

where we have to consider the following charge-transport components:

1. Penetration and scattering of primary electrons and backscattering results in a current density $j_{\mathrm{PE}}(z, E)$.
2. Generation, scattering, recombination or emission of SE ($E_{\mathrm{SE}} < 50$ eV). Their straggling is affected by the electric field component E_z across the insulating layer resulting in $j_{\mathrm{SE}}(z, E, E_z)$.
3. Creation of holes as a consequence of the SE excitation from the valence band and their drift current $j_{\mathrm{h}}(z, E, E_z)$ in the internal electric field.
4. Poole–Frenkel current $j_{\mathrm{PF}}(z, E_{\mathrm{t}}, T, E_z)$ of field-enhanced thermal emission from traps with activation energy E_{t} at temperature T.
5. Fowler–Nordheim tunnelling current

$$j_{\mathrm{FN}}(z, E_z(d)) = AE_z^2 \exp(-B/E_z) \tag{8.19}$$

   caused by the electric field component $E_z(d)$ of the order of a few 10^6 V/cm at the Si–SiO_2 boundary at a depth $z = d$ ($A = 8.77 \times 10^{-8}$ A/V^2 and $B = 2.26 \times 10^8$ V/cm).
6. Retarding field current $j_{\mathrm{R}}(z, E_{z,ext})$ of emitted SE caused by an external field component $E_{z,ext}(z < 0)$.

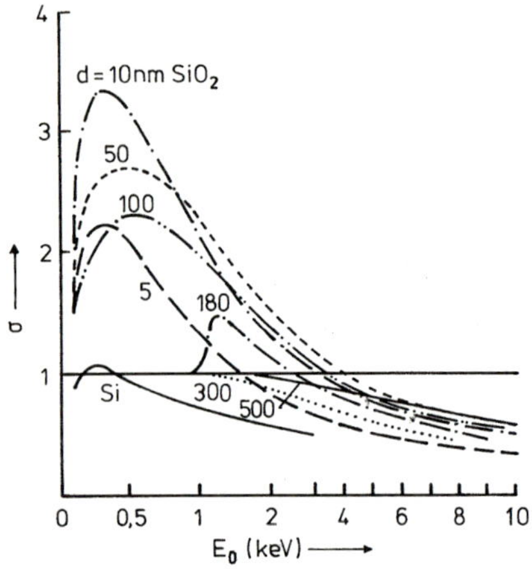

Fig. 8.11

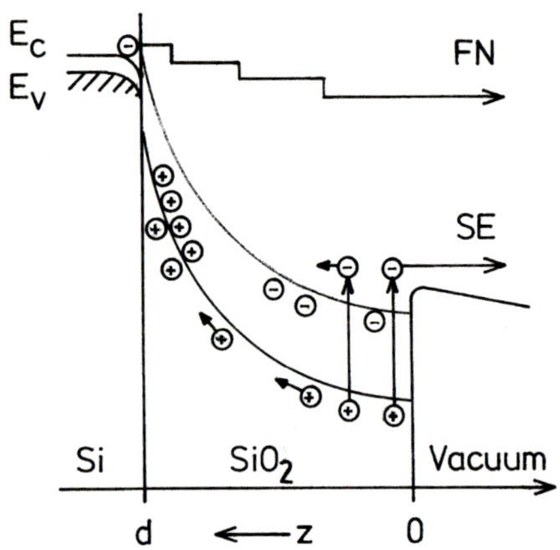

Fig. 8.12

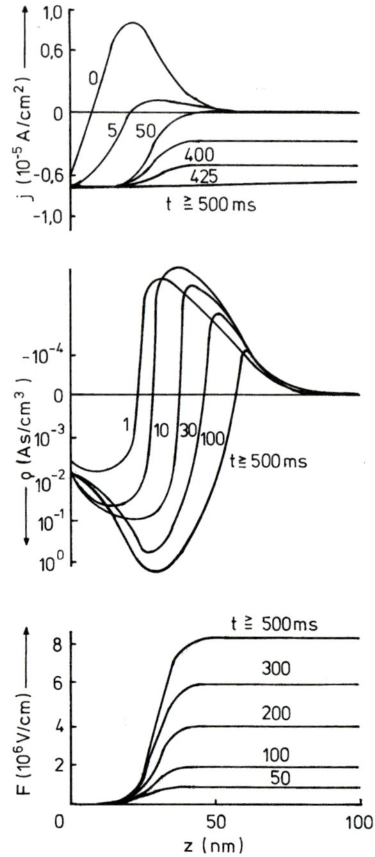

Fig. 8.13

Fig. 8.11. Measurements of the dependence of the total yield $\sigma = \eta + \delta$ on electon energy E for different thicknesses d of SiO_2 on a Si substrate [8.92]

Fig. 8.12. Potential distribution of a positively-charged 100 nm SiO_2 layer on a Si substrate with the conventional SE emission and FN as the Fowler-Nordheim part of SE

Fig. 8.13. Depth distributions of current density j, charge density ρ and electric field strength E_z within an insulating 100 nm SiO_2 film on Si irradiated with a current density $j_0 = 10^{-5}$ A/cm^2 of 1 keV electrons for increasing irradiation times t [8.92]

The equation of continuity for a planar geometry

$$\nabla j^{+} = -\frac{\mathrm{d}}{\mathrm{d}z} j = -\frac{\mathrm{d}}{\mathrm{d}t}\rho \tag{8.20}$$

results in the charge density $\rho(z)$ and with the Poisson equation

$$-\Delta U = \nabla \boldsymbol{E} = \frac{\mathrm{d}}{\mathrm{d}z} E_z = \rho/\epsilon_0\epsilon_r \tag{8.21}$$

we get the electric field strength $E_z(z)$. Substitution for ρ in (8.20) from (8.21) yields the partial differential equation

$$\epsilon_0\epsilon_r \frac{\partial}{\partial t} E_z(z) = j(z, t, E_z(z')) \tag{8.22}$$

Substitution of j from (8.18) results in a partial integro-differential equation which can only be solved by computer iteration with the boundary condition $E_z(z = 0, t) = 0$. Figures 8.13a–c show $j(z,t)$, $\rho(z,t)$ and $E_z(z,t)$ for a $d =$ 100 nm SiO_2 film on Si irradiated with $E = 1$ keV electrons and a current density $j_0 = 10^{-5}$ A/cm^2 for increasing irradiation times t. The total yield, given by

$$\sigma = 1 - \frac{j(z = 0, t)}{j_0} \tag{8.23}$$

is found to be $\simeq$1.75 for the steady state and this is in good agreement with the experimental values shown in Fig. 8.11 [8.92].

The kinetics of charges in localized traps of spatial concentration n_{t0}, capture cross-section σ_{t} and thermal activation energy E_{t} can in first-order be described by

$$\frac{\mathrm{d}n_{\mathrm{t}}}{\mathrm{d}t} = \frac{j_{\mathrm{i}}}{e}\sigma_{\mathrm{t}}(n_{\mathrm{t0}} - n_{\mathrm{t}}) - f \exp\left(-\frac{E_{\mathrm{t}}}{kT}\right) \tag{8.24}$$

where $n_{\mathrm{t}}(t,T)$ denotes the actual concentration of filled traps, j_{i}/e the injection charge density and f a nearly constant factor. The first term on the right-hand side of (8.24) describes the charge storage and dominates at low temperatures resulting in

$$n_{\mathrm{t}} = n_{\mathrm{t0}}\left[1 - \exp\left(-\frac{t}{\tau}\right)\right] \tag{8.25}$$

with the mean filling time $\tau = e/j_{\mathrm{i}}\sigma_{\mathrm{t}}$.

In the case of SiO_2, electron traps dominate and filling the traps results in a decrease of $E_z(z = d)$ and of the total yield σ with increasing irradiation time t as shown in the left-hand part of Fig. 8.13 with an irradiation at constant room temperature. Otherwise, hole traps dominate in Ta_2O_5 and Al_2O_3 layers and σ decreases [8.92]. Analysis of these curves allow n_{t0} and σ_{t} to be determined.

When the temperature is increased, the second right-hand term of (8.24) dominates and a heating rate $\mu = \mathrm{d}T/\mathrm{d}t$ results in a decrease of the occupied traps in the following heating cycle shown on the right-hand part of Fig. 8.13. In front of this maximum a slight hole emission is detectable.

$$n_\mathrm{t} = n_{\mathrm{t}0} \left[-\frac{f}{\mu} \int_{T_0}^{T} \exp\left(-\frac{E_\mathrm{t}}{kT} \right) \right] \mathrm{d}T \tag{8.26}$$

This release of trapped electrons results in a maximum of σ between $T = 250°$ and $350°$. By applying different heating rates the activation energy E_t can be obtained.

When the time–temperature cycle is stored pixel per pixel, this mode of electron trap microscopy [8.94] allows $n_{\mathrm{t}0}$, σ_t and E_t to be mapped in order to image hidden defects, inhomogeneities and pre-treatments. A spot irradiation with 30 keV electrons followed by the SEFE technique at 1 keV allows the halo of charge trapped by backscattered electrons to be imaged, for example.

8.4 Transmitted Electron Mode

If the specimen is thin enough for 10–50 keV electrons to be transmitted in a SEM, it will be more reasonable to observe it in a 100 keV-TEM where the resolution is much better. In a TEM, only electrons scattered through angles of a few tens of milliradians are used for image formation because of the high spherical aberration of the objective lens. A scanning transmission electron microscope (STEM) with a field-emission gun or a STEM attachment to a TEM works in the same conditions and similar contrast is seen owing to the theorem of reciprocity [8.95].

However, it is sometimes of interest to use transmitted electrons in SEM, even though this type of microscope is designed for the investigation of solid specimens. The maximum specimen thickness in TEM is limited by the decrease of transmission through the small objective aperture and by the energy losses, owing to the high chromatic aberration of the objective lens. In a SEM, on the other hand, the transmitted electrons can be collected with a large aperture and even with a solid angle 2π corresponding to the entire lower hemisphere because there is no imaging lens below the specimen. Even for thick specimens, the intensity and signal-to-noise ratio are often sufficient. The transmission follows an exponential law

$$T = \exp(-x/x_\mathrm{k}) \tag{8.27}$$

with the mass–thickness $x = \rho t$ and the contrast thickness x_k depending on aperture and Z. Measurements of x_k in a TEM between 20 and 100 keV [8.96] coincide with measurements in a LVSEM between 2 and 30 keV [8.97].

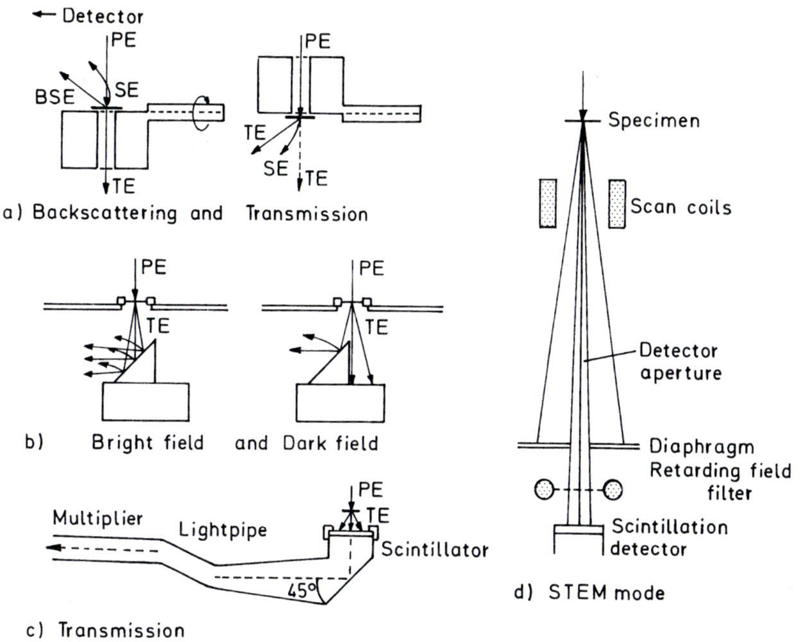

Fig. 8.14. Different specimen stubs and detector arrangements used for recording transmitted electrons in a SEM: (**a**) specimen stub for backscattered and transmitted electrons, (**b**) TE/SE conversion for bright- and dark-field images, (**c**) scintillator and light-pipe below a transparent specimen, (**d**) STEM mode for recording diffraction patterns and electron energy filtering

As shown in Fig. 3.21, the transmission decreases continuously with increasing thickness and can be used to estimate the local specimen thickness [8.98]. The resolution will be of the order of the electron-probe diameter for structures at the entrance side and decreases with increasing depth because of the electron-beam broadening caused by multiple scattering (Sect. 3.3.3) and electron diffusion. In special cases it will be possible to image precipitates in alloys, for example [8.99]. The transmission mode can also be used if a specimen is to be investigated both in TEM and SEM in order to identify the same specimen area. Similar types of contrast can be produced by using a small detector aperture (see below).

Figures 8.14a,b show two simple attachments, which use the conversion of transmitted electrons (TE) to SE (a) at the top or bottom of the specimen [8.100] and (b) at a metal surface below the specimen [8.101]. The SE are detected by the conventional Everhart–Thornley detector. In (c) the TE are directly recorded by a scintillator on a light-pipe coupled to a photomultiplier [8.102]. The arrangement (d) with a larger specimen–detector distance can be used for observing a transmission electron diffraction pattern on the screen, for limiting the detector aperture α_d to get contrast conditions comparable with TEM or when a retarding-field filter is required in front of the scintillator. When a transparent specimen, a biological section, for example, is mounted on a glass substrate with a conductive and light-reflecting metal film, the cathodoluminescence of transmitted electrons can be used as a signal [8.103].

8.5 Scanning Electron Mirror Microscopy

In electron mirror microscopy, the specimen is biased a few volts more negative than the cathode, so that the incident electrons are retarded and reflected just in front of the specimen. The electrons are deflected by distortions of the equipotentials near the surface caused by surface topography and by differences in the surface bias, at p-n junctions and biased conductive pads on an IC, for example. Because of the low electron energy at the point at which the electron is reflected, the trajectories are also affected by magnetic stray fields analogous to the type–1 magnetic contrast of the SE (Sect. 8.2.1).

In conventional electron mirror microscopy with comparatively broad illumination, the electron beam is formed just as in a TEM by a condenser lens system. First, the incident electrons are turned through 90° by a magnetic prism and, after being reflected at the specimen and again turned through 90°, the electron beam is on-axis again. Smaller prism angles have also been used to separate the incident and the reflected beam in such a magnetic bridge. An enlarged image can be formed by means of an electron lens system, the final image plane of which is conjugate to the plane of reflection in front of the specimen. The image can be observed on a fluorescent screen or recorded on a photographic emulsion by direct electron exposure as in TEM [8.104, 105]. Owing to the deflections mentioned above, the image intensity will be non-uniform. Electron deflection causes a redistribution of the image brightness, which is increased at certain image points not conjugate to the specimen point. This overlap of undeflected and deflected electrons leads to a modulation of the image brightness, which can only be used quantitatively by comparison with model calculations. Another way of creating contrast is to intercept the deflected electrons at a diaphragm (bright field) or to intercept the undeflected electrons (dark field).

When the specimen is explored with an electron probe instead of an electron beam, however, we have an unambiguous relation between image point and specimen point. This is the idea behind scanning electron mirror microscopy (SEMM) [8.106–108]. Contrast can be generated by using a detector that is sensitive to deflections at the deformed equipotentials.

SEMM can be realized in a SEM as shown in the three configurations of Fig. 8.15: (a) the specimen is placed normal to the electron beam and the deflected electrons are recorded by an annular detector (dark-field mode);

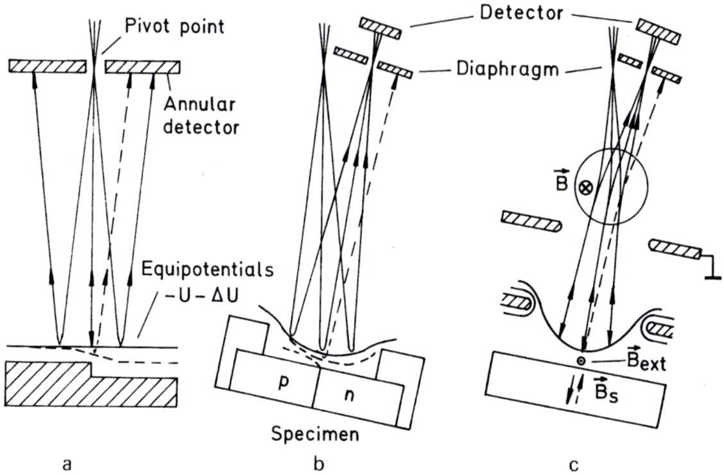

Fig. 8.15. Different specimen and detector arrangements and additional electrodes for scanning electron mirror microscopy. The full lines represent equipotentials in front of a uniform surface and the dashed lines show the equipotentials caused by (**a**) a surface step and (**b**) a p-n junction; (**c**) shows the effect of an external magnetic stray field

(b) the specimen is inclined and the reflected electrons that have undergone no deflection can be separated by means of a diaphragm from the deflected electrons (bright-field mode); (c) incident and undeflected beams are separated by a magnetic prism as in conventional mirror microscopy.

As shown in (a), when the scanning probe is far from the centre, the component of electron momentum parallel to the surface may be quite large, with the result that the undeflected electrons also hit the detector system at off-axis positions while the deflected electrons return to the centre. This effect can be decreased by using a concave mirror surface in front of the cup-shaped mirror electrode (b), but this modification inevitably suffers from high-order aberrations and distortions. The same effect can be achieved by inserting an additional electrode (c) in front of the specimen. Geometries for generating topographic, voltage and magnetic contrast are illustrated in Figs. 8.15a-c, respectively. The full equipotentials correspond to a plane specimen and the dashed one to distortions caused by a surface step, a p-n junction and an external magnetic stray field, respectively.

The main advantages of such SEMM modes are:

1. There is no contamination and no radiation damage at the specimen because no electrons hit the surface.
2. The sensitivity to electric and magnetic surface fields is high.
3. No disturbing specimen current or EBIC is produced when observing ICs in operation.

The following inconveniences have to be taken into account:

4. The specimen has to be insulated to apply the negative bias of the cathode and an additional retarding bias; the power supply for an IC thus has to work at the high negative bias.
5. In order to prevent electrical breakdown, the working distance of the final probe-forming lens must be long.
6. During the retarding of the electron beam, the gun brightness is decreased because it depends on the local electron energy E, see (2.7). The electron-probe size therefore increases or the resolution decreases as compared with a conventional non-retarding SEM mode.

This discussion of the basic electron optics and contrast mechanisms of SEMM shows that the main application will be in the field of inspection of electric microfields in front of semiconductors. Other types of contrast caused by topography or external magnetic stray fields can be better studied in the conventional SE mode. SEMM has also been used [8.109] to examine the action of a retarding grid. Owing to the potential drop at the mesh centre, a stronger negative bias has to be applied to reflect the electron probe as the latter is scanned across a mesh centre.

8.6 Environmental Scanning Electron Microscopy (ESEM)

The high vacuum at the electron gun and in the electron optical column of a SEM has to be less than 10^{-3} Pa = 10^{-5} mbar. Though some biological specimens with low water content can be observed directly for a short time at low magnifications, high resolution demands dehydration, which must be preceded by chemical fixation to preserve the structure. Shrinkage and swelling of biological specimens accompany all preparation processes during fixation, dehydration, critical point or freeze-drying and embedding [8.110–112]. One way of observing hydrated specimens is shock-freezing and transfer through a special system to the cooled specimen stage [8.113]. Another way to observe biological specimen surfaces in a SEM in their native state is to increase the partial pressure of water vapour near the specimen in order to suppress evaporation. The necessary partial pressure of liquid H_2O at 0°C is 609 Pa = 6.09 mbar. This principle can also be applied to the direct examination of other liquids. Environmental cells with reactive gases and specimen heating are of interest for observing in-situ crystal growth or oxydation and corrosion processes, for example. Some such cell designs incorporate a gas flux through a capillary tube [8.114, 115].

In such an environmental SEM (ESEM) [8.116–119], two aligned pressure-limiting diaphragms of 50–300 μm at the end of the probe-forming lens allow the high vacuum in the SEM column to be maintained with a higher working pressure of 0–50 mbar between the specimen and the last diaphragm. The space between the two diaphragms separated by a few millimetres is evacuated by an additional rotary pump to a vacuum of the order of 0.1–0.5 mbar. One of the diaphragms can be the final-lens diaphragm. The construction and design of this vacuum system has been discussed in detail in [8.118, 121]. A bibliography of ESEM applications is published in [8.120].

The pressure of 0–50 mbar between the specimen and the last pressure-limiting diaphragm causes additional electron scattering and electron–gas interactions. We start the discussion of these interactions with a recapitulation of some laws of gas kinetics. At this low pressure each gas behaves as an ideal gas with the law

$$pV = \nu RT = \frac{m}{M} N_{\mathrm{A}} kT \tag{8.28}$$

where $\nu = m/M$ is the number of moles and $R = kN_{\mathrm{A}} = 8314$ J/K kmol is the universal gas constant. Avogadro's law tells us that one mole of a gas with $N_{\mathrm{A}} = 6.023 \times 10^{23}$ molecules has the same mole volume $V_{\mathrm{m}} = 22.414$ l at the normal conditions of $p_0 = 1.013 \times 10^5$ Pa and $T_0 = 273.15$ K irrespective of the composition of the gas. In a gas mixture, the total pressure is the sum of the partial pressures. The density of a gas becomes

$$\rho = \frac{M}{V_{\mathrm{m}}} \frac{pT_0}{p_0 T} \,. \tag{8.29}$$

The gas molecules move randomly with a mean kinetic energy

$$\frac{1}{2} m\overline{v^2} = \frac{3}{2} kT \quad \rightarrow \quad v = 146\,(T/M)^{1/2}\ \mathrm{m/s}\,. \tag{8.30}$$

The number of molecules striking a specimen per unit area and unit time is

$$n = \frac{Nv}{4} = 2.6 \times 10^{24} \frac{p}{MT^{1/2}}\ \mathrm{m^{-2}s^{-1}}\,. \tag{8.31}$$

The mean-free path Λ between collisions of gas molecules is inversely proportional to the pressure p and is about 120 μm for H_2, 45 μm for H_2O and 65 μm for N_2 and air at $p = 1$ mbar and $T = 293$ K. It is interesting to note that the mean distance between molecules in these conditions is about 35 nm.

The gas flow through the lowest pressure-limiting diaphragm can be treated as viscous effusion flow resulting in a jet of gas in contrast to free molecular (Knudsen) flow through a thin capillary. This can increase the effective concentration of gas molecules for the electron beam when crossing the space between the two diaphragms. Of the order of 10% of the primary beam electrons will be scattered in this region. Near the entrance side of the diaphragm the pressure will be somewhat decreased but this can, however, obviously be neglected [8.118].

The gas atoms between the lowest diaphragm and the specimen at a distance $t = 0.5$–5 mm corresponding to a mass–thickness $x = \rho t$ will scatter the electron beam. A H_2O vapour layer of 1 mm in thickness at a pressure of 10 mbar corresponds to $x = 0.7\ \mu\mathrm{g/cm}^2$. At the specimen the electron beam profile consists of the unscattered part with a small diameter equal to the electron-optical electron-probe diameter d_p decreasing in amplitude with increasing x and a very broad distribution (skirt) of electrons scattered elastically and inelastically through small angles. The mean-free-path Λ_t or $x_\mathrm{t} = \rho\Lambda_\mathrm{t}$ in units of mass–thickness has been introduced in Sect. 3.3.1. The decrease of the unscattered part can be described by the exponential law $I = I_0 \mathrm{e}^{-p_c}$ (3.103) where $p_c = x/x_\mathrm{t}$ is the mean number of collisions. From Table 3.1, we see that for carbon and approximately for air or N_2, the total mean-free-path length $x_\mathrm{t} = \rho\Lambda_\mathrm{t}$ is about 2 μg cm^{-2} for $E = 20$ keV. We have 37% unscattered electrons for $p_c = 1$, 18% for $p_c = 2$ and 6% for $p_c = 3$. This means that ESEM should work in the region $p_c = 1$–2. Substituting $2x_\mathrm{t}$ for x in (3.103) (p_c=2), we obtain a maximum value of $p \cdot t = 4 \times 10^3$ Pa mm, which means that we can work with 18% of the unscattered primary electrons at $p = 10$ mbar and $t = 4$ mm.

Though the mean elastic and inelastic scattering angles are small, the scattered part is distributed in a skirt orders of magnitude larger in diameter than the unscattered electron probe. This beam broadening is larger than in bulk films (Sect. 3.3.3) because of the long distance t. For a calculation of this diameter, the elastic and inelastic differential cross-sections discussed in Sects. 3.1.5 and 3.2.4 can be used. The multiple-scattering theories of Sects. 3.3.2 and 3.4 assume larger thicknesses up to $p_c \simeq 20$. For the range $p_c = 1$–2 the spatial distribution after passing a layer of thickness t can be calculated by evaluating the multiple-scattering integral [8.118, 122] or by Monte Carlo simulations. This skirt distribution can be described in terms of a normalized radial distribution function

$$R(r) = \int_0^r 2\pi r n(r)\mathrm{d}r \bigg/ \int_0^\infty 2\pi r n(r)\mathrm{d}r \tag{8.32}$$

with the distribution $n(r)$ of scattered electrons only or of a normalized step contrast function [8.123]

$$S(x) = \int_{-\infty}^{+\infty}\int_{-\infty}^{x} n(x,y)\mathrm{d}x\mathrm{d}y \quad \text{where} \quad \int_{-\infty}^{+\infty}\int_{-\infty}^{+\infty} n(x,y)\mathrm{d}x\mathrm{d}y = 1 \tag{8.33}$$

over all electrons. Figure 8.16 shows $R(r)$ and $S(x)$ for $E = 20$ keV, $p_{\mathrm{H_2O}}$= 10 mbar and $t = 2$ mm. $S(x)$ also contains the step at $x = 0$ caused by

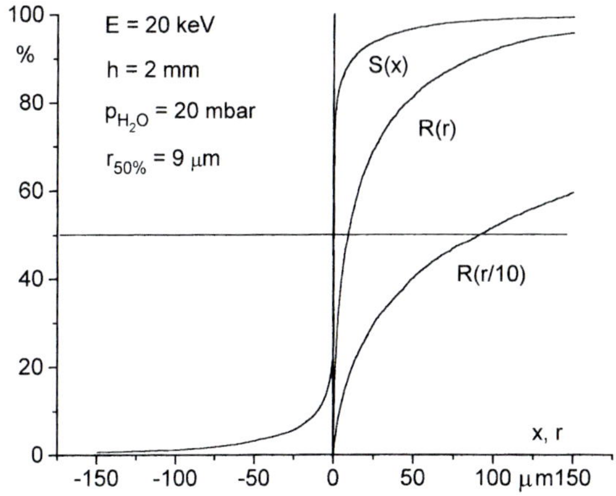

Fig. 8.16. Radial distribution function $R(r)$ and step distribution function $S(x)$ of the skirt of scattered electrons in an environmental atmosphere with $p_{\mathrm{H_2O}} = 10$ mbar and a depth $t = 2$ mm (calculated with the Monte Carlo simulation program MOCASIM [3.135])

the unscattered electrons. This plot is analoguous to $S(x)$ for the exit distribution of secondary electrons (Fig. 4.28). To characterize the size of the skirt distribution, the radius r_{m} of a circle that contains $m\%$ of the scattered electrons can be used (excluding the unscattered part). In Fig. 8.16 $r_{50} = 9$ μm is found from $R(r_{50}) = 0.5$.

Because $r_{\mathrm{m}} \gg d_{\mathrm{p}}$, the skirt of scattered electrons is distributed over such a large area that the averaging over specimen structures will result in a nearly constant background, which can be subtracted by adjustment of the black level (contrast) knob of the SEM; a highly resolved image can thus be obtained from the variation of the unscattered electrons.

The BSE of the specimen can penetrate the gas atmosphere without large energy losses and the small-angle scattering does not disturb the image when using a detector of large solid angle. Figure 8.17a shows two semi-annular scintillation detectors for BSE around the last pressure-limiting diaphragm with which it is possible to work with $t = 1$ mm. When t is further decreased for higher p the solid angle of collection will decrease. A signal from the charge carriers generated by the primary and backscattered electrons can be collected by electrodes biased at $\simeq$10 V a distance 1–5 mm from the specimen [8.118, 123]. The detection of SE emitted at the specimen will hardly be possible with a conventional detector. Though their energy of 1–10 eV is too low to ionize the gas atoms, they can be captured forming negative ions, especially in electronegative gases, which move about a thousand times slower than free electrons in a collecting field. However, an interesting alternative is to detect SE by initiating a Townsend discharge in the gas [8.123, 124]. In such a gaseous secondary electron detector (Fig. 8.17b), the conical detection electrode is biased by a few hundred volts. The SE emitted are accelerated and can produce further SE in collisions with the gas molecules resulting in proportional cascade multiplication. The probability of ionization and the

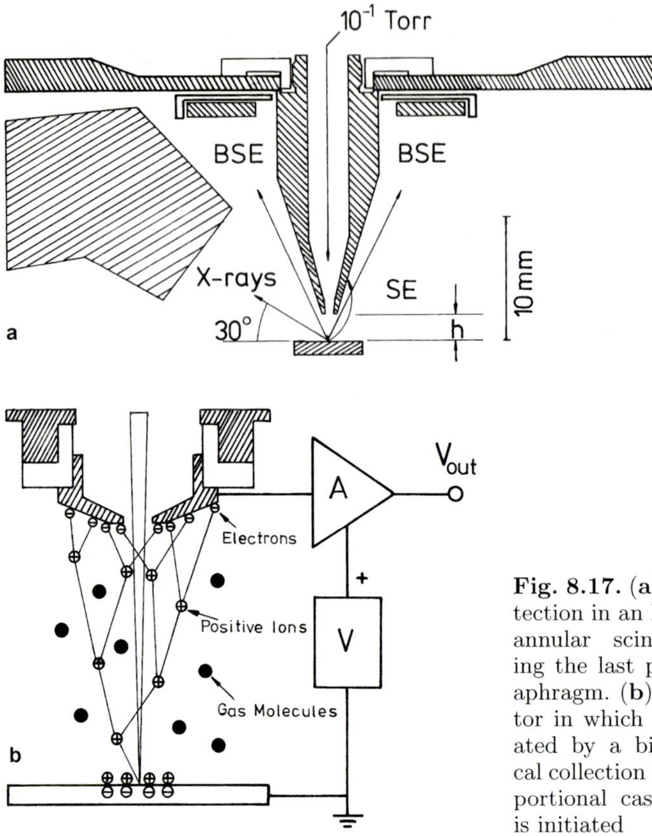

Fig. 8.17. (**a**) BSE and x-ray detection in an ESEM by two semi-annular scintillators surrounding the last pressure-limiting diaphragm. (**b**) Gaseous SE detector in which the SE are accelerated by a bias V on the conical collection electrode and a proportional cascade multiplication is initiated

gain depend on the composition of the gas. Water vapour, the most commonly used environmental gas, provides good imaging performance.

It is also possible to record x-rays (Fig. 8.17a), which are not affected by the low gas pressure, by placing a specially designed energy-dispersive detector near the specimen. Increasing the H_2O pressure increases linearly the peak-to-background ratio of oxygen by direct ionization of O atoms in the gas atmosphere and introduction of argon gas results in clear argon peaks. A handicap for spatial resolution is again the skirting of the beam, which can be reduced by using high electron energies and as low a gas pressure as possible [8.125].

Any positive ions generated in the gas by PE, BSE and SE will be attracted by the negative surface potentials on insulators for electron energies beyond E_2 (Sect. 3.6.2) and can compensate this charging.

8.7 Scanning Electron Acoustic Microscopy

Photoacoustic microscopy was first achieved by irradiating a specimen with a periodic pulsed laser beam [8.126]. An electron beam chopped at frequencies in the range 100 kHz to 5 MHz also produces periodic specimen heating, which propagates into the material as a damped thermal wave. The periodic changes in thermal expansion caused by the temperature oscillation excite an acoustic wave, which can be picked up by a piezoelectric crystal transducer of lead zirconate titanate (PZT) [8.127–130]. This technique has, therefore, also been called acoustic thermal-wave microscopy.

Figure 8.18 shows the experimental set-up, which can be mounted on the specimen stub. The cable has to be shielded carefully to avoid spurious signal pick-up. The periodic signal of the transducer can be used directly for high primary electron currents; alternatively, the signal may be selectively amplified by a lock-in amplifier with signal amplitude and phase as different sources of information. The lock-in amplifier can also select the second harmonic. Another possibility is to record the acoustic wave by means of a capacitive microphone [8.131]. The capacitance C is formed by two metal plates a distance 5–10 μm apart. One of the plates is in acoustic contact with the specimen. An applied voltage creates induced currents since $\mathrm{d}Q/\mathrm{d}t = U\,\mathrm{d}C/\mathrm{d}t$. This device has the advantage that it has no resonance frequency and can be heated or cooled. Using an InSb photodiode, sensitive to radiation of wavelengths 1–5 μm, and an elliptical mirror, the surface temperature can be recorded directly [8.132, 133] and non-adherent zones of metal layers can be imaged as in the acoustic mode.

In Sect. 3.5.1, we have studied the steady-state equation of heat conduction $P = -\lambda S \nabla T$, which can also be written $\boldsymbol{j} = -\lambda \nabla T$ where $\boldsymbol{j}$ is the heat current density. Here, we must use the transient equation for the diffusion of heat

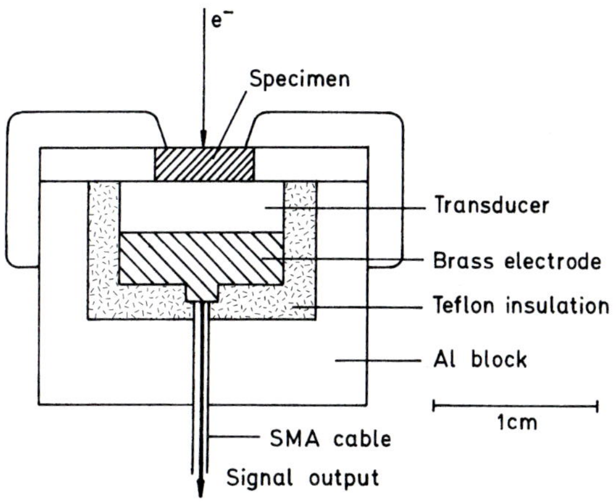

Fig. 8.18. Experimental setup for recording thermally induced acoustic waves (SMA: metal shielded coaxial cable)

$$\rho C \partial T/\partial t = \lambda \nabla^2 T + g(\boldsymbol{r}, t) \tag{8.34}$$

where the left-hand side is the change of heat energy per unit mass and time where ρ denotes the density [g cm^{-3}] and C heat capacity [J g^{-1} K^{-1}]; this must be equal to the change of energy density $-\nabla \cdot \boldsymbol{j}$ plus the local generation of heat $g(\boldsymbol{r}, t)$. The latter will be zero for $r \geq R$ outside the electron diffusion cloud and we assume

$$g(r, t) = g_0 \exp(\mathrm{i}\omega t) \tag{8.35}$$

inside the region $r < R$. This will create a temperature variation

$$T(r, t) = T(r) \exp(\mathrm{i}\omega t) \; . \tag{8.36}$$

For a source of radius R in an infinite homogeneous medium, substitution of (8.35) in (8.34) results in the radially-symmetric equation

$$\frac{1}{r^2} \frac{\partial}{\partial r} \left(r^2 \frac{\partial T}{\partial r} \right) = \frac{\mathrm{i}\rho c \omega}{\lambda} T(r) - g_0 \tag{8.37}$$

with the solution

$$T(r) \propto \frac{1}{\omega \rho c} \frac{\exp[-r(1 + \mathrm{i})/d_{\mathrm{t}}]}{r} g_0 \exp(\mathrm{i}\omega t) \quad \text{for} \quad r > R \, , \tag{8.38}$$

which is an exponentially damped spherical wave, the amplitude of which decays by a factor 1/e in a propagation distance

$$d_{\mathrm{t}} = \left(\frac{2\lambda}{\omega \rho c} \right)^{1/2} \; . \tag{8.39}$$

The thermal wave generates a local internal pressure

$$p = K \alpha(r) T(r) \, , \tag{8.40}$$

where K is the bulk modulus of elasticity and α the coefficient of thermal volume expansion. Assuming that the material is isotropic with Poisson's ratio ν, the displacement $\boldsymbol{u}(\boldsymbol{r}) = \exp(\mathrm{i}\omega t)$ of the excited acoustic wave must satisfy the equation

$$\nabla(\nabla \cdot \boldsymbol{u}) - \frac{1 - 2\nu}{2(1 - \nu)} \nabla \times (\nabla \times \boldsymbol{u}) + \frac{1 + \nu}{3E_{\mathrm{el}}(1 - \nu)} \omega^2 \boldsymbol{u} = \frac{1 + \nu}{3(1 - \nu)} \alpha \nabla T \, , \tag{8.41}$$

which is not easy to solve, especially when we take into account that most materials and even cubic crystals are elastically anisotropic. From (8.40), we see that the amplitude of the acoustic wave is proportional to

$$\int \alpha(r) T(r) \mathrm{d}^3 r = -\frac{\mathrm{i}}{\omega} \frac{\alpha}{\rho c} g_0 \frac{2\pi R^3}{3} = \frac{\mathrm{i}}{\omega} \frac{\alpha}{\rho c} f I_{\mathrm{p}} U \; . \tag{8.42}$$

The presence of the factor i means that the phase of the signal is shifted by $\pi/2$ relative to the electron-beam chopping. This estimate shows that the amplitude will depend on the power $I_{\mathrm{p}} U$ of the electron beam (U: acceleration voltage) and on the factor f (Sect. 3.5.1) which depends on backscattering. The acoustic-wave amplitude depends on the combination $\alpha/\rho c$ of the material constants but not on the thermal conductivity λ. The wavelength of the acoustic wave is

$$\lambda_{ac} = \frac{2\pi v_{ac}}{\omega} = \frac{2\pi}{\omega}\left(\frac{E_{el}}{\rho}\right)^{1/2} \tag{8.43}$$

(E_{el}: Young's modulus), and is of the order of centimetres for 100 kHz and of millimetres for 1 MHz. The specimen and the pick-up device of Fig. 8.18, therefore, act as an acoustic resonator. The output signal of the transducer will be greatest at the resonance frequency of this system. For frequencies higher than a few hundred kHz, the signal amplitude also depends on the position of the electron probe and in particular, on whether there is a node or an antinode of vibration at the electron impact point. A typical node pattern can hence be imaged at high frequencies and low magnification.

This brief summary of the excitation mechanism of acoustic waves shows that the interpretation of acoustic images will not be easy but that this method offers much new information not available from other imaging modes. For example, aluminium conductive pads on integrated circuits are imaged with a bright contrast because the thermal expansion coefficient is much higher for Al than for Si or SiO_2. Areas with no SiO_2 layer also appear brighter [8.127]. The imaging of doped regions in Si has also been observed and interpreted in terms of phase shifts caused by the differences in thermal conductivity between doped and undoped Si [8.128, 134]. As a result of elastic anisotropy, grains of polycrystalline material can produce different intensity levels and specific contrast is seen at grain boundaries. The absence of such contrast in elastically isotropic tungsten confirms this model. It must also be recalled that the thermally induced pressure will mainly be relieved by an outward movement of the surface when the excitation of flexural modes is weak because the relevant modulus has a low value along the surface normal. Grain and grain boundary contrast are even visible for rough surfaces, which show only topographic contrast in SE and BSE micrographs. An increase in the sharpness of grain boundary contrast has been observed in polycrystalline Si when the image is formed with the second harmonic of the chopping frequency [8.128]. The acoustic signal from silicon shows a strong dependence on temperature [8.131]. The contrast obtained from $BaTiO_3$ [8.128, 135] can be attributed to a direct electron-piezoelectric coupling mechanism.

The resolution of the mode is limited by the volume of heat generation, the radius of which is of the order of the electron range R, caused by the electron probe; the diameter of the latter may be comparable with R to provide a large probe current $I_p \geq 10^{-7}$ A because the output signal will be proportional to I_p. The resolution is also limited by the damping distance d_t (8.39) of the thermal wave. However, the largest contribution comes from a volume of the order of R in diameter so that a resolution of the order of 0.1–0.3 μm is possible [8.128]. For one or two metal layers on a semiconductor substrate calculated electron acoustic signals are compared with experiments in [8.136].

8.8 Electron-Beam Lithography and Metrology

This technique is used for the manufacture of large scale and very large scale integrated circuits (LSI and VLSI). A photoresist in the form of a polymer film of 0.1–1 μm in thickness is illuminated by ultraviolet light, x-rays, electrons or ions. It is not the aim of this Section to discuss electron-beam lithography (EBL) in detail. Special electron optical designs and instruments are essential. The imaging mode in these instruments is mainly used for searching and alignment of markers on wafers. In comparison with exposure by ultraviolet light, EBL has the disadvantage of being much slower because of the pixel by pixel exposure, though methods have been developed to expose all the pixels in a rectangular area simultaneously. The field of EBL is mainly the production of structures of the order of 10 nm and the production of masks for soft x-ray exposure. For laboratory work any SEM with a beam blanker can be used for home-made structures.

The same radiation mechanism that cause damage to organic material and contamination (Sect.3.6.3) can be used for the exposure of resists in EBL. In exposed areas, a positive resist becomes soluble because the principal effect of irradiation is scission of macromolecular chains whereas a positive resist becomes insoluble because of the dominance of cross-linking processes. The necessary doses for exposure are in the range 10^{-8}–10^{-4} C cm^{-2} (see [8.137] for references). This exposure is orders of magnitude lower than that necessary for the total loss of mass by radiation damage (Sect. 3.5.3). The most widely used positive resist is polymethylmethacrylate (PMMA) because of its superior resolution, though its sensitivity of about 10^{-4} C cm^{-2} is very low. The long chains with a molecular weight of $\simeq$100000 are broken by scission. The exposed areas are developed (dissolved) in isopropyl alcohol and methyl ethyl (or isobutyl) ketone at 4:3 (or 1:1), for example. The substrate material can be processed through this thin film mask by chemical etching or ion milling.

Unlike light and x-ray exposure, EBL does not need a rigid mask and the pattern can be exposed by deflection of an electron probe and by beam blanking using a digitally stored pattern. An important point during fabrication is the alignment of the pattern to be exposed with a pattern already processed. This can be achieved by writing fiducial marks near the edges of a chip, which can be recognized in the SE or BSE mode [8.138].

The resolution of EBL is limited by the electron probe diameter and the lateral spread of primary electrons (σ_f in (8.44)) in the $\simeq$0.5 μm resist layer and the much broader contribution (σ_b) of electrons backscattered from the substrate to the resist. The exposure intensity distribution can be calculated by Monte Carlo simulations [8.139, 140] and determined experimentally by examining cross-sectional profiles [8.139] or by light-optical observation [8.141] of developed resists with increasing exposure.

Assuming that the electron probe is focused to a point, the energy density $\mathrm{d}E/\mathrm{d}V$ deposited in the resist can be represented by a radial point spread function, which can be approximated by two or three Gaussians [8.142–143]

$$\frac{\mathrm{d}E}{\mathrm{d}V} \propto \frac{1}{\pi\sigma_\mathrm{f}^2} \exp(-r^2/\sigma_\mathrm{f}^2) + \eta_\mathrm{e} \frac{1}{\pi\sigma_\mathrm{b}^2} \exp(-r^2/\sigma_\mathrm{b}^2) \ . \tag{8.44}$$

When a single line of 10 nm in width, for example, is exposed the contribution from the second term can be neglected because $\sigma_\mathrm{b} \gg \sigma_\mathrm{f}$ and the energy density is spread over a large area. However, when writing a large number of parallel lines with a separation much smaller than σ_b, the background from the second term is summed up and causes a continuous background below the single line exposures by the first term so that it becomes more difficult to find a threshold exposure between the line-maxima and the background exposure. This is one typical example of the so-called proximity effect [8.145].

The normalized Fourier transform of (8.44) is the modulation transfer function

$$M(q) = \frac{1}{1+\eta_\mathrm{e}} [\exp(-\pi^2\sigma_\mathrm{f}^2/\Lambda^2) + \eta_\mathrm{e} \exp(-\pi^2\sigma_\mathrm{b}^2/\Lambda^2)] \tag{8.45}$$

where $q = 1/\Lambda$ is the spatial frequency (number of period lengths Λ per unit length). The modulation transfer function $M(q)$ is plotted in Fig. 8.19 for the numerical examples discussed below. For 20 keV electrons and a resist layer of 0.5 μm the 1/e-radius σ_f resulting from the lateral spread of primary electrons in the resist is about 0.12 μm and σ_b caused by electrons backscattered from the substrate is about 2 μm. The ratio $\eta_\mathrm{e} = 0.74$ for a silicon substrate describes the relative fraction of the last term in (8.44) and tells us how much more energy is deposited in the resist per BSE than per primary electron. As in secondary electron emission (Sect. 4.2.3), we have $\eta_\mathrm{e} \simeq \beta\eta$ though the real value can be larger if the BSE trajectory inside the resist layer is increased by multiple scattering.

The proximity effect can be partially reduced by correction of the deposited dose for backscattering or by an equalization of the background (GHOST method) [8.146–151]. With increasing energy, σ_f decreases and σ_b increases. The tendency is therefore to use $\simeq$50 keV for sub-100 nm lithography. Otherwise it makes no sense to use low energies to exposure a 0.5 μm resist layer. However, σ_f decreases as $t^{1.75}$ with increasing resist thickness t. Figure 8.19 also shows the modulation transfer function for a resist layer of 35 nm exposed by 5 keV electrons. Low-energy electrons of 1 keV can also be used for the exposure of ultrathin films of PMMA of the order of 8.5 nm [8.152].

Another way of reducing proximity effects is by the use of multilayer resists. For example, the latent image for exposure is formed in an upper

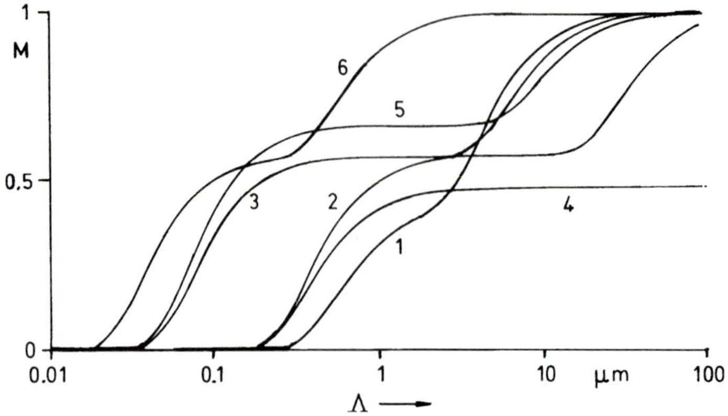

Fig. 8.19. Examples of the dependence of the contrast transfer function M on resist thickness and electron beam energy: 0.5 μm resist film on silicon for beam energies of (1) 5 keV, (2) 20 keV, (3) 50 keV and (4) 20 keV with a correction exposure to reduce the proximity effect; (5) double-layer structure consisting of a 0.2 μm resist film on 3 μm lower film for 20 keV and (6) 35 nm resist film on silicon for 5 keV

thin layer of 0.2 μm, which reduces σ_f to 0.02 μm for a 20 keV exposure. A lower thick layer of 2 μm has no image-forming function but reduces the backscattering factor to $\eta_e = 0.5$ and σ_b increases to 3 μm [8.144]. The upper layer is developed as usual and the lower layer with a much lower etch rate can be etched by reactive ion etching. To decrease charging, a suitably conductive intermediate layer can be used [8.153, 154].

Owing to the increased ionization density of low-energy electrons and the increased rate of contamination, direct deposition of carbon can be used for writing and storage. For example, conductive palladium acetate films of 0.05–0.5 μm in thickness can be deposited from solutions in chloroform and exposed by 5 keV electrons [8.155]. Using a differentially pumped subchamber around the specimen, metals such as gold can be deposited by 2 keV electrons with a resolution of $\simeq$ 50 nm from organometallic compounds in the gas phase at $\simeq 10^{-4}$ mbar [8.156]; an increased halo caused by gas scattering can decrease the resolution at low energies. For writing with beams of high intensity resist heating effects have also to be taken into account [8.157].

The production of sub-100 nm structures by electron-beam lithography requires an exact measurement of the *critical dimensions*. SEM will of course also be a suitable method for metrology.

As shown in Sect. 6.1.2, edges are imaged with characteristic imaging artifacts in the BSE and SE signals, an increase of image intensity with a width of the order of the electron range, for example. Comparison with Monte Carlo simulations can help to analyse which points of the signal profiles can be used for measuring a critical dimension [8.158–164]. These calculations and the decrease of electron range with decreasing energy shows that LVSEM is very important for accurate metrology of submicron VLSI structures [8.165–

167], especially when using specially designed magnification standards for LVSEM [8.168] or laser-interferometrically controlled specimen stages. Other reasons for preferring LVSEM are the reduced device damage, resist exposure and specimen charging, though residual charges can influence the primary trajectories [8.169–173].

A better recognition of edge position was observed by imaging BSE with a microchannel plate [8.165], which gives symmetrical line profiles across bars and trenches, whereas the SE image can be affected by specimen charging and can show asymmetric effects due to shadowing contrast. The latter can be avoided in SE imaging by using a system of four Everhart–Thornley detectors [8.174] or a through-lens detector system (Sect. 5.4.1). The latter requires careful attention to the symmetry. Otherwise the detector system still shows a preferential collection.

The following procedures can be used for the measurement of critical dimensions from linescans with an accuracy of the order of 10 to 20 nm:

1. Threshold method, by measuring the distance between the two cross-points in a signal and a preset threshold between 25–75%.
2. Peak detection method, by computing the points of maximum signal intensity.
3. Maximum slope method, by finding the maximum in a differentiated signal.

The maximum slope method gives better values of the critical dimensions of spaces in resist, whereas the peak-to-peak method can giver better values for lines [8.166], for example. When deeply etched trenches have to be measured, the trench walls strongly limit the collection angle of BSE and SE and the critical dimension at the bottom cannot be measured accurately. The best procedure will depend on the structure of lines or spaces and on the slope of the edge and the existence of undercuts, for example. The critical dimension and the shape of edges can therefore measured more accurately in a sideview of a cross-section. When measuring resist overlayers, charging effects can influence the critical dimension, which can be tested by making a second measurement after coating the specimen with a 5 nm Au/Pd film, for example, whereas often no charging effect is observed when the electron range is larger than the thickness of the overlay [8.167]. In all cases, the accuracy of SEM metrology is limited by the lack of accurate standards, which only can be used for measuring the magnification.

Alternatives for sub-100 nm metrology are scanning optical microscopy (SOM) [8.166] and scanning tunnelling microscopy (STM) [8.175, 176] in the constant current and atomic force modes. Although the signal profile of lines and trenches recorded in a SOM is strongly influenced by diffraction effects, critical dimensions can be measured with an accuracy of about 20 nm. The confocal mode allows us to measure top and bottom critical dimensions separately by correct focusing. A comparison with SEM metrology and cross-section images is more essential than for SEM alone, to detect systematic errors that depend sensitively on the shape of the feature.

The STM allows height differences normal to the surface to be detected and measured with atomic resolution. The shape of the conical probe tip limits the lateral resolution and the measurement of deep trenches, and STM and atomic force microscopy suffer from just as many problems as the SEM metrology. When scanning a step with undercut side walls, the step will be convolved with the shape of the tip, which can change during scanning. For the measurement of trench depths of 1 μm or more with submicron accuracy, conductive needles of about 0.1 μm diameter can be grown on top of normal probe tips by electron irradiation in a SEM.

8.9 Combination of SEM and STM

Scanning tunnelling microscopy (STM) has the advantage of high lateral and depth resolution of the order of 0.1 nm and 0.01 nm, respectively, and atomic force microscopy (AFM) allows us to investigate insulating specimens [8.177–180]. The disadvantage is that the scanned area is relatively small, with an edge length of 1–10 μm. This makes it difficult to search for interesting details on large areas and the shape of the tip often does not allow rough specimens to be imaged without artifacts.

Some missing information can be recovered when the same specimen is also investigated by SEM or TEM, which should be done more frequently. To combine the advantages of these imaging modes, STM or AFM units have been installed inside a SEM in which the specimen is usually inclined by $\simeq 45°$ [8.181–186]. This allows us to search over a large specimen and to look at an interesting detail with the atomic or near atomic resolution of the STM. With small electron probe currents, the scanning of the STM tip can be observed directly.

9. Crystal Structure Analysis by Diffraction

Electrons are Bragg diffracted at lattice planes. The geometry of a diffraction pattern can be described by the kinematical theory. For the discussion of intensities it is necessary to use the dynamical theory of electron diffraction and the Bloch-wave model. Because a Bloch-wave field has nodes and antinodes at the nuclei and the probability density at the nuclei depends sensitively on the tilt relative to the Bragg position, the backscattering coefficient shows an anisotropy. When an electron probe is rocked, this anisotropy of the backscattering results in the electron channelling pattern (ECP). For a stationary electron probe, the angular distribution of backscattered electrons is modulated by excess and defect Kikuchi bands, leading to the formation of an electron backscattering pattern (EBSP), which can be observed on a fluorescent screen beside the specimen and recorded on a photographic emulsion or via a TV camera. At oblique incidence of the electron beam, the reflection high-energy electron diffraction (RHEED) pattern may contain Bragg diffraction spots and Kikuchi lines. ECP and EBSP are related by the theorem of reciprocity. These patterns contain information about the crystal structure, orientation and distortion.

X-rays excited by electron irradiation inside a single crystal are also Bragg reflected on their path through the crystal and result in x-ray Kossel patterns with excess and defect Kossel lines; these contain high resolution information from which the lattice constants and any changes caused by mechanical strains can be obtained.

9.1 Theory of Electron Diffraction

9.1.1 Kinematical Theory and Bragg Condition

The kinematical and dynamical theory of electron diffraction have been discussed in detail in [9.1–3]. Here we merely summarize the most important facts that are needed to understand the diffraction effects that can occur in bulk materials in SEM.

Crystals can be described in terms of the seven Bravais translation lattices [9.4, 5], which consist of unit cells defined by three non-coplanar vectors $\boldsymbol{a}_i$ (i = 1,2,3). The origins of the unit cells are at the positions

$$\boldsymbol{r}_g = m\boldsymbol{a}_1 + n\boldsymbol{a}_2 + o\boldsymbol{a}_3 \quad (m, n, o\,\text{integers})\,. \tag{9.1}$$

Each unit cell can contain $p \geq 1$ atoms at the positions

$$\boldsymbol{r}_k = u_k\boldsymbol{a}_1 + v_k\boldsymbol{a}_2 + w_k\boldsymbol{a}_3 \quad (u_k, v_k, w_k \leq 1, k = 1, ..., p)\,. \tag{9.2}$$

For example, one atom at the cube corner, $\boldsymbol{r}_1 = (0,0,0)$ and a second one at the cube centre $\boldsymbol{r}_2 = (\frac{1}{2}, \frac{1}{2}, \frac{1}{2})$ form a body-centred cubic lattice or one atom at the cube corner and three atoms at the centres of the neighbouring faces form a face-centred cubic lattice.

In the kinematical theory of electron diffraction, each atom scatters the incident electron wave of wavevector $\boldsymbol{k}_0 = \boldsymbol{u}_0/\lambda$ elastically through an angle θ with a scattering amplitude $f(\theta)$ (Sect. 3.1.5) into a direction with wavevector $\boldsymbol{k} = \boldsymbol{u}\lambda$, where $|\boldsymbol{k}_0| = |\boldsymbol{k}| = 1/\lambda$ and $\boldsymbol{u}$ is a unit vector. Waves scattered at two atoms separated by a distance $\boldsymbol{r}$ have a path difference $\Delta = \boldsymbol{u}\cdot\boldsymbol{r} - \boldsymbol{u}_0\cdot\boldsymbol{r}$ resulting in a phase shift (Fig. 9.1)

$$\phi = \frac{2\pi}{\lambda}\Delta = \frac{2\pi}{\lambda}(\boldsymbol{u} - \boldsymbol{u}_0)\cdot\boldsymbol{r} = 2\pi(\boldsymbol{k} - \boldsymbol{k}_0)\cdot\boldsymbol{r}\,. \tag{9.3}$$

Superposition of the waves scattered at all the atoms of a lattice into the direction $\boldsymbol{k}$ gives the wave amplitude

$$\psi = \psi_0 \sum_{m=1}^{N_1}\sum_{n=1}^{N_2}\sum_{o=1}^{N_3}\sum_{k=1}^{p} f_k(\theta)\exp[2\pi i(\boldsymbol{k} - \boldsymbol{k}_0)\cdot(\boldsymbol{r}_g + \boldsymbol{r}_k)] \tag{9.4}$$

where we assume that the crystal forms a parallelepiped with $N = N_1N_2N_3$ unit cells and $f_k(\theta)$ denotes the elastic scattering amplitude of the kth atom in the unit cell at the position $\boldsymbol{r}_g + \boldsymbol{r}_k$. From now on, we set the amplitude ψ_0 of the incident wave equal to unity. Equation (9.4) can be split into the factors

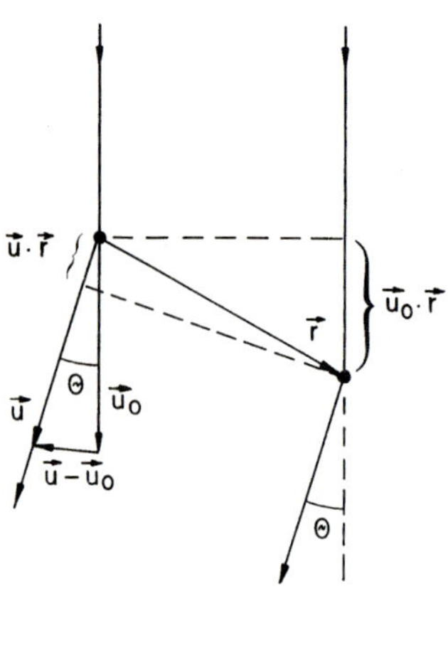

Fig. 9.1. Scattering of an incident plane wave of wavevector $\boldsymbol{k}_0 = \boldsymbol{u}_0/\lambda$ at two atoms separated by a distance $\boldsymbol{r}$ through an angle θ into a plane wave of wavevector $\boldsymbol{k} = \boldsymbol{u}/\lambda$ from which the path difference $\Delta = (\boldsymbol{u} - \boldsymbol{u}_0)\cdot\boldsymbol{r}$ may be calculated ($\boldsymbol{u}_0$, $\boldsymbol{u}$ = unit vectors)

$$\psi = \underbrace{\sum_{k=1}^{p} f_k(\theta)\exp[2\pi\mathrm{i}(\boldsymbol{k}-\boldsymbol{k}_0)\cdot\boldsymbol{r}_k]}_{F} \times \underbrace{\sum_{m=1}^{N_1}\sum_{n=1}^{N_2}\sum_{o=1}^{N_3}\exp[2\pi\mathrm{i}(\boldsymbol{k}-\boldsymbol{k}_0)\cdot\boldsymbol{r}_g]}_{G} \tag{9.5}$$

where F and G denote the structure and lattice amplitude, respectively. The waves will interfere constructively only if the arguments in the terms $\exp[2\pi\mathrm{i}(\boldsymbol{k}-\boldsymbol{k}_0)\cdot\boldsymbol{r}_g]$ of G are all integral multiples of 2π. Substitution of (9.1) yields the Laue equations

$$(\boldsymbol{k}-\boldsymbol{k}_0)\cdot\boldsymbol{a}_i = h_i \quad (h_i \text{ are integers}, i = 1,2,3). \tag{9.6}$$

Only those combinations of $\boldsymbol{k}_0$ and $\boldsymbol{k}$ for which this system of Laue equations is satisfied result in a scattered amplitude corresponding to constructive interference. The solution is that $\boldsymbol{k}-\boldsymbol{k}_0$ has to be equal to a reciprocal lattice vector $\boldsymbol{g}$:

$$\boldsymbol{k}-\boldsymbol{k}_0 = \boldsymbol{g} = h_1\boldsymbol{a}_1^* + h_2\boldsymbol{a}_2^* + h_3\boldsymbol{a}_3^* \tag{9.7}$$

where the elementary reciprocal lattice vectors are defined by

$$\boldsymbol{a}_1^* = \frac{\boldsymbol{a}_2\times\boldsymbol{a}_3}{V_\mathrm{e}}; \quad \boldsymbol{a}_2^* = \frac{\boldsymbol{a}_3\times\boldsymbol{a}_1}{V_\mathrm{e}}; \quad \boldsymbol{a}_3^* = \frac{\boldsymbol{a}_1\times\boldsymbol{a}_2}{V_\mathrm{e}} \tag{9.8}$$

and $V_\mathrm{e} = \boldsymbol{a}_1\cdot(\boldsymbol{a}_2\times\boldsymbol{a}_3)$ ist the volume of the unit cell. The vectors $\boldsymbol{a}_i^*$ of a reciprocal lattice fulfil the relations

$$\boldsymbol{a}_i\cdot\boldsymbol{a}_j^* = \delta_{ij} = \begin{cases} 1 & \text{if} \quad i=j \\ 0 & \text{if} \quad i\neq j \end{cases} . \tag{9.9}$$

Substitution of the solution (9.7) into (9.6) confirms that the Laue equations are satisfied by (9.7). Each reciprocal lattice vector $\boldsymbol{g}$ corresponds to a set of parallel and equidistant lattice planes with Miller indices $(h_1h_2h_3)$ or (hkl) and $\boldsymbol{g}$ is perpendicular to the lattice planes with a lattice-plane spacing $d_{hkl} = 1/|\boldsymbol{g}|$ (Fig. 9.2a).

Equation (9.7) is the Bragg law in vector notation. In the Ewald construction (Fig. 9.2a) the wavevector $\boldsymbol{k}_0$ of the incident wave ends at the origin O of the reciprocal lattice and starts at the excitation point M ($\overline{\mathrm{MO}} = |\boldsymbol{k}_0| = 1/\lambda$). The Bragg condition (9.7) is satisfied when the Ewald sphere of radius $|\boldsymbol{k}| = 1/\lambda$ with the excitation point M as centre passes through a reciprocal lattice point $\boldsymbol{g}$. This expresses the phenomenon of Bragg reflection. The vectors $\boldsymbol{k}_0$ and $\boldsymbol{k}$ are inclined at equal angles (Bragg angles) θ to the lattice planes and $\boldsymbol{k}_0$, $\boldsymbol{k}$ and $\boldsymbol{g}$ are coplanar. With $|\boldsymbol{g}| = 1/d_{hkl}$ and $|\boldsymbol{k}| = 1/\lambda$, simple geometry applied to the triangle MOH in Fig. 9.2a yields the Bragg equation

$$2d_{hkl}\sin\theta_\mathrm{B} = \lambda . \tag{9.10}$$

Not all possible lattice planes with Miller indices hkl contribute to diffraction. Using the Bragg condition (9.7), we can substitute $\boldsymbol{k}-\boldsymbol{k}_0 = \boldsymbol{g}$ in the structure amplitude F of (9.5) and find

$$F = \sum_{k=1}^{p} f_k(\theta) \exp(2\pi i \boldsymbol{g} \cdot \boldsymbol{r}_k) . \quad (9.11)$$

which depends on the structure of the unit cell and can vanish for certain values of hkl. These extinction rules and the dependence of the magnitude of F on hkl allows us to distinguish different crystals with unit cells of the same external shape but with different arrangements of atoms inside. As an example, we list the structure amplitudes F of several different cubic lattices in Table 9.1.

We have seen that all scattered waves interfere constructively when $\boldsymbol{k} - \boldsymbol{k}_0$ takes the value $\boldsymbol{g}$ in (9.5). It is important to know how the intensity of a Bragg spot decreases for small deviations from the Bragg condition, and for this we introduce an excitation error $\boldsymbol{s}$ (Fig. 9.2a) and substitute $\boldsymbol{k} - \boldsymbol{k}_0 = \boldsymbol{g} + \boldsymbol{s}$ in (9.5). This has no influence on the first term in (9.5) because $\boldsymbol{r}_g$ is only a vector inside the unit cell and the product $\boldsymbol{s} \cdot \boldsymbol{r}_k$ is negligible. This is not the case for the last term G, where $\boldsymbol{r}_g$ extends over the whole diffracting crystal. The variation of $\boldsymbol{s} \cdot \boldsymbol{r}_g$ from unit cell to unit cell is again small and the sum over all N unit cells can be replaced by an integral over the crystal dimension with a continuously varying phase factor $\exp(2\pi i s z)$. This yields the following expression for the intensity of the Bragg diffraction spot for a single-crystal film of thickness t

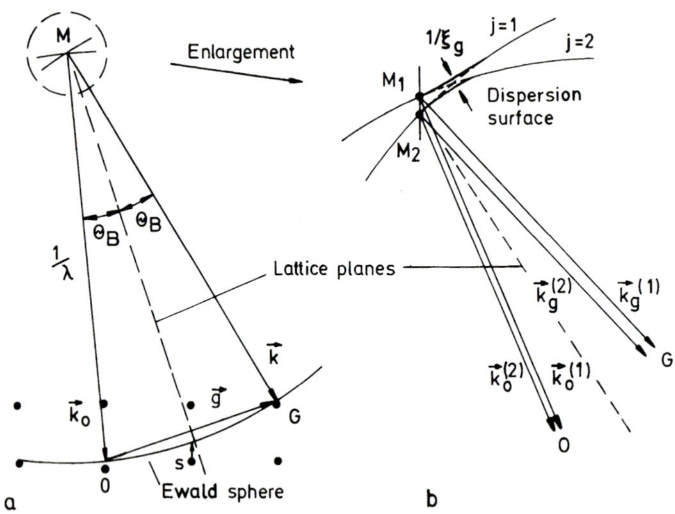

Fig. 9.2. (**a**) Construction of the Ewald sphere defined by the wavevector $\boldsymbol{k}_0$ of length $1/\lambda$ from the excitation point M to the origin O of the reciprocal lattice. The Bragg condition is satisfied when $\boldsymbol{k} - \boldsymbol{k}_0 = \boldsymbol{g}$ (s = excitation error). (**b**) Enlargement around the excitation point M showing the splitting into the dispersion surface and the formation of four waves from the excitation points M_1 and M_2. The waves with wavevectors $\boldsymbol{k}_0^{(i)}$ and $\boldsymbol{k}_g^{(i)}$ (i=1,2) form the primary and reflected waves, respectively; $\boldsymbol{k}_0^{(1)}$, $\boldsymbol{k}_g^{(1)}$ and $\boldsymbol{k}_0^{(2)}$, $\boldsymbol{k}_g^{(2)}$ form the Bloch waves of types 1 and 2

Table 9.1. Structure amplitude of some cubic lattices

---

1) Body-Centred Cubic Lattice

Atoms of atomic scattering amplitude f at $\boldsymbol{r}_1$=(0,0,0) and $\boldsymbol{r}_2=(\frac{1}{2},\frac{1}{2},\frac{1}{2})$

$F_{\text{bbc}} = f\{1+\exp[\pi\text{i}(h+k+l)]\}$

$F = 2f$	for $h+k+l$ even
$F = 0$	for $h+k+l$ odd

2) Face-Centred Cubic Lattice

Atoms at $\boldsymbol{r}_1$=(0,0,0), $\boldsymbol{r}_2=(\frac{1}{2},\frac{1}{2},0)$, $\boldsymbol{r}_3=(\frac{1}{2},0,\frac{1}{2})$, $\boldsymbol{r}_4=(0,\frac{1}{2},\frac{1}{2})$

$F_{\text{fcc}} = f\{1+\exp[\pi\text{i}(h+k)]+\exp[\pi\text{i}(h+l)]+\exp[\pi\text{i}(k+l)]\}$

$F = 4f$	for h,k,l all even or all odd
$F = 0$	for h,k,l mixed

3) CsCl Structure

Two primitive cubic lattices of Cs and Cl shifted by half the body diagonal

$F_{\text{CsCl}} = F_{\text{Cs}} + F - \text{Cl}\exp[\pi\text{i}(h+k+l)]$

$F = f_{\text{Cs}} + f_{\text{Cl}}$	for $h+k+l$ even
$F = f_{\text{Cs}} - f_{\text{Cl}}$	for $h+k+l$ odd

4) NaCl Structure

Two face-centred lattices of Na and Cl shifted by half the body diagonal

$F_{\text{NaCl}} = \{f_{\text{Na}} + f_{\text{Cl}}\exp[\pi\text{i}(h+k+l)]\}F_{\text{fcc}}$

$F = 4(f_{\text{Na}} + f_{\text{Cl}})$	for h,k,l even
$F = 4(f_{\text{Na}} - f_{\text{Cl}})$	for h,k,l odd
$F = 0$	for h,k,l mixed

5) ZnS (Adamantine) Structure

Two face-centred lattices of Zn and S shifted by a quarter of the body diagonal

$F_{\text{ZnS}} = \{f_{\text{Zn}} + f_{\text{S}}\exp[\pi\text{i}(h+k+l)/2]\}F_{\text{fcc}}$

$F = 4(f_{\text{Zn}} \pm \text{i}f_{\text{S}})$	for h,k,l odd
$F = 4(f_{\text{Zn}} + f_{\text{S}})$	for h,k,l even and $h+k+l$=4n
$F = 4(f_{\text{Zn}} - f_{\text{S}})$	for h,k,l even and $h+k+l$=4n+2
$F = 0$	for h,k,l mixed

The structure amplitudes of the *diamond structure* are obtained by writing $f_{\text{Zn}} = f_{\text{S}} = f$.

$$I_g = I_0\frac{\pi^2}{\xi_g^2}\frac{\sin^2(\pi ts)}{(\pi s)^2}\,, \tag{9.12}$$

where I_0 is the incident intensity and

$$\xi_g = \frac{\pi V_{\text{e}}}{\lambda F(\theta)} = \frac{\pi m v V_{\text{e}}}{hF(\theta)} \tag{9.13}$$

is known as the extinction distance. Some values of ξ_g are listed in Table 9.2 for E = 10, 20 and 30 keV. Calculated and measured values of ξ_g agree reasonably well at E = 100 keV [9.3]. The values listed in Table 9.2 on page 345 may be extrapolated by using the proportionality between ξ_g and the electron velocity v. Figure 9.6 shows how the intensity varies with s. The

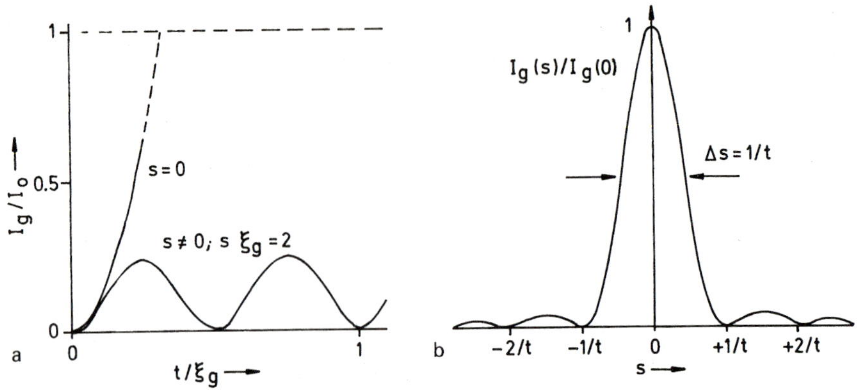

Fig. 9.3. Dependence of the kinematical intensity I_g of a Bragg reflection on (**a**) the foil thickness t and (**b**) the excitation error s

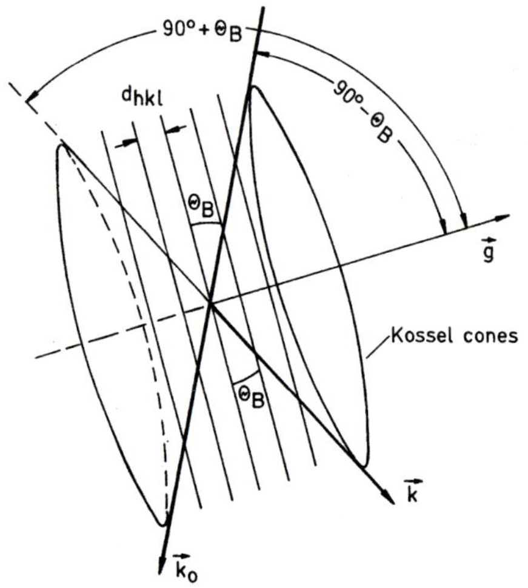

Fig. 9.4. Orientation of the two Kossel cones with $\boldsymbol{g}$ as zone axis parallel to the normal to the lattice planes of spacing d_{hkl} (θ_B = Bragg angle)

maximum of this rocking curve has a width $\Delta s = 1/t$ (Fig. 9.3b). For $s \neq 0$, the intensity I_g oscillates with increasing film thickness t (Fig. 9.3a). For $s = 0$ (exact Bragg position), I_g increases as t^2 and we leave the domain in which the kinematical theory is valid when I_g approaches I_0.

9.1.2 Kossel Cones and Kossel Maps

The Bragg condition (9.7) or (9.10) will be satisfied for all directions of incidence for which the wavevector $\boldsymbol{k}_0$ lies on a so-called Kossel cone with a semi-apex angle $90° - \theta_\mathrm{B}$ and $\boldsymbol{g}$ as the cone axis. The corresponding Bragg-reflected beam with wavevector $\boldsymbol{k}$ lies in the complimentary cone with semi-apex angle $90° + \theta_\mathrm{B}$ (Fig. 9.4), so that $\boldsymbol{k}_0$, $\boldsymbol{k}$ and $\boldsymbol{g}$ are coplanar.

The geometry of electron channelling patterns, electron backscattering patterns and x-ray Kossel patterns represents none other than the intersection of these Kossel cones with the plane of observation or recording and we therefore describe how to construct a map of Kossel lines, the curves in which the Kossel cones intersect the plane of observation.

The reciprocal lattice vectors $\boldsymbol{g}$ (9.7) consist of all possible linear combinations of the $\boldsymbol{a}_i^*$ with integral values of hkl, including zero and negative integers. Those values of hkl for which the structure amplitude becomes zero (Table 9.1, for example) have to be omitted. A unit vector $\boldsymbol{u}_g = \boldsymbol{g}/|\boldsymbol{g}|$ becomes a distinct point P (pole) on a unit sphere and characterizes the direction of the lattice-plane normal. The lattice planes are parallel to a plane that is perpendicular to $\boldsymbol{u}_g$ and intersects the unit sphere in a great circle (Fig. 9.5). Depending on the de Broglie wavelength λ (2.23) of the electrons or the x-ray wavelength (10.2) of a characteristic line, we can calculate θ_{B} from the Bragg condition (9.10). The lattice-plane spacing of the (h_1, h_2, h_3) planes can be calculated from

$$1/d_{h_1,h_2,h_3} = |\boldsymbol{g}| = \sum_{i,j} h_i h_j \boldsymbol{a}_i^* \cdot \boldsymbol{a}_j^* \quad (i,j = 1,2,3) . \tag{9.14}$$

Two cones of semi-apex angles $90° \pm \theta_{\mathrm{B}}$ with $\boldsymbol{u}_g$ as axis form a pair of Kossel cones. After repeating this procedure for all those hkl for which the structure amplitude is not zero, the intersections of the Kossel cones with the unit sphere form a characteristic line pattern, which also indicates the symmetry of the crystal.

Two possible ways of projecting the unit sphere onto a plane, which are commonly used, will now be discussed. The first is the stereographic projection, in which a point P on the upper half of the sphere is projected onto a point P' on the equatorial plane by connecting P to the south pole (the projection centre) (Fig. 9.5a). In the second procedure, known as the gnomonic

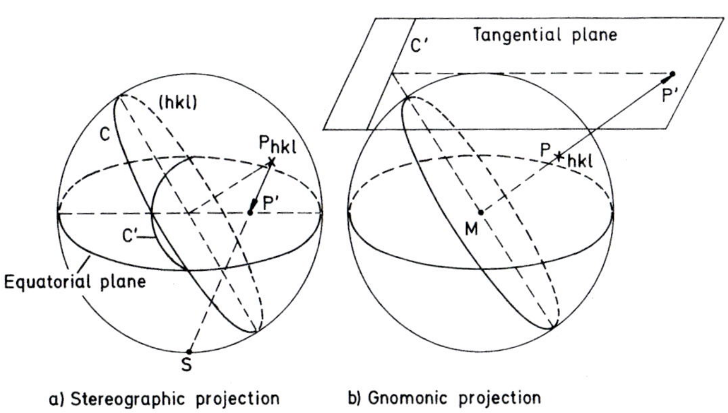

Fig. 9.5. Principle of (**a**) the stereographic projection and (**b**) the gnomonic projection methods of mapping a unit sphere onto a plane

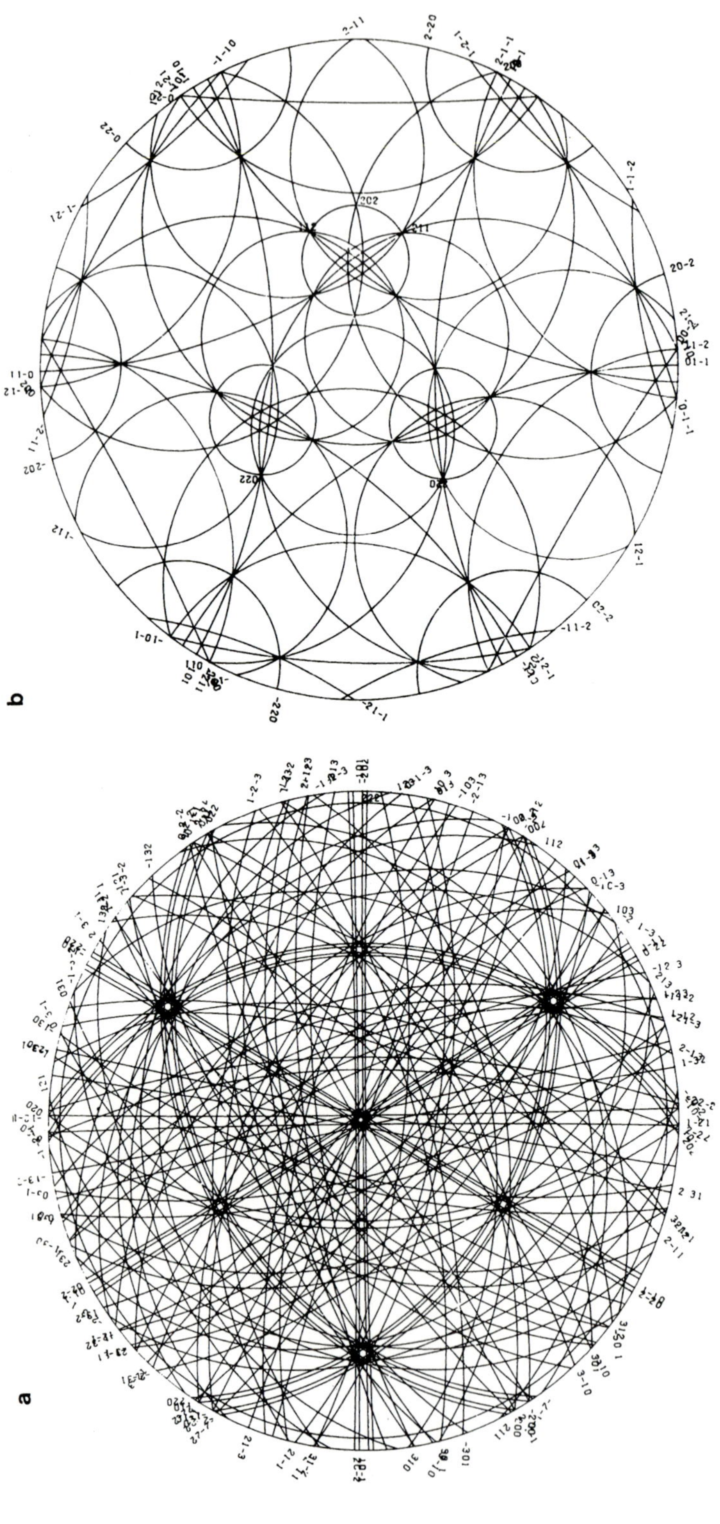

Fig. 9.6. Kossel map for the diffraction of (**a**) 20 keV electrons ($\lambda = 8.586$ pm) and (**b**) Cu Kα x-rays ($\lambda_x = 0.154$ nm) at a copper crystal of 111 orientation

projection, the projection centre is situated at the centre of the unit sphere and points on the sphere are projected onto a tangential plane of arbitrary orientation (Fig. 9.5b), which in practice is parallel to the plane of observation.

The projections have the following fundamental properties and differences. The stereographic projection transforms great circles on the sphere into circles and preserves all angles between great circles, that is, the angles between the tangents to the great circles at their points of intersection. Wulff's net, which contains great circles equivalent to lines of equal geographic longitude and lines which correspond to equal latitude can be used to measure angles on a stereographic projection (see textbooks on crystallography for details). Conversely, the gnomonic projection transforms all great circles into straight lines but does not preserve the angles between the great circles.

Since the Bragg deflection angles $2\theta_B$ are smaller than 5° in electron diffraction, the two Kossel cones corresponding to semi-apex angles $90° \pm \theta_B$ form approximately parallel lines on either side of the intersection of the lattice planes, though the intersection of a cone becomes strictly hyperbolic in the gnomonic projection. In x-ray diffraction, Bragg deflection angles $2\theta_B$ cover the whole range 0 to 180° and the projection of Kossel cones looks quite different from that observed in the case of electron diffraction, as can be seen by comparison of Figs. 9.6a,b.

Given these basic laws, it is possible to write a computer program for plotting Kossel maps [9.6]. A stereographic plot for all possible hkl can be reduced to a spherical triangle on the unit sphere limited by great circles, and the whole Kossel map is then obtained by invoking the symmetries of the corresponding crystal system. These unit triangles are shown in Fig. 9.7

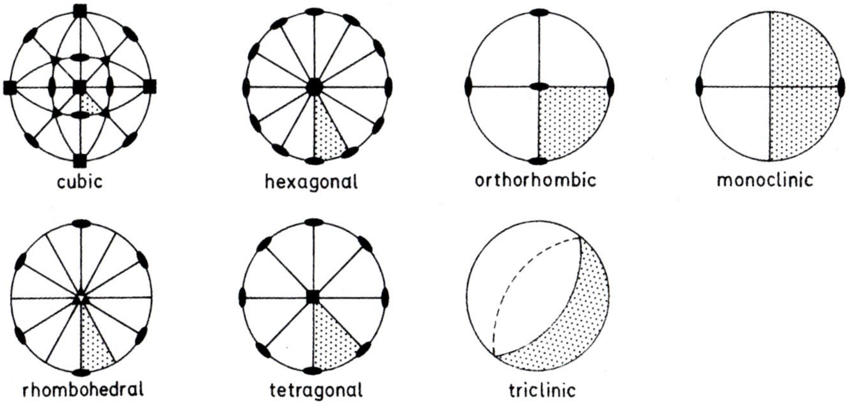

Fig. 9.7. Symmetry elements of the seven Bravais lattices with twofold, threefold and fourfold rotational axes showing the minimum unit triangles (shaded) necessary for the computation of a Kossel map

together with the multiplicities of rotational symmetry. The size of the unit triangle increases with decreasing symmetry from the cubic to the triclinic lattice. For the cubic system, it is only necessary to calculate the Kossel map inside the shaded triangle between the [100], [110] and [111] directions after which the whole map can be obtained by successively reflecting the map inside the triangle at each of the great circles that limit the unit triangle. In other cases, the symmetries observed in the Kossel maps can be used to determine the crystal symmetry.

9.1.3 Dynamical Theory of Electron Diffraction and Bloch Waves

The kinematical approach (9.12) discussed in Sect. 9.1.1 can only be used for very thin films with thicknesses t of the order of a few nanometres because the diffracted intensity increases as t^2 at the Bragg position ($s = 0$). The dynamical theory [9.1–3] sets out from the Schrödinger equation (3.21) and provides a solution to the following problem: if a plane incident wave satisfies the boundary conditions at the interface crystal–vacuum, how will it be transformed into a wave-field $\Psi(\boldsymbol{r})$ having the same periodicity as the crystal? This is analysed by writing the periodic, overlapping Coulomb potentials of the single atoms as a Fourier series over all reciprocal lattice points $\boldsymbol{h}$:

$$\nabla^2\Psi + \frac{2m}{\hbar^2}\left[E - \sum_h V_h \exp(2\pi \mathrm{i}\boldsymbol{h}\cdot\boldsymbol{r})\right]\Psi = 0\,. \tag{9.15}$$

The Fourier coefficients $V_h = V_h + \mathrm{i}V_h'$ are formally complex. The real parts are related by

$$V_h = \frac{h^2}{2\pi m V_\mathrm{e}} F(\theta_h) \tag{9.16}$$

to the structure amplitude $F(\theta_h)$ in (9.5). The imaginary part V_g' represents the scattering into directions between the primary and the Bragg reflected beams. The solution of (9.15) must also exhibit the same periodicity as the crystal and can be written as a sum of Bloch waves $\psi^{(j)}$

$$\Psi = \sum_j \epsilon^{(j)}\psi^{(j)} = \sum_j \epsilon^{(j)} \sum_g C_g^{(j)} \exp[2\pi\mathrm{i}(\boldsymbol{k}_0^{(j)} + \boldsymbol{g})\cdot\boldsymbol{r}] \exp(-2\pi q^{(j)} z) \tag{9.17}$$

where the $\epsilon^{(j)}$ are the excitation amplitudes of the Bloch waves $\psi^{(j)}$ and the $C_g^{(j)}$ are the amplitudes of the partial waves with wavevectors $\boldsymbol{k}_g^{(j)} = \boldsymbol{k}_0^{(j)} + \boldsymbol{g}$. The last term, containing the absorption parameters $q^{(j)}$, expresses the exponential decrease of the wave amplitudes with increasing depth z. The indices j and g run over an infinite number of reciprocal lattice points, including the origin ($\boldsymbol{g} = 0$). In practice, only a limited number n of strongly excited beams will be taken into account (n-beam case), which in the notation of the Ewald sphere only show small excitation errors s_g (Fig. 9.8), the latter being the distance of the end point of $\boldsymbol{g}$ from the Ewald sphere. This means that the plane, incident wave splits into n^2 partial waves with wavevectors $\boldsymbol{k}_g^{(j)}$ as it enters the crystal.

Before substituting $\psi^{(j)}$ from (9.17) into (9.15) – at first without the last exponential damping term – we introduce the wavevector

$$K = \frac{1}{h}[2m(E + V_0)]^{1/2}, \tag{9.18}$$

which respresents the reciprocal wavelength inside the crystal characterized by the mean value V_0 (inner potential) associated with the $\boldsymbol{h} = 0$ term of $V(r)$ in (9.15). This results in a system of simultaneous differential equations:

$$4\pi^2[K^2 - (\boldsymbol{k}_0^{(j)} + \boldsymbol{g})^2 + \frac{2m}{h^2}\sum_{h\neq 0} V_h \exp(2\pi\mathrm{i}\boldsymbol{h}\cdot\boldsymbol{r})] \times C_g^{(j)} \exp[2\pi\mathrm{i}(\boldsymbol{k}_0^{(j)} + \boldsymbol{g})\cdot\boldsymbol{r}] = 0 \tag{9.19}$$

for all $\boldsymbol{g}$. This system of equations can only be satisfied if the coefficients of identical exponential terms become simultaneously zero. Collecting up terms containing the factor $\exp[2\pi\mathrm{i}(\boldsymbol{k}_0^{(j)} + \boldsymbol{g})\cdot\boldsymbol{r}]$, we obtain the fundamental equations of dynamical theory:

$$[K^2 - (\boldsymbol{k}_0^{(j)} + \boldsymbol{g})^2]C_g^{(j)} + \frac{2m}{h^2}\sum_{h\neq 0} V_h C_{g-h}^{(j)} = 0 \quad \text{for} \quad \boldsymbol{h} = \boldsymbol{g}_1, \ldots, \boldsymbol{g}_n\,. \tag{9.20}$$

The solution of these equations gives a set of n values of $\boldsymbol{k}_0^{(j)}$ and n^2 values of $C_g^{(j)}$ for a preset direction of incidence characterized by the wavevector $\boldsymbol{K}$. For a more detailed discussion of the solution, the reader is referred to [9.3]. The first factor in (9.20) can be written (Fig. 9.8)

$$[K^2 - (\boldsymbol{k}_0^{(j)} + \boldsymbol{g})^2] = (K + |\boldsymbol{k}_0 + \boldsymbol{g}|)\cdot(K - |\boldsymbol{k}_0 + \boldsymbol{g}|) \simeq 2K(s_g - \gamma^{(j)})\,. \tag{9.21}$$

The $\gamma^{(j)}$ values are much smaller than the $k_0^{(j)}$ and determine the position of the excitation points M_j. In the kinematical theory, the excitation point M (Fig. 9.2a), the starting point of the wavevector $\boldsymbol{k}_0$ ending at O, lies on a circle of radius $1/\lambda$ around O. In the dynamical two-beam case, the primary and the excited Bragg waves are equivalent and we have two excitation points $\mathrm{M}_{1,2}$, which however do not lie on spheres around O and G in Figs. 9.2b and 9.8, respectively; on the branches j=1,2 of the so-called dispersion surface the spheres do not intersect and the smallest distance between these branches is $1/\xi_g$ at the Bragg position, where ξ_g is the extinction distance introduced in (9.13). The values of $\gamma^{(j)}$, which are obtained when (9.21) is substituted in (9.20), describe the splitting of the dispersion surface into n branches for the n-beam case. A knowledge of the dispersion surface allows us to construct n^2 partial waves with wavevectors $\boldsymbol{k}_g^{(j)} = \boldsymbol{k}_0^{(j)} + \boldsymbol{g}$ (Fig. 9.8) from the n excitation points M_j to the n excited reciprocal lattice points $\boldsymbol{g}_n$.

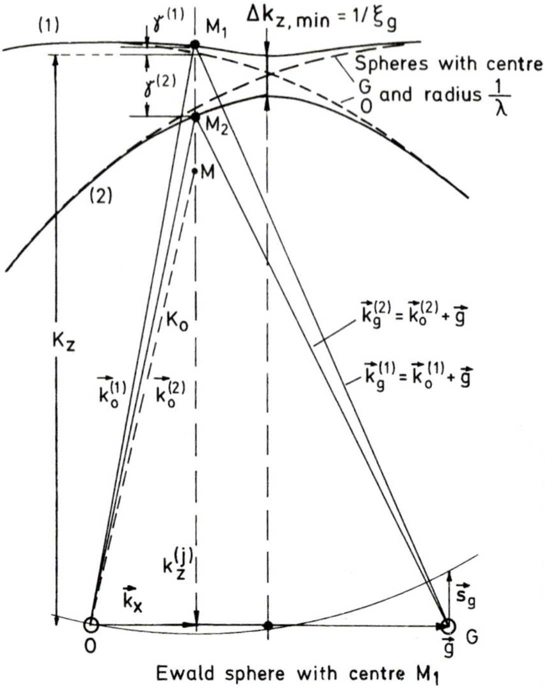

Fig. 9.8. Splitting of the excitation point M into M_1 and M_2 on the dispersion surface with branches $j = 1$ and 2 and the four possible wavevectors $\boldsymbol{k}_0^{(j)} + \boldsymbol{g}$ ($\boldsymbol{g} = 0, \boldsymbol{g}$) for the dynamical two-beam case. $\boldsymbol{s}_g$ = excitation error, $\boldsymbol{k}_x$ = x-component of $\boldsymbol{k}_0^{(j)}$

Substitution of (9.21) in (9.20) and division by $2K$ leads to an eigenvalue problem

$$\begin{pmatrix} A_{11} & A_{12} & \dots & A_{1n} \\ \dots & \dots & \dots & \dots \\ A_{21} & A_{22} & \dots & A_{2n} \\ A_{n1} & A_{n2} & \dots & A_{nn} \end{pmatrix} \begin{pmatrix} C_1^{(j)} \\ C_2^{(j)} \\ \dots \\ C_3^{(j)} \end{pmatrix} = \gamma^{(j)} \begin{pmatrix} C_1^{(j)} \\ C_2^{(j)} \\ \dots \\ C_3^{(j)} \end{pmatrix} \quad \text{for} \quad j = 1, \dots, n \tag{9.22}$$

with the matrix elements

$$A_{11} = 0,\ A_{gg} = s_g,\ A_{hg} = A_{gh} = \frac{1}{2\xi_{g-h}}\ .$$

ξ_{g-h} is the extinction length given by (9.13) when θ_{g-h} in $F(\theta_{g-h})$ is the angle between $\boldsymbol{g}$ and $\boldsymbol{h}$. From the Born approximation (3.47), we find that $F(\theta)$ and the Fourier coefficients V_g of the periodic lattice potential are related by $F(\theta) = (2\pi m/h^2)V_\mathrm{e}V_g$ and we have

$$\xi_{g-h} = \frac{\pi V_\mathrm{e}}{\lambda F(\theta_{g-h})} = \frac{h^2}{2m\lambda V_{g-h}}\ . \tag{9.23}$$

Algorithms and computer subroutines exist for calculating the n eigenvalues $\gamma^{(j)}$ and the corresponding eigenvectors with components $C_g^{(j)}$. The

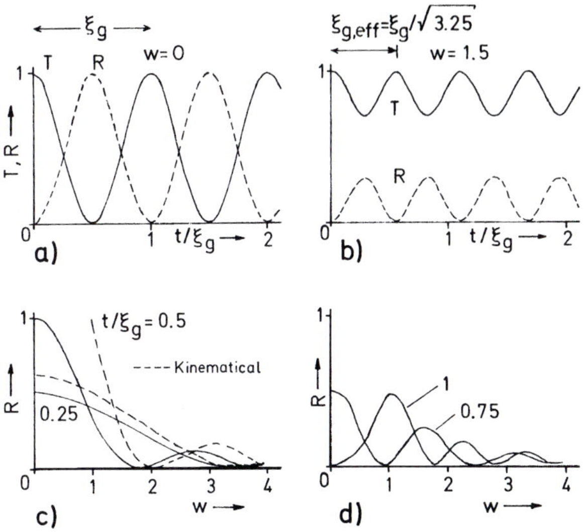

Fig. 9.10. Transmission T and Bragg-reflected intensity R for the two-beam case without absorption, (**a**) at $w = 0$, (**b**) at $w \neq 0$ for increasing film thickness t and (**c**), (**d**) variation of R for constant values of $t/\xi_g = 0.25$, 0.5, 0.75 and 1.0 as a function of the tilt parameter $w = s\xi_g$ (rocking curve)

$$|\psi_g|^2 = 1 - |\psi_0|^2 = \frac{1}{1+w^2}\sin^2\left(\pi\sqrt{1+w^2}\frac{t}{\xi_g}\right) . \tag{9.30}$$

Figures 9.10a,b show how the transmission $T = |\psi_0|^2$ and the reflected intensity $R = |\psi_g|^2$ vary with increasing foil thickness t for $w = 0$ and $w \neq 0$, respectively, and Figs. 9.10c,d show rocking curves for R at constant t and varying tilt parameter w similar to Fig. 9.3b of the kinematical theory.

The other way of grouping the partial waves is as Bloch waves (9.17), which can be written together with their excitation amplitudes $\epsilon^{(j)} = C_0^{(j)}$ for normal incidence:

$$\psi^{(j)} = \sum_g C_0^{(j)} C_g^{(j)} \exp[2\pi\mathrm{i}(\boldsymbol{k}_0^{(j)} + \boldsymbol{g})\cdot\boldsymbol{r}]\exp(-2\pi q^{(j)} z) . \tag{9.31}$$

We consider the Bragg condition (w=0) of the two-beam case (Figs. 9.9c,d) for which all products of the coefficients $C_0^{(j)} C_g^{(j)}$ take the absolute value 1/2. Again neglecting the final absorption term, substitution with the correct sign in (9.31) results in

$$\psi^{(1,2)} = \exp(2\pi\mathrm{i}\boldsymbol{k}_0^{(1,2)}\cdot\boldsymbol{r})\exp(\pi\mathrm{i}\boldsymbol{g}\cdot\boldsymbol{r})[\exp(\pi\mathrm{i}\boldsymbol{g}\cdot\boldsymbol{r}) \pm \exp(-\pi\mathrm{i}\boldsymbol{g}\cdot\boldsymbol{r})]/2 \tag{9.32}$$

with the positive and negative sign for j = 1,2, respectively. Hence, we find

$$\begin{aligned} |\psi^{(1)}|^2 &\propto \cos^2(\pi\boldsymbol{g}\cdot\boldsymbol{r}) = \cos^2(\pi x/d_{hkl}) , \\ |\psi^{(2)}|^2 &\propto \sin^2(\pi\boldsymbol{g}\cdot\boldsymbol{r}) = \sin^2(\pi x/d_{hkl}) , \end{aligned} \tag{9.33}$$

because the scalar product $\boldsymbol{g} \cdot \boldsymbol{r}$ contains only the x component of $\boldsymbol{r}$, which is parallel to $\boldsymbol{g}$ and normal to the z axis, the latter being parallel to the electron beam and the foil normal. At the lattice planes at $x/d_{hkl} = 0, 1 \ldots$ (integer), the probability density $|\psi^{(1)}|^2$ of Bloch wave 1 shows maxima (antinodes) and $|\psi^{(2)}|^2$ shows minima (nodes). This is simply the interference pattern of two waves with wavevectors $\boldsymbol{k}_g^{(j)} = \boldsymbol{k}_0^{(j)} + \boldsymbol{g}$ where j = const and $g = 0, \boldsymbol{g}$. These wavevectors have approximately the same length (same excitation point M_j at the branch j of the dispersion surface) but are inclined at an angle $2\theta_\mathrm{B}$, which results in interference maxima distance d_{hkl} apart perpendicular to the bisector of $\boldsymbol{k}_0^{(j)}$ and $\boldsymbol{k}_g^{(j)}$, that is, parallel to the lattice planes (Fig. 9.8). Midway between the lattice planes, the situation is reversed, nodes for Bloch waves 1 and antinodes for Bloch waves 2. Because most of the nuclei lie on the lattice planes, all interaction processes that are concentrated at the nuclei are excited more strongly when there are antinodes at the lattice planes. These interactions are the elastic large-angle scattering (Mott scattering, Sect. 3.1.6), the ionization of inner shells and as a consequence the emission of characteristic x-ray quanta and Auger electrons.

The higher probability for these effects results in a stronger exponential attenuation of Bloch wave 1 whereas Bloch wave 2, with nodes at the lattice planes, shows a lower attenuation; this effect can be described in terms of the different absorption parameters $q^{(j)}$ that arise when an exponential damping term $\exp(-2\pi q^{(j)} z)$ is introduced in (9.17), (9.29) and (9.31).

We mentioned previously that this attenuation can be considered by formally introducing complex coefficients $V_g + \mathrm{i}V_g'$ in the Schrödinger equation (9.15), with the result that the eigenvalues $\gamma^{(j)} + \mathrm{i}\,q^{(j)}$ become complex. If we assume that $V_g' < V_g$, the eigenvalue problem can first be solved with the real part V_g only after which the $q^{(j)}$ are calculated by perturbation theory

$$q^{(j)} = \frac{m\lambda}{h^2} \sum_g \sum_h V_{g-h}' C_g^{(j)} C_h^{(j)} = \frac{1}{2} \sum_g \sum_h \frac{1}{\xi_{g-h}'} C_g^{(j)} C_h^{(j)} \tag{9.34}$$

where the absorption lengths

$$\xi_g' = \frac{h^2}{2m\lambda V_g'} \tag{9.35}$$

are introduced by analogy with the extinction lengths ξ_g in (9.23).

Substitution of the two-beam values of $C_g^{(j)}$ from (9.28) in (9.34) gives

$$q^{(j)} = \frac{1}{2}\left(\frac{1}{\xi_0'} - \frac{(-1)^j}{\xi_g'\sqrt{1+w^2}}\right) \tag{9.36}$$

which is plotted in Fig. 9.9b and confirms the stronger absorption of Bloch wave 1 and the reduced absorption of Bloch wave 2 for tilt angles near the Bragg condition.

Table 9.2. Extinction distances ξ_g and absorption lengths ξ'_g in nanometres for some substances at $E = 20$ keV. $\xi_g(E)/\xi_g(20\text{keV}) = (E/20)^{1/2}$; $\xi'_g(E)/\xi'_g(20\text{keV}) = E/20$ with E in keV

			$g = 111$	220	311
Al		$\xi_g=$	25	50	95
	$\xi'_0=140$	$\xi'_g=$	420	600	650
Si		$\xi_g=$	26	33	60
	$\xi'_0=124$	$\xi'_g=$	760	1050	1050
Cu		$\xi_g=$	13	14	25
	$\xi'_0=57$	$\xi'_g=$	100	106	140
Ge		$\xi_g=$	19	20	33
	$\xi'_0=64$	$\xi'_g=$	160	260	–
Au		$\xi_g=$	8	12	15
	$\xi'_0=32$	$\xi'_g=$	42	52	56

The absorption lengths ξ'_0 and ξ'_g vary as v^2 while the extinction lengths ξ_g are proportional to v. The imaginary potential coefficient V'_g results from elastic electron–phonon scattering and hence ξ'_0 and ξ'_g vary with temperature. In the limit of large scattering angles, the electron–phonon scattering resembles Rutherford or Mott scattering. Table 9.2 contains values extrapolated to $E = 20$ keV from measurements in a TEM at $E = 100$ keV, see also Tables 7.2 and 7.3 of [9.3]. The low values of ξ'_0 and ξ'_g at $E = 10$–30 keV demonstrate the strong attenuation of the Bloch-wave field at these low electron energies.

This characteristic difference in the attenuation of the various Bloch waves seems to suggest that the Bloch waves can be treated independently. In such an independent Bloch-wave model, the intensities of the Bloch waves have to be squared and added to get the local probability density $P_{\text{ind}}(\boldsymbol{r})$ for the interaction processes

$$P_{\text{ind}}(\boldsymbol{r}) = \sum_j |\psi^{(j)}|^2 = \sum_j \Big| \sum_g C_0^{(j)} C_g^{(j)} \exp[2\pi \mathrm{i}(\boldsymbol{k}_0^{(j)} + \boldsymbol{g}) \cdot \boldsymbol{r}] \exp[-2\pi q^{(j)} z] \Big|^2 \tag{9.37}$$

Figure 9.11a shows how the contributions corresponding to $j = 1$ and 2 are summed to obtain P_{ind} and their decrease in amplitude after traversing a thickness $z = \xi_g$. It can be seen that the intensity profile between two lattice planes at $\boldsymbol{g} \cdot \boldsymbol{r} = 0$ and 1 changes owing to the difference between the absorption parameters $q^{(1)}$ and $q^{(2)}$. However, this independent Bloch-wave model does not satisfy the boundary condition, which requires that at the

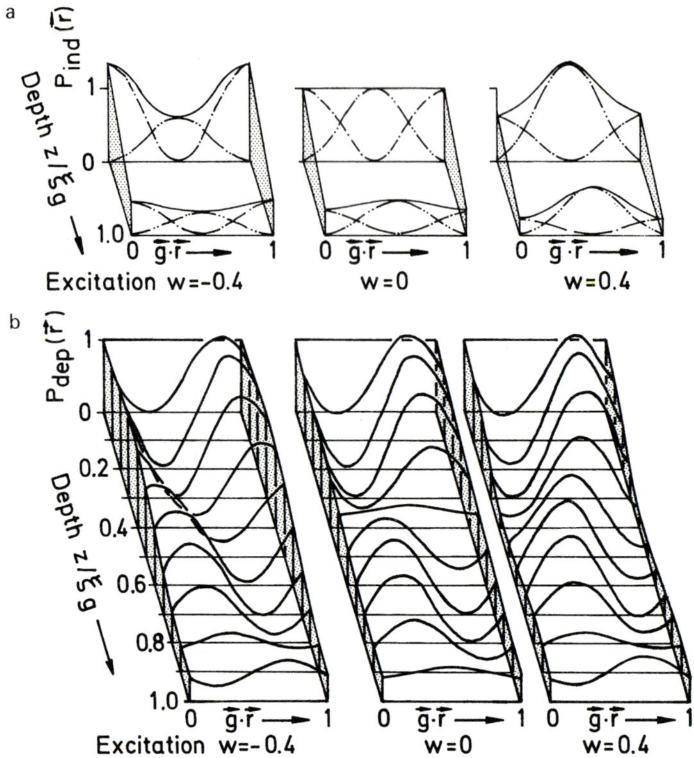

Fig. 9.11. Variation of the probability density $P(x, z)$ for (**a**) the independent and (**b**) the dependent Bloch-wave model with increasing depth z below the surface for various values of the tilt parameter w; the x direction is normal to the lattice planes which lie at $\boldsymbol{g} \cdot \boldsymbol{r} = 0$ and 1

surface ($z = 0$) both the external plane wave and the internal Bloch-wave field have constant normalized unit amplitude. This is the case for $w = 0$ but not for $w \neq 0$ in Fig. 9.11a.

In reality, we have to use the dependent Bloch-wave model [9.9], which allows the different types of Bloch waves to interfere and we have first to sum the amplitudes and then square to get the dependent probability density

$$\begin{aligned} P_{\text{dep}}(\boldsymbol{r}) &= \Big|\sum_j \psi^{(j)}\Big|^2 \\ &= \Big|\sum_j \sum_g C_0^{(j)} C_g^{(j)} \exp[2\pi\mathrm{i}(\boldsymbol{k}_0^{(j)} + \boldsymbol{g}) \cdot \boldsymbol{r}] \exp(-2\pi q^{(j)} z)\Big|^2 . \end{aligned} \tag{9.38}$$

This probability density is plotted in Fig. 9.11b, which shows that there are additional oscillations at the lattice planes and in the space between them, which are not found with the independent Bloch-wave model (Fig. 9.11a). This dependent Bloch-wave model also satisfies the boundary condition of

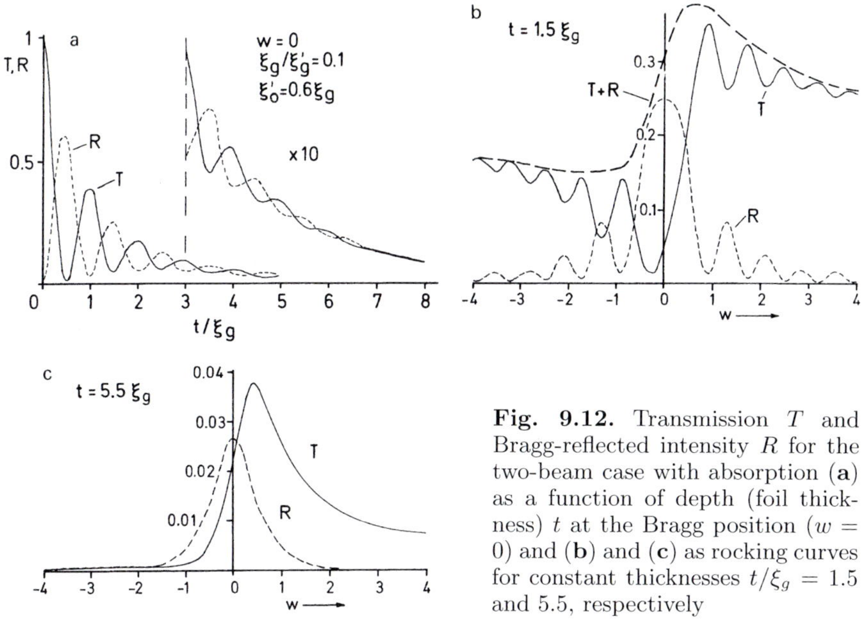

Fig. 9.12. Transmission T and Bragg-reflected intensity R for the two-beam case with absorption (**a**) as a function of depth (foil thickness) t at the Bragg position ($w = 0$) and (**b**) and (**c**) as rocking curves for constant thicknesses $t/\xi_g = 1.5$ and 5.5, respectively

constant amplitude at $z = 0$ for all tilt parameters w. Nevertheless, use of the independent Bloch-wave model often simplifies the discussion of the dependence of interaction processes on the tilt parameter. Differences between the results given by the dependent and independent Bloch-wave models are only perceptible for very thin films, as discussed in the next section.

Finally, we consider the modification of T and R in Fig. 9.10 introduced by the Bloch-wave absorption represented by the exponential factors $\exp(-2\pi q^{(j)}z)$ in (9.29). The dependence on the specimen thickness t is shown in Fig. 9.12a and the rocking curve is drawn in Fig. 9.12b,c. For negative w, Fig. 9.9c shows that the Bloch-wave excitation intensity $|C_0^{(1)}|^2$ is strong and the transmitted intensity in Fig. 9.12b shows anomalous absorption. For positive w, $|C_0^{(2)}|^2$ is strong, which results in anomalous transmission.

Figure 9.13 shows the rocking curve for a 3-beam case with the reflections $-\boldsymbol{g}$, 0 and $+\boldsymbol{g}$ excited. Instead of the tilt parameter w, we here use the paramater k_x/g, which becomes +0.5 at $+\boldsymbol{g}$ and −0.5 at the $-\boldsymbol{g}$ Bragg position (Fig. 9.8). The rocking curve for the $-\boldsymbol{g}$ reflection is not shown, being merely a shifted version of that for $+\boldsymbol{g}$. The dynamical theory without absorption, (9.30) and Fig. 9.10a, results in $T + R = 1$ for two beams. This is no longer true when absorption is considered. The sum of the intensities of the primary beam and of the Bragg reflections is the same as the total intensity surviving in the Bloch-wave field at a depth z. With (9.29) we have for n beams

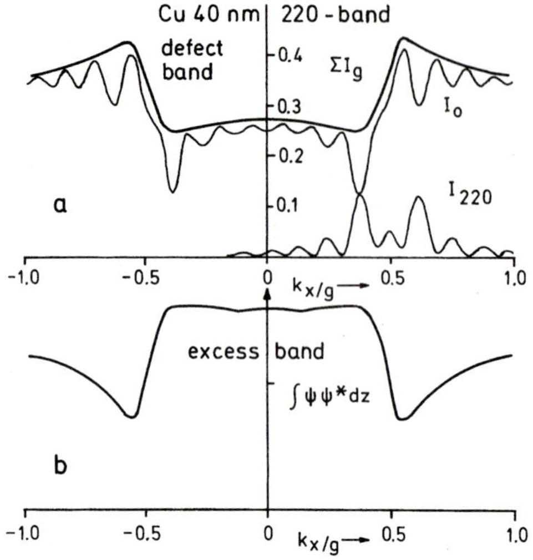

Fig. 9.13. (**a**) Rocking curve showing $I_0 = \psi\psi_0^*$ and $I_g = \psi_g\psi_g^*$ for the three-beam case of dynamical theory with the excitation of $-\boldsymbol{g}$, 0, $+\boldsymbol{g}$ and $\boldsymbol{g} = 220$. ΣI_g represents the defect band profile in a Kossel pattern. (**b**) Probability of electron backscattering resulting in an excess band profile in an electron channelling pattern

$$
\begin{aligned}
\sum_g I_g &= \sum_g \psi_g \psi_g^* \\
&= \sum_g \Big(\sum_{i,j} C_0^{(i)} C_g^{(i)} C_0^{(j)} C_g^{(j)} \exp[2\pi \mathrm{i}(\boldsymbol{k}_0^{(i)} - \boldsymbol{k}_0^{(j)}) \cdot \boldsymbol{r}] \exp[-2\pi(q^{(i)} - q^{(j)})z] \Big) \\
&= \sum_{i,j} C_0^{(i)} C_0^{(j)} \underbrace{\sum_g C_g^{(i)} C_g^{(j)}}_{\delta_{ij}} \exp[2\pi \mathrm{i}(\boldsymbol{k}_0^{(i)} - \boldsymbol{k}_0^{(j)}) \cdot \boldsymbol{r}] \exp[-2\pi(q^{(i)} - q^{(j)})z] \\
&= \sum_i |C_0^{(i)}|^2 \exp(-4\pi q^{(i)} z)\,. \qquad (9.39)
\end{aligned}
$$

where the order of summation has been interchanged and the orthonormality of the eigenvector components (9.24) implies that non-zero values are obtained only for $i = j$. This sum over the reflected intensities including the transmitted beam intensity does not show the subsidiary maxima of T and R but a deficient or defect band due to anomalous absorption (Figs. 9.12b and 9.13a).

This intensity distribution can be observed in transmission electron diffraction patterns by producing the pattern with conical illumination: the latter can be realized either externally, by using the convergent rays of an electron probe of large aperture or by diffuse scattering in the top layer of a thick crystal foil. In the former irradiation mode, the band is called a Kossel band and in the latter, a Kikuchi band. The rays of the illumination cone (Fig. 9.14b) that contribute to the intensity in the direction OP of the diffraction pattern are those transmitted by the foil ($\overline{\mathrm{P}}_0\mathrm{O}$) or Bragg diffracted from $\overline{\mathrm{P}}_g\mathrm{O}$ into OP and the intensity $\Sigma_g I_g$ of (9.39) will then be observed at P. The result is the so-called Kossel pattern, with defect Kossel bands (Fig. 9.14a). The Kossel lines of a Kossel map lie at the edges of these defect bands, that is, at $k_x/g = -0.5$ and $+0.5$ in the intensity profile across a band shown in Fig. 9.13a.

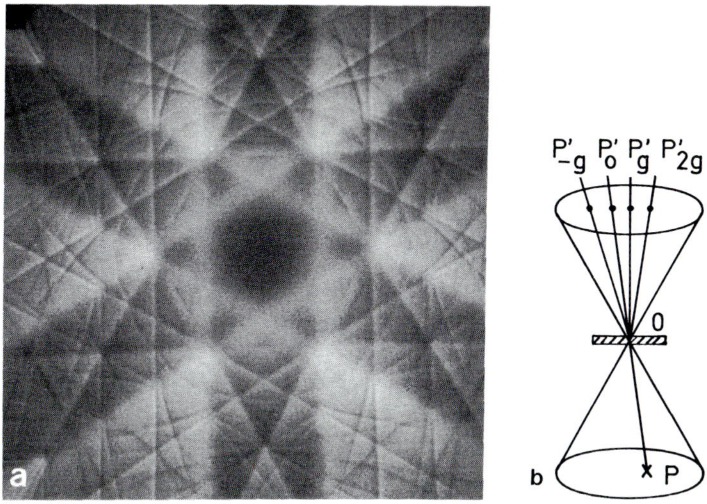

Fig. 9.14. (**a**) Transmission Kossel pattern of a 111-oriented Si foil recorded with 100 keV electrons [9.10]. (**b**) Generation of the Kossel pattern by a convergent illumination cone and superposition of the transmission and Bragg reflections from the directions $\overline{\mathrm{P}}_g\mathrm{O}$, $\overline{\mathrm{P}}_0\mathrm{O}$, $\overline{\mathrm{P}}_g\mathrm{O}$ and $\overline{\mathrm{P}}_{2g}\mathrm{O}$ into the direction OP

Defect Kikuchi bands of this type will also be observed in electron backscattering patterns (Sect. 9.3.3) when electrons, diffusely scattered at large depths, penetrate through the surface as backscattered electrons; see also the discussion of contrast reversal effects in Sect. 9.3.3.

9.1.4 Orientation Anisotropy of Electron Backscattering

We discuss the theory of the dependence of the backscattering coefficient on the tilt angle [9.11–17] for the simple case of single large-angle scattering in a film of thickness t, which is larger than the absorption lengths of the dynamical theory. This means that the primary Bloch-wave field is exponentially damped in the top layer of the film or single crystal. We further assume that electrons are scattered out of the Bloch-wave field as plane waves. We shall see, in Sect. 9.1.5, that this is not strictly correct and that each incident or scattered electron has to propagate in a Bloch-wave field inside a crystal. However, the presence of different scattering angles causes an averaging with the result that the plane-wave approximation can be used because we are not interested in the variation of $\mathrm{d}\eta/\mathrm{d}\Omega$ with the take-off angle (as shown in Fig. 9.18) but only in the influence of channelling on the total backscattering coefficient.

When the exponential attenuation terms are retained, the total Bloch-wave field can be described by (9.17) with $\epsilon^{(j)} = C_0^{(j)}$ for normal incidence. The probability density $P_{\mathrm{dep}}(\boldsymbol{r})$, (9.38), of the Bloch-wave field at the lattice planes can be obtained from $\Psi\Psi^*$ with $(\boldsymbol{g}-\boldsymbol{h})\cdot\boldsymbol{r}=$ integer, where we assume that the large-angle scattering processes are localized at the nuclei on the lattice planes. We also substitute $k_{0,z}^{(i)}-k_{0,z}^{(j)}=\gamma^{(i)}-\gamma^{(j)}$ and (9.17) can then be written

$$\begin{aligned}\Psi\Psi^* &= \sum_i\sum_j C_0^{(i)}C_0^{(j)}\left(\sum_g\sum_h C_g^{(i)}C_h^{(j)}\right)\\ &\times \exp[2\pi\mathrm{i}(\gamma^{(i)}-\gamma^{(j)})z]\exp[-2\pi(q^{(i)}-q^{(j)})z]\,. \end{aligned}\tag{9.40}$$

For the terms for which $g=h$, the double sum over g and h becomes δ_{ij} from the orthonormality (9.28) of the eigenvectors and we obtain

$$\begin{aligned}\Psi\Psi^* &= \sum_i\sum_j C_0^{(i)}C_0^{(j)}\sum_g\left(\delta_{ij}+\sum_{h\neq g} C_g^{(i)}C_h^{(j)}\right)\\ &\times \exp[2\pi\mathrm{i}(\gamma^{(i)}-\gamma^{(j)})z]\exp[-2\pi(q^{(i)}-q^{(j)})z]\,. \end{aligned}\tag{9.41}$$

At a depth z, the fraction (9.39) is still inside the Bloch-wave field and the remainder is scattered out of the Bloch-wave field as plane waves, the amplitudes of which are assumed for simplicity to be independent of $\boldsymbol{r}$. With $N=N_\mathrm{A}\rho/A$ atoms per unit volume and the scattering cross-section σ_B for backscattering through angles larger than 90°, a thin layer of thickness $\mathrm{d}z$ at depth z contributes to the backscattering coefficient with

$$\mathrm{d}\eta/\mathrm{d}z = N\sigma_\mathrm{B}\{\Psi\Psi^* + [1-\sum_j |C_0^{(j)}|^2\exp(-4\pi q^{(j)})z]\}\,. \tag{9.42}$$

When we integrate (9.42) over z to get the total backscattering coefficient, the unity term in the square bracket results in a linear increase of the backscattering coefficient, $\eta = N\sigma_\mathrm{B}t$, with increasing t, the solution to be expected for an amorphous or polycrystalline specimen. All the other terms in (9.42) cause deviations from this linear increase. If we subtract this linearly increasing term from (9.42), integration between $z=0$ and t gives the variation Δz of the backscattering coefficient or its orientation anisotropy. The second term in the square bracket of (9.42) cancels with the term of (9.41) that contains δ_{ij} and we obtain

$$\begin{aligned}\Delta\eta &= N\sigma_\mathrm{B}\sum_i\sum_j C_0^{(i)}C_0^{(j)}\sum_{g\neq h} C_g^{(i)}C_h^{(j)}\int_0^\infty \cos[2\pi(\gamma^{(i)}-\gamma^{(j)})z]\\ &\times \exp[-2\pi(q^{(i)}-q^{(j)})z]\mathrm{d}z\,. \end{aligned}\tag{9.43}$$

Formally, we may take $z=\infty$ as the upper limit of integration because we assume that the Bloch wave amplitudes become negligible at the depth t. As there are terms j,i for each i,j, the exponential functions containing the eigenvalues combine to give a cosine function. Integrating (9.43) and splitting the expression into terms for which $i=j$ and those for which $i\neq j$ gives in

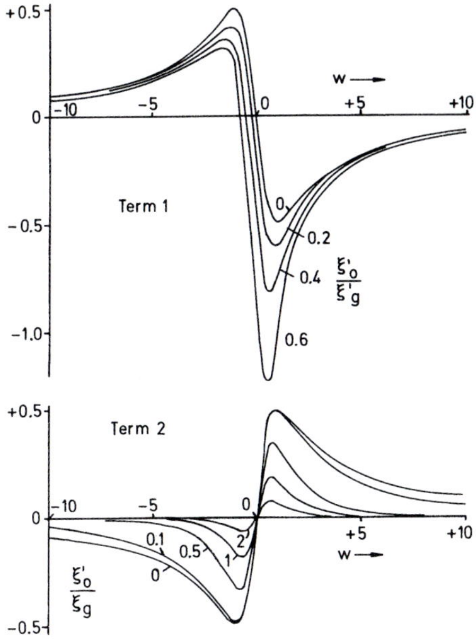

Fig. 9.15. Terms 1 and 2 of equation (9.45); these contribute to the anisotropy of electron backscattering in the two-beam case of the dynamical theory [9.14]

$$
\begin{aligned}
\Delta\eta &= N\sigma_{\mathrm{B}} \sum_i |C_0^{(i)}|^2 \sum_{g\neq h} C_g^{(i)} C_h^{(i)} / 4\pi q^{(i)} \\
&+ \frac{N\sigma_{\mathrm{B}}}{2\pi} \sum_{i\neq j} C_0^{(i)} C_0^{(j)} \sum_{g\neq h} C_g^{(i)} C_h^{(j)} (q^{(i)} + q^{(j)}) / [\Delta\gamma_{ij}^2 + (q^{(i)} + q^{(j)})^2] \,.
\end{aligned} \tag{9.44}
$$

The first term corresponds to the independent Bloch-wave model discussed in Sect. 8.1.3 while if we use both terms, we obtain the solution for the dependent Bloch-wave model [9.14].

All the quantities occuring in (9.44) are known once the eigenvalue problem of the dynamical theory has been solved. To get an idea of how $\Delta\eta$ varies as a function of the tilt parameter w, we substitute the values for the two-beam case from (9.27), (9.28 and (9.36) in (9.44) and find [9.14]

$$
\Delta\eta = \frac{N\sigma_{\mathrm{B}}}{2\pi}\xi_0' \left(\underbrace{-\frac{w + \xi_0'/\xi_g'}{1 + w^2 - (\xi_0'/\xi_g')^2}}_{\text{1st term for } i = j} + \underbrace{\frac{w}{1 + w^2 + [(1 + w^2)\xi_0'/\xi_g)]^2}}_{\text{2nd term for } i \neq j} \right) . \tag{9.45}
$$

The two terms are plotted against w in Fig. 9.15a,b for various values of the parameters ξ_0'/ξ_g' and ξ_0'/ξ_g, respectively. In practice, we always have $\xi_0'/\xi_g' > 1$ and $\xi_0'/\xi_g > 1$ (Table 8.2). For the latter, we see in Fig. 9.15 that the second term can be neglected and that the independent Bloch-wave model represented by the first term is a good approximation. In Fig. 9.15a this term shows a typical anisotropy with an increase of the backscattering

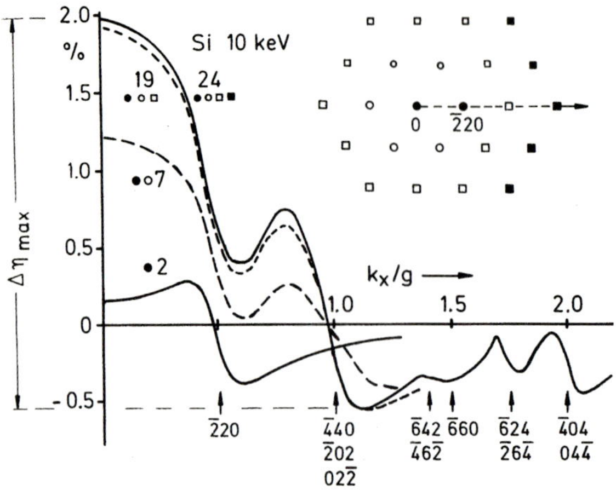

Fig. 9.16. Calculated variation $\Delta\eta$ of the backscattering coefficient from the centre of the 111 pole along the indicated line normal to the $\bar{2}20$ Kikuchi band and parallel to $\boldsymbol{g} = \bar{2}20$ (see also Fig. 9.23a) for an increasing number $n = 2$–24 of beams in the dynamical theory

coefficient ($\Delta\eta$ positive) for negative w (reciprocal lattice point outside the Ewald sphere) and a decrease ($\Delta\eta$ negative) for positive w.

Equation (9.45) and Fig. 9.15 can also be used to estimate the width of a channelling profile in the two-beam approximation corresponding to a high-order reflection. A width Δw in Fig. 9.15 corresponds to

$$\Delta w = \Delta s \xi_g = g \Delta\theta \xi_g \to \Delta\theta = \frac{\Delta w}{g s_g} = \frac{d_{hkl} \Delta w}{\xi_g} . \tag{9.46}$$

The angular width $\Delta\theta$ therefore decreases with increasing hkl owing to the decrease of d_{hkl} and the increase of ξ_g and high-order Kikuchi lines become sharper than the low-indexed Kikuchi lines, which overlap as Kikuchi bands. This can be seen in theoretical calculations (Fig. 9.16) and experiments (Figs. 9.14a, 9.23a,b), especially for the fine lines from high-order Laue zones near the 111 pole (Fig. 9.23c), which are discussed in Sect. 9.2.4. The problem is to define Δw in terms of some width occurring in the dynamical two-beam theory of Fig. 9.15. A value $\Delta w = 2$ may be too optimistic [9.18].

Across a systematic row of Bragg reflections $-ng, \ldots, -g, 0, +g, \ldots, +ng$, the $-g$ and $+g$ anisotropies overlap to form a bright excess band (Fig. 9.13b). The Kossel cones cut the diagram at tilts $k_x/g = -0.5$ and $+0.5$. Three of the 220 bands that cross the 111 pole of an electron channelling pattern (Sect. 9.2) can be seen in Fig. 9.23. At the centre of the 111 pole of silicon, for example, the contributions from all the Bragg reflections sum up to a very bright maximum. Fig. 9.16 shows the intensity distribution along the line starting at the 111 pole and normal to the 220 Kikuchi band also indicated in Fig. 9.23a and calculated by the dynamical n-beam theory. This shows that the calculated intensity settles down for $n > 20$ and that the two-beam case is a very bad approximation.

The maximum variation $\Delta\eta_{\max}$ between the extrema in Fig. 9.16 has been measured in electron channelling patterns at different primary electron ener-

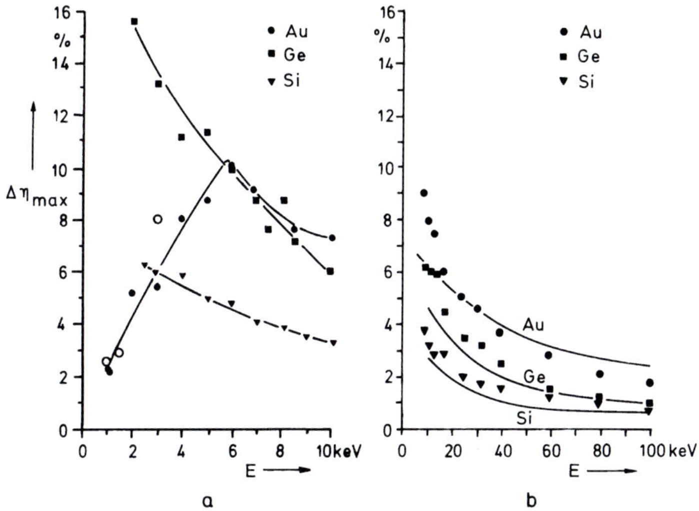

Fig. 9.17. Dependence of $\Delta\eta_{\max}$ measured at the 111 pole of Si, Ge and Au for (**a**) low and (**b**) high electron energies

gies [9.13–15, 19] and is plotted in Fig. 9.17. The decrease of the orientatioin anisotropy $\Delta\eta_{\max}$ with increasing E can be explained by the prefactor of (9.45). When we assume Rutherford scattering, σ_B decreases as E^{-2} with increasing E. The absorption length ξ_0' increases as $v^2 \propto E$, with the result that $\Delta\eta_{\max} \propto E^{-1}$, in rather good agreement with experiments for energies in the range $E = 10$–100 keV.

When $\Delta\eta_{\max}$ is calculated from (9.44) and only single scattering through $\theta \geq 90°$ is used for the calculation of σ_B, the observed values of $\Delta\eta_{\max}$ are larger than the calculated ones by a factor 1.5–4. One reason for this disagreement is that Mott scattering cross-sections (Sect. 3.1.6) should be used instead of Rutherford cross-sections for calculating σ_B [9.14, 15]. For example, the decrease of $\Delta\eta_{\max}$ that is observed for gold at low energies (Fig. 9.17a) can only be explained by Mott scattering theory [9.15]. Furthermore, the theory has to be modified to take into account multiple scattering and diffusion. Owing to the small absorption length of the primary Bloch-wave field, that part of σ_B which is directly scattered out of the Bloch-wave field through angles larger than 90° will be correct. With increasing backscattering, at negative w for example, there is also more scattering through angles smaller than 90°, however, and these electrons have a larger probability of escaping as BSE after multiple scattering. Likewise, the low-angle scattering is also decreased for positive w. This can be allowed for by increasing the effective value of σ_B, either by using values obtained by Monte Carlo simulations together with experimental results about backscattering [9.14, 15] or by employing a multiple scattering theory [9.16, 17].

For non-normal incidence of the primary beam on the single-crystal surface, the relation $\epsilon^{(j)} = C_0^{(j)}$ for the excitation amplitude of the Bloch waves in (9.17) has to be modified and the excitation points M_j on the branches of the dispersion surface (Fig. 9.8) lie on straight line parallel to the surface normal. This leads to asymmetry of the Kikuchi band profiles of a systematic row, for example [9.8, 12, 16].

9.1.5 Angular Anisotropy of Electron Backscattering

We have seen in the last section that rocking the incident electron beam causes the probability density $P(\boldsymbol{r})$ of the Bloch-wave field at the nuclei to vary: there is a high probability of back- and forward scattering when $P(\boldsymbol{r})$ is large or when there are antinodes at the nuclei and a low probability of scattering when there are nodes at the nuclei. These scattered electrons contribute to the attenuation of the Bloch-wave field. Inside the crystal, however, they have also to propagate as Bloch waves in the scattering direction to conform to the crystal periodicity. The consequence is that the angular distribution of the scattered electrons will not vary monotonically with increasing θ, in proportion to $\mathrm{d}\eta/\mathrm{d}\Omega$; instead, this angular distribution will be modulated by excess or defect Kikuchi bands. The behaviour of the differential backscattering coefficient $\mathrm{d}\eta/\mathrm{d}\Omega$ that is shown for polycrystalline films in Fig. 4.6 results in a strongly modulated variation for a single-crystal Au film (Fig. 9.18).

We first explain the excess band contrast, which can be observed in pure form for very thin films for which strong multiple scattering can be ignored. The intensity scattered into the direction Q from an incident plane wave from P (source) (Fig. 9.19a) will be proportional to $\Psi_{\mathrm{in}}\Psi_{\mathrm{in}}^*$ at the nuclei. Due to the

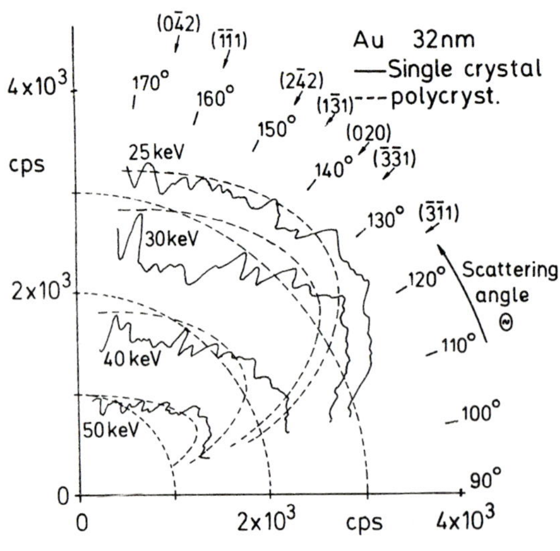

Fig. 9.18. Angular distribution of BSE from a 30 nm Au single-crystal film plotted as a polar diagram [9.20]

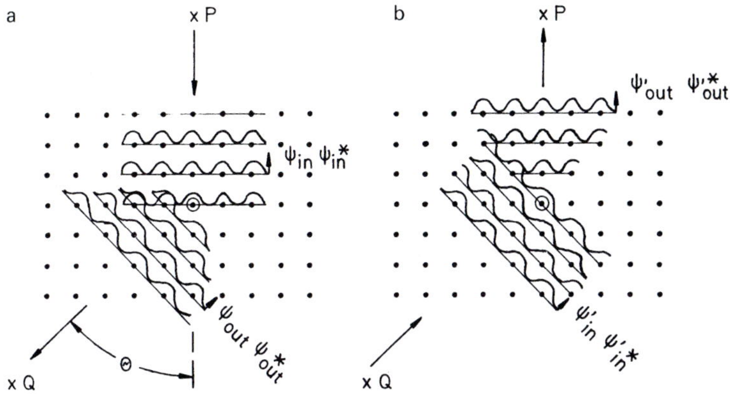

Fig. 9.19. Theorem of reciprocity for the large-angle scattering of electrons from (**a**) a primary Bloch-wave field Ψ_{in} to a secondary Bloch-wave field Ψ_{out} and (**b**) in the reverse direction from Ψ'_{in} to Ψ'_{out}

localization of large-angle scattering near the nuclei, the scattered wave will also show a large probability density at the nuclei; the condition is fulfilled by a Bloch wave Ψ_{out} that also has antinodes at the nuclei. In other words, there can only be scattering to Q with a large probability when the orientation OQ relative to the crystal allows a Bloch-wave field that shows strong antinodes at the nuclei to propagate in this direction. The question whether there are antinodes or nodes can be solved by considering a plane incident wave with Q as source propagating in the reverse direction QO (Fig. 9.19b). A consequence is the theorem of reciprocity [9.21, 22], which states that the intensity at Q with P as source is the same as that at P with Q as source:

$$I_{\mathrm{P}} = I_{\mathrm{Q}} = \Psi_{\mathrm{in}}\Psi_{\mathrm{in}}^{*}\frac{\mathrm{d}\sigma_{\mathrm{el}}(\theta)}{\mathrm{d}\Omega}\Psi_{\mathrm{out}}\Psi_{\mathrm{out}}^{*} = \Psi'_{\mathrm{in}}\Psi'^{*}_{\mathrm{in}}\frac{\mathrm{d}\sigma_{\mathrm{el}}(\theta)}{\mathrm{d}\Omega}\Psi'_{\mathrm{out}}\Psi'^{*}_{\mathrm{out}} \,. \tag{9.47}$$

A similar effect can be observed in the emission of β-particles from a nucleus in a single crystal; these too can be emitted and propagate only as a Bloch-wave field with antinodes at the emitting nucleus [9.23, 24]. All these mechanisms result in excess Kikuchi bands with a band profile as shown in Fig. 9.13b. The theorem of reciprocity will be further discussed in Sect. 9.3.3 in connection with the reciprocity between electron channelling patterns (ECP) and electron backscattering patterns (EBSP).

Electrons that are scattered in a transparent film through small angles form a cone of diffusely scattered electrons and result in defect Kikuchi bands, as discussed at the end of Sect. 9.1.3; such bands can be seen at the centre of Fig. 9.20. With increasing foil thickness, the angular width of the cone of diffusely scattered electrons increases and the central area with defect bands extends to larger angles [9.25–28].

At the transition from the inner defect to the outer excess region of the Kikuchi band pattern, the cancellation of the different terms of ΣI_g in (9.39),

Fig. 9.20. 100 keV electron transmission pattern from an 80 nm thick silicon crystal foil with excess Kikuchi bands at high scattering angles and defect bands at low scattering angles and bright Kikuchi lines at intermediate angles [9.10]

which leads to defect bands, is not complete. The Kossel cone that has more electrons will be Bragg-scattered to the opposite cone (Fig. 9.4). In other words, the small number of electrons scattered from the opposite cone to the first one result in a net contrast, which is seen as a dark or a bright Kikuchi line where the first or opposite cone, respectively, intersects the observation plane. These sharp dark and bright Kikuchi lines can be seen in Fig. 9.20 at medium scattering angles.

In the case of a tilted bulk single crystal, those electrons which are scattered out of the primary Bloch-wave field by single scattering will contribute to the excess bands (Fig. 9.21). In the forward direction of the reflection-like angular distribution $d\eta/d\Omega$ of the BSE (Fig. 4.12), electrons that are multiply scattered at large depths penetrate the surface as a diffuse cone and result in the transmission defect bands discussed above. For increasing energy loss, the width of these bands increases and the band contrast will be blurred. This mechanism explains satisfactorily the observed appearance of excess and defect bands [9.10, 28, 29]. For quantitative calculations, it will be necessary to know the dependence on take-off angle of both contributions, in order to see which contribution dominates.

9.1.6 Transmission and Reflection Electron Diffraction Patterns

Electron diffraction is a powerful analytical method in TEM for determining the crystal structure and orientation. The cone of $\simeq 5°$ aperture around the primary beam with an illumination aperture $\alpha_i \leq 1$ mrad contains Bragg diffraction spots and the electrons scattered diffusely between the diffraction spots by electron–phonon (thermal diffuse) scattering and by core and valence electron excitations generate Kikuchi lines and bands (Fig. 9.20). When using an electron probe of small diameter ($\leq 0.1\ \mu$m) with a larger illumination

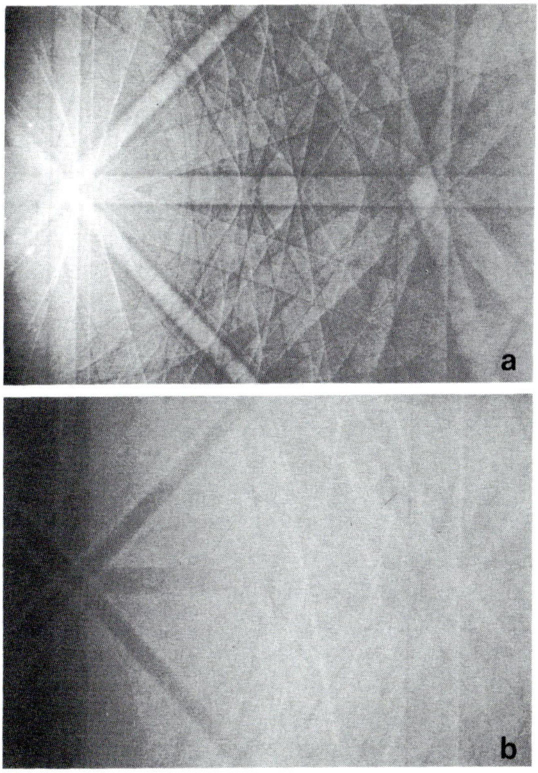

Fig. 9.21. 100 keV electron backscattering pattern (EBSP) from a bulk Ge single crystal tilted through (**a**) $\phi = 75°$ and (**b**) $\phi = 65°$ and recorded on a photographic emulsion situated below the crystal and normal to the electron beam (see Fig. 9.27a). The pattern (**b**) shows the contrast reversal from defect to excess bands with increasing take-off angle from left to right [9.10]

aperture (1–10 mrad), the primary beam and the diffraction spots spread out to circles and this convergent-beam electron diffraction (CBED) pattern contains an intensity modulation, which is a direct image of the rocking curves of dynamical theory. If the aperture is further increased ($\geq$ 10 mrad), the circles of the primary beam and the Bragg spots overlap. The result is a cancellation of the pendellösung fringes. Nevertheless, such a Kossel diagram (Fig. 9.14) still contains an intensity modulation in the form of defect Kikuchi bands, caused by anomalous absorption of the Bloch-wave field and therefore provides information about crystal symmetry and orientation, the latter because the system of Kikuchi bands or Kossel cones is fixed to the crystal. Such a pattern can also be observed behind foils which would normally be too thick for TEM because of the chromatic aberration of the objective lens. In this case, the upper part of the foil acts as a diffuser for the electron beam and the Kossel pattern represents the lower part of the foil, from a depth of the order of the mean absorption length of dynamical theory.

The stationary diffraction patterns produced by a stationary beam are normally observed on a fluorescent screen and recorded on photographic emulsions or by a CCD. Alternatively, deflection coils, often known as Grigson coils, situated below the specimen may be used to scan the diffraction pattern across a small detector diaphragm [9.30, 31]. This method allows the diffraction intensities to be recorded directly and presented in Y-modulation on a TV screen. Besides these techniques with stationary beams or electron probes and stationary diffraction patterns, techniques using a rocking beam can also be used in transmission (see [9.10] for further details).

In SEM, electron diffraction in transmission is uncommon because it is mostly thick specimens that are investigated and if the specimen can be traversed by electrons, the useful thickness must be less than in TEM owing to the lower electron energy, typically 10–50 keV. Normally, such specimens will be investigated more successfully in a 100–300 keV TEM.

Nevertheless, this possibility should not be forgotten in SEM. For example, a transmission electron diffraction tube with scanning coils can be mounted below the specimen chamber (Fig. 8.14d). However, it is not possible to produce electron probes having the small aperture ($\leq$0.1 mrad) needed to produce a conventional diffraction pattern without inreasing the illuminated area on the specimen. The electron-probe diffraction techniques mentioned above can, however, be used and, in particular, a Kossel pattern can be observed with much thicker specimens than those used for conventional electron diffraction.

Reflection high-energy electron diffraction (RHEED) for the investigation of bulk single-crystal specimens is another possibility, which can also be employed in SEM when the latter is equipped with a fluorescent screen, a photographic emulsion or a CCD camera. Bragg diffraction spots from bulk material, which can be seen at the right of Fig. 9.21a, can only be obtained from lattice planes nearly parallel to the surface and at tilt angles of the surface normal very close to 90°. As in conventional diffraction, Bragg diffraction spots will be seen only if the illumination aperture is small, it should not exceed 0.1 mrad. The RHEED pattern also contains Kikuchi lines and bands. Such a RHEED pattern is very sensitive to contamination layers because only the first few lattice planes contribute to the pattern.

When the tilt angle of the surface normal becomes smaller than 80°, Bragg diffraction spots can no longer be observed (Fig. 9.21b). The diffraction pattern becomes an electron backscattering pattern (EBSP), which will be discussed in Sect. 9.3. Unlike the case of RHEED patterns, the illumination aperture can be larger, which means that the electron beam can be focused to a small electron probe at the specimen surface.

Decreasing the electron energy increases the Bragg angles. At $E = 1$ keV, 220 lattice planes of Si ($d_{220} = 1.6$ nm) result in $2\theta_{\mathrm{B}} = 45°$. A low-energy electron diffraction (LEED) pattern can be observed on a fluorescent screen after enhancing the contrast by energy-filtering of elastically scattered electrons using retarding grids in front of the screen. This diffraction pattern contains information about the two-dimensional lattice of surface reconstructions caused by the dangling bonds, which will be different from the bulk crystal structure. Such a pattern can be observed in a LVSEM with ultra-high vacuum on a fluorescent screen coupled to a CCD camera [9.32, 33].

9.2 Electron Channelling Patterns (ECP)

9.2.1 Electron-Optical Parameters for the Recording of ECP

We have seen in Sect. 9.1.4 that the channelling of Bloch waves results in a variation of the backscattering coefficient η (orientation anisotropy) when the specimen is tilted or the incident beam rocked. In order to record an electron channelling pattern (ECP) (Figs. 9.23 and 9.24), the angle of incidence or rocking angle γ must be varied in a raster [9.34], so that each point (x, y) of the synchronously scanned CRT corresponds to an angle of incidence with components (γ_x, γ_y) where $\gamma = (\gamma_x^2 + \gamma_y^2)^{1/2}$ is the rocking angle relative to optic axis. For reviews about ECP see [9.35–38].

The following optical parameters are important when producing an ECP and must be chosen correctly to obtain the best results with the various methods described in Sect. 8.2.2:

d_p	=	electron-beam diameter,
α_p	=	electron-probe aperture,
I_p	=	electron-probe current,
$2\gamma_\mathrm{max}$	=	maximum rocking angle, and
D_f	=	diameter of the scanned area contributing to the ECP.

The electron-beam diameter d_p determines the spatial resolution of the surface topography when rocking is combined with specimen scanning in the large-area mode (standard method). In this case the diameter of the scanned area D_f can be a few millimetres but even in the selected-area modes D_f will be larger than d_p because of the spherical aberration. The probe aperture α_p limits the angular resolution of the ECP. Owing to the sensitivity of the orientation anisotropy to small tilts of the electron beam relative to the lattice planes, the aperture angles must be small enough to provide the desired angular resolution of the ECP, which implies $\alpha_\mathrm{p} \leq 1$ mrad for medium and ≤ 0.1 mrad for high angular resolution. The normal electron-probe aperture in the imaging mode, of the order of a few milliradians, is not small enough to resolve the band structure of the ECP. For example, the Bragg angle θ_B for a 220 Bragg reflection in Si ($a = 0.543$ nm, $d_{220} = 0.192$ nm) at $E = 20$ keV ($\lambda = 8.586$ pm) is 22 mrad.

The electron-probe current is important, for it must be high enough to provide the signal-to-noise ratio needed for detecting small variations of $\Delta\eta/\eta$ of the order of 1–5% (Fig. 9.17). With N incident electrons per pixel, the statistical variation due to the shot noise is $N^{1/2}$. The signal-to-noise ratio $\propto N^{1/2}$ should be larger than 1000, which corresponds to $N > 10^6$ recorded

BSE or 5×10^6 primary electrons per pixel if we assume $\eta = 20\%$, for example. This in turn requires 5×10^{12} electrons per frame with 10^6 pixels. This implies that the electron-probe current must be $I_p = 8 \times 10^{-9}$ A for a frame time $T_f = 100$ s.

However, the quantities I_p, d_p and α_p cannot be varied independently and are related by the gun brightness (2.27):

$$\beta = 0.4 I_p / \alpha_p^2 d_p^2 \,. \tag{9.48}$$

For a thermionic cathode, $\beta = 5 \times 10^4$ A cm^{-2} sr^{-1} at $E = 20$ keV. For the typical conditions listed above, $I_p \simeq 10^{-8}$ A, $T_f = 100$ s, this results in $\alpha_p d_p \simeq 3$ mrad μm, which means that only combinations such as

$$\begin{array}{lllll} \alpha_p & = & 0.1\,\text{mrad} & 1\,\text{mrad} & 10\,\text{mrad} \\ d_p & = & 30\,\mu\text{m} & 3\,\mu\text{m} & 0.3\,\mu\text{m} \end{array} \tag{9.49}$$

are attainable. The first combination results in an excellent angular resolution of the ECP but modest spatial resolution. The last gives a better spatial resolution but completely inadequate angular resolution.

The lens excitations that furnish aperture angles $\alpha_p \leq 1$ mrad are those that provide as parallel a beam as possible (collimated beam condition) between the final lens and the specimen [9.39]. Not all present-day SEMs provide the necessary independent variation of the lens excitations over a large range

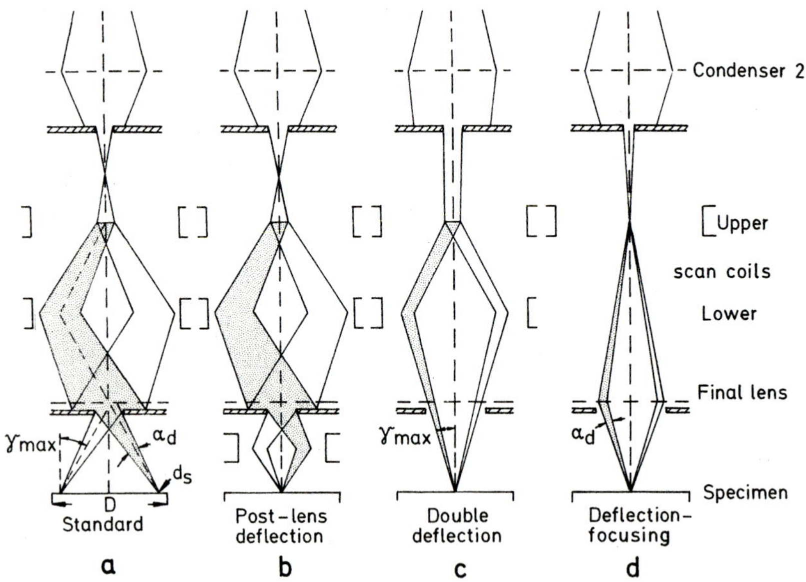

Fig. 9.22. Four different ways of recording electron channelling patterns by (**a**) the standard, (**b**) the post-lens deflection, (**c**) the double-deflection and (**d**) the deflection-focusing methods ($\alpha_d = \alpha_p$ = electron-probe aperture)

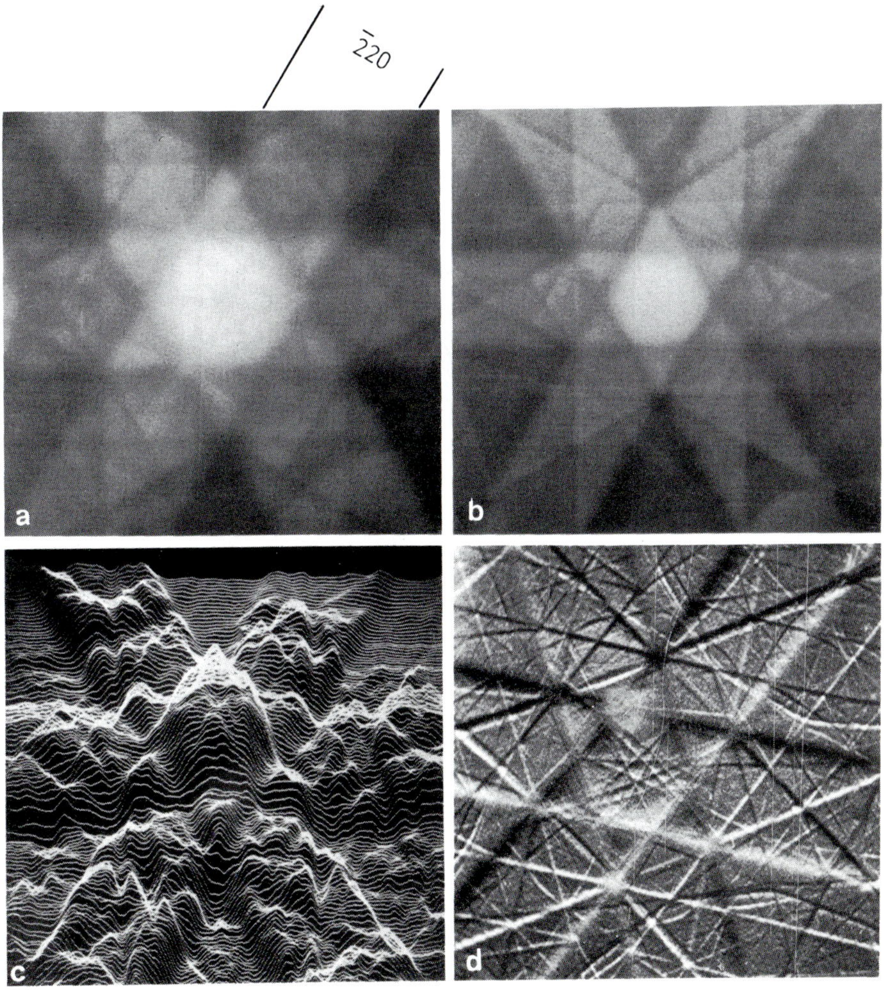

Fig. 9.23. Electron channelling pattern (ECP) of a silicon single crystal with the 111 pole parallel to to the electron beam recorded at (**a**) 10 keV and (**b**) 20 keV. (**c**) Pattern (a) displayed in the Y-modulation mode and (**d**) the differentiated signal of (b) showing the fine structure inside the pole with better contrast

of excitation currents. However, this is not a real handicap because the tendency is to use electron backscattering patterns (EBSP) for crystallographic information whenever possible (see the advantages discussed in Sect. 9.3.2).

9.2.2 The Standard and Tilt Method of Generating ECP

Standard Method. The simplest way of recording an ECP is the standard or large-area method [9.34], in which a large single crystal is scanned at

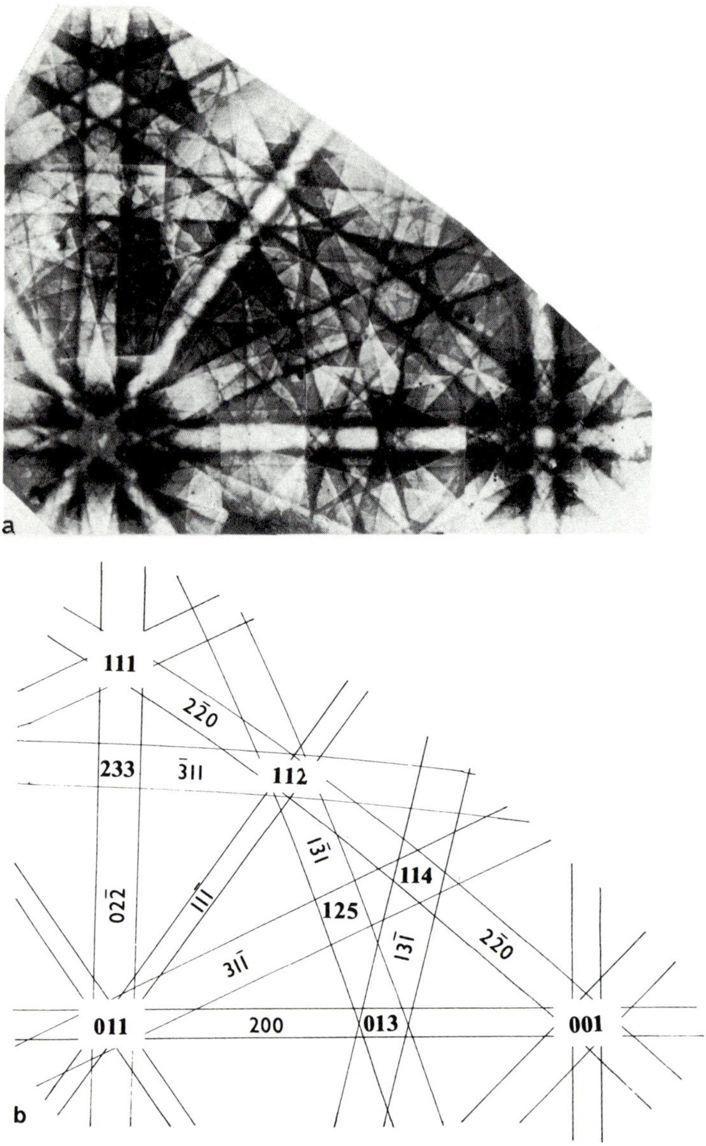

Fig. 9.24. (**a**) Panorama diagram of copper ($E = 20$ keV) composed of successively tilted and overlapping ECPs. The reduced computer map (**b**) of the low-indexed Kikuchi bands is useful for identifying the poles at the intersection of two bands. The indices of the poles are obtained by forming the vector product of the *hkl* of two intersecting bands [9.50]

low magnification (Figs. 9.22a and 9.23a,b). The angle of incidence varies across the scanned area and each point of the crystal surface corresponds to a particular pair of values of γ_x, γ_y. The maximum rocking angle γ_{max} can be estimated from the relation

$$\gamma_{max} = \frac{D}{2w} = \frac{1}{2w}\frac{B}{M} \tag{9.50}$$

where w denotes the working distance, D the width of the scanned area, B the width of the CRT screen and M the magnification. A working distance $w = 1$ cm and $M = 20$ give $\gamma_{max} = \pm 14° \simeq \pm 0.25$ rad and a width $D = 5$ mm of the scanned area contributing to the ECP. This method can therefore be employed only with large perfect single crystals such as Si, Ge or GaAs wafer crystals with and without epitaxial layers. D and γ_{max} are inversely proportional to M. If the surface is flat and if the BSE signal is uniform, the video signal is modulated by the orientation anisotropy of the backscattering coefficient.

In this method (Fig. 9.22a), the scan coil system is excited as in the conventional imaging mode and the plane of the final-lens diaphragm is a pivot point of rocking. An angular resolution of the order of 0.1 mrad can be achieved [9.40–42], which permits the fine structure from the high-order Laue zone reflections inside the 111 pole of Si, for example, to be resolved (Fig. 9.23d). In this mode, the diameter of the electron beam at the specimen is of the order of the diameter of the final-lens diaphragm and topographical resolution will not be possible when the angular resolution is high. By increasing the beam divergence, the electron-probe diameter d_p can be reduced and it then becomes possible to observe topographic contrast superimposed on the ECP.

Specimen Tilt Method. A disadvantage of the standard method is obviously the large scanned area of a few millimetres. One solution is to use a stationary electron probe and tilt the specimen mechanically in synchronism with the CRT beam [9.41]. Such specimen stages have the advantage that the maximum rocking angle can be increased and tilt ranges of $120° \times 90°$ have been achieved [9.43, 44]. Two scintillation detectors opposite one another and an electronic circuit that keeps the main signal constant have to be used to avoid shadowing effects and a strong signal gradient when the specimen is tilted through such large angles. The ECP was recorded not by a pen but by a small light-emitting diode mounted on an x-y recorder, which was scanned over a photographic emulsion.

9.2.3 Small-Area Methods for ECP

We now describe various electron-optical methods of rocking the electron probe on as small a specimen area as possible; the pivot point of rocking is in the specimen plane. They are again limited in their maximum rocking angles γ_{max}.

Post-Lens Deflection Method. An additional scan coil system is placed between the polepiece of the final lens and the specimen and used to deflect the scanning beam of the standard method on axis at the specimen (Fig. 9.22b). This method has the disadvantage that the working distance has to be large and is adversely affected by the aberrations of the deflection unit. Similar problems arise when a post-lens double-deflection system is used for generating stereo pairs without mechanical tilting of the specimen [9.45, 46].

Double-Deflection Method. Both pairs of scan coils inside the final lens are used, as in the conventional imaging modes, but the last coils and the final lens are less strongly excited so that the pivot point, which lies at the plane of the last diaphragm in the imaging mode and the standard method, now moves down to the specimen (Fig. 9.22c) [9.42, 47]. The aperture diaphragm has to be removed or to be of the order of a few millimetres so that the full range of deflected electrons can pass through. The electron-probe aperture or beam divergence, which is normally limited by the final-aperture diaphragm, must be controlled by a diaphragm in the upper part of the column. Figure 9.22c shows the ray diagram for the case in which the final lens is switched off.

Though the double-deflection method has the advantage that the scanned area is much smaller than in the standard method, the electron beam cannot be rocked at a fixed position because the deflection systems suffers from spherical aberration. At the plane of least confusion, the electron beam at $\gamma_{\max} = \pm 5° \simeq 0.1$ rad is deflected through a distance $d_s = 0.5 C_s \gamma_{\max}^3$ (2.16) and the minimum selected area becomes 50–100 μm in diameter. Unlike the standard method. where the electron probe scans in a raster across the specimen, the electron beam here scans across a caustic cross-section. A decrease of $\gamma_{\max}$ reduces the minimum selected area since the third power of $\gamma_{\max}$ is involved. However, if the rocking angle is too small, the identification and indexing of an ECP becomes more difficult. For this reason, the double deflection method has not been widely used.

Deflection-Focusing Method. In this method, the aperture or beam divergence is again limited by a small diaphragm in the upper column. In contrast to the standard and double-deflection methods, the lower scan coils are switched off. The final lens deflects the beam on axis when the specimen plane and the plane of the upper scan coils are conjugate (Fig. 9.22d) [9.48]. As in the double-deflection method, the final-aperture diaphragm must be of the order of a few millimetres in diameter. If the excitation of the condenser lenses is chosen so that a demagnified image of the crossover is formed at the plane of the upper scan coils, the electron probe diameter at the specimen will be optimum. Furthermore, it is very easy to switch from the rocking mode to the imaging mode merely by switching on the lower scan-coil system. In this mode, condenser lens 2 determines the focus while the excitation of condenser lens 1 governs the electron-probe size. Instruments in which the first two lenses are ganged together are not flexible enough to permit this technique to be used in optimal conditions.

The minimum selected area is smaller than in the double-deflection mode. In the deflection-focusing mode, the spherical aberration of the final lens rather than that of the scan coils is the limiting factor. At a working distance $w = 10$ mm, values of $D \simeq 10$ μm are obtainable, which can be reduced to $\simeq 3\mu$m by using the smallest possible working distance [9.49]. In the latter case, the detection of BSE becomes difficult but it is possible to record the ECP with the specimen current mode (Sect. 6.2.8) because relatively large values of I_{p} are used to increase the signal-to-noise ratio as discussed in Sect. 9.2.1.

The area selection can be performed by using a dynamic correction method [9.48]. Unlike the situation in the normal imaging mode, only a fine pencil of rays passes through the lens at any one time. The final lens current and the focal length can therefore be changed to suit the deflection angle γ in such a way that the shift of the electron probe by spherical aberration is compensated. When using the conventional raster, the focal length has to be changed by $\Delta f = C_{\text{s}}\gamma^2 = C_{\text{s}}(\gamma_x^2 + \gamma_y^2)$ along each line. This needs a special lens design capable of following such rapid changes without hysteresis. When a spiral raster is used, the total change of defocus is spread over a complete frame. Selected areas of the order of 1 μm can be obtained at $E = 30$ keV [9.51].

9.2.4 Recording of Electron Channelling Patterns

The recording of ECP is hampered by the shallow information depth, of the order of a few nanometres, and by the fact that the variation of $\Delta\eta/\eta$ is so small, of the order of a few percent (Fig. 9.17).

The contrast of ECP is often perturbed by thin surface layers in the form of oxide and contamination layers and by strong deformations of the surface layer after mechanical polishing (Beilby layer). Surface layers cause a broadening of the angular distribution by multiple scattering and destroy all the effort made to achieve a small electron aperture or beam divergence discussed in Sect. 9.2.2,3. This increase of the effective aperture when the electron beam reaches the signal–crystal surface blurs the excess band contrast and worsens with decreasing electron energy [9.52–54]. The contrast of ECP also decreases with increasing dislocation density (Sect. 9.2.4) and the distortion after mechanical polishing can be so high that no channelling contrast can be seen. Careful cleaning of the surface is therefore necessary, by chemical etching, electrolytical polishing or ion-beam etching. The first of these can be applied to Si and Ge using 90% HNO_3, 10% HF, for example. Electropolishing is a good method for metals because it also smooths the surface thereby eliminating overlapping topographic contrast. However, nearly every metal needs a different electrolyte. Ion-beam etching with argon ions of a few keV, for example, has the advantage of being applicable to any material although the sputtering rate can differ by orders of magnitude. It is necessary to bombard the specimen with a low angle of 10°–20° between ion beam and surface and to rotate the specimen around its normal to avoid the formation of hillocks and other preferential etching effects typical of ion-beam etching with a fixed beam direction. For Si and Ge, care must be taken to ensure that the ion-beam etching does not render the surface amorphous and for ionic crystals, that point defects are not created. The amorphization of sapphire single crystals, for example, by ion implantation can be investigated by varying the electron energy to get information about the depth of the amorphized zone [9.55]. Ion-beam etching can also be successfully employed for observing the crystal orientation or channelling contrast (Sect. 6.2.3) when imaging individual grains in polycrystalline materials [9.56].

The electron bombardment of ionic crystals, alkali halides for example, leads to the formation of colour centres and defect clusters and eventually to a degradation of the ECP contrast. The contrast reappears after annealing, NaCl at 300°C and KCl at 210°C, for example [9.57, 58].

As we have already discussed in Sect. 9.2.1, small variations of $\Delta\eta/\eta$ of a few percent need a large electron-probe current. It is also important to collect as many BSE as possible. Usually, ECP are recorded with normal incidence of the electron beam. The most effective type of detector will therefore be an annular semiconductor detector [5.31] or a scintillator sheet below the polepiece. Such detectors allow ECP to be recorded in the standard method without any signal gradient across the scanned area. The number of excited electron-hole pairs or light quanta is proportional to the BSE energy, which increases the contrast because BSE that contribute to the channelling contrast are directly backscattered inside the primary Bloch-wave field with low energy losses. Thus the channelling contrast can also be increased by energy filtering of the BSE (Sect. 6.2.6).

The signal amplitude $\Delta I/I \propto \Delta\eta/\eta$ being low, electronic suppression of the constant background by black level control is necessary (Sect. 5.5.2) (Fig. 9.23a,b). For this to be effective, overlapping topographic and other contrast and any signal gradient across the scanned area must be small. Another efficient method is to record the differentiated signal [9.59] (Fig. 9.23d). This has the advantage that sharp lines of low contrast, from high-order Laue zones (Sect. 9.2.4) inside the 111 pole, for example, can be imaged with a better contrast, while all slowly varying contrast is suppressed. Another way of displaying ECP is to use the Y-modulation mode, which allows rocking curves to be recorded directly as linescans or as a two-dimensional rocking map as in Fig. 9.23c.

9.2.5 Information in Electron Channelling Patterns

Crystal Structure. Determination of crystal structure requires a knowledge of the symmetry elements at the very least. The angular size of the ECP is too small to extract such information from a single pattern. It is necessary to assemble a panorama diagram (Fig. 9.24) based on successive tilts of the specimen, looking at the poles showing 2-, 3- or 4-fold axes and at mirror planes. Obtaining a panorama diagram of cubic crystals is straightforward because of the high symmetry and the small size of the unit triangle (Figs. 9.7 and 9.24). Since the ECP are gnomonic projections, it may be difficult to join tilt series over too large an angular range.

Crystal Orientation. The orientation of a crystal is determined once the crystallographic direction of the surface normal and one direction in the specimen plane are known. The centre of the ECP corresponds to the direction in the specimen plane without rocking. For accurate measurements, this direction has to be established by means of a Si wafer that is optically adjusted exactly normal to the electron beam, for example. Such a pattern also allows every point on the CRT to be associated with a rocking direction γ_x, γ_y and so distances on the pattern to be calibrated in radians. It is also necessary to check the linearity of the CRT or digital display, which can introduce errors of the order of 10% [9.60], by imaging a cross-ruled grating [9.61]. Any rotation of the ECP relative to the imaged area when changing from the imaging to the diffraction mode has also to be taken into account [9.62–64].

The width of a Kikuchi band corresponds to $2\theta_B$ and allows the lattice-plane distance d_{hkl} to be determined from the Bragg condition (9.10). This value can be used to index the band by comparing the measured value with a list of possible hkl. However, the measurement of a width at the positions $k_x/g = \pm 0.5$ in Fig. 9.13b is not very accurate. For bands of low-order indices, parallel high-order lines should be used whenever possible to increase the accuracy. Lines and bands of high-order indices normally do not appear as pairs inside the limited range of rocking angles of an ECP. Indexing and the determination of orientation therefore become difficult if the orientation is not near a low-indexed pole. With a specimen goniometer, the crystal can be tilted to a more favourable orientations and the tilt angle can be taken into account when determining the orientation. The best method for a known substance is to use a panorama diagram (Fig. 9.24a) (see [9.38] for further examples of panorama diagrams). The observed ECP can be compared with this diagram by shifting and rotating. It is also possible to use a computed map of Kossel cones as shown in Figs. 9.6a or 9.26b, though the identification of corresponding areas is more difficult without the characteristic grey level distribution of a panorama diagram. In both cases the diagrams have to be taken for the same electron energy because a change of electron energy changes the widths of the bands (Figs. 9.23a,b), the position of high-indexed lines (Fig. 9.25) and the grey-level distribution.

For routine applications, a computer program can be used, which overcomes the problem of indexing by comparing ECP at two different electron energies [9.65]. The accuracy of the orientation will be of the order of $\pm 1°$ [9.66]. In a comparison of orientations on the same specimen, between twinned

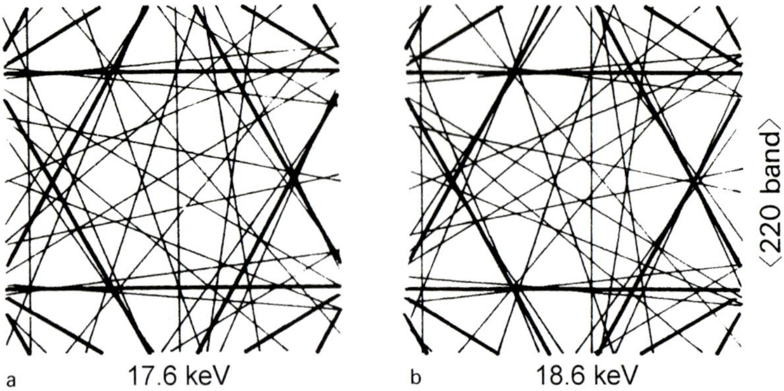

Fig. 9.25. Computed maps of high-order Kikuchi lines from high-order Laue zones inside the 111 pole of silicon formed by the 220 lines (*thick lines*) from the zero Laue zone at (**a**) $E = 17.6$ keV and (**b**) 18.6 keV (the latter agrees with the line structure in Fig. 9.23d inside the 111 pole pattern)

crystals or between matrix and precipitate [9.67, 68], for example, it is possible to reach an accuracy of the order of $\pm 0.1°$.

A crystal that has been bent by plastic deformation or showing a mosaic structure can form curved Kikuchi lines [9.69]. The magnetic stray field above unidirectional ferromagnetics such as cobalt, for example, can likewise cause distortions of the pattern [9.70]. The acceleration voltage needed to calculate the wavelength λ in the Bragg condition can be determined accurately from the Duane-Hunt limit of continuous x-rays at $E_x = eU$, recorded with an energy-dispersive x-ray detector [9.71]. The accuracy is ± 40 V at $U = 20$ kV. Another method uses the channeling pattern of a substance such as silicon, for which the lattice constant $a = 0.192017$ nm at 25°C is exactly known. Comparison of a recorded pattern (Fig. 9.25d) with a series of calculated patterns (Fig. 9.25) of the fine structure caused by high-order Kikuchi lines also allows us to reach an accuracy ± 50 V for U=20 kV [9.72].

Dislocation Density. High dislocation densities D of the order of 10^9–10^{12} cm^{-2} blur the widths (9.46) of high-indexed Kikuchi lines from initial values of the order of 1 mrad to $\simeq$10 mrad for $D \simeq 10^{12}$ cm^{-2} or a 10–20% tensile elongation [9.38, 73–76]. The width $\Delta\theta$ in mrad and D are related by the empirical formula $\Delta\theta = 3\log D - 24$. The imaging of single dislocations has been discussed in Sect. 6.2.2,5.

9.3 Electron Backscattering Patterns (EBSP)

9.3.1 Recording of EBSP

We learnt in Sects. 9.1.4,5 that channelling effects result not only in the orientation anisotropy of the total backscattering coefficient η and in proportional

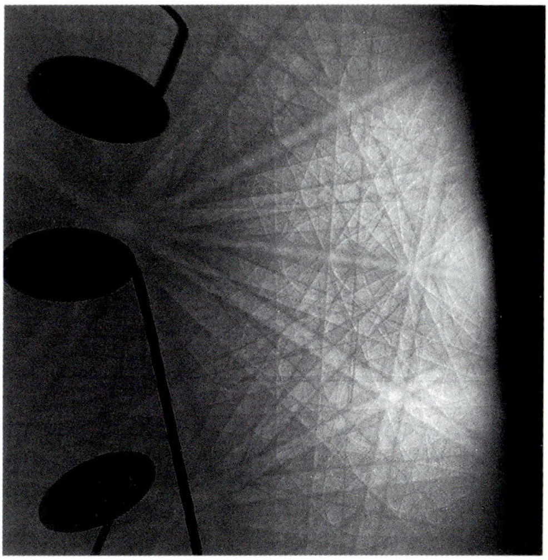

Fig. 9.26. Electron backscattering pattern (EBSP) of Si recorded with 20 keV electrons by the method of Fig. 9.27a and the shadows of three balls used for determining the coordinates x_M, y_M and z_M of the projection centre

variations of $\Delta\eta/\Delta\Omega$ when the electron probe is rocked for the recording of an ECP but also in a modulation of the angular distribution $d\eta/d\Omega$ of Kikuchi bands when the electron beam is stationary. Such an electron backscattering pattern (EBSP) can be observed on a fluorescent screen, with the aid of a video camera or by direct exposure of a photographic emulsion (Fig. 9.26) [9.77–79]. Figure 9.27 shows two possible positions of the screen or film. In the lower position (a) with oblique incidence of the electron beam and a low take-off direction, the maximum of the angular distribution of BSE (Fig. 4.12) falls on the screen. The resulting EBSP is very bright but there is a strong brightness gradient across the screen (Fig. 9.21). In the case of $\phi \simeq 90°$ it is even possible to see the Bragg spots of the RHEED pattern (Sect. 9.1.6 and Fig. 9.21a). In the arrangement of Fig. 9.27b, the tilt angles of the specimen normal are of the order of $\phi = 50° - 70°$. The brightness on the screen is lower but more uniform.

The EBSP recording methods shown schematically in Fig. 9.27 have the following characteristics. A film camera can be mounted inside the specimen chamber [9.79, 80]. The EBSP can first be adjusted by observing it on a fluorescent screen situated on the cover plate of the film camera. When this plate is insulated, the absorbed electron current can be measured by an electrometer (Fig. 9.27b) and used to estimate the exposure time of the photographic emulsion ($\simeq$20 s for $I_p \simeq 10^{-9}$ A). Recording on a photographic emulsion has the advantage of high resolution of the order of 20–50 μm and a good contrast but the disadvantage of time-consuming dark-room work. A current $I_p \simeq 10^{-7}$ A is needed for external photography of a fluorescent screen [9.83]. Direct recording methods have therefore been proposed in which a channel plate is employed [9.81, 82]; alternatively the fluorescent screen may be observed through a lead-glass window by an external low-light level TV camera

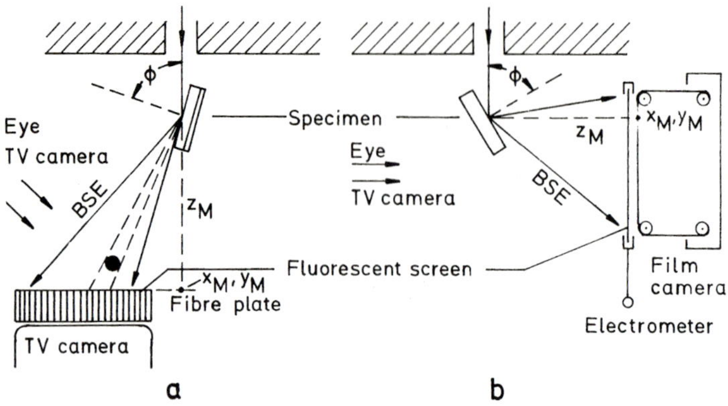

Fig. 9.27. Possible arrangements of fluorescent screen, photographic emulsion or CCD camera for the observation of electron backscattering patterns (EBSP) in (**a**) a low take-off and (**b**) a high take-off position

[9.84] or internally by means of a transparent fluorescent screen to a Peltier-cooled slow-scan CCD camera by means of a fibre plate [9.85]. Observation of the screen by direct recording methods has the advantage that channelling contrast (Sect. 6.2.3) and an EBSP can be seen simultaneous when scanning the specimen. The EBSP changes when the electron probe crosses the boundary between crystals of different orientations.

9.3.2 Advantages of EBSP over ECP

Whereas the sector of the orientation sphere that can be observed on an ECP with a single pattern is limited by the maximum rocking angle $\gamma_{\max} = \pm 5° - 10°$ for the standard technique, the sector on an EBSP increases to $\gamma = \pm 25°$ when a screen of 50 mm in edge-length and situated at a distance of 50 mm from the specimen is used (Figs. 9.27 and 9.26), which would cover a panorama map of 25 ECPs. No problem arises in identifying the characteristic pole patterns and indexing the Kikuchi bands. In the case of cubic crystals, this angular sector can cover the whole unit triangle of Fig. 9.7a or 9.24 between the [100], [110] and [111] directions.

A particular attraction of the EBSP is that it is produced by a stationary electron probe. The problem of rocking the electron probe and keeping it within a small area, as discussed for ECP in Sect. 9.2.3, do not arise. The spatial resolution of an EBSP is limited only by the electron-probe diameter. As shown by the theorem of reciprocity in the next section, there is no limitation on the electron-probe aperture α_p for EBSP, whereas α_p must be smaller than 1 mrad for recording an ECP. The optimum aperture for producing the minimum probe diameter can hence be employed. As a result, an EBSP can be recorded with a probe diameter smaller than 1 μm at $I_p \simeq 10^{-9}$ A when a thermionic cathode is used and of the order of a few tens of nanometres with a field-emission gun [9.83]. This has the further advantage that, inside this smaller information volume which is contributing to the EBSP, the lattice distortions will often be less severe than in a larger area. Because the system of Kossel cones is fixed to the crystal lattice, a tilt of the crystal results in a shift and statistical distortions in a blurring of both ECP and EBSP.

The sensitivity to surface contamiation is the same as that discused for ECP. By using the surface cleaning methods discussed for ECP in Sect. 9.2.3, it is possible to record EBSP under normal vacuum conditions. However, it is far better to use a field-emission gun and an ultra-high vacuum chamber [9.83].

9.3.3 The Reciprocity Theorem Between ECP and EBSP

To understand in more detail the characteristic differences between ECP and EBSP, it is useful to discuss the reciprocity of these techniques with the aid of reciprocal ray diagrams [9.10]. Figure 9.28a shows the arrangement of the screen on which the EBSP is recorded, as in Fig. 9.27b. Figure 9.28b shows the reciprocal system used to record an ECP, where the incident electron beam comes from the right.

For ECP recording, the electron beam of small aperture $\alpha_p \leq 1$ mrad, indispensable to obtain good angular resolution, rocks through an angle $\pm\gamma$. The solid angle $\Delta\Omega$ of detection with an aperture α_d can be arbitrary: it does not matter whether the solid angle of detection is small or large. The rocking of the incident electron beam causes variations $\Delta\eta$ at all angles; these are proportional to $\Psi_{in}^*\Psi_{in}$ of the primary Bloch-wave field, see Fig. 9.19a and (9.48). The solid angle of collection influences only the signal intensity and

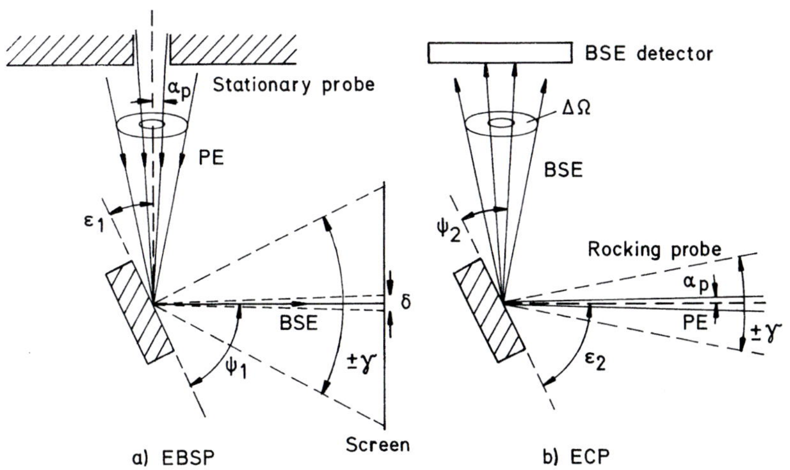

Fig. 9.28. Demonstration of the reciprocity of the ray diagrams representing (**a**) EBSP and (**b**) ECP formation

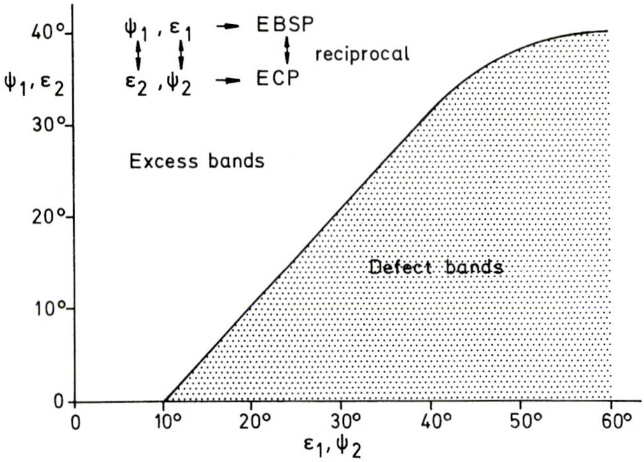

Fig. 9.29. Regions of ϵ_1, ψ_1 and ϵ_2, ψ_2 space showing predominant excess and defect Kikuchi bands in the EBSP and ECP modes, respectively (see Fig. 9.28 for the definition of ϵ and ψ)

the signal-to-noise ratio of the ECP. It is therefore best to use as large a solid angle of detection as possible.. However, the contrast reversals discussed below have to be taken into account for tilted specimens and the direction of detection should coincide with the angular range covering excess Kikuchi bands only.

In the EBSP ray diagram (Fig. 9.28a), the rocking of the beam in ECP corresponds to scanning with the eye across the screen or with a densitometer across an exposed emulsion. The angular resolution δ of the EBSP is related to the resolution of the fluorescent screen ($\simeq$0.1 mm) or of the photographic emulsion or CCD ($\simeq$10–50 μm). The incident electron-probe aperture α_p of the EBSP, which corresponds to the detector aperture α_d of the ECP mode, can be arbitrary because the EBSP is produced by scattering into the secondary Bloch-wave field, proportional to $\Psi^*_\mathrm{out}\Psi_\mathrm{out}$ in Fig. 9.19a and (9.48), and not by the primary Bloch-wave field.

The theorem of reciprocity can also be invoked to interpret the contrast reversals of Kikuchi bands, as shown in Fig. 9.21b for an EBSP with an angle of incidence $\epsilon_1 = 90° - \phi = 25°$ at low take-off angles ψ_1 (see Fig. 9.28 for the definition of ψ and ϵ). Figure 9.29 shows the regions of defect (dark) and excess (bright) band contrast in a $\psi_1 - \epsilon_1$ plane. Contrast reversals are also observed in ECP [9.12, 28, 92]. From Fig. 9.28, we see that ϵ_2 and ψ_2 in the ECP mode are reciprocal to ψ_1 and ϵ_1 in the EBSP mode, respectively. Measurements of the critical angle of contrast reversal confirm that the coordinates in Fig. 9.29 have to be interchanged to use the same diagram. Energy-filtering of ECP results in excess bands for all angles of incidence and take-off [9.93]. This confirms that low-loss electrons are directly scattered out of the Bloch-wave field and thus show excess contrast.

9.3.4 Determination of Crystal Structure and Orientation

The point-symmetry group can be read from the symmetry of low-indexed pole patterns and Kikuchi-line intersections [9.86–88]; the failure of Friedel's law in case of non-centrosymmetric crystals can also be turned to account though differences in intensity cannot always be recognized [9.87, 89]. The profiles of low-indexed Kikuchi bands obtained from dynamical n-beam calculations depend on crystal structure and composition [9.94] in accordance with experiments on silicon and alkali halides.

The accurate determination of orientation requires a knowledge of the coordinates x_{M} and y_{M} in the recording plane of the EBSP or in an x-ray Kossel pattern (Sect. 9.5.1) and of the height z_{M} of the electron impact on the crystal (Fig. 9.27a,b) that corresponds to the projection centre M of a gnomonic projection at the centre of the orientation sphere (Fig. 9.5b). The origin of the coordinates x_{M} and y_{M} is arbitrary and can be at the pattern centre or at one corner. The orientation of the specimen plane and its normal is known from the geometry of the tilt stage or can be seen in the arrangement of Fig. 9.27a as the shadow of an edge (right side of Fig. 9.26). The position of the projection centre M can be determined by recording the shadow of a mask, or more accurately, of three small steel balls placed in front of the screen or film (Figs. 9.27a and 9.26), the exact position of which is important [9.78, 81]. The shadows of the balls become elliptical and the point of intersection of their long axes, continued if necessary, is $x_{\mathrm{M}}, y_{\mathrm{M}}$. By reading about 20 coordinate pairs of the outlines of the ellipses, a least-squares computation routine finds the best-fit ellipses. The height z_{M} can be obtained from the positions of the centres of the ellipses and the ratios of their long and short axes by simple geometry [9.78]. Another possibility, which does not require shadow-casting structures, is to determine the positions of three or four low-indexed poles, from which x_{M}, y_{M} and z_{M} can be deduced [9.95].

Once the coordinates x_{M}, y_{M} and z_{M} of M are known, the direction cosines of $\overline{\mathrm{PM}}$ can be calculated from the coordinates $x_{\mathrm{P}}, y_{\mathrm{P}}$ of a point P on the pattern. From the cosines of the angles between M and two points on opposite edges of a Kikuchi band, the angle $2\theta_{\mathrm{B}}$ between these two directions and the lattice-plane spacings d_{hkl} or the hkl can be calculated, though this will be not very accurate for low-indexed bands. However, this method is important because in the gnomonic projection, which does not preserve angles, the widths of the Kikuchi band are different depending on the orientation of the band relative to the pattern axes. The coordinates of Kikuchi bands can be read by a cursor [9.90] or by digital image processing, making use of a Hough transformation to localize the bands [9.91].

When two Kikuchi bands with Miller indices or reciprocal lattice vectors $\boldsymbol{g}_1 = (h_1, k_1, l_1)$ and $\boldsymbol{g}_2 = (h_2, k_2, l_2)$ cross each other, the intersection centre will be a zone axis or pole the crystallographic direction of which is given by $[uvw] = \boldsymbol{g}_1 \times \boldsymbol{g}_2$. Once the direction cosines of two poles and of the surface normal and one direction parallel to the surface are known, the problem of determining the orientation is solved. In practice it is easier to determine directly the coordinates of a low-indexed pole, the indices $[uvw]$ of which can be recognized from its characteristic symmetry or obtained by comparison with a computer map. The use of more than two poles increases the accuracy when a least-squares routine is employed but ambiguity in the signs of the $[uvw]$ or hkl has to be taken into account and the values deduced must be checked.

The Kikuchi bands are about 2° wide and are somewhat asymmetrical in intensity for tilted crystals, wich causes an inaccuracy in the determination of the pole direction of the order of a few tenths of a degree; the accuracy in the determination of small differences in orientation will be of the order of ±0.5° [9.78].

9.4 X-Ray Kossel Patterns

9.4.1 Generation and Recording of X-Ray Kossel Patterns

The divergent-beam x-ray diffraction method was first investigated by *Kossel* [9.96, 97], who produced a source of divergent characteristic x-rays by electron bombardment of an anticathode in the form of a single crystal in an x-ray tube (see reviews [9.98–104] for applications in x-ray microprobes and SEM).

The x-rays emitted are Bragg reflected at the lattice planes with spacings d_{hkl} when the Bragg condition

$$2d_{hkl}\sin\theta_{\rm B} = \lambda_{\rm x} \qquad (9.51)$$

is satisfied. We have seen in Sect. 9.1.2 that this will be the case for all directions of incidence that lie on a Kossel cone of semi-apex angle $90° - \theta_{\rm B}$. Unlike electron diffraction, where the Bragg angles $\theta_{\rm B}$ are very small owing to the small de Broglie wavelength and the rapid decrease of the elastic scattering amplitude $f(\theta)$ with increasing θ (3.56), the angular dependence of the scattering intensity of x-rays is proportional to $(1+\cos^2\theta)$, thus allowing Bragg angles $\theta_{\rm B}$ between zero and 90° to be observed.

Figure 9.30a demonstrates how x-rays from a source inside the electron diffusion cloud are diffracted at a set of lattice planes. The resulting reflection cones intersect a photographic film plane in ellipses or hyperbolae as shown in the example of Fig. 9.31. In the case of electron diffraction, the Kossel cones lie near great circles of the unit sphere, which are the intersections of the lattice planes with the sphere (Figs. 9.6a and 9.23). In x-ray Kossel patterns, on the contrary, much smaller circles corresponding to intersections of the Kossel cones with the unit sphere or with an observation plane can be observed (Figs. 9.6b and 9.31).

From the Bragg condition (9.51), we see that the inequality

$$\sin\theta_{\rm B} = \frac{\lambda}{2d_{hkl}} < 1 \quad \text{or} \quad d_{hkl} > \lambda_{\rm x}/2 \qquad (9.52)$$

has to be satisfied, so that x-rays generated inside the specimen with the characteristic wavelength λ_x can be diffracted at the lattice planes with deflection angles $2\,\theta_B$ between zero and 180°. No Bragg diffraction will be observed if $d_{hkl} < \lambda_x/2$. This means that the K radiation of an element cannot produce a Kossel pattern for all atomic numbers. Only for $Z = 20$–30, that is for Ti, V, Cr, Fe, Co, Ni, Cu and Zn, can the Kα radiation be used and for Mo, Ag, Pb and Sn the Lα radiation. For other elements a pseudo-Kossel technique has to be applied in which the surface is coated with a film or dusted with a powdered element whose characteristic radiation is of suitable wavelength, Ni on Si for example, to satisfy (9.52). X-ray Kossel patterns can be recorded by mounting a photographic emulsion with a central hole below the polepiece or by means of the arrangement of Fig. 9.27b also used for EBSP. Kossel patterns need longer exposure times (2–10 min) than EBSPs: the backscattered electrons have to be absorbed in an organic film about 0.1 mm thick and the emulsion has to be protected against exposure to light. When a large beryllium window is used, the photographic emulsion can be outside the vacuum [9.104, 107]. When the divergent x-ray source is generated in a separate foil at the end of a capillary a distance 1–2 mm in front of the specimen, the spatial resolution will be reduced but so too will the diffuse background generated by BSE in the metal foil in front of the film (Fig. 9.31) [9.105, 106].

This model describes the principal features of the geometry of the observed Kossel pattern. However, for details of Kossel line intensities, the dynamical theory of x-ray diffraction has to be taken into account [9.103, 104]. An absorption length ξ_p of the primary x-ray wave analogous to the absorption length ξ'_g of the dynamical theory of electron diffraction is introduced. The consequence is that a perfect crystal shows sharp lines but of very low intensity, which can hardly be discerned against the strong background of the pattern. The total diffracted intensity increases considerably when the crystal consists of mosaic blocks of the order of one micrometre in size, created by annealing and lightly working (Fig. 9.30b), corresponding to dislocation densities between 10^6 and 10^9 cm^{-2}. At densities 10^9–10^{10} cm^{-2}, the Kossel lines are broadened and the precision of the measurement of lattice constants (Sect. 9.4.2) is lost. Even higher dislocation densities result in complete disappearance of the Kossel pattern.

9.4.2 Information from X-Ray Kossel Patterns

Just as for the quantitative exploitation of an EBSP, it is necessary to determine the position x_M, y_M, z_M of the x-ray source point by using the three-ball method, for example, as described in Sect. 9.3.4. Since the lines in x-ray Kossel patterns are sharper and a precision of 1/3000 in the x-ray source point coordinates can be attained, lattice spacings can be measured with an accuracy of 1/7000 for high-order reflection and orientation relationships can be accurate to ±0.2° [9.108]. For the determination of orientation, the necessary line indexing can be done with the aid of computer-drawn maps in gnomonic

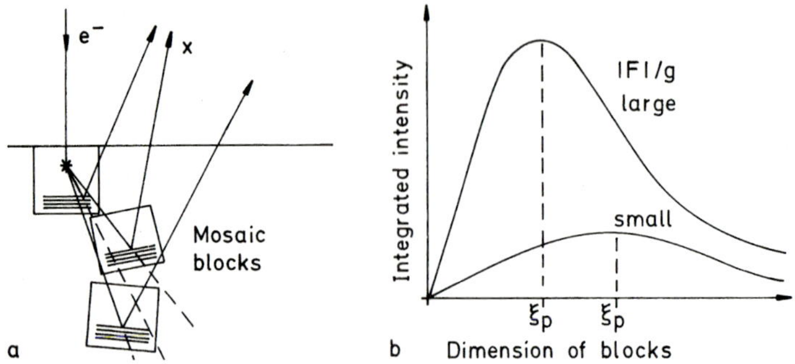

Fig. 9.30. (**a**) Schematic representation showing how slightly misoriented regions of a crystal with mosaic structure lead to the production of stronger intensities on Kossel cones. (**b**) Dependence of Kossel line intensity on the size of the mosaic blocks (ξ_p = absorption length of the primary x-ray wave)

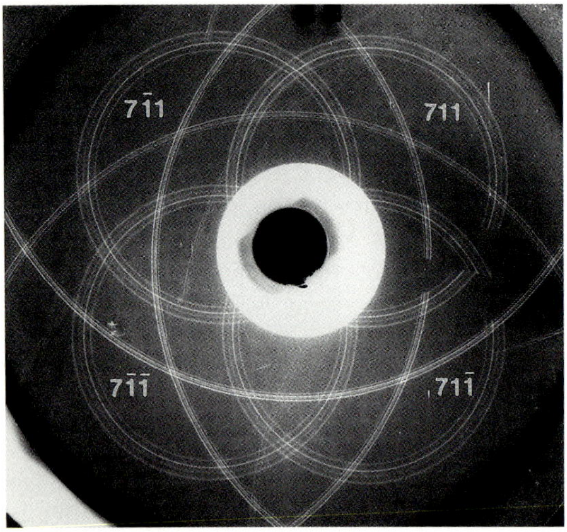

Fig. 9.31. X-ray Kossel pattern of epitaxially grown GaAsSb layers on a GaAs substrate. The shift of the $K\alpha_1$–$K\alpha_2$ doublet of the 711 lines of layer and substrate result in a relative difference $\Delta d/d = (43.58 \pm 0.35) \times 10^{-4}$ of lattice-plane distances [9.106]

or stereographic projection (Fig. 9.6b); alternatively, the direction of the axis of a Kossel cone may be calculated from a large number of points on the cone. This allows the semi-apex angle $90° - \theta_B$ and hence d_{hkl} to be calculated.

A high accuracy in the determination of lattice parameters can be attained by considering the lens-shaped regions of two overlapping cones or the triangle formed by three overlapping cones of different *hkl*. The accuracy increases with decreasing width of the lens or triangle [9.100, 104, 108]. The accuracy of the relative lattice parameter $\Delta a/a \simeq 10^{-5} - 10^{-4}$ is sufficiently high for the lattice distortion caused by elastic strains and the distribution of strains during and after deformation to be measured with a lateral resolution of the order of one micrometre. Figure 9.31 shows as an example the measurement of differences in the lattice parameter a of a GaAsSb layer on a GaAs substrate obtained from the shift of Kossel lines from layer and substrate [9.105, 106]. The dislocation density must not exceed 10^{10} cm^{-2}, as higher values result in a broadening of the Kossel lines. For further details and applications we refer to [9.100, 104, 108].

10. Elemental Analysis and Imaging with X-Rays

The deceleration of fast electrons in the Coulomb field of a nucleus can result in the emission of a background x-ray quantum. The de-excitation of an inner-shell ionization results either in the emission of a characteristic x-ray quantum or in the emission of an Auger electron, both of which can be used for elemental analysis.

Either a wavelength- (WDS) or an energy-dispersive spectrometer (EDS) can be used to measure the wavelength or quantum energy $E_x = h\nu$ of characteristic x-ray lines for elemental analysis, Though the WDS has a better spectroscopic resolution, the resolution $\Delta E_x \simeq 150$–180 eV of an EDS is sufficient for routine work and the whole x-ray spectrum can be recorded simultaneously by means of a multichannel analyser.

For quantitative x-ray microanalysis, the number of x-ray quanta in characteristic lines produced in the specimen and in a pure element standard or one of known concentration are counted. The concentration of the element in the specimen can be calculated from the ratio of these counts. However, several corrections are necessary since various factors are different for specimen and standard: the backscattering and the stopping power, which depend on atomic number (Z); the x-ray absorption (A) and the x-ray fluorescence (F). The ZAF correction programs developed for x-ray microprobes at normal incidence can also be used quantitatively in SEM for normal and tilted specimens. For small particles and films on substrates and for biological specimens, special correction problems have to be taken into account. X-ray fluorescence analysis can be employed in SEM at some sacrifice of spatial resolution.

Imaging with x-rays can be realized by using the non-dispersive signal (total rate imaging with x-rays) or the dispersive signal in the x-ray mapping mode, in which the number of characteristic counts is used to modulate the brightness of the cathode-ray tube during a scan. X-ray projection microscopy and x-ray topography can also be performed in a SEM and usefully complement the other imaging and analysing modes.

Auger electron spectroscopy is a surface-sensitive method which, however, needs an ultra-high vacuum. It can in principle be combined with SEM but the use of special Auger electron microprobes is much more satisfactory.

10.1 X-Ray and Auger Electron Emission

10.1.1 X-Ray Spectra

X-rays are electromagnetic waves of quantum energy $E_x = h\nu$; they are generated by the deceleration of charged particles (x-ray continuum) or by electron transitions from a filled electron state into a vacancy in a lower atomic shell (characteristic x-ray lines). The vacancy may be the result of ionization by a charged particle or of photoionization (fluorescence) by a quantum of sufficiently high energy. The quantum energies normally used in electron-probe microanalysis lie in the range 100 eV to 50 keV, corresponding to wavelengths from 10 nm to 0.02 nm. The low-energy tail of the x-ray spectrum excited by electron bombardment extends continuously into the ultraviolet and visible part of the electromagnetic spectrum (see transition radiation and Lilienfeld radiation in Sect. 7.2.1).

X-ray emission spectra are plotted versus the quantum energy E_x or versus the wavelength λ_x, depending on whether wavelength- or energy-dispersive spectrometers (Sect. 5.4) are used. We use the suffix x to distinguish these quantities from the electron energy $E = eU$ and electron wavelength $\lambda = h/mv$, respectively. They are related by

$$E_x = h\nu = hc/\lambda_x \tag{10.1}$$

where $h = 6.6256 \times 10^{-34}$ J s denotes Planck's constant and $c = 2.99793 \times 10^8$ m s^{-1} the speed of light. These values give the numerical relation

$$E_x \lambda_x = 1.23989 \tag{10.2}$$

where E_x and λ_x are measured in units of keV and nm, respectively.

The following point has to be considered when plotting and using x-ray spectra. As a function of quantum energy E_x or wavelength λ_x, either the radiation power P in watts (1 W = 1 J s^{-1}) or the number N_x of quanta in the ranges $E_x, E_x + dE_x$ or $\lambda_x, \lambda_x + d\lambda_x$ can be plotted. A plot of radiation power will be of interest when recording the absorbed x-ray energy per unit time by means of an ionization chamber or a scintillator without single photon counting, for example. A plot of the number of quanta will be used when the x-ray quanta are counted individually by a proportional counter, a scintillator or a semiconductor detector. For the widely used energy-dispersive spectrometry, a plot of the number $N(E_x)dE_x$ of x-ray quanta versus quantum energy E_x is of most interest because such a spectrum is recorded directly with a multichannel analyser (Sect. 5.4.2).

These spectra are related by

$$\begin{aligned} P(\lambda_x)|d\lambda_x| &= P(E_x)|dE_x| \rightarrow P(\lambda_x) = (hc/\lambda_x^2)P(E_x) \\ N(E_x)dE_x &= P(E_x)dE_x/E_x\,. \end{aligned} \tag{10.3}$$

Figures 10.1a-c show schematically for the x-ray continuum excited by 10, 20 and 30 keV electrons how these different ways of plotting influence the shape of the spectra.

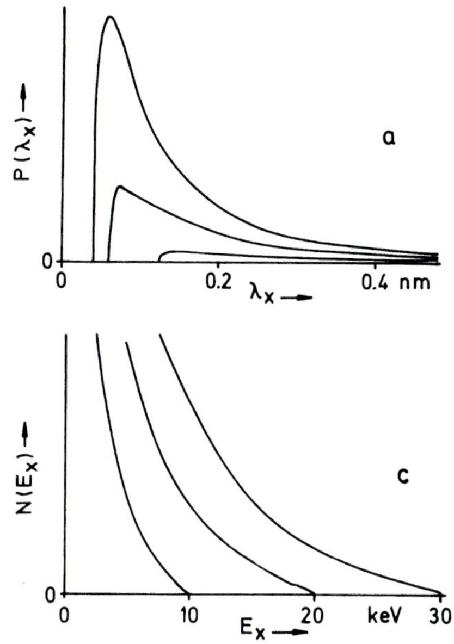

Fig. 10.1. Different ways of plotting x-ray spectra (x-ray continuum only) as the radiation power P in watts as a function of (**a**) wavelength λ_x and (**b**) quantum energy E_x. In (**c**) the number $N(E_x)$ dE_x of x-ray quanta in the interval $E_x, E_x + dE_x$ is plotted against E_x

10.1.2 The X-Ray Continuum

Electrodynamics shows that an accelerated charged particle of charge e with an acceleration $\ddot{x}$ emits an electromagnetic wave with an intensity I, defined as the energy flux through unit area per unit time, proportional to the Poynting vector $\boldsymbol{E} \times \boldsymbol{H}$:

$$I(\theta) = \frac{c}{16\pi^2\epsilon_0} \left(\frac{e\ddot{x}}{rc^2}\right)^2 \sin^2\theta \tag{10.4}$$

where θ is the angle between the directions of emission and acceleration. Deviations resulting from relativity will be discussed below. A typical application of (10.4) is the radiation of a dipole (antenna), which emits a harmonic, monochromatic wave when excited by an ac current; the latter generates an oscillating dipole moment $\boldsymbol{p} = e\boldsymbol{x} = \boldsymbol{p}_0 \sin\omega t$. Substitution in (10.4), using $\langle p^2 \rangle = p_0^2\omega^4 \langle \sin^2 \omega t \rangle = p_0^2\omega^4/2$, leads to

$$I(\theta) = \frac{p_0^2\omega^4}{32\pi^2\epsilon_0 c^3 r^2} \sin^2\theta \, . \tag{10.5}$$

In the case of continuous x-ray emission, the acceleration is caused by the screened Coulomb potential of the nucleus, which normally leads to elastic scattering (Sect. 3.1). With a low probability (cross-section), this interaction can also result in the emission of an x-ray quantum of energy $0 \le E_x \le E = eU$ and the incident electron is then decelerated from E to $E - E_x$. Unlike the periodic oscillation of a dipole, the acceleration is a δ-function peak in time.

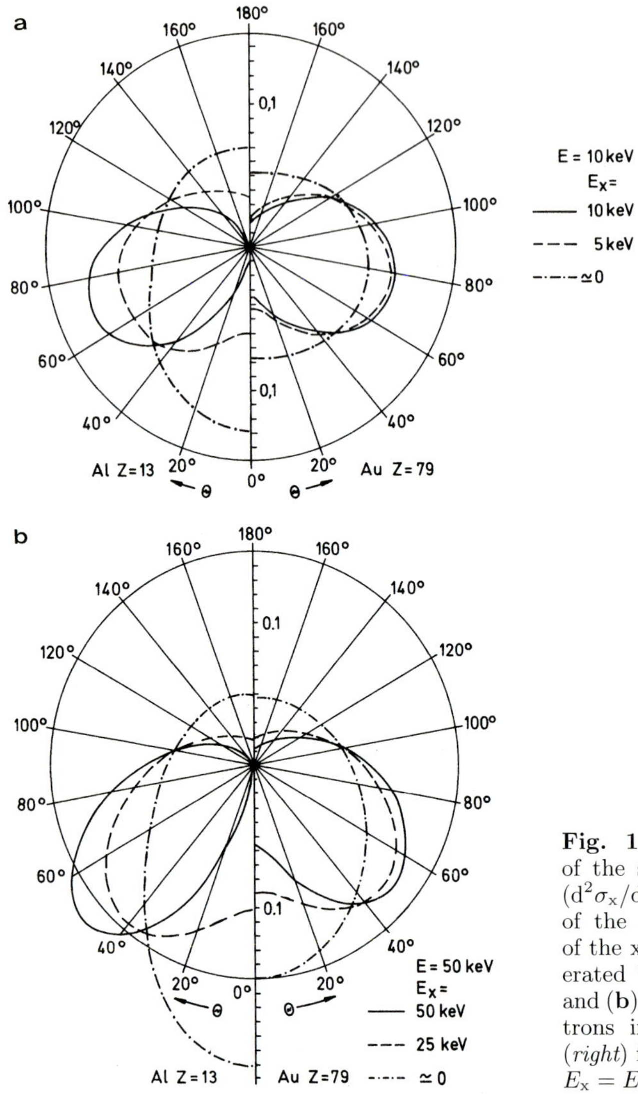

Fig. 10.2. Polar diagram of the shape function $S = (\mathrm{d}^2\sigma_\mathrm{x}/\mathrm{d}E_\mathrm{x}\mathrm{d}\Omega)/(\mathrm{d}\sigma_\mathrm{x}/\mathrm{d}E_\mathrm{x})$ of the angular distribution of the x-ray continuum generated by (**a**) $E = 10$ keV and (**b**) 50 keV primary electrons in Al (*left*) and Au (*right*) for quantum energies $E_\mathrm{x} = E$, $E/2$ and 0 [10.3]

These analogies as well as differences between an oscillating dipole and radiative deceleration in a Coulomb field allows us to deduce in the following properties of the continuous x-ray spectrum.

a) Angular Distribution. The proportionality $I(\theta) \propto \sin^2\theta$ in (10.4) and (10.5) results in the typical dipole characteristics with a maximum of emission perpendicular ($\theta = 90°$) and zero emission parallel ($\theta = 0, 180°$) to the dipole. Conversely, the maximum emission of the x-ray continuum is shifted to smaller angles θ because the Coulomb field of a moving charge is not exactly radially-symmetric owing to relativistic effects and because retardation has to be taken into account because of the finite velocity of wave propagation [10.1]. Though the relativistic increase of mass can be neglected for 10–50 keV electrons, this shift of the maximum is a relativistic effect. The calculated angular characteristics in Fig. 10.2 show this shift for $E_x = E$. These values of the differential cross-section $d^2\sigma_x(E, Z, E_x, \theta)/\mathrm{d}E_x\mathrm{d}\Omega$ per atom are normalized by division by $\mathrm{d}\sigma_x(E, Z, E_x)/\mathrm{d}E_x$ [10.2, 3] and the resulting shape function $S(E, Z, E_x, \theta) = (\mathrm{d}^2\sigma_x/\mathrm{d}E_x\mathrm{d}\Omega)/(\mathrm{d}\sigma_x/\mathrm{d}E_x)$ is plotted in Fig. 10.2. In these calculations, the screening of the Coulomb potential of the nucleus by the orbital electrons is considered, whereas the older theories [10.4, 5] neglected screening and the resulting deceleration in the unscreened Coulomb potential of the nucleus. Figure 10.2 also shows the changes in the angular characteristics for $E_x < E$; the maximum emission is observed for low $E_x \simeq 0$ in the forward direction.

b) Energy Distribution. An oscillating dipole emits quanta of energy $h\nu$. The δ-peak of acceleration has to be Fourier transformed to get the frequency or energy distribution. The Fourier transform of a δ-function is a constant for all ν. However, the energy $E_x = h\nu$ of the x-ray quanta cannot exceed the electron energy $E = eU$ since the total energy must be conserved. There is thus a maximum frequency (Fig. 10.1b) or minimum wavelength (Fig. 10.1a) (Duane–Hunt limit):

$$\nu_{\max} = \frac{eU}{h}; \quad \lambda_{\min} = \frac{c}{\nu_{\max}} = \frac{ch}{eU} . \tag{10.6}$$

Figure 10.3 shows energy distributions calculated by integrating the cross-section $\mathrm{d}^2\sigma_x/\mathrm{d}E_x\mathrm{d}\Omega$ over all emission angles and multiplying the resulting $\mathrm{d}\sigma_x/\mathrm{d}E_x$ by $E_x\beta^2/Z^2$ where $\beta = v/c$; these curves show only small deviations from the constancy estimated above. The proportionality $E_x\mathrm{d}\sigma_x/\mathrm{d}E_x \propto Z^2/\beta^2$ and the independence of E_x was already obtained by *Kramers* [10.6] in a semi-classical treatment in which the electron trajectories were assumed to be parabolic and screening was neglected:

$$E_x\frac{\mathrm{d}\sigma_x}{\mathrm{d}E_x} = \frac{32\pi^2e^6}{3\sqrt{3}(4\pi\epsilon_0)^3hc^5m^2}\frac{Z^2}{\beta^2} = a_K\frac{Z^2}{\beta^2} \quad \text{for} \quad E_x \le E . \tag{10.7}$$

The Kramers constant $a_K = 5.54 \times 10^{-31}$ m^2 is also plotted in Fig. 10.3 for comparison. The examples of exact calculations in Fig. 10.3 show no serious departures from the constant value a_K obtained by Kramers' approach. This explains the success of this formula in estimating the continuous x-ray intensity. The number of x-ray quanta per atom and per incident electron in the energy interval $E_x, E_x + \mathrm{d}E_x$ is obtained by division of (10.7) by E_x, see (10.3).

c) Polarization. The radiation of an oscillating dipole is linearly polarized with the $\boldsymbol{E}$ vector in the plane containing $\boldsymbol{p}$ and the direction of emission. The x-ray emission is also polarized but this is not of direct interest in SEM.

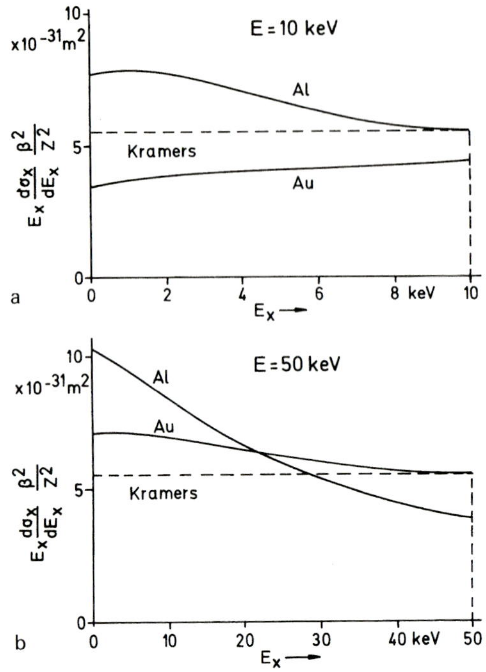

Fig. 10.3. Angular integrated continuous x-ray spectra plotted as $E_{\mathrm{x}}(\mathrm{d}\sigma_{\mathrm{x}}/\mathrm{d}E_{\mathrm{x}})\beta^2/Z^2$ excited by (**a**) 10 keV and (**b**) 50 keV electrons in Al and Au. The constant value (– – –) resulting from Kramers' approach is shown for comparison

d) Continuous X-Ray Emission from a Solid Target. We assume that E_{m} is the mean energy of electrons at depth z. The number n_{c} of continuous x-ray quanta emitted per incident electron in the interval $E_{\mathrm{x}}, E_{\mathrm{x}}+\mathrm{d}E_{\mathrm{x}}$ in a layer $z, z+\mathrm{d}z$ with $N_{\mathrm{A}}\rho\mathrm{d}z/A$ atoms per unit area becomes, with the Kramers' cross-section (10.7),

$$n_{\mathrm{c}}\mathrm{d}(\rho z)\mathrm{d}E_{\mathrm{x}} = \frac{\mathrm{d}\sigma_{\mathrm{x}}}{\mathrm{d}E_{\mathrm{x}}}\mathrm{d}E_{\mathrm{x}}\frac{N_{\mathrm{A}}\rho\mathrm{d}z}{A} = \frac{16\pi^2}{3\sqrt{3}}\frac{e^6 Z^2}{(4\pi\epsilon_0)^3 hc^5 m E_{\mathrm{m}} E_{\mathrm{x}}}\frac{N_{\mathrm{A}}\rho\mathrm{d}z}{A}\mathrm{d}E_{\mathrm{x}}\,. \tag{10.8}$$

The mean energy E_{m} decreases with the Bethe stopping power $S = |\mathrm{d}E_{\mathrm{m}}/\mathrm{d}x|$ (Sect. 3.3.4). *Kramers* [10.6] adopted for simplicity the Thomson–Whiddington law (3.135). The first part of (3.135) can be solved for $\mathrm{d}x/E_{\mathrm{m}}$ and substituted in (10.8) where $x = \rho z$:

$$n_{\mathrm{c}}\mathrm{d}x\mathrm{d}E_{\mathrm{x}} = \frac{8\pi e^2}{3\sqrt{3}4\pi\epsilon_0 hc^5 ml}\frac{Z}{E_{\mathrm{x}}}|\mathrm{d}E_{\mathrm{m}}|\mathrm{d}E_{\mathrm{x}} = k\frac{Z}{E_{\mathrm{x}}}|\mathrm{d}E_{\mathrm{m}}|\mathrm{d}E_{\mathrm{x}} \tag{10.9}$$

which has to be integrated from $E_{\mathrm{m}} = E$ (initial energy) to $E_{\mathrm{m}} = E_{\mathrm{x}}$ because only electrons with $E_{\mathrm{m}} \geq E_{\mathrm{x}}$ can generate continuous x-ray quanta of energy E_{x}. This results in

$$N_{\mathrm{c}}(E_{\mathrm{x}})\mathrm{d}E_{\mathrm{x}} = kZ\frac{E - E_{\mathrm{x}}}{E_{\mathrm{x}}}\mathrm{d}E_{\mathrm{x}} \tag{10.10}$$

or using (10.3)

$$P(E_x)\mathrm{d}E_x = kZ(E - E_x)\mathrm{d}E_x \ . \tag{10.11}$$

Both dependences on E_x are plotted schematically in Figs. 10.1b and c.

The angular characteristics of the x-ray continuum shown in Fig. 10.2 have only to be taken into account for quantum energies near the Duane–Hunt limit. The angular characteristics of lower quantum energies excited by more strongly decelerated electrons are more randomized owing to the increase of angular spread by electron diffusion. The number $N_c(E_x)\mathrm{d}E_x$ of continuous x-ray quanta generated can therefore be treated as isotropic in a first-order approximation. However, the influence of electron backscattering and x-ray absorption decreases the observed number $N_c\mathrm{d}E_x\Delta\Omega/4\pi$ of quanta where $\Delta\Omega$ is the solid angle of collection. This can be taken into account by applying the ZAF correction technique (Sect. 9.1) to the continuum as well, which gives acceptable agreement with experimental values obtained with a Si(Li) x-ray detector [10.7].

Further details of continuous x-ray emission are summarized in the reviews [10.8, 9].

10.1.3 The Characteristic X-Ray Spectrum

We assume that a vacancy is formed by inner-shell ionization (Sect. 3.2.3). Figure 10.4 shows the nomenclature of the subshells with the corresponding quantum numbers n, l and j. When an atomic electron from an outer shell fills the vacancy, a characteristic x-ray quantum can be emitted whose quantum energy E_x is equal to the de-excitation energy, which is the difference between the energies of the initial and final states (Fig. 10.5a). The x-ray lines observed can be classified in series according to the shell in which ionization took place; thus the first letter is K for an initially ionized K shell, L for an ionized L shell, etc. Within a series, the second label is traditionally a greek letter with an arabic subscript, Kα_1, for example (Fig. 10.4). Some of the possible transitions are forbidden. The allowed transitions have to fulfil the selection rules $\Delta l = \pm 1, \Delta j = 0, \pm 1$.

To a first approximation, in which the subshell structure is disregarded, the quantum energies $E_{x,K}$ of the K series can be estimated from modified terms of the Bohr model ($E_I = E_K$: ionization energy of the K shell):

$$E_{x,K} = E_n - E_1 = -R(Z-1)^2\left(\frac{1}{n^2} - \frac{1}{1^2}\right);\ Z \geq 3;\ n = 2 : \mathrm{K}\alpha;\ n = 3 : \mathrm{K}\beta \tag{10.12}$$

where $R = 2\pi^2 e^4 m/(4\pi\epsilon_0)^2 h^2 = 13.6\mathrm{eV}$ denotes the ionization (Rydberg) energy of the hydrogen atom. The reduction of Z by unity represents the screening of the nuclear charge $+Ze$ by the remaining electron in the K shell. Likewise for the L series ($E_2 = E_L$):

$$E_{x,L} = E_n - E_2 = -R(Z-7.4)^2\left(\frac{1}{n^2} - \frac{1}{2^2}\right);\ Z \geq 11;\ n = 3, 4, \ldots \tag{10.13}$$

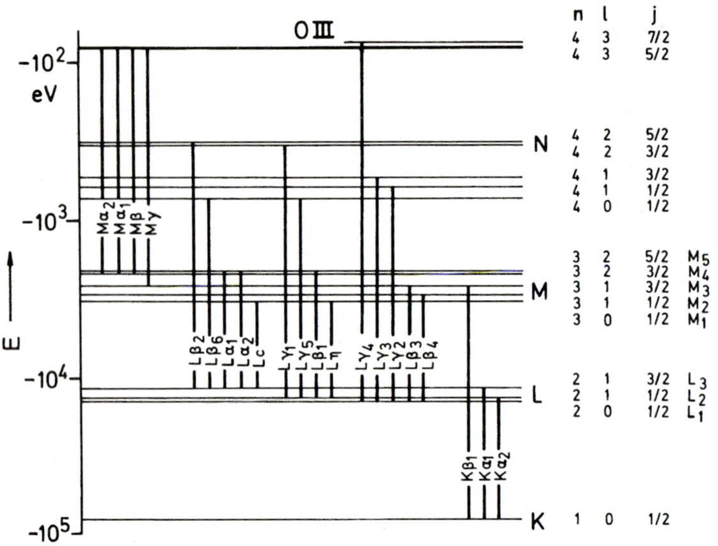

Fig. 10.4. Energy levels of atomic shells in Au (logarithmic scale) and nomenclature of the subshells with quantum numbers n, l and j and possible transitions resulting in the emission of characteristic x-ray lines

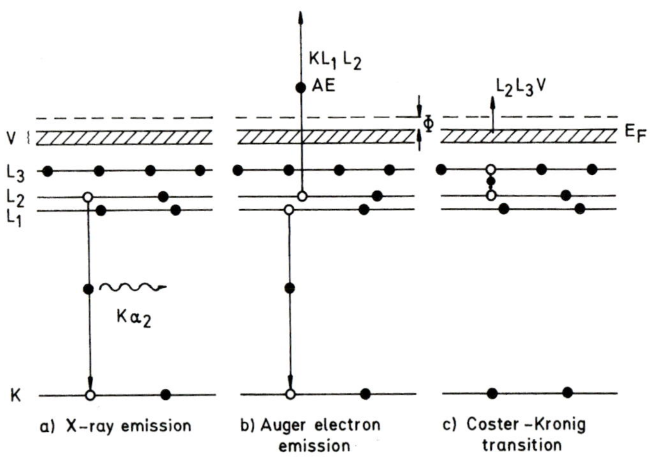

Fig. 10.5. (**a**) X-ray emission, (**b**) Auger electron emisssion and (**c**) Coster–Kronig transitions as possible de-excitation processes after ionization of the K or L shell

A consequence is that the quantum energies of a series satisfy Moseley's law

$$E_{\mathrm{x}} \propto (Z - \sigma)^2 \tag{10.14}$$

where σ is the screening constant of the corresponding shell. The validity of this law is demonstrated in Fig. 10.6 where $E_{\mathrm{x}}^{1/2}$ is plotted against Z.

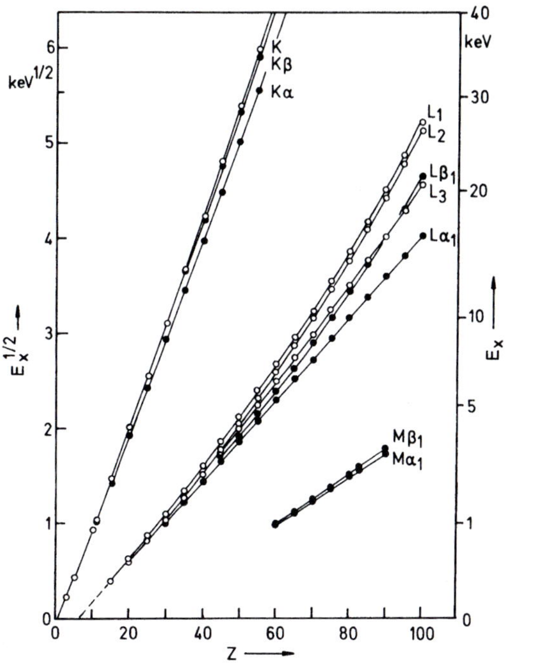

Fig. 10.6. Demonstration of the Moseley law – linear relation between $E_x^{1/2}$ and Z – by plotting the ionization (edge) energies E_I, (I = K, L$_1$, L$_2$, L$_3$) (o o o) and the quantum energies $E_x^{1/2}$ of the x-ray lines (• • •) against the atomic number Z (values for E_x in keV on the left-hand ordinate)

A splitting of the energy levels of subshells (Fig. 10.4) is caused by differences in the spin-orbit coupling

$$E_{nl} = -R\left[\frac{(Z-\sigma)^2}{n^2} + \frac{\alpha^2(Z-\sigma')^2}{n^4}\left(\frac{n}{j+1/2} - \frac{3}{4}\right)\right] , \qquad (10.15)$$

where $\alpha = 1/137$ and σ' is a spin-relativity screening constant. Both σ and σ' also depend on l. Because two values of j, $j = l \pm 1/2$, belong to each quantum number l, transitions to or from such a pair of levels will give rise to a doublet. A typical example is the Kα_1, Kα_2 doublet which arises from the transitions $L_3 \rightarrow K$ and $L_2 \rightarrow K$, respectively ($L_1 \rightarrow K$ is forbidden).

It is fortunate for elemental analysis based on characteristic x-ray lines that the energy states of inner shells are not strongly influenced by the nature and strength of the chemical bonds. However, chemical effects on x-ray emission will be observed if transitions occur from the valence electron states, which are involved in chemical bonds [10.1, 10]. Changes in shape and position less than 1 eV can be observed in the Kβ line for $Z < 20$ and in the L spectrum for $Z < 30$. For example, the observed shifts of sulphur S Kα in a large number of compounds depend on the effective charge: S^{2-}(–0.14 eV), S^{2+}(+0.31 eV), S^{4+}(+0.95 eV), S^{6+}(+1.19 eV) [10.11]. This chemical shift can be neglected in energy-dispersive spectrometry because of the poorer energy resolution but can be of interest in wavelength-dispersive spectrometry.

An x-ray quantum is only emitted with a probability ω, the fluorescence yield, when the vacancy in an inner shell is filled by an atomic electron from the outer shells. When the vacancy has been generated by electron bombardment, the emission has nothing in common with fluorescence. The same fluorescence yield does occur when the vacancy is generated by the photoelectric effect, which means absorption of an x-ray quantum of energy E_{x} larger than the ionization energy (Sect. 10.1.4). With the complementary probability $a = 1-\omega$, the Auger electron yield, the energy difference between the initial and final states may be transferred in a radiationless transition to another atomic electron, which leaves the atom as an Auger electron (Fig. 10.5b) (see Sect. 10.1.5 for details). It is sometimes wrongly stated in the literature that the Auger electron is generated by the internal conversion of an x-ray quantum. Auger electron emission is an alternative process to x-ray emission and no x-ray quantum is excited, not even as an intermediate state.

The K fluorescence yield ω_{K} increases with increasing atomic number Z (Fig. 10.7a) and can be approximated by the expression [10.12]

$$\omega_{\mathrm{K}} = \frac{d^4}{1+d^4}; \quad d = -0.044 + 0.0346Z - 1.35 \times 10^{-6} Z^3 \,. \tag{10.16}$$

This means that for low atomic numbers, the probability of emission of an x-ray quantum is very low, whereas the Auger electron yield is near unity.

For the L shells we have three yields ω_{L1}, ω_{L2} and $\omega_{\mathrm{L_3}}$ but only the $\omega_{\mathrm{L_3}}$ yield (Fig. 10.7b) is important because the $\mathrm{L}\alpha_1$ line, the strongest line of the L series, is generated by de-excitation of a hole in the $\mathrm{L_3}$ subshell. The so-called Coster–Kronig transitions affect the number of vacancies in the subshells, being primarily produced by electron collision. These are radiationless transitions from electrons in higher states of a subshell to vacancies in a lower one. The de-excitation energy is transferred to an electron at the atomic periphery (Fig. 10.5c). The energy difference $E_{\mathrm{L3}} - E_{\mathrm{L2}}$, for example, can be transferred to a valence electron, which is called an $\mathrm{L_2L_3V}$ transition. If the probabilities of direct ionization of the L subshells are n_1, n_2 and n_3 and the Coster–Kronig transition probabilities are f_{ij} then the total probabilities of x-ray emission from the subshells are

$$\begin{aligned} n_1\omega_{\mathrm{L1}}(\mathrm{eff}) &= n_1\omega_{\mathrm{L1}} \\ n_2\omega_{\mathrm{L2}}(\mathrm{eff}) &= n_2(1+n_1 f_{12}/n_2)\omega_{\mathrm{L2}} \\ n_3\omega_{\mathrm{L3}}(\mathrm{eff}) &= n_3[1+n_1 f_{12}/n_2 + n_1(f_{13}+f_{12}f_{23})/n_3]\omega_{\mathrm{L3}} \,. \end{aligned} \tag{10.17}$$

This shows that the effective values of the fluorescence yields are increased by the Coster–Kronig transitions. An extensive discussion of fluorescence yields and Coster–Kronig transition probabilities is to be found in [10.13].

Since the fluorescence yield, ω_{K} for example, is only the probability that filling a vacancy in the K shell results in the emission of an x-ray quantum, we need the relative transition probabilities p for the appearance of single lines of a series or for the subshell from which the electron that fills the vacancy

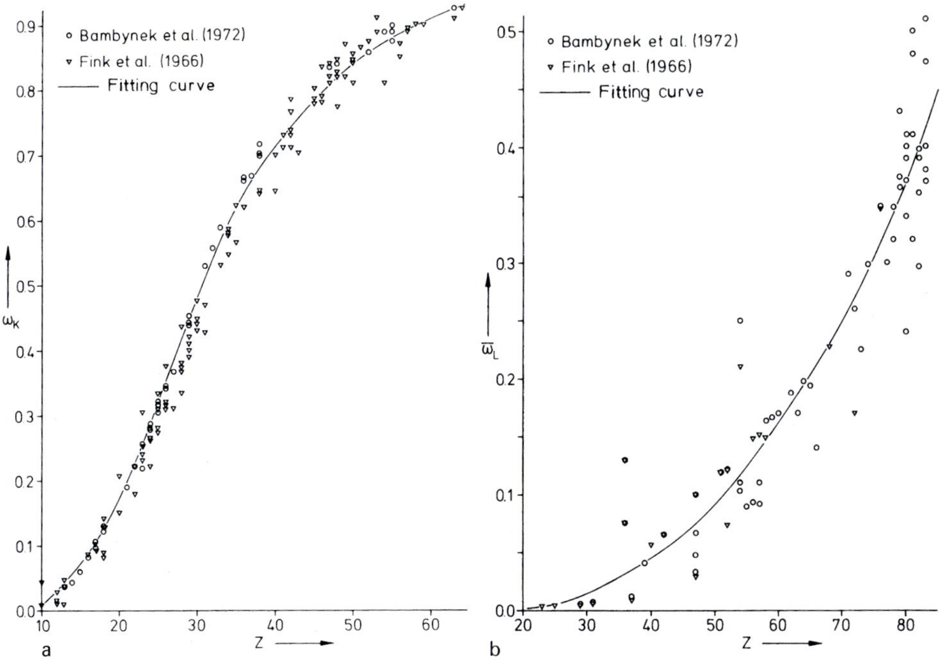

Fig. 10.7. (**a**) X-ray fluorescence yield ω_K of the K shell and (**b**) ω_L of the L shell versus the atomic number Z

will come. The sum of the transition probabilities of the K$\alpha_{1,2}$ and K$\beta_{1,2,3}$ lines has to be unity

$$p_{K\alpha} + p_{K\beta} = 1 \, . \tag{10.18}$$

The ratio $p_{K\alpha_1}/p_{K\alpha_2}$ of the intensities of the Kα_1 line ($L_3 \to$K) to that of the Kα_2 line ($L_2 \to$K) is in a first-order approximation proportional to the number of electrons in the subshells, which is $4/2 = 2$ (sum rule). The probabilities p_K are plotted in Fig. 10.8a. The reason for this dependence on Z is the gradual emptying of the N shell as Z falls towards 30 and of the M shell as Z falls towards 10. Experimental p_L values are plotted in Fig. 10.8b. Experiments and theories for these probabilities are described in [10.13].

It will be of interest for x-ray spectroscopy to know the width ΔE_x of x-ray lines, which are related to the mean lifetime τ of the energy states by Heisenberg's uncertainty relation $\tau \Delta E = \hbar$. The lifetime τ is inversely proportional to the sum of the transition probabilities for the radiationless Auger and the x-radiative transitions; this sum is inversely proportional to the denominator in (10.16), so that

$$\Delta E_x \propto \tau^{-1} \propto A + BZ^4 \, . \tag{10.19}$$

For example, the width of Kα lines increases from $\Delta E_x \simeq 1$ eV for $Z = 20$, $\Delta E_x = 5$ eV for $Z = 40$ to $\Delta E_x = 54$ eV for $Z = 79$ (see also Fig. 10.15).

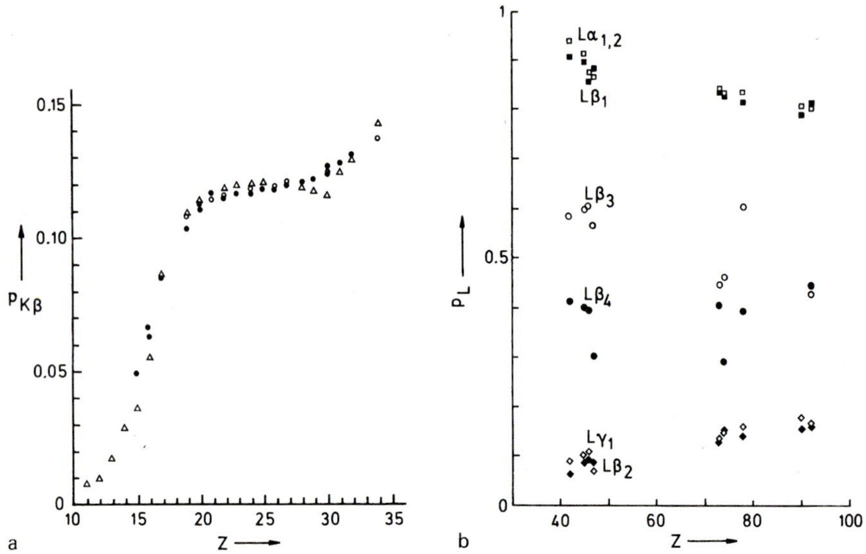

Fig. 10.8. (**a**) Relative transition probability p_K for the appearance of a Kβ line in the K series and (**b**) of the most important L lines [10.13]

The latter corresponds to a lifetime $\tau = 12 \times 10^{-18}$ s. For L lines the situation becomes more complex because of the additional possibility of radiationless Coster–Kronig transitions, so that the line-width increases in those regions where the Coster–Kronig transitions are most probable.

For SEM applications, this shows that the energy resolution of a Si(Li) detector $\Delta E_x > 100$ eV will be larger than the natural line width whereas the resolution of a wavelength-dispersive spectrometer can be of the order of $\Delta E_x = 1$–50 eV and is more comparable with the natural width (Fig. 10.15).

The dependence of the x-ray line intensity of solid targets on atomic number and electron energy will be discussed in detail in Sect. 10.3. For further details about characteristic x-ray emission see the reviews [10.1, 14, 15].

10.1.4 Absorption of X-Rays

The following interactions of x-rays with matter have to be taken into account and contribute to x-ray absorption:

a) Elastic or Thomson Scattering. The electric field of the electromagnetic x-ray wave can excite the atomic electrons into forced oscillations, which can in turn emit dipole radiation of the same frequency. This effect causes elastic scattering of x-rays (without changing their energy) and is responsible for the atomic scattering amplitude in x-ray diffraction and for the elastic peak in x-ray fluorescence (Sect. 10.5.3).

b) Inelastic or Compton Scattering. The x-ray quantum of energy $E_x = h\nu$ and momentum $p_x = h\nu/c$ hits an atomic electron at rest energy $E_0 = m_0c^2$ and zero momentum. The scattered quantum has a lower energy E'_x and an energy $E_x - E'_x$ is transferred to the electron. Conservation of energy and momentum results in

$$E_x + m_0c^2 = E'_x + mc^2 = E'_x + c(m_0c^2 + p_e^2)^{1/2} \tag{10.20}$$

$$\boldsymbol{p}_x = \boldsymbol{p}'_x + \boldsymbol{p}_e \,. \tag{10.21}$$

Solving (10.20) for p_e^2 gives

$$p_e^2 = \frac{(E_x + m_0c^2 - E'_x)^2}{c^2} - m_0c^2 = \frac{1}{c^2}[E_x^2 + E'^2_x + 2(E_x - E'_x)m_0c^2 - 2eE_xE'_x] \,. \tag{10.22}$$

Forming p_e from (10.21), we obtain

$$p_e^2 = p_x^2 + p'^2_x - 2\boldsymbol{p}_x \cdot \boldsymbol{p}'_x = \frac{1}{c^2}(E_x^2 + E'^2_x - 2E_xE'_x \cos\theta) \tag{10.23}$$

where θ is the angle through which the x-ray quantum is scattered. Equating the two results for p_e^2, cancelling common terms and solving for $E_x - E'_x$ gives an expression for the energy loss of the x-ray quantum, which is equal to the kinetic energy transferred to the electron:

$$\Delta E_x = E_x - E'_x = \frac{E_x^2(1-\cos\theta)}{m_0c^2 + E_x(1-\cos\theta)} \simeq \frac{E_x^2}{E_0}(1-\cos\theta) \tag{10.24}$$

where the electron rest energy $E_0 = m_0c^2 = 511$ keV. The term $(1 - \cos\theta)$ varies between 0 and 2 for $\theta = 0$ and 180°, respectively. Such a Compton loss can be seen as a shifted diffuse peak in front of an elastically scattered line in an x-ray fluorescence spectrum (Fig. 10.39). For $E_x = 17.5$ keV (Mo Kα), we find $\Delta E_x = 600$ eV for $\theta = 90°$, which can be resolved by an energy-dispersive Si(Li) detector. This approximate calculation of Compton scattering assumed quasi-free electrons with binding energies less than the energy transferred.

c) Photoelectric Absorption. X-rays with energies $E_x \geq E_{nl}$ can ionize the corresponding shell of ionization energy E_{nl}. The quantum energy is completely absorbed and the excess energy is converted into kinetic energy of a photoelectron. Unlike electron excitation of inner shells, where the maximum of the cross-section is at an overvoltage ratio of the order of $u = 2$–3 (Fig. 3.9), the cross-section σ_{ph} for photoabsorption shows a steep increase (absorption edge in Figs. 10.9 and 10.32) when E_x exceeds the ionization energy E_{nl}. Behind this edge, σ_{ph} decreases with a negative power of E_x

$$\sigma_{ph} \propto Z^4 E_x^{-n} \tag{10.25}$$

where the exponent n varies between 2.5 and 3.5. The absorption edges of outer shells are split owing to the differences in the ionization energies of

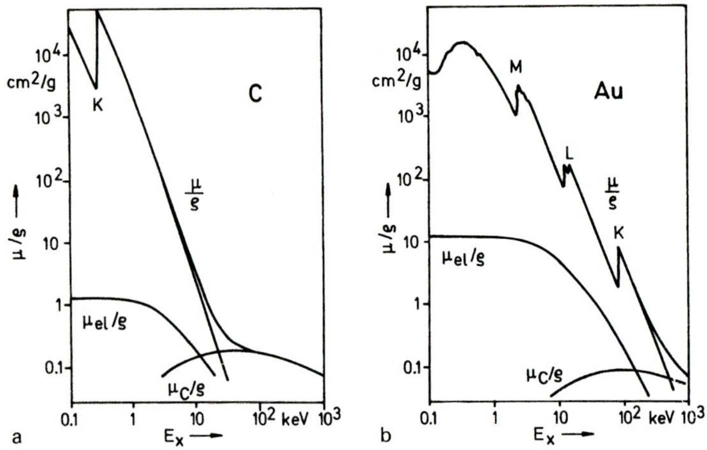

Fig. 10.9. X-ray absorption (attenuation) coefficients μ/ρ for (**a**) C (**b**) Au as a function of quantum energy E_x shown as the sum of elastic scattering (μ_el), inelastic Compton scattering (μ_c) and photoionization (μ_ph). The latter shows the characteristic absorption edges when E_x exceeds E_I

subshells; there are three L and five M edges, for example (Figs. 10.9 and 10.32).

With $N_\mathrm{A}\mathrm{d}(\rho z)/A$ atoms per unit area of a film of mass–thickness $\mathrm{d}(\rho z)$ and the total x-ray absorption cross-section $\sigma_\mathrm{x} = \sigma_\mathrm{el} + \sigma_\mathrm{C} + \sigma_\mathrm{ph}$, the dependence of the number of x-ray quanta passing through this layer becomes

$$\mathrm{d}N = -N\frac{\sigma_\mathrm{x} N_\mathrm{A}}{A}\mathrm{d}(\rho z) = -N\frac{\mu}{\rho}\mathrm{d}(\rho z) \qquad (10.26)$$

where $\mu = \mu_\mathrm{el} + \mu_\mathrm{C} + \mu_\mathrm{ph}$ is the linear and μ/ρ the mass-attenuation coefficient in units of cm^{-1} and cm^2/g, respectively. Integration of (10.26) results in the exponential absorption law with the mass–thickness $x = \rho z$

$$N = N_0 \exp[-(\mu/\rho)x] = N_0 \exp(-\mu z) \; . \qquad (10.27)$$

For a multi-component target, we have the sum rule

$$\frac{\overline{\mu}}{\rho} = \sum_i c_i \frac{\mu_i}{\rho} \; . \qquad (10.28)$$

The dependence of σ_ph on Z and E in (10.25) gives

$$\frac{\mu}{\rho} \propto \frac{Z^4}{A} E_\mathrm{x}^{-n} \simeq Z^3 E_\mathrm{x}^{-n} \; . \qquad (10.29)$$

Such a formula has been used [10.16–18] to fit experimental and tabulated values of μ/ρ [10.19]. Photon absorption cross-sections from $E_\mathrm{x} = 0.1$–1000 keV are tabulated in [10.20, 21].

In a double-logarithmic plot of μ/ρ versus E_x the power law in E_x produces a straight line between absorption edges with a negative slope $-n$. The contributions μ_C/ρ and μ_{el}/ρ and the total mass-attenuation coefficient μ/ρ are plotted in Fig. 10.9 against the quantum energy, which shows that in the energy range under consideration, μ_{ph}/ρ is the dominant term.

10.1.5 Auger Electron Emission

As mentioned in Sect. 10.1.3, the de-excitation energy produced when an inner-shell vacancy is filled with an electron from an upper shell can be transferred without radiation directly to an atomic electron, which then leaves the atom as an Auger electron [10.22–24]. For de-excitation energies below 2 keV, the probability $a = 1 - \omega$ for Auger electron emission is near unity. In x-ray emission, only two atomic levels, that of the inner-shell vacancy and that from which an electron fills the hole, are involved, whereas in Auger electron emission a third level (subshell) has to be taken into account, that from which the Auger electron is ejected. These three subshells are used to characterize the Auger electron. In the Auger process shown in Fig. 10.5b, ionization occurs in the K shell, the hole is filled with an electron from the L_1 shell and the Auger electron is emitted from the L_2 shell. The transition is therefore called KL_1L_2 and the symbol V is used when the last or the two last levels lie in the valence band. This simple spectroscopic notation will in general be sufficient to describe and identify an Auger peak in an energy spectrum.

For the example of a KL_1L_2 Auger electron, the exit energy becomes

$$\begin{aligned} E_{AE} &= E_K(Z) - E_{L_1}(Z) - E_{L_2}(Z+\Delta) - \phi_w \\ &= E_K(Z) - E_{L_1}(Z) - \Delta[E_{L_2}(Z+1) - E_{L_2}(Z)] - \phi_w \,. \end{aligned} \tag{10.30}$$

The first energy difference will be emitted as an x-ray quantum when the transition is radiative with a probability ω. The subtraction of $E_{L_2}(Z)$ (third term) considers in a first-order approach that the Auger electron has to overcome the ionization energy of the L_2 subshell. The term containing Δ, which varies between 0.5 and 1.5, descibes relaxation effects. In the case of x-ray emission, we finally have one and in the case of Auger-electron emission, two holes in the L shell, which influences the position of the effective energy levels. The last term represents the influence of the work function ϕ_w of the material used in the construction of the energy spectrometer. Figure 10.10 shows the position of E_{AE} versus atomic number Z for different series of Auger electrons.

The shape and position of the Auger peak are influenced by the chemical environment and type of binding [10.25].

The Auger electrons with discrete exit energies and with an energy width of the order of a few electronvolts are superposed on the low energy tail of the BSE spectrum (Figs. 1.5 and 10.12a). For a better identification of the

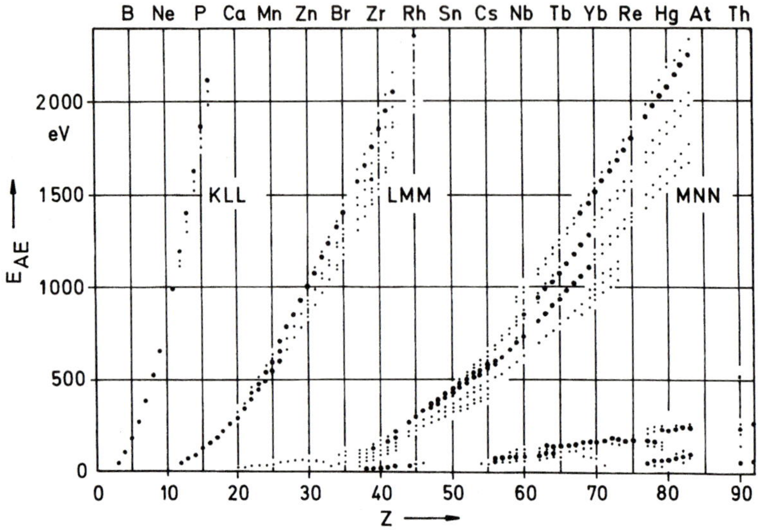

Fig. 10.10. Dependence of the Auger electron energies E_{AE} of the different series on atomic number Z

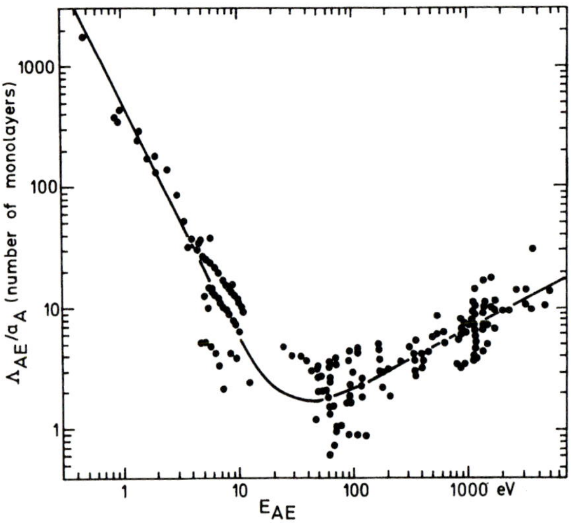

Fig. 10.11. Dependence of inelastic mean-free-path length Λ_{AE} in multiples of atomic monolayers on the Auger electron energy E_{AE} for a variety of materials [10.26]

peaks against the BSE background, the Auger spectrum is often recorded as a differentiated spectrum (Fig. 10.12b).

Energy losses caused by inelastic scattering of Auger electrons lead to a decrease of the no-loss Auger peak proportional to $\exp(-x/\Lambda_{\mathrm{AE}})$ where x is the path length inside the solid before leaving the specimen and Λ_{AE} is the

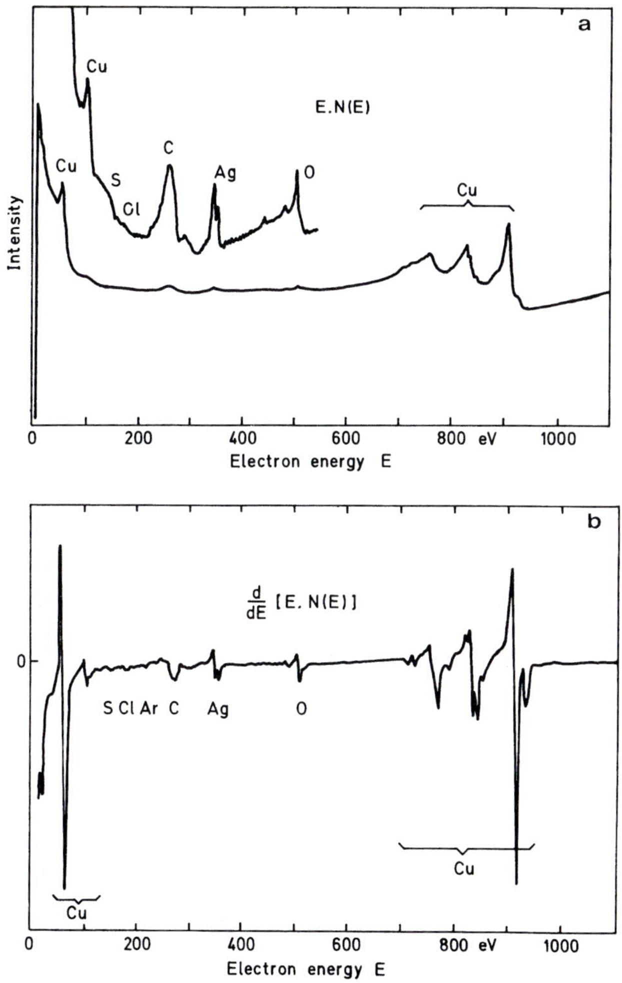

Fig. 10.12. (**a**) Directly recorded Auger electron spectrum $E \cdot N(E)$ of contaminated copper and (**b**) its derivative [10.28]

mean-free-path length between the plasmon losses and/or intra- and interband transitions that contribute to the energy-loss spectrum (Sect. 3.2.1). Because most of these energy losses show maxima broader than 1 eV, these plurally inelastically scattered Auger electrons can hardly be identified in the BSE background. For this reason, only atoms in a thin surface layer of the order of $\Lambda_{\rm AE}$ can contribute to the sharp peaks of no-loss Auger electrons. This exit depth $\Lambda_{\rm AE}$ can be measured by examining increasingly thick overlayers on a substrate [10.27], for example. The experimental values of $\Lambda_{\rm AE}$ lie in the range of one to ten monolayers for $E_{\rm AE}$ = 10–2000 eV with a shallow minimum at 30–50 eV (Fig. 10.11) and can be fitted [10.28] by

$$\Lambda_{\mathrm{AE}} = [538 E_{\mathrm{AE}}^{-2} + 0.41 (a_{\mathrm{A}} E_{\mathrm{AE}})^{1/2}] a_{\mathrm{A}} \tag{10.31}$$

where E_{AE} is in eV and Λ_{AE} in nm, and $a_{\mathrm{A}}^3 = A/N_{\mathrm{A}}\rho$ is the volume per atom in nm^3 . The ratio $\Lambda_{\mathrm{AE}}/a_{\mathrm{A}}$ can be interpreted as the number of monolayers. The question of whether there exists a universal mean-free-path curve for inelastic scattering remains open [10.29, 30]. The integral peak intensity of Auger electrons can be estimated from the formula

$$n_{\mathrm{AE}} = n_0 \sec\phi\, \sigma_{\mathrm{I}} a_{\mathrm{I}} (1 + r)(N_{\mathrm{A}}\rho/A) \Lambda_{\mathrm{AE}} T_{\mathrm{s}} \epsilon_{\mathrm{s}} \tag{10.32}$$

where n_0 denotes the number of incident electrons with a tilt angle ϕ of the specimen, σ_{I} and a_{I} are the ionization cross-section and the Auger electron yield of the shell involved and r = 0–0.6 is a backscatter factor included because backscattered electrons on their trajectories through the surface layer can also ionize the corresponding shell and contribute to the Auger electron emission [10.31, 32]. This means, for example, that the number of Auger electrons from a thin Al film on a Au substrate becomes of the order of a factor $(1+r)$ = 1.5 larger than for a bulk Al specimen. The AE signal, therefore, shows a matrix effect similar to that for SE though both are emitted only from a very thin surface layer. The last factors in (10.32) represent the transmittance (collection efficiency) T_{s} and the detection efficiency ϵ_{s} of the spectrometer used. Electron channelling effects also influence the number of Auger electrons emitted, which is thus dependent on crystal orientation (Sect. 8.1.4).

10.2 X-Ray Spectrometers

10.2.1 Wavelength-Dispersive X-Ray Spectrometers

A wavelength-dispersive spectrometer selects a distinct x-ray wavelength λ_{x} from an x-ray spectrum by Bragg reflection at a single crystal with lattice-plane spacings d parallel to the surface. Bragg's law

$$2d \sin\theta_{\mathrm{B}} = n\lambda_{\mathrm{x}} \tag{10.33}$$

requires that the incident and emergent angles θ_{B}, should be equal and so the incident and emergent directions include an angle $180°-2\theta_{\mathrm{B}}$. A spectrum can be scanned by rotating the crystal through an annular range and rotating the detector at twice the angular speed.

Optimum focusing with a larger solid angle of collection can be achieved with a Johansson spectrometer (Fig. 10.13). The analysing crystal and the two foci, where the electrons strike the specimen and on the detector slit, are located on a focal circle of radius r_{f} (Rowland circle). The crystal-lattice planes are cylindrically bent to twice the radius of the focal circle ($r_{\mathrm{d}} = 2r_{\mathrm{f}}$) and the crystal is ground to the radius r_{f} of the Rowland circle.

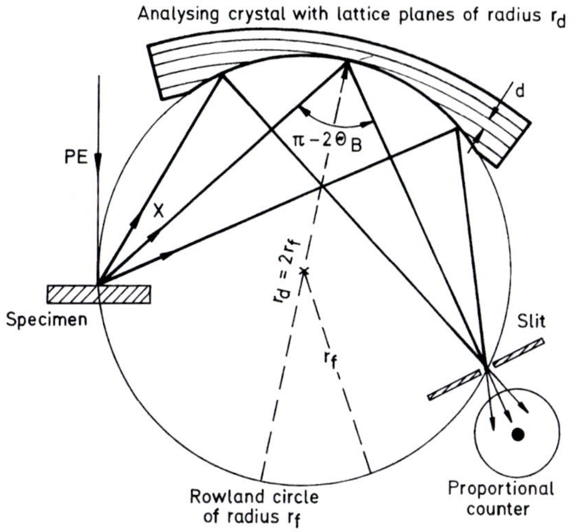

Fig. 10.13. Focusing of x-rays with a Johannsson spectrometer where the source on the specimen, the ground crystal surface and the slit in front of the proportional counter lie on a Rowland circle of radius r_f. The lattice planes of the crystal are bent to a radius $r_d = 2r_f$

Figure 10.14 shows how the crystal has to be moved along a straight line C and to be rotated to satisfy Bragg's law. The detector slit moves on a cloverleaf figure D and its axis must always face the crystal. The centre of the focusing circle moves on the circle R. Owing to the finite sizes of the crystal and the detector, ranges of Bragg angles near 0 and 90° have to be excluded and one crystal can analyse only a limited range of wavelengths inside the angular range of the order of $\theta = 15° - 70°$. The analyser crystals have to be changed from a material with a small lattice-plane spacing such as (200) LiF suitable for short wavelengths to another with a large spacing such as lead stearate for long wavelengths. Figure 10.15 shows the range of quantum energies E_x which can be analysed and the resolution obtainable as the full width ΔE_{FWHM} at half maximum. This figure also contains for comparison the ranges covered by a Si(Li) energy-dispersive detector and a proportional counter.

A crystal spectrometer can record only one wavelength at a time (characteristic line or background) inside its range. X-ray microanalysers are therefore equipped with two or three spectrometers in which the crystals can be changed without breaking the vacuum.

The x-ray quanta passing the analyser slit are counted by a gas-filled proportional counter, which has the familiar features of a Geiger counter, namely, a central wire at a positive bias of a few hundreds or thousands of volts and an earthed cylindrical electrode. The x-rays enter the counter through a window in the radial or axial direction. The latter has the advantage that the absorption path is longer, which is necessary to absorb high-energy x-ray quanta. The lowest detectable quantum energy is limited by the absorption of x-rays in the window, which therefore consists of a thin mylar foil a few micrometres in thickness supported by a metal grid. With ultrathin windows, quantum

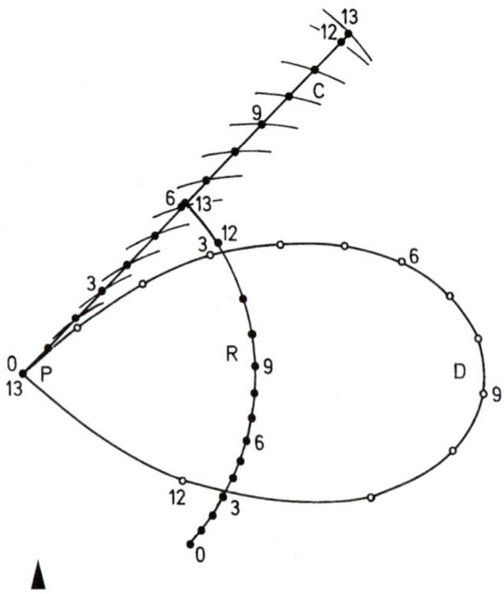

Fig. 10.14. Movement of the analysing crystal on a straight line C, of the centre of the Rowland circle on the circle R and of the detector slit on the cloverleaf figure D to permit variation of the x-ray wavelength analysed while fulfilling the Rowland condition

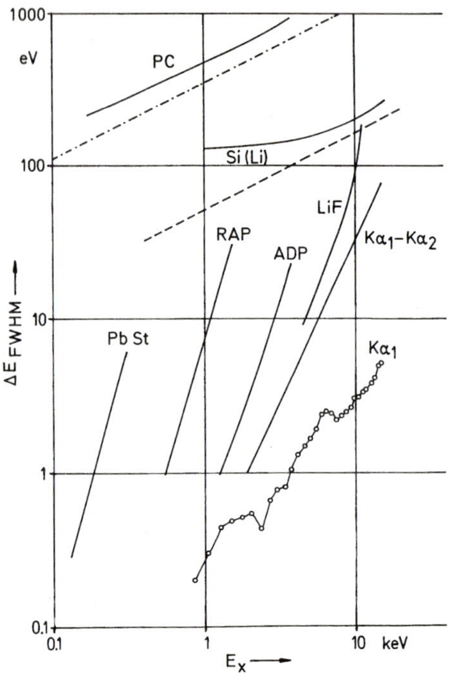

Fig. 10.15

Fig. 10.15. Range of quantum energies E_x that can be analysed by crystals of LiF. ADP = ammonium dihydrogen phosphate, RAP = rubidium acid phthalate and PbSt = lead stearate and the full widths at half maximum ΔE_{FWHM} obtainable, as a measure of resolution. Values for an energy-dispersive Si(Li) detector and a proportional counter are included for comparison (– – –: theoretical lower limits). $K\alpha_1 - K\alpha_2$ = separation of the $K\alpha_{1,2}$ doublet and $K\alpha_1$ = natural width of the $K\alpha_1$ lines

energies down to Be $K\alpha$ (E_K = 108 eV, λ_x = 11.6 nm) can be analysed. Gas-flow proportional counters, which have a lower leak rate at the windows and a longer lifetime, are also in use.

The x-ray quanta absorbed by the gas atoms by photoabsorption (Sect. 10.1.4) create a pulse with a mean number $E_x/\overline{E}_i$ of primary electron–ion pairs ($\overline{E}_i$ = 20–30 eV is the mean ionization energy of the detector gas). The electrons are accelerated by the high electric field near the wire and ionize further gas atoms, so that an avalanche of electrons is produced. The internal amplification factor is between 10^3 and 10^5 and the time constant of one pulse is of the order of 10^{-5} s. The pulse height is proportional to the quantum energy E_x and a pulse-height discriminator can be used to eliminate the high-order Bragg reflections with $n > 1$ in (10.33), for example.

The proportional counter is, therefore, an example of an energy-dispersive detector, but because of the high mean ionization energy $\overline{E}_i$ and the large

Fano factor (Sects. 5.4.2 and 10.2.2), the energy resolution is not comparable with that of a lithium-drifted silicon detector (Fig. 10.15). Pulses of Kα lines can only be separated for differences in the atomic number $\Delta Z > 3$ and for a weak line near a strong one, even $\Delta Z = 3$ will not be sufficient [10.33].

10.2.2 Energy-Dispersive X-Ray Spectrometers

An energy-dispersive spectrometer [10.34, 35] measures the energy of x-ray quanta from the number of electron-hole pairs generated in a semiconductor. The pulse height of the charge collected will be proportional to the quantum energy E_x of the x-rays absorbed. The x-ray quanta are absorbed by the photoelectric effect (Sect. 10.1.4) and the excess energy of an excited electron is dissipated in the crystal and creates electron–hole pairs, the mean energy required being $\overline{E}_i = 3.8$ eV per electron–hole pair in silicon. If the K shell of a Si atom is ionized, the vacancy will be filled by an electron from an outer shell and a Si Kα quantum of energy $E_{K,Si} = 1.739$ keV can either be reabsorbed by the photoelectric effect or leave the crystal. In the former case, the whole primary quantum energy is converted to produce $N = E_x/\overline{E}_i$ electron–hole pairs, but in the latter case, the number is reduced to $N' = (E_x - E_{K,Si})/\overline{E}_i$ and an escape peak can be observed in the recorded spectrum at an energy of 1.739 keV below the strongly excited characteristic peak (Fig. 10.22b) [10.36].

Because of the high penetrating power of x-rays, it is necessary to use 3–5 mm thick silicon crystals and the charge carriers generated have to be separated and collected irrespective of where the Auger and photoelectrons are formed. For this, a depletion layer of high electrical resistance with a thickness of 3–5 mm is needed. This can be obtained by the diffusion of lithium atoms at 500°C into a p-type silicon crystal. After the diffusion, the exponential concentration profile of Li atoms divides the crystal into a p- and an n-region. The p-n junction is at the concentration c_i where the Li atoms compensate the acceptor atoms of the p-type crystal and the crystal becomes intrinsic. The action of a reverse-biased electric field at 150°C

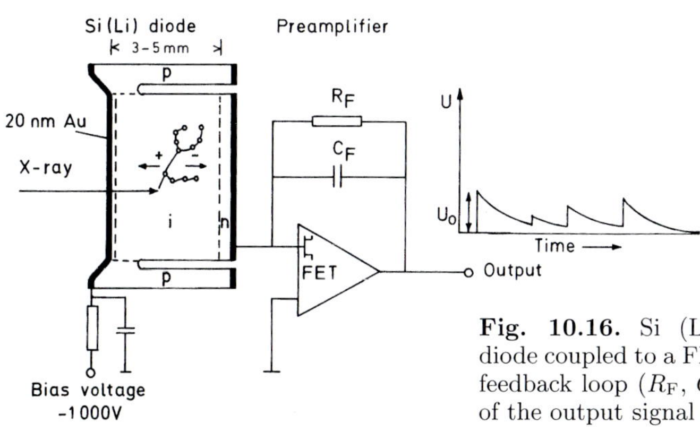

Fig. 10.16. Si (Li) x-ray detector diode coupled to a FET with a resistive feedback loop (R_F, C_F) and the shape of the output signal

increases the width of the intrinsic zone to 3–5 mm and only thin p- and n-type zones remain at the left and right-hand sides, respectively. The result is a p-i-n diode, which possesses the desired properties. Figure 10.16 shows a cross-section through such a Si(Li) detector crystal with a conductive Au coating of the order of 20 nm and a dead layer of p-type Si of the order of 100 nm. Cooling the crystal with liquid nitrogen reduces the leakage current from 500 nA at room temperature to 0.1 pA. This allows us to apply a reverse bias of the order of 1000 V across the intrinsic zone of the crystal so that, depending on the position at which the charge carriers are generated, the collection time ranges from 20 to 100 ns. The Si(Li) crystal should be permanently cooled but a short stay at room temperature does not alter the performance of the detector. However, application of a bias while the crystal is at room temperature causes diffusion of the Li atoms and destroys the intrinsic zone. The cooling to liquid nitrogen temperature requires a high vacuum to avoid contamination and a 7.5–25 μm thick beryllium or some other ultrathin window (see below) is situated in front of the detector crystal to separate the vacua of the microscope and the detector.

The energy resolution of an energy-dispersive spectrometer is limited by the statistics of electron–hole pair production and the noise amplitude of the preamplifier. A monochromatic x-ray line thus becomes Gaussian in shape and the full width at half maximum (FWHM) of the observed pulse-height statistics can be used as a measure of resolution. For the commercial specification of detector resolution, the FWHM of the 5.894 kev Mn Kα line is used (dashed line in Fig. 10.17). The limitation by the statistics can be calculated from the mean number $N = E_x/\overline{E}_i$ of electron–hole pairs with a statistical variation $\Delta N = [\text{var}(N)]^{1/2} = (NF)^{1/2}$. The Fano factor F would be unity for statistically independent processes. In this mechanism of electron–hole formation, this factor is much less than unity because the events causing electron–hole pair creation are not statistically independent. The same Fano factor also had to be taken into account in the discussion of semiconductor detectors in Sect. 5.4.2. The Fano factor of silicon is of the order of 0.09 for present-day detectors and 0.05 is the theoretical lower limit. The statistical variation $\Delta N = [\text{var}(N)]^{1/2}$ means that 68.3% of the counts are within the interval $N \pm \Delta N$ and 95% within $N \pm 2\Delta N$. The FWHM is a factor 2.35 broader and the contribution of the statistics to the FWHM becomes

$$\Delta E_{\text{st}} = 2.35\overline{E}_i\Delta N = 2.35(E_x\overline{E}_iF)^{1/2} . \tag{10.34}$$

The contribution ΔE_n of the preamplifier noise is independent of the x-ray energy but depends on the pulse-shaping time-constant of the main amplifier. For a detector 12 mm^2 in area, the optimum time-constant is about 14 μs with ΔE_n = 80–120 eV. The total resolution becomes

$$\Delta E_{\text{FWHM}} = [(\Delta E_n)^2 + (\Delta E_{\text{st}})^2]^{1/2} \tag{10.35}$$

Figure 10.17 shows the variation of resolution as a function of quantum energy E_x for different noise levels of the amplifier.

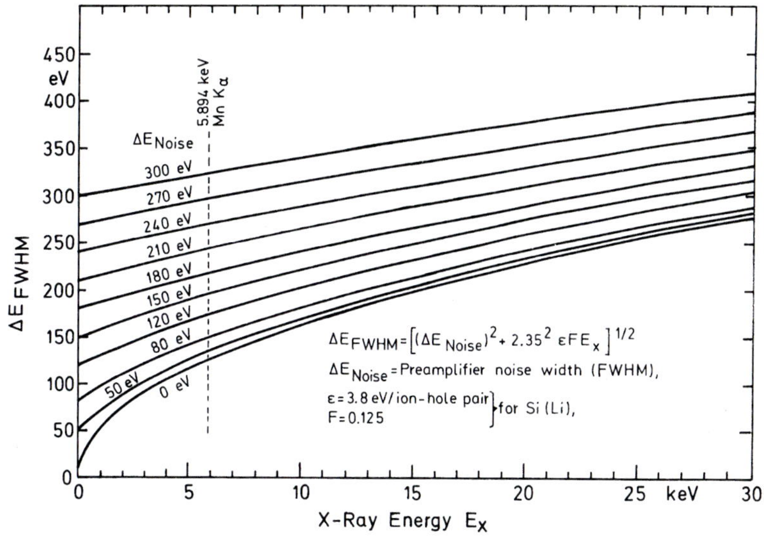

Fig. 10.17. Resolution ΔE_{FWHM} as the full width at half maximum of the Gaussian peak x-ray line of quantum energy E_{x} for different levels of amplifier noise

The efficiency $\eta(E_{\mathrm{x}})$ of recording an x-ray quantum of energy E_{x} depends on the transmission through the Be window and the probability of absorption inside the intrinsic zone of the Si(Li) crystal ($t_{\mathrm{Si}} = 3$–5 mm):

$$\eta(E_{\mathrm{x}}) = \exp[-\mu_{\mathrm{Be}}(E_{\mathrm{x}})t_{\mathrm{Be}}]\{1 - \exp[-\mu_{\mathrm{Si}}(E_{\mathrm{x}})t_{\mathrm{Si}}]\} \tag{10.36}$$

where the μ are the corresponding x-ray absorption coefficients (Sect. 10.1.4). Curves for different thicknesses t_{Be} of the window and t_{Si} of the crystal are plotted in Fig. 10.18. When using a Be window, only Kα lines of elements down to $Z = 11$ (Na) can be analysed (see positions of the K lines for various elements at the top of Fig. 10.18). In principle, it is also possible to analyse the C, N and O Kα lines with quantum energies of 277, 392 and 524 eV, respectively, if a windowless detector is used; the window is removed when there is high vacuum on both sides [10.37, 38]. Nowadays, ultrathin windows of $\simeq 500$ nm of diamond-like carbon or boron compounds, or metal-coated plastic films (e.g. 300 nm pyrolene + 20–40 nm Al) capable of withstanding atmospheric pressure are in use. Even so, the lines of these elements are not well resolved since the FWHM is of the order of 100 eV and the absorption of x-rays in the 20 nm Au coating and the 100 nm dead layer also have to be taken into account (see dashed transmission curves in Fig. 10.18). The formation of ice and CO_2 and contamination layers on the cooled detector reduces the efficiency especially for low quantum energies [10.39]. The efficiency and the action of conditioning should be controlled frequently, therefore, by recording a standard spectrum [10.40]. $CaCO_3$ can be used as a standard by controlling the ratio of the C K (277 eV) and O K (525 eV) intensities with

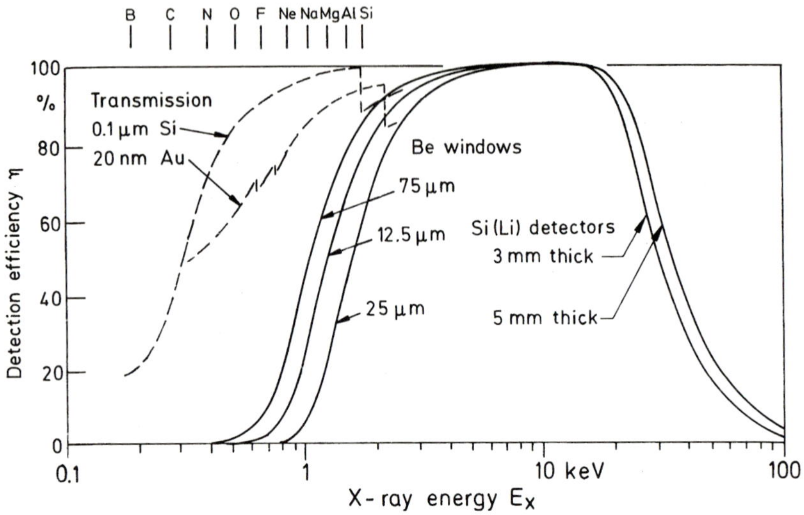

Fig. 10.18. Dependence of the detection efficiency $\eta(E_x)$ of a Si (Li) x-ray detector on quantum energy E_x and on the thicknesses of the Be window and the detector crystal. The transmission through a 20 nm Au film and a 0.1 μm dead-layer in Si decreases the efficiency of a windowless detectors. Energies of the K lines of low Z elements are drawn at the top of the figure

the Ca K (3.692 keV) intensity; the latter is less affected by contamination [10.41]. Another possibility is a control by the continuous background below $E_x = 1$ keV in Ti, Si or Ge spectra [10.42]. This contamination problem and the different efficiencies of commercial detectors are a mayor handicap for quantitative analysis by low-energy x-ray quanta [10.43]. The absorption of weak x-rays in contamination layers on the specimen makes cleaning, e.g. by ion-beam sputtering [10.44] important, as it is for more frequently used standards.

Backscattered electrons with energies higher than 25 keV can penetrate the 7.5 μm Be window and increase and falsify the continuous background. It is therefore absolutely necessary to work with a permanent magnet deflection system in front of the Si(Li) detector when the primary electron energies exceed 25 keV (Fig. 10.22a) [10.45, 46].

Nowadays high-purity intrinsic germanium detectors (HPGe or IG for short) are also in use [10.50–52]. Because of the higher absorption of x-rays in Ge, the last term of (10.36) in curly brackets extends to higher x-ray energies. For example, a 3 mm thick crystal shows 10% transmission at 23 keV in a Si- and at 74 keV in a Ge-detector. A somewhat better energy resolution is observed thanks to the lower Fano factor $F = 0.06$. The stronger and numerous escape peaks are a disadvantage. Therefore, such a detector is more suitable for TEM where the K x-ray lines can be excited for nearly all elements.

10.2.3 Electronic Circuit of an Energy-Dispersive Spectrometer

The silicon detector diode and the preamplifier work at liquid nitrogen temperature to avoid lithium diffusion in the intrinsic zone and to minimize electronic noise. The preamplifier operates as a charge-sensitive amplifier with a feedback loop. The idea of continuous feedback is illustrated in Fig. 10.16 with resistive feedback. The integrating capacitor C_F in the feedback loop of the preamplifier with a FET collects the charge of one pulse giving a voltage step

$$V_x = eE_x/C_F \tag{10.37}$$

with a pulse rise-time shorter than 100 ns. The presence of the feedback resistor R_F in Fig. 10.16 leads to an exponential decay of the step pulse:

$$V(t) = V_x \exp(-t/R_F C_F) \tag{10.38}$$

where the time constant $\tau_F = R_F C_F$ is typically in the range 1–10 ms.

Because the feedback resistor is a source of noise, pulsed feedback methods are in use, which work without R_F and correspond to $\tau = \infty$. When, by

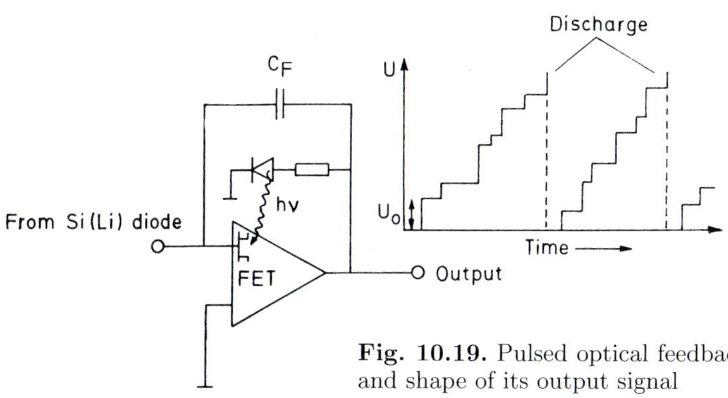

Fig. 10.19. Pulsed optical feedback preamplifier and shape of its output signal

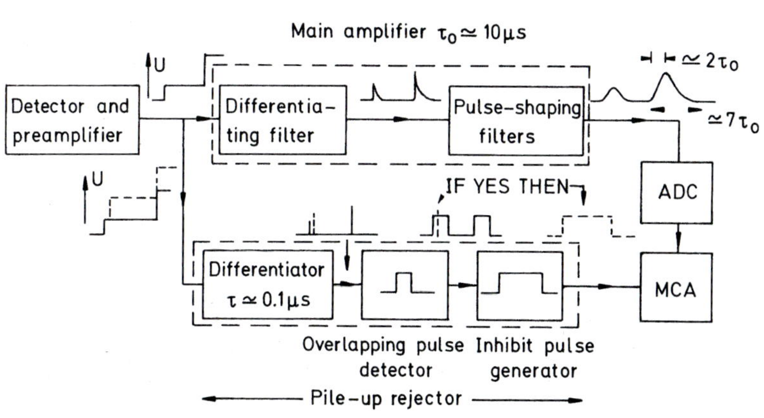

Fig. 10.20. Scheme of pulse-shaping and pulse-pileup rejection

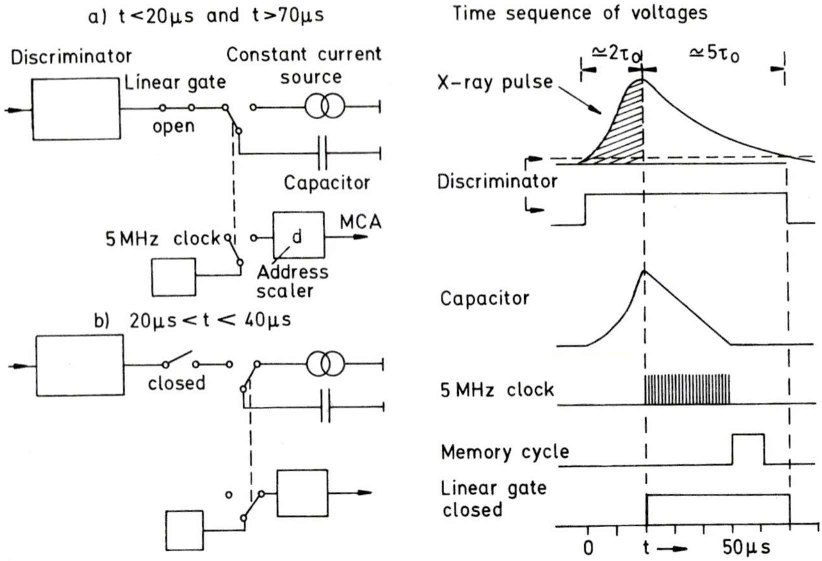

Fig. 10.21. Different stages in the analogue-to-digital conversion (ADC) of shaped x-ray pulses

integration of single pulses, the voltage U at C_F increases step by step and reaches the amplifier power-supply voltage (Fig. 10.19), the light-emitting diode is pulsed on and the light emission is optically coupled to the FET and generates a large leakage current across the gate-to-drain junction which rapidly discharges the capacitor C_F. During this short time interval, the amplifier and the multichannel analyser are gated off, because a large negative pulse is generated by this discharge.

The function of the main amplifier is to convert the signal from the preamplifier into a form that can be processed by the analogue-to-digital converter (ADC) of the multichannel analyser (MCA) with a high signal-to-noise ratio. This is achieved by means of an differentiator filter followed by active integrator filters, which produce pulses of a semi-Gaussian form (Fig. 10.20). These are low-pass filters, which cut off high-frequency noise. The shaped pulses reach their peak amplitudes, which are proportional to E_x, after about $2\tau_0$, where τ_0 is the pulse-shape time constant and the total pulse width at the baseline is of the order of $7\tau_0$. The optimum value, which finally gives the lowest value of the FWHM of the x-ray line, is of the order of 10 μs. In the case of a resistive feedback loop, the differentiation also produces negative values behind the pulse and a baseline restoration unit, which is not necessary for pulsed optical feedback, has to be used.

The principle of operation of a Wilkinson-type ADC is shown in Fig. 10.21. A discriminator just above the noise level of the baseline indicates a pulse and a capacitor is charged by the pulse (Fig. 10.21a). When the pulse reaches its peak amplitude, which is indicated by its change in slope, the capacitor is first disconnected from the input, a constant current source is then connected, the linear gate is closed and a 50 MHz clock is connected to the address scaler (Fig. 10.21b). The constant-current source discharges the capacitor linearly. The time of discharge is proportional to the x-ray quantum energy E_x. During this discharge period, the 50 MHz clock produces in the address scaler a binary number $N = 1$–1024 in the appropriate channel, increasing the previously collected number of x-ray quanta by one. During the time interval between the start of the discharging period and the end of the discriminator pulse or the end of the memory pulse, whichever is later, no input signal is accepted by the ADC. After this time interval the linear gate is open for a new pulse.

With increasing count rate, there is an increasing probability that two pulses will lie within a time interval of the order of $2\tau_0$, which means that a second pulse contributes to the peak amplitude of the first. This results in pulse pile-up, which distorts the spectrum. This can be partly avoided by means of a pile-up rejector (Fig. 10.20). The latter works with a smaller time constant, of the order of 1 μs, than the main amplifier with a pulse-shaping time constant τ_0 because it will only be necessary to detect a second pulse inside an interval of the order of $2\tau_0$, after the first pulse but not to measure its peak amplitude exactly. When a second pulse is detected inside this time interval, an inhibit pulse is fed to the MCA which prevents the counting of both x-ray pulses from the main amplifier circuit. For example, the additive lines 2Fe Kα and Fe Kα + Fe Kβ at the end of the spectrum, which occur when two characteristic x-ray quanta are counted with a time difference less than $2\tau_0$ are decreased in magnitude when using pulse pile-up rejection (Fig. 10.22b, note the logarithmic scale). Though this technique for pile-up rejection is available in most energy-dispersive detector systems, count rates higher than 2000–5000 cps should be avoided for quantitative work when using a ZAF correction program (Sect. 9.1). In another way of avoiding pulse pile-up, an electron beam blanker is triggered to blank the electron beam for a certain period each time an x-ray pulse is detected by the fast discriminator of the pulse pile-up rejection circuit. This can increase the useful linear range to 10^4 cps [10.48, 49].

In a ZAF correction program, two measurements of the counts of a characteristic line are necessary, from the substance under investigation and from a standard of known elemental concentration, and both measurements should be made at or reduced to the same value of the product $I_p \cdot t$ where I_p is the electron-probe current and t the counting time. The measurement time has to be prolonged by the dead-times during the memory period of the ADC and MCA by using a life-time corrector, which can be operated by the 50 MHz clock of the ADC and the pulse that closes the linear gate behind the discriminator.

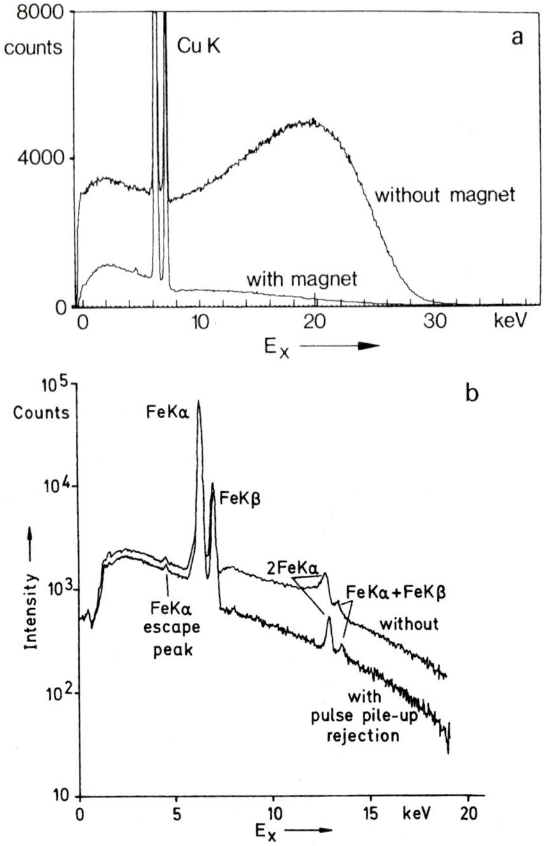

Fig. 10.22. Artifacts in energy-dispersive x-ray spectrometry. (**a**) Increase of the continuous background by recording of electron energies $\geq$ 25 keV and their avoidance by placing a permanent magnet in front of the detector [10.46]. (**b**) Pulse pile-up of additive 2 Fe Kα and Fe Kα + Fe Kβ peaks when using a high count rate and decrease of these pulses by a pulse pile-up rejector (note logarithmic scale) [10.47]. In front of the Fe Kα peak, an escape peak can be observed

10.2.4 Comparison of Wavelength- and Energy-Dispersive Spectrometers

The operation of wavelength- and energy-dispersive spectrometers, WDS and EDS, respectively, will now be summarized to bring out the advantages and disadvantages of the two types of x-ray spectrometers.

The main advantage of WDS lies in their energy resolution, which is one order of magnitude better than that of EDS: $\Delta E_x = 6$ eV (quartz) –30 eV (LiF) for a WDS compared to $\Delta E_x \simeq 150$ eV for an EDS for $E_x = 5.9$ keV (Mn Kα) (Fig. 10.17). This enables us to resolve narrower lines and to separate overlaps of L and M lines of high Z elements with K lines of low Z elements. A smaller fraction of the x-ray continuum, by a factor of about 0.1, is counted inside the smaller energy window of WDS. This increases considerably the analytical sensitivity (Sect. 10.5.2).

Owing to the preselection of the spectrometer crystal and the high count rates of the order of 10^5 cps that are possible with a proportional counter, large electron-probe currents of the order of $10^{-9} - 10^{-7}$ A can be used. By contrast, EDS have to work with count rates smaller than 10000 cps distributed over the whole spectrum. The electron-probe current cannot therefore exceed $10^{-12} - 10^{-10}$ A, which decreases the count rate for a single x-ray line. In this context, it is also important to recall the difference in the solid angle of collection, which is of the order of $\Delta\Omega \leq 10^{-3}$ sr for WDS and $10^{-3} - 10^{-2}$ sr for EDS.

An advantage of EDS is that the whole spectrum can be recorded simultaneously. Nowadays, therefore, x-ray microprobes are also equipped with an EDS because it permits a rapid survey of the main elements present in the target. With a WDS, only one line is measured per spectrometer crystal, and a spectrum has to be recorded sequentially.

The lowest quantum energy is limited to the K lines of C, N and O when using a thin window in front of the EDS but the resolution is much worser than in WDS, which can even analyse the K radiation of Be if a special spectrometer crystal and a proportional counter with an ultrathin window are used.

The focusing condition of a WDS needs exact adjustment of the specimen position on axis and at the correct height with an accuracy of the order of a few tens of micrometres. These conditions are normally satisfied by adjusting the specimen position with a light microscope. In contrast, there is no focusing condition for EDS and specimen position and height can vary over several millimetres without changing the solid angle of collection of the Si(Li) crystal. It is possible to perform qualitative x-ray microanalysis even with a rough specimen. For more quantitative analysis and application of a ZAF correction program (Sect. 10.3), the specimen has to be flat over a circle of the order of the electron range around the electron impact point and the orientation of the surface facet has to be known. ZAF correction formulae for non-normal incidence (Sect. 10.4.1) can then be applied.

X-ray mapping based on the counts of a single characteristic line (Sect. 10.6.1) can be performed with an EDS by scanning the electron beam whereas in the case of a WDS, the specimen has to be moved mechanically to satisfy the focusing condition. Otherwise, an x-ray map recorded with a WDS would contain a lower contribution of continuous x-ray quanta as discussed above.

10.3 Correction Methods for X-Ray Microanalysis

10.3.1 Problems of Quantitative Analysis

The aim of quantitative x-ray microanalysis is the measurement of atomic fractions m_i or mass fractions c_i of the elements i = a, b, c,... in a multicomponent target; for the definition of these fractions see (3.132). Because the number of 'a' atoms is proportional to c_a, we expect that the x-ray signal will be proportional to c_a in a first-order approach and an analytic calibration curve should be a straight line (Fig. 10.23). For quantitative analysis of

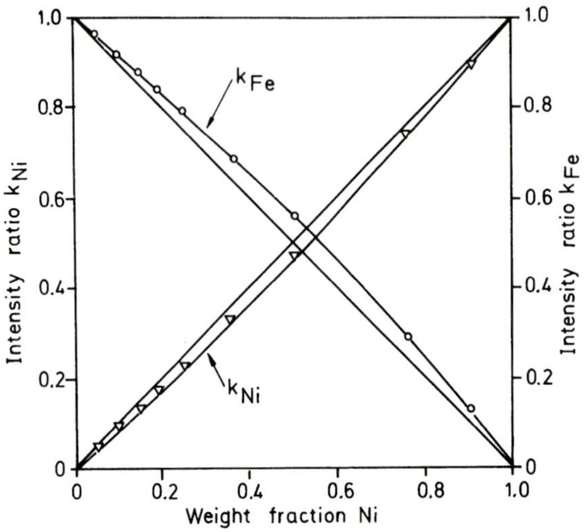

Fig. 10.23. Analytical calibration curves of the x-ray intensity ratio k for Fe-Ni binary alloys showing the dominant influences of absorption for k_{Ni} (concave shape) and of x-ray fluorescence of Fe atoms excited by Ni $K\alpha$ for k_{Fe} (convex shape of the calibration curve [10.53])

an element, two sets of x-ray counts of a characteristic line will be necessary, one, n_{a}, from the specimen under investigation and another, n_{s}, from a pure element standard ($c_{\text{s}} = 1$) of the same element or from a standard of known concentration c_{s}. These counts have to be recorded with an equal number $n_{\text{e}} = I_{\text{p}}\tau/e$ of incident electrons or have to be matched to the same product $I_{\text{p}}\tau$ (τ: counting time). This method has the advantage that all inaccurately known quantities such as the collection efficiency, the ionization cross-sections and the fluorescence yield cancel out when the x-ray intensity ratio is formed, i.e.,

$$k = n_{\text{a}}/n_{\text{s}} \simeq c_{\text{a}}/c_{\text{s}} \,. \tag{10.39}$$

However, a linear calibration curve is hardly ever found in practice; instead either a convex (Fe Kα) or concave (Ni Kα) calibration curve is observed as shown for the example of Fe-Ni binary alloys in Fig. 10.23. The following effects (1–4 in Fig. 10.24) are responsible for this lack of linearity:

1. The Bethe stopping power $S = |\mathrm{d}E_{\text{m}}/\mathrm{d}s|$ (Sect. 3.3.4) depends on the composition, and the Bethe range R_{B}, which is the maximum path length of electron trajectories, has an effect on the probability of ionization. No further x-ray quanta can be excited once the electron energy has decreased below the ionization energy E_{nl} of the corresponding shell with quantum numbers n and l.
2. The fraction of backscattered electrons with exit energies E_{B} larger than the ionization energy E_{nl} will decrease the number of x-ray quanta generated. This decrease depends on the backscattering coefficient and on the energy spectrum $\mathrm{d}\eta/\mathrm{d}E_{\text{B}}$ of the BSE of the specimen being investigated. These two effects will be considered in the atomic number (Z) correction discussed in Sect. 10.3.2.

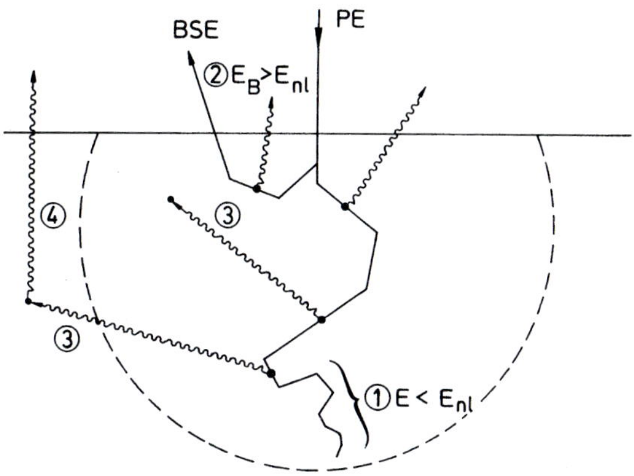

Fig. 10.24. Effects influencing the x-ray emission from a solid target (explanation of processes 1–4 in the text)

3. X-rays can be absorbed inside the specimen, depending on the depth of x-ray generation and on the mean mass-attenuation coefficient $\overline{\mu}/\rho$ of the specimen. This is taken into account by the absorption (A) correction discussed in Sect. 10.3.3.
4. Characteristic x-ray quanta of an element 'b' and that part of the continuous x-ray spectrum with quantum energies beyond the ionization energy $E_{nl,\mathrm{a}}$ can ionize atoms of element 'a' by photo-ionization (fluorescence), resulting in an increase in the number of x-ray quanta of element 'a' generated. This fluorescence (F) correction will be described in Sect. 10.3.4.

In the case of the binary Fe-Ni system in Fig. 10.23, we can neglect the influence of the first two effects because of the small difference between their atomic numbers. The concave shape of the Ni Kα calibration curve results from the strong absorption of Ni Kα quanta by Fe atoms (Phenomenon 3) whereas the Fe Kα emission is enhanced by x-ray fluorescence excited in the Fe atoms by the Ni Kα radiation (Phenomenon 4) resulting in the convex shape in Fig. 10.23.

If a sufficient number of well-known concentrations of a binary system are available, a calibration curve like that of Fig. 10.23 can be obtained; this curve can be fitted to a hyperbola to a good approximation [10.54]:

$$c = \frac{\alpha k}{1 + k(\alpha - 1)} \quad \text{or} \quad \frac{c}{1-c} = \alpha \frac{k}{1-k} . \tag{10.40}$$

If $\alpha < 1$ or $k > c$, the curve is convex like the curve for Fe in Fig. 10.23 for which $\alpha = 0.79$. If $\alpha > 1$ or $k < c$, the curve is concave like that for Ni for which $\alpha = 1.15$.

In the case of a multicomponent specimen, (10.39) must be modified: for the element 'a', we have a formula analogous to (10.40) for binary systems:

$$k_\mathrm{a} = \frac{\alpha_\mathrm{a} c_\mathrm{a}}{\sum_i \alpha_i c_i} . \tag{10.41}$$

The α_i can be calculated for the analysis of minerals using the theory developed in Sects. 10.3.2–4 [10.55, 56].

In practice, it will be easier to prepare a single standard, which in general need not be a pure-element standard. The unknown concentration c_{a} is then found by correcting the x-ray intensity ratio k (10.39) by three multiplicative factors

$$c_{\mathrm{a}} = k c_{\mathrm{s}} k_{\mathrm{Z}} k_{\mathrm{A}} k_{\mathrm{F}} \tag{10.42}$$

corresponding to the atomic number (Z), absorption (A) and fluorescence (F) corrections. However, these factors depend on the unknown concentrations themselves and the ZAF correction procedure has to be iterated. A number of ZAF correction programs have been proposed, MAGIC [10.57], MULTI [10.58], COR [10.59], FRAME [10.60] and BICEP for biological specimens [10.61, 62], for example. Nowadays, correction programs are offered together with x-ray detectors though often the source code and the model used is not stated. Controls with specimens of known composition should therefore not be neglected. Standards available commercially are listed in [10.63].

All the programs contain approximate expressions for the correction terms. The physical background and some examples will be presented in Sects. 10.3.2–4. A detailed and critical discussion of the ZAF correction methods has been published in [10.14, 64]. The user of such a program should know what approximations are being used for the different correction terms and should understand the physics behind them to enable him to decide where the correction fails or runs into uncertainty.

For quantitative analysis, the specimen composition must be homogeneous inside the x-ray excitation volume and the specimen must be plane. The ZAF correction procedure was developed for x-ray microprobes, which work with polished specimens and normal electron incidence. For microanalysis in a SEM, it can be of interest to work with tilted specimens to optimize the take-off direction though x-ray detectors with a high take-off angle at normal electron incidence with a conical polepiece are becoming more and more usual. Otherwise the influence of surface tilt must be considered for a rough specimen. Modifications of the correction procedure for tilted specimens are discussed in Sect. 10.4.1. As a special case of an inhomogeneous specimen, we mention coating films on a bulk substrate (Sect. 10.4.3); other types of problem arise when small particles on a substrate are being analysed (Sect. 10.4.4).

10.3.2 X-Ray Excitation and Atomic Number Correction

The first step is to calculate the number of characteristic x-ray quanta of element 'a' generated along the electron trajectory. The decrease of electron energy along the trajectory is described by the stopping power $S = |\mathrm{d}E_{\mathrm{m}}/\mathrm{d}x|$ (3.134) of the Bethe continuous-slowing-down approximation, where the path length element $\mathrm{d}x = \rho \mathrm{d}s$ is measured in units of g cm^{-2}. In this approach, the electron energy loss is assumed to occur continuously in small steps with energy losses mainly below 100 eV: this is not strictly true because the electrons can lose energy in larger discrete amounts if they ionize an inner shell. However, K and L shell ionizations are comparatively rare events and hence this approximation is reasonable.

If a thin layer of mass–thickness $\mathrm{d}x$ with $N_\mathrm{a} = c_\mathrm{a} N_\mathrm{A}/A_\mathrm{a}$ atoms of type 'a' and of mass concentration c_a per unit mass [in grams] is crossed by $n_\mathrm{e} = I_\mathrm{p}\tau/e$ electrons, the number $\mathrm{d}N_\mathrm{xa}$ of x-ray quanta generated of one characteristic line of the n = K, L or M series will be

$$\mathrm{d}N_\mathrm{xa} = n_\mathrm{e}\sigma_n\omega_n p_{nm} N_\mathrm{a}\mathrm{d}x = n_\mathrm{e}c_\mathrm{a}(N_\mathrm{A}/A_\mathrm{a})\sigma_n\omega_n p_{nm}\mathrm{d}x \tag{10.43}$$

where σ_n = ionization cross-section (Sect. 3.2.3, Fig. 3.9), ω_n = fluorescence yield of the n-shell of element 'a' and p_{nm} = transition probability in the n series (Fig. 10.6).

This contribution from a thin layer has to be integrated from zero to the Bethe range R_B (Sect. 3.4.1) to find the total number N_xa of x-ray quanta generated in a solid target. From the relation $\mathrm{d}x = -\mathrm{d}E_\mathrm{m}/S$, which connects the path length to a mean energy E_m, and remembering that electrons with an energy loss less than the ionization energy E_{nl} cannot ionize the nl shell, we obtain the number N_xa of x-ray quanta of element 'a' emitted from a target of mass concentration c_a

$$N_\mathrm{xa} = n_\mathrm{e}\frac{N_\mathrm{A}}{A_\mathrm{a}}\omega_n p_{nm} R_\mathrm{x,a}\int\limits_{E_{nl}}^{E}\frac{\sigma_n}{\overline{S}_\mathrm{a}}\mathrm{d}E_\mathrm{m} \tag{10.44}$$

where $\overline{S_\mathrm{a}}$ is the average stopping power of the multicomponent target, which can be calculated from (3.134), and $R_\mathrm{x,a}$ is the backscatter correction factor, which will be discussed below. Equation (10.44) already shows that N_xa is not proportional to c_a but depends on both $R_\mathrm{x,a}$ and the stopping power $\overline{S_\mathrm{a}}$ of the target.

We now consider the influence of backscattering. Backscattered electrons with exit energies $E_{nl} \le E_\mathrm{B} \le E$ cannot produce further x-ray quanta. Thus, the relative loss of ionizations by BSE characterized by the backscattering correction factor R_x depends on the energy distribution $\mathrm{d}\eta/\mathrm{d}E_\mathrm{B}$ [10.65], namely

$$R_\mathrm{x} = 1 - \left[\int\limits_{E_{nl}}^{E}\frac{\mathrm{d}\eta}{\mathrm{d}E_\mathrm{B}}\left(\int\limits_{E_{nl}}^{E_\mathrm{B}}\frac{\sigma_n}{\overline{S}}\mathrm{d}E_\mathrm{m}\right)\mathrm{d}E_\mathrm{B}\Big/\int\limits_{E_{nl}}^{E}\frac{\sigma_n}{\overline{S}}\mathrm{d}E_\mathrm{m}\right] . \tag{10.45}$$

When forming the ratio k (10.39), the average stopping power $\overline{S}$ as well as the backscatter factor R_x will be different in the specimen and the standard (subscripts 'a' and 's', respectively). Both effects depend on the atomic number. They are combined, therefore, in the atomic number correction factor k_Z of (10.42)

$$k_{\mathrm{Z}} = \frac{R_{\mathrm{x,s}}}{R_{\mathrm{x,a}}} \frac{\int_{E_{nl}}^{E} (\sigma_n/\overline{S}_{\mathrm{s}})\mathrm{d}E_{\mathrm{m}}}{\int_{E_{nl}}^{E} (\sigma_n/\overline{S}_{\mathrm{a}})\mathrm{d}E_{\mathrm{m}}} . \tag{10.46}$$

The constants ω_n and p_{nm} and a prefactor $\eta(E_{\mathrm{x}})\Delta\Omega/4\pi$ representing the detector and collection efficiencies cancel out in this ratio, which is one advantage of a standard-based method over a standardless method.

To derive analytical expressions for the integrals in (10.44) and (10.45), we use the simple formulae (3.96) for σ_n and (3.134) for $\overline{S}$ and introduce the overvoltage ratio $u = E_{\mathrm{m}}/E_{nl}$ or $u_0 = E/E_{\mathrm{nl}}$. Omitting the factors ω, p and R_{x} in (10.44) and setting $n_{\mathrm{e}} = 1$, we get the total mean number N_n of n-shell ionizations per electron [10.66]

$$\begin{aligned} N_n &= c_{\mathrm{a}} \frac{N_{\mathrm{A}}}{A_{\mathrm{a}}} \int_{E_{nl}}^{E} \frac{\sigma_n}{\overline{S}} \mathrm{d}E_{\mathrm{m}} = \frac{c_{\mathrm{a}} b_n z_n}{2A_{\mathrm{a}}} \int_{E_{nl}}^{E} \frac{(E_{\mathrm{m}}/E_{nl}^2)(\ln u/u)}{\sum_j c_j (Z_j/A_j) \ln(bE_{nl}u/J_j)} \mathrm{d}E_{\mathrm{m}} \\ &= \frac{c_{\mathrm{a}} b_n z_n}{2A_{\mathrm{a}}} \int_{1}^{u_0} \frac{\ln u}{\sum_j c_j (Z_j/A_j) \ln(bE_{nl}u/J_j)} \mathrm{d}u . \end{aligned} \tag{10.47}$$

We introduce the auxiliary variables

$$L = \sum_j c_j Z_j/A_j \quad \text{and} \quad \ln M = \frac{1}{L} \sum_j c_j \frac{Z_j}{A_j} \ln(bE_{nl}/J_j) \tag{10.48}$$

so that the denominator in the integral of (10.47) becomes $L\ln(uM)$ and (10.47) transforms to

$$N_n = \frac{c_{\mathrm{a}} b_n z_n}{2A_{\mathrm{a}} L} \int_{1}^{u_0} \frac{\ln u}{\ln(uM)} \mathrm{d}u . \tag{10.49}$$

We write $t = \ln(uM)$ and get $\ln u = t - \ln M$ and $M\mathrm{d}u = \mathrm{e}^t \mathrm{d}t$. Substitution in (10.49) results in

$$\begin{aligned} N_n &= \frac{c_{\mathrm{a}} b_n z_n}{2A_{\mathrm{a}} L} \frac{1}{M} \int_{\ln M}^{\ln(u_0 M)} \frac{t - \ln M}{t} \mathrm{e}^t \mathrm{d}t \\ &= \frac{c_{\mathrm{a}} b_n z_n}{2A_{\mathrm{a}} L} \left\{ u_0 - 1 - \frac{\ln M}{M} [\mathrm{li}(u_0 M) - \mathrm{li}(M)] \right\} \end{aligned} \tag{10.50}$$

where $\mathrm{li}(x) = \int_{-\infty}^{\ln x} (\mathrm{e}^t/t)\mathrm{d}t$ is the tabulated logarithmic integral. In the case of a pure element ($c_{\mathrm{a}} = 1$) and K shell ionization, (10.50) simplifies to

$$N_{\mathrm{K}} = \frac{b_{\mathrm{K}}}{Z} \left\{ u_0 - 1 - \frac{\ln M}{M} [\mathrm{li}(u_0 M) - \mathrm{li}(M)] \right\} \tag{10.51}$$

where $M = b_{\mathrm{K}} E_{\mathrm{K}}/J$. The function $ZN_{\mathrm{K}}/b_{\mathrm{K}}$ inside the braces is plotted in Fig. 10.25 as a function of u_0 for different values of M.

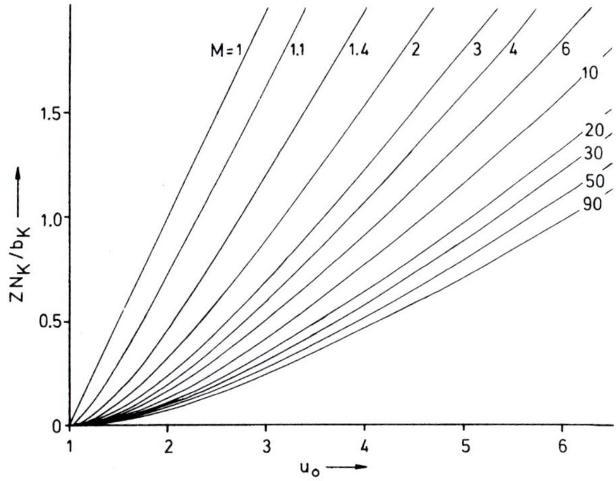

Fig. 10.25. Plot of ZN_K/b_K (10.51) versus the overvoltage ratio $u_0 = E/E_{nl}$ where N_K is the number of x-ray K quanta excited and M a parameter defined in (10.48)

We have seen in Sect. 3.3.4 that by neglecting the logarithmic term in the Bethe stopping power S ($\mathrm{d}E_x/\mathrm{d}x = -c/E$), we obtain the Thomson–Whiddington law $E^2 - E_m^2 = cx$ (3.135). We can make this approximation for S in our present calculation to obtain a simpler formula for the number of x-ray quanta emitted per electron [10.67]. This is valid if we work with modest overvoltage ratios, so that the slowly varying logarithmic term can be ignored. The cross-section σ_K will then be given approximately by (3.96), $\sigma_K = aE_K^{-2}(\ln u/u)$, and the following formula corresponding to (10.49) is obtained (c_a=1, z_K=2)

$$N_K = \frac{2b_K}{A_a}\frac{a}{c}\int_1^{u_0} \ln u\, du\,. \tag{10.52}$$

Since $\int \ln u\, du = u \ln u - u$, (10.52) becomes

$$N_K = \frac{2b_K}{A_a}\frac{a}{c}[u_0 \ln u_0 - (u_0 - 1)]\,. \tag{10.53}$$

The approximation $[u_0 \ln u_0 - (u_0 - 1)] \simeq 0.365(u_0 - 1)^{1.67}$ is applicable for values of the overvoltage ratio between 1.5 and 16 with less than 10% error. This shows that the number of x-ray quanta generated increases as

$$N_K \propto (E - E_K)^{1.67} \tag{10.54}$$

with increasing electron energy E; a similar expression was proposed long ago [10.65] with the exponent 1.5 and confirmed experimentally [10.67, 68]. In reality, N_K can take lower values at higher electron energies as a result of the increase of the electron range and stronger x-ray absorption; alternatively

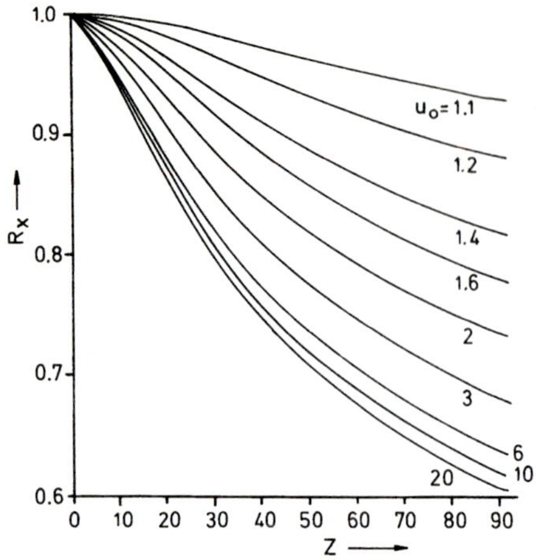

Fig. 10.26. Backscatter correction factor R_x for normal electron incidence as a function of atomic number Z and overvoltage ratio $u_0 = E/E_{nl}$ calculated by *Duncumb* and *Reed* [10.69] using experimental energy spectra of backscattered electrons obtained by *Bishop* (Fig. 4.13b)

it can take larger values if the increasing x-ray generation of the continuum contributes to the corresponding line by x-ray fluorescence.

In order to calculate the backscatter correction factor R_x, we have to know the energy distribution $d\eta/dE_B$ of the BSE; the integration in (10.45) can only be performed numerically with the aid of energy spectra obtained experimentally or by Monte Carlo simulation. Calculated values of R_x using the energy spectra of *Bishop* (Fig. 4.13b) are plotted in Fig. 10.26 as a function of the overvoltage ratio u and the atomic number [10.69]. The following power series in u_0 and Z fits these numerical results for $E = 20$ keV and can be used in a computer program

$$R_x = 1 + \sum_{j=0}^{5} \sum_{k=1}^{5} a_{jk} u_0^j 10^{-2k} Z^k \tag{10.55}$$

with the matrix elements a_{jk} listed in Table. 10.1

It must not be forgotten that this series was obtained with the values of η and $d\eta/dE_B$ corresponding to $E = 20$ keV. It is of doubtful validity in light element analysis when low primary electron energies are employed.

Table 10.1. Matrix elements a_{jk} for calculating R_x from (10.55)

	$j = 0$	1	2	3	4	5
$k = 1$	−0.581	+2.162	−5.137	+9.213	−8.619	+2.962
2	−1.609	−8.298	+28.791	−47.744	+46.744	−17.676
3	+5.400	+19.184	−75.733	+120.050	+110.700	+41.792
4	−5.725	−21.645	+88.128	−130.060	+117.750	−42.445
5	+2.095	+8.947	−36.510	+55.694	−46.079	+15.851

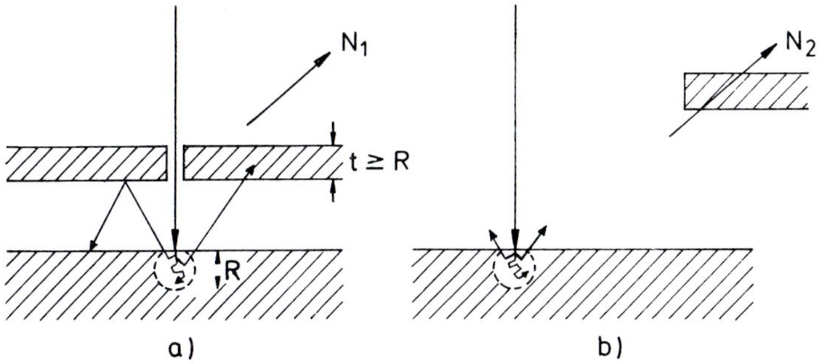

Fig. 10.27. Method of *Dérian* [10.76] for measuring the backscatter correction factor R_x

A rule similar to the mixing rule (4.18) for the backscattering coefficient of a multicomponent target

$$R_\mathrm{x} = \sum_i c_i R_{\mathrm{x},i} \tag{10.56}$$

has been proposed [10.64, 10.69]. The quantity R_x can hence be represented better in terms of the backscattering coefficient η than of Z, as proposed and confirmed by Monte Carlo simulations [10.64, 70, 71]:

$$R_\mathrm{x} = 1 - \eta \left[\sum_{n=1}^{4} \left(b_n + \frac{\eta}{u_0} \right) (\ln u_0)^n \right]^{1.67} \tag{10.57}$$

with the coefficients

b_1=+0.33148 b_2=+0.05596 b_3=−0.06339 b_4=+0.00947
c_1=+2.87898 c_2=−1.51307 c_3=+0.81312 c_4=−0.08241

Different formulae for R_x are compared in [10.72–75].

For a multicomponent target, the backscattering coefficient η (4.18) can be substituted. For a tilted specimen surface, a formula describing the dependence of η on the tilt angle ϕ such as (4.14–4.16) can be used [10.77].

A method of *Dérian* [10.76] allows R_x to be determined experimentally by using a bulk specimen with a foil of the same material in front of it (Fig. 10.27). The thickness of the foil should be larger than the electron range of the incident electrons but not much larger to avoid x-ray absorption. The electron beam first passes through a small hole in the foil (Fig. 10.27a) and we measure the number N_1 of counts equal to the mean number of x-ray quanta emitted multiplied by an absorption factor representing the penetration through the thin foil. In a second experiment (Fig. 10.27b), the number N_2 that corresponds to the number of x-ray quanta when we consider the backscatter correction is counted. A foil in front of the detector absorbs the same fraction of x-rays from the bulk as in the first experiment. The backscatter factor is given by $R_\mathrm{x} = N_2/N_1$, in which we can also take into account the fact that x-rays generated inside the foil by the BSE have to penetrate a smaller effective foil thickness for absorption.

10.3.3 Depth Distribution of X-Ray Production and the Absorption Correction

The x-ray quanta are generated at different depths, which can be described by a depth distribution function $\Phi_n(\rho z)$, normalized so that $\int_0^\infty \Phi_n(\rho z)\mathrm{d}\rho z = 1$. Usually, the depth is characterized by the mass–thickness ρz in units of mg/cm^2. X-rays generated at a depth z and collected at a take-off angle ψ with solid angle $\Delta\Omega$ are absorbed along the path length $z\,\mathrm{cosec}\psi$ (Fig. 10.28) and the number of characteristic x-ray quanta collected becomes

$$n_\mathrm{a} = N_\mathrm{xa}\frac{\Delta\Omega}{4\pi}\int_0^\infty \Phi_n(\rho z)\exp\left(-\frac{\overline{\mu}}{\rho}\,\mathrm{cosec}\psi\cdot\rho z\right)\mathrm{d}\rho z = N_\mathrm{xa}\frac{\Delta\Omega}{4\pi}f(\chi) \quad (10.58)$$

in which

$$\chi = (\overline{\mu}/\rho)\,\mathrm{cosec}\psi \quad (10.59)$$

and N_xa denotes the total number of x-rays of element 'a' generated, as in (10.44). The mean mass-attenuation coefficient $\overline{\mu}/\rho$ is the average of the coefficients for all the elements of the specimen (10.16). The absorption correction function $f(\chi)$ in (10.58) has the mathematical form of the Laplace transform of $\Phi(\rho z)$ and can be calculated from either experimental or calculated depth distribution functions. The factors $f_\mathrm{a}(\chi)$ and $f_\mathrm{s}(\chi)$ are different for the specimen and the standard and the absorption correction factor k_A in (10.42) becomes

$$k_\mathrm{A} = f_\mathrm{a}(\chi)/f_\mathrm{s}(\chi)\,. \quad (10.60)$$

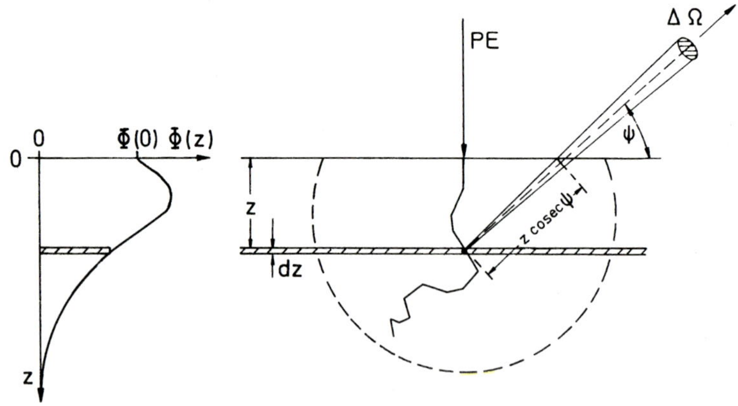

Fig. 10.28. Geometrical relations for the absorption of x-rays excited at a depth z with a depth distribution function $\Phi(z)$ and recorded with a take-off angle ψ

The depth distribution function $\Phi(\rho z)$ can be measured by the tracer method [10.78]. A thin layer of element B of the order of a few $\mu g/cm^2$ is evaporated on a substrate of element A. By evaporating additional layers of element A of various thicknesses z_A, targets are obtained for which the intensity of characteristic radiation of element B is a measure of the ionization probability, proportional to $\Phi(\rho z_A)$ at this depth. The measured intensity has to be corrected for exponential absorption by the factor $\exp[(\mu/\rho)\mathrm{cosec}\psi\, \rho z_A]$. If the atomic numbers of A and B are not very different, the diffusion will not be disturbed by the tracer layer. In this case, the target can also consist of increasing thicknesses z_B of element B on a substrate [10.79, 80]. The depth distribution is derived from the difference between the signals from two successive thicknesses. These methods can also be employed to measure $\Phi(\rho z)$ for x-ray energies below 1 keV. Comparisons with Monte Carlo simulations show that for characteristic x-ray energies $E_x < 1$ keV the generation by fast secondary electrons have to be taken into account [10.81, 82].

Various sets of $\Phi(\rho z)$ curves have been measured by these methods [10.78, 79, 83–85] and examples are shown in Figs. 10.29 and 10.30. The fact that $\Phi(\rho z)$ passes through a maximum for normal incidence is a consequence of backscattering inside the target (see discussion of the depth distribution of energy dissipation in Sect. 3.4.2). The only difference is that the maximum depth z_x of x-ray generation is smaller than the electron range R because electron energies smaller than the ionization energy E_{nl} cannot ionize the corresponding shell. Just as the electron range can be represented approximately by a power law $R = aE^n$ (Sect. 3.4.1), so can the range of x-ray excitation be approximated by

$$\rho z_x = a(E^n - E_{nl}^n) , \tag{10.61}$$

where the constants a and n can be deduced from one of the range laws in Table 3.2. Equation (10.61) expresses the fact that the depth z_x decreases with increasing ionization energy E_{nl} and can be used to estimate the information volume of x-ray analysis (see also lateral dimensions of x-ray excitation in Sect. 10.4.2).

A direct procedure for measuring $f(\chi)$ is the variable take-off method. At normal incidence the take-off angle ψ is varied, for which a detector that can be pivoted about the electron impact point is required [10.86]. For non-normal incidence and energy-dispersive analysis, a cone-shaped specimen can be used [10.87]. As the electron beam moves around the cone, the take-off angle can be varied from a maximum value depending on the cone angle to zero. Alternatively, a flat inclined specimen can be rotated about an axis parallel to the electron beam. The number of counts recorded for different take-off angles ψ is plotted as $N(\chi)$ against χ (10.59) and extrapolated to $\chi = 0$, intersecting the ordinate axis at $N(0)$. The absorption correction is

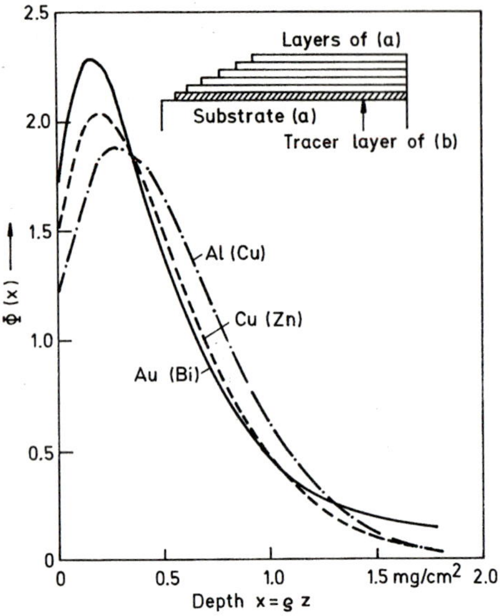

Fig. 10.29. Depth distribution function $\Phi(\rho z)$ measured by the tracer method at $E = 29$ keV [10.78] [Au(Bi) = Bi tracer in Au target, for example]. Unity at $x = 0$ means a normalization to the number $dN_a dx$ (10.43) of x-ray quanta per mass-thickness excited by primary electrons at normal incidence. The surface ionization $\Phi(0) > 1$ results from x-ray quanta excited by BSE

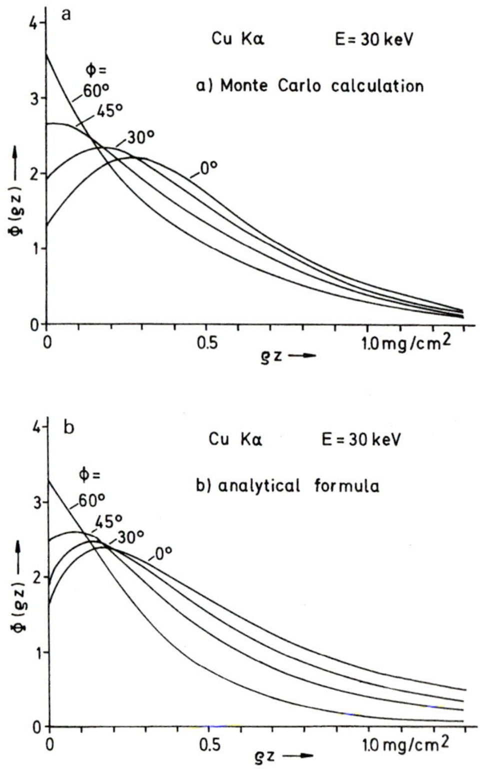

Fig. 10.30. Depth distribution function $\Phi(\rho z)$ of Cu at $E = 30$ keV for tilt angles $\phi = 0$–$60°$ (**a**) from Monte Carlo simulations and (**b**) calculated by (10.85)

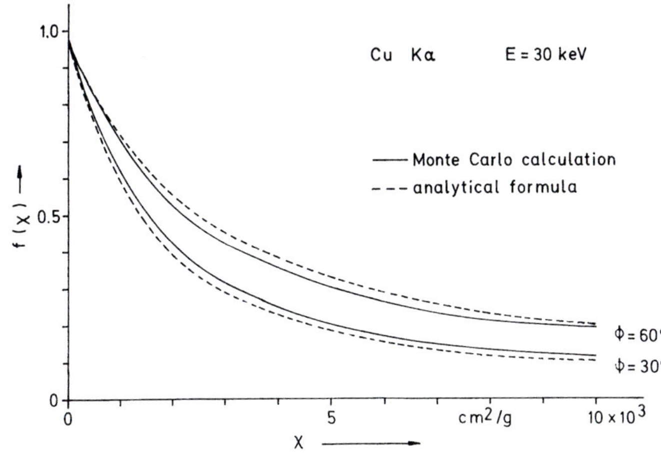

Fig. 10.31. Absorption correction factor $f(\chi)$ for copper at $E = 30$ keV for tilt angles $\phi = 30°$ and $60°$ (——: experiments [10.77], - - -: calculated by (10.86))

given by $f(\chi) = N(\chi)/N(0)$ (Fig. 10.31). The weak point of this method lies in the mode of extrapolation.

If absorption correction factors $f(\chi)$ are to be used in a computer program, an analytical approach is required. However, the depth distribution function $\Phi(\rho z)$ cannot be evaluated from the fundamental laws of electron scattering without considerable oversimplification and, in practice, the parameters occuring in the formulae obtained have to be fitted to experimental values of $f(\chi)$. In the following we describe some approximate representations of $\Phi(\rho z)$ and the corresponding $f(\chi)$, the Laplace transform of $\Phi(\rho z)$. All the $\Phi(\rho z)$ will be normalized.

In an early approximation, *Castaing* assumed that at large depths z the normalized function $\Phi(\rho z)$ can be described by a single exponential decrease

$$\Phi_n(\rho z) = \sigma \exp(-\sigma \rho z) \tag{10.62}$$

which results in

$$f(\chi) = \int_0^\infty \sigma \exp[-(\chi + \sigma)\rho z] \mathrm{d}\rho z = 1/[1 + (\chi/\sigma)] \,. \tag{10.63}$$

The expression (10.62) does not represent the maximum of emission below the surface shown in Fig. 10.30. However, it can be a good approximation to $\Phi_n(\rho z)$ for tilted specimens (Fig. 10.30).

The model of *Philibert* [10.88] uses two exponential terms

$$\Phi_n(\rho z) = \exp(-\sigma \rho z) \frac{\sigma \rho (1 + h)}{\Phi_1 + h\Phi_0} [\Phi_1 - (\Phi_1 - \Phi_0) \exp(-\sigma \rho z/h)] \,. \tag{10.64}$$

to represent the maximum. For $z = 0$, the square bracket reduces to Ψ_0, which is therefore proportional to the surface ionization. The Laplace transform of (10.64) has the form

$$f(\chi) = \frac{1 + [h\Psi_0/(\Psi_1 + h\Psi_0)]\dfrac{\chi}{\sigma}}{\left(1 + \dfrac{\chi}{\sigma}\right)\left(1 + \dfrac{h}{1+h}\dfrac{\chi}{\sigma}\right)} . \tag{10.65}$$

In a simplified model, it is assumed that $\Phi_0 = 0$; substituting this in (10.64) results in $\Phi(0) = 0$. This model thus neglects the surface ionization, which is excited not only by the primary beam but also by the BSE on their return trajectories through the surface (see also Sect. 10.4.3). Equation (10.65) simplifies to

$$f(\chi) = \frac{1}{\left(1 + \frac{\chi}{\sigma}\right)\left(1 + \frac{h}{1+h}\frac{\chi}{\sigma}\right)} . \tag{10.66}$$

a formula that is often used in correction programs, though it fails for strong absorption when the surface ionization $\Phi(0)$ becomes important. The best fit of measured values of $f(\chi)$ is obtained with the following values proposed by *Heinrich* [10.89]

$$h = \sum_i c_i \frac{1.2A_i}{Z_i} ; \quad \sigma = \frac{4.5 \times 10^5}{E^{1.65} - E_{nl}^{1.65}} \tag{10.67}$$

with σ in cm^2 g^{-1} and E, E_{nl} in keV.

In further simplied approximations, $\Phi(\rho z)$ is represented either by a rectangular distribution [10.90] or by straight lines (quadrilateral model) connecting the points $(0,\Phi(0))$, $(z_{\rm max}, \Phi_{\rm max})$ $(z_{\rm m},0)$ [10.71, 91]. The former results in

$$\Phi_n(\rho z) = \frac{1}{2\rho\bar{z}} \quad \text{for} \quad z \leq 2\bar{z} \quad \text{and zero for} \quad z > 2\bar{z} \tag{10.68}$$

where $\bar{z}$ denotes a mean depth of x-ray production. This gives

$$f(\chi) = \frac{1 - \exp(-2\chi\rho\bar{z})}{2\chi\rho\bar{z}} . \tag{10.69}$$

The maximum below the surface can also be approximated by a Gaussian function $\Phi(\rho z) = \gamma \exp[-\alpha(\rho z)^2]$ [10.92, 93]. In the PAP model (Sect. 10.3.5) $\Phi(\rho z)$ is approximated by two parabolae [10.94].

10.3.4 Fluorescence Correction

The photoelectric absorption of characteristic x-rays and of the x-ray continuum in the target causes inner-shell ionizations and the subsequent emission of secondary or fluorescent radiation. The photo-ionization cross-section

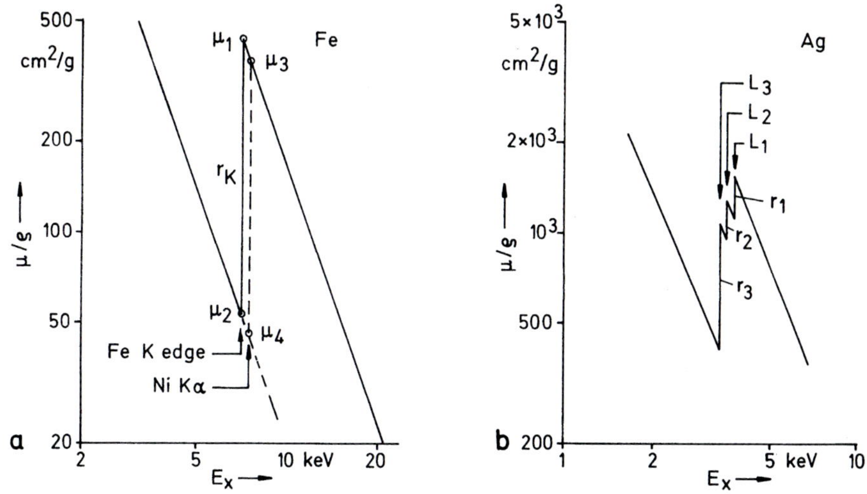

Fig. 10.32. Double-logarithmic plot of the mass-attenuation coefficient μ/ρ versus x-ray quantum energy E_x and definition of the jump ratio r for (**a**) the Fe K edge and (**b**) the Ag L edges

shows a sudden rise when the quantum energy E_x exceeds the ionization energy E_{nl} of an element after which it decreases with increasing E_x following a power law $\propto E^{-n}$ where $n \simeq 2.5$–3.5 (Sect. 10.1.4). This can be seen from the mass-attenuation coefficient μ/ρ in Figs. 10.8 and 10.32. Photo-ionization and x-ray fluorescence therefore shows a maximum for $E_x = E_{nl}$ unlike inner-shell ionization by electron impact, for which the maximum cross-section is observed at overvoltage ratios $u = E/E_{nl} \simeq 3 - 4$ (Fig. 3.9). X-ray fluorescence will, therefore, be strongest when the characteristic x-ray energy E_{xb} of an element 'b' just exceeds the ionization energy $E_{nl,a}$ of element 'a'.

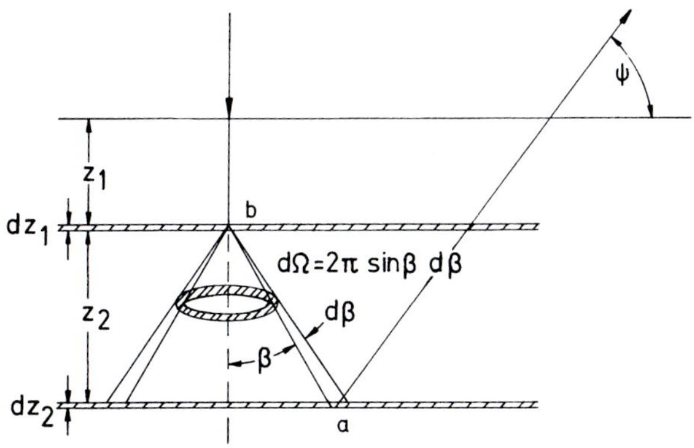

Fig. 10.33. Geometrical relations for calculating the fluorescence correction. Characteristic x-ray quanta of element 'b' excited at depth z_1 are absorbed by atoms of element 'a' at a depth $z_1 + z_2$ and excite fluorescence in 'a' that is collected with a take-off angle ψ

This process will be discussed with the aid of Fig. 10.33, where a characteristic x-ray line of element 'b' is excited by the primary electron beam in a layer of thickness $\mathrm{d}z_1$ at a depth z_1 and we wish to know how many fluorescent quanta of element 'a' are generated in a layer of thickness $\mathrm{d}z_2$ at a depth $z_1 + z_2$ [10.95, 96]. Given the depth distribution function $\Phi_b(z)$ of x-ray generation, the number of x-rays of element 'b' emitted into a solid angle $\Delta\Omega$ will be

$$\mathrm{d}^2 N_{\mathrm{pb}} = N_{\mathrm{pb}} \Phi_b(z_1) \frac{\mathrm{d}\Omega}{4\pi} \mathrm{d}z_1 = N_{\mathrm{pb}} \Phi_b(z_1) \frac{1}{2} \sin\beta \, \mathrm{d}\beta \, \mathrm{d}z_1 \tag{10.70}$$

where N_{pb} (10.44) is the total number of x-ray quanta emitted by element 'b' excited by the primary electrons. We use $\mathrm{d}\Omega = 2\pi \sin\beta \, \mathrm{d}\beta$ (Fig. 10.33) for directions of emission for which the path length of absorption $z \sec\beta$ between the layers at depth z_1 and $z_1 + z_2$ is the same. The fraction of primary radiation of element 'b' absorbed by atoms 'a' of mass concentration c_{a} in the layer $\mathrm{d}z_2$ becomes

$$\mathrm{d}^3 N_{\mathrm{ab}} = \mathrm{d}^2 N_{\mathrm{pb}} \exp(-\mu_{\mathrm{b}}^* z_2 \sec\beta) c_{\mathrm{a}} \mu_{\mathrm{a,b}} \sec\beta \, \mathrm{d}z_2 \tag{10.71}$$

where $\mu_{\mathrm{a,b}}^*/\rho$ is the mass-attenuation coefficient for the radiation of element 'b' in the matrix and $\mu_{\mathrm{a,b}}/\rho$ that of element 'b' by atoms 'a'.

Not all absorbed x-ray quanta of element 'b' will cause a photo-ionization in atoms of element 'a', but only those that correspond to the jump of the mass-attenuation coefficient. If μ_1 and μ_2 are the absorption coefficients on either side of the jump (Fig. 10.32a), we call $r = \mu_1/\mu_2$ the jump ratio; the fraction $(\mu_1 - \mu_2)/\mu_1 = (r_K - 1)/r_K$ of absorbed quanta of element 'b' in the K shell of atoms of 'a' at the depth $z_1 + z_2$ will cause ionization of atoms 'a' with the fluorescence yield ω_{a} and the transition probability p_{a} (10.43). If the quantum energy of element 'b' is somewhat larger than the ionization energy of element 'a', it can be seen from Fig. 10.32a that for a K shell ionization

$$\frac{\mu_3 - \mu_4}{\mu_3} \simeq \frac{\mu_1 - \mu_2}{\mu_1} = \frac{r_K - 1}{r_K} \simeq 0.294 - 0.00144Z \tag{10.72}$$

provided that the exponents n of the decrease of μ with increasing E_{x} on either side of the jump are of the same order. The last term in (10.72) comes from an empirical approach [10.99].

The situation becomes more complex when the L shell of element 'a' is excited by fluorescence (Fig. 10.32b). The ratio depends on where the line of element 'b' is situated relative to the ionization energies of the L subshells of 'a' and which subshell is ionized. If r_1, r_2 and r_3 are the jump ratios of the subshells, the ratios corresponding to (10.72) are listed in Table 10.2

The latter ratio can be approximated by [10.99]

$$\frac{r_3 - 1}{r_1 r_2 r_3} \simeq 0.548 - 0.00231Z \,. \tag{10.73}$$

Table 10.2. Ratios for the L subshells analoguous to (10.72) for calculating x-ray fluorescence

Subshell ionized	L_1	L_2	L_3
for $E_{L_3} < E_{\mathrm{xb}} < E_{L_2}$	—	—	$(r_3-1)/r_3$
$E_{L_2} < E_{\mathrm{xb}} < E_{L_1}$	—	$(r_2-1)/r_2$	$(r_3-1)/r_2r_3$
$E_{L_1} < E_{\mathrm{xb}}$	$(r_1-1)/r_1$	$(r_2-1)/r_1r_2$	$(r_3-1)/r_1r_2r_3$

Similar relations can be written down for the M shell but with five absorption edges (Fig. 10.9).

For the example of K radiation, the x-ray fluorescence of element 'a' recorded by a detector of solid angle $\Delta\Omega$ at a take-off angle ψ is given by

$$\mathrm{d}^3N_{\mathrm{fa}} = \mathrm{d}^3N_{\mathrm{ab}}\frac{r_K-1}{r_K}\exp[-\mu_a^*(z_1+z_2)\mathrm{cosec}\psi]\frac{\Delta\Omega}{4\pi} \qquad (10.74)$$

which we have to integrate over z_1, β and z_2 to get the total number N_{fa} of x-ray fluorescence quanta. Introducing the conversion ratio

$$G = c_{\mathrm{a}}\mu_{\mathrm{a,b}}\frac{r_k-1}{r_K}\omega_{\mathrm{a}}p_{\mathrm{a}} \qquad (10.75)$$

from the primary to the secondary intensities and substituting (10.69) and (10.70) into (10.72), we find

$$\begin{aligned} N_{\mathrm{fa}} &= \frac{\Delta\Omega}{8\pi}N_{\mathrm{pb}}G\int_0^{\infty}\int_0^{\pi}\int_{-z_1}^{\infty}\Phi_{\mathrm{b}}(z_1)\exp(-\mu_{\mathrm{a}}^*z_1\mathrm{cosec}\psi) \\ &\times \exp[-(\mu_{\mathrm{b}}^*\sec\beta+\mu_{\mathrm{a}}^*\mathrm{cosec}\psi)z_2]\tan\beta\,\mathrm{d}z_2\,\mathrm{d}\beta\,\mathrm{d}z_1 \,. \end{aligned} \qquad (10.76)$$

A knowledge of $\Phi_{\mathrm{b}}(z_1)$ is needed to perform this integration. When using the Castaing approach (10.62), (10.76) gives

$$N_{\mathrm{fa}} = \frac{\Delta\Omega}{8\pi}N_{\mathrm{pb}}\frac{G}{\mu_{\mathrm{b}}^*}f(E_{nl}^{(b)},\chi_{\mathrm{a}}^*)\left(\frac{\ln(1+\mu_{\mathrm{a}}^*/\mu_{\mathrm{b}}^*)}{\mu_{\mathrm{a}}^*/\mu_{rmb}^*}+\frac{\ln(1+\sigma/\mu_{\mathrm{b}}^*)}{\sigma/\mu_{\mathrm{b}}^*}\right) . \qquad (10.77)$$

The first term in the square bracket results from integration over z_2 in the range $z_1 \le z_2 < \infty$ and the last term from integration in the range $-z_1 \le z_2 \le 0$. The function $f(E_{nl}^{(b)},\chi_{\mathrm{a}}^*)$ is the $f(\chi)$ function introduced in Sect. 10.3.3 for the consideration of absorption but with the ionization energy $E_{nl}^{(b)}$ of element 'b' and the parameter $\chi_{\mathrm{a}} = (\mu_{\mathrm{a}}^*/\rho)\mathrm{cosec}\psi$ of element 'a'.

In most cases only K-K and L-L fluorescence have to be taken into account where the first letter stands for the line of element 'a' and the other for that of element 'b'; x-ray fluorescence only becomes significant for small differences in Z of a few units. However, whether or not K-L or L-K fluorescence has to be taken into account for neighbouring lines must be checked.

For the correction of the total measured number of x-ray quanta $N_{pa}+N_{fa}$, the multiplicative fluorescence correction factor

$$k_F = \frac{N_{pa}}{N_{pa}+N_{fa}} = \frac{1}{1+N_{fa}/N_{pa}} \tag{10.78}$$

must be substituted in (10.42) assuming that there is no fluorescence in the pure element standard. Otherwise and notably when considering fluorescence excited by the continuum, k_F has to be replaced by the ratio k_{Fa}/k_{Fs}.

The ratio N_{fa}/N_{pa} can be obtained by taking N_{fa} from (10.77) and N_{pb} from (10.53–10.54), which results in

$$\begin{aligned} \frac{N_{fa}}{N_{pa}} &= \frac{c_b}{2} P_{ij} \frac{A_a}{A_b} \left(\frac{\mu_b - 1}{\mu_a - 1}\right)^{1.67} \omega_b \frac{r_{ia}-1}{r_{ia}} \frac{\mu_{a,b}}{\mu_b^*} \\ &\times \left(\frac{\ln(1+\mu_a^*/\mu_b^*)}{1+\mu_a^*/\mu_b^*} + \frac{\ln(1+\sigma/\mu_b^*)}{1+\sigma/\mu_b^*} \right) . \end{aligned} \tag{10.79}$$

The factor P_{ij} where i, j stand for the shells involved of the elements 'a' and 'b', respectively, takes the value $P_{ij} = 1$ for K-K and L-L, 0.24 for K-L and 4.2 for L-K fluorescence.

Figure 10.34 shows the enhancement factor $1/k_F$ for the typical example of Fe-Ni binary alloys. This factor takes the value 1.4 for low Fe concentrations c_a, which means that 40% of the Fe $K\alpha$ emission is caused by fluorescence excited by the Ni $K\alpha$ radiation. The factor varies nonlinearly with c_a because μ_b^* in (10.79) decreases as $c_b = 1 - c_a$ increases owing to the reduced absorption by element 'b' (Ni) of its own characteristic radiation.

A similar treatment is necessary for the x-ray fluorescence excited by the continuum, from which all quantum energies $E_{nl,a} \leq E_x \leq eU$ can contribute

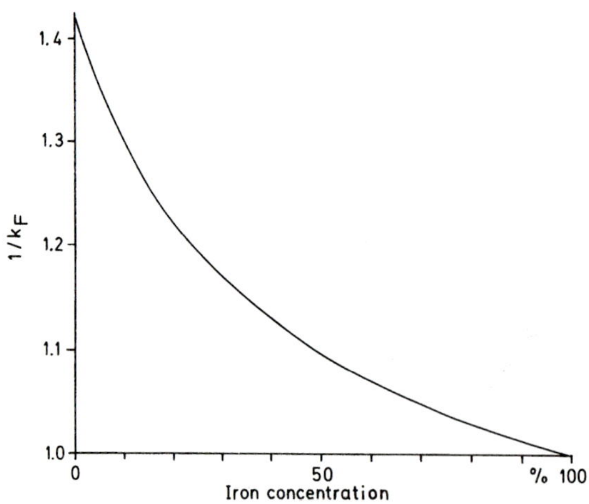

Fig. 10.34. Fluorescence enhancement factor $1/k_F = 1 + N_{fa}/N_{pa}$ for iron of concentratiopn c_{Fe} in Fe-Ni binary alloys ($E = 25$ keV, $\psi = 40°$) [10.97]

to the fluorescence. This needs a further integration over E_x which is difficult because little is known about the depth distribution of the continuous radiation, and the emission of the continuum is not isotropic.

A simple model due to *Hénoc* [10.98] assumes that all continuous radiation is emitted isotropically from the point of electron impact at the surface. This is justified in so far as the high-energy tail of the continuum is excited only at shallow depths. *Springer* [10.99, 100] assumed that the depth distribution of the continuum is the same as that of characteristic quanta of equal energy. However, for simplicity, only those isotropically emitted continuous quanta that are emitted downwards are taken into account, which is equivalent to an integration over $z_1 \le z_2 \le \infty$ in Fig. 10.33.

10.3.5 Refined Methods and Standardless X-Ray Microanalysis

The question whether the multiplicative ZAF correction scheme (10.42) is correct has often be discussed [10.101]; in particular, is it permissible to consider the backscatter factor R and the stopping power separately. We saw in the last Sections that the depth distribution of ionization $\Phi(\rho z)$ is the key for calculating both the atomic number and absorption correction and it will also be of interest in the calculation of the fluorescence correction. In case of the ZAF method, we can summarize the number of emitted x-rays for specimen or standard by

$$N = n_e c \frac{\Delta\Omega}{4\pi} \eta(E_x) \omega_{\text{eff}}\, p_{nm} \frac{1}{k_F} f(\chi) \frac{N_A}{A} R \int_E^{E_n} \frac{\sigma_n(E')}{\mathrm{d}E'/\mathrm{d}\rho s} \mathrm{d}E' \tag{10.80}$$

in which the influences of detection efficiency, fluorescence, absorption and atomic number corrections are shown sequentially. In a $\Phi(\rho z)$ model this equation can be written

$$N = n_e c \frac{\Delta\Omega}{4\pi} \eta(E_x) \omega_{\text{eff}}\, p_{nm} \frac{1}{k_F} \frac{N_A}{A} \sigma(E) \int_0^{z_{\max}} \Phi(\rho z) \exp(-\chi \rho z) \mathrm{d}\rho z\ . \tag{10.81}$$

Different modifications have therefore been proposed based on experimental, calculated and Monte-Carlo-simulated $\Phi(\rho z)$: the proposal of *Brown* and *Packwood* [10.102], the *Love* and *Scott* model [10.71] using quadrilateral $\Phi(\rho z)$ profiles (Sect. 10.3.3), the *Pouchou* and *Pichoir* (PAP) model using parabolic fits of $\Phi(\rho z)$ [10.94] and modifications by *Bastin* and *Heijligers* [10.93] especially developed for ultra-light elements. For soft x-ray lines of light elements Monte Carlo simulations of $\Phi(\rho z)$ are in excellent agreement with experiment [10.81, 82, 103] and also allow depth distributions in multi-layer (stratified) specimens to be calculated in 10–30 s [10.104]. These simulations are therefore becoming of growing interest and in principle an iterative correction with Monte Carlo simulations should be possible. The use of Monte Carlo simulations for optimum electron energies in the analysis of multi-layer structures is also of importance.

The sequential record of x-ray spectra for specimen and standard requires high constancy of the electron probe current. The ZAF method using a standard can be applied to plane but not to rough specimens. Though an EDS is less sensitive to the position of specimen and standard, deviations in the surface tilt can influence the absorption correction. A standardless method [10.105–107] makes use of the fact that the whole x-ray spectrum is a fingerprint of the analysed specimen. It includes the continuum and the peak-to-background ratios and is particularly applicable to the analysis of low Z elements.

By using (10.44), (10.58) and (10.78) the number of x-ray quanta recorded of a characteristic line (ch) can be written

$$N_a^{\mathrm{ch}} = \frac{\Delta\Omega}{4\pi}\epsilon(E_a)\frac{It}{e}\frac{N_\mathrm{A}}{A}\omega_n\, p_{nm}\, c_a R_{x,a}^{\mathrm{ch}} f(\chi)_a^{\mathrm{ch}}(1 + N_{\mathrm{fa}}/N_{\mathrm{pa}})\int_{nl}^{E}\frac{\sigma_n}{S_a}\mathrm{d}E_\mathrm{m}\,. \tag{10.82}$$

By using (10.10) the number of recorded continuous (con) x-ray quanta in the energy interval ΔE_x below the characteristic line becomes

$$N_a^{\mathrm{con}} = \frac{\Delta\Omega}{4\pi}\epsilon(E_a)\frac{It}{e}R_{x,a}^{\mathrm{con}} f(\chi)_a^{\mathrm{con}}\overline{kZ}\frac{E - E_x}{E_x}\Delta E_x\,. \tag{10.83}$$

In the peak-to-background ratio of (10.82) and (10.83), the solid angle $\Delta\Omega$, the detection efficiency $\epsilon(E_a)$ and the number of incident electrons It/e are cancelled. It is further assumed that peak and background radiation shows the same absorption and backscatter correction. The other quantities have to be calculated from fitting formula. When calculating the concentrations from the peak-to-background ratio of any element, Σc_i should be unity. In case of L lines, the strong influence of Coster–Kronig transitions on the effective fluorescence yield (10.17) has to be considered more carefully [10.106].

10.3.6 X-Ray Microanalysis of Nonconducting Specimens

The following effects can occur during the analysis of nonconducting specimens such as glass, ceramics or mineralogical specimens [10.108]. As shown in Sect. 3.5.2, strong negative charging of the surface results for electron energies larger than E_2, which reduces the landing energy of the primary electrons, retarded in the external electric field in the vacuum. This landing energy can be measured directly from the short-wavelength or high-energy cut-off of the x-ray continuum in an energy-dispersive x-ray spectrum thanks to Duane–Hunt's law (10.6). The reduced electron energy can either increase or decrease the intensity of an x-ray line depending on whether the electron energy is higher or lower than the energy corresponding to the maximum cross-section of inner-shell ionization (Fig. 3.9).

Thin conductive metal or carbon films can prevent this charging, and the quality criterion should be that the high-energy cut-off is shifted to $E_\mathrm{x} = eU$.

However, coating films influence the intensity of soft x-ray lines, of O K radiation for carbon films, for example. Instead, copper layers have been proposed since these have no interfering lines at low quantum energies. Nevertheless, each coating film influences the intensity ratio of different x-ray lines. After burning a hole in a carbon layer by increasing the oxygen partial pressure with a gas jet, the correct high-energy cut-off is observed.

Another problem is the internal electric field caused by trapped charges. The analysis of SE emission from thin SiO_2 films on Si in Sect. 8.2.7 indicate that high internal field strengths of the order of 10^7 V/cm $= 10^3$ V/μm can be built up, which can internally retard the electrons. This may shift the ionization density distribution $\Phi(\rho z)$ to lower depths. Because of the influence of charge carrier mobilities such an effect cannot be considered quantitatively. The fact that the composition can be changed by radiation damage and by a drift of mobile ions in this external electric field (Sect. 3.5.3) also has to be taken into account. It is difficult to decide which effect dominates in any given material.

10.4 X-Ray Microanalysis for Special Geometries

10.4.1 Tilted Specimens

The ZAF correction scheme described in Sect. 9.1 has been developed for plane specimens and normal electron incidence in x-ray microprobes. The correction methods have been extended to tilted specimens [10.91, 109–113] for two reasons. If the last probe-forming lens of the SEM has a flat polepiece, the take-off direction of an energy-dispersive spectrometer relative to the electron beam will be nearly 90° though this angle can be decreased by using a conically shaped polepiece or a specially designed spectrometer. To avoid low x-ray take-off angles ψ, for which the χ parameter of the absorption correction is high, it may be advantageous to tilt the specimen towards the detector thereby increasing the take-off angle. The other reason is that an energy-dispersive spectrometer has the advantage that rough surfaces can also be investigated. Usually, this only provides qualitative information about elemental composition. However, in the case of surface zones that are plane within the dimension of the electron diffusion cloud, the ZAF method can be extended to make the analysis more quantitative if the local tilt angle and azimuth angle of the surface normal are known.

The backscatter factor R_x decreases owing to the increase of η with increasing tilt angle ϕ, which can be described by (4.16), for example. Equation (10.55) expresses R_x approximately as a function of η and the overvoltage ratio u_0. Substitution of (4.16) extends this formula to the case $\phi > 0$.

Because the absorption correction function $f(\chi)$ is simply the Laplace transform (10.58) of the depth distribution function $\Phi(\rho z)$, we first consider the variation of $\Phi(\rho z)$ with increasing tilt angle ϕ. The maximum of $\Phi(\rho z)$

below the surface (Fig. 10.30) shifts towards the surface as ϕ increases and for $\phi \geq 60°$ the maximum is at $z = 0$. Experiments in our laboratory show that the surface ionization $\Phi(0, \phi)$ increases with ϕ and can be approximated by

$$\Phi(0, \phi) = \Phi_0 \sec\phi \,. \tag{10.84}$$

Furthermore, a depth ρz for normal incidence ($\phi = 0$) corresponds to a path length $\rho z \sec\phi$ for non-normal incidence. This relation and (10.84) can be substituted in (10.64) to give the modified depth distribution

$$\begin{aligned}\Phi(z, \phi) &= \frac{\sigma\rho(1+h)\sec\phi}{\Phi_1 + h\Phi_0 \sec\phi} \exp(-\sigma\rho z \sec\phi) \\ &\times \left[\Phi_1 - (\Phi_1 - \Phi_0 \sec\phi)\exp(-\sigma\rho z \sec\phi/h)\right] .\end{aligned} \tag{10.85}$$

Figure 10.30 shows a comparison of $\Phi(z, \phi)$ calculated (a) by Monte Carlo simulation and (b) using (10.85). This demonstrates that (10.85) describes the modification of the depth distribution with increasing ϕ rather well. The Laplace transform of (10.85) is

$$f(\chi) = \frac{1 + [h\Phi_0/(\Phi_1 + h\Phi_0 \sec\phi))]\frac{\chi}{\sigma}}{\left(1 + \frac{\chi}{\sigma}\sec\phi\right)\left(1 + \frac{h}{1+h}\frac{\chi}{\sigma}\sec\phi\right)} \tag{10.86}$$

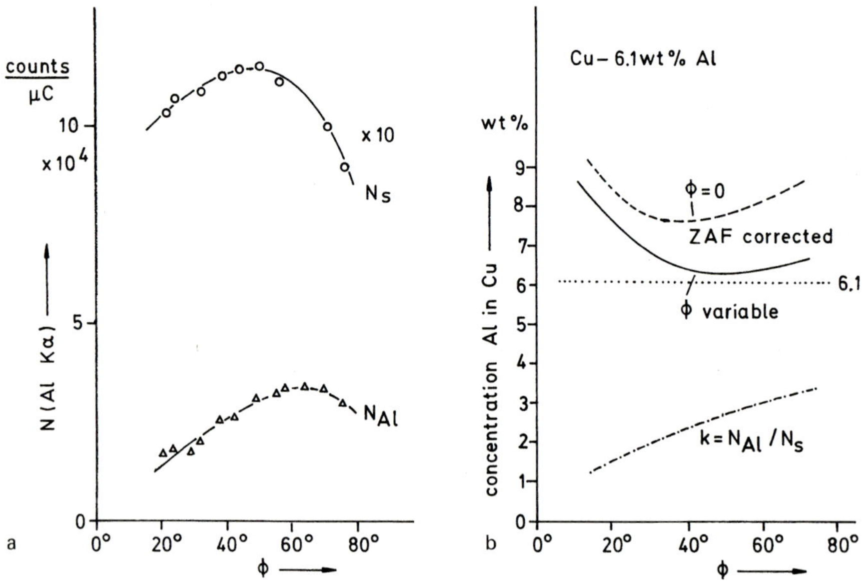

Fig. 10.35. X-ray microanalysis of a Cu-6.1 *wt*% Al alloy for different tilt angles ϕ and $E = 20$ keV. (**a**) Counts per unit incident charge [μC] for the Al K line in the alloy (N_{Al}) and a pure Al standard (N_{s}). (**b**) $k = N_{\mathrm{Al}}/N_{\mathrm{s}}$ and application of a ZAF correction for normal incidence (- - -) and of a modified program for tilted specimens (—) [10.113]

and this is the extension of (10.63) to the case $\phi > 0$. In the parameter χ, (10.59), the effective take-off angle $\psi_{\rm eff}$ has to be substituted. The tilt angle ϕ is defined to be the angle between surface normal and electron beam. We introduce the angle δ between detector take-off direction and the electron beam and the azimuthal angle α of the surface normal with the electron beam as axis ($\alpha = 0$: normal in the plane containing the beam and the take-off direction). From spherical trigonometry we find

$$\sin\psi_{\rm eff} = \cos\phi\cos\delta + \sin\phi\sin\delta\cos\alpha\ . \qquad (10.87)$$

Figure 10.35a shows measurements of the number of Al $K\alpha$ counts from a Cu-6.1wt%Al alloy and from a pure Al standard for different tilt angles ϕ ($\delta = 90°$). The decrease of the number of counts as ϕ is decreased below the maximum can be attributed to the decrease of $\psi_{\rm eff}$ and the increase of χ. The decrease beyond the maximum is caused by a decrease of x-ray excitations due to the increase of electron backscattering. The uncorrected k-ratio in Fig. 10.35b (dash-dotted line) predicts too low concentrations, whereas the correction with a ZAF program modified for specimen tilt (full line) agrees well in the range $\phi = 40° - 60°$. The poorer agreement for low ϕ can be attributed to an inaccuracy in the μ/ρ value. This example and others [10.113] demonstrate that quantitative analysis is possible for medium tilt angles.

10.4.2 Perpendicular Interfaces and Spatial Distribution of X-Ray Emission

The spatial distribution $\Phi(r,z)$ of x-ray emission plays no role for semi-infinite specimens but it is of interest for the analysis of concentration profiles near perpendicular interfaces and other simple geometries. If the electron-probe diameter is much smaller than the electron range, the lateral resolution will be limited by the diameter of the excitation volume of the corresponding x-ray line. If the atomic numbers of the materials 'a' and 'b' on the left and right sides of the interface and their densities do not differ significantly, the electron diffusion will be only slightly perturbed by the interface, and the x-ray emission of element 'a' with the electron probe at a distance x from the interface will be proportional to

$$N_{\rm a} \propto \int\limits_{-\infty}^{x} \Phi(x)\,{\rm d}x \quad \text{with} \quad \Phi(x) = \int\limits_{-\infty}^{+\infty}\left[\int\limits_{0}^{\infty}\Phi(\sqrt{x^2+y^2},z)\,{\rm d}z\right]{\rm d}y\ . \qquad (10.88)$$

A typical profile of k-ratios across an artificial, perpendicular Fe/Cu interface is shown in Fig. 10.36 [10.114]. The longer tail of Fe $K\alpha$ in copper reveals the influence of x-ray fluorescence because Fe $K\alpha$ can be excited by absorption of Cu $K\alpha$. The total electron range of 40 keV electrons is of the order of $R = 1.2\ \mu$m. The K shell ionization energies of Fe and Cu are 7.1 and 8.9 keV, respectively, and using (10.61), we see that the depth of x-ray emission is only of the order of 10% below the electron range R. The measurements in

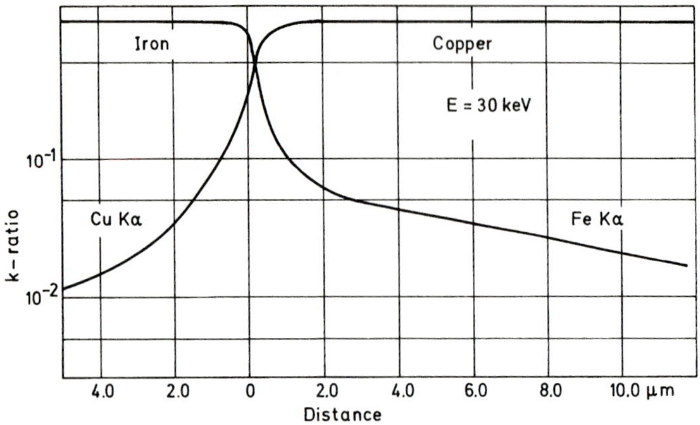

Fig. 10.36. Relative intensities of Fe $K\alpha$ and Cu $K\alpha$ when scanning in the x direction across a Fe/Cu interface perpendicular to the surface [10.114]

Fig. 10.36 result in a half-width of $\Phi(x)$ for copper of 0.2 μm and a width at 5% of the peak intensity of the order of 2 μm. Such lateral extensions of the x-ray emission have to be taken into account when the concentration profiles near interfaces are being investigated, the true profiles being convolved with the lateral distribution $\Phi(x)$.

The lateral distribution can only be decreased by reducing the electron energy or by using free-supported, electron transparent foils for which (3.120) can be used to estimate the lateral electron-beam broadening and the volume of x-ray excitation. Segregations at interfaces, for example, can therefore be analysed with a higher lateral resolution only as thin films, which will be one advantage of performing x-ray microanalysis in a TEM.

10.4.3 Thin Films on Substrates

For a film on a bulk substrate the ideal of microanalysis is to determine its composition and mass–thickness ρt. An appropiate modification of the ZAF correction is incorporated in the MAGIC program [10.57]. We discuss the correction procedures, making use of the depth distribution $\Phi(\rho z)$ of x-ray emission (Sect. 10.3.3). First we assume that an element 'a' of interest in the film is not present in the substrate. The intensity ratio k_a of counts from the film (f) and a pure element standard (s) becomes [10.115, 116]

$$k_a = \frac{N_{af}}{N_{as}} = c_a \int_0^{\rho t} \Psi_a^*(\rho z)\exp(-\chi_{af} z)\mathrm{d}(\rho z) \Big/ \int_0^{\infty} \Phi_a(\rho z)\exp(-\chi_{as} z)\,\mathrm{d}(\rho z) \tag{10.89}$$

when fluorescence excitation is neglected, where c_a is the concentration of element 'a' in the film, and χ_{af} and χ_{as} are the absorption parameters $\chi = (\mu/\rho)\mathrm{cosec}\psi$ for radiation of element 'a' in the film and in the standard, respectively. The depth distribution $\Phi^*(\rho z)$ of the film-substrate combination will normally be different from the depth distribution $\Phi(\rho z)$ of the standard. Only if the mean atomic numbers of the coating and the substrate do not differ by more than five units can we assume $\Phi^* = \Phi$.

The number of quanta of the substrate element 'b' will decrease with increasing ρt and the equation corresponding to (10.89) becomes

$$\begin{aligned} k_\mathrm{b} &= \frac{N_\mathrm{b,sub}}{N_\mathrm{bs}} = c_\mathrm{b}\exp[-(\chi_\mathrm{bf} - \chi_\mathrm{b,sub})] \\ &\times \int_{\rho t}^{\infty} \Phi_\mathrm{b}^*(\rho z)\exp(-\chi_\mathrm{b,sub} z)\mathrm{d}(\rho z) \Big/ \int_0^\infty \Phi_\mathrm{b}(\rho z)\exp(-\chi_\mathrm{bd} z)\mathrm{d}(\rho z)\,, \end{aligned} \tag{10.90}$$

where the index 'sub' stands for substrate. In the simplest case, absorption can be neglected and the atomic number of the pure element film is close to that of the pure element standard so that

$$k_\mathrm{a} = p_\mathrm{a} t;\quad k_\mathrm{b} = 1 - p_\mathrm{b} t \quad \text{with} \quad p(t) = \int_0^t \Phi(z)\mathrm{d}z \Big/ \int_0^\infty \Phi(z)\mathrm{d}z\,. \tag{10.91}$$

The thickness can be determined directly from the ratio k_a and/or k_b once the depth distributions are known.

For thin films $p_\mathrm{a}(t)$ can be approximated by $\Phi_\mathrm{a}^*(0)\rho t$ and becomes proportional to the film thickness for $p_\mathrm{a} > 20\%$. Hence, experimental calibration curves with films of known thickness can be used. The primary beam (p) will generate

$$n_\mathrm{p}(\rho t) = \frac{N_\mathrm{A}}{A_\mathrm{a}} c_\mathrm{a}\, \omega \mathrm{a}\, p_{nm} \rho t = \Phi_\mathrm{pa}(0)\rho t \tag{10.92}$$

x-ray quanta per electron if electron backscattering and x-ray absorption can be neglected. Electrons backscattered from the substrate can generate x-ray quanta in the film when their exit energy E_B exceeds E_{nl}. Thus (10.92) has to be multiplied by the enhancement factor $\Phi_a^*(0)/\Phi_\mathrm{pa}(0)$ which can be approximated by the relation [10.117]

$$\Phi_\mathrm{a}^*(0)/\Phi_\mathrm{pa}(0) = 1 + 2.8\left(1 - \frac{0.9}{u}\right)\eta \tag{10.93}$$

where η is the backscattering coefficient of the substrate and $u = E/E_{nl}$ the overvoltage ratio. The factor (10.93) also tells us how many more x-ray quanta will be emitted from a film on a substrate than from a free-supported film of equal mass–thickness.

Absorption can be considered in a first-order approximation by writing

$$\begin{aligned} k_\mathrm{a} &= p_\mathrm{a}(t)\frac{\exp(-\chi_\mathrm{af} t/2)}{\exp(-\chi_\mathrm{as}\bar{z}_\mathrm{s})}\,, \\ k_\mathrm{b} &= p_\mathrm{b}(t)\exp(-\chi_\mathrm{bf} t)\frac{\exp[-\chi_\mathrm{b,sub}(\bar{z}_\mathrm{sub} - t)]}{\exp(-\chi_\mathrm{bs}\bar{z}_\mathrm{s})};, \end{aligned} \tag{10.94}$$

where the $\bar{z}$ are the corresponding mean depths of x-ray excitation.

If the element 'a' of interest is present in both the film and the substrate with concentrations c_1 and c_2, respectively, (10.91) becomes

$$k_a = c_1 p_a(t) + c_2[1 - p_a(t)] \,. \tag{10.95}$$

Even for $c_2 = 0$, we have two unknowns c_1 and t. In the general case, a series of measurements with different electron energies allows c_1, c_2 and t to be deduced.

This brief summary of the basic concepts show that the key to quantitative analysis of thin films is a knowledge of $\Phi^*(\rho z)$ and $\Phi(\rho z)$. These functions cannot all be stored in a computer program and analytical fits have to be used [10.118]. Modern PCs allow these functions to be calculated for the film–substrate combination of interest by Monte Carlo simulation [10.119] or by solving a transport equation. It may be necessary to make a further fluorescence correction to allow for the radiation of a substrate element and this must be routinely checked [10.120].

For the analysis of submicron films and multilayers, the use of electron energies $E \leq 5$ keV and low x-ray energies can be advantageous. The depth and lateral resolutions fall below 0.3 μm and the minimum detectable concentration or film thickness may likewise decrease though the influence of any contamination has to be taken into account. Depth profiling of thin film structures can also be investigated by bevelled (angle-lapped) specimens [10.122].

10.4.4 Free-Supporting Films and Particles

First we discuss the x-ray microanalysis of thin free-supporting films because some of the results from TEM [10.121] can be transferred to the analysis of small particles on a supporting film or on a bulk substrate of low x-ray emission.

The number of x-ray quanta emitted by a film of mass–thickness $x = \rho t$ can be obtained from (10.43) if we assume that the x-ray emission is uniform through the film, which will be the case for such small x that the ionization cross-section σ_n is not increased by a reduction in the mean electron energy by energy losses and that the path lengths inside the film are not increased by multiple scattering. From Sect. 3.3. we know that these conditions will be satisfied for thicknesses smaller than 10–30 times the mean-free-path length Λ_t. A backscatter correction will not be necessary, because the number of BSE from a thin film is small (Sect. 4.1.2). The absorption correction can be simplified because we are assuming that the x-rays are generated uniformly across the film. Consequently, we can use (10.69) with $2\rho\bar{z} = x$. From (10.43) and (10.69), we obtain for an element 'a':

$$N_{xa}(\rho t) = n_e c_a \frac{N_A}{A_a} \omega_a \, p_a \, x \, \frac{1 - \exp(-\chi_a x)}{\chi_a x} \,. \tag{10.96}$$

Because x is small, the exponential in the absorption correction factor can be expanded in a Taylor series; retaining only the first three terms the last term of (10.96) becomes$(1 - \chi_a x/2 + \ldots)$. Normally, the absorption correction will be negligible but this approximation can be used to estimate the influence of absorption if an element 'b' that strongly absorbs the radiation of element 'a' is present in the film.

Even without knowing the thickness, which will only be needed for the absorption correction, the concentration c_a/c_b of two elements can be obtained from the ratio of their counts (ratio method) [10.123]

$$\frac{N_{xa}}{N_{xb}} = \frac{\omega_a \sigma_a p_a \eta(E_a) A_b}{\omega_b \sigma_b p_b \eta(E_b) A_a} \frac{c_a}{c_b} \frac{1 - \chi_a x/2}{1 - \chi_b x/2} = k_{ab}[1 - (\chi_a - \chi_b)x/2] \frac{c_a}{c_b} , \quad (10.97)$$

where $\eta(E_x)$ takes into account the dependence of the detector efficiency (10.36) on x-ray energy. The *Cliff–Lorimer* correction factor k_{ab} depends on the electron energy and cannot be calculated very accurately, because values of ω and σ are not known with sufficient accuracy. However, $k_{a,Si}$ values, for example, can be experimentally determined [10.123-125].

This ratio method will also be applicable to small particles with dimensions smaller than 10–30 mean-free-path lengths. Whether an absorption correction will be necessary can be estimated by replacing x by the mean diameter of the particle.

However, this is an ideal situation; in reality, particles have irregular shapes and are non-uniform in composition. For the determination of the concentration it will be necessary to know the total mass of the particle. Another extreme situation arises when the dimension of the particle is larger than the electron range. The most complicated case will be that in which the particle dimension is of the order of the electron range [10.126–130]. A higher backscatter correction factor R_x is then to be expected because more electrons leave the sides of the particle by multiple scattering. Monte Carlo simulations can also be used to consider this effect [10.131].

To avoid any influence of the substrate, the particles should be mounted on thin supporting films or substrates of low atomic number such as beryllium or graphite. The particles should be scanned in a raster because the uniformity of the electron current density cannot be guaranteed when electron probes larger than or comparible with the particle dimension are used.

There is no signal that gives the particle mass or its volume and density directly. However, the total signal of BSE (Sect 6.2.1), of the x-ray continuum [10. 132, 133] and of the absorbed signal of x-ray emission from the substrate [10.134] are quantities that are proportional to the mass of the particles but depend in different ways on its composition.

10.4.5 X-Ray Microanalysis of Biological Specimens

In principle, elemental concentrations in biological material are of interest as physiological concentrations, typically in units of mmol/l or mmol/kg. When the specimen is not analysed in the frozen-hydrated state, it must be dehydrated after fixation and is usually though not always embedded in a resin before electron irradiation. Only mass fractions per unit dry mass can be obtained therefore and the dry mass is reduced by electron bombardment because of mass loss caused by radiation damage (Sect. 3.5.3). As a rule of thumb, the mass loss is of the order of the non-carbon mass fraction though some non-carbon atoms remain in the carbonized material and some carbon atoms can leave the specimen as volatile fragments. For x-ray microanalysis, a large charge density of the order of 1 C/cm^2 is needed to produce a sufficient number of x-ray counts; these values are much larger than the charge densities, of the order of 10^{-2} and 10^{-1} C/cm^2, that are sufficient to saturate the mass loss of aliphatic and aromatic compounds, respectively, at E = 40 keV. Another problem will be whether the element under investigation remains in the specimen during the steps of preparation and during the mass loss caused by electron irradiation [10.135]. Even in the ideal case, in which the atoms analysed do not leave the specimen, the concentration relative to the dry mass will increase until the mass loss saturates and will show a slow decrease with further irradiation time with the growth of a contamination layer.

The local distribution of elements and of the dry mass inside a biological specimen is mostly very inhomogeneous and measurements of the local variation of elemental concentration per dry mass becomes unreliable because the mass loss can differ locally. This has to be taken into account for sections and for bulk material. For the latter, the analytical depth can even increase during irradiation with the gradual mass loss until a fully carbonized layer stops all electrons. The mass loss can be reduced by specimen cooling [10. 136, 137]. The question is, whether the volatile fragments still are present in the specimen at liquid nitrogen temperature. Experiments at liquid helium temperature with TEM specimens show that only the mass loss can be halted but not the ionization damage of the specimen.

X-ray analysis in the frozen state is also not without problems because ice suffers radiolysis under electron bombardment. The accuracy of quantitative analysis of biological specimens is only of the order of 10–20% therefore and can never reach the accuracy of elemental analysis of inorganic material.

The dry mass and the mass of a frozen-hydrated specimen cannot be measured by elemental analysis with an energy-dispersive spectrometer. Though C, N, or O can be detected when using an ultra-thin window, the fluorescence yield ω is too low for an accurate measurement. Information about the mass is therefore drawn from measurements of the x-ray continuum or from transmisison experiments in TEM and STEM.

The relative concentration of two elements c_a/c_b can be obtained by the ratio method as in (10.97); x-ray absorption can be neglected in organic material up to thicknesses of 1–4 μm. The ratio method can be applied to bulk and sectioned material. The correction factor k_{ab} has to be determined by standard specimens [10.138, 139] which should have a comparable composition of the organic matrix and a known concentration of the element, which is mostly incorporated in epoxy resin [10.136, 140, 141] or aminoplastic resins [10.142, 143] in the form of inorganic salts. Another possibility is to use 20–30 μm embedded and sectioned Chelex beads with calibrated concentrations [10.144]. When using a resin-embedded specimen, the standard will be embedded in the same resin and when using frozen-hydrated specimens the standard can be a frozen solution. Problems arising from differences in thickness of specimen and standard can be avoided by the use of peripheral standards. The tissue is surrounded by an albumine solution containing inorganic salts before shock-freezing [10.145–147].

Absolute concentration per unit dry mass or wet mass can be obtained by measuring the x-ray counts N_{xc} of the continuum just below the x-ray line of interest [10.148] with N_{xa} counts or in a continuous part of the spectrum with no further lines [10.149]. The concentration is given by

$$c_{\text{a}} = k_{\text{ac}} N_{\text{xa}} / N_{\text{xc}} , \tag{10.98}$$

where the calibration constant k_{ac} also has to be measured with the aid of standards. The idea of the normalization by the continuum is that the x-ray counts of equal quantum energy will be produced at the same depth and will show the same absorption and that the continuum intensity will be the same for material of equal mean atomic number. Equation (10.98) can also be used when the section thicknesses of the specimen and standard are different because both N_{xa} and N_{xc} increase as the thickness

Equation (10.98) is also the basis of the *Hall* method [10.137, 150, 151] for thin sections. Because the tissue can contain additional elements and/or different concentrations of H, C, O, N, P, S or Ca, for example, relative to the standard, the ratio $N_{\text{xa}}/N_{\text{xc}}$ may be different and N_{xc} depends also on the concentration of all elements. Kramers formula (10.8) shows that the continuum from a thin layer increases as Z^2/A. With the constants of proportionality k_1 and k_2, we can write

$$N_{\text{xa}} = k_1 m_{\text{a}} \quad \text{and} \quad N_{\text{xc}} = k_2 m G \tag{10.99}$$

with

$$G = \overline{Z^2/A} = \frac{1}{m} \Big(\sum_a m_a Z_a^2 / A_a + \sum_i m_i Z_i^2 / A_i \Big) \tag{10.100}$$

where m_i are the masses of the elements not recorded in the x-ray spectrum (especially light elements) and $m = \sum m_a + \sum m_i$ is the total mass. Thus the concentration becomes

$$c_a = \frac{m_a}{m} = c_s \frac{(N_{\text{xa}}/N_{\text{xc}})_{\text{Specimen}}}{(N_{\text{xa}}/N_{\text{xc}})_{\text{Standard}}} \frac{G_{\text{Standard}}}{G_{\text{Specimen}}} . \tag{10.101}$$

This method can be used iteratively by first neglecting the first term in (10.100) and then calculating a new value of G with the obtained concentration c_a. The peak-to-background method can also be applied to small particles [10.132, 133] and bulk organic specimens [10.149, 152].

For further details on the x-ray microanalysis of biological specimens the reader is referred to [10.14, 153].

10.5 Data Processing and Sensitivity in X-Ray and Auger Electron Analysis

10.5.1 Background and Peak-Overlap Correction

For quantitative analysis, the continuous background must be subtracted to obtain the number of counts in a characteristic peak. As shown in Sect. 4.3.2, Kramer's formula (10.10) is only an approximation. The easiest method is to measure the background counts at energies below and beyond the characteristic line and use a linear interpolation. For background subtraction over a larger range of quantum energy, the continuum can be fitted, for example, by an expression of the form [10.154]

$$N_{\mathrm{xc}} \propto f(\chi, E_{\mathrm{x}})\eta(E_{\mathrm{x}})\left(a\,\frac{E-E_{\mathrm{x}}}{E_{\mathrm{x}}}+b\,\frac{(E-E_{\mathrm{x}})^2}{E_{\mathrm{x}}}\right), \tag{10.102}$$

where $f(\chi, E_{\mathrm{x}})$ is the absorption correction for the quantum energy E_{x} of the continuum, $\eta(E_{\mathrm{x}})$ the detection efficiency (10.36) and the constants a and b have to be determined by least-squares fitting of experimental background spectra. In the case of high concentration of an element, the background shows a discontinuity characterized by the absorption jump ratio (Sect. 9.1.4), which lies at the absorption edge of the corresponding shell and often near below the characteristic line because of the limited energy width $\Delta E_{\mathrm{FWHM}} \simeq 20$–200 eV of energy-dispersive spectrometers. Characteristic lines and absorption edges of the continuum can be more satisfactorily separated by a wavelength-dispersive spectrometer.

If possible, peak overlaps of characteristic lines should be avoided by looking for a single line of the element of interest. Unavoidable overlaps can be separated [10.155] by fitting the coefficients A_n for Gaussian profiles

$$N_i = \sum_n A_n \exp\left[-\ln 2\left(\frac{E_i - E_n}{\Delta E_{\mathrm{FWHM}}}\right)^2\right], \tag{10.103}$$

where i is the number of the channel and E_i its quantum energy, E_n are the peak energies of the characteristic lines and ΔE_{FWHM} the full-width at half maximum which varies with E_{x} as shown in Fig. 10.17.

10.5.2 Counting Statistics and Sensitivity

The number of counts measured when the same experiment is repeated with equal numbers of incident electrons or incident charge $I_{\mathrm{p}}\tau$ follows Poisson statistics. A plot of the probability of measuring N_i counts becomes a Gaussian distribution with a standard deviation $\sigma = \overline{N}^{1/2}$. This means that 68.3% of the measurements lie inside the interval $(\overline{N} \pm \sigma)$, 95% in $(\overline{N} \pm 2\sigma)$ and 99.7% in $(\overline{N} \pm 3\sigma)$.

We employ this statistical model to estimate the analytical sensitivity, which is the smallest detectable difference $\Delta c = c_2 - c_1$ between two concentrations c_1 and c_2 with counts N_1 and N_2, respectively. For small Δc, we have $N_1 \simeq N_2 \simeq N$ with a standard deviation $\sigma_1 \simeq \sigma_2 = N^{1/2}$. The peak-to-continuum ratio should be large and because the x-ray continuum can be measured repeatedly beside the x-ray line, its standard deviation becomes smaller than $N_c^{1/2}$. According to the law of error propagation, N_2 will be different from N_1 with a 99.7% level of confidence, if

$$N_2 - N_1 > 3(\sigma_1^2 + \sigma_2^2)^{1/2} \simeq 3\sqrt{2}N^{1/2} \simeq 4N^{1/2} . \tag{10.104}$$

This results in

$$\frac{\Delta c}{c} = \frac{N_2 - N_1}{N} = \frac{4}{N^{1/2}} . \tag{10.105}$$

For example, for an analytical sensitivity $\Delta c/c = 1\%$ or for testing the homogeneity of a specimen to within $\Delta c/c = \pm\, 0.5\%$, we need $N = 1.6 \times 10^5$ counts in the x-ray peak of interest.

To obtain an estimate of the minimum detectable concentration $c_{\min}$, we assume that the number of counts N_p at the x-ray peak is only slightly larger than the number N_c of the continuous background counts, or $N_p - N_c \ll N_c$. Recalling (10.104), we find

$$N_p - N_c > 3(\sigma_p^2 + \sigma_c^2)^{1/2} \simeq 3N_p^{1/2} \simeq N_c^{1/2} \tag{10.106}$$

in which we have neglected σ_c because it is possible to measure the continuum repeatedly beside the x-ray line and to extrapolate for the background level N_c below the line. Hence

$$c_{\min} = \frac{N_p - N_c}{N_s} = \frac{3N_c^{1/2}}{N_s} = \frac{3}{(I_p\tau)^{1/2}} \frac{n_c^{1/2}}{n_s} \tag{10.107}$$

where n_c and n_s are the counts of the continuous background and of a pure element standard per unit charge $I_p\tau$ of the incident electron beam ($N_c = n_c I_p \tau$). Because the magnitude of the electron-probe current I_p is limited by the electron optics, $c_{\min}$ can only be reduced by increasing the counting time τ but this is also limited by contamination and long-term drift of the electron probe. The number n_c is given by the matrix and the number n_s by the trace element. Therefore, $c_{\min}$ increases with the atomic number of the matrix.

The last factor in (10.107) can be optimized by using a suitably chosen width ΔE_W of the energy window for counting N_p and N_c (Fig. 10.37). For a small energy window $\Delta E_W < \Delta E_{FWHM}$, only a few channels near the peak are measured and the statistics is poor. With increasing ΔE_W, the number N_c increases in proportion. For large $\Delta E_W \gg \Delta E_{FWHM}$, the number of peak counts $N_p - N_c$ saturates. An optimum is therefore to be expected, and occurs at $\Delta E_W/\Delta E_{FWHM} = 1.2$ [10.156].

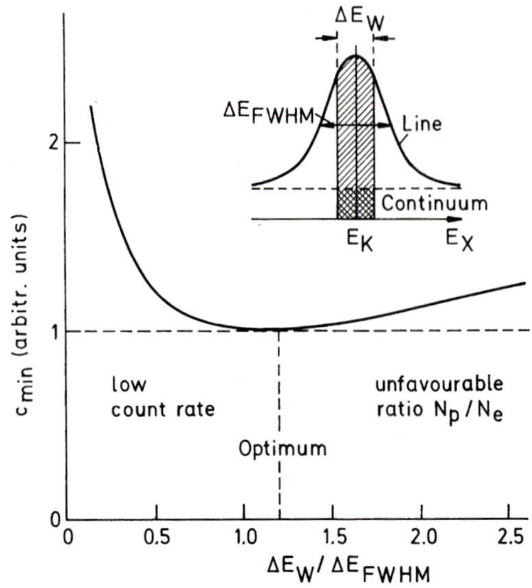

Fig. 10.37. Dependence of the minimum detectable concentration c_{min} on the width ΔE_{W} of the energy window in energy-dispersive x-ray microanalysis (ΔE_{FWHM} = full-width at half-maximum of the analysed characteristic x-ray line) [10.156]

10.5.3 X-Ray Fluorescence Spectroscopy in SEM

As shown in Sect. 9.2.3, the analytical sensitivity for trace elements is limited by the relatively large background of the x-ray continuum when the x-ray emission is excited by electron bombardment. This background is considerably decreased when x-ray fluorescence is excited by externally produced x-ray quanta. With an x-ray tube and a crystal or wavelength-dispersive spectrometer, the technique is widely used for elemental analysis. The technique allows low concentrations to be determined and the thickness of thin films on substrates to be measured down to monolayers.

The presence of an electron beam and an energy-dispersive detector makes it possible to perform x-ray fluorescence spectroscopy in a SEM [10.157–162]. The electron beam strikes not the specimen but a thin-foil target of Mo, for example, which is thick enough to absorb all the primary electrons but not too thick to absorb the x-rays (Fig. 10.38), or a massive target in front of a polymer film. It is also possible to use a thin metal film, the thickness of which just saturates the x-ray emission and which is backed by a polymer film. The x-rays are emitted isotropically and a diaphragm has to be used to limit the the x-ray irradiated area on the specimen to 0.1–1 mm in diameter. The use of smaller diameters will drastically reduce the x-ray fluorescence count rate. It will therefore not be necessary to use a small electron-probe diameter as is the case for x-ray projection microscopy (Sect. 10.6.3); the electron current at the x-ray target can be increased to the maximum available valuc by exciting only one condenser lens and removing the final-lens diaphragm. Since the irradiated area is large, this technique is by no means a microanalytical tool but it has the advantage of detecting trace elements with a high sensitivity.

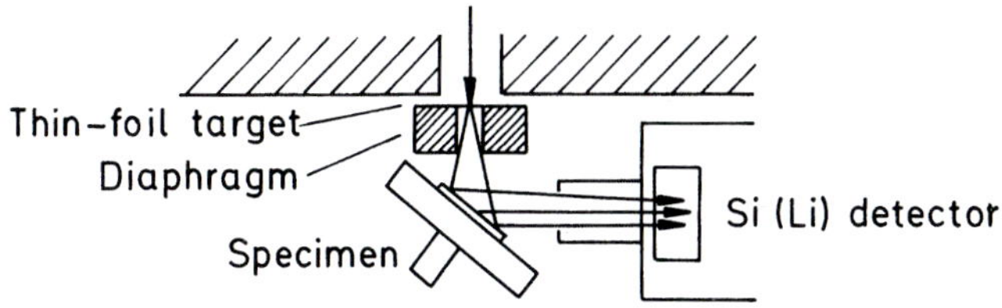

Fig. 10.38. Device for x-ray fluorescence analysis in a SEM where the thin foil is non-transparent to electrons

A comparison of typical x-ray spectra excited by direct electron bombardment and by x-ray fluorescence is shown in Fig. 10.39; the number of counts per channel is measured on a logarithmic scale. The specimen is a Si wafer doped with 100 ppm As. The spectra are recorded with the indicated currents I_p and counting times τ. The lower background in the fluorescence spectrum, which allows weaker concentrations to be detected, can be seen. The background in fluorescence spectra is caused by scattering of the continuum x-rays produced in the target foil. The x-ray fluorescence spectra contain the elastically scattered exciting line (Mo $K\alpha$) of the Mo target foil and the broader maximum of Compton scattered x-rays below the elastic peak. These elastic scattering processes have been discussed in Sect. 10.1.4. The electron-beam excited spectrum shows pulse pile-up beyond the Si $K\alpha$ peak.

For x-ray fluorescence to be excited by inner-shell ionization caused by the photoelectric effect, the quantum energy E_x of the exciting x-rays must reach a value equal to or larger than the ionization energy E_{nl} of the corresponding shell. The photoelectric cross-section reaches a maximum just at $E_x = E_{nl}$ and then decreases as E_x^{-n} ($n = 2.5$–3.5), corresponding to the absorption edge seen in Figs. 10.9 and 10.32. This means that the condition shown in Fig. 10.39 for exciting As K with Mo K radiation is not optimum and the highest probability of exciting x-ray fluorescence occurs if E_x of the target just exceeds E_{nl} of the tracer element with an associated increase in sensitivity. The Si $K\alpha$ peak in Fig. 10.39 is excited with a low intensity because the the energy of Mo K is much larger than the Si ionization energy. It will be convenient, therefore, to mount several target foils on a pivot.

The quantitative interpretation of x-ray fluorescence spectra is more difficult than in the case of excitation by electron bombardment and calibration standards of similar composition have to be used.

This discussion of the x-ray fluorescence mode shows that it is very easy to implement in a SEM. It allows trace elements to be surveyed. SEM equipment cannot, however, attain the sensitivity of commercial fluorescence apparatus.

10.5.4 Auger Electron and X-Ray Photoelectron Spectroscopy

The basic mechanism of Auger electron emission has been described in Sect. 10.1.5. X-ray photoelectrons can be excited by Mg or Al $K\alpha$ radiation and are

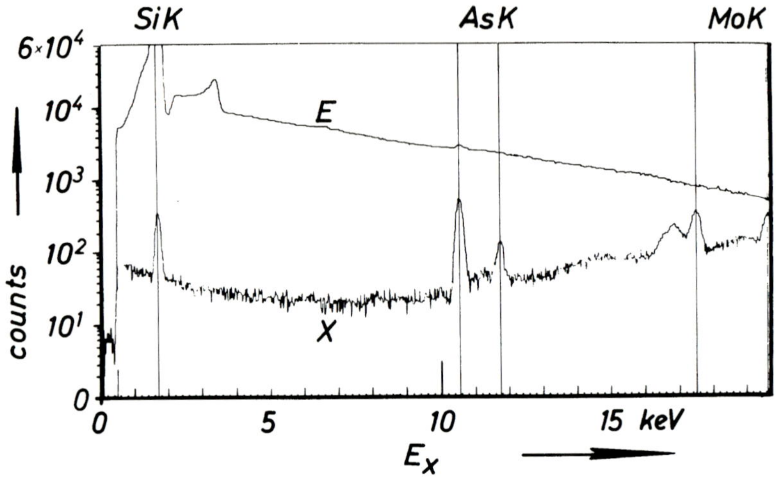

Fig. 10.39. Comparison of x-ray spectra obtained by electron bombardment (E) with $I_p = 2 \times 10^{-10}$ A, $\tau = 400$ s from 100 ppm As in silicon and by x-ray fluorescence (X) with $I_p = 4.7 \times 10^{-7}$ A, $\tau = 70$ s ($E = 30$ keV) using a Mo foil target [10.159]

then analysed by means of an electron spectrometer. The photoelectrons and Auger electrons generated show a peak shift and change of peak shape caused by differences in chemical bonding, which is why XPS (x-ray photoelectron spectroscopy) is sometimes called ESCA (electron spectroscopy for chemical analysis) [10.163]. Both techniques are sensitive to surface films a few monolayers thick and are standard methods in surface physics. The full capability of these methods can be attained only with specially designed ultra-high vacuum Auger electron microprobes or x-ray photoelectron spectrometers, which are most frequently equipped with cylindrical mirror analysers (Fig. 5.19d) or hemispherical spectrometers [10.164–166]. The combination of AES, XPS and SIMS (secondary ion mass spectroscopy) [10.167] in one instrument [10.168] offers the additional advantage of depth profiling by ion-beam etching. Otherwise, SE imaging of the specimen in an Auger electron microprobe will be inferior to that attainable in a SEM. The use of these techniques in SEM [10.169–171] can only be a compromise with a loss of sensitivity. To increase the probability of Auger electron excitation inside the surface layer, Auger electron probes normally work with electron energies below 5 keV. Electron probes of 0.1 μm and $I_p = 6 \times 10^9$ A can be produced. The probe diameter can be decreased by the use of a field-emission gun. AES can be used for local analysis or for elemental mapping.

XPS is normally not used as a microanalytical method. However, when the x-ray source is produced in a 5 μm Al foil in front of a thin specimen, resolutions below 20 μm are possible [10.172, 173].

10.6 Imaging Methods with X-Rays

10.6.1 X-Ray Mapping and Linescans

The counts from characteristic lines of the x-ray spectrum can be selected in a multichannel analyser by setting energy windows at these lines in order to produce distribution maps of the corresponding elements. As the specimen area is scanned, all the pulses of the energy-dispersive detector that fall inside the selected energy interval are used to switch on the beam of the synchronously scanned CRT to produce a bright spot at the point corresponding to the position of the beam on the specimen. This single-pulse technique does not need more than one recorded x-ray quantum per pixel and the concentration of an element is indicated by the concentration of bright spots per unit area. This is illustrated in Fig. 10.40 with a larvicite sample [10.174]. The energy-dispersive spectrum is shown in (a) and the BSE micrograph of the surface in (b). Figures 10.40c–g are x-ray maps recorded with Si, K, Fe, Ti and Ca $K\alpha$ lines, respectively. The concentration of bright spots shows that the grains mainly contain Fe and Ti and the matrix Si and K. Because of the continuous background inside the selected energy windows, spurious bright dots can also be produced in regions not containing the element of interest. This becomes a serious problem for small concentrations. X-ray peaks of small height can only be used for x-ray mapping if the element is concentrated within small particles, for example. The x-ray maps of different elements can be recorded simultaneously by setting energy windows (channels) at the corresponding x-ray lines and presenting each channel as a different colour on a colour TV display [10.175, 176]. X-ray maps should show at least 2×10^4 counts per frame. The number of x-ray counts necessary to form a two-dimensional map with many grey levels will be very large, of the order of $10^5 - 10^6$ counts per frame. We have to bear in mind that a Si(Li) detector can only operate with a count rate of $\simeq 10^5$ cps and recording times of several minutes will hence be necessary. The single-pulse technique has the disadvantage that only a few grey levels (counts per unit area) – of the order of eight – can be distinguished on a CRT [10.176].

Because the resolution is limited by the excitation volume of the x-rays, of the order of 1–3 μm, only magnifications of a few hundreds will be useful and it is no problem to work with electron-probe currents of 10^{-10} A to increase the counting rate. At a 100-fold magnification, the frame of the CRT corresponds to a specimen area of 1 mm^2. In the case of a wavelength-dispersive spectrometer the focusing condition of the spectrometer cannot be satisfied for shifts of the electron beam position of the order of 0.2 mm but no such problem arises in energy-dispersive spectrometry. For the former, the specimen is therefore scanned by translating the stage under a fixed electron beam. Such an arrangement can also be used with an EDS when larger areas are to be scanned with high accuracy. Otherwise a WDS has the advantage of better resolution. The fraction of spurious dots caused by the continuum below the selected line can therefore be reduced to of the order of one tenth of that recorded with an EDS. The proportional counter of the WDS can operate at higher count rates and one spectrometer of the WDS only counts one line. The electron-probe current should hence be increased up to 10^{-8} A to get a better statistics.

Two-dimensional maps containing a number of grey levels can be presented in the Y-modulation mode, by shading each grey level differently or by colour coding [10.176]. Digital storage and image processing are very advantageous because it is then possible to average over larger areas around particular points to get a compromise between signal averaging and resolution.

When the x-ray intensity is recorded by a slow single linescan, the statistics will be good enough to generate an analogue signal by integrating the pulses in a preamplifier with a time constant smaller than the scan time between two pixels; otherwise the signal will not follow abrupt changes of the x-ray intensity. For example, the Fe and Ti maps of the Fe-Ti oxide grains in Fig. 10.40e,f show striations, which cannot be clearly resolved due to the inadequate counting statistics of the two-dimensional map. A smaller area inside a grain is shown in Fig. 10.40h,i, on which linescans representing the signal from the Fe and Ti $K\alpha$ quanta are superposed. The complementarity of the Fe and Ti concentrations in the striations can be recognized.

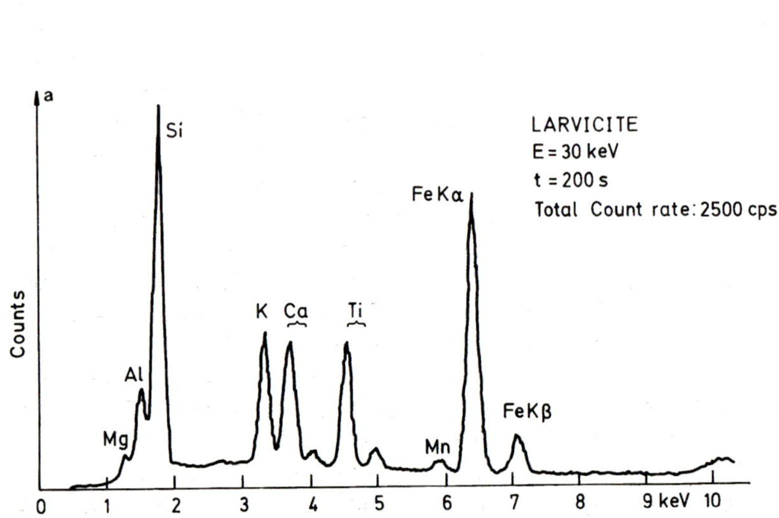

Fig. 10.40. X-ray microanalysis of larvicite with Fe-Ti oxide inclusions. (**a**) Total x-ray spectrum of the specimen. (**b**) BSE micrograph of an oxide grain surrounded by feldspath and mica. Elemental distribution maps recorded with the (**c**) Si, (**d**) K, (**e**) Fe, (**f**) Ti and (**g**) Ca $K\alpha$ lines. Lamellar structure of the Fe-Ti phase with superposed x-ray linescans of (**h**) Fe and (**i**) Ti $K\alpha$ [10.174]

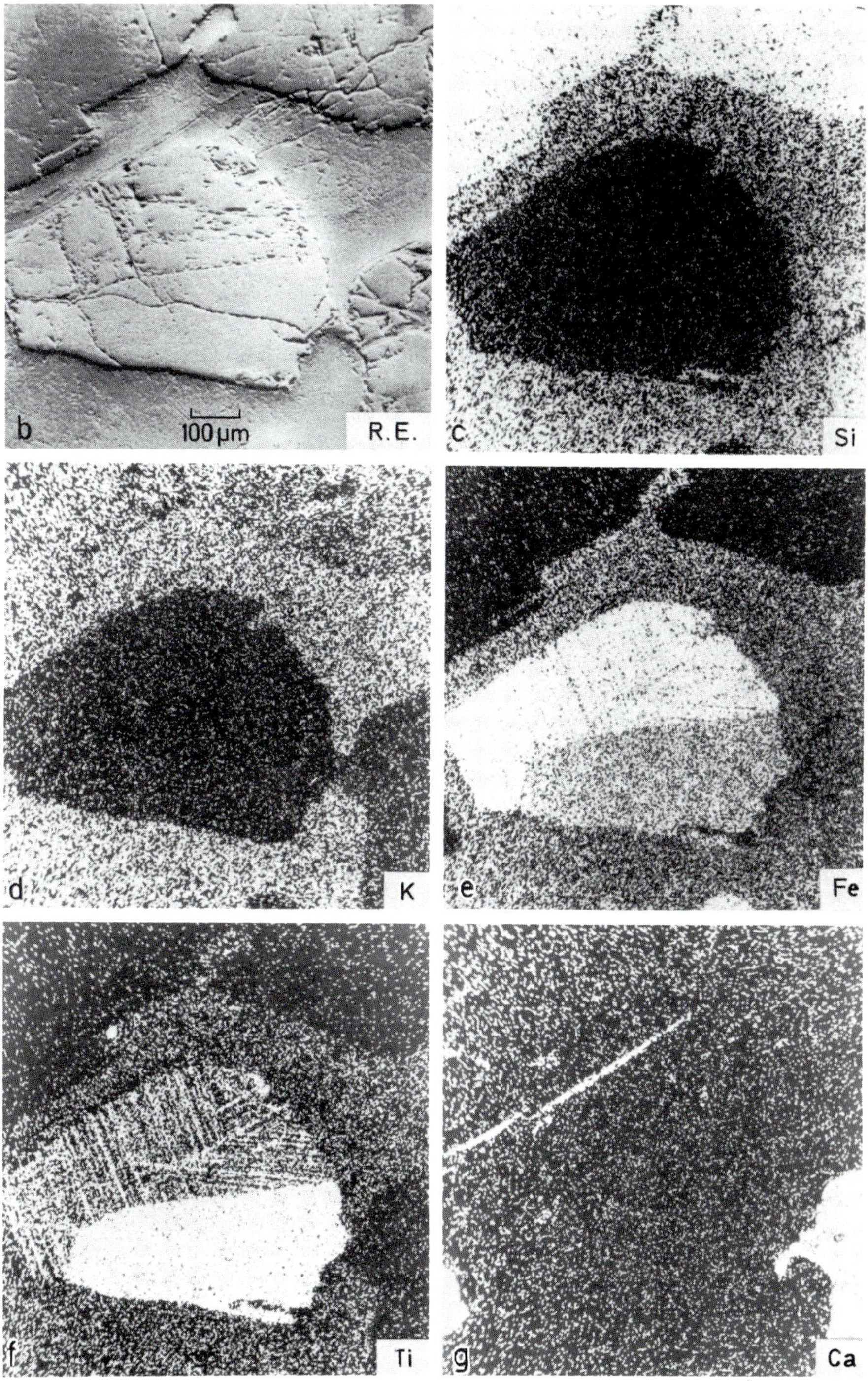

Fig. 10.40b–g

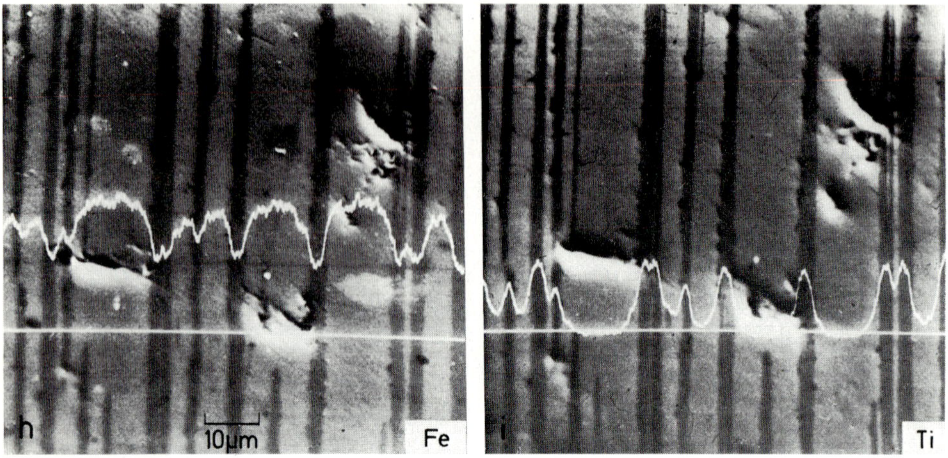

Fig. 10.40h,i

10.6.2 Total Rate Imaging with X-Rays (TRIX)

The TRIX technique was devised specially for imaging heavy metal inclusions in an organic matrix with better contrast than that of a BSE image [10.177]; in this method an analogue signal from the Si(Li) detector is recorded with all x-ray quanta. The method can be improved by recording the x-ray intensity with the aid of a thick scintillator [10.178].

The measured TRIX signal intensities reproduced in Fig. 10.41 show the increase with increasing atomic number at an electron energy $E = 45$ keV. The intensity of the continuum increases monotonically with increasing Z, whereas the additional contribution from the characteristic x-ray emission shows maxima. The calculations (full lines) agree rather well with the experiments and are not fitted. The intensity of the K radiation first increases, subsequently dropping to zero when the ionization energy of the element exceeds the incident electron energy and K shell ionization is no longer possible. This fall in the K contribution shifts to higher Z as E is increased [10.178]. The same effect occurs for the L and M contributions but these overlap and do not show a pronounced maximum followed by a minimum of the TRIX signal.

10.6.3 X-Ray Projection Microscopy and Tomography

A stationary electron probe of diameter d_{p} that strikes a metal target acts like a point source of x-rays; the latter can penetrate a small specimen at a distance a and produce an enlarged projection on a photographic emulsion at a distance A from the source. The magnification will be $M = A/a$ and is typically of the order of 4–50. The contrast is caused by x-ray absorption

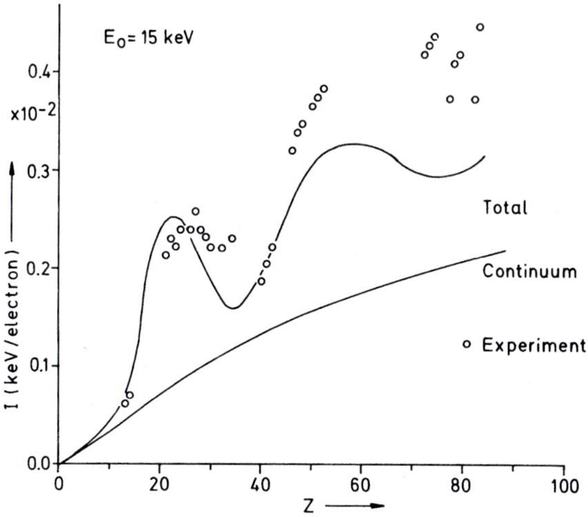

Fig. 10.41. Intensity of the total x-ray signal recorded with a scintillator and comparison with calculations of the x-ray continuum intensity and of the superposed characteristic x-ray lines as a function of atomic number at $E = 45$ keV [10.178]

in the specimen and can be altered by varying the electron energy and the target material.

When the target consists of a thin metal foil separating the vacuum of the electron-optical column from the atmosphere, such an x-ray projection microscope can be used with the specimen and emulsion at atmospheric pressure. Such special instruments were constructed in the years 1940–1960 [10.180–183] but have not been followed up.

A stationary electron probe is readily available inside a SEM and x-ray projection microscopy with a film camera inside the specimen chamber can therefore be added to the modes of operation of the instrument [10.170, 184, 185]. Digital recording is possible by using an x-ray sensitive scintillator coupled by a fibre plate to a cooled CCD array [10.186]. The resolution will be of the order of the electron range R or the electron-probe diameter d_p if $d_p > R$. The resolution can be improved by using a thin target of high atomic number, e.g. a 0.05–0.1 μm evaporated Au film on a 1 μm polymer film, which stops the 10 keV electrons [10.179]. Fresnel diffraction of the x-rays limits the resolution, the first Fresnel fringe maximum occurring at a distance $d_\lambda = (a\lambda_x)^{1/2}$ [10.180]. The problem is to combine a large electron-probe current I_p to reduce the exposure time with a small electron-probe size to increase the resolution.

The best design [10.184, 185] will allow the foil target to be shifted into the electron beam and the distance a between foil and specimen to be varied. If there is a small gap between foil amd polepiece, the focusing of the electron probe on the foil can be controlled by using the conventional SE mode. Before or after taking an x-ray projection micrograph, a SE or BSE image can be formed at the same position after withdrawing the foil from the beam, which allows the surface topography and the internal structure of the specimen to be recorded. Stereo x-ray projection micrographs can be obtained by shifting the specimen through a small distance x with the x-y translation stage. Specimen details with a difference in height h will show a parallax (6.19). When recording a series by rotating a specimen in steps around an axis perpendicular to the electron beam also x-ray tomography can be realized with a resolution of $\simeq$10 μm and 1 μm should be possible in principle [10.186, 187]. This will be the main advantages of combining SEM and x-ray projection microscopy. Typical specimens which can be investigated by this technique are small living organisms with sizes of a few millimetres and below and metal and alloy foils.

The image contrast is generated by x-ray absorption described by the exponential law (10.27) and differences in the mass-attenuation coefficient μ/ρ times the mass–thickness can be recorded

$$I = I_0 \exp\left[-\int_{-\infty}^{+\infty} (\mu/\rho)\rho \mathrm{d}z\right] . \qquad (10.108)$$

The dependence of μ/ρ on x-ray quantum energy E_x is plotted in Figs. 10.9 and 10.32. This shows absorption edges, when the quantum energy exceeds the ionization energy of an atomic shell. By using the characteristic $K\alpha$ radiation of different targets, it is possible to produce contrast differences [10.180, 183, 185]. For example a Cu target results in a better contrast of alloy constituents that contain Fe than an Fe target because the Cu $K\alpha$ radiation ($\mu/\rho = 307$ cm^2/g) is absorbed more strongly by iron than the Fe $K\alpha$ radiation ($\mu/\rho = 68$ cm^2/g) [10.185]. Absorption also sets an upper limit on the useful thickness, which is of the order of 100 μm for minerals and metals and a few millimetres for biological specimens. The lower limit is set by the minimum detectable contrast.

The characteristic radiation of an x-ray point source can be Bragg reflected at a single crystal for all points on a line such that Bragg's law (9.51) is satisfied. If the x-ray point source is shifted a small distance on a foil target, the line that satisfies Bragg's law also shifts on the specimen and on the photographic emulsion. In this way, x-ray topographic images that are sensitive to crystal lattice distortions can be recorded in the SEM [10.188].

Scanning x-ray microscopy is a reciprocal technique. The image is now not recorded on a photographic emulsion with a stationary x-ray source but instead the x-ray signal is recorded by an energy-dispersive Si(Li) detector with a small diaphragm and the electron beam (point source) is scanned over the target foil. The image of the specimen between target and detector is formed on a CRT using the signal from the x-ray detector. This method also allows x-ray projection images to be recorded with different windows of the multi-channel analyser [10.189]. A disadvantage is the low signal intensity due to the necessarily small solid angle of detection with a consequently large increase in the recording time.

A very different version of x-ray microscopy is soft x-ray contact microscopy [10.190, 191]. A 1–2 μm thick photoresist layer of PMMA on a silicon wafer is the support for organic material, which is exposed to soft x-rays (C $K\alpha$, for example). The synchrotron radiation of an electron accelerator furnishes an intense beam of soft x-rays, thereby decreasing the exposure time. After the exposure (analogue to electron-beam lithography but here with the soft x-rays), the organic material is removed by washing and the resist is developed, the resulting surface profile being dependent on the exposure and absorption of soft x-rays. The developed resist is coated with a 1–4 nm Au-Pd film and observed in a high-resolution SEM. Resolutions of the order of 5 nm are possible.

References

Chapter 1

[1.1] L. Reimer: *Transmission Electron Microscopy, Physics of Image Formation and Microanalysis*, 4th edn., Springer Ser. Opt. Sci., Vol.36 (Springer, Berlin, Heidelberg 1997)

[1.2] P.R. Thornton: *Scanning Electron Microscopy, Application to Materials and Device Science* (Chapman and Hall, London 1972)

[1.3] C.W. Oatley: *Scanning Electron Microscopy 1. The Instrument* (University Press, Cambridge 1972)

[1.4] D.B. Holt, M.D. Muir, P.R. Grant, I.M. Boswarva (eds.): *Quantitative Scanning Electron Microscopy* (Academic, London 1974)

[1.5] O.C. Wells: *Scanning Electron Microscopy* (McGraw-Hill, New York 1974)

[1.6] J.I. Goldstein, H. Yakowitz: *Practical Scanning Electron Microscopy* (Plenum, New York 1975)

[1.7] L. Reimer, G. Pfefferkorn: *Raster-Elektronenmikroskopie*, 2. Aufl. (Springer, Berlin, Heidelberg 1977)

[1.8] F. Maurice, L. Meny, R. Tixier (eds): *Microanalyse et Microscopie Electronique à Balayage* (Les Editions de Physique, Orsay 1978) Engl. Transl.: *Microanalysis and Scanning Electron Microscopy*, 1979

[1.9] D.A. Erasmus: *Electron Microprobe Analysis in Biology* (Chapman and Hall, London 1978)

[1.10] P. Lechene, R.R. Warner (eds.): *Microbeam Analysis in Biology* (Academic, New York 1979)

[1.11] J.I. Goldstein, D.E. Newbury, P. Echlin, D.C. Joy, C. Fiori, E. Lifshin: *Scanning Electron Microscopy and X-Ray Microanalysis* (Plenum, New York 1981)

[1.12] B.L. Gabriel: *Biological Scanning Electron Microscopy* (Van Nostrand, New York 1982)

[1.13] S.J.B. Reed: *Electron Microprobe Analysis* (Cambridge Univ. Press, Cambridge 1975)

[1.14] T.E. Hutchinson, A.B. Somlyo: *Microprobe Analyis of Biological Systems* (Academic, New York 1981)

[1.15] K.F.J. Heinrich: *Electron Beam X-Ray Microanalysis* (Van Nostrand, New York 1981)

[1.16] M. A. Hayat (ed.): *X-Ray Microanalysis in Biology* (McMillan, London 1981)

[1.17] C. Quintana, S. Halpern: *Microanalyse X en Biologie* (Soc. Française de Micr. Electronique, Paris 1983)

[1.18] H. Bethge, J. Heydenreich (eds.): *Elektronenmikroskopie in der Festkörperphysik* (VEB Deutscher Verlag der Wissenschaften, Berlin 1980)

[1.19] P. Echlin: *Analysis of Organic and Biological Surfaces* (Wiley, New York 1984)

[1.20] D.B. Holt, D.C. Joy: *SEM Microcharcterization of Semiconductors* (Academic, London 1989)

[1.21] K.F.J. Heinrich: *Electron Beam Quantitation* (Plenum, New York 1991)

[1.22] J.B. Pawley: LVSEM for high resolution topography and density contrast imaging. Adv. Electr. Electron Phys. **83**, 203 (1992)

[1.23] L. Reimer: *Image Formation in Low-Voltage Scanning Electron Microscopy*, Tutorial Texts in Optical Engineering, Vol. TT 12 (SPIE, Bellingham, WA 1993)

[1.24] J.L. Thong (ed.): *Electron Beam Testing Technology* (Plenum, New York 1993)

[1.25] P. Schmidt et al.: *Praxis der Rasterelektronenmikroskopie und Mikrobereichsanalyse* (expert verlag, Remmingen-Marheim 1994)

[1.26] G.M. Roomans, B.L. Gupta, R.D. Leapman, T. von Zylinicki (eds.): *The Science of Biological Microanalysis*. Scanning Microscopy Suppl. **8** (SEM Inc., AMF O'Hare, IL 1994)

[1.27] V.D. Scott, G. Love, S.J.B. Reed: *Quantitative Electron-Probe Microanalysis* (Ellis Hoswood, New York 1995)

[1.28] D.B. Williams, J.I. Goldstein, D.E. Newbury (eds.): *X-Ray Spectrometry in Electron Beam Instruments* (Plenum, New York 1995)

Chapter 2

[2.1] M.E. Haine, P.A. Einstein: Characteristics of the hot cathode electron microscope gun. Brit. J. Appl. Phys. **3**, 40 (1952)

[2.2] J.A. Swift, A.C. Brown: SEM electron source: pointed tungsten filaments with long-life and high brightness. Scanning **2**, 42 (1979)

[2.3] A.N. Broers: Some experimental and estimated characteristics of the LaB_6 rod cathode electron gun. J. Phys. E **2**, 273 (1969)

[2.4] S.F. Vogel: Pyrolithic graphite in the design of a compact inert heater of a LaB_6 cathode. Rev. Sci. Instr. **41**, 585 (1970)

[2.5] R. Vogt: Richtstrahlwert und Energieverteilung der Elektronen aus einem Elektronenstrahlerzeuger mit LaB_6-Kathode. Optik **36**, 262 (1972)

[2.6] S.D. Ferris, D.C. Joy, H.J. Leamy, C.K. Crawford: A directly heated LaB_6 electron source. SEM 1975 (ITTRI, Chicago 1975) p.11

[2.7] S. Nakagawa, T. Yanaka: A highly stable electron probe obtained with LaB_6 cathode electron gun. SEM 1975 (ITTRI, Chicago 1975) p.19

[2.8] C.K. Crawford: Mounting methods and operating characteristics for LaB_6 cathodes. SEM 1979/I (SEM Inc., AMF O'Hare, IL 1979) p.19

[2.9] M. Gesley, F. Hohn: Emission distribution, brightness, and mechanical stability of the LaB_6 triode electron gun. J. Appl. Phys. **64**, 3380 (1988)

[2.10] P.H. Schmidt, D.C. Joy, L.D. Longinotti, H.J. Leamy, S.D. Ferris, Z. Fisk: Anisotropy of thermionic electron emission values for LaB_6 single-crystal emitter cathodes. Appl. Phys. Lett. **29**, 400 (1976)

[2.11] J.D. Verhoeven: On the problem of obtaining optimum brightness from your LaB_6 gun. SEM 1977/I (ITTRI, Chicago 1977) p.581

[2.12] P.B. Sewell: High brightness thermionic electron guns for electron microscopes. SEM 1980/I (SEM Inc., AMF O'Hare, IL 1980) p.11

[2.13] H. Hagiwara, H. Hiraoka, T. Terasaki, M. Ishii, R. Shimizu: Crystallographic and geometrical effects on thermionic emission change of single crystal LaB_6 cathodes. SEM 1982/II (SEM Inc., AMF O'Hare, IL 1982) p.473

[2.14] H. Boersch: Experimentelle Bestimmung der Energieverteilung in thermisch ausgelösten Elektronenstrahlen. Z. Physik **139**, 115 (1954)

[2.15] K.H. Loeffler: Energy-spread generation in electron-optical instruments. Z. angew. Phys. **27**, 145 (1969)

[2.16] H. Rose, R. Spehr: On the theory of the Boersch effect. Optik **57**, 339 (1980)

[2.17] D.B. Langmuir: Theoretical limitations of cathode ray tubes. Proc. IRE **25**, 977 (1937)

[2.18] J. Dosse: Theoretische und experimentelle Untersuchungen über Elektronenstrahler. Z. Physik **115**, 530 (1940)

[2.19] D.W. Tuggle, J.Z. Li, L.W. Swanson: Point cathodes for use in virtual source electron optics. J. Microsc. **140**, 293 (1985)

[2.20] D.W. Tuggle, L.W. Swanson: Emission characteristics of the ZrO/W thermal field electron source. J. Vac. Sci. Techn. B **3**, 220 (1985)

[2.21] S. Saito, Y. Nakaizumi, H. Mori, T. Nagatani: A field emission SEM controlled by microprocessor. In *Electron Microscopy 1982*, Vol.1 (Deutsche Ges. für Elektronenmikroskopie, Frankfurt 1982) p.379

[2.22] M. Ohi, K. Harasawa, T. Niikura, H. Okazaki, Y. Ishimori, T. Miyokawa, S. Nakagawa: Development of a new digital FE SEM. In *Electron Microscopy 1990*, Vol.1, ed. by L.D. Peachey, D.B. Williams (San Francisco Press, San Francisco 1990) p. 432

[2.23] A.N. Broers: Electron sources for SEM. SEM 1975 (ITTRI, Chicago 1975) p.661

[2.24] L. Reimer, B. Volbert, P. Bracker: STEM semiconductor detector for testing SEM quality parameters. Scanning **2**, 96 (1979)

[2.25] P.B. Sewell, K.N. Ramachandran: A source imaging detector for the SEM. SEM 1977/I (ITTRI, Chicago 1977) p.17

[2.26] O.C. Wells: Experimental method for measuring the electron optical parameters of the SEM. SEM 1977/I (ITTRI, Chicago 1977) p.25

[2.27] W. Glaser: *Grundlagen der Elektronenoptik* (Springer, Wien 1952)

[2.28] V.E. Cosslett: Probe size and probe current in the STEM. Optik **36**, 85 (1972)

[2.29] W. Glaser: Strenge Berechnung magnetischer Linsen der Feldform $H = H_o/[1+(z/a)^2]$. Z. Physik **117**, 285 (1941)

[2.30] R.L. Barnes, I.K. Openshaw: A comparison of experimental and theoretical C_s values for some probe-forming lenses. J. Phys. E **1**, 628 (1968)

[2.31] S. Nakagawa: A method for measuring the spherical and chromatic aberration coefficients of an objective lens. SEM 1977/I (ITTRI, Chicago 1977) p.33

[2.32] O. Scherzer: Sphärische und chromatische Korrektur von Elektronenlinsen. Optik **2**, 114 (1947)

[2.33] H. Rose: Abbildungseigenschaften sphärisch korrigierter elektronenoptischer Achromate. Optik **33**, 1 (1971)

[2.34] H. Rose: Inhomogeneous Wien filter as a corrector compensating for the chromatic and spherical aberration of low-voltage electron microscopes. Optik **84**, 91 (1990)

[2.35] J. Zach: Design of a high-resolution low-voltage SEM. Optik **83**, 30 (1989)

[2.36] J. Zach, M. Haider: Correction of spherical and chromatic aberrations in a LVSEM. In *Electron Microscopy 1994*, Vol.1 (Les Editions de Physique, Les Ulis 1994) p.199

[2.37] J. Pawley, J. Wall: A low voltage SEM optimized for high-resolution topographical imaging. In *Electron Microscopy 1982*, Vol.1 (Deutsche Ges. für Elektronenmikroskopie, Frankfurt 1992) p.383

[2.38] T. Nagatani, S. Saito, M. Sato, M. Yamada: Development of an ultrahigh resolution SEM by means of a field emission source and in-lens system. Scanning Microscopy **1**, 901 (1987)

[2.39] T. Nagatani, M. Sato, M. Osumi: Development of an ultra-high-resolution low voltage SEM with an optimized in-lens design. In *Electron Microscopy 1990*, Vol.1, ed. by L.D. Peachey, D.B. Williams (San Francisco Press, San Francisco 1990), p.388

[2.40] H. Nakagawa, N. Nomura, T. Koizumi, N. Anazawa, K. Harafuji: A novel high-resolution SEM for the surface analysis of high-aspect-ratio three-dimensional structures. Jpn. J. Appl. Phys. **30**, 2112 (1991)

[2.41] O.C. Wells, F.K. LeGoues, R.T. Hodgson: In-lens low-loss electron detector for the upper specimen stage in the SEM. In *Electron Microscopy 1990*, Vol.1, ed. by L.D. Peachey, D.B. Williams (San Francisco Press, San Francisco 1990) p.382

[2.42] Z. Shao: Extraction of secondary electrons in a newly proposed immersion lens. Rev. Sci. Instr. **60**, 693 (1989)

[2.43] K. Ogura, Y. Ono, H. Kazumori, S. Nakagawa, J. Critchell: Advantages of low voltage imaging with the JSM-6320F semi-in-lens FESEM. In *Electron Microscopy 1994*, Vol.1 (Les Editions de Physique, Les Ulis 1994) p.63

[2.44] T. Miyokawa, H. Kazumori, S. Nakagawa, R.C. Hertsens: Strong magnetic field lens with built-in secondary electron detector. In *Electron Microscopy 1994*, Vol.1 (Les Editions de Physique, Les Ulis 1994) p.255

[2.45] T. Mulvey: Unconventional lens design. in *Magnetic Electron Lenses*, ed. by P.W. Hawkes, Topics Current Phys., Vol.18 (Springer, Berlin, Heidelberg 1982) p.359

[2.46] A. Yonezawa, Y. Takeuchi,, T. Kano, H. Ishijima: Single-pole objective lens for low-voltage SEM high resolution wafer observation. In *Electron Microscopy 1990*, Vol.1, ed. by L.D. Peachey, D.B. Williams (San Francisco Press, San Francisco 1990) p.396

[2.47] M. Bode, L. Reimer: Detector strategy for a single-polepiece lens. Scanning **7**, 125 (1985)

[2.48] V. Kolarik, I. Müllerová, M. Lenc: SEM with single-polepice lens. Scanning Microscopy **3**, 1003 (1989)

[2.49] I. Müllerová, M. Lenc: The scanning very low-energy electron microscope. Mikrochim. Acta **12** (Suppl.), 173 (1992)

[2.50] Y.W. Yau, R.F.W. Pease, A.A. Iranmanesh, K.J. Polasko: Generation and applications of finely focused beams of low-energy electrons. J. Vac. Sci. Techn. **19**, 1048 (1981)

[2.51] C. Kinalidis, J.C. Wolfe: Characterization of a low voltage, high current density electron probe. J. Vac. Sci. Techn. B **5**, 150 (1987)

[2.52] I. Müllerová, M. Lenc: Some approaches to low-voltage SEM. Ultramicroscopy **41**, 399 (1992)

[2.53] J. Frosien, E. Plies, K. Anger: Compound magnetic and electrostatic lenses for low-voltage applications. J. Vac. Sci. Techn. B **7**, 1874 (1989)

[2.54] E. Weimer, J.P. Martin: Development of a new ultra-high performance SEM. In *Electron Microscopy 1994*, Vol.1 (Les Editions de Physique, Les Ulis 1994) p.67

[2.55] J. Frosien, S. Lanio, H.P. Feuerbaum: High precision electron optical system for absolute and CD-measurements on large substrates. Nucl. Instr. Meth. Phys. Res. A **363**, 25 (1995)

[2.56] M. Lenc, I. Müllerová: Electron optical properties of a cathode lens. Ultramicroscopy **41**, 411 (1992)

[2.57] I. Müllerová, L. Frank: Very low energy microscopy in commercial SEMs. Scanning **15**, 193 (1993)

[2.58] E. Bauer: Low energy electron microscopy. In *Chemistry and Physics of Solid Surfaces VIII*, ed. by R. Vanselov, R. Howe, Springer Ser. Surf. Sci., Vol.22 (Springer, Berlin, Heidelberg 1990) p.267

[2.59] L.H. Veneklasen: The continuing development of low-energy electron microscopy for characterizing surfaces. Rev. Sci. Instr. **63**, 5513 (1992)

[2.60] M. Fukuoka, Y. Sakai, K. Tsunoda, T. Ichinokawa: A microscanning electron microscope in ultrahigh vacuum for surface microanalysis. Rev. Sci. Instr. **65**, 2844 (1994)

[2.61] T.H.P. Chang, D.P. Kern, L.P. Muray: Microminiaturization of electron-optical systems. J. Vac. Sci. Techn. B **8**, 1698 (1990)

[2.62] A.D. Feinerman, D.A. Crewe, D.C. Perng, S.E. Shoaf: Sub-centimeter micromachined electron microscope. J. Vac. Sci. Techn. A **10**, 611 (1992)

[2.63] L.P. Muray, U. Staufer, D.P. Kern, T.H.P. Chang: Performance measurements of a 1-keV electron-beam microcolumn. J. Vac. Sci. Techn. B **10**, 2749 (1992)

[2.64] D.P. Kern, T.H.P. Chang: Miniaturized electron optics: Basis and applications. In *Electron Microscopy 1994*, Vol.1 (Les Editions de Physique, Les Ulis 1994) p.149

[2.65] C.W. Oatley, W.C. Nixon, R.F.W. Pease: *Scanning Electron Microscopy*, Adv. Electr. Electron Phys. **21**, 181 (1965)

[2.66] J. Ximen, Z. Shao, P.S.D. Lin: Theoretical calculation of probe size of low-voltage SEM. J. Microsc. **170**, 119 (1993)

[2.67] P. Gentsch, P. Hagemann, L. Reimer: Comparison of the resolution using a 100 keV electron microscope in the conventional mode and with a scanning device. In *Electron Microscopy 1974*, Vol.1, ed. by J.V. Sanders and D.J. Goodchild (Austral. Acad. Sci., Canberra 1974) p.256

[2.68] J.R.A. Cleaver, K.C.A. Smith: Two-lens probe forming system employing field emission guns. SEM 1973 (ITTRI, Chicago 1973) p.49

[2.69] A.V. Crewe: Optimization of small electron probes. Ultramicroscopy **23**, 159 (1987)

[2.70] C. Colliex, C. Mory: Quantitative aspects of STEM. In *Quantitative Electron Microscopy*, ed. by J.N. Chapman and A.J. Craven (Scottish Univ. Summer School in Physics, Edinburgh 1984) p.149

[2.71] J.E. Barth, P. Kruit: Addition of different contributions to the charged particle probe size. Optik **101**, 101 (1996)

[2.72] O. Scherzer: The theoretical resolution limit of the electron microscope. J. Appl. Phys. **20**, 20 (1949)

[2.73] H. Hantsche: Entwicklung einer verbesserten Sondenstromstabilisierung für Rasterelektronenmikroskope. Scanning **2**, 20 (1979)

[2.74] S.J.B. Reed: Probe current stability in electron-probe microanalysis. J. Phys. E **1**, 136 (1968)

[2.75] H.F. Wellenstein, R.E. Ensman: A regulated filament temperature power supply for electron guns. Rev. Sci. Instr. **44**, 922 (1973)

[2.76] J. Arndt, H. Hantsche, D. Schmidt: A simple beam current stabilizing unit for SEMs. Scanning **1**, 125 (1978)

[2.77] L.J. Balk, K. Elbern, E. Kubalek: Elektronische Zusätze zum Rasterelektronenmikroskop. Beitr. elektr. mikr. Direktabb. Oberfl. **8**, 313 (1975)

[2.78] J.R.A. Cleaver, K.C.A. Smith: Optical characteristics of a field emission scanning microscope. In *Scanning Electron Microscopy: systems and applications* (Inst. of Physics, London 1973) p.6

[2.79] M.T. Postek, W.J. Keery, A.E. Vladar: Modification of a commercial SEM with a computer controlled stabilized power supply. Scanning **15**, 208 (1993)

[2.80] G.F. Bahr, E. Zeitler: The determination of magnification in the electron microscope. Lab. Invest. **14**, 880 (1965)

[2.81] A. Higgs: A digital rotating system. J. Phys. E **15**, 266 (1982)

[2.82] W. Bröcker, G. Hauck, H. Weigelt, G. Pfefferkorn: Lock-in technique applied to cathodoluminescence of biomedical specimens in the SEM. Scanning **4**, 165 (1981)

[2.83] L.J. Balk, E. Kubalek: Use of phase sensitive-(lock-in)-amplification with SEM. Beitr. elektr. mikr. Direktabb. Oberfl. **6**, 551 (1973)

[2.84] A.E. Lukianov, G.V. Spivak: Electron mirror microscopy of transient phenomena in semiconductor diodes. In *Electron Microscopy 1966*, Vol.I, ed. by R. Uyeda (Maruzen, Tokyo 1966) p.611

[2.85] O.I. Szentesi: Stroboscopic electron mirror microscopy at frequencies up to 100 MHz. J. Phys. E **5**, 563 (1972)

[2.86] M. Weinfeld, A. Bouchoule: Electron gun for generation of subnanosecond electron packets at very high repetition rate. Rev. Sci. Instr. **47**, 412 (1976)

[2.87] S.M. Davidson: Wehnelt modulation beam blanking in the SEM. In *Electron Microscopy and Analysis 1981*, ed. by M.J. Goringe (Inst. of Physics, Bristol 1982) p.39

[2.88] R. Schief, M. Steiner: Energieverbreiterung eines durch hochfrequente Wehneltspannung gepulsten Elektronenstrahles. Optik **38**, 261 (1973)

[2.89] A.J. Gonzales, M.W. Powell: Internal waveform measurements of the MOS three transistor, dynamic RAM using SEM stroboscopic techniques. Techn. Digest of IEDM, IEEE, New York 1975, p.119

[2.90] G.S. Plows, W.C. Nixon: Stroboscopic SEM. J. Phys. E **1**, 595 (1968)

[2.91] E. Menzel, E. Kubalek: Electron beam chopping in the SEM. SEM 1979/I (SEM Inc., AMF O'Hare, IL 1979) p.305

[2.92] J.T.L. Thong, B.C. Breton, W.C. Nixon: High repetition rate electron beam chopping system for electron beam testing at microwave frequencies. J. Vac. Sci. Techn. B **8**, 2048 (1990)

[2.93] K. Ozaki, A. Ito, Y. Goto, Y. Furukawa, T. Inagaki: Electron beam probing system with ultrahigh time resolution. J. Vac. Sci. Techn. B **5**, 84 (1987)

[2.94] A. Gopinath, M.S. Hill: SEM stroboscopy at 9 GHz. SEM 1973 (ITTRI, Chicago) p.197

[2.95] G.Y. Robinson: Stroboscopic SEM at GHz frequencies. Rev. Sci. Instr. **42**, 251 (1971)

[2.96] K. Ura, H. Hujioka, T. Hosokawa: Electron optical design of picosecond pulse stroboscopic SEM. SEM 1978/I (SEM Inc., AMF O'Hare, IL 1978) p.747

[2.97] T. Hosokawa, H. Fujioka, K. Ura: Generation and measurement of subpicosecond electron beam pulses. Rev. Sci. Instr. **49**, 624 (1978)

[2.98] A. Gopinath, M.S. Hill: Deflection beam chopping in the SEM. J. Phys. E **10**, 229 (1977)

[2.99] J. Stabenow: Herstellung dünnwandiger Objektivaperturblenden für die Elektronenmikroskopie. Naturwiss. **54**, 163 (1967)

[2.100] E. Schablach: A method for the fabrication of thin foil apertures for electron microscopy. J. Microsc. **101**, 121 (1974)

[2.101] J.P. Martin, R. Speidel: Zur Verwendung von Dünnschicht-Aperturblenden im Elektronen-Rastermikroskop. Beitr. elektr. mikr. Direktabb. Oberfl. **4/2**, 345 (1971)

[2.102] N.C. Yew: Dynamic focusing technique for tilted samples in SEM. SEM 1971 (ITTRI, Chicago 1971) p.33

[2.103] J. Hersener, Th. Ricker: Eine automatische Fokussierungseinrichtung für Rasterelektronenmikroskope. Beitr. elektr. mikr. Direktabb. Oberfl. **5**, 377 (1972)

[2.104] S. Shirai, A. Onoguchi, T. Ichinogawa: An automatic focusing system for the SEM. In *Proc. 6th Int'l. Conf. on X-Ray Optics and Microanalysis*, ed. by G. Shinoda et al. (Univ. Tokyo Press, Tokyo 1972) p.511

[2.105] W.J. Tee, K.C.A. Smith, D.M. Holburn: An automatic focusing and stigmating system for the SEM. J. Phys. E **12**, 35 (1979)

[2.106] S.J. Erasmus, K.C.A. Smith: An automatic focusing and astigmatism correction system for the SEM and CTEM. J. Microsc. **127**, 185 (1982)

[2.107] P.E. Batson: Imaging of bulk inter-band excitations in small structures with inelastically scattered electrons, in *Electron Microscopy 1986*, Vol.1, ed. by T. Imura et al. (Jpn. Soc. Electron Microscopy, Tokyo 1986) p.95

[2.108] T.E. Everhart, N. Saeki, R. Shimizu, T. Koshikawa: Measurement of structure in the energy distribution of slow secondary electrons from aluminium. J. Appl. Phys. **47**, 2941 (1976)

[2.109] R.F.W. Pease, W.C. Nixon: High resolution SEM. J. Sci. Instr. **42**, 81 (1965)

[2.110] A.N. Broers: A new high resolution reflection SEM. Rev. Sci. Instr. **40**, 1040 (1969)

[2.111] R. Simon: Resolving power of the SEM. J. Appl. Phys. **40**, 2851 (1969)

[2.112] P. Gentsch, H. Gilde, L. Reimer: Measurement of the top-bottom effect in STEM of thick amorphous specimens. J. Microsc. **100**, 81 (1974)

[2.113] S.A. Rishton, S.P. Beaumont, C.D.W. Wilkinson: Measurement of the profile of finely focused electron beams in a SEM. J. Phys. E **17**, 296 (1984)

[2.114] K. Schur, Ch. Schulte, L. Reimer: Auflösungsvermögen und Kontrast von Oberflächenstufen bei der Abbildung mit einem Raster-Elektronenmikroskop. Z. Angew. Phys. **23**, 405 (1967)

[2.115] J. Bentley, N.D. Evans, E.A. Kenik: Resolution measurement for SEM. In *Electron Microscopy 1994*, Vol.1, ed. by B. Jouffrey, C. Colliex (Les Éditions de Physique, Les Ulis, France) p. 109

[2.116] R. Speidel, J.P. Martin, B. Bauer: Testobjekte für die Bestimmung des Punktauflösungsvermögens im Elektronen-Rastermikroskop. Optik **34**, 321 (1971)

[2.117] L. Reimer, B. Volbert, P. Bracker: Quality control of SEM micrographs by laser diffractometry. Scanning **1**, 233 (1978)

[2.118] S.J. Erasmus, D.M. Holburn, K.C.A. Smith: On-line computation of diffractograms for the analysis of SEM images. Scanning **3**, 273 (1980)

Chapter 3

[3.1] L. Reimer: *Transmission Electron Microscopy, Physics of Image Formation and Microanalysis*, 4th ed., Springer Ser. Opt. Sci., Vol.36 (Springer, Berlin, Heidelberg 1997)

[3.2] R.D. Birkhoff: The Passage of Fast Electrons through Matter. *Encyclopedia of Physics*, Vol.34, ed. by S. Flügge (Springer, Berlin 1958) p.53

[3.3] J.M. Lévy-Leblond: Nonrelativistic particles and wave equations. Comm. Math. Phys. **6**, 286 (1967)

[3.4] W.Greiner: *Theoretische Physik*, Bd.4: *Quantenmechanik* (Harry Deutsch, Thun 1979) p.300

[3.5] H.S.W. Massey, E.H.S. Burhop, H.B. Gilbody: *Electronic and Ionic Impact Phenomena. Vol.I. Collison of Electrons with Atoms* (Clarendon, Oxford 1965)

[3.6] H.L. Cox, R.A. Bonham: Elastic electron scattering amplitudes for neutral atoms calculated using the partial wave method at 10, 40, 70 and 100 kV for $Z = 1$ to $Z = 54$. J. Chem. Phys. **47**, 2599 (1967)

[3.7] F. Lenz: Zur Streuung mittelschneller Elektronen in kleinste Winkel. Z. Naturforschg. **9a**, 185 (1954)

[3.8] H. Raith: Komplexe Streuamplituden für die elastische Elektronenstreuung an Festkörperatomen. Acta Cryst. A **24**, 85 (1968)

[3.9] L. Reimer, K.H. Sommer: Messungen und Berechnungen zum elektronenmikroskopischen Streukontrast für 17 bis 1200 keV-Elektronen. Z. Naturforschg. **23a**, 1569 (1968)

[3.10] F. Salvat, R. Mayol: Elastic scattering of electrons and positrons by atoms. Schrödinger and Dirac partial wave analysis. Comp. Phys. Comm. **74**, 358 (1993)

[3.11] N.F. Mott, H.S.W. Massey: *The Theory of Atomic Collisions*, 3rd edn. (Oxford Univ. Press, London 1965)

[3.12] D.W. Walker: Relativistic effects in low energy electron scattering from atoms. Adv. Physics **20**, 257 (1971)

[3.13] J. Kessler: *Polarized Electrons*, 2nd edn., Springer Ser. Atoms Plasmas, Vol.1 (Springer, Berlin, Heidelberg 1985)

[3.14] W. Bühring: Computational improvement in phase shift calculations of elastic electron scattering. Z. Physik **187**, 180 (1965)

[3.15] A.C. Yates: Spin polarization of low-energy electrons scattered elastically from atoms and molecules. Phys. Rev. **176**, 173 (1968)

[3.16] L. Reimer, B. Lödding: Calculation and tabulation of Mott cross-sections for large-angle electron scattering. Scanning **6**, 128 (1984)

[3.17] A. Jablonski, G. Gergely: Mott factors for P, V, Fe, Ga, As, Pd, In, Ta, and W for 500–3000 eV electrons. Scanning **11**, 29 (1989)

[3.18] Z. Czyzewski, D.O. MacCallum, A. Romig, D.C. Joy: Calculations of Mott scattering cross section. J. Appl. Phys. **68**, 3066 (1990)

[3.19] R. Browning: Universal elastic scattering cross sections for electrons in the range 1–100 keV. Appl. Phys. Lett. **58**, 2845 (1991)

[3.20] R. Browning: T. Eimori, E.P. Traut, B. Chui, R.F.W. Pease: An elastic cross sections model for use with Monte Carlo simulations of low energy scattering from high atomic number targets. J. Vac. Sci. Techn. B **9**, 3578 (1991)

[3.21] D. Drouin, R. Gauvin, D.C. Joy: Computation of polar angle of collisions from partial elastic Mott cross-sections. Scanning **16**, 67 (1994)

[3.22] G. Gergely: Elastic peak electron spectroscopy. Scanning **8**, 203 (1986)

[3.23] H. Raether: *Excitations of Plasmons and Interband Transitions by Electrons*. Springer Tracts Mod. Phys., Vol. 88 (Springer, Berlin, Heidelberg 1980)

[3.24] J. Daniels, C. von Festenberg, H. Raether, K. Zeppenfeld: Optical constants of solids by electron spectroscopy. Springer Tracts Mod. Phys., Vol. 54 (Springer, Berlin, Heidelberg 1970) p.77

[3.25] R.F. Egerton: *Electron Energy-Loss Spectroscopy in the Electron Microscope*, 2nd edn. (Plenum, New York, London 1996)

[3.26] F. Hofer: Inner-shell ionization. In *Energy-Filtering Transmission Electron Microscopy*, ed. by L. Reimer, Springer Ser. Opt. Sci., Vol. 71 (Springer, Berlin, Heidelberg 1995) p.225

[3.27] C.C. Ahn, O.L. Krivanek: *EELS Atlas* (Gatan Inc., Warrendale, PA 1983)

[3.28] L. Reimer, U. Zepke, J. Moesch, St. Schulze-Hillert, M. Ross-Messemer, W. Probst, E. Weimer: *EELSpectroscopy. A Reference Handbook of Standard Data for Identification and Interpretation of Electron Energy Loss Spectra and for Generation of Electron Spectroscopic Images*, Carl Zeiss, Electron Optics Division (Oberkochen, Germany 1992)

[3.29] M. Creuzburg, H. Raether: Über die charakteristischen Energieverluste bei Elektronenstreuung an Si-Spaltflächen. Z. Physik **171**, 436 (1963)

[3.30] J. Schilling: Energieverlustmessungen von schnellen Elektronen an Oberflächen von Ga, In, Al und Si. Z. Physik B **25**, 61 (1976)

[3.31] H. Froitzheim: Electron energy loss spectroscopy, In *Electron Spectroscopy of Surface Analysis*, ed. by H. Ibach, Topics Current Phys., Vol.4 (Springer, Berlin, Heidelberg 1977) p.205

[3.32] C. Møller: Zur Theorie des Durchgangs schneller Elektronen durch Materie. Ann.Physik **14**, 531 (1932)

[3.33] M. Gryziński: Classical theory of atomic collisions. I. Theory of inelastic collisions. Phys.Rev. A **138**, 336 (1965)

[3.34] C.R. Worthington, S.G. Tomlin: The intensity of emission of characteristic x-radiation. Proc. Phys. Soc. A **69**, 401 (1956)

[3.35] C.J. Powell: Cross-sections for ionization of inner-shell electrons by electrons. Rev. Mod. Phys. **48**, 33 (1976)

[3.36] C.J. Powell: Inelastic scattering of electrons in solids. In *Electron Beam Interactions with Solids*, ed. by D.F. Kyser et al. (SEM Inc., AMF O'Hare, IL 1982) p.19

[3.37] J. Philibert, R. Tixier: Some problems with quantitative electron probe microanalysis. in *Quantitative Electron Probe Microanalysis*, ed. by K.F.J. Heinrich, Nat. Bur. Stand. Spec.Publ. 298 (Washington 1968) p.13

[3.38] R.F. Egerton: Measurement of inelastic/elastic scattering ratio for fast electrons and its use in the study of radiation damage. Phys. Stat. Solidi A **37**, 663 (1976)

[3.39] L. Reimer, M. Ross-Messemer: Contrast in the electron spectroscopic imaging mode of a TEM. II. Z-ratio, structure-sensitive and phase contrast. J. Microsc. **159**, 143 (1990)

[3.40] M. Kotera, R. Ijichi, T. Fujiwara, H. Suga, D.B. Wittry: A simulation of electron scattering in metals. Jpn. J. Appl. Phys. **29**, 2277 (1990)

[3.41] S. Goudsmit, J.L. Saunderson: Multiple scattering of electrons. Phys. Rev. **57**, 24 (1940); **58**, 36 (1940)

[3.42] H.W. Lewis: Multiple scattering in an infinite medium. Phys. Rev. **78**, 526 (1950)

[3.43] G. Molière: Theorie der Streuung schneller geladener Teilchen II. Mehrfachstreuung und Vielfachstreuung. Z. Naturforschg A **3**, 78 (1948)

[3.44] H.S. Snyder, W.T. Scott: Multiple scattering of fast charged particles. Phys. Rev. **76**, 220 (1949)

[3.45] H.A. Bethe: Molière's theory of multiple scattering. Phys. Rev. **89**, 1256 (1953)

[3.46] W. Bothe: Durchgang von Elektronen durch Materie. In *Handbuch der Physik*, Vol.22/2, hersg. v. H. Geiger and K. Scheel (Springer, Berlin 1933) p.1

[3.47] V.E. Cosslett, R.N. Thomas: Multiple scattering of 5–30 keV electrons in evaporated metal films. I. Total transmission and angular distribution. Brit. J. Appl. Phys. **15**, 883 (1964)

[3.48] P. Gentsch, H. Gilde, L. Reimer: Measurement of the top-bottom effect in STEM of thick amorphous specimens. J. Microsc. **100**, 81 (1974)

[3.49] J.I. Goldstein, J.L. Costley, G.W. Lorimer, S.J.B. Reed: Quantitative x-ray analysis in the electron microscope. SEM 1977/I (ITTRI, Chicago 1977) p.315

[3.50] K. Jost, J. Kessler: Die Ortverteilung mittelschneller Elektronen bei Mehrfachstreuung. Z. Physik **176**, 126 (1963)

[3.51] L. Reimer, H. Gilde, K.H. Sommer: Die Verbreiterung eines Elektronenstrahles (17–1200 keV) durch Mehrfachstreuung. Optik **30**, 590 (1970)

[3.52] T. Groves: Thick specimens in the CEM and STEM. Resolution and image formation. Ultramicroscopy **1**, 15 and 170 (1975)

[3.53] H. Rose: The influence of plural scattering on the limit of resolution in electron microscopy. Ultramicroscopy **1**, 167 (1975)

[3.54] D.F. Kyser, K. Murata: Application of Monte-Carlo simulation to electron microprobe analysis of thin films on substrates. In *Use of Monte Carlo Calculations in Electron Probe Microanalysis and Scanning Electron Microscopy.* NBS Spec. Publ. 460 (U.S. Dep. of Commerce, Washington, DC 1976) p.129

[3.55] D.F. Kyser: Monte Carlo calculations for electron microscopy, microanalysis, and microlithography. In *Electron Beam Interactions with Solids*, ed. by D.F. Kyser et al. (SEM Inc., AMF O'Hare, IL 1982) p.119

[3.56] H. Bethe: Zur Theorie des Durchgangs schneller Korpuskularstrahlen durch Materie. Ann. Phys. **5**, 325 (1930)

[3.57] H. Bethe: Quantenmechanik der Ein- und Zwei-Elektronenprobleme. In *Handbuch der Physik 24*, hersg. v. J. Geiger and K. Scheel (Springer, Berlin 1933) p.273

[3.58] R.D. Evans: *The Atomic Nucleus.* Int'l. Ser. in Pure and Appl. Phys. (McGraw Hill, New York 1955)

[3.59] N. Bohr: The penetration of atomic particles through matter. Kgl. Danske Videnskabernes. Selskat, Matematisk-fysike Medd. **18**, No.8 (1948)

[3.60] H.A. Bethe, J. Ashkin: Passage of radiation through matter. In *Experimental Nuclear Physics*, Vol.1, ed. by E. Segrè (Wiley, New York 1953) p.166

[3.61] M.J. Berger, S.M. Seltzer: Tables of energy loss and ranges of electrons and positrons. Nat. Acad. Sci., Nat. Res. Council Publ. 1133, Washington DC (1964) p.205

[3.62] R. Whiddington: The transmission of cathode rays through matter. Proc. Roy. Soc. A **86**, 360 (1912); **89**, 554 (1914)

[3.63] H.M. Terrill: Loss of velocity of cathode rays in matter. Phys. Rev. **22**, 101 (1923)

[3.64] L. Reimer, K. Brockmann, U. Rhein: Energy losses of 20–40 keV electrons in 150–650 μg/cm^2 metal films. J. Phys. D **11**, 2151 (1978)

[3.65] D.C. Joy, S. Luo: An empirical stopping power relationship for low-energy electrons. Scanning **11**, 176 (1989)

[3.66] D.C. Joy, S. Luo, X. Zhang: Electron stopping powers and their experimental determination. In *Electron Microscopy 1990*, Vol.2 (San Francisco Press, San Francisco 1990) p.8

[3.67] S. Luo, X. Zhang, D.C. Joy: Experimental determination of electron stopping power at low energies. Rad. Effects and Defects in Solids **117**, 235 (1991)

[3.68] D.C. Joy, S. Luo, J.R. Dunlap, D. Williams, S. Cao: Stopping-power determination for compounds by EELS. In *Proc. 52th Ann. Meeting Micr. Soc. of America* (San Francisco Press, San Francisco 1994) p.948

[3.69] T.S. Rao-Sahib, D.B. Wittry: X-ray continuum from thick element targets for 10–50 keV electrons, J. Appl. Phys. **45**, 5060 (1974)

[3.70] C.J. Tung, J.C. Ashley, R.H. Ritchie: Electron inelastic mean free paths and energy losses in solids II. Surf. Sci. **81**, 427 (1979)
[3.71] R.M. Nieminen: Stopping power for low-energy electrons. Scanning Microscopy **2**, 1917 (1988)
[3.72] L. Landau: On the energy loss of fast electrons by ionization. J. Phys. USSR **8**, 201 (1944)
[3.73] O. Blunck, S. Leisegang: Zum Energieverlust schneller Elektronen in dünnen Schichten. Z. Physik **128**, 500 (1950)
[3.74] D.F. Hebbard, P.R. Wilson: The effect of multiple scattering on energy loss distributions. Austral. J. Phys. **8**, 90 (1955)
[3.75] C.N. Yang: Actual path length of electrons in foils. Phys. Rev. **84**, 599 (1951)
[3.76] L. Reimer, R. Senkel: Calculation of enery spectra of electrons transmitted through thin aluminium foils. J. Phys. D **25**, 1371 (1992)
[3.77] L. Reimer: Calculation of the angular and energy distribution of multiple scattered electrons using Fourier transforms. Ultramicroscopy **31**, 169 (1989)
[3.78] H. Drescher, L. Reimer, H. Seidel: Rückstreukoeffizient und Sekundärelektronen-Ausbeute von 10–100 keV Elektronen und Beziehungen zur Raster-Elektronenmikroskopie. Z. angew. Phys. **29**, 331 (1970)
[3.79] A.F. Makhov: The penetration of electrons into solids. Sov. Phys. Solid State **2**, 1934. 1942 and 1945 (1961)
[3.80] A. Ya. Vyatskin, A.N. Pilyankevich: Some energy characteristics of electron passage through a solid. Sov. Phys. Solid State **5**, 1662 (1964)
[3.81] R. Böngeler, U. Golla, M. Kässens, L. Reimer, B. Schindler, R. Senkel, M. Spranck: Electron-specimen interactions in LVSEM. Scanning **15**, 1 (1993)
[3.82] L. Reimer: Electron-specimen interactions and image formation in LVSEM. In *Electron Microscopy 1992*, Vol. 1, ed. by A. Rios et al. (Secr. Publ. Univers. Granada, Spain 1992) p.9
[3.83] M. Kotera, K. Murata, K. Nagami: Monte Carlo simulation of 1–10 keV electron scattering in a gold target. J. Appl. Phys. **52**, 997 (1981)
[3.84] M. Kotera, K. Murata, K. Nagami: Monte Carlo simulation of 1–10 keV. electron scattering in an aluminium target. J. Appl. Phys. **52**, 7403 (1981)
[3.85] H. Kanter, E.J. Sternglass: Interpretation of range measurements for kilovolt electrons in solids. Phys. Rev. **126**, 620 (1962)
[3.86] J.E. Holliday, E.J. Sternglass: New method for range measurements of low energy ellectrons in solids. J. Appl. Phys. **30**, 1428 (1959)
[3.87] V.E. Cosslett, R.N. Thomas: Multiple scattering of 5–30 keV electrons in evaporated metal films. II Range-energy relations. Brit. J. Appl. Phys. **15**, 1283 (1964)
[3.88] V.V. Makarov: Spatial distribution of excitation density in a solid bombarded by electrons at 0.5-500 eV. Sov. Phys. Techn. Phys. **23**, 324 (1978)
[3.89] H. Kanter: Electron scattering by thin foils for energies below 10 keV. Phys. Rev. **121**, 461 (1961)
[3.90] K.O. Al-Ahmad, D.E. Watt: Stopping powers and extrapolated ranges for electrons (1–10 keV) in metals. J. Phys. D **16**, 2257 (1983)
[3.91] I.R. Kanicheva,, V.V. Burtsev: Investigation of transmission of 0.5–16 keV electrons through collodion and gold films. Sov. Phys. Solid State **1**, 1146 (1960)
[3.92] J.R. Young: Penetration of electrons and ions in aluminium. J. Appl. Phys. **27**, 1 (1956)
[3.93] W.F. Libby: Measurement of radioactive tracers. Anal. Chem. **19**, 2 (1947)

[3.94] R.O. Lane, D.J. Zaffarano: Transmission of 0–40 keV electrons by thin films with application to beta-ray spectroscopy. Phys. Rev. **94**, 960 (1954)

[3.95] T.E. Everhart, P.H. Hoff: Determination of kilovolt electron energy dissipation vs penetration distance in solid materials. J. Appl. Phys. **42**, 5837 (1971)

[3.96] A.E. Grün: Lumineszenz-photometrische Messungen der Energieabsorption im Strahlungsfeld von Elektronenquellen, Eindimensionaler Fall in Luft. Z. Naturforschg. A **12**, 89 (1957)

[3.97] L. Reimer, H. Seidel, H. Gilde: Einfluß der Elektronendiffusion auf die Bildentstehung im Rasterelektronenmikroskop. Beitr. elektr. mikr. Direktabb. Oberfl. **1**, 53 (1968)

[3.98] A. Cohn, G. Caledonia: Spatial distribution of the fluorescent radiation emission caused by an electron beam. J. Appl. Phys. **41**, 3767 (1970)

[3.99] W. Ehrenberg, D.E.N. King: The penetration of electrons into luminescent materials. Proc. Phys. Soc. **81**, 751 (1963)

[3.100] M. Hatzakis: New method of observing electron penetration profiles in solids. Appl. Phys. Lett. **18**, 7 (1971)

[3.101] B. Linnemann, L. Reimer: Electron flux inside a gas target. Scanning **4**, 199 (1981)

[3.102] L. Reimer: Monte-Carlo-Rechnungen zur Elektronendiffusion. Optik **27**, 86 (1968)

[3.103] D.F. Kyser, D.B. Wittry: Spatial distribution of excess carriers in electron-beam excited semiconductors. Proc. IEEE **55**, 733 (1967)

[3.104] C. Donolato: On the theory of SEM charge-collection imaging of localized defects in semiconductors. Optik **52**, 19 (1978)

[3.105] J.F. Bresse: Electron beam induced current in silicon planar p-n junctions: physical model of carrier generation. Determination of some physical parameters in silicon. SEM 1972 (ITTRI, Chicago 1972) p.105

[3.106] T.E. Everhart: Simple theory concerning the reflection of electrons from solids. J. Appl. Phys. **31**, 1483 (1960)

[3.107] G.D. Archard: Backscattering of electrons. J. Appl. Phys. **32**, 1505 (1961)

[3.108] H.A. Bethe, M.E. Rose, L.P. Smith: The multiple scattering of electrons. Proc. Amer. Phil. Soc. **78**, 573 (1938)

[3.109] K. Kanaya, S. Ono: The energy dependence of a diffusion model for an electron probe into solid targets. J. Phys. D **11**, 1495 (1978)

[3.110] H.W. Thümmel: *Durchgang von Elektronen und Betastrahlung durch Materieschichten. Streuabsorptionsmodelle* (Akad. Verlag, Berlin 1974)

[3.111] H. Niedrig: Ein kombiniertes Einfachstreu- und Diffusionsmodell für die Elektronen-Rückstreuung dünner Schichten. Beitr. elektr. mikr. Direktabb. Oberfl. **14**, 291 (1981)

[3.112] H. Niedrig: Analytical models in electron backscattering. In *Electron Beam Interactions with Solids*, ed. by D.F. Kyser et al. (SEM Inc., AMF O'Hare, IL 1982) p.51

[3.113] U. Werner, H. Bethge, J. Heydenreich: An analytical model of electron backscattering for the energy range 10–100 keV. Ultramicroscopy **8**, 417 (1982)

[3.114] H.J. Dudek: Ein Rechenmodell zur räumlichen Beschreibung der Elektronenstrahl-Materie-Wechselwirkung. Optik **56**, 149 (1980)

[3.115] Z.T. Bödy: On the backscattering of electrons from solids. Brit. J. Appl. Phys. **13**, 483 (1962)

[3.116] L.V. Spencer: Theory of electron penetration. Phys. Rev. **98**, 1597 (1955)

[3.117] D.J. Fathers, P. Rez: A transport equation theory of electron. In *Electron Beam Interactions with Solids*, ed. by D.F. Kyser et al. (SEM Inc., AMF O'Hare, IL 1982) p.193

[3.118] K.E. Hoffmann, H. Schmoranzer: Inelastic and elastic multiple scattering of fast electrons described by the transport equation. In *Electron Beam Interactions with Solids*, ed. by D.F. Kyser et al. (SEM Inc., AMF O'Hare, IL 1982) p.209

[3.119] E.R. Krefting, L. Reimer: Monte-Carlo Rechnungen zur Elektronendiffusion. In *Quantitative Analysis with Electron Microprobes and Microanalysis*, ed. by E. Preuss (Kernforschungsanlage Jülich 1973) Jül-Conf.-8, p.114

[3.120] K.F.J. Heinrich, D.E. Newbury, H. Yakowitz (eds.): In *Use of Monte Carlo Calculations in Electron Probe Microanalysis and Scanning Electron Microscopy.* NBS Spec. Publ. 460 (U.S. Dep. of Commerce, Washington, DC 1976)

[3.121] J. Hénoc, F. Maurice: Characteristics of a Monte Carlo program for microanalysis study of energy loss. In *Use of Monte Carlo Calculations in Electron Probe Microanalysis and Scanning Electron Microscopy.* NBS Spec. Publ. 460 (U.S. Dep. of Commerce, Washington, DC 1976) p.61

[3.122] R.L. Myklebust, D.E. Newbury, H. Yakowitz: NBS Monte Carlo electron trajectory calculation program. In *Use of Monte Carlo Calculations in Electron Probe Microanalysis and Scanning Electron Microscopy.* NBS Spec. Publ. 460 (U.S. Dep. of Commerce, Washington, DC 1976) p.105

[3.123] R. Shimizu, Y. Kataoka, T. Matsukawa, T. Ikuta, M. Murata, H. Hashimoto: Energy distribution measurement of transmitted electrons and Monte Carlo simulation for kilovolt electron. J. Phys. D **8**, 820 (1975)

[3.124] R. Shimizu, Y. Kataoka, T. Ikuta, T. Koshikawa, H. Hashimoto: A Monte Carlo approach to the direct simulation of electron penetration in solids. J. Phys. D **9**, 101 (1976)

[3.125] D.C. Joy: An introduction to Monte Carlo simulations. Scanning Electron Microscopy **5**, 329 (1991)

[3.126] D.C. Joy: *Monte Carlo Modeling for Electron Microscopy and Microanalysis* (Oxford Univ. Press, New York, Oxford 1995)

[3.127] L. Reimer, D. Stelter: FORTRAN 77 Monte Carlo program for minicomputers using Mott cross-sections. Scanning **8**, 265 (1986)

[3.128] L. Reimer, R. Senkel: Monte Carlo simulations in low voltage electron microscopy. Optik **98**, 85 (1995)

[3.129] L. Reimer: Monte Carlo simulation techniques for quantitative x-ray microanalysis. In *Microbeam and Nanobeam Analysis*, ed. by D. Benoit et al., Mikrochim. Acta [Suppl.] **13**, 1 (1996)

[3.130] J.J. Quinn: Range of excited electrons in metals. Phys. Rev. **126**, 1453 (1962)

[3.131] H.W. Streitwolf: Zur Theorie der Sekundärelektronenemission von Metallen: der Anregungsprozeß. Ann. Physik **3**, 183 (1959)

[3.132] Z.J. Ding, R. Shimizu: A Monte Carlo modeling of electron interaction with solids including cascade secondary electron production. Scanning **18**, 92 (1996)

[3.133] L. Reimer, M. Kässens, L. Wiese: Monte Carlo simulation program with a free configuration of specimen and detector geometries. In *Microbeam and Nanobeam Analysis*, ed. by D. Benoit et al., Mikrochim. Acta [Suppl.] **13**, 485 (1996)

[3.134] C.O. Schiebl, A. Pfeiffer, H.W. Wagner, W.S.M. Werner, H. Stippel: Monte Carlo simulation of electron scattering for arbitrary 2D structures using a modified quadtree geometry discretization. In *Microbeam and Nanobeam Analysis*, ed. by D. Benoit et al., Mikrochim. Acta [Suppl.] **13**, 533 (1996)

[3.135] L. Reimer: MOCASIM – Ein Monte Carlo Programm für Forschung und Lehre. Beitr. elektr. mikr. Direktabb. Oberfl. **29**, 1 (1996)

[3.136] H.J. Dudek: Zeitlicher Temperaturanstieg im Material bei Elektronenbestrahlung. Beitr. elektr. mikr. Direktabb. Oberfl. **3**, 179 (1970)

[3.137] R. Christenhuß, L. Reimer: Schichtdickenabhängigkeit der Wärmeerzeugung durch Elektronenbestrahlung im Energiebereich zwischen 9 und 100 keV. Z. angew. Phys. **23**, 397 (1967)

[3.138] L.G. Pittaway: The temperature distribution in thin foil and semi-infinite targets bombarded by an electron beam. Brit. J. Appl. Phys. **15**, 967 (1964)

[3.139] H. Kohl, H. Rose, H. Schnabl: Dose-rate effect at low temperatures in FBEM and STEM due to object-heating. Optik **58**, 11 (1981)

[3.140] C.W. Price, P.L. McCarthy: Low-voltage SEM of low-density materials. Scanning **10**, 29 (1988)

[3.141] J. Cazaux: Some considerations on the electric field induced in insulators by electron bombardment. J. Appl. Phys. **59**, 1418 (1986)

[3.142] J. Cazaux, C. LeGressus: Phenomena relating charge in insulators: macroscopic effects and microscopic causes. Scanning Microscopy **5**, 17 (1991)

[3.143] K. Kotera, H. Suga: A simulation of keV electron scattering in a charged-up specimen. J. Appl. Phys. **63**, 261 (1988)

[3.144] V.I. Arkhipov, A.I. Rudenko, G.M. Sessler: Space-charge distribution in electron-beam charged dielectrics. J. Phys. D **24**, 731 (1991)

[3.145] R. Rettig, M. Kässens, L. Reimer: Measurement of specimen charging in SEM with a Kelvin probe. Scanning **16**, 221 (1994)

[3.146] R.D. van Veld, T.J. Shaffner: Charging effects in SEM. SEM 1971 (ITTRI, Chicago 1971) p.17

[3.147] H. Fujioka, K, Nakamae, K. Ura: Core model for potential distribution on insulator surfaces in the SEM. In *Electron Microscopy 1986*, Vol.1, ed. by T. Imura et al. (Jpn. Soc. of Electron Microscopy, Tokyo 1986) p.643

[3.148] S.C.J. Garth: Electron beam testing of ultra large scale ICs. Microelectr. Eng. **4**, 121 (1986)

[3.149] M.H. Ying, J.T.L. Thong: Insulator charging under irradiation with a stationary electron probe. Meas. Sci. Techn. **5**, 1089 (1994)

[3.150] L. Reimer, U. Golla, R. Böngeler, M. Kässens, B. Schindler, R. Senkel: Charging of bulk specimens, insulating layers and free-supporting films in SEM. Optik **92**, 14 (1992)

[3.151] D.M. Taylor: The effect of passivation on the observation of voltage contrast in the SEM. J. Phys. D **11**, 2443 (1978)

[3.152] W.W. Adams, G. Price, S.J. Krause: Imaging of polymer single crystals in low-voltage, high-resolution SEM. In *Electron Microscopy 1990*, Vol.4 (San Francisco Press, San Francisco 1994) p.1106

[3.153] T. Ichinokawa, M. Iijama, A. Onoguchi, T. Kobayashi: Charging effect of specimen in SEM. Jpn. J. Appl. Phys. **13**, 1272 (1974)

[3.154] D.C. Joy: Control of charging in low-voltage SEM. Scanning **11**, 1 (1989)

[3.155] M. Knoll: Aufladepotential und Sekundäremission elektronenbestrahlter Körper. Z. Techn. Phys. **16**, 467 (1935)

[3.156] H. Fujioka, K. Miyaji, K. Ura: A new method for measuring charging characteristics of an electrically floating target under electron beam irradiation. J. Phys. E **21**, 583 (1988)

[3.157] H.E. Bauer, H. Seiler: Determination of the non-charging electron beam energies of electrically floating metal samples. SEM 1984/III (SEM Inc. AMF O'Hare, IL 1984) p.1081

[3.158] M. Ostrow, E. Menzel, E. Postulka, S. Görlich, E. Kubalek: IC-internal electron beam logic state analysis. SEM 1982/II (SEM Inc. AMF O'Hare, IL 1982) p.563

[3.159] M. Brunner, E. Menzel: Surface potential measurements on floating targets with a parallel beam technique. J. Vac. Sci. Techn. B **1**, 1344 (1983)

[3.160] W. Liu, J. Ingino, R.F. Pease: Resist charging in electron beam lithography. J. Vac. Sci. Techn. B **13**, 1979 (1995)

[3.161] M. Brunner, D. Winkler, B. Lischke: Crucial parameters in electron beam short/open testing. In *Microcircuit Engineering 1984* (Academic, London 1985)

[3.162] J. Chang, S. Krause, R. Gorur: Application of low-voltage SEM to studies of polymer near-surface diffusion and hydrophobicity. In *Electron Microscopy 1990*, Vol.4 (San Francisco Press, San Francisco 1990) p.1108

[3.163] H.J. Fitting, P. Magdanz, W. Mehnert, D. Hecht, Th. Hingst: Charge trap spectroscopy in single and multiple layer dielectrics. Phys. Stat. Solidi (a) **122**, 297 (1990)

[3.164] N. Sugiyama, S. Ikeda, J. Uchikawa: Low voltage SEM inspection of microelectronic devices. J. Electron Microsc. **35**, 9 (1986)

[3.165] E.A. Flinn, M. Salehi: Accuracy of the theoretical predictions concerning the location of the cross-over points on the secondary-electron emission yield curve. J. Appl. Phys. **52**, 5800 (1981)

[3.166] L.A. Weitzenkamp: Measurement of fibre potentials in a SEM. J. Phys. E **2**, 561 (1969)

[3.167] D.R. Clarke, P.R. Stuart: An anomalous contrast effect in the SEM. J. Phys. E **3**, 705 (1970)

[3.168] M. Brunner: Simulation of BSE by reflection of primary electrons applied to optimization of detector designs. Appl. Phys. Lett. **43**, 391 (1983)

[3.169] J. Hejna, L. Reimer: Backscattered electron multidetector systems for improved quantitative topographic contrast. Scanning **9**, 162 (1987)

[3.170] R. Autrata, J. Hejna: Detectors for LVSEM. Scanning **13**, 275 (1991)

[3.171] P. Echlin: Sputter coating techniques for SEM. SEM 1975 (ITTRI, Chicago 1975) p.217

[3.172] P.N. Panayi, D.C. Cheshire, P. Echlin: A cool sputtering system for coating heat-sensitive specimens. SEM 1977/I (ITTRI, Chicag 1977) p.463

[3.173] G. Pfefferkorn: Specimen preparation techniques. SEM 1970 (ITTRI, Chicago 1970) p. 89

[3.174] R.O. Kelley, R.A.F. Dekker, J.G. Bluemink: Ligand-mediated osmium binding: its application in coating biological specimens for SEM. J. Ultrastruct. Res. **45**, 254 (1973)

[3.175] L.E. Malick, R.B. Wilson: Evaluation of a modified technique for SEM examination of vertebrate specimens without evaporated metal layers. SEM 1975 (ITTRI, Chicago 1975) p.259

[3.176] G.V. Spivak, E.I. Rau, A.E. Lukianov, V.I. Petrov, M.V. Bicov: Des images non altérées des isolants dans un microscope électronique à balayage. In *Electron Microscopy 1972* (Inst. of Physics, London 1972) p.492

[3.177] P. Morin, M. Pitaval, E. Vicario: Direct observation of insulators by SEM. In *Developments in Electron Microscopy and Analysis*, ed. by J.A. Venables (Academic, London 1976) p.115

[3.178] G. Pfefferkorn, H. Grüter, M. Pfautsch: Observations on the prevention of specimen charging. SEM 1972 (ITTRI, Chicago 1972) p.147

[3.179] P.R. Thornton: *Scanning Electron Microscopy: Applications to Materials and Devices* (Chapman and Hall, London 1968) p.179

[3.180] C.K. Crawford: Ion charge neutralization effects in SEM. SEM 1980/IV (SEM Inc., AMF O'Hare, IL 1980) p.11

[3.181] G.D. Danilatos, R. Postle: The environmental SEM and its applications. SEM 1982/I (SEM Inc., AMF O'Hare, IL 1982) p.1

[3.182] L. Reimer: Irradiation changes in organic and inorganic objects. Lab. Invest. **14**, 1082 (1965)

[3.183] K. Stenn, G.F. Bahr: Specimen damage caused by the beam of the TEM, a correlative consideration. J. Ultrastruct. Res. **31**, 526 (1970)

[3.184] B.M. Siegel, D.R. Beaman (eds.): *Physical Aspects of Electron Microscopy and Microbeam Analysis* (Wiley, New York 1975)

[3.185] M.S. Isaacson: Specimen damage in the electron microscope. In *Principles and Techniques of Electron Microscopy*, Vol.7, ed. by M.A. Hayat (Van Nostrand-Reinhold, New York 1977) p.1

[3.186] R.M. Glaeser, K.A. Taylor: Radiation damage relative to TEM of biological specimens at low temperatures: a review. J. Microsc. **122**, 127 (1978)

[3.187] L. Reimer, A. Schmidt: The shrinkage of bulk polymers by radiation damage in an SEM. Scanning **7**, 47 (1985)

[3.188] S.J. Erasmus: Damage to resist structures during SEM inspection. J. Vac. Sci. Techn. B **5**, 409 (1987)

[3.189] W. Bröcker, E.R. Krefting, L. Reimer: Beobachtung der Strahlenschädigung während des Abrastvorganges im Raster-Elektronenmikroskop mit Hilfe der Kathodolumineszenz. Beitr. elektr. mikr. Direktabb. Oberfl. **7**, 75 (1974)

[3.190] J.B. Pawley, P. Walther, S.J. Shih, M. Malecki: Early results using high-resolution, low-voltage low-temperature SEM. J. Microsc. **161**, 327 (1991)

[3.191] P. Walther, E. Wehrli, R. Hermann, M. Müller: Double-layer coating for high-resolution low-temperature SEM. J. Microsc. **179**, 229 (1995)

[3.192] L.W. Hobbs: Radiation effects in analysis of inorganic specimens by TEM, in *Introduction to Analytical Electron Microscopy*, ed. by J.J. Hren et al. (Plenum, New York 1979) p.437

[3.193] D.L. Medlin, D.G. Howitt: Radiation damage processes affecting electron beam lithography of inorganic materials. Scanning **14**, 86 (1992)

[3.194] A.K. Varshneya, A.R. Cooper, M.J. Cable: Changes in composition during electron microprobe analysis of K_2O-SrO-SiO_2 glass. J. Appl. Phys. **37**, 2199 (1966)

[3.195] M.P. Boron, R.E. Hanneman: Local compositional changes in alkali silicate glasses during electron microprobe analysis. J. Appl. Phys. **38**, 2406 (1967)

[3.196] K. Jurek, V. Hulinský, O. Gedeon: Electron beam induced migration of alkaline ions in silica glass. Mikrochim. Acta, Suppl. 13, 339 (1996)

[3.197] W. Bröcker, L. Reimer: Specimen damage by negative oxygen ions from the SEM cathode detected by CdS cathodoluminescence. Scanning **1**, 60 (1978)

[3.198] O.C. Wells: Ion damage to the specimen in the SEM studied by the EBIC technique. Scanning **1**, 182 (1978)

[3.199] J.J. Hren: Barriers to AEM: Contamination and etching. In *Introduction to Analytical Electron Microscopy*, ed. by J.J. Hren et al. (Plenum, New York 1979) p.481

[3.200] J.P. Martin, R. Speidel: Zur Verwendung von Dünnschicht-Aperturblenden im Elektronen-Rastermikroskop. Beitr. elektr. mikr. Direktabb. Oberfl. **4/2**, 345 (1971)

[3.201] J.B. Pawley: LVSEM for high resolution topographic and density contrast imaging. Adv. Electr. Electron Phys. **83**, 203 (1992)

[3.202] P. Hirsch, M. Kässens, M. Püttmann, L. Reimer: Contamination in a SEM and the influence of specimen cooling. Scanning **16**, 101 (1994)

[3.203] K.H. Müller: Elektronen-Mikroschreiber mit geschwindigkeits-gesteuerter Strahlführung. Optik **33**, 296 (1971)

[3.204] J.T. Fourie: Contamination phenomena in cryopumped TEM and ultrahigh vacuum field-emission STEM systems. SEM 1976/I (ITTRI, Chicago IL 1976) p.53

[3.205] G. Love, V.D. Scott, N.M.T. Dennis, L. Laurenson: Sources of contamination in electron optical equipment. Scanning **4**, 32 (1981)

[3.206] E.K. Brandis, F.W. Anderson, R. Hoover: Reduction of carbon contamination in the SEM. SEM 1971 (ITTRI, Chicago 1971) p.505

[3.207] A.T. Marshall: Residual gas analysis in a SEM. J. Microsc. **133**, 119 (1984)

[3.208] H.G. Heide: Die Objektverschmutzung im Elektronenmikroskop und das Problem der Strahlenschädigung durch Kohlenstoffabbau. Z. angew. Phys. **15**, 116 (1963)

[3.209] D. Wang, P.C. Hoyle, J.R.A. Cleaver, G.A. Porkolab, N.C. MacDonald: Lithography using electron beam induced etching of a carbon film. J. Vac. Sci. Techn. B **13**, 1984 (1995)

[3.210] L. Reimer: *Image Formation in Low-Voltage Scanning Electron Microscopy*, Tutorial Texts in Optical Engineering, Vol. TT 12 (SPIE, Optical Engineering Press, Bellingham WA 1993)

[3.211] H.G. Heide: Die Objektraumkühlung im Elektronenmikroskop. Z. angew. Phys. **17**, 73 (1964)

[3.212] V. Harada, T. Tomita, T. Watabe, H. Watanabe, T. Etoh: Reduction of contamination in analytical electron microscopy. SEM 1979/II (SEM Inc., AMF O'Hare, IL 1979) p.103

[3.213] L. Reimer, M. Wächter: Contribution to the contamination problem in TEM. Ultramicroscopy **3**, 169 (1978)

[3.214] R. Buhl: Verringerung der Kontamination durch Ionen-Bombardement der das Präparat umgebenden Wände. Optik **19**, 122 (1962)

Chapter 4

[4.1] H. Drescher, L. Reimer, H. Seidel: Rückstreukoeffizient und Sekundärelektronen-Ausbeute von 10–100 keV Elektronen und Beziehungen zur Raster-Elektronenmikroskopie. Z. Angew. Phys. **29**, 331 (1970)

[4.2] E. Weinryb, J. Philibert: Mesure du coefficient de la rétrodiffusion des électrons de 5 à 30 keV. Compt. rend. Acad. Sci. (Paris) **258**, 4535 (1964)

[4.3] K.F.J. Heinrich, Interrelationships of sample composition, backscattering coefficient, and target current measurements. Adv. X-Ray Analysis, Vol.7, ed. by W.M. Mueller et al. (Plenum, New York 1964) p.235

[4.4] J.W. Colby: Backscattered and secondary electron emission as ancillary techniques in electron probe analysis. In *Electron Probe Microanalysis*, ed. by A.J. Tousimis and L. Marton (Academic, New York 1969) p.177

[4.5] L. Reimer, C. Tollkamp: Measuring the backscattering coefficient and secondary electron yield inside a SEM. Scanning **3**, 35 (1980)

[4.6] H. Niedrig, P. Sieber: Rückstreuung mittelschneller Elektronen an dünnen Schichten. Z. angew. Phys. **31**, 27 (1971)

[4.7] V.E. Cosslett, R.N. Thomas: Multiple scattering of 5–30 keV electrons in evaporated metal films. III Backscattering and absorption. Brit. J. Appl. Phys. **16**, 779 (1965)

[4.8] T. Just, H. Niedrig, H. Yersin: Schichtdickenbestimmung mittels Elektronen-Rückstreuung. Z. angew. Phys. **25**, 89 (1968)

[4.9] F.J. Hohn, H. Niedrig: Elektronenrückstreuung an dünnen Metall- und Isolatorschichten. Optik **35**, 290 (1972)

[4.10] H. Niedrig: Film thickness determination in electron microscopy; the electron backscattering method. Optica Acta **24**, 679 (1977)

[4.11] H. Niedrig: Electron backscattering from thin films. J. Appl. Phys. **53**, R15 (1982)

[4.12] G. Kuhnle, J.P. Martin, H. Seiler: Schichtdickenuntersuchungen im Elektronen-Rastermikroskop. Beitr. elektr. mikr. Direktabb. Oberfl. **2**, 101 (1969)

[4.13] H. Seiler: Determination of the "information depth" in the SEM. SEM 1976/I (ITTRI, Chicago 1976) p.9

[4.14] H.G. Badde, H. Drescher, E.R. Krefting, L. Reimer, H. Seidel, W. Bühring: Use of Mott scattering cross sections for calculating backscattering of 10–100 keV electrons. In *Proc. of 25th Anniv. Meeting of EMAG* (IoP, Bristol 1971) p.74

[4.15] M. Baciocchi, E. Di Fabricio, M. Gentili, L. Grella, R. Maggiore, L. Mastrogiacomo, D. Peschiaroli: High accuracy thickness measurements by means of BSE metrology. J. Vac. Sci. Techn. B **13**, 2676 (1995)

[4.16] F.J. Hohn, M. Kindt, H. Niedrig, B. Stuth: Elektronenrückstreumessungen an dünnen Schichten auf massiven Trägersubstanzen. Optik **46**, 491 (1976)

[4.17] H.J. Hunger, S. Rogaschewski: A study of electron backscattering of thin films on substrates. Scanning **8**, 257 (1986)

[4.18] W. Reuter: The ionization function and its application to the electron probe analysis of thin films. In *Proc. 6th Int'l. Conf. on X-Ray Optics and Microanalysis*, ed. by G. Shinoda et al., (Univ. Tokyo Press, Tokyo 1972) p.121

[4.19] P.B. DeNee: Measurement of mass and thickness of respirable size dust particles by SEM backscattered electron imaging. SEM 1978/I (SEM Inc., AMF O'Hare, IL 1978) p.741

[4.20] M.R. Sogard: An empirical study of electron backscattering from thin films. J. Appl. Phys. **51**, 4417 (1980)

[4.21] N.G. Nakhodkin, A.A. Ostroukhov, V.A. Romanovski: Electron inelastic scattering in thin films. Sov. Phys. Solid State **4**, 1112 (1962)

[4.22] H. Niedrig: Ein kombiniertes Einfachstreu- und Diffusionsmodell für die Elektronen-Rückstreuung dünner Schichten. Beitr. elektr. mikr. Direktabb. Oberfl. **14**, 291 (1981)

[4.23] H.J. August, J. Wernisch: Theoretical prediction of the electron backscattering coefficient for multilayer structures. J. Microsc. **157**, 247 (1990)

[4.24] D.F. Kyser, K. Murata: Application of Monte Carlo simulation to electron microprobe analysis of thin films on substrates. NBS Spec. Publ. 460 (US Dep. of Commerce, Washington, DC 1976) p.129

[4.25] K. Murata, H. Kawata, K. Nagami: Electron scattering in LVSEM targets. Scanning Microscopy, Suppl. 1 (SEM Inc., AMF O'Hare, IL 1987) p.83

[4.26] G. Neubert, S. Rogaschewski: Measurement of the backscattering of 20 to 60 keV electrons from double layers at various angles of incidence. Phys. Stat. Solidi (a) **87**, 451 (1985)

[4.27] J. Hejna: A study of BSE signals of double layers on substrates at various detector take-off angles. Scanning **14**, 256 (1992)

[4.28] H. Schmoranzer, H. Grabe: Absolute differential cross sections for scattering of 40 keV electrons from thin gold and carbon foils. In *Electron Microscopy 1976*, Vol.I, ed. by D.G. Brandon (Tal International, Jerusalem 1976) p.293

[4.29] F.J. Hohn: Angular dependence of electron intensities backscattered by carbon films. Optik **47**, 491 (1977)

[4.30] K.E. Hoffmann, H. Schmoranzer: Inelastic and elastic multiple scattering of fast electrons described by the transport equation. In *Electron Beam Interactions with Solids*, ed. by D.F. Kyser et al. (SEM Inc., AMF O'Hare, IL 1982) p.209

[4.31] H. Kanter: Zur Rückstreuung von Elektronen im Energiebereich von 10–100 keV. Ann. Phys. **20**, 144 (1957)

[4.32] F. Arnal, P. Verdier, P.D. Vinsensini: Coefficient de rétrodiffusion dans de cas d'électrons monocinétiques arrivant sur la cible sous une incidence oblique. Compt. rend. Acad. Sci. (Paris) **268**, 1526 (1969)

[4.33] E.H. Darlington, V.E. Cosslett: Backscattering of 0.5–10 keV electrons from solid targets. J. Phys. D **5**, 1969 (1972)

[4.34] E.H. Darlington: Backscattering of 10–100 keV electrons from thick targets. J. Phys. D **8**, 85 (1975)

[4.35] H. Bruining: Secondary electron emission. Physica **5**, 901 (1938)

[4.36] B. Lödding, L. Reimer: Monte Carlo Rechnungen im Energiebereich 1–20 keV. Beitr. elektr. mikr. Direktabb. Oberfl. **14**, 315 (1981)

[4.37] R. Castaing: Electron-probe microanalysis. Adv. Electr. Electron Phys. **13**, 317 (1960)

[4.38] R. Herrmann, L. Reimer: Backscattering coefficient of multicomponent specimens. Scanning **6**, 20 (1984)

[4.39] H.J. Fitting, R. Technow: Electron backscattering at various angles of incidence. Phys. Stat. Solidi (a) **76**, K151 (1984)

[4.40] R. Böngeler, U. Golla, M. Kässens, L. Reimer, B. Schindler, R. Senkel, M. Spranck: Electron-specimen interactions in LVSEM. Scanning **15**, 1 (1993)

[4.41] A. Darliński: Measurements of angular distribution of the backscattered electrons in the energy range 5 to 30 keV. Phys. Stat. Solidi (a) **63**, 663 (1981)

[4.42] L. Reimer, M. Riepenhausen: Detector strategy for secondary and backscattered electrons using multiple detector systems. Scanning **7**, 221 (1985)

[4.43] M. Spranck, M. Kässens, L. Reimer: Influence of the angular distribution of backscattered electrons on signals at different take-off angles in LVSEM. Scanning **17**, 97 (1995)

[4.44] H. Jahrreiss, W. Oppel: Angular distribution of secondary electrons originating from thin films of different metals in re-emission and transmission. J. Vac. Sci. Techn. **9**, 173 (1972)

[4.45] Z.J. Ding, H. Yoshikawa, R. Shimizu: Angular distribution of elastically reflected electrons from Au. Phys. Stat. Solidi (b) **161**, 257 (1990)

[4.46] H. Yoshikawa, Z.J. Ding, R. Shimizu: On the radial distribution function for sub-kV backscattered electron diffraction from disordered surface. Surf. Sci. **242**, 434 (1991)

[4.47] J.O. Brand: Über die Energieverteilung rückdiffundierter Elektronenstrahlen. Ann. Phys. **26**, 609 (1936)

[4.48] E.J. Sternglass: Backscattering of kilovolt electrons from solids. Phys. Rev. **95**, 345 (1954)

[4.49] H. Kulenkampff, W. Spyra: Energieverteilung rückdiffundierter Elektronen. Z. Physik **137**, 416 (1954)

[4.50] W. Bothe: Zur Rückdiffusion schneller Elektronen. Z. Naturforschg. A **4**, 542 (1949)

[4.51] H.E. Bishop: Electron scattering in thin targets. Brit. J. Appl. Phys. **18**, 703 (1967)

[4.52] T. Matsukawa, R. Shimizu, H. Hashimoto: Measurements of the energy distribution of backscattered kilovolt electrons with a spherical retarding-field energy analyser. J. Phys D **7**, 695 (1974)

[4.53] L. Reimer, R. Böngeler, M. Kässens, F.F. Liebscher, R. Senkel: Calculation of energy spectra from layered structures for backscattered electron spectrometry and relations to Rutherford backscattering spectrometry by ions. Scanning **13**, 381 (1991)

[4.54] V.P. Afanas'ev, D. Naujoks: Energy spectra of electrons reflected from layered structures. Z. Physik B **84**, 397 (1991)

[4.55] H. Boersch, R. Wolter, H. Schoenebeck: Elastische Energieverluste kristallgestreuter Elektronen. Z. Physik **199**, 124 (1967)

[4.56] P. Sommerkamp: Elektronen-Rückstreumessungen an Ta und Ni, ein Beitrag zur Leistungsbilanz des Elektronenstrahlschmelzens. Z. angew. Phys. **28**, 220 (1970)

[4.57] H.D. Bauer: Messungen zur Energieverteilung von Rückstreuelektronen an polykristallinen Festkörpern. Exp. Techn. Phys. **27**, 331 (1979)

[4.58] O.C. Wells: Explanation of the low-loss image in the SEM in terms of electron scattering theory. SEM 1972 (ITTRI, Cicago 1972) p.169

[4.59] H. Bruining: *Physics and Application of Secondary Electron Emission* (Pergamon, London 1964)

[4.60] A.J. Dekker: Secondary electron emisison. Solid State Phys. **6**, ed. by F. Seitz and D. Turnbull (Academic, New York 1958) p.251

[4.61] R. Kollath: Sekundärelektronen-Emission fester Körper bei Bestrahlung mit Elektronen. In *Encyclopedia of Physics*, Vol. 21, ed. by S. Flügge (Springer, Berlin, Göttingen 1956) p.232

[4.62] C. Hachenberg, W. Brauer: Secondary electron emission from solids. Adv. Electr. Electron Phys. **11**, 413 (1959)

[4.63] H. Seiler: Einige aktuelle Probleme der Sekundärelektronenemission. Z. angew. Phys. **22**, 249 (1967)

[4.64] H. Seiler: Secondary electron emission. In *Electron Beam Interactions with Solids*, ed. by D.F. Kyser et al. (SEM Inc., AMF O'Hare, IL 1982) p.33

[4.65] J. Schou: SE emission from solids by electron and proton bombardment. Scanning Microscopy **2**, 607 (1988)

[4.66] W. Dietrich, H. Seiler: Energieverteilung von Elektronen, die durch Ionen und Elektronen in Durchstrahlung an dünnen Folien ausgelöst werden. Z. Physik **157**, 576 (1960)

[4.67] J. Schäfer, J. Hölzl: A contribution to the dependence of secondary electron emission from the work function and Fermi energy. Thin Solid Films **18**, 81 (1972)

[4.68] V.E. Henrich: Role of bulk and surface plasmons in the emission of slow secondary electrons: polycrystalline aluminium. Phys. Rev. B **7**, 3512 (1973)

[4.69] T.E. Everhart, N. Saeki, R. Shimizu, T. Koshikawa: Measurement of structure in the energy distribution of slow secondary electrons from aluminium. J. Appl. Phys. **47**, 2941 (1976)

[4.70] R.E. Chase, W.L. Gordon, R.W. Hoffman: Measurement of low energy secondary electron emission using a double-pass cylindrical mirror analyzer. Appl. Surf. Sci. **4**, 271 (1980)

[4.71] F.J. Pijper, P. Kruit: Detection of energy-selected SE in coincidence with energy-loss events in thin carbon foils. Phys. Rev. B **44**, 9192 (1991)

[4.72] H. Mullejans, A.L. Bleloch, A. Howie, M. Tomita: SE coincidence detection and time of flight spectroscopy. Ultramicroscopy **52**, 360 (1993)

[4.73] A. Howie: Recent developments in SE imaging. J. Microsc. **180**, 192 (1995)

[4.74] K. Kanaya, H. Kawakatsu: Secondary electron emission due to primary and backscattered electrons. J. Phys. D **5**, 1727 (1972)

[4.75] V.V. Makarov, N.N. Petrov: Regularities of secondary electron emission of the elements of the periodic table. Sov. Phys. Solid State **23**, 1028 (1981)

[4.76] P.W. Palmberg: Secondary emission studies on Ge and Na-covered Ge. J. Appl. Phys. **38**, 2137 (1967)

[4.77] K. Kanaya, S. Ono: Interaction of the electron beam with the target in SEM. In *Electron Beam Interactions with Solids*, ed. by D.F. Kyser et al. (SEM Inc., AMF O'Hare, IL 1982) p.69

[4.78] H. Mayer, J. Hölzl: Experimentelle Bestimmung der maximalen Austrittstiefen monoenergetischer Sekundärelektronen. Phys. Stat. Solidi **18**, 779 (1966)

[4.79] I.M. Bronshtein, V.A. Dolinin: The secondary electron emission of solids at large angles of incidence of the primary beam. Sov. Phys. Solid State **9**, 2133 (1968)

[4.80] W. Oppel, H. Jahrreiss: Messungen der Winkelverteilung von Sekundärelektronen an dünnen freitragenden Al- und Au-Schichten. Z. Physik **252**, 107 (1972)

[4.81] J. Burns: Angular distribution of secondary electrons from (100) faces of Cu and Ni. Phys. Rev. **119**, 102 (1960)

[4.82] G. Appelt: Fine structure measurements in the energy angular distribution of secondary electrons from a (110) face of copper. Phys. Stat. Solidi **27**, 657 (1968)

[4.83] R.F. Willis, N.E. Christensen. Secondary-electron-emission spectroscopy of tungsten: angular dependence and phenomenology. Phys. Rev. B **18**, 5140 (1978)

[4.84] J. Schäfer, R. Schoppe, J. Hölzl, R. Feder: Experimental and theoretical study of the angular resolved secondary electron spectroscopy (ARSES) for W(100) in the energy range $0 \leq E \leq 20$ eV. Surf. Sci. **107**, 290 (1981)

[4.85] J.P. Martin: Zum Einfluß der Objektkontamination auf das rastermikroskopische Bild. Beitr. elektr. mikr. Direktabb. Oberfl. **4/2**, 387 (1971)

[4.86] A.G. Knapp: The effect of electron bombardment on the secondary electron emission from Na_3AlF_6. J. Appl. Phys. **50**, 5961 (1979)

[4.87] D.C. Joy: A model for calculating secondary and backscattered electron yields. J. Microsc. **147**, 51 (1987)

[4.88] S. Luo, D.C. Joy: Monte Carlo calculations of SE emission. Scanning Microscopy **2**, 1901 (1988)

[4.89] M. Kotera: A Monte Carlo simulation of primary and secondary electron trajectories in a specimen. J. Appl. Phys. **65**, 3991 (1989)

[4.90] M.S. Chung, T.E. Everhart: Simple calculation of energy distribution of low-energy secondary electrons emitted from metals under electron bombardment. J. Appl. Phys. **45**, 707 (1974)
[4.91] H.W. Streitwolf: Zur Theorie der Sekundärelektronenemission von Metallen: der Anregungsprozeß. Ann. Phys. **3**, 183 (1959)
[4.92] J.J. Quinn: Range of excited electrons in metals. Phys. Rev. **126**, 1453 (1962)
[4.93] R. Bindi, H. Lanteri, P. Rostaing: A new approach and resolution method of the Boltzmann equation applied to secondary electron emission by reflection from polycrystalline aluminium. J. Phys. D **13**, 267 (1980)
[4.94] J. Schou: Transport theory for kinetic emission of secondary electrons from solids. Phys. Rev. B **22**, 2141 (1980)
[4.95] M. Rösler, W. Brauer: Theory of secondary electron emission. Phys. Stat. Solidi (b) **104**, 161 and 575 (1981)
[4.96] F. Hasselbach: The emission microscope: a valuable tool for investigating the fundamentals of the SEM. Scanning Microscopy **2**, 41 (1988)
[4.97] L. Reimer, H. Drescher: Secondary electron emission of 10–100 keV electrons from transparent films of Al and Au. J. Phys. D **10**, 805 (1977)
[4.98] H. Jahrreiß: Secondary electron emission from thin films. Thin Solid Films **12**, 187 (1972)
[4.99] J. Kadlec, L. Eckertová: Zur Bestimmung der Sekundärelektronenemission dünner Schichten. Z. angew. Phys. **30**, 141 (1970)
[4.100] H.J. Fitting: Transmission, energy distibution, and SE excitation of fast electrons in thin solid films. Phys. Stat. Solidi (a) **26**, 525 (1974)
[4.101] V.N.E. Robinson: The dependence of emitted secondary electrons upon the direction of travel of the exciting electron. J. Phys. D **8**, L74 (1975)
[4.102] K. Murata, M. Yasuda, H. Kawata: The spatial distribution of BSE revisited with a new Monte Carlo simulation. Scanning Microscopy **6**, 943 (1992)
[4.103] A. van der Zeil: *Noise* (Prentice Hall, Englewood Cliffs, NJ 1970)
[4.104] W. Shockley, J.R. Pierce: A theory of noise for electron multipliers. Proc. Inst. Radio Eng. **26**, 321 (1938)
[4.105] B. Kurrelmeyer, L.J. Hayner: Shot effect of secondary electrons from nickel and beryllium. Phys. Rev. **52**, 952 (1937)
[4.106] T.E. Everhart, O.C. Wells, C.W. Oatley: Factors affecting contrast and resolution in the SEM. J. Electron. Contr. **7**, 97 (1959)
[4.107] L. Reimer: Rauschen der Sekundärelektronenemission. Beitr. elektr. mikr. Direktabb. Oberfl. **4/2**, 299 (1971)

Chapter 5

[5.1] C.W. Oatley: Detectors for the SEM. J. Phys. E **14**, 971 (1981)
[5.2] M. Maussion: High resolution SEM with absorbed current at low intensity primary beams. J. Phys. E **18**, 11 (1985)
[5.3] H. Vitturro, C. Peez, R.D. Bonetto, A.G. Alvarez: A simple BSE and specimen current detection device for the SEM. Scanning **15**, 232 (1993)
[5.4] M. Hatzakis: A new method of forming scintillators for electron collectors. Rev. Sci. Instr. **41**, 128 (1970)
[5.5] M.E. Taylor: An improved light pipe for the SEM. Rev. Sci. Instr. **43**, 1846 (1972)
[5.6] D.C. Marshall, J. Stephen: Secondary electron detector with an extended life for use in a SEM. J. Phys. E **5**, 1046 (1972)

[5.7] R.J. Woolf, D.C. Joy, D.W. Tansley: A transmission stage for the SEM. J. Phys. E **5**, 230 (1972)

[5.8] J.B. Pawley: Performance of SEM scintillation materials. SEM 1974 (IT-TRI, Chicago 1974) p.27

[5.9] N.R. Comins, M.M.E. Hengstberger, J.T. Thirlwall: Preparation and evaluation of P-47 scintillators for a SEM. J. Phys E **11**, 1041 (1978)

[5.10] G. Mustoe: Preparation of P-47 crystalline phosphor scintillator discs. Scanning **2**, 41 (1979)

[5.11] W. Baumann, A. Niemietz, L. Reimer, B. Volbert: Preparation of P-47 scintillators for STEM. J. Microsc. **122**, 181 (1981)

[5.12] N.R. Comins, J.T. Thirlwall; Quantitative studies and theoretical analysis of the performance of the scintillation electron-detector. J. Microsc. **124**, 119 (1981)

[5.13] R. Autrata, P. Schauer, J. Kvapil, J. Kvapil: A single crystal of YAG – new fast scintiilator in SEM. J. Phys. E **11**, 707 (1978)

[5.14] R. Autrata, P. Schauer, J. Kvapil, J. Kvapil: A single crystal of $YAlO_3:Ce^{3+}$ as a fast scintillator in SEM. Scanning **5**, 91 (1983)

[5.15] R. Autrata, P. Schauer, J. Kvapil, J. Kvapil: Single-crystal aluminates – a new generation of scintillators for SEM and transparent screens in electron optical devices. SEM 1983/II (SEM Inc., AMF O'Hare, IL 1983) p.489

[5.16] P. Schauer, R. Autrata: Time response of single crystal scintillation detectors for SEM/STEM. In *Electron Microscopy 1994*, Vol.1, ed. by B. Jouffrey and C. Colliex (Les Editions de Physique, Les Ulis, France 1994), p.227

[5.17] T.E. Everhart, R.F.M. Thornley: Wideband detector for micro-microampere low-energy electron currents: J. Sci. Instr. **37**, 246 (1960)

[5.18] R. Autrata, R. Hermann, M. Müller: An efficient single crystal BSE detector in SEM. Scanning **14**, 127 (1992)

[5.19] P. Schauer, R. Autrata: Light transport in single-crystal scintillation detectors in SEM. Scanning **14**, 325 (1992)

[5.20] S. Yamada, T. Ito, K. Gouhara, Y. Uchikawa: Electron count imaging in SEM. Scanning **13**, 165 (1991)

[5.21] Y. Uchikawa, K. Gouhara, S. Yamada, T. Ito, T. Kodama, P. Sardeshmukh: Comparative study of electron counting and conventional analogue detection of SE electrons in SEM. J. Electr. Microsc. **41**, 253 (1992)

[5.22] E. Breitenberger: Scintillation spectrometer statistics. Progr. Nucl. Phys. **4**, 56 (1955)

[5.23] B. Saleh: *Photoelectron Statistics*, Springer Ser. Opt. Sci., Vol.6 (Springer, Berlin, Heidelberg 1978)

[5.24] P.R. Evrard, C. Gazier: Propriétés statistiques de certains photo-multiplicateurs. Journal de Physique **26**, 37 A (1965)

[5.25] A. Williams, D. Smith: Afterpulses in liquid scintillation counters. Nucl. Instr. Meth. **112**, 131 (1973)

[5.26] K.A. Hughes, D.V. Sulway, R.C. Wayte, P.R. Thornton: Application of secondary-electron channel multipliers to SEM. J. Appl. Phys. **38**, 4922 (1967)

[5.27] H. Hantsche, G. Schreiber: Zur Verwendung eines Channeltrons als Elektronendetektor im Raster-Elektronenmikroskop. Beitr. elektr. mikr. Direktabb. Oberfl. **3**, 167 (1970)

[5.28] P.E. Russel, J.F. Mancuso: Microchannel plate detector for LVSEM. J. Microsc. **140**, 323 (1985)

[5.29] M.T. Postek, W.J. Keery, N.V. Frederick: Low-profile, high-efficiency microchannel-plate detector system for SEM applications. Rev. Sci. Instr. **61**, 1648 (1990)

[5.30] J.B. Elsbrock, L.J. Balk: Quantitative evaluation of micromagnetic fields by means of a SEM. SEM 1984/I (SEM Inc., AMF O'Hare, IL 1984), p.131

[5.31] E.D. Wolf, T.E. Everhart: Annular diode detector for high angular resolution pseudo-Kikuchi patterns. SEM 1969 (ITTRI, Chicago 1969) p.41

[5.32] E. Miyazaki, H. Maeda, K. Miyaji: The Evoscope – a fixed pattern generator using a Au-Si diode. Adv. Electr. Electron Phys. A **22**, 331 (1966)

[5.33] W. Czaja: Response of Si and GaP p-n junctions to a 5 to 40 keV electron beam. J. Appl. Phys. **37**, 4236 (1966)

[5.34] O.C. Wells: Effect of collector position on type-2 magnetic contrast in the SEM. SEM 1978/I (SEM Inc., AMF O'Hare, IL 1978) p.293

[5.35] J. Stephen, B.J. Smith, D.C. Marshall, E.M. Wittam: Application of a semiconductor backscattered electron detector in a SEM. J. Phys. E **8**, 607 (1975)

[5.36] M. Kikuchi, S. Takashima: Multi-purpose backscattered electron detector. In *Electron Microscopy 1978*, Vol.1, ed. by J.M. Sturgess (Microsc. Soc. of Canada, Toronto 1978) p.82

[5.37] D.A. Gedcke, J.B. Ayers, P.B. DeNee: A solid state backscattered electron detector capable of operating at TV scan rates. SEM 1978/I (SEM Inc., AMF O'Hare, IL 1978) p.581

[5.38] A.V. Crewe, M. Isaacson, D. Johnson: Secondary electron detection in a field emission scanning microscope. Rev. Sci. Instr. **41**, 20 (1970)

[5.39] H. Drescher, L. Reimer, H. Seidel: Rückstreukoeffizient und Sekundärelektronen-Ausbeute von 10–100 keV Elektronen und Beziehungen zur Raster-Elektronenmikroskopie. Z. angew. Phys. **29**, 331 (1970)

[5.40] C.W. Oatley: Electron currents in the specimen chamber of a SEM. J. Phys. E **16**, 308 (1983)

[5.41] K.R. Peters: Generation, collection and properties of an SE-I enriched signal suitable for high resolution SEM of bulk specimens. In *Electron Beam Interactions with Solids*, ed. by D.F. Kyser et al. (SEM Inc., AMF O'Hare, IL 1984) p.363

[5.42] T. Nagantani, A. Okura: Enhanced secondary electron detection at small working distances in the field emission SEM. SEM 1977/I (SEM Inc., AMF O'Hare, IL 1977) p.695

[5.43] H. Koike, K. Ueno, M. Suzuki: Scanning device combined with conventional electron microscope. Proc. EMSA 1971 (Claytor's Publ. Div., Baton Rouge) p.28

[5.44] J.R. Banbury, W.C. Nixon: A high-contrast directional detector for the SEM. J. Phys. E **2**, 1055 (1969)

[5.45] J. Zach, H. Rose: Efficient detection of SE in LVSEM. Scanning **8**, 285 (1986)

[5.46] R. Schmid, M. Brunner: Design and application of a quadrupole detector for LVSEM. Scanning **8**, 294 (1986)

[5.47] M. Brunner, R. Schmid: Characteristics of an electric/magnetic quadrupole detector for LVSEM. Scanning Microscopy **1**, 1501 (1987)

[5.48] J. Zach, H. Rose: High-resolution low-voltage electron microprobe with large SE detection efficiency. Inst. of Phys. Conf. Ser. No. 93 (IoP, Bristol 1988) p.81

[5.49] J. Zach: Design of a high-resolution low-voltage SEM. Optik **83**, 30 (1989)

[5.50] R. Autrata, P. Schauer: Collection of low energy signal electrons in the rotationally symmetric electrostatic field of a detector. Beitr. elektr. mikr. Direktabb. Oberfl. **26**, 19 (1993)

[5.51] V. Kolařik, J. Mejzlik: A design of a new axially symmetric SE detector for the TEM. Meas. Sci. Techn. **1**, 391 (1990)

[5.52] S.H. Moll, F. Healey, B. Sullivan, W. Johnson: A high efficiency, nondirectional backscattered electron detection mode for SEM. SEM 1978/I (SEM Inc., AMF O'Hare, IL 1978) p.303

[5.53] L. Reimer, B. Volbert: Detector system for backscattered electrons by conversion to secondary electrons. Scanning **2**, 238 (1979)

[5.54] A. Boyde, M.J. Cowham: An alternative method for obtaining converted BSE images and other uses for specimen biasing in biological SEM. SEM 1980/I (SEM Inc., AMF O'Hare, IL 1980) p.227

[5.55] B. Volbert, L. Reimer: Advantages of two opposite Everhart–Thornley detectors in SEM. SEM 1980/IV (SEM Inc., AMF O'Hare, IL 1980) p.1

[5.56] A.R. Walker, G.R. Booker: A simple energy-filtering backscattered electron detector. In *Developments in Electron Microscopy and Analysis*, ed. by J.A. Venables (Academic, London 1976) p.119

[5.57] O.C. Wells: New contrast mechanism for SEM. Appl. Phys. Lett. **16**, 151 (1970)

[5.58] O.C. Wells, C.G. Bremer: Collector turret for SEM. Rev. Sci. Instr. **41**, 1034 (1970)

[5.59] R. Blaschke, K. Schur: Der Informationsgehalt des Rückstreubildes im Raster-Elektronenmikroskop. Beitr. elektr. mikr. Direktabb. Oberfl. **7**, 33 (1974)

[5.60] K. Schur, R. Blaschke, G. Pfefferkorn: Improved conditions for backscattered electron SEM micrographs on polished sections using a modified scintillator detector. SEM 1974 (ITTRI, Chicago 1974) p.1003

[5.61] V.N.E. Robinson: The construction and uses of an efficient backscattered electron detector for SEM. J. Phys. E **7**, 650 (1974)

[5.62] V.N.E. Robinson: Imaging with backscattered electrons in a SEM. Scanning **3**, 15 (1980)

[5.63] R. Autrata: BSE imaging using single crystal scintillation detectors. Scanning Microscopy **3**, 739 (1989)

[5.64] R. Autrata, J. Hejna: Detectors for LVSEM. Scanning **13**, 275 (1991)

[5.65] R. Autrata: New configurations of single-crystal scintillation detectors in SEM. In *Electron Microscopy 1990*, Vol. 1, (San Francisco Press, San Francisco 1990) p.376

[5.66] V.N.E. Robinson: BSE imaging at low accelerating voltages. Hitachi Instr. News **19**, 32 (1990)

[5.67] L. Reimer: Electron signal and detector strategy. In *Electron Beam Interactions with Solids*, ed. by D.F. Kyser et al. (SEM Inc., AMF O'Hare, IL 1982) p.299

[5.68] L. Reimer: SEM of surfaces. In *Electron Microscopy 1982*, Vol.1 (Deutsche Ges. für Elektronenmikroskopie, Frankfurt 1982) p.79

[5.69] L. Reimer, M. Riepenhausen: Detector strategy for secondary and backscattered electrons using multiple detector systems. Scanning **7**, 221 (1985)

[5.70] S. Kimoto, H. Hashimoto: Stereoscopic observation in SEM using multiple detectors. In *The Electron Microprobe*, ed. by T.D. McKinley et al (Wiley, New York 1966), p.480

[5.71] J. Hejna: Detection of topographic contrast in the SEM at low and medium resolution by different detectors and detector systems. Scanning Microscopy **8**, 143 (1994)

[5.72] J. Hejna: A ring scintillation detector for detection of BSE in the SEM. Scanning Microscopy **1**, 983 (1987)

[5.73] J. Hejna: Optimization of the ring scintillation detector for BSE in the SEM. In *EUREM 88* (IoP, Bristol 1988) p.119

[5.74] J. Lebiedzik, E.W. White: Multiple detector method for quantitative determination of microtopography in the SEM. SEM 1975 (ITTRI, Chicago 1975) p.181

[5.75] J. Lebiedzik, J. Lebiedzik, R. Edwards, B. Phillips: Use of microtopography capability in the SEM for analysing fracture surfaces. SEM 1979/II (SEM Inc., AMF O'Hare, IL 1979) p.61

[5.76] M. Lange, L. Reimer, C. Tollkamp: Testing of detector strategies in SEM by isodensities. J. Microsc. **134**, 1 (1984)

[5.77] J. Hejna, L. Reimer: Backscattered electron multidetector systems for improved quantitative topographic contrast. Scanning **9**, 162 (1987)

[5.78] W. Kuypers, S. Lichtenegger: Universal detector system for BSE, transmitted electrons and cathodoluminescence. In *Electron Microscopy 1980*, Vol.1, ed. by P. Brederoo and G. Boom (7th European Congr. on Electron Microscopy Foundation, Leiden 1980) p.522

[5.79] L. Reimer, W. Pöpper, W. Bröcker: Experiments with a small solid angle detector for BSE. SEM 1978/I (SEM Inc., AMF O'Hare, IL 1978) p.705

[5.80] T.E. Everhart, O.C. Wells, C.W. Oatley: Factors affecting contrast and resolution in the SEM. J. Electron. and Contr. **7**, 97 (1959)

[5.81] W. Baumann, L. Reimer: Comparison of the noise of different electron detection systems using a scintillator–photomultiplier combination. Scanning **4**, 141 (1981)

[5.82] R.C. Jones: Quantum efficiency of detectors for visible and infrared radiation. Adv. Electr. Electron Phys. **11**, 87 (1959)

[5.83] F.J. Maher, C.J. Rossouw: Design and performance of an amplifier for EBIC imaging in a SEM. J. Phys. E **16**, 1238 (1983)

[5.84] W. Steckelmacher: Energy analysers for charged particle beams. J. Phys. E **6**, 1061 (1973)

[5.85] H.T. Pearce-Percy: The design of spectrometers for energy loss spectroscopy. SEM 1978/I (SEM Inc., AMF O'Hare, IL 1978) p.41

[5.86] L. Reimer, K. Brockmann, U. Rhein: Energy losses of 20–40 keV electrons in 150–650 μg/cm^2 metal films. J. Phys. D **11**, 2151 (1978)

[5.87] A. Rusterholz: *Elektronenoptik I: Grundlagen der theoretischen Elektronenoptik.* (Birkhäuser, Basel 1950)

[5.88] E. Menzel, E. Kubalek: Secondary electron detection systems for quantitative voltage measurements. Scanning **5**, 151 (1983)

[5.89] O.C. Wells, C.G. Bremer: Improved energy analyser for the SEM. J. Phys. E **2**, 1120 (1969)

[5.90] L.J. Balk, H.P. Feuerbaum, E. Kubalek, E. Menzel: Quantitative voltage contrast at high frequencies in the SEM. SEM 1976/I (ITTRI, Chicago 1976) p.615

[5.91] Y. Goto, A. Ito, Y. Furukawa, T. Inagaki: Hemispherical retarding type energy analyser for IC testing by electron beam. J. Vac. Sci. Techn. **19**, 1030 (1981)

[5.92] K. Nakamae, H. Fujioka, K. Ura, T. Takagi, S. Takashima: A hemispherical retarding-field analyser with a microchannel plate detector and high extraction field for voltage measurement in the SEM. J. Phys. E **19**, 847 (1986)

[5.93] H.P. Feuerbaum: VLSI testing using the electron probe. SEM 1979/I (SEM Inc., AMF O'Hare, IL 1979) p.285

[5.94] H. Todokara, S. Yoneda, K. Yamaguchi, S. Fukuhara, T. Komoda: Stroboscopic testing of LSIs with LVSEM. J. Microsc. **140**, 313 (1985)

[5.95] J. Frosien, E. Plies, K. Anger: Compound magnetic and electrostatic lenses for low-voltage applications. J. Vac. Sci. Techn. B **7**, 1874 (1989)

[5.96] A.R. Dinnis, A. Khursheed: A voltage-contrast analyzer using radial magnetic field exctraction. J. Vac. Sci. Techn. B **6**, 2003 (1988)

[5.97] T. Aton, S.C.J. Garth, J.N. Sackett, D.F. Spicer: Characteristics of a virtual immersion lens spectrometer for electron beam testing. J. Vac. Sci. Techn. B **6**, 1953 (1988)

[5.98] P. Kruit, F.H. Read: Magnetic field paralleliser for 2π electron-spectrometer and electron-image magnifier. J. Phys. E **16**, 313 (1983)

[5.99] W.J. Tee, A. Gopinath: A voltage measurement scheme for the SEM using a hemispherical retarding analyser. SEM 1976/I (ITTRI, Chicago 1976) p.595

[5.100] W. Hylla, H. Niedrig, T. Wittich: The use of silicon avalanche diodes as energy sensitive electron detectors. In *Electron Microscopy 1986*, Vol. 1, ed. by T. Imura et al. (Jpn. Soc. of Electron Microscopy, Tokyo 1986) p.443

[5.101] P. Gerard, J.L. Balladore, H. Pinna, J.P. Martinez, A. Ouabbou, G. Ball: Method for the analysis of backscattered electron energy. In *Electron Microscopy 1992*, Vol.1, ed. by A. Rios et al. (Secr. Publ. Universidad de Granada, Spain 1992) p.99

[5.102] L. Reimer, R. Böngeler, M. Kässens, F.F. Liebscher, R. Senkel: Calculation of energy spectra from layered structures for backscattered electron spectrometry and relations to Rutherford backscattering spectrometry by ions. Scanning **13**, 381 (1991)

[5.103] O.C. Wells: Low-loss image for surface SEM. Appl. Phys. Lett. **19**, 232 (1971)

[5.104] P. Morin, M. Pitaval, D. Besnard, G. Fontaine: Electron-channelling imaging in SEM. Phil. Mag. A **40**, 511 (1979)

[5.105] O.C. Wells, A.N. Broers, C.G. Bremer: Method for examining solid specimens with improved resolution in the SEM. Appl. Phys. Lett. **23**, 353 (1973)

[5.106] N.N. Dremova, E.I. Rau, V.N.E. Robinson: Energy analysers for SEM. Instr. Exp. Techn. **38**, 95 (1995)

[5.107] E.I. Rau, V.N.E. Robinson: An annular toroidal BSE energy analyser for use in in SEM. Scanning **18**, 556 (1996)

Chapter 6

[6.1] R. Böngeler, U. Golla, M. Kässens, L. Reimer, B. Schindler, R. Senkel, M. Spranck: Electron-specimen interactions in LVSEM. Scanning **15**, 1 (1993)

[6.2] M. Lange. L. Reimer, C. Tollkamp: Testing of detector strategies in SEM by isodensities. J.Microsc. **134**, 1 (1984)

[6.3] D.K. Hindermann, R.H. Davis: SEM techniques for the examination of blind and through holes. SEM 1974 (ITTRI, Chicago 1974) p.183

[6.4] A.E. Lukianov, G.V. Spivak, E.I. Rau, D.D. Gorodsky: The secondary electron SEM-collector with magnetic field. In *Electron Microscopy 1972* (IoP, London 1972) p.186

[6.5] K. Saito, M. Yoshizawa, K. Wada: A SEM for trench observation. J. Vac. Sci. Techn. B **8**, 1152 (1990)

[6.6] H. Nakagawa, N. Nomura, T. Koizumi, N. Anazawa, K. Harafuji: A novel high-resolution SEM for the surface analysis of high-aspect-ratio three-dimensional structures. Jpn. J. Appl. Phys. **30**, 2112 (1991)

[6.7] K. Schur, Ch. Schulte, L. Reimer: Auflösungsvermögen und Kontrast von Oberflächenstufen bei der Abbildung mit einem Raster-Elektronenmikroskop. Z. Angew. Phys. **23**, 405 (1967)

[6.8] G. Pfefferkorn, R. Blaschke: Der Informationsgehalt rasterelektronenmikroskopischer Aufnahmen. Beitr. elektr. mikr. Direktabb. Oberfl. **1**, 1 (1968)

[6.8] N.S. Griffin: Enhanced topography imaging in the SEM. Meas. Sci. Techn. **5**, 1403 (1994)

[6.9] R.H. Milne: Surface steps imaged by SE. Ultramicroscopy **27**, 433 (1989)

[6.10] L. Reimer, M. Riepenhausen, C. Tollkamp: Detector strategy for improvement of image contrast analogous to light illumination. Scanning **6**, 155 (1984)

[6.11] C. Le Gressus, H. Okuzumi, D. Massignon: Changes of SEI brightness under electron irradiation as studied by electron spectroscopy. SEM 1981/I (SEM Inc., AMF O'Hare, IL 1981) p.251

[6.12] O. Lee-Deacon, C. Le Gressus, D. Massignon: Analytical SEM for surface science. In *Electron Beam Interactions with Solids*, ed. by D.F. Kyser et al. (SEM Inc., AMF O'Hare, IL 1982) p.271

[6.13] A. Howie: Recent developments in SE imaging. J. Microsc. **180**, 192 (1995)

[6.14] C.A. Walsh: Effect of specimen bias on SE images in the STEM. Ultramicroscopy **45**, 85 (1992)

[6.15] R. Persaud, H. Moro, M. Azim, R.H. Milne, J.A. Venables: Recent surface studies using biassed SE imaging. Scanning Microscopy **8**, 803 (1994)

[6.16] Y. Homma, M. Suzuki, M. Tomita: Atomic configuration dependent SE emission from reconstructed silicon surfaces. Appl. Phys. Lett. **62**, 3276 (1993)

[6.17] H.W. Ren, M. Tanaka, T. Nishinaga: Real time observation of reconstruction transitions on GaAs(111)B surface by SEM. Appl. Phys. Lett. **69**, 565 (1996)

[6.18] M. Aven, J.Z. Devine, R.B. Bolon. G.W. Ludwig: SEM and cathodoluminescence of $ZnSe_xTe_{1-x}$ p-n junctions. J. Appl. Phys. **43**, 4136 (1972)

[6.19] G.R. Sawyer, T.F. Page: Microstructural characterisation of 'REFEL' (reaction bonded) silicon carbides. J. Mater. Sci. **13**, 885 (1978)

[6.20] D.D. Perovic, M.R. Castell, A. Howie, C. Lavoie, T. Tiedje, J.S.W. Cole: Field-emisison SEM imaging of compositional and doping layer semiconductor superlattices. Ultramicroscopy **58**, 104 (1995)

[6.21] D. Venables, D.M. Maher: Quantitative two-dimensional dopant profiles obtained directly from SE images. J. Vac. Sci. Techn. B **14**, 421 (1996)

[6.22] L. Reimer: Electron signal and detector strategy. In *Electron Beam Interactions with Solids*, ed. by D.F. Kyser et al. (SEM Inc., AMF O'Hare, IL 1982) p.299

[6.23] L. Reimer: SEM of surfaces. In *Electron Microscopy 1982*, Vol.1 (Deutsche Ges. für Elektronenmikroskopie, Frankfurt (1982) p.79

[6.24] L. Reimer, B. Volbert: Detector system for backscattered electrons by conversion to secondary electrons. Scanning **2**, 238 (1979)

[6.25] K.H. Peters: Conditions required for high quality high magnification images in secondary electron-I SEM. SEM 1982/IV (SEM Inc., AMF O'Hare, IL 1982) p.1359

[6.26] B. Volbert, L. Reimer: Advantages of two opposite Everhart–Thornley detectors in SEM. SEM 1980/IV (SEM Inc., AMF O'Hare, IL 1980) p.1

[6.27] M. Kässens, L. Reimer: Contrast effects using a two-detector system in LVSEM. J. Microsc. **181**, 277 (1996)

[6.28] L. Reimer, C. Tollkamp: Recording of topography by secondary electrons with a two-detector system. In *Electron Microscopy 1982*, Vol.2 (Deutsche Ges. für Elektronenmikroskopie, Frankfurt 1982) p.543

[6.29] L. Reimer, R. Böngeler, V. Desai: Shape from shading using multiple detector signals in SEM. Scanning Microscopy **1**, 963 (1987)

[6.30] B. Volbert: True surface topography: the need for signal mixing. In *Electron Microscopy 1982*, Vol.2 (Deutsche Ges. für Elektronenmikroskopie, Frankfurt 1982) p.233

[6.31] P.S.D. Lin, R.P. Becker: Detection of BSE with high resolution. SEM 1975 (ITTRI, Chicago 1975) p.61

[6.32] J.L. Abraham, P.B. DeNee: Biomedical applications of BSE imaging – one year's experience with SEM histochemistry. SEM 1974 (ITTRI, Chicago 1974) p.251

[6.33] R.P. Becker, M. Sogard: Visualization of subsurface structures in cells and tissues by BSE imaging. SEM 1979/II (SEM Inc., AMF O'Hare, IL 1979) p. 835

[6.34] R.G. Richards, I. A. Gwynn: BSE imaging of the undersurface of resin-embedded cells by field-emission SEM. J. Microsc. **177**, 43 (1995)

[6.35] J. Brostin: Compositional imaging of polymers using a field emission SEM with a microchannel plate BSE detector. Scanning **17**, 327 (1995)

[6.36] B. Ocker, R. Wurster, H. Seiler: Investigation of nanoparticles in high resolution SEM and LVSEM by digital image analysis. Scanning Microscopy **9**, 63 (1995)

[6.37] P. Hirsch, M. Kässens, L. Reimer, R. Senkel, M. Spranck: Contrast of colloidal gold particles and thin films on a silicon substrate oberved by BSE in a LVSEM. Ultramicroscopy **50**, 263 (1993)

[6.38] L. Reimer, W. Pöpper, W. Bröcker: Experiments with a small solid angle detector for BSE. SEM 1978/I (SEM Inc., AMF O'Hare, IL 1978) p.705

[6.39] O.C. Wells: Effect of collector position on type–2 magnetic contrast in the SEM. SEM 1978/I (SEM Inc., AMF O'Hare, IL 1978) p.293

[6.40] O.C. Wells: Effects of collector take-off angle and energy filtering on the BSE image in the SEM. Scanning **2**, 199 (1979)

[6.41] R. Christenhuß: Zur Darstellbarkeit kristalliner Objekte in der Auflicht-Elektronenmikroskopie. Beitr. elektr. mikr. Direktabb. Oberfl. **1**, 67 (1968)

[6.42] J. Philibert, R. Tixier: Effects of crystal contrast in SEM. Micron **1**, 174 (1969)

[6.43] D.C. Joy, D.E. Newbury, P.M. Hazzledine: Anomalous crystallographic contrast on rolled and annealed specimens. SEM 1972 (ITTRI, Chicago 1972) p.97

[6.44] P.F. Schmidt, H.G. Grewe, G. Pfefferkorn: SEM imaging of the cell structure of dislocation networks in cold worked Cu single crystals. Scanning **1**, 174 (1978)

[6.45] H. Seiler: Determination of the "information depth" in the SEM. SEM 1976/I (ITTRI, Chicago 1976) p.9

[6.46] D.J. Prior, P.W. Trimby, U.D. Weber: Orientation contrast imaging of microstructures in rocks using forescatter detectors in the SEM. Mineral. Mag. **60**, 859 (1996)

[6.47] L. Reimer, B. Volbert: The origin and correction of SEM imaging artifacts arising from the use of the difference signal of two detectors. Philips Electron Optics Bulletin No. 118 (Philips, Eindhoven 1982)

[6.48] M.D. Ball, D.G. McCartney: The measurement of atomic number and composition in an SEM using BSE detectors. J. Microsc. **124**, 57 (1981)

[6.49] D. Weirauch: Quantitative Auswertung der Intensität der Rückstreuelektronen am Rasterelektronenmikroskop. Beitr. elektr. mikr. Direktabb. Oberfl. **26**, 7 (1993)

[6.50] P. Roschger, H. Plenk, K. Klaushofer, J. Eschberger: A new SEM approach to the quantification of bone mineral distribution – BSE image grey levels correlated to calcium Kα-line intensities. Scanning Microscopy **9**, 75 (1995)

[6.51] H. Niedrig: BSE as a tool for film thickness determination. SEM 1978/I (SEM Inc., AMF O'Hare, IL 1978) p.841

[6.52] O.C. Wells, R.J. Savoy, P.J. Bailey: BSE imaging in the SEM – measurement of surface-layer mass–thickness. In *Electron Beam Interaction with Solids*, ed. by D.F. Kyser et al. (SEM Inc., AMF O'Hare, IL 1982) p.287

[6.53] P.B. DeNee: Measurement of mass and thickness of respirable size dust particles by SEM BSE imaging. SEM 1978/I (SEM Inc., AMF O'Hare, IL 1978) p.741

[6.54] J. Lebiedzik, E.W. White: Multiple detector method for quantitative determination of microtopography in the SEM. SEM 1975 (ITTRI, Chicago 1975) p.181

[6.55] J. Lebiedzik, J. Lebiedzik, R. Edwards, B. Phillips: Use of microtopography capability in the SEM for analysing fracture surfaces. SEM 1979/II (SEM Inc., AMF O'Hare, IL 1979) p.61

[6.56] I.C. Carlsen: Reconstruction of true surface-topographies in SEM using BSE. Scanning **7**, 169 (1985)

[6.57] W. Bell, I.C. Carlsen, V. Desai, L. Reimer: Three-dimensional digital surface reconstruction in SEM using signals of a multiple detector system. Europ. J. Cell Biology (Suppl. 25) **48**, 13 (1989)

[6.58] D.A. Wassink, J.Z. Raski, J.A. Levitt, D. Hildreth, K.C. Ludema: Surface topographical and compositional characterization using BSE methods. Scanning Microscopy **5**, 919 (1991)

[6.59] A.R. Walker, G.R. Booker: A simple energy filtering BSE detector. In *Developments in Electron Microscopy and Analysis*, ed. by J.A. Venables (Academic, London 1976), p.119

[6.60] O.C. Wells: Low-loss image for surface SEM. Appl. Phys. Lett. **19**, 232 (1971)

[6.61] P. Morin, M. Pitaval, D. Besnard, G. Fontaine: Electron-channelling imaging in SEM. Phil. Mag. A **40**, 511 (1979)

[6.62] O.C. Wells, A.N. Broers, C.G. Bremer: Method for examining solid specimens with improved resolution in the SEM. Appl. Phys. Lett. **23**, 353 (1973)

[6.63] O.C. Wells: Explanation of the low-loss image in the SEM in terms of electron scattering theory. SEM 1972 (ITTRI, Chicago 1972) p.169

[6.64] J.P. Spencer, C.J. Humphreys, P.B. Hirsch: A dynamical theory for the contrast of perfect and and imperfect crystals in the SEM using BSE. Phil. Mag. **26**, 193 (1972)

[6.65] R.M. Stern, T. Ichinokawa, S. Takashima, H. Hashimoto, S. Kimoto: Dislocation images in the high resolution SEM. Phil. Mag. **26**, 1495 (1972)

[6.66] G.R. Booker, D.C. Joy, J.P. Spencer, C.J. Humphreys: Imaging of crystal defects in the SEM. SEM 1973 (ITTRI, Chicago 1973) p.251

[6.67] P. Morin, M. Pitaval, E. Vicario, G. Fontaine: SEM observation of single defects in solid crystalline materials. Scanning **2**, 217 (1979)

[6.68] D.E. Newbury: The utility of specimen current imaging in the SEM. SEM 1976/I (ITTRI, Chicago 1976) p.111

[6.69] P. Echlin, G. Kaye: Thin films for high resolution conventional SEM. SEM 1979/II (SEM Inc., AMF O'Hare, IL 1979) p.21

[6.70] J.D. Geller, T. Yoshioka, D.A. Hurd: Coating by ion sputtering for ultrahigh resolution SEM. SEM 1979/II (SEM Inc., AMF O'Hare, IL 1979) p. 355

[6.71] K.R. Peters: SEM at macromolecular resolution in low energy mode on biological specimens coated with ultra thin metal films. SEM 1979/II (SEM Inc., AMF O'Hare, IL 1979) p.133

[6.72] K.R. Peters: Metal deposition by high energy sputtering for high magnification electron microscopy. In *Advanced Techniques in Biological Electron Microscopy III*, ed. by J.K. Koehler (Springer, Berlin, Heidelberg 1986) p.101

[6.73] R. Hermann, M. Müller: High resolution biological SEM: a comparative study of low temperature coating techniques. J. Electron Microsc. Techn. **18**, 440 (1991)

[6.74] R. Wepf, M. Amrein, U. Bürkli, H. Gross: Platinum/iridium/carbon: a high-resolution shadowing material for TEM, STM and SEM of biological macromolecular structure. J. Microsc. **163**, 51 (1991)

[6.75] P. Walther, E. Wehrli, R. Hermann, M. Müller: Double-layer coating for high-resolution low-temperature SEM. J. Microsc. **179**, 229 (1995)

[6.76] T. Nakadera, A. Mitsushima, K. Tanaka: Application of high-resolution SEM to biological macromolecules. J. Microsc. **163**, 43 (1991)

[6.77] O.C. Wells: BSE image in the SEM. SEM 1977/I (ITTRI, Chicago 1977) p.747

[6.78] P.G. Merli, M. Nacucci: Resolution of superlattice structures with BSE in a SEM. Ultramicroscopy **50**, 83 (1993)

[6.79] L. Reimer, P. Wahlbring: Monte-Carlo simulation of imaging cross-section of layered structures by BSE. In *Electron Microscopy*, Vol.1, ed. by B. Jouffrey and C. Colliex (Les éditions de Physique, Les Ulis, France 1994) p.355

[6.80] L. Reimer, M. Kässens, L. Wiese: Monte Carlo simulation program with a free configuration of specimen and detector geometries. In *Microbeam and Nanobeam Analysis*, ed. by D. Benoit et al., Mikrochim. Acta [Suppl.] **13**, 485 (1996)

[6.81] B.H. Fishbine, R.J. Macy: First results with a high-imaging-speed SEM. Rev. Sci. Instr. **61**, 2534 (1990)

[6.82] B. Dunger, D. Schmidt: Zum Einfluß der Bildröhre auf den Kontrast in REM-Bildern. Beitr. elektr. mikr. Direktabb. Oberfl. **4/2**, 381 (1971)

[6.83] H. Hantsche: Systematische Fehler bei Rasterabbildungen als Folge endlichen Strahldurchmessers. Beitr. elektr. mikr. Direktabb. Oberfl. **3**, 371 (1970)

[6.84] J. Bahr, B. Dunger, W. Schwarz: Zum Einfluß von Rastergrößen auf die Schärfentiefe beim Raster-Elektronenmikroskop. Beitr. elektr. mikr. Direktabb. Oberfl. **3**, 379 (1970)

[6.85] E. Oho. N. Ichise, T. Ogashiwa: Proper acquisition and handling of SEM images using a high-performance personal computer. Scanning **18**, 72 (1996)

[6.86] G.A.C. Jones, H. Ahmed, W.C. Nixon: Large field SEM. In *Developments in Electron Microscopy and Analysis*, ed. by J.A. Venables (Academic, London 1976) p.65

[6.87] K. Schur: Ein Moiré-Effekt im Raster-Elektronenmikroskop. Beitr. elektr. mikr. Direktabb. Oberfl. **3**, 143 (1970)

[6.88] C.E. Fiori, H. Yakowitz, D.E. Newbury: Some techniques of signal processing in SEM. SEM 1974 (ITTRI, Chicago 1974) p.167

[6.89] L. Reimer, P. Hagemann: Recording of mass thickness in STEM. Ultramicroscopy **2**, 297 (1977)

[6.90] J.P. Flemming: The display of information from scanned measuring systems by contour mapping. J. Phys. E **2**, 93 (1969)

[6.91] P.R. Thornton, D.V. Sulway, D.A. Shaw: SEM in device diagnostics and reliability. IEEE Trans. ED-**16**, 360 (1969)

[6.92] M.C. Bagget, L.H. Glassman: SEM image processing by analog homomorphic filtering techniques. SEM 1974 (ITTRI, Chicago 1974) p.199

[6.93] L.J. Balk, E. Kubalek, E. Menzel: Untersuchung von GaAlAs-Elektrolumineszenzdioden im Rasterelektronenmikroskop. Beitr. elektr. mikr. Direktabb. Oberfl. **7**, 245 (1974)

[6.94] M.T. Postek, A.E. Vladár: Digital imaging for SEM. Scanning **18**, 1 (1996)

[6.95] T.J. Pitt: Application of array processor to image processing in electron microscopy. J. Microsc. **127**, 85 (1982)

[6.96] B. Jähne: Digital Image Processing (Springer, Berlin, Göttingen Heidelberg 1993)

[6.97] T.S. Huang: *Picture Processing and Digital Filtering*, 2nd edn., Topics Appl. Phys. Vol.6 (Springer, Berlin, Heidelberg 1979)

[6.98] W.K. Pratt: *Digital Image Processing* (Wiley, New York 1991)

[6.99] A. Rosenfeld, A.C. Kak: *Digital Picture Processing*, 2nd edn., Vol.1,2 (Academic, New York 1982)

[6.100] J.C. Russ: *The Image Processing Handbook*, 2nd edn. (CRC Press, London 1995)

[6.101] A.V. Jones, K.C.A. Smith: Image processing for scanning microscopists. SEM 1978/I (SEM Inc., AMF O'Hare, IL 1978) p.13

[6.102] K.C.A. Smith: On-line digital computer techniques in electron microscopy : general introduction. J. Microsc. **127**, 3 (1982)

[6.103] P.W. Hawkes, E. Kasper: *Principles of Electron Optics*, Vol. 3, *Wave Optics* (Academic, London 1994)

[6.104] P.W. Hawkes: Processing electron images. In *Quantitative Electron Microscopy*, ed. by J.N. Chapman and A.J. Craven (Scottish Univ. Summer School in Physics, Edinburgh 1984) p.351

[6.105] E. Oho, N. Ichise, W.H. Martin, K.R. Peters: Practical method for noise removal in SEM. Scanning **18**, 50 (1996)

[6.106] H.J. Nussbaumer: *Fast Fourier Transform and Convolution Algorithms*, 2nd edn., Springer Ser. Inf. Sci., Vol.2 (Springer, Berlin, Heidelberg 1982)

[6.107] L. Reimer, B. Volbert, P. Bracker: Quality control of SEM micrographs by laser diffractometry. Scanning **1**, 233 (1978)

[6.108] S.J. Erasmus, D.M. Holburn, K.C.A. Smith: On-line computation of diffractograms for the analysis of SEM images. Scanning **3**, 273 (1980)

[6.109] A. Niemitz, L. Reimer: Digital image processing of multiple detector signals in SEM. Ultramicroscopy **16**, 161 (1985)

[6.110] F. Yano, S. Nomura: Deconvolution of SEM images. Scanning **15**, 19 (1993)

[6.111] E.R. Weibel,, G.S. Kistler, W.F. Schobe: Practial stereological methods for morphometric cytology. J. Cell. Biol. **30** 23 (1966)

[6.112] H. Giger: Grundgleichungen der Stereologie. Metrika **16**, 43 (1970)

[6.113] E.E. Underwood: *Quantitative Stereology* (Addison-Wesley, Reading, MA 1972)

[6.114] C.G. Amstutz, H. Giger: Stereological methods applied to mineralogy, petrology, mineral deposits and ceramics. J. Microsc. **95**, 145 (1972)

[6.115] E.R. Weibel: Stereological techniques for electron microscopic morphometry. In *Principles and Techniques of Electron Microscopy*, Vol.3, ed. by M.A. Hayat (Van Nostrand-Reinhold, New York 1973) p.237

[6.116] J. Lebiedzik, K.G. Burke, S. Troutman, G.G. Johnson, E.W. White: New methods for quantitative characterization of multiphase particulate materials including thickness measurement. SEM 1973 (ITTRI, Chicago 1973) p.121

[6.117] W.R. Stott, E.J. Chatfield: A precision SEM image analysis system with full-feature EDXA characterization. SEM 1979/II (SEM Inc., AMF O'Hare, IL 1979) p.53

[6.118] R.J. Lee, J.F. Kelly: Overview of SEM-based automated image analysis. SEM 1980/I (SEM Inc., AMF O'Hare, IL 1980) p.303

[6.119] D.L. Johnson: Automated SEM characterization of particulate inclusions in biological tissues. SEM 1983/III (SEM Inc., AMF O'Hare, IL 1983) p.1211

[6.120] T. Mattfeldt (Guest Editor): Journal of Microscopy **186**, Pt.2, 91–209 (May 1997) (papers from the 9th Intl' Congr. for Stereology)

[6.121] A. Boyde: Quantitative photogrammetric analysis and qualitative stereoscopic analysis of SEM images. J. Microsc. **98**, 452 (1973)

[6.122] A. Boyde: Photogrammetry of stereopair SEM images using separate measurements from the two images. SEM 1974 (ITTRO, Chicago 1974) p.101

[6.123] P.G.T. Howell: The derivation of working formulae for SEM photogrammetry. Scanning **1**, 230 (1978)

[6.124] G. Koenig, W. Nickel, J. Storl, D. Meyer, J. Stange: Digital stereophotogrammetry for processing SEM data. Scanning **9**, 185 (1987)

[6.125] J.T.L. Thong, B.C. Breton: A topographic measurement instrument based on the SEM. Rev. Sci. Instr. **63**, 131 (1992)

[6.126] A.R. Dinnis: After-lens deflection and its use. In *Scanning Electron Microscopy: systems and applications* (IoP, London 1973) p.76

[6.127] A. Boyde: A stereo-plotting device for SEM micrographs; and a real time 3-D system for the SEM. SEM 1974 (ITTRI, Chicago 1974) p.93

[6.128] E.J. Chatfield, J. More, V.H. Nielsen: Stereoscopic SEM at T.V. scan rates. SEM 1974 (ITTRI, Chicago 1974) p.117

[6.129] J.B. Pawley: Design and performance of presently available TV rate stereo SEM systems. SEM 1978/I (SEM Inc., AMF O'Hare, IL 1978) p.157

Chapter 7

[7.1] J.F. Bresse: Quantitative investigation in semiconductor devices by electron beam induced current mode: a review. SEM 1982/IV (SEM Inc., AMF O'Hare, IL 1983) p.1487

[7.2] S.M. Davidson: Semiconductor material assessment by SEM. J. Microsc. **110**, 177 (1977)

[7.3] D.B. Holt: Quantitative conduction mode in SEM. In *Quantitative Scanning Electron Microscopy*, ed. by D.B. Holt et al. (Academic, London 1974) p.213

[7.4] H.J. Deamy: Charge collection scanning electron microscopy. J. Appl. Phys. **53**, R51 (1982)

[7.5] D.B. Holt, M. Lesniak: Recent developments in electrical microcharacterization using the charge collection mode of SEM. SEM 1985/I (SEM Inc., AMF O'Hare, IL 1985) p.67

[7.6] L.C. Kimerling, H.J. Leamy, J.R. Patel: The electrical properties of stacking faults and precipitates in heat-treated dislocation-free Czochralski silicon. Appl. Phys. Lett. **30**, 217 (1977)

[7.7] D.E. Ioannou, S.M. Davidson: SEM observation of dislocations in boron implanted silicon using Schottky barrier EBIC technique. Phys. Stat. Solidi (a) **48**, K1 (1978)

[7.8] J. Palm, H. Alexander: Local investigation of grain boundaries by grain boundary EBIC. Phys. Stat. Solidi (a) **138**, 639 (1993)

[7.9] L.J. Balk, E. Kubalek, E. Menzel: Time-resolved and temperature dependent measurements of electron beam induced current (EBIC), voltage (EBIV) and cathodoluminescence (CL) in the SEM. SEM 1975 (ITTRI, Chicago 1975) p.447

[7.10] A.E. Grün: Lumineszenz-photometrische Messungen der Energieabsorption im Strahlungsfeld von Elektronenquellen, Eindimensionaler Fall in Luft. Z. Naturforschg. A **12**, 89 (1957)

[7.11] L. Reimer, H. Seidel, H. Gilde: Einfluß der Elektronendiffusion auf die Bildentstehung im Raster-Elektronenmikroskop. Beitr. elektr. mikr. Direktabb. Oberfl. **1**, 53 (1968)

[7.12] M. Hatzakis: New method of observing electron penetration profiles in solids. Appl. Phys. Lett. **18**, 7 (1971)

[7.13] R. Shimizu, T. Ikuta, T.E. Everhart, W.J. DeVore: Experimental and theoretical study of energy dissipation profiles of keV electrons in PMMA. J. Appl. Phys. **46**, 1581 (1975)

[7.14] R. Shimizu, T.E. Everhart: Monte-Carlo simulation of the energy dissipation of an electron beam in an organic specimen. Optik **36**, 59 (1972)

[7.15] J.F. Bresse: Electron beam induced current in silicon planar p-n junctions: physical model of carrier generation. Determination of some physical parameters in Si. SEM 1972 (ITTRI, Chicago 1972) p.105

[7.16] S.M. Davidson, C.A. Dimitriadis: Advances in the electrical assessment of semiconductors using the SEM. J. Microsc. **118**, 275 (1980)

[7.17] C.A. Dimitriadis: Determination of bulk diffusion length in thin semiconductors layers by SEM-EBIC. J. Phys. D **14**, 2269 (1981)

[7.18] W. van Roosbroeck: Injected current carrier transport in a semi-infinite semiconductor and the determination of lifetimes and surface recombination velocities. J. Appl. Phys. **26**, 380 (1955)

[7.19] J.F. Bresse, D. Lafeuille: SEM beam-induced current in planar p-n junctions, diffusion length and generation factor measurements. In *Proc. 25th Ann. Meeting of EMAG*, ed. by W.C. Nixon (IoP, London 1971) p.220

[7.20] C.J. Wu, D.B. Wittry: Investigation of minority-carrier diffusion length by electron bombardment of Schottky barriers. J. Appl. Phys. **49**, 2827 (1978)

[7.21] W.H. Hackett, R.H. Saul, R.W. Dixon, G.W. Kammlott: SEM characterization of GaP red-emitting diodes. J. Appl. Phys. **43**, 2857 (1972)

[7.22] W.H. Hackett: Electron beam excited minority carrier diffusion profiles in semiconductors. J. Appl. Phys. **43**, 1649 (1972)

[7.23] F. Berz, H.K. Kuiken: Theory of lifetime measurements with the SEM: steady state. Solid State Electron. **19**, 437 (1976)

[7.24] C. van Opdorp: Methods of evaluating diffusion lengths and near-junction luminescence-efficiency from SEM scans. Philips Res. Rep. **32**, 192 (1977)

[7.25] O. van Roos: On the determination of diffusion lengths by means of angle lapped p-n junctions. Solid State Electron. **22**, 113 (1978)

[7.26] D.E. Ioannou, S.M. Davidson: Diffusion length evaluation of boron-implanted silicon using the SEM-EBIC Schottky diode technique. J. Phys. D **12**, 1339 (1979)

[7.27] L. Chernyak, A. Osinsky, H. Temkin, J.W. Yang, Q. Chen, M. Asif Khan: EBIC measurements of minority carrier diffusion length in gallium nitride. Appl. Phys. Lett. **69**, 2531 (1996)

[7.28] J.D. Kamm: A method for investigation of fluctuations in doping concentration and minority carrier diffusion length in semiconductors by SEM. Solid State Electron. **19**, 921 (1976)

[7.29] P.H. Hoff, T.E. Everhart: Carrier profiles and collection efficiencies in Gaussian p-n junctions under electron bombardment. IEEE Trans. ED-**17**, 452 (1970)

[7.30] G.E. Possin: C.G. Kirkpatrick: Electron beam depth profiling in semiconductors. SEM 1979/I (SEM Inc., AMF O'Hare, IL 1979) p.245

[7.31] G.E. Possin: C.G. Kirkpatrick: Electron-beam measurements of minority carrier lifetime distributions in ion-beam damaged silicon. J. Appl. Phys. **50**, 4033 (1979)

[7.32] M. Kittler, W. Seifert, H. Raith: A new technique for the determination of minority-carrier diffusion length by EBIC. Scanning **11**, 24 (1989)

[7.33] J.Y. Chi, H.C. Gatos: Non-destructive determination of the depth of planar p-n junctions by SEM. IEEE Trans. ED-**24**, 1366 (1977)

[7.34] N.C. MacDonald, T.E. Everhart: Direct measurement of the depletion layer width variation vs applied bias for a p-n junction. Appl. Phys. Lett. **7**, 267 (1965)

[7.35] M. Aven, J.Z. Devine, R.B. Bolon, G.W. Ludwig: SEM and cathodoluminescence of $ZnSe_xTe_{1-x}$ p-n junctions. J. Appl. Phys. **43**, 4136 (1972)

[7.36] A. Castaldini, A. Cavallini, C. del Papa, M. Alietti, C. Canali, F. Nava, G. Lanzieri: Bias dependence of the depletion layer width in semi-insulating GaAs by charge collection scanning microscopy. Scanning Microscopy **8**, 969 (1994)

[7.37] A. Georges, J.M. Fournier, D. Bois: Time resolved EBIC: a non destructive method technique for an accurate determination of p-n junction depth. SEM 1982/I (SEM Inc., AMF O'Hare, IL 1982) p.147

[7.38] W. Zimmermann: Measurement of spatial variations of the carrier lifetime in silicon power devices. Phys. Stat. Solidi (a) **12**, 671 (1972)

[7.39] D.R. Hunter, D.H. Paxman, M. Burgess, G.R. Booker: Use of the SEM for measuring minority carrier lifetimes and diffusion lengths in semiconductor devices. In *Scanning Electron Microscopy: systems and applications* (IoP, London 1973) p. 208

[7.40] H.H. Kuiken: Theory of lifetime measurements with the SEM: transient analysis. Solid State Electron. **19**, 447 (1976)

[7.41] S. Othmer: Determination of the diffusion length and drift mobility in silicon by use of a modulated SEM beam. SEM 1978/I (SEM Inc., AMF O'Hare, IL 1978) p.727

[7.42] M. Bode, A. Jakubowics. H.U. Habermeier: A computerized signal acquisition system for the quantitative evaluation of EBIC and CL data in a SEM. Scanning **10**, 169 (1988)

[7.43] T. Fuyuki, H. Matsunami: Determination of lifetime and diffusion constant of minority carriers by a phase-shift technique using EBIC. J. Appl. Phys. **52**, 3428 (1981)

[7.44] A. Romanowski, L. Kordas, A. Mulak: Measurement of local minority carrier diffusion length and lifetime by an AC-EBIC method. Scanning **11**, 207 (1989)

[7.45] T.E. Everhart, O.C. Wells, T.K. Matta: A novel method of semiconductor device measurement. Proc. IEEE **52** 1642 (1964)

[7.46] A.J. Gonzales: On the electron beam induced current analysis of semiconductor devices. SEM 1974 (ITTRI, Chicago 1974) p.941

[7.47] J.D. Schick: Failure analysis of integrated circuits with SEM beam induced currents. SEM 1974 (ITTRI, Chicago 1974) p.949

[7.48] H. Raith: Spezielle Anwendungen des JEOL-Raster-Elektronenmikroskopes insbesondere auf dem Gebiet der Halbleitertechnik. Beitr. elektr. mikr. Direktabb. Oberfl. **1**, 275 (1968)

[7.49] G.V. Lukianoff: Electrical junction delineation by SEM beam technique. Sol. State Technol. **14**, 39 (1971)

[7.50] H.J. Leamy, L.C. Kimerling, S.D. Ferris: Silicon single crystal characterization by SEM. SEM 1976/I (ITTRI, Chicago 1976) p.529

[7.51] A.J.R. de Kock, S.D. Ferris, L.C. Kimerling, H.J. Leamy: SEM observation of dopant striae in silicon. J. Appl. Phys. **48**, 301 (1977)

[7.52] J.Y. Chi, H.C. Gatos: Determination of dopant-concentration diffusion length and lifetime variations in silicon by SEM. J. Appl. Phys. **50**, 3433 (1979)

[7.53] A.A. Patrin, A.E. Luk'yanov: Anomalous EBIC contrast in Si wafers. Scanning **12**, 334 (1990)

[7.54] A.J.R. de Kock, S.D. Ferris, L.C. Kimerling, H.J. Leamy: Investigation of defects and striations in as-grown Si crystals by SEM using Schottky diodes. Appl. Phys. Lett. **27**, 313 (1975)

[7.55] S. Kawado, Y. Hayafuji, T. Adachi: Observation of lattice defects in Si by SEM utilizing beam induced current generated in Schottky barriers. Jpn. J. Appl. Phys. **14**, 407 (1975)

[7.56] V.I. Petrov, R.S. Gvozdover: The influence of electrophysical parameters on spatial resolution of p-n-junctions with EBIC measurements. Scanning **15**, 322 (1993)

[7.57] M. Kittler, W. Seifert: On the origin of EBIC contrast of extended defects in Si. Scanning Microscopy **6**, 979 (1992)

[7.58] C. Donolato: On the theory of SEM charge-collection imaging of localized defects in semiconductors. Optik **52**, 19 (1978)

[7.59] C. Donolato: Contrast formation in SEM charge-collection images of semiconductor defects. SEM 1979/I (SEM Inc., AMF O'Hare, IL 1979) p.257

[7.60] C. Donolato, H. Klann: Computer simulation of SEM EBIC images of dislocations and stacking faults. J. Appl. Phys. **51**, 1624 (1980)

[7.61] M. Kittler, W. Seifert: On the sensitivity of the EBIC technique as applied to defect investigations in Si. Phys. Stat. Solidi (a) **66**, 573 (1981)

[7.62] A. Ourmazd, G.R. Booker: The electrical recombination efficiency of individual dislocations and stacking faults in n-type Si. Phys. Stat. Solidi (a) **55**, 771 (1979)

[7.63] L. Pasemann, H. Blumtritt, R. Gleichmann: Interpretation of the EBIC contrast of dislocations in Si. Phys. Stat. Solidi (a) **70**, 197 (1982)

[7.64] A. Jakubowicz: Theory of EBIC and CL contrasts from structural defects of semiconductor crystals; steady state and time-resolved problems. Scanning Microscopy **1**, 515 (1987)

[7.65] K.V. Ravi, C.J. Varker, C.E. Volk: Electrically active stacking faults in silicon. J. Electrochem. Soc. **120**, 533 (1975)

[7.66] C.J. Varker, G. Ehlenberger: Investigation of pre-breakdown sites in shallow diffused structures with the SEM. SEM 1971 (ITTRI, Chicago 1971) p.441

[7.67] E. Ress, G. Ress: REM-Darstellung von Sperrschichten in der Fehleranalyse elektronischer Halbleiterbauteile. Beitr. elektr. mikr. Direktabb. Oberfl. **6**, 363 (1973)

[7.68] S. Kusanagi, T. Sekiguchi, K. Sumino: Difference of the electrical properties of screw and 60° dislocations in Si as detected with temperature-dependent EBIC technique. Appl. Phys. Lett. **61**, 792 (1992)

[7.69] I.E. Bondarenko, E.B. Yakimov: Effect of dislocations on the local electrical properties of n-Si. In *Defect Control in Semiconductors*, ed. by K. Sumino (North-Holland, Amsterdam 1990) p.1443

[7.70] A. Cavallini, A. Castaldini: Recombination activity of individual extended defects in Si. MRS Proc. **262**, 223 (1992)

[7.71] M.Eckstein, H.U. Habermeier: Numerical analysis of the tempertaure dependence of EBIC and CL contrasts. J. de Physique **IV 1**, C6-23 (1991)

[7.72] J.I. Hanoka, R.O. Bell: EBIC in semiconductors. Ann. Rev. Mater. Sci. **11**, 353 (1981)

[7.73] V. Higgs, C.E. Norman, E.C. Lightowlers, P. Kightley: Characterization of dislocations in the presence of transition metal contamination. Inst. of Physics Conf. Ser. 117 (IoP, London 1991) p.737

[7.74] C.T. Ho, D.B. Sandstrom, C.E. Dube: Classification of macroscopic defects contained in p-type EFG ribbon Si. Solid-State Electronics **29**, 495 (1986)

[7.75] A. Jakubowicz, H.U. Habermeier, A. Eisenbeiss, D. Käss: EBIC versus temperature investigations of localized dislocations in heat-treated Czochralski Si. Phys. Stat. Solidi (a) **104**, 635 (1987)

[7.76] M. Kittler, W. Seifert,, K.W. Schröder: Temperature-dependent EBIC diffusion-length measurements in Si. Phys. Stat. Solidi (a) **93**, K101 (1986)

[7.77] A. Ourmadz, P.R. Wilshaw, G.R. Booker: The electrical behaviour of individual dislocations, Shockley partials and stacking fault ribbons in Si. Physica B **116**, 600 (1983)

[7.78] Z.J. Radzimski, T.Q. Zhou, A.B. Buczkowski, G.A. Rozgonyi: Electrical activity of dislocations. Prospects for practical utilization. Appl. Physics A **53**, 189 (1991)

[7.79] B. Raza, D.B. Holt: EBIC contrast of grain boundaries in polycrystalline solar cells. In *Polycrystalline Semiconductors II*, ed. by J.H. Werner, H.P. Streunk (Springer, Berlin, Göttingen, Heidelberg 1991) p.72

[7.80] B. Sieber: Experiments and theory of dislocations in GaAs. In *Gettering and Defect Engineering in Semiconductor Technology Solid State Phenomena* Vol. 19 and 20, ed. by M. Kittler, H. Richter (Trans. Tech. Publications, Zürich 1991) p.353

[7.81] P.R. Wilshaw, T.S. Fell, G.R. Booker: Recombination at dislocations in Si and GaAs. In *Point and Extended Defects in Semiconductors*, ed. by G. Benedek, A. Cavallini, W. Schröter (Plenum, New York 1989) p.243

[7.82] P.R. Wilshaw, T.S. Fell, M.D. Coteau: EBIC contrast of defect in semiconductors. J. de Physique IV **1**, C6-3 (1991)

[7.83] M. Kittler, W. Seifert: On the origin of EBIC defect contrast in Si. A reflection on injection and temperature dependent investigations. Phys. Stat. Solidi. (a) **138**, 687 (1993)

[7.84] M. Kittler, W. Seifert: Two classes of defect recombination behaviour as studied by SEM-EBIC. Scanning **15**, 316 (1993)

[7.85] M. Kittler, W. Seifert, Z.J. Radzimski: Two classes of recombination behavior as studied by the technique of the EBIC: $NiSi_2$ partcles and misfir dislocations in Ni contaminated n-type Si. Appl. Phys. Lett. **62**, 2513 (1993)

[7.86] T.S. Fell, P.R. Wilshaw, M.D. Coteau: EBIC investigation of dislocations and their interactions with impurities in Si. Phys. Stat. Solidi (a) **138**, 695 (1993)

[7.87] P.M. Petroff, D.V. Lang, J.L. Strudel, R.A. Logan: STEM techniques for simultaneous electronic analysis and observation of defects in semiconductors. SEM 1978/I (SEM Inc., AMF O'Hare, IL 1978) p.325

[7.88] T.G. Sparrow, U. Valdrè: Application of STEM to semiconductor devices. Phil. Mag. **36**, 1517 (1977)

[7.89] P.M. Petroff, R.A. Logan, A. Savage: Nonradiative recombination at dislocations in III-V compound semiconductors. J. Microsc. **118**, 255 (1980)

[7.90] D. Fathy, T.G. Sparrow, U. Valdrè: Observation of dislocations and microplasm sites in semiconductors by direct correlations of STEBIC, STEM and ELS. J. Microsc. **118**, 263 (1980)

[7.91] P.M. Petroff, D.V. Lang: A new spectroscopic technique for imaging the spatial distribution of nonradiative defects in a STEM. Appl. Phys. Lett. **31**, 60 (1977)

[7.92] O. Breitenstein: A capacitance meter of high absolute sensitivity suitable for scanning DLTS application. Phys. Stat. Solidi (a) **71**, 159 (1982)

[7.93] J. Heydenreich, O. Breitenstein: Characterization of defects in semiconductors by combined application of SEM(EBIC) and SDLTS. J. Microsc. **141**, 129 (1986)

[7.94] R.G. Woodham, G.R. Booker: Development of a SDLTS system and application to well-characterized dislocations in silicon. Inst. of Physics Conf. Ser **87** (IoP, London 1987) p.781

[7.95] J. Kreis, H. Alexander: Application of current-scanning deep level transient spectroscopy to deformed silicon crystals – first results. Scanning **15**, 338 (1993)

[7.96] A.S. Bruk, A.V. Govorkov, M.G. Milvidskij, A.Y. Polyakov: Determination of local deep-levels homogeneity using SEM. Scanning **15**, 345 (1993)

[7.97] R.C. Hughes: Hole mobility and transport in thin SiO_2 films. Appl. Phys. Lett. **26**, 436 (1975)

[7.98] G.H. Bernstein, S.W. Polchlopek, R. Kamath, W. Porod: Determination of fixed electron-beam-induced positive oxide charge. Scanning **14**, 345 (1992)

[7.99] N.C. MacDonald, T.E. Everhart: Selective electron-beam irradiation of metal–oxide–semiconductor structures. J. Appl. Phys. **39**, 2433 (1968)

[7.100] R. Hetzel: Electron-beam induced-current investigations on MOS NMOS devices. Sol. State Electron. **22**, 735 (1979)

[7.101] T.P. Ma, G. Scoggan, R. Leone: Comparison of interface-state generation by 25-keV electron beam irradiation in p-type and n-type MOS capacitors. Appl. Phys. Lett. **27**, 61 (1975)

[7.102] D. Green, J.E. Sandor, T.W. O'Keeffe, R.K. Matta: Reversible changes in transistor characteristics caused by SEM examination. Appl. Phys. Lett. **6**, 3 (1965)

[7.103] M. Miyoshi, M. Ishikawa, K. Okumura: Effects of electron beam testing on short channel metal oxide semiconductor characteristics. SEM 1982/IV (SEM Inc., AMF O'Hare, IL 1982) p.1507

[7.104] W.J. Keery, K.O. Leedy, K.F. Galloway: Electron beam effects on microelectronic devices. SEM 1976/I (ITTRI, Chicago 1976) p.507

[7.105] N.C. MacDonald: Quantitative SEM: solid state applications. SEM 1969 (ITTRI, Chicago 1969) p.431

[7.106] W. Reiners, R. Görlich, E. Kubalek: On the primary electron energy dependence of radiation damage in passivated NMOS transistors. Inst. of Physics Conf. Ser. No. 76 (IoP, Bristol 1985) p.507

[7.107] D.W. Ramasinghe, D.J. Machin, G. Proctor: Electron beam irradiation effects on MOS-transistors and its significance to e-beam testing. Microelectr. Eng. **7**, 379 (1987)

[7.108] C. Munakata: On the voltage induced by an electron beam in a bulk semiconductor crystal. Jpn. J. Appl. Phys. **5**, 756 (1966)

[7.109] C. Munakata: Measurement of potential distribution in a semiconductor crystal with an electron beam. Jpn. J. Appl. Phys. **6**, 548 (1967)

[7.110] C. Munakata: An electron beam method of measuring diffusion voltage in semiconductors. Jpn. J. Appl. Phys. **6**, 274 (1967)

[7.111] C. Munakata: An application of beta conductivity to measurement of resistivity distribution. J. Phys. E **1**, 639 (1968)

[7.112] A. Gopinath: On SEM conduction mode signals in bulk semiconductor devices. J. Phys. D **3**, 467 (1970)

[7.113] A. Gopinath, T. de Monts de Savasse: On SEM conduction-mode signals in bulk semiconductor devices: annular geometry. J. Phys. D **4**, 2031 (1971)

[7.114] J. Tauc: *Photo and Thermoelectric Effects in Semiconductors* (Pergamon, Oxford 1962) p.89

[7.115] A.M.B. Shaw, G.R. Booker: A new method for obtaining SEM beam-induced conductivity images. SEM 1969 (ITTRI, Chicago 1969) p.459

[7.116] L. Pensak: Conductivity induced by electron bombardment in thin insulating films. Phys. Rev. **75**, 472 (1949)

[7.117] F. Ansbacher, W. Ehrenberg: Electron bombardment conductivity of dielectric films. Proc. Phys. Soc. A **64**, 362 (1951)

[7.118] L. Reimer, J. Senss: Change of conductivity in amorphous selenium films by electron bombardment (10 to 100 keV). Phys. Stat. Solidi (a) **2**, 809 (1970)

[7.119] H.S. Fresser, F.E. Prins, D.P. Kern: Low-energy electron detection in microcolumns. J. Vac. Sci. Techn. B **13**, 2553 (1995)

[7.120] S. Zolgharnain, K.Y. Lee, S.A. Rishton, D. Kister, T.H.P. Chang: Characterization of a GaAs metal–semiconductor–metal low-energy electron detector. J. Vac. Sci. Techn. B **13**, 2556 (1995)

[7.121] R. Godehardt, M. Füting: Untersuchung von Stromfilamenten in Sandwichobjekten mittels Potentialkontrast und EDX. Beitr. elektr. mikr. Direktabb. Oberfl. **23**, 67 (1990)

[7.122] H. Baumann, T. Pioch, H. Dahmen, D. Jäger: Current filament formation in gold compensated silicon p-i-n diodes. SEM 1986/II (SEM Inc., AMF O'Hare, IL 1986) p.441

[7.123] K.M. Mayer, J. Parisi, R.P. Huebener: Imaging of self-generated multifilamentary current patterns in Ga As. Z. Physik B **71**, 171 (1988)

[7.124] A. Wierschem, F.J. Niedernostheide, A. Gorbatyuk, H.G. Purwins: Observation of current-density filamentation in multilayer structures by EBIC measurements. Scanning **17**, 106 (1995)

[7.125] R. Gross, T. Doderer, R.P. Huebener, F. Kober, D. Koelle, C. Kruelle, J. Mannhart, B. Mayer, D. Quenter, A. Ustinov: Low-temperature SEM studies of superconducting thin films and Josephson junctions. Physica B **169**, 415 (1991)

[7.126] M.D. Muir, P.R. Grant: Cathodoluminescence. In *Quantitative Electron Microscopy*, ed. by D.B. Holt et al. (Academic, London 1974) p.287

[7.127] D.B. Holt: Quantitative SEM studies of cathodoluminescence in adamantine semiconductors. In *Quantitative Electron Microscopy*, ed. by D.B. Holt et al. (Academic, London 1974) p.335

[7.128] G.F.J. Garlick: Cathodo- and radioluminescence. In *Luminescence of Inorganic Solids*, ed. by P. Goldberg (Academic, New York 1966) p.685

[7.129] D.B. Holt: New directions in SEM CL microcharacterization. Scanning Microscopy **6**, 1 (1992)

[7.130] G. Remond, F. Cesbron, R. Chapoulie, D. Ohnenstetter, C. Roques-Carmes, M. Schvoerer: CL applied to the microcharacterization of mineral materials. - A present status in experimentation and interpretation. Scanning Microscopy **6**, 23 (1992)

[7.131] W. Bröcker, G. Pfefferkorn: Bibliography on cathodoluminescence. SEM 1975/I, p.725; SEM 1977/I, p.455 (ITTRI, Chicago 1975 and 1977); SEM 1978/I, p.333; SEM 1980/I. p. 298 (SEM Inc., AMF O'Hare, IL 1978 and 1980)

[7.132] S.M. Davidson, A.W. Vaidya: High resolution temperature measurements in microwave devices. In *Proc. Int'l. Symp. on GaAs and Related Compounds* (IoP, Bristol 1977) p. 287

[7.133] P.M. Williams, A.D. Yoffe: Monochromatic cathodoluminescence image in SEM. Nature **221**, 952 (1969)

[7.134] J.B. Steyn, P. Giles, P.B. Holt: An efficient spectroscopic detection system for cathodoluminescence mode SEM. J. Microsc. **107**, 107 (1976)

[7.135] S.M. Davidson, A. Rasul: Application of a high performance SEM-based CL analysis system to compound semiconductors devices. SEM 1977/I (ITTRI, Chicago 1977) p.225

[7.136] H.C. Casey, R.H. Kaiser: Analysis of n-type GaAs with electron-beam-excited radiative recombination. J. Electrochem. Soc. **114**, 149 (1967)

[7.137] D.A. Cusano: Radiative recombination from GaAs directly excited by electron beams. Solid State Commun. **2**, 353 (1964)

[7.138] A. Rasul, S.M. Davidson: Recombination around dislocations in GaP. In *Proc. Int'l. Symp. on GaAs and Related Compounds* (IoP, Bristol 1977) p.306

[7.139] H. Boersch, C. Radeloff, G. Sauerbrey: Über die an Metallen durch Elektronen ausgelöste sichtbare und ultraviolette Strahlung. Z. Physik **165**, 464 (1961)

[7.140] H. Boersch, P. Dobberstein, D. Fritzsche, G. Sauerbrey: Transition radiation, Bremsstrahlung und Plasmastrahlung. Z. Physik **187**, 97 (1965)

[7.141] D.F. Kyser, D.B. Wittry: Cathodoluminescence in GaAs. In *The Electron Microprobe*, ed. by T.D. McKinley et al. (Wiley, New York 1964), p.691

[7.142] P.R. Harmer, S.D. Ford, R.A. Dugdale: Electron bombardment luminescence stage for optical microscopy. J. Phys. E **1**, 59 (1968)

[7.143] S.H. Roberts, J.W. Steeds: Cathodoluminescence from ZnS and CdSe. Scanning **3**, 165 (1980)

[7.144] B.H. Vale, R.T. Greer: SEM cathodoluminescence detection system for transmission optical fluorescence analysis. SEM 1977/I (ITTRI, Chicago 1977) p.241

[7.145] F.J. Judge, J.M. Stubbs, J. Philp: A concave mirror, light pipe photon collecting system for cathodoluminescent studies on biological specimens in the JSM2 SEM. J. Phys E **7**, 173 (1974)

[7.146] H.C. Casey, R.H. Kaiser: Analysis of n-type GaAs with electron beam excited radiative recombination. J. Electrochem. Soc. **114**, 149 (1967)

[7.147] E.J. Korda, I.C. Pruden, J.P. Williams: SEM of a P-16 phosphor – cathodoluminescent and secondary electron emission modes. Appl. Phys. Lett. **10**, 205 (1967)

[7.148] A. Boyde, S.A. Reid: New methods for cathodoluminescence in the SEM. SEM 1983/IV (SEM Inc., AMF O'Hare, IL 1983) p.1803

[7.149] J.B. Steyn, D.B. Holt: Monochromator-photomultiplier-photon counter detection in luminescent mode SEM. In *Scanning Electron Microscopy: systems and applications*, ed. by W.C. Nixon (IoP, London 1973) p.272

[7.150] E.F. Bond, D. Beresford, G.H. Haggis: Improved cathodoluminescence microscopy. J. Microsc. **100**, 271 (1974)

[7.151] G.A.C. Jones, B.R. Nag, A. Gopinath: Temperature dependence of cathodoluminescence in n-type GaAs. J. Phys. D **7**, 183 (1974)

[7.152] W.R. McKinney, P.V.C. Hough: A new detector system for cathodoluminescence microscopy. SEM 1977/I (ITTRI, Chicago 1977) p.251

[7.153] A. Rasul, S.M. Davidson: SEM measurements of minority carrier lifetimes at dislocations in GaP, employing photon counting. SEM 1977/I (ITTRI, Chicago 1977) p.233

[7.154] L.J. Balk, E. Kubalek: Cathodoluminescence studies of semiconductors in SEM. Beitr. elektr. mikr. Direktabb. Oberfl. **6**, 559 (1973)

[7.155] E.M. Hörl: Verbessertes Ellipsenspiegel-Detektorsystem für die Kathodolumineszenz-Rasterelektronenmikroskopie. Beitr. elektr. mikr. Direktabb. Oberfl. **8**, 369 (1975)

[7.156] L. Carlsson, C.G. van Essen: An efficient apparatus for studying cathodoluminescence in the SEM. J. Phys. E **7**, 98 (1974)

[7.157] A. Ishikawa, F. Mizuno, Y. Uchikawa, S. Maruse: High resolution and spectroscopic cathodoluminescent images in SEM. Jpn. J. Appl. Phys. **12**, 286 (1973)

[7.158] S. Myhajlenko, M. Wong, J.L. Edwards, G.N. Maracas, R.J. Roedel: Spectrally resolved transmission CL evaluation of vertical cavity surface emitting lasers. Scanning Microscopy **6**, 955 (1992)

[7.159] D. Hoder, R. Herbst, A.M. Multier-Lajous: Ein einfaches Verfahren zur Herstellung von Farb-Kathodolumineszenz-Aufnahmen am Rasterelektronenmikroskop. Beitr. elektr. mikr. Direktabb. Oberfl. **12/1**, 273 (1979)

[7.160] G.V. Spivak, G.V. Saparin, M.K. Antoshin: Colour contrast in SEM. (in Russian) Usp. Fiz. Nauk **113**, 695 (1974)

[7.161] E.M. Hörl, F. Buschbeck: Rasterelektronenmikroskopie unter Verwendung eines Farbmonitors. Beitr. elektr. mikr. Direktabb. Oberfl. **8**, 233 (1975)

[7.162] F. Buschbeck, E.M. Hörl: Electronic adding-up and storing of SEM colour images. SEM 1978/I (SEM Inc., AMF O'Hare, IL 1978) p.835

[7.163] S.K. Obyden, G.V. Saparin, G.V. Spivak: Observation of long persistance luminescent materials using colour TV SEM. Scanning **3**, 181 (1980)

[7.164] W. Bröcker, G. Hauck, H. Weigelt, G. Pfefferkorn: Lock-in technique applied to cathodoluminescence of biomedical specimens in the SEM. Scanning **4**, 165 (1981)

[7.165] G.V. Saparin, S.K. Obyden: Colour display of video information in SEM: Principles and applications to physics, geology, soil science, biology and medicine. Scanning **10**, 87 (1988)

[7.166] G.V. Saparin: Cathodoluminescence in SEM. In *Electron Microscopy in Materials Science*, ed. by P.G. Merli and M.V. Antisari (World Scientific, Singapore 1996) p.547

[7.167] K. Löhnert, M. Hastenrath, L. Balk, E. Kubalek: Optische Vielkanalanalyse der Kathodolumineszenz im Rasterelektronenmikroskop. Beitr. elektr. mikr. Direktabb. Oberfl. **11**, 95 (1978)

[7.168] D. Bimberg, H. Münzel, A. Steckenborn, J. Christen: Kinetics of relaxation and recombination of nonequilibrium carriers in GaAs: Carrier capture by impurities. Phys. Rev. B **31**, 7788 (1985)

[7.169] M.D. Muir, P.R. Grant, G. Hubbard, J. Mundell: Cathodoluminescence spectra. SEM 1971 (ITTRI, Chicago 1971) p.403

[7.170] D.B. Wittry, D.F. Kyser: Measurement of diffusion lengths in direct-gap semiconductors by electron-beam excitation. J. Appl. Phys. **38**, 375 (1967)

[7.171] B. Akamatsu, P. Hénoc, A.C. Papadopoulo: Diffusion length measurement in InP and GaAs by filtered cathodoluminescence in a SEM. SEM 1983/IV (SEM Inc., AMF O'Hare, IL 1983) p. 1579

[7.172] T.S. Rao-Sahib, D.B. Wittry: Measurement of diffusion lengths in p-type GaAs by electron beam excitation. J. Appl. Phys. **40**, 3745 (1969)

[7.173] H.C. Casey, J.S. Jayson: Cathodoluminescent measurements in GaP(Zn,O). J. Appl. Phys. **42**, 2774 (1971)

[7.174] G.A.C. Jones, B.R. Nag, A. Gopinath: Temperature variation of cathodoluminescence in direct gap semiconductors. SEM 1973 (ITTRI, Chicago 1973) p.309

[7.175] M. Hastenrath, E. Kubalek: Time-resolved CL in SEM. SEM 1982/I (SEM Inc., AMF O'Hare, IL 1982) p.157

[7.176] M. Hastenrath, L.J. Balk, K. Löhnert, E. Kubalek: Time rsolved CL in the SEM by use of a streak technique. J. Microsc. **118**, 303 (1980)

[7.177] H.C. Casey: Investigation of inhomogeneities in GaAs by electron beam excitation. J. Electrochem. Soc. **114**, 153 (1967)

[7.178] D.B. Holt, B.D. Chase: Scanning-electron-beam-excited charge collection micrography of GaAs lasers. J. Mater. Sci. **3**, 178 (1968)

[7.179] D.A. Shaw, P.R. Thornton: Cathodoluminescent studies of laser quality GaAs. J. Mat. Sci. **3**, 507 (1968)

[7.180] A.L. Esquivel, W.N. Lin, D.B. Wittry: Cathodoluminescence study of plastically deformed GaAs. Appl. Phys. Lett. **22**, 414 (1973)

[7.181] P.R. Grant, S.H. White: Cathodoluminescence and microstructure of quartz overgrowth on quartz. SEM 1978/I (SEM Inc., AMF O'Hare, IL 1978) p.789

[7.182] A. Steckenborn: Minority-carrier lifetime mapping in the SEM. J. Microsc. **118**, 297 (1980)

[7.183] D. Krinsley, N.K. Tovey: Cathodoluminescence in quartz sand grains. SEM 1978/I (SEM Inc., AMF O'Hare, IL 1978) p.887

[7.184] K.L. Pey, J.C.H. Phang, D.S.H. Chan: Investigation of dislocations in GaAs using CL in the SEM. Scanning Microscopy **7**, 1195 (1993)

[7.185] K.L. Pey, J.C.H. Phang, D.S.H. Chan: CL contrast of localized defects. Scanning Microscopy **9**, 355 and 367 (1995)

[7.186] J. Hersener, Th. Ricker: Lumineszenzuntersuchungen an Leuchtstoffen für Farbbildröhren. Beitr. elektr. mikr. Direktabb. Oberfl. **4/2**, 523 (1971)

[7.187] W. Bröcker, E.R. Krefting, L. Reimer: Abhängigkeit des Kathodolumineszenzsignals vom Kippwinkel der Probe im Raster-Elektronenmikroskop. Beitr. elektr. mikr. Direktabb. Oberfl. **10**, 647 (1977)

[7.188] W. Bröcker, E.R. Krefting, L. Reimer: Beobachtung der Strahlenschädigung während des Abrastvorganges im Raster-Elektronenmikroskop mit Hilfe der Kathodolumineszenz. Beitr. elektr. mikr. Direktabb. Oberfl. **7**, 75 (1974)

[7.189] G.V. Saparin, G.V. Spivak: Applications of stroboscopic cathodoluminescence microscopy. SEM 1979/I (SEM Inc., AMF O'Hare, IL 1979) p.267

[7.190] V.I. Petrov, R.S. Gvozdover: Spatial resolution of CL SEM of semiconductors. Scanning **13**, 410 (1991)
[7.191] S. Myhajlenko: LVSEM CL observation of GaAs. Scanning Microscopy **5**, 603 (1991)
[7.192] M. De Mets, A. Lagasse: An investigation of some organic chemicals as cathodoluminescent dyes using the SEM. J. Microsc. **94**, 151 (1971)
[7.193] M. De Mets, K.J. Howlett, A.O. Yoffe: Cathodoluminescent spectra of organic compounds. J. Microsc. **102**, 125 (1974)
[7.194] M. DeMets: Relationship between cathodoluminescence and molecular structure of organic compounds. Microscopica Acta **76**, 405 (1975)
[7.195] R. Herbst, D. Hoder: Cathodoluminescence in biological studies. Scanning **1**, 35 (1978)
[7.196] W. Bröcker, G. Pfefferkorn: Applications of the cathodoluminescence method in biology and medicine. SEM 1979/II (SEM Inc., AMF O'Hare, IL 1979) p.125
[7.197] E.M. Hörl, P. Roschger: CL SEM investigations of biological material at liquid helium and liquid nitrogen temperatures. SEM 1980/II (SEM Inc., AMF O'Hare, IL 1980) p.285

Chapter 8

[8.1] O. Kitakami, T. Sakurai, Y. Shimada: High density recorded patterns observed by high-resolution Bitter SEM method. J. Appl. Phys. **79**, 6074 (1996)
[8.2] O.C. Wells: Some theoretical aspects of type-1 magnetic contrast in the SEM. J. Microsc. **139**, 187 (1985)
[8.3] J.R. Dorsey: Scanning electron probe measurements of magnetic fields. in *Electron Probe Microanalysis*, ed. by A.J. Tousimis and L. Marton: Adv. Electr. Electron Physics, Suppl.6 (Academic, New York 1969) p.291
[8.4] D.C. Joy, J.P. Jacubovics: Direct observation of magnetic domains by SEM. Phil. Mag. **17**, 61 (1968); J. Phys. D **2**, 1367 (1969)
[8.5] G.W. Kammlott: Observation of ferromagnetic domains with the SEM. J. Appl. Phys. **42**, 5156 (1971)
[8.6] J.R. Banbury, W.C. Nixon: The direct observation of domain structure and magnetic fields in the SEM. J. Sci. Instr. **44**, 889 (1967)
[8.7] G.A. Jones: On the quality of type 1 magnetic contrast obtained in the SEM. Phys. Stat. Solidi (a) **36**, 647 (1976)
[8.8] J.R. Banbury, W.C. Nixon: A high-contrast directional detector for the SEM. J. Phys. E **2**, 1055 (1969)
[8.9] P. Gentsch, L. Reimer: Messungen zum magnetischen Kontrast im Rasterelektronenmikroskop. Beitr. elektr. mikr. Direktabb. Oberfl. **5**, 299 (1972)
[8.10] Ji Yuan, R. Senkel, L. Reimer: Recording of magnetic contrast type I by a two-detector system. Scanning **9**, 249 (1987)
[8.11] G.A. Wardly: Magnetic contrast in the SEM. J. Appl. Phys. **42**, 376 (1971)
[8.12] W.K. Chim: An analytical model for SEM type–1 magnetic contrast with energy filtering. Rev. Sci. Instr. **65**, 374 (1994)
[8.13] T. Yamamoto, K. Tsuno: Magnetic contrast in SE images of uniaxial ferromagnetic materials obtained by SEM. Phys. Stat. Solidi (a) **28**, 479 (1975)
[8.14] D.J. Fathers, J.P. Jacubovics, D.C. Joy, D.E. Newbury, H. Yakowitz: A new method of observing magnetic domains by SEM. Phys. Stat. Solidi (a) **20**, 535 (1973); **22**, 609 (1974)

[8.15] D.E. Newbury, H. Yakowitz, R.L. Myklebust: Monte Carlo calculations of magnetic contrast from cubic materials in the SEM. Appl. Phys. Lett. **23**, 488 (1973)
[8.16] T. Yamamoto. H. Nishizawa, K. Tsuno: Magnetic domain contrast in BSE images obtained with a SEM. Phil. Mag. **34**, 311 (1976)
[8.17] K. Tsuno, T. Yamamoto: Observed depths of magnetic domains in high-voltage SEM. Phys. Stat. Solidi (a) **35**, 437 (1976)
[8.18] O.C. Wells: Effect of collector position on type–2 magnetic contrast in the SEM. SEM 1978/I (SEM Inc., AMF O'Hare, IL 1978) p.293
[8.19] O.C. Wells: Effects of collector take-off angle and energy filtering on the BSE image in the SEM. Scanning **2**, 199 (1979)
[8.20] D.E. Newbury: The utility of specimen current imaging in the SEM. SEM 1976/I (ITTRI, Chicago 1976) p.111
[8.21] A.R. Walker, G.R. Booker: A simple energy filtering BSE detector. In *Developments in Electron Microscopy and Analysis*, ed. by J.A. Venables (Academic, London 1976), p.119
[8.22] L. Pogany, K. Ramstock, A. Hubert: Quantitative magnetic contrast – Part I: Experiment. Scanning **14**, 263 (1992)
[8.23] P.J. Grundy, R.S. Tebble: Lorentz electron microscopy. Adv. Physics **17**, 153 (1968)
[8.24] J.N. Chapman, E.M. Waddell, P.E. Batson, R.P. Ferrier: The Fresnel mode of Lorentz microscopy using a STEM. Ultramicroscopy **4**, 283 (1979)
[8.25] L. Marton, S.H. Lachenbruch: Electron optical mapping of electromagnetic fields. J. Appl. Phys. **20**, 1171 (1949)
[8.26] Ch. Schwink: Über neue quantitative Verfahren der elektronenoptischen Schattenmethode. Optik **12**, 481 (1955)
[8.27] R.F.M. Thornley, J.D. Hutchison: Magnetic field measurements in the SEM. Appl. Phys. Lett. **13**, 249 (1968)
[8.28] T. Ishiba, H. Suzuki: Measurements of magnetic field of magnetic recording head by a SEM. Jpn. J. Appl. Phys, **13**, 457 (1974)
[8.29] M.D. Coutts, E.R. Levin: Examination of local magnetic fields by SEM. In *Microscopie Electronique 1970*, Vol.1, ed. by P. Favard (Soc. Française Micr. Electronique, Paris 1970) p.261
[8.30] E.I. Rau, G.V. Spivak: SEM of two-dimensional magnetic stray fields. Scanning **3**, 27 (1980)
[8.31] J.B. Elsbrock, L.J. Balk: Quantitative evaluation of micromagnetic fields by means of a SEM. SEM 1984/I (SEM Inc., AMF O'Hare, IL 1984) p.131
[8.32] J. Kessler: *Polarized Electrons*, 2nd edn., Springer Ser. Atoms Plasmas, Vol.1 (Springer, Berlin, Heidelberg 1985)
[8.33] N. Müller, W. Eckstein, W. Heiland: Electron spin polarization in field emission from EuS-coated tungsten tip. Phys. Rev. Lett. **29**, 1651 (1972)
[8.34] D.T. Pierce, R.J. Celotta, G.C. Wang, W.N. Unertl, A. Galejs, C.E. Kuyatt, S.R. Mielczarek: GaAs spin polarized electron source. Rev. Sci. Instr. **51**, 478 (1980)
[8.35] E. Kisker, W. Gudat, K. Schröder: Observation of a high spin polarization of secondary electrons from single crystal Fe and Co. Sol. State Commun. **44**, 591 (1982)
[8.36] K. Koike, K. Hayakawa: Scanning electron microscope observation of magnetic domains using spin-polarized secondary electrons. Jpn. J. Appl. Phys. **23**, L187 (1984)
[8.37] H.P. Oepen, J. Kirschner: Imaging of magnetic microstructures at surfaces: The scanning electron microscope with spin polarization analysis. Scanning Microscopy **5**, 1 (1991)

[8.38] J. Kirschner, R. Feder: Spin polarization in double diffraction of low-energy electrons from W(001): experiment and theory. Phys. Rev. Lett. **42**, 1008 (1979)

[8.39] J. Kirschner: *Polarized electrons at surfaces* (Springer, Berlin, Heidelberg 1985)

[8.40] D.T. Pierce, S.M. Girvin, J. Unguris, R.J. Celotta: Absorbed current spin polarization detector. Rev. Sci. Instr. **52**, 1437 (1981)

[8.41] H.C. Siegmann, D.T. Pierce, R.J. Celotta: Spin-dependent absorption of electrons in a ferromagnetic metal. Phys. Rev. Lett. **46**, 452 (1981)

[8.42] K. Koike, H. Matsuyama, K. Hayakawa: Spin-polarized SEM equipped with a thumb-sized spin detector. Jpn. J. Appl. Phys. **27**, L1352 (1988)

[8.43] J. Unguris, D.T. Pierce, R.J. Celotta: Low-energy diffuse scattering electron-spin polarization analyzer. Rev. Sci. Instr. **57**, 1314 (1986)

[8.44] J. Unguris, D.T. Pierce, A. Galejs, R.J. Celotta: Spin and energy analysed secondary electron emission from a ferromagnet. Phys. Rev. Lett. **49**, 72 (1982)

[8.45] T. Kohashi, H. Matsuyama, K. Koike: A spin rotator for detecting all three magnetization vector components by spin-polarized SEM. Rev. Sci. Instr. **66**, 5537 (1995)

[8.46] D.T. Pierce, R.J. Celotta: Spin polarization in electron scattering from surfaces. Adv. Electr. Electron Phys. **56**, 219 (1981)

[8.47] K. Koike, K. Hayakawa: Spin polarization due to low-energy electron diffraction at the W(001) surface. Jpn. J. Appl. Phys. **22**, 1332 (1983)

[8.48] S. Kimoto, H. Hashimoto, K. Mase: Voltage contrast in SEM. In *Electron Microscopy 1968*, Vol.1, ed. by D.S. Bocciarelli (Tipografia Poliglotta Vaticana, Rome 1968) p.83

[8.49] P. Gentsch: Potentialkontrast und magnetischer Kontrast im Rasterelektronenmikroskop. Diplomarbeit, Münster 1970

[8.50] J.R. Banbury, W.C. Nixon: Voltage measurement in the SEM. SEM 1970 (ITTRI, Chicago 1970) p.473

[8.51] H. Yakowitz, J.P. Ballantyne, E. Munro, W.C. Nixon: The cylindrical SE detector as a voltage measuring device in the SEM. SEM 1972 (ITTRI, Chicago 1972) p.33

[8.52] G.Y. Robinson, R.M. White: SEM of ferroelectric domains in barium titanate. Appl. Phys. Lett. **10**, 320 (1967)

[8.53] D.G. Coates, N. Shaw: Direct observation of ferroelectric domains in triglycine sulphate using the SEM. In *Microscopie Electronique 1970*, Vol.1, ed. by P. Favard (Soc. Française Micr. Electronique, Paris 1970) p.259

[8.54] R. Le Bihan, M. Maussion: Observation des domaines ferroélectriques au microscope électronique à balayage. Rev. Phys. Appl. **9**, 427 (1974)

[8.55] R. Le Bihan, M. Maussion: Study of the surface of ferroelectric crystals with the SEM. Ferroelectrics **7**, 307 (1974)

[8.56] M. Maussion, R. Le Bihan: Study of ferroelectric domains on KD_2PO_4 and $BaTiO_3$ crystals with the SEM. Ferroelectrics **13**, 465 (1976)

[8.57] Y. Uchikawa, S. Ikeda: Application of SEM to analysis of surface domain structure of ferroelectrics. SEM 1981/I (SEM Inc., AMF O'Hare, IL 1981) p.209

[8.58] R. Godehardt, J. Landgraf: SEM investigations of charge distributions and stray fields of ferroelectric samples. Beitr. elektr. mikr. Direktabb. Oberfl. **25**, 19 (1992)

[8.59] H. Bahadur, R. Parshad: SEM of vibrating quartz crystals – a review. SEM 1980/I (SEM Inc., AMF O'Hare, IL 1980) p.509

[8.60] H.P. Feuerbaum, G. Eberharter, G. Tobolka: Visualization of traveling surface acoustic waves using a SEM. SEM 1980/I (SEM Inc., AMF O'Hare, IL 1980) p.503

[8.61] D.V. Roshchupkin, M. Brunel: SEM visualization of surface acoustic wave propagation in a $LiNbO_3$ crystal. Acustica **81**, 173 (1995)

[8.62] D.V. Roshchupkin, M. Brunel: SEM observation of the voltage contrast image of the ferroelectric domain structure in the $LiNbO_3$ crystal. Scanning Microscopy **7**, 543 (1993)

[8.63] D.V. Roshchupkin, Th. Fournier, M. Brunel, O.A. Plotitsyna, N.G. Sorokin: SEM observation of the interaction between the surface acoustic waves and regular domain structures in the $LiNbO_3$ crystals. Scanning Microscopy **6**, 367 (1992)

[8.64] N.C. MacDonald: Auger electron spectroscopy in SEM: potential measurements. Appl. Phys. Lett. **16**, 76 (1970)

[8.65] W.R. Hardy, S.K. Behera, D. Cavan: A voltage contrast deetctor for the SEM. J. Phys. E **8**, 789 (1975)

[8.66] E. Menzel, E. Kubalek: SE detection systems for quantitative voltage measurements. Scanning **5**, 151 (1983)

[8.67] A.P. Janssen, P. Akhter, C.J. Harland, J.A. Venables: High spatial resolution surface potential measurements using SE. Surf. Sci. **93**, 453 (1980)

[8.68] A. Gopinath, C.C. Sanger: A technique for the linearization of voltage contrast in the SEM. J. Phys. E **4**, 334 (1971)

[8.69] A. Khursheed, A.R. Dinnis: A comparison of voltage contrast detectors. Scanning **6**, 85 (1984)

[8.70] A. Gopinath: Estimate of minimum measurable voltage in the SEM. J. Phys. E **10**, 911 (1977)

[8.71] E. Menzel, E. Kubalek: Fundamentals of electron beam testing of integrated circuits. Scanning **5**, 103 (1983)

[8.72] Y. Petit-Clerc, J.D. Carette: Effect of temperature on surface charges caused by an incident electron beam on a metallic surface. Appl. Phys. Lett. **12**, 227 (1968)

[8.73] D.M. Taylor: The effect of passivation on the observation of voltage contrast in the SEM. J. Phys. D **11**, 2443 (1978)

[8.74] D.L. Crosthwait, F.W. Ivy: Voltage contrast methods for semiconductor device failure analysis. SEM 1974 (ITTRI, Chicago 1974) p.935

[8.75] G.V. Lukianoff, T.R. Touw: Voltage coding : temporal versus spatial frequencies. SEM 1975 (ITTRI, Chicago 1975) p.465

[8.76] K. Nakamae, H. Fujioka, K. Ura: Accurate measurement of the operating frequency in ICs with the SEM. J. Phys. E **21**, 913 (1988)

[8.77] H.P. Feuerbaum, D. Kantz, E. Wolfgang, E. Kubalek: Quantitative measurement with high time resolution of internal waveform on MOS RAMs using a modified SEM. IEEE J. Solid State Circuits **SC-13**, 319 (1978)

[8.78] H. Fujioka, K. Nakamae, K. Ura: Function testing of bipolar ICs and LSIs with the stroboscopic SEM. IEEE J. Solid State Circuits **SC-15**, 177 (1980)

[8.79] H. Fujioka, K. Ura: Waveform measurements on gigahertz semiconductor devices by SEM stroboscopy. Appl. Phys. Lett. **39**, 81 (1981)

[8.80] H. Fujioka, H. Kunizawa, K. Ura: 0.1 ps resolution delay circuit for waveform measurements in the stroboscopic SEM. J. Phys. E **19**, 1025 (1986)

[8.81] F.M. Boland, E.R. Lynch: Analysis of the stroboscopic waveform mode in SEM. J. Phys. E **20**, 1011 (1987)

[8.82] E. Menzel, E. Kubalek: Electron beam test techniques for integrated circuits. SEM 1981/I (SEM Inc., AMF O'Hare, IL 1981) p.305

[8.83] E. Wolfgang: Electron beam testing: problems in practice. Scanning **5**, 71 (1983)

[8.84] Y. Watanabe, Y. Fukuda, T. Jinno: Analysis of capacitive coupling voltage contrast in SEM. Jpn. J. Appl. Phys. **24**, 1294 (1985)

[8.85] E.I. Cole, C.R. Bagnell, B. Davies, A. Neascu, W. Oxford, S. Roy, R.H. Propst: A novel method for depth profiling and imaging of semiconductor devices using capacitive coupling voltage contrast. J. Appl. Phys. **62**, 4909 (1987)

[8.86] W. Reiners, K.D. Herrmann, E. Kubalek: Electron beam testing of passivated devices via capacitive coupling voltage contrast. Scanning Microscopy **2**, 161 (1988)

[8.87] W. Reiners: Fundamentals of electron beam testing via capacitive coupling voltage contrast. Microelect. Eng. **12**, 325 (1990)

[8.88] M. Batinic, S. Görlich, K.D. Herrmann: On geometrical dependencies of capacitive coupling voltage contrast. Microelectr. Eng. **12**, 341 (1990)

[8.89] W. Mertin, K.D. Herrmann, E. Kubalek: The capacitive coupling voltage error and the capacitive coupling cross talk in electron beam testing of passivated IC and measures for their reduction. Microelectr. Eng. **12**, 349 (1990)

[8.90] V.V. Aristov, O. Kononchuk, E.I. Rau, E.B. Yakimov: SEM investigation of semiconductors by the capacitance technique. Microelectr. Eng. **12**, 179 (1990)

[8.91] J. Kirschner: Threshold spectroscopies and prospects for their application in the SEM. SEM 1983/IV (SEM Inc., AMF O'Hare, IL 1983) p.1665

[8.92] H.J. Fitting, D. Hecht: Secondary electron field emission. Phys. Stat. Solidi (a) **108**, 265 (1988)

[8.93] H.J. Fitting, P. Magdanz, W. Mehnert, D. Hecht, Th. Hingst: Charge trap spectroscopy in single and multiple layer dielectrics. Phys. Stat. Solidi (a) **122**, 297 (1990)

[8.94] H.J. Fitting, Th. Hingst, R. Franz, E. Schreiber: Electron trap microscopy – a new mode for SEM. Scanning Microscopy **8**, 165 (1994)

[8.95] E. Zeitler, M.G.R. Thomson: Scanning transmission electron microscopy. Optik **31**, 258 and 359 (1970)

[8.96] L. Reimer, K.H. Sommer: Messungen und Berechnungen zum elektronenmikroskopischen Streukontrast für 17–1200 keV Elektronen. Z. Naturforschg. **23a**, 1569 (1968)

[8.97] U. Golla, B. Schindler, L. Reimer: Contrast in the transmission mode of a LVSEM. J. Microsc. **173**, 219 (1994)

[8.98] J. Weise: Messung des Materialtransportes in dünnen Al-Filmen mit dem Raster-Elektronenmikroskop. Beitr. elektr. mikr. Direktabb. Oberfl. **4/2**, 477 (1971)

[8.99] P. Furrer: Verbindung von Raster- und Durchstrahlungselektronenmikroskopie zur Untersuchung des Ausscheidungsverlaufs in dünnen Folien. Beitr. elektr. mikr. Direktabb. Oberfl. **4/2**, 463 (1971)

[8.100] R. Blaschke: Ein Präparathalter für Durchstrahlungsexperimente und für Stereobildpaare. Beitr. elektr. mikr. Direktabb. Oberfl. **3**, 161 (1970)

[8.101] B.J. Crawford, C.R.W. Liley: A simple transmission stage using the standard collection system in the SEM. J. Phys. E **3**, 461 (1970)

[8.102] J.A. Swift, A.C. Brown, C.A. Saxton: STEM with the Cambridge Stereoscan Mk II. J. Phys. E **2**, 744 (1969)

[8.103] A. Ishikawa, F. Mizuno, Y. Uchikawa, S. Maruse: High resolution and spectroscopic cathodoluminescent images in SEM. Jpn. J. Appl. Phys. **12**, 286 (1973)

[8.104] A.B. Bok: Mirror electron microscopy: theory and applications. In *Modern Diffraction and Imaging Methods in Material Science*, ed. by S. Amelinckx et al. (North-Holland, Amsterdam 1978) p.655

[8.105] A.B. Bok, J.B. LePoole, J. Roos, H. De Lang, H. Bethge, J. Heydenreich, N.E. Barnett: Mirror Electron Microscopy. In *Adv. in Optical and Electron Microscopy*, Vol.4, ed. by R. Barer and V.E. Cosslett (Academic, London 1971) p.161

[8.106] R.E. Ogilvie, M.A. Schippert, S.H. Moll, D.M. Koffman: Scanning electron mirror microscopy. SEM 1969 (ITTRI, Chicago 1969) p.425

[8.107] G.V. Spivak, V.P. Ivannikov, A.E. Luk'yanov, E.I. Rau: Development of scanning mirror electron microscopy for quantitative evaluation of electric microfields. J. Micr. Spectr. Electron. **3**, 89 (1978)

[8.108] J. Witzani, E.M. Hörl: Scanning electron mirror microscopy. Scanning **4**, 53 (1980)

[8.109] R.S. Paden, W.C. Nixon: Retarding field SEM. J. Phys. E **1**, 1073 (1968)

[8.110] A. Boyde, E. Maconnachie: Volume changes during preparation of mouse embryonic tissue for SEM. Scanning **2**, 149 (1979)

[8.111] A. Boyde, F. Franc, E. Maconnachie: Measurements of critical point shrinkage of glutaraldehyde fixed mouse liver. Scanning **4**, 69 (1981)

[8.112] G.J. Campbell, M.R. Roach: Dimensional changes associated with freeze-drying of the internal elastic lamina from cerebral arteries. Scanning **5**, 137 (1983)

[8.113] J.B. Pawley, J.T. Norton: A chamber attached to the SEM for fracturing and coating frozen biological samples. J. Microsc. **112**, 169 (1977)

[8.114] A. Maas: Direct observation and analysis of crystal growth processes in a SEM. In *Electron Microscopy 1974*, Vol.1, ed. by J.V. Sanders and D.J. Goodchild (Australian Acad. of Science, Canberra 1974) p.162

[8.115] U. Finnström: Dynamic studies of the reduction of iron oxides in the SEM. In *Electron Microscopy 1974*, Vol.1, ed. by J.V. Sanders and D.J. Goodchild (Australian Acad. of Science, Canberra 1974) p.164

[8.116] G.D. Danilatos, V.N.E. Robinson: Principles of SEM at high specimen chamber pressures. Scanning **2**, 72 (1979)

[8.117] G.D. Danilatos, R. Postle: The environmental SEM and its application. SEM 1982/I (SEM Inc., AMF O'Hare, IL 1982) p.1

[8.118] G.D. Danilatos: Foundations of environmental SEM. Adv. Electr. Electron Physics **71** 109 (1988)

[8.119] A.N. Farley, J.S. Shah: High-pressure SEM of insulating materials: a new approach. J. Microsc. **164**, 107 (1991)

[8.120] G.D. Danilatos: Bibliography of environmental SEM. Microsc. Res. Techn. **25**, 529 (1993)

[8.121] G.D. Danilatos: Design and construction of an atmospheric or environmental SEM. Scanning **4**, 9 (1981)

[8.122] K. Jost, J. Kessler: Die Ortverteilung mittelschneller Elektronen bei Mehrfachstreuung. Z. Physik **176**, 126 (1963)

[8.123] A.N. Farley, J.S. Shah: Primary considerations for image enhancement in high-pressure SEM. J. Microsc. **158**, 379 and 389 (1990)

[8.124] G.D. Danilatos: Theory of the gaseous detector device in the environmental SEM Adv. Electr. Electron Physics **78** 1 (1990)

[8.125] C. Gilpin, D.C. Sigee: X-ray microanalysis of wet biological specimens in the ESEM. J. Microsc. **179**, 22 (1995)

[8.126] A. Rosencwaig, G. Busse: High resolution photoacoustic thermal-wave microscopy. Appl. Phys. Lett. **36**, 257 (1980)

[8.127] D.G. Davies: Scanning electron acoustic microscopy. SEM 1983/III (SEM Inc., AMF O'Hare, IL 1983) p.1163

[8.128] L.J. Balk, N. Kultscher: Scanning electron acoustic microscopy. Beitr. elektr. mikr. Direktabb. Oberfl. **16**, 107 (1983)

[8.129] G.S. Cargill: Ultrasonic imaging in SEM. Nature **286**, 691 (1980)

[8.130] N. Kultscher, L.J. Balk: Signal generation and contrast mechanisms in scanning electron acoustic microscopy. SEM 1986/I (SEM Inc., AMF O'Hare, IL 1986) p.33

[8.131] M. Domnik, L.J. Balk: Quantitative scanning electron acoustic microscopy of silicon. Scanning Microscopy **7**, 37 (1993)

[8.132] H. Ermert, F.H. Dacol, R.L. Melcher, T. Baumann: Noncontact thermal-wave imaging of subsurface structure with infrared detection. Appl. Phys. Lett. **44**, 1136 (1984)

[8.133] J.F. Bresse: Scanning electron acoustic microscopy and SEM imaging of III-V compound devices. Scanning Microscopy **5**, 961 (1991)

[8.134] A. Rosencwaig, R.M. White: Imaging of dopant regions in silicon with thermal-wave electron microcopy. Appl. Phys. Lett. **38**, 165 (1981)

[8.135] B.Y. Zhang, F.M. Jiang, Y. Yang, Q.R. Yin, S. Kojima: Electron acoustic imaging of $BaTiO_3$ single crystals. J. Appl. Phys. **80**, 1916 (1996)

[8.136] J.F. Bresse: Electron acoustic signal of metallic layers over a semiconductor substrate. Scanning Microscopy **7**, 523 (1993)

[8.137] M. Hatzakis: Lithographic processes in VLSI circuit fabrication. SEM 1979/I (SEM Inc., AMF O'Hare, IL 1979) p.275

[8.138] D. Stephani: Monte Carlo calculation of backscattered electrons at registration marks. J. Vac. Sci. Techn. **16**, 1739 (1979)

[8.139] K. Murata: Monte Carlo simulation of electron scattering in resist film/substrate targets. In *Electron Beam Interactions with Solids*, ed. by D.F. Kyser et al. (SEM Inc., AMF O'Hare, IL 1982) p.311

[8.140] D.F. Kyser: Monte Carlo simulation of spatial resolution limits in electron beam lithography. In *Electron Beam Interactions with Solids*, ed. by D.F. Kyser et al. (SEM Inc., AMF O'Hare, IL 1982) p.331

[8.141] M. Kisza, Z. Maternia, Z. Radzimski: Backscattering of electrons from complex structures. In *Electron Beam Interactions with Solids*, ed. by D.F. Kyser et al. (SEM Inc., AMF O'Hare, IL 1982) p.109

[8.142] S.A. Rishton, D.P. Kern: Point exposure distribution measurements for proximity correction in electron beam lithography on a sub-100 nm scale. J. Vac. Sci. Techn. B **5**, 135 (1987)

[8.143] S.J. Wind, M.G. Rosenfield, G. Pepper, W.W. Molzen, P.D. Gerber: Proximity correction for electron beam lithography using a three-Gaussian model of the electron energy distribution. J. Vac. Sci. Techn. B **7**, 1507 (1989)

[8.144] G. Owen: Methods for proximity effect correction in electron lithography. J. Vac. Sci. Techn. B **8** 1889 (1990)

[8.145] T.H.P. Chang: Proximity effect in electron-beam lithography. J. Vac. Sci. Techn. **12**, 1271 (1975)

[8.146] M. Parikh: Corrections to proximity effects in electron beam lithography. J. Appl. Phys. **50**, 4371 (1979)

[8.147] G. Owen, P. Rissman: Proximity effect correction in electron beam lithography by equalization of background dose. J. Appl. Phys. **54**, 3573 (1983)

[8.148] M.G.R. Thomson: Incident dose modification for proximity effect correction. J. Vac. Sci. Techn. B **11**, 2768 (1993)

[8.149] H. Eisenmann, Th. Waas, H. Hartmann: PROXECCO – proximity effect correction by convolution. J. Vac. Sci. Techn. B **11**, 2741 (1993)

[8.150] T.R. Groves: Efficiency of electron-beam proximity effect correction. J. Vac. Sci. Techn. B **11**, 2746 (1993)

[8.151] B.D. Cook, S.Y. Lee: Fast proximity effect correction. An extension of PYRAMID for thicker resists. J. Vac. Sci. Techn. B **11**, 2762 (1993)

[8.152] S.W.J. Kuan, C.W. Frank, Y.H. Yen Lee, T. Eimori, D.R. Allee, R.F.W. Pease, R. Browning: Ultrathin poly(methylmethacrylate) resist films for microlithography. J.Vac. Sci. Techn. B **7**, 1745 (1989)

[8.153] J.B. Kruger, P. Rissman, M.S. Chang: Silicon transfer layer for multi-layer resist systems. J. Vac. Sci. Techn. **19**, 1320 (1981)

[8.154] M. Angelopoulos, J.M. Shaw, R.D. Kaplan, S. Perreault: Conducting polyanilines: Discharge layers for electron-beam lithography. J. Vac. Sci. Techn. B **7**, 1519 (1989)

[8.155] T.J. Stark, T.M. Mayer, D.P. Griffis, P.E. Russell: Electron beam induced metalization of palladium acetate. J. Vac. Sci. Techn. B **9**, 3475 (1991)

[8.156] K.L. Lee, M. Hatzakis: Direct electron-beam patterning for nanolithography. J. Vac. Sci. Techn. B **7** 1941 (1989)

[8.157] M. Yasuda, H. Kawata, K. Murata, K. Hashimoto, Y. Hirai, N. Nomura: Resist heating effect in electron beam lithography. J. Vac. Sci. Techn. B **12**, 1362 (1994)

[8.158] L. Reimer, D. Stelter: Monte Carlo calculations of electron emission at surface edges. Scanning Microscopy **1**, 951 (1987)

[8.159] S. Hasegawa, T. Hidaka: Three-dimensional Monte Carlo calculation by a supercomputer. J. Vac. Sci. Techn. B **5**, 142 (1987)

[8.160] D.C. Joy: Image simulation for secondary electron micrographs in the SEM. Scanning Microscopy **2**, 57 (1988)

[8.161] T.E. Allen, R.R. Kunz, T.M. Mayer: Monte Carlo calculation of low-energy electron emission from surfaces. J. Vac. Sci. Techn. B **6**, 2057 (1988)

[8.162] S. Takeuchi, H. Nakamura, Y. Watakabe: A study of electron beam metrology using computer simulation. J. Vac. Sci. Techn. B **7**, 73 (1989)

[8.163] X. Wang, D.C. Joy: A new high-speed simulation method for electron-beam critical dimension metrology profile modeling. J. Vac. Sci. Techn. B **9**, 3573 (1991)

[8.164] K.S. Maher: Techniques for LVSEM linewidth measurements. Scanning Microscopy **7**, 65 (1993)

[8.165] M.T. Postek: Low accelerating voltage SEM imaging and metrology using backscattered electrons. Rev. Sci. Instr. **61**, 3750 (1990)

[8.166] M.G. Rosenfield: Measurement techniques for submicron resist images. J. Vac. Sci. Techn. B **6**, 1944 (1988)

[8.167] M.G. Rosenfield: Overlay measurement using the LVSEM. Microelectr. Eng. **17**, 439 (1992)

[8.168] M.T. Postek: SEM-based metrological SEM magnification standard. Scanning Microscopy **3**, 1087 (1989)

[8.169] L. Grella, E. Di Fabricio, M. Goutili, M. Baciocchi, L. Mastrogiacomo, R. Maggiora, L. Capodicci: Secondary electron line scans over high resolution resist images: theoretical and experimental investigation of induced local electric field effects. J. Vac. Sci. Techn. B **12**, 3555 (1994)

[8.170] W. Liu, J. Ingino, R.F. Pease: Resist charging in electron beam lithography. J. Vac. Sci. Techn. B **13**, 1979 (1995)

[8.171] J. Kato, H. Morita, K. Saito, N. Shimazu: Beam position stabilization by suppression of electrons reentering the electron-beam column. J. Vac. Sci. Techn. B **13**, 2450 (1995)

[8.172] F. Mizuno, S. Yamada: Effect of electron-beam parameters on critical-dimension measurements. J. Vac. Sci. Techn. B **13**, 2682 (1995)

[8.173] E. Di Fabricio, L. Grella, M. Gentili, M. Baciocchi, L. Mastrogiacomo, R. Maggiora: Nanometer metrology by means of BSE. J. Vac. Sci. Techn. B **13**, 321 (1995)

[8.174] G. Matsuoka, M. Ichihasi, H. Marukoshi, K. Yamamoto: Automatic electron beam metrology system for development of very large-scale integrated devices. J. Vac. Sci. Techn. B **5**, 79 (1987)

[8.175] J.E. Griffith. D.A, Grigg, M.J. Vasile, P.E. Russell, E.A. Fitzgerald: Characterization of scanning probe microscope tips for linewidth measurement. J. Vac. Sci. Techn. B **9**, 3586 (1991)

[8.176] K.L. Lee, D.W. Abraham, F. Secord, L. Landstein: Submicron Si trench profiling with an electron-beam fabricated atomic force microscope tip. J. Vac. Sci. Techn. B **9**, 3562 (1991)

[8.177] R.J. Behm, N. Garcia, H. Rohrer (eds.): *Scanning Tunneling Microscopy and Related Methods*, NATO ASI Series E: Applied Sciences, Vol.184 (Kluwer, Dordrecht 1990)

[8.178] H.J. Güntherodt, R. Wiesendanger (eds.): *Scanning Tunneling Electron Microscopy I*, 2nd edn., Springer Ser. in Surface Sciencs, Vol. 20 (Springer, Berlin, Heidelberg 1994); R. Wiesendanger, H.J. Güntherodt (Eds.): *Scanning Tunneling Microscopy* II and III, 2nd edn., Springer Surf. Sci., Vols. 28 and 29 (Springer, Berlin, Heidelberg 1995 and 1996)

[8.179] D.A. Bonnell (ed.): *Scanning Tunneling Microscopy and Spectroscopy* (VCH Publishers, New York 1993)

[8.180] R. Wiesendanger: *Scanning Probe Microscopy and Spectroscopy* (Cambridge Univ. Press, Cambridge 1994)

[8.181] M. Anders, M. Mück, C. Heiden: SEM/STM combination for STM tip guidance. Ultramicroscopy **25**, 123 (1988)

[8.182] L. Vázquez, A. Bartolomé, R. Garcia, A. Buendia, A.M. Baró: Combination of a STM with a SEM. Rev. Sci. Instr. **59**, 1286 (1988)

[8.183] H. Fuchs, R. Laschinski: Surface investigations with a combined scanning electron – scanning tunneling microscope. Scanning **12**, 126 (1990)

[8.184] A.O. Golubok, V.A. Timofeev: STM combined with SEM without SEM capability limitations. Ultramicroscopy **42–44**, 1558 (1992)

[8.185] M. Troyon, H.N. Lei, A. Bourhettar: Integration of an STM in an SEM. Ultramicroscopy **42–44**, 1564 (1992)

[8.186] K. Nakamoto, K. Uozumi: A compact STM compatible with conventional SEM. Ultramicroscopy **42–44**, 1569 (1992)

Chapter 9

[9.1] P.B. Hirsch, A.Howie, R.B. Nicholson, D.W. Pashley, M.J. Whelan: *Electron Microscopy of Thin Crystals* (Butterworths, London 1965)

[9.2] S. Amelinckx, R. Gevers, G. Remaut, J. Van Landuyt: *Modern Diffraction and Imaging Techniques in Material Science* (North-Holland, Amsterdam 1970)

[9.3] L. Reimer: *Transmission Electron Microscopy. Physics of Image Formation and Microanalysis.* Springer Ser. Opt. Sci., Vol. 36, 4th ed. (Springer, Berlin, Heidelberg 1997)

[9.4] B. Vainshtein: *Fundamanetals of Crystals*, 2nd edn., Modern Crystallography 1, (Springer, Berlin, Heidelberg 1994)

[9.5] B. Vainshtein, V.M. Fridkin, V.L. Indenbom: *Modern Crystallography II.* Springer Ser. Solid-State Sci., Vol. 21 (Springer, Berlin, Heidelberg 1982)

[9.6] C.T. Young, J.L. Lytton: Computer generation and identification of Kikuchi projections. J. Appl. Phys. **43**, 1408 (1972)

[9.7] J.P. Spencer, C.J. Humphreys: Electron diffraction from tilted specimens and its application in SEM. In *Electron Microscopy and Analysis*, ed. by W.C. Nixon (IoP, London 1971) p.310

[9.8] P. Pirouz, L.M. Boswarva: Pseudo-Kikuchi pattern contrast from tilted specimens. in *Scanning Electron Microscopy: systems and applications* (IoP, London 1973) p.238

[9.9] P. Hagemann, L. Reimer: An experimental proof of the dependent Bloch wave model by large-angle electron scattering from thin crystals. Phil. Mag. A 40, 367 (1979)

[9.10] L. Reimer: Electron diffraction methods in TEM, STEM and SEM. Scanning **2**, 3 (1979)

[9.11] G.R. Booker, A.M.B. Shaw, M.J. Whelan, P.B. Hirsch: Some comments on the interpretation of the Kikuchi-like reflection patterns observed by SEM. Phil. Mag. **16**, 1185 (1967)

[9.12] E. Vicario, M. Pitaval, G. Fontaine: Etude des pseudo-lignes de Kikuchi observeés en microscopie électronique à balayage. Acta Cryst. A **27**, 1 (1971)

[9.13] L. Reimer, H.G. Badde, H. Seidel: Orientierungsanisotropie des Rückstreukoeffizienten und der Sekundärelektronenausbeute von 10–100 keV Elektronen. Z. Angew. Phys. **31**, 145 (1971)

[9.14] H.G. Badde, H. Drescher, E.R. Krefting, L. Reimer, H. Seidel, W. Bühring: Use of Mott scattering cross sections for calculating backscattering of 10–100 keV electrons. In *Proc. 25th Anniv. Meeting EMAG* (IoP, London 1971) p.74

[9.15] H. Drescher, E.R. Krefting, L. Reimer, H. Seidel: The orientation dependence of the electron backscattering coefficient of gold single crystal films. Z. Naturforschg. **29a**, 833 (1974)

[9.16] J.P. Spencer, C.J. Humphreys, P.B. Hirsch: A dynamical theory for the contrast of perfect and and imperfect crystals in the SEM using backscattered electrons. Phil. Mag **26**, 193 (1972)

[9.17] R. Sandström, J.F. Spencer, C.J. Humphreys: A theoretical model for the energy dependence of electron channelling patterns in SEM. J. Phys. D 7, 1030 (1974)

[9.18] E.M. Schulson: Interpretation of the widths of SEM electron channelling lines. Phys. Stat. Solidi (b) **46**, 95 (1971)

[9.19] H. Seiler, G. Kuhnle: Zur Anisotropie der Elektronenausbeute in Abhängigkeit von der Energie der auslösenden Primärelektronen von 5–50 keV. Z. Angew. Phys. **29**, 254 (1970)

[9.20] H. Niedrig: Electron backscattering from thin films. J. Appl. Phys. **53**, R15 (1982)

[9.21] M. von Laue: *Materiewellen und ihre Interferenzen* (Akad. Verlagsges., Leipzig 1948)

[9.22] Y. Kainuma: The theory of Kikuchi pattern. Acta Cryst. **8**, 247 (1955)

[9.23] R.E. De Wames, W.F. Hall, G.W. Lehman: Mass dependence of the angular dependence of charged particle emission from crystals: transition to the classical limit. Phys. Rev. **174**, 392 (1968)

[9.24] D.S. Gemmell: Channelling and related effects in the motion of charged particles through crystals. Rev. Mod. Phys. **46**, 129 (1974)

[9.25] H. Boersch: Über Bänder bei Elektronenbeugung. Z. Physik **38**, 1000 (1937)

[9.26] C.R. Hall: On the thickness dependence of Kikuchi band contrast. Phil. Mag. **22**, 63 (1970)

[9.27] M. Komura, S. Kojima, T. Ichinokawa: Contrast reversals of Kikuchi bands in transmission electron diffraction. J. Phys. Soc. Jpn. **33**, 1415 (1972)

[9.28] L. Reimer, W. Pöpper, B. Volbert: Contrast reversals in the Kikuchi bands of backscattered and transmitted electron diffraction patterns. in *Developments in Electron Microscopy and Analysis*, ed. by D.L. Misell (IoP, London 1977) p.259

[9.29] M.N. Alam, M. Blackman, D.W. Pashley: High-angle Kikuchi patterns. Proc. Roy. Soc. A **221**, 224 (1954)

[9.30] C.W.B. Grigson: Improved scanning electron diffraction system. Rev. Sci. Instr. **36**, 1587 (1965)

[9.31] M.F. Tompsett: Scanning high-energy electron diffraction (SHEED) in materials science. J. Mater. Sci. **7**, 1069 (1972)

[9.32] T. Ichinokawa, Y. Ishikawa, M. Kemmochi, N. Ikeda, Y. Hosokawa, J. Kirschner: Low energy SEM combined with LEED. Surf. Sci. **176**, 397 (1986)

[9.33] J. Kirschner, T. Ichinokawa, Y. Ishikawa, M. Kemmochi, N. Ikeda, Y. Hosokawa: LEED with microscopic resolution. SEM 1986/II (SEM Inc., AMF O'Hare, IL 1986) p.331

[9.34] D.G. Coates: Kikuchi-like reflection patterns obtained with the SEM. Phil. Mag. **16**, 1179 (1967)

[9.35] G.R. Booker: Scanning electron microscopy: electron channelling effects, In *Modern Diffraction and Imaging Techniques in Material Science* ed. by S. Amelinckx et al. (North-Holland, Amsterdam 1970) p.613

[9.36] D.C. Joy: Electron channelling patterns in the SEM. In *Quantitative Electron Microscopy*, ed. by D.B. Holt et al. (Academic, London 1974) p.131

[9.37] D.E. Newbury: The origin, detection, and uses of electron channelling contrast. SEM 1974 (ITTRI, Chicago 1974) p.1047

[9.38] D.C. Joy, D.E. Newbury, D.L. Davidson: Electron channelling pattern in the SEM. J. Appl. Phys. **53**, R81 (1982)

[9.39] E.M. Schulson, C.G. van Essen: Optimum conditions for generating channelling patterns in the SEM. J. Phys. E **2**, 247 (1969)

[9.40] E.D. Wolf, T.E. Everhart: Annular diode detector for high annular resolution pseudo-Kikuchi patterns. SEM 1969 (ITTRI, Chicago 1969) p.41

[9.41] D.G. Coates: Pseudo-Kikuchi orientation analysis in the SEM. SEM 1969 (ITTRI, Chicago 1969) p.27

[9.42] E.M. Schulson, C.G. van Essen, D.C. Joy: The generation and application of SEM electron channeling patterns. SEM 1969 (ITTRI, Chicago 1969) p.45

[9.43] J. Frosien, W. Gaebler, H. Niedrig: New display and specimen stage for large angle Kikuchi-like patterns. in *Electron Microscopy 1974*, Vol.1 ed. by J.V. Sanders and D.J. Goodchild (Australian Acad. of Science, Canberra 1974) p.158

[9.44] M. Brunner, H.J. Kohl, H. Niedrig: Großwinkel-Elektronen-Channeling-Diagramme zur Untersuchung epitaktisch hergestellter Schichten. Optik **49**, 477 (1978)

[9.45] A.R. Dinnis: Limiting factors in direct stereo viewing. In *Scanning Electron Microscopy: systems and applications* (IoP, London 1973) p.76

[9.46] A.R. Dinnis: Stereoscopic viewing in the SEM. In *Developments in Electron Microscopy and Analysis*, ed. by D.L. Misell (IoP, London 1977) p.87

[9.47] C.G. van Essen, E.M. Schulson: Selected area channelling patterns in the SEM. J. Mater. Sci. **4**, 336 (1969)

[9.48] C.G. van Essen, E.M. Schulson, R.H. Donaghay: The generation and identification of SEM channelling patterns from 10 μm areas. J. Mater. Sci. **6**, 213 (1971)

[9.49] G.R. Booker, R. Stickler: SEM selected-area channelling patterns: dependence of area on rocking angle and working distance. J. Mater. Sci. **7**, 712 (1972)

[9.50] C.G. van Essen: Selected area diffraction in the SEM – towards 1 micron. In *Proc. 25th Anniv. Meeting of EMAG* (IoP, London 1971) p.314

[9.51] D.C. Joy, D.E. Newbury: SEM selected area channelling patterns from 1 micron specimen areas. J. Mater. Sci. **7**, 714 (1972)

[9.52] H. Seiler: Determination of the "information depth" in the SEM. SEM 1967/I (ITTRI, Chicago 1967) p.9

[9.53] E.D. Wolf, M. Braunstein, A.I. Braunstein: Pseudo-Kikuchi pattern degradation by a thin amorphous silicon film. Appl.Phys. Lett. **15**, 389 (1969)

[9.54] S.M. Davidson, G.R. Booker: Decollimation of a parallel electron beam by thin surface films and its effect on SEM channelling patterns. In *Microscopie Electronique 1970*, Vol.1, ed. by P. Favard (Soc. Française Micr. Electronique, Paris 1970) p.235

[9.55] T.F. Page, C.J. McHargue, C.W. White: SEM electron channelling patterns as a technique for the characterization of ion-implantation damage. J. Microsc. **163**, 245 (1991)

[9.56] M. Hoffmann, L. Reimer: Channelling contrast on metal surfaces after ion beam etching. Scanning **4**, 91 (1981)

[9.57] E.M. Schulson: A SEM study of the degradation of electron channelling effects in alkali halide crystals during electron irradiation. J. Mater. Sci. **6**, 377 (1971)

[9.58] E.M. Schulson: SEM electron channelling line width (broadening) and pattern degradation in alkali halide crystals. SEM 1971 (ITTRI, Chicago 1971) p.489

[9.59] A.D.G. Stewart: Recent developments in SEM. Beitr. elektr. mikr. Direktabb. Oberfl. 1, 283 (1968)

[9.60] A. Boyde: Practical problems and methods in the 3D analysis of SEM images. SEM 1970 (ITTRI, Chicago 1970) p.105

[9.61] S. Murray, A.H. Windle: Characterisation and correction of distortions in SEM micrographs. In *Scanning Electron Microscopy: systems and applications* (IoP, London 1973) p.88

[9.62] J.D. Verhoeven, E.D. Gibson: Rotation between SEM micrograph and electron channelling patterns. J. Phys. E **8**, 15 (1975)

[9.63] D.C. Joy, C.M. Maruszewski: The rotation between selected area channelling patterns and micrographs in the SEM. J. Mater. Sci. **10**, 178 (1975)

[9.64] D.L. Davidson: Rotation between SEM micrographs and electron channelling patterns. J. Phys. E **9**, 341 (1976)

[9.65] D.E. Newbury, D.C. Joy: A computer technique for the analysis of electron channelling patterns. In *Proc. 25th Anniv. Meeting of EMAG* (IoP, London 1971) p.306

[9.66] D.C. Joy, G.R. Booker, E.O. Fearon, M. Bevis: Quantitative crystallographic orientation determinations of microcrystals present on solid specimens using the SEM. SEM 1971 (ITTRI, Chicago 1971) p.497

[9.67] G.R. Booker: Electron channelling effects using the SEM. SEM 1970 (ITTRI, Chicago 1970) p.489

[9.68] J.D. Ayers, D.C. Joy: A crystallographic study of massive precipitates in Cu-Zn and Ag-Zn alloys utilizing selected area electron channelling. Acta Met. **20**, 1371 (1972)

[9.69] D.E. Newbury, D.C. Joy: SEM dynamical studies of the deformation of Pb-Sn superplastic alloys. In *Proc. 25th Anniv. Meeting of EMAG* (IoP, London 1971) p.216

[9.70] D.C. Joy, E.M. Schulson, J.P. Jacubovics, C.G. van Essen: Electron channelling patterns from ferromagnetic crystals in the SEM. Phil. Mag. **20**, 843 (1969)

[9.71] L.F. Solovsky, D.R. Beaman: A simple method for determining the acceleration potential in electron probes and SEM. Rev. Sci. Instr. **43**, 1100 (1972)

[9.72] L. Reimer, H. Seidel, R. Blaschke: Energieabhängigkeit der Feinstruktur eines Channelling-Diagrammes am Beispiel des 111-Poles von Silizium. Beitr. elektr. mikr. Direktabb. Oberfl. **4/2**, 289 (1971)

[9.73] R. Stickler, C.W. Hughes, G.R. Booker: Application of the SA-ECP method to deformation studies. SEM 1971 (ITTRI, Chicago 1971) p.473

[9.74] J.P. Spencer, G.R. Booker, C.J. Humphreys, D.C. Joy: Electron channelling patterns from deformed crystals. SEM 1974 (ITTRI, Chicago 1974) p.919

[9.75] D.L. Davidson: A method for quantifying electron channelling pattern degradation due to material deformation. SEM 1974 (ITTRI, Chicago 1974) p.927

[9.76] R.C. Farrow, D.C. Joy: Measurements of electron channelling pattern linewidths in silicon. Scanning **2**, 249 (1979)

[9.77] J.A. Venables, C.J. Harland: Electron back-scattering patterns – a new technique for obtaining crystallographic information in the SEM. Phil. Mag. **27**, 1193 (1973)

[9.78] J.A. Venables, R. Bin-Jaya: Accurate microcrystallography using electron back-scattering patterns. Phil. Mag. **35**, 1317 (1977)

[9.79] D.J. Dingley: Diffraction from sub-micron areas using electron backscattering in a SEM. SEM 1984/II (SEM Inc., AMF O'Hare, IL 1984) p.569

[9.80] L. Reimer, D. Grün: Zur Aufzeichnung von Elektronen-Rückstreudiagrammen (EBSP) in einem SEM. Beitr. elektr. mikr. Direktabb. Oberfl. **19**, 105 (1986)

[9.81] J.A. Venables, C.J. Harland, R. bin-Jaya: Crystallographic orientation determination in the SEM using electron backscattering patterns and channel plates. In *Developments in Electron Microscopy and Analysis*, ed. by J.A. Venables (IoP, London 1976) p.101

[9.82] D.L. Barr, W.L. Brown: A channel plate detector for electron backscatter diffraction. Rev. Sci. Instr. **66**, 3480 (1995)

[9.83] C.J. Harland, P. Akhter, J.A. Venables: Accurate microcrystallography at high spatial resolution using electron back-scattering patterns in a field emission gun SEM. J. Phys. E **14**, 175 (1981)

[9.84] K.Z. Baba-Kishi: A study of directly recorded RHEED and BKD patterns in the SEM. Ultramicroscopy **34**, 205 (1990)

[9.85] F. Ouyang, O.L. Krivanek: Slow-scan CCD observation of backscattering patterns in SEM. In *Proc. 50th Ann. Meeting EMSA* (San Francisco Press, San Francisco 1982) p. 1308

[9.86] D.J. Dingley, K. Baba-Kishi: Use of electron backscatter diffraction patterns for determination of crystal symmetry elements. SEM 1986/II (SEM Inc., AMF O'Hare, IL 1986) p.383

[9.87] K. Baba-Kishi, D.J. Dingley: Backscatter Kikuchi diffraction in the SEM for identification of crystallographic point groups. Scanning **11**, 305 (1989)

[9.88] K.Z. Baba-Kishi: Use of Kikuchi line intersections in crystal symmetry determination : application to chalcopyrite structure. Ultramicroscopy **36**, 355 (1991)

[9.89] K. Marthinsen, R. Høier. On the breakdown of Friedel's law in electron backscattering channelling patterns. Acta Cryst. A **44**, 700 (1988)

[9.90] N.H. Schmidt, J.B. Bilde Sørensen, P.J. Jensen: Band position used for on-line crystallographic orientation determination from EBSP. Scanning Microscopy **5**, 637 (1991)

[9.91] N.C.K Lassen, D.J. Jensen, K. Conradsen: Image processing procedure for analysis of EBSP. Scanning Microscopy **6**, 115 (1992)

[9.92] T. Ichinokawa, M. Nishimura, H. Wada: Contrast reversals of pseudo-Kikuchi bands and lines due to detector position in SEM. J. Phys. Soc. Jpn. **36**, 221 (1974)

[9.93] M. Pitaval, P. Morin, J. Baudry, E. Vicario, G. Fontaine: Advances in crystalline contrast from defects. SEM 1977/I (ITTRI, Chicago 1977) p.439

[9.94] L. Reimer, U. Heilers, G. Saliger: Kikuchi band contrast in diffraction patterns recorded by transmitted and backscattered electrons. Scanning **8**, 101 (1986)

[9.95] N.C. Krieger Lassen, J.B. Bilde-Sørensen: Calibration of an electron backscattering pattern set-up. J. Microsc. **170**, 125 (1993)

[9.96] W. Kossel, H. Voges: Röntgeninterferenzen an der Einkristallantikathode. Ann. Phys. **23**, 677 (1935)

[9.97] W. Kossel: Zur Systematik der Röntgenreflexe eines Raumgitters. Ann. Phys. **25**, 512 (1936); Messungen am vollständigen Reflexsystem eines Kristallgitters. Ann. Phys. **26**, 533 (1936)

[9.98] M. Bevis, N. Swindells: The determination of the orientation of microcrystals using back-reflection Kossel technique and an electron probe microanalyser. Phys. Stat. Solidi **20**, 197 (1967)

[9.99] M. Bevis, E.O. Fearon, P.C. Rowlands: The accurate determination of lattice parameters and crystal orientations from Kossel patterns. Phys. Stat. Solidi (a) **1**, 653 (1970)

[9.100] H. Yakowitz: The divergent beam x-ray technique. In *Electron Probe Microanalysis*, Suppl.IV, Adv. Electr. Electron Phys. (Academic, New York 1969) p.361

[9.101] R. Tixier, C. Wachè: Kossel patterns. J. Appl. Cryst. **3**, 466 (1970)

[9.102] H. Yakowitz: Role of divergent beam (Kossel) x-ray technique in SEM. In *Quantitative Scanning Electron Microscopy*, ed. by D.B. Holt et al. (Academic, London 1974) p.451

[9.103] D.J. Dingley, J.W. Steeds: Application of the Kossel x-ray back reflection technique in the SEM. In *Quantitative Scanning Electron Microscopy*, ed. by D.B. Holt et al. (Academic, London 1974) p.487

[9.104] D.J. Dingley: Theory and application of Kossel x-ray diffraction in the SEM. Scanning **1**, 79 (1978)

[9.105] J. Hejna: Divergent-beam x-ray diffraction in the SEM and its use for the study of the semiconductor epitaxial layers. SEM 1985/III (SEM Inc., AMF O'Hare, IL 1985) p.1103

[9.106] J. Hejna, E.B. Radojewska, H. Szymánski, M. Wolcyrz: Determination of the lattice mismatch in heterostructures by x-ray diffraction in the SEM. Scanning **8**, 177 (1986)

[9.107] N. Swindells, J.C. Ruckman: A new Kossel camera design concept for the SEM. In *Scanning Electron Microscopy: systems and applications* (IoP, London 1973) p.303

[9.108] S. Biggin, D.J. Dingley: A general method for locating the x-ray source point in Kossel diffraction. J. Appl. Cryst. **10**, 376 (1977)

Chapter 10

[10.1] B.K. Arawal: *X-Ray Spectroscopy, an Introduction*, 2nd edn., Springer Ser. Opt. Sci. Vol. 15 (Springer, Berlin, Heidelberg 1991)

[10.2] R.H. Pratt, H.K. Tseng, C.M. Lee, L. Kissel: Bremsstrahlung energy spectra from electrons of kinetic energy 1 keV $\leq T_1 \leq$ 2000 keV incident on neutral atoms $2 \leq Z \leq 92$. Atomic Data and Nuclear Data Tables **20**, 175 (1977)

[10.3] H.K. Tseng, R.H. Pratt, C.M. Lee: Electron bremsstrahlung angular distributions in the 1–500 keV energy range. Phys. Rev. A **19**, 187 (1979)

[10.4] A. Sommerfeld: Über die Beugung und Bremsung der Elektronen. Ann. Phys. **11**, 257 (1931)

[10.5] P. Kirkpatrick, L. Wiedmann: Theoretical continuous x-ray energy and polarization. Phys. Rev. **67**, 321 (1945)

[10.6] H.H. Kramers: On the theory of x-ray absorption and of the continuous x-ray spectrum. Phil. Mag. **46**, 836 (1923)

[10.7] L. Reimer, P. Bernsen: Total rate imaging with x-rays in a SEM. SEM 1984/IV (SEM Inc., AMF O'Hare, IL 1984) p.1707

[10.8] H. Kulenkampff: Das kontinuierliche Röntgenspektrum. In *Handbuch der Physik*, Bd. 23/2, ed. by H. Geiger and K. Scheel (Springer, Berlin 1933) p.142

[10.9] S.T. Stephenson: The continuous x-ray spectrum. In *Encyclopedia of Physics*, Vol.30, ed. by S. Flügge (Springer, Berlin 1957) p.337

[10.10] W.L. Baun: Changes in x-ray emission spectra observed between pure elements in combination with others to form compounds or alloys. Adv. Electr. Electron Phys. Suppl. **6**, 155 (1969)

[10.11] A. Faessler, M. Goehring: Röntgenspektrum und Bindungszustand. Die Kα-Fluoreszenzstrahlung des Schwefels. Naturwiss. **39**, 169 (1952)

[10.12] E.H.S. Burhop: Le rendement de fluorescence. J. Phys. Radium **16**, 625 (1965)

[10.13] W. Bambynek, B. Crasemann, R.W. Fink, H.U. Freund, H. Mark, C.D. Swift, R.E. Price, P.V. Rao: X-ray fluorescence yields, Auger, and Coster–Kronig transition probabilities. Rev. Mod. Phys. **44**, 716 (1972)

[10.14] K.F.J. Heinrich: *Electron Beam X-Ray Microanalysis* (Van Nostrand, New York 1981)

[10.15] N.A. Dyson: *X-Rays in Atomic and Nuclear Physics* (Longman, London 1973)

[10.16] K.F.J. Heinrich: X-ray absorption uncertainty. In *The Electron Microprobe*, ed. by T.D. McKinley et al. (Wiley, New York) p.296

[10.17] J.Z. Frazer: A computer fit to mass absorption coefficient data. Rep. S.I.O. 67–29, Inst. for the Study of Matter (Univ. of California, La Jolla 1967)

[10.18] G. Springer, B. Nolan: Mathematical expression for the evaluation of x-ray emission and critical energies, and of mass absorption coefficients. Canad. J. Spectr. **21**, 134 (1976)

[10.19] R. Theisen: *Quantitative Electron Microprobe Analysis* (Springer, Berlin, Heidelberg 1965)

[10.20] W.J. Veigele: Photon cross-sections from 0.1 keV to 1 MeV for elements Z=1 to Z=94. Atomic Data Tables **5**, 51 (1973)

[10.21] B.L. Henke, E.M. Gullikwa, J.C. Davies: X-ray interactions: photoabsorption, scattering, transmission and reflection at E=50–30000 eV, Z=1–92. Atomic Data and Nucl. Tables **54**, 181 (1993)

[10.22] E.H.S. Burhop: *The Auger Effect* (Cambridge Univ. Press, Cambridge 1952)

[10.23] T.A. Carlson: *Photoelectron and Auger Electron Spectroscopy* (Plenum, New York 1974)

[10.24] T. Åberg, G. Howart: Theory of the Auger Effect. In *Encyclopedia of Physics*, Vol.32, ed. by S. Flügge (Springer, Berlin, Heidelberg 1982) p. 469

[10.25] H.H. Madden: Chemical information from Auger electron spectroscopy. J. Vac. Sci. Techn. **18**, 677 (1981)

[10.26] M.P. Seah: A review of quantitative Auger electron spectrocopy. SEM 1983/II (SEM Inc., AMF O'Hare, IL 1983) p.521

[10.27] M.L. Tarng, G.K. Wehner: Escape length of Auger electrons. J. Appl. Phys. **44**, 1534 (1973)

[10.28] M.P. Seah, W.A.D. Dench: Quantitative electron spectroscopy of surfaces. A standard base for electron inelastic mean free paths in solids. Surf. Interface Anal. **1**, 2 (1979)

[10.29] J. Szajman, R.C.G. Leckey: An analytical expression for the calculation of electron mean free paths in solids. J. Electron Spectr. **23**, 83 (1981)

[10.30] J. Szajman, J. Liesegang, J.G. Jenkin, R.C.G. Leckey: Is there a universal mean-free path curve for electron inelastic scattering in solids? J. Electron Spectr. **23**, 97 (1981)

[10.31] H.E. Bishop, J.C. Rivière: Estimates of the efficiencies of production and detection of electron-excited Auger emission. J. Appl. Phys. **40**, 1740 (1969)

[10.32] J. Kirschner: The role of backscattered electrons in scanning Auger microscopy. SEM 1976/I (ITTRI, Chicago 1976) p.215

[10.33] J. Weihrauch: Nichtdispersive Röntgenmikroanalyse am Raster-Elektronenmikroskop. Beitr. elektr. mikr. Direktabb. Oberfl. **1**, 121 (1968)

[10.34] D.A. Gedcke: The Si(Li) x-ray energy spectrometer for x-ray microanalysis. In *Quantitative Scanning Electron Microscopy*, ed. by D.B. Holt et al. (Academic, London 1974) p.403

[10.35] E. Fiori, D.E. Newbury: Artifacts observed in energy-dispersive x-ray spectrometry in the SEM. SEM 1978/I (SEM Inc., AMF O'Hare, IL 1978) p.401

[10.36] S.J.B. Reed, N.G. Ware: Escape peaks and internal fluorescence in x-ray spectra recorded with lithium-drifted silicon detectors. J. Phys. E **5**, 582 (1972)

[10.37] N.C. Barbi, A.O. Sandborg, J.C. Russ, C.E. Soderquist: Light element analysis on the SEM using windowless energy dispersive x-ray spectrometer. SEM 1974 (ITTRI, Chicago 1974) p.151

[10.38] J.C. Russ, G.C. Baerwaldt, W.R. McMillan: Routine use of a second generation windowless detector for energy dispersive ultra-light element x-ray analysis. X-Ray Spectrometry **5**, 212 (1976)

[10.39] D.G. Rickerby: Barriers to energy dispersive spectrometry with low energy x-rays. Mikrochim. Acta (Suppl.) **13**, 493 (1996)

[10.40] P. Hovington, G. L'Espérance, E. Baril, M. Rigaud: Monitoring the performance of energy dispersive spectrometer detectors at low energy. Scanning **17**, 136 (1995)

[10.41] F. Eggert: Effektivität energiedispersiver Röntgenspektrometer im Energiebereich kleiner 1 keV. Beitr. elektr. mikr. Direktabb. Oberfl. **29**, 31 (1996)

[10.42] M. Procop: Abschätzung der Dicke absorbierender Schichten eines EDX-Detektorsystems mit Hilfe von Bremsstrahlungsspektren. Beitr. elektr. mikr. Direktabb. Oberfl. **29**, 47 (1996)

[10.43] W.P. Rehbach: Vergleich von energiedispersiven Röntgenspektren im Energiebereich <1 keV, gemessen mit unterschiedlichen EDX-Detektoren. Beitr. elektr. mikr. Direktabb. Oberfl. **29**, 37 (1996)

[10.44] M. Procop: Über Probleme bei der EDX-Analyse im Energiebereich unterhalb 1 keV. Beitr. elektr. mikr. Direktabb. Oberfl. **27**, 1 (1994)

[10.45] T.A. Hall: Reduction of background due to backscattered electrons in energy-dispersive x-ray microanalysis. J. Microsc. **110**, 103 (1977)

[10.46] B. Neumann, L. Reimer, B. Wellmanns: A permant magnet system for electron deflection in front of an energy-dispersive x-ray spectrometer. Scanning **1**, 130 (1978)

[10.47] E. Lifshin, M.F. Ciccarelli: Present trends in x-ray analysis with the SEM. SEM 1973 (ITTRI, Chicago 1973) p.89

[10.48] P.J. Statham, J.V.P. Long, G. White, K. Kandiah: Quantitative analysis with an energy-dispersive detector using a pulsed electron probe and active signal processing. X-ray Spectrometry **3**, 153 (1974)

[10.49] M.H. Peeters: An electron-beam blanker mounted on a SEM to improve the counting efficiency of an energy dispersive x-ray analyser, in *EUREM 88*, Inst. of Physics Conf. Ser. 93, Vol.1 (IoP, London 1988) p.137

[10.50] R. Schmidt, M. Feller-Kniepmeier: Investigation of system-induced background radiation using a 0–160 keV high-purity germanium detector. Ultramicroscopy **34**, 229 (1990)

[10.51] T.J. White, D.R. Cousens, G.J. Auchterlonie: Preliminary characterization of an intrinsic germanium detector on a 400 keV microscope. J. Microsc. **162**, 379 (1991)

[10.52] G. Clooth, C. Paggen, W.G. Burchard: Quantifizierung des Sauerstoffs mittels der energiedispersiven Röntgenanalyse. Beitr. elektr. mikr. Direktabb. Oberfl. **27**, 7 (1994)

[10.53] J.I. Goldstein, R.E. Ganneman, R.E. Ogilvie: Diffusion in the Fe-Ni system at 1 atm and 40 kbar pressure. Trans. Met. Soc. AIME **233**, 812 (1965)

[10.54] T.O. Ziebold, R.E. Ogilvie: An empirical method for electron microanalysis. Anal. Chem. **36**, 322 (1964)

[10.55] A.E. Bence, A.L. Albee: Empirical correction factors for the electron microanalysis of silicates and oxides. J. Geology **76**, 382 (1968)

[10.56] D. Laguitton, R. Rousseau, F. Claisse: Computed alpha coefficients for electron microprobe analysis. Anal. Chem. **47**, 2174 (1975)

[10.57] J.W. Colby: Quantitative microprobe analysis of thin insulating films. Adv. X-Ray Analysis **11**, 287 (1968)

[10.58] K.F.J. Heinrich, R.L. Myklebust, H. Yakowitz, S.D. Rasberry: A simple correction procedure for quantitative electron probe microanalysis. NBS Spec. Techn. Note 719 (U.S. Dep. of Commerce, Washington, DC 1972)

[10.59] J. Hénoc, K.F.J. Heinrich, R.L. Myklebust: A rigorous correction procedure for quantitative electron probe microanalysis (COR 2). NBS Techn. Note 769 (U.S. Dep. of Commerce, Washington D.C. 1973)

[10.60] H. Yakowitz, R.L. Myklebust, K.F.J. Heinrich: An on-line correction procedure for quantitative electron probe microanalysis. NBS Techn. Note 796 (U.S. Dep. of Commerce, Washington, DC 1973)

[10.61] R.R. Warner, J.R. Coleman: A procedure for quantitative electron probe microanalysis of biological material. Micron **4**, 61 (1973)

[10.62] R.R. Warner, J.R. Coleman: A biological thin specimen microprobe quantitation method that calculates composition and ρx. Micron **6**, 79 (1975)

[10.63] R.B. Marinenko: Standards for electron probe microanalysis. In *Electron Probe Quantitation*, ed. by K.F.J. Heinrich, D.E. Newbury (Plenum, New York 1991) p.251

[10.64] G. Love, V.D. Scott: Updating correction procedures in quantitative electron probe microanalysis. Scanning **4**, 111 (1981)

[10.65] D.L. Webster, H. Clark, W.W. Hansen: Effects of cathode-ray diffusion on intensities in x-ray spectra: Phys. Rev. **37**, 115 (1931)

[10.66] J. Philibert, R. Tixier: Some problems with quantitative electron probe microanalysis. In *Quantitative Electron Probe Microanalysis*, ed. by K.F.J. Heinrich, NBS Spec.Publ. 298 (U.S. Dep. of Commerce, Washington, DC 1968) p.13

[10.67] M. Green, V.E. Cosslett: The efficiency of production of characteristic x-radiation in thick targets of pure elements. Proc. Phys. Soc. **78**, 1206 (1961)

[10.68] M. Green, V.E. Cosslett: Measurement of K, L and M shell x-ray production efficiencies. J. Phys. D **1**, 425 (1968)

[10.69] P. Duncumb, S.J.B. Reed: The calculation of stopping power and backscatter effects in electron probe microanalysis. In *Quantitative Electron Probe Microanalysis*, ed. by K.F.J. Heinrich, NBS Spec. Publ. 298 (U.S. Dep. of Commerce, Washington, DC 1968) p.133

[10.70] G. Love, M.G.C. Cox, V.D. Scott: A simple Monte Carlo method for simulating electron-solid interactions and its application to electron probe microanalysis. J. Phys. D **10**, 7 (1977)

[10.71] V.D. Scott, G. Love: An EPMA correction method based upon quadrilateral $\Phi(\rho z)$ profile. In *Electron Probe Quantitation*, ed. by K.F.J. Heinrich, D.E. Newbury (Plenum, New York 1991) p. 19

[10.72] J.L. Labar: Comparison of backscatter loss calculations in electron microanalysis. Scanning **8**, 188 (1986)

[10.73] H.J. August, R. Razka, J. Wernisch: Calculation and comparison of the backscattering factor R for characteristic x-ray emission. Scanning **10**, 107 (1988)

[10.74] R.L. Myklebust, D.E. Newbury: The R factor: The x-ray loss due to electron backscatter. In *Electron Probe Quantitation*, ed. by K.F.J. Heinrich, D.E. Newbury (Plenum, New York 1991) p. 177

[10.75] JG. Springer: The loss of x-ray intensity due to backscattering in microanalyser targets. Microchim. Acta 1966, p.587

[10.76] J.C. Dérian: PhD thesis, CEA Rep. R 3052, Univ.Paris 1966

[10.77] B. Lödding, L. Reimer: An experimental test of a ZAF correction program for tilted specimens and energy dispersive spectrometry. In *10th Int'l Congr. on X-Ray Optics and Microanalysis*. J. Physique **45**, C2, 37 (1984)

[10.78] R. Castaing, J. Descamps: Sur le bases physique de l'analyse ponctuelle par spectrographie X. J. Phys. Rad. **16**, 304 (1955)

[10.79] U. Schmitz, P.L. Ryder, W. Pitsch: An experimental method for determining the depth distribution of characteristic x-rays in electron microprobe specimens. In *5th Int'l.Conf. on X-Ray Optics and Microanalyis*, ed. by G. Möllenstedt and K.H. Gaukler (Springer, Berlin, Heidelberg 1969) p.104

[10.80] A.R. Büchner, W. Pitsch: A new correction for absorption and for atomic number in quantitative microprobe analysis of metals. Z. Metallkd. **62**, 392 (1971)

[10.81] P. Karduck, W. Rehbach: Experimental determination of the depth distribution of x-ray production $\Phi(\rho z)$ for x-ray energies below 1 keV. Mikrochim. Acta (Suppl.) **11**, 289 (1985)

[10.82] W. Rehbach, P. Karduck: Measurement of $\Phi(\rho z)$ depth distributions for low energy characteristic x-rays. In *11th Int'l. Congr. in X-Ray Optics and Microanalysis*, ed. by J.D. Brown and R.H. Packwood (London, Ontario 1986) p.244

[10.83] A. Vignes, G. Dez: Distribution depth of the primary x-ray emission in anticathodes of titanium and lead. J. Phys. D **1**, 1309 (1968)

[10.84] J.D. Brown, L. Parobek: X-ray production as a function of depth for low electron energies. X-Ray Spectrometry **5**, 36 (1976)

[10.85] J.D. Brown, L. Parobek: The sandwich sample technique applied to the atomic number effect. In *6th Int'l. Conf. on X-Ray Optics and Microanalysis*, ed. by G. Shinoda et al. (Univ. of Tokyo Press, Tokyo 1972) p.163

[10.86] M. Green: The target absorption correction in x-ray microanalysis In *Proc. 3rd Int'l.Conf on X-Ray Optics and Microanalysis*, ed. by H.H. Pattee et al. (Academic, New York 1963) p.361

[10.87] B. Neumann, L. Reimer: Method for measuring the absorption correction $f(\chi)$ with an energy dispersive x-ray detector. Scanning **1**, 243 (1978)

[10.88] J. Philibert: A method for calculating the absorption correction in electron probe microanalysis. In *Proc 3rd. Int'l.Conf. on X-Ray Optics and Microanalysis*, ed. by H.H. Pattee et al. (Academic, New York 1963) p.379

[10.89] K.F.J. Heinrich: Present state of the classical theory of quantitative electron probe microanalysis. NBS Techn. Note 521 (U.S. Dep. of Commerce, Washington, DC 1970)

[10.90] H.E. Bishop: The prospects for an improved absorption correction in electron probe microanalysis. J. Phys. D **7**, 2009 (1974)

[10.91] G. Love, V.D. Scott: Evaluation of a new correction procedure for quantitativce electron probe microanalysis. J. Phys. D **11**, 1369 (1978)

[10.92] R.H. Packwood, J.D. Brown: A Gaussian expression to describe $\Phi(z)$ curves for quantitative electron probe microanalysis. X-Ray Spectrometry **10**, 138 (1981)

[10.93] G.F. Bastin, H.J.M. Heijligers, F.J.J. van Loo: A further improvement in the Gaussian $\Phi(\rho z)$ approach for matrix correction in quantitative electron probe microanalysis. Scanning **8**, 45 (1986); see also *Electron Probe Quantitation*, ed. by K.F. J. Heinrich, and D.E. Newbury (Plenum, New York 1991) p.145

[10.94] J.L. Pouchou, F. Pichoir: Quantitative analysis of homogeneous or stratified microvolumes applying the model PAP. In *Electron Probe Quantitation* ed. by K.F.J. Heinrich and D.E. Newbury (Plenum, New York 1991) p.31

[10.95] S.J.B. Reed: Characteristic fluorescence corrections in electron-probe microanalysis. Brit. J. Appl. Phys. **16**, 913 (1967)

[10.96] G. Springer: Die Berechnung von Korrekturen für die quantitative Elektronenstrahl-Mikroanalyse. Fortschr. Miner. **45**, 103 (1967)

[10.97] S.J.B. Reed: *Electron Probe Microanalysis* (Univ. Press, Cambridge, 1975)

[10.98] J. Hénoc: Fluorescence excited by the continuum. In *Quantitative Electron Probe Microanalysis*, ed. by K.F.J. Heinrich, NBS Spec. Publ. 298 (U.S. Dep. of Commerce, Washington D.C., 1968) p.197

[10.99] G. Springer: The correction for continuum fluorescence in electron probe microanalysis. Neues Jahrbuch Miner. Abh. **106**, 241 (1967)

[10.100] G. Springer: Fluorescence by continuum radiation in multi-element targets. In *6th Int'l. Conf. on X-Ray Optics and Microanalysis*, ed. by G. Shinoda et al. (Univ. of Tokyo Press, Tokyo 1972) p.141

[10.101] K.F.J. Heinrich: Strategies of electron probe data reduction. In *Electron Probe Quantitation*, ed. by K.F.J. Heinrich, D.E. Newbury (Plenum, New York 1991) p. 9

[10.102] R. Packwood: A comprehensive theory of electron probe microanalysis. In *Electron Probe Quantitation*, ed. by K.F.J. Heinrich, D.E. Newbury (Plenum, New York 1991) p. 83

[10.103] Karduck, W. Rehbach: Progress in the measurement and calculation of the depth distribution of low-energy x-ray generation. In *Microbeam Analysis 1990*, ed. by D.E. Newbury (San Francisco Press, San Francisco 1990) p.277

[10.104] L. Reimer: Monte Carlo simulation techniques for quantitative x-ray microanalysis. Mikrochim. Acta (Suppl.) **13**, 1 (1996)

[10.105] P.J. Statham: Measurement and use of peak-to-background ratios in x-ray analysis, Mikrochim. Acta (Suppl.8), 229 (1979)

[10.106] J.L. Pouchou: Standardless x-ray analysis of bulk specimens. Mikrochim. Acta **114/115**, 33 (1994)

[10.107] F. Eggert: Standardfreie Elektronenstrahl-Mikroanalyse. Beitr. elektr. mikr. Direktabb. Oberfl. **27**, 15 (1994)

[10.108] G.F. Bastin, H.J.M. Heijligers: Nonconductive specimens in the electron probe microanalyzer – a hitherto poorly discussed problem. In *Electron Probe Quantitation*, ed. by K.F. J. Heinrich, D.E. Newbury (Plenum, New York 1991) p. 163

[10.109] M. Green: The angular distribution of characteristic x-radiation and its origin within a solid target. Proc. Phys. Soc. **83**, 435 (1964)

[10.110] H.E. Bishop: The absorption and atomic number correction in electron-probe x-ray microanalysis. J. Phys. D **1**, 673 (1968)

[10.111] J.C. Russ: Microanalysis of thin sections, coatings and rough surfaces. SEM 1973 (ITTRI, Chicago 1973) p.113

[10.112] G. Love, M.G. Cox, V.D. Scott: A versatile atomic number correction for electron-probe microanalysis. J. Phys. D **11**, 7 (1978)

[10.113] B. Lödding, L. Reimer: Energy dispersive x-ray microanalysis of tilted specimens using a modified ZAF correction. Scanning **1**, 225 (1978)

[10.114] H.J. Dudek, R. Borath: Preparation of a sharply defined boundary between two elements. Scanning **2**, 39 (1979)

[10.115] W.E. Sweeney, R.E. Seebold, L.S. Birks: Electron probe measurements of evaporated metal films. J. Appl. Phys. **31**, 1061 (1960)

[10.116] G.H. Cockett, C.D. Davies: Coating thickness measurement by electron probe microanalysis. Brit. J. Appl. Phys. **14**, 813 (1963)

[10.117] W. Reuter: The ionization function and its application to the electron probe analysis of thin films. In *6th Int'l. Conf. on X-Ray Optics and Microanalysis*, ed. by G. Shinoda et al. (Univ. of Tokyo Press, Tokyo 1972) p.121

[10.118] H. Yakowitz, D.E. Newbury: A simple analytical method for thin film analysis with massive pure element standards. SEM 1976/I (ITTRI, Chicago 1976) p.151

[10.119] H.E. Bishop, D.M. Poole: A simple method of thin film analysis in the electron probe microanalyser. J. Phys. D **6**, 1142 (1973)

[10.120] M.G.C. Cox, G. Love, V.D. Scott: A characteristic fluorescence correction for electron-probe microanalysis of thin coatings. J. Phys. D **12**, 1441 (1979)

[10.121] L. Reimer: *Transmission Electron Microscopy. Physics of Image Formation and Microanalysis*, Springer Ser. Opt. Sci., Vol.36, 4th edn. (Springer, Berlin, Heidelberg 1997)

[10.122] P. Willich, R. Bethke: Practical aspects and applications of EPMA at low electron energies. Mikrochim. Acta (Suppl.) **13**, 631 (1996)

[10.123] G. Cliff, G.W. Lorimer: The quantitative analysis of thin specimens. J. Microsc. **103**, 203 (1975)

[10.124] J.I. Goldstein, J.L. Costley, G.W. Lorimer, S.J.B. Reed: Quantitative x-ray analysis in the electron microscope. SEM 1977/I ITTRI, Chicago 1977) p.315

[10.125] T.P. Schreiber, A.M. Wims: A quantitative x-ray microanalysis thin film method using K-, L- and M-lines. Ultramicroscopy **6**, 323 (1981)

[10.126] H.J. Hoffmann, J.H. Weihrauch, H. Fechting: Eine empirische Methode zur quantitativen chemischen Analyse von Mikroteilchen mit der Mikrosonde. In *5th Int'l. Conf. on X-Ray Optics and Microanalyis*, ed. by G. Möllenstedt and K.H. Gaukler (Springer, Berlin, Heidelberg 1969) p.166

[10.127] J.T. Armstrong, P.R. Buseck: Quantitative chemical analysis of individual microparticles using the electron microprobe. Anal. Chem. **47**, 2178 (1975)

[10.128] N.C. Barbi, M.A. Giles, D.P. Skinner: Estimating elemental concentrations in small particles using x-ray analysis in the electron microscope. SEM 1978/I (SEM Inc., AMF O'Hare, IL 1978) p.193

[10.129] J.T. Armstrong: Methods of quantitative analysis of individual microparticles with electron beam instruments. SEM 1978/I (SEM Inc., AMF O'Hare, IL 1978) p.455

[10.130] K.F.J. Heinrich: Characterization of particles. NBS Spec.Publ. 533 (Nat. Bur. of Standards, Washington, DC 1980)

[10.131] R.L. Myklebust, D.E. Newbury, K.F.J. Heinrich, J.A. Small, C.E. Fiori: Monte Carlo electron trajectory simulation – an aid for particle analysis. In *Proc. 13th Ann. Conf. Microbeam Anal. Soc.* (Ann Arbor 1978) p.61A

[10.132] P.J. Statham, J.B. Pawley: A new method for particle x-ray microanalysis based on peak-to-background measurements. SEM 1978/I (SEM Inc., AMF O'Hare, IL 1978) p.469

[10.133] J.A. Small, K.F.J. Heinrich, D.E. Newbury, R.L. Myklebust: Progress in the development of the peak-to-background method for the quantitative analysis of single particles with the electron probe. SEM 1979/II (SEM Inc., AMF O'Hare, IL 1979) p.807

[10.134] P. Wieser, R. Wurster: Some remarks about quantitative characterization of small particles by the electron microprobe. Scanning **2**, 29 (1979)

[10.135] A.J. Morgan, T.W. Davies, D.A. Erasmus: Specimen preparation. In *Electron Probe Microanalysis in Biology*, ed. by D.A. Erasmus (Chapman and Hall, London 1978) p.94

[10.136] H. Shuman, A.V. Somlyo, A.P. Somlyo: Quantitative electron probe microanalysis of biological thin sections: methods and validity. Ultramicroscopy **1**, 317 (1976)

[10.137] T.A. Hall: Problems of the continuum-normalization method for the quantitative analysis of sections of soft tissue. In *Microbeam Analysis in Biology*, ed. by C. Lechéne, R.R. Warner (Academic, New York 1979) p.185

[10.138] A. Warley: Standards for the application of x-ray microanalysis to biological specimens. J. Microsc. **157**, 135 (1990)

[10.139] A. Patak, A. Wright, A.T. Marshall: Evaluation of several common standards for the x-ray microanalysis of thin biological specimens. J. Microsc. **170**, 265 (1993)

[10.140] G.M. Roomans: Standards for x-ray microanalysis of biological specimens. SEM 1979/II (SEM Inc., AMF O'Hare, IL 1979) p.649

[10.141] M.J. Ingram, F.D. Ingram: Electron microprobe calibration for measurements of intracellular water. SEM 1983/III (SEM Inc., AMF O'Hare, IL 1983) p.1249

[10.142] N. Roos. T. Barnard: Aminoplastic standards for quantitative x-ray microanalysis of thin sections of plastic-embedded biological material. Ultramicroscopy **15**, 277 (1984)

[10.143] A.J. Morgan, C. Winters: Practical notes on the production of thin aminoplast standards for quantitative x-ray microanalysis. Micron Microsc. Acta **20**, 209 (1989)

[10.144] W.C. de Bruin, M.I. Cleton-Soeteman: Application of Chelex standard beads in integrated morphometrical and x-ray microanalysis. SEM 1985/II (SEM Inc., AMF O'Hare, IL 1985) p.715

[10.145] A. Dörge, R. Rick, K. Gehring, K. Thuraus: Preparation of frozen-dried cryosections for quantitative x-ray microanalysis of electrolytes in biological soft tissue. Pflügers Arch. **373**, 85 (1978)

[10.146] T. von Zglinicki, M. Bimmler, W. Krause: Estimation of organelle water fractions from frozen-dried cryosections. J. Microsc. **146**, 67 (1987)

[10.147] K.E. Tvedt, G. Kopstad, J. Halgunset, O.A. Haugen: Rapid freezing of small biopsies and standard for cryosectioning and x-ray microanalysis. Am. J. Clin. Pathol. **92**, 51 (1989)

[10.148] A. Boekestein, F. Thiel, A.L.H. Stols, E. Bouw, A.M. Stadhouders: Surface roughness and the use of peak to background ratio in the x-ray microanalysis of bio-organic bulk specimens. J. Microsc. **134**, 327 (1984)

[10.149] I. Zs-Nagy, C. Pieri, C. Giuli, C. Bertoni-Freddari, V. Zs.-Nagy: Energy-dispersive x-ray microanalysis of the electrolytes in biological bulk specimens. J. Ultrastruct. Res. **58**, 22 (1977)

[10.150] T.A. Hall: Quantitative electron probe x-ray microanalysis in biology. Scanning Microscopy **3**, 461 (1989)

[10.151] G.M. Roomans: The Hall method in the quantitative x-ray microanalysis of biological specimens: a review. Scanning Microscopy **4**, 1055 (1990)

[10.152] G.M. Roomans: Quantitative electron probe x-ray microanalysis of biological bulk specimens. SEM 1981/II (SEM Inc., AMF O'Hare, IL 1981) p.345

[10.153] J.I. Goldstein, D.E. Newbury, P. Echlin, D.C. Joy, C. Fiori, E. Lifshin: *Scanning Electron Microscopy and X-Ray Microanalysis* (Plenum, New York 1981)

[10.154] C.E. Fiori, R.L. Myklebust, K.F.J. Heinrich, H. Yakowitz: Prediction of continuum intensity in energy-dispersive x-ray microanalysis. Anal. Chem. **48**, 172 (1976)

[10.155] C.E. Fiori, R.L. Myklebust, K.F.J. Heinrich: A method for resolving overlapping energy dispersive peaks of an x-ray spectrometer. In *Conf. Microbeam Anal. Soc.* (Ann Arbor 1976) p.12A

[10.156] P. Ryder, S. Baumgartl: Die Eignung eines energiedispersiven Röntgenspektrometers für die Elektronenstrahl-Mikroanalyse. Arch. Eisenhüttenwesen **42**, 635 (1971)

[10.157] R.W. Gould, J.T. Healey: Secondary fluorescent excitation in the SEM: improved sensitivity of energy dispersive analysis. Rev. Sci. Instr. **46**, 1427 (1975)

[10.158] L.M. Middleman, J.D. Geller: Trace element analysis using x-ray excitation with an energy dispersive spectrometer on a SEM. SEM 1976/I (ITTRI, Chicago 1976) p.171

[10.159] B. Linnemann, L. Reimer: Comparison of x-ray elemental analysis by electron excitation and x-ray fluorescence. Scanning **1**, 109 (1978)

[10.160] J.B. Warren, H.W. Kraner: Optimized stage design for x-ray fluorescence analysis in the SEM. SEM 1982/IV (SEM Inc., AMF O'Hare, IL 1982) p.1373

[10.161] R. Eckert: X-ray fluorescence analysis in the SEM with a massive anode. SEM 1983/IV (SEM Inc., AMF O'Hare, IL 1983) p.1535

[10.162] E. Valamontes, A. G. Nassiopoulos: Comparison of back-foil scanning x-ray microfluorescence and electron probe microanalysis for elemental characterisation of thin coatings. Mikrochim. Acta, Suppl. **13**, 597 (1996)

[10.163] K. Siegbahn et al. *ESCA: Atomic, Molecular and Solid State Structure studied by Means of Electron Spectroscopy* (Almqvist and Wiksells, Uppsala 1967)

[10.164] C.C. Chang: Auger electron spectroscopy. In *Characterization of Solid Surfaces*, ed. by P.F. Kane and G.B. Larrabee (Plenum, New York 1974) p.509

[10.165] M.B. Chamberlain: Instrumentation and methods for scanning Auger microscopy. SEM 1982/III (SEM Inc., AMF O'Hare, IL 1982) p.937

[10.166] L.L. Levenson: Fundamentals of Auger electron spectroscopy. SEM 1983/IV (SEM Inc., AMF O'Hare, IL 1983) p.1643

[10.167] A. Benninghoven: Surface investigation of solids by the statical method of secondary ion mass spectroscopy (SIMS). Surf. Sci. **35**, 427 (1973)

[10.168] P.W. Palmberg, W.M. Riggs: Unique instrument for multiple surface characterization by ESCA, scanning Auger, UPS and SIMS. J. Vac. Sci. Techn. **15**, 786 (1978)

[10.169] N.C. MacDonald: Auger electron spectroscopy for SEM. SEM 1971 (ITTRI, Chicago 1971) p.89

[10.170] E.K. Brandis: High spatial resolution Auger electron spectroscopy in an ordinary diffusion pumped SEM. SEM 1975 (ITTRI, Chicago 1975) p.141

[10.171] R. Holm, B. Reinfandt: Auger microanalysis in a conventional SEM. Scanning **1**, 42 (1978)

[10.172] J. Cazaux: X-ray probe microanalyser and scanning x-ray microscopies. Ultramicroscopy **12**, 321 (1984)

[10.173] C.T. Hovland: Scanning ESCA: a new dimension for electron spectroscopy. Appl. Phys. Lett. **30**, 274 (1977)

[10.174] R. Plattner, D. Schünemann: Das energiedispersive Röntgenspektrometer – gegenwärtiger Leistungsstand und zukünftige Entwicklung. Beitr. elektr. mikr. Direktabb. Oberfl. **4/2**, 77 (1971)

[10.175] J.B. Pawley, T. Hayes, R.H. Falk: Simultaneous three-element x-ray mapping using color TV. SEM 1976/I (ITTRI, Chicago 1976) p.187

[10.176] K.F.J. Heinrich: Elemental mapping in the microscope domain. SEM 1977/I (ITTRI, Chicago 1977) p.605

[10.177] P. Ingram, J.D. Shelburne: Total rate imaging with x-rays (TRIX) – a simple method of forming a non-projective x-ray image in the SEM using an energy dispersive detector and its application to biological specimens. SEM 1980/II (SEM Inc., AMF O'Hare, IL 1980) p.285

[10.178] P. Bernsen, L. Reimer: Total rate imaging with x-ray in a SEM. J. Physique **45**, C2-297 (1984)

[10.179] B. Neumann, L. Reimer: Versuche zur Röntgenprojektionsmikroskopie im Rasterlektronenmikroskop. Beitr. elektr. mikr. Direktabb. Oberfl. **9**, 147 (1976)

[10.180] V.E. Cosslett, W.C. Nixon: *X-Ray Microscopy* (Cambridge Univ. Press, Cambridge 1960)

[10.181] M. von Ardenne: Zur Leistungsfähigkeit des Elektronen-Schattenmikroskops und über ein Röntgenstrahl-Schattenmikroskop. Naturwiss. **27**, 485 (1939)

[10.182] V.E. Cosslett, W.C. Nixon: The x-ray shadow microscope. J. Appl. Phys. **24**, 616 (1953)

[10.183] S.P. Ong: *Microprojection with X-Rays*. Martinus Nijhoff, The Hague 1959

[10.184] H.R.F. Horne, H.G. Waltinger: Röntgenmikrographie und Röntgenabsorptionsanalyse im REM. Beitr. elektr. mikr. Direktabb. Oberfl. **6**, 163 (1974)

[10.185] H.R.F. Horne, H.G. Waltinger: How to obtain and use x-ray projection microscopy in the SEM. Scanning **1**, 100 (1978)

[10.186] G. Fuhrmann, H. Halling, R. Möller, J. Vell, Z. Shuan-Ren, E. Wallura: Mikrotomographiesystem als Zusatz für Rasterelektronenmikroskope. Beitr. elektr. mikr. Direktabb. Oberfl. **25**, 95 (1992)

[10.187] R.F. de Paiva, M. Bisiaux, J. Lyach, E. Rosenberg: High-resolution x-ray tomography in an electron microprobe. Rev. Sci. Instr. **67**, 2251 (1996)

[10.188] H.R.F. Horne: X-ray reflection-topography in the SEM. Scanning **6**, 69 (1984)

[10.189] W. Brünger: Scanning x-ray projection microscopy using an energy-dispersive spectrometer. SEM 1978/I (SEM Inc., AMF O'Hare, IL 1978) p.423

[10.190] R. Feder, E. Spiller, J. Topalian, A.N. Broers, W. Gudat, B. Panessa: High resolution soft x-ray microscopy: Science **197**, 259 (1977)

[10.191] B.J. Panessa, J.B. Warren, P. Hoffman, R. Feder: Imaging unstained proteoglycan aggregates by soft x-ray contact microscopy. Ultramicroscopy **5**, 267 (1980)

Index

Printing: Mercedesdruck, Berlin
Binding: Buchbinderei Lüderitz & Bauer, Berlin

Springer Series in Optical Sciences

1 **Solid-State Laser Engineering**
By W. Koechner 4th Edition

2 **Table of Laser Lines in Gases and Vapors**
By R. Beck, W. Englisch, and K. Gurs
3rd Edition

3 **Tunable Lasers and Applications**
Editors: A. Mooradian, T. Jaeger, and P. Stokseth

4 **Nonlinear Laser Spectroscopy**
By V. S. Letokhov and V. P. Chebotayev
2nd Edition

5 **Optics and Lasers** Including Fibers and Optical Waveguides By M. Young
4th Edition (available as a textbook)

6 **Photoelectron Statistics**
With Applications to Spectroscopy and Optical Communication By B. Saleh

7 **Laser Spectroscopy III**
Editors: J. L. Hall and J. L. Carlsten

8 **Frontiers in Visual Science**
Editors: S. J. Cool and E. J. Smith III

9 **High-Power Lasers and Applications**
Editors: K.-L. Kompa and H. Walther

10 **Detection of Optical and Infrared Radiation**
By R. H. Kingston

11 **Matrix Theory of Photoelasticity**
By P. S . Theocaris and E. E. Gdoutos

12 **The Monte Carlo Method in Atmospheric Optics**
By G. I. Marchuk, G. A. Mikhailov, M. A. Nazaraliev, R. A. Darbinian, B. A. Kargin, and B. S. Elepov

13 **Physiological Optics**
By Y. Le Grand and S. G. El Hage

14 **Laser Crystals** Physics and Properties
By A. A. Kaminskii 2nd Edition

15 **X-Ray Spectroscopy** By B. K. Agarwal
2nd Edition

16 **Holographic Interferometry**
From the Scope of Deformation Analysis of Opaque Bodies
By W. Schumann and M. Dubas

17 **Nonlinear Optics of Free Atoms and Molecules**
By D. C. Hanna, M. A. Yuratich, and D. Cotter

18 **Holography in Medicine and Biology**
Editor: G. von Bally

19 **Color Theory and Its Application in Art and Design** By G. A. Agoston 2nd Edition

20 **Interferometry by Holography**
By Yu. I. Ostrovsky, M. M. Butusov, and G. V. Ostrovskaya

21 **Laser Spectroscopy IV**
Editors: H. Walther and K. W. Rothe

22 **Lasers in Photomedicine and Photobiology**
Editors: R. Pratesi and C. A. Sacchi

23 **Vertebrate Photoreceptor Optics**
Editors: J. M. Enoch and F. L. Tobey, Jr.

24 **Optical Fiber Systems and Their Components**
An Introduction By A. B. Sharma, S. J. Halme, and M. M. Butusov

25 **High Peak Power Nd: Glass Laser Systems**
By D. C. Brown

26 **Lasers and Applications**
Editors: W. O. N. Guimaraes, C. T. Lin, and A. Mooradian

27 **Color Measurement**
Theme and Variations
By D. L. MacAdam 2nd Edition

28 **Modular Opticsl Design**
By O. N. Stavroudis

29 **Inverse Problems of Lidar Sensing of the Atmosphere** By V. E. Zuev and I. E. Naats

30 **Laser Spectroscopy V**
Editors: A. R. W. McKellar, T. Oka, and B. P. Stoicheff

31 **Optics in Biomedical Sciences**
Editors: G. von Bally and P. Greguss

32 **Fiber-Optic Rotation Sensors** and Related Technologies
Editors: S. Ezekiel and H. J. Arditty

33 **Integrated Optics:** Theory and Technology
By R. G. Hunsperger
4th Edition (available as a textbook)

34 **The High-Power Iodine Laser**
By G. Brederlow, E. Fill, and K. J. Witte

35 **Engineering Optics**
By K. Iizuka 2nd Edition

36 **Transmission Electron Microscopy**
Physics of Image Formation and Microanalysis
By L. Reimer 4th Edition

37 **Opto-Acoustic Molecular Spectroscopy**
By V. S. Letokhov and V. P. Zharov

38 **Photon Correlation Techniques**
Editor: E. O. Schulz-DuBois

39 **Optical and Laser Remote Sensing**
Editors: D. K. Killinger and A. Mooradian

40 **Laser Spectroscopy VI**
Editors: H. P. Weber and W. Lüthy

41 **Advances in Diagnostic Visual Optics**
Editors: G. M. Breinin and I. M. Siegel

Springer and the environment

At Springer we firmly believe that an international science publisher has a special obligation to the environment, and our corporate policies consistently reflect this conviction.

We also expect our business partners – paper mills, printers, packaging manufacturers, etc. – to commit themselves to using materials and production processes that do not harm the environment. The paper in this book is made from low- or no-chlorine pulp and is acid free, in conformance with international standards for paper permanency.